The San Andreas Fault System:
Displacement, Palinspastic Reconstruction, and Geologic Evolution

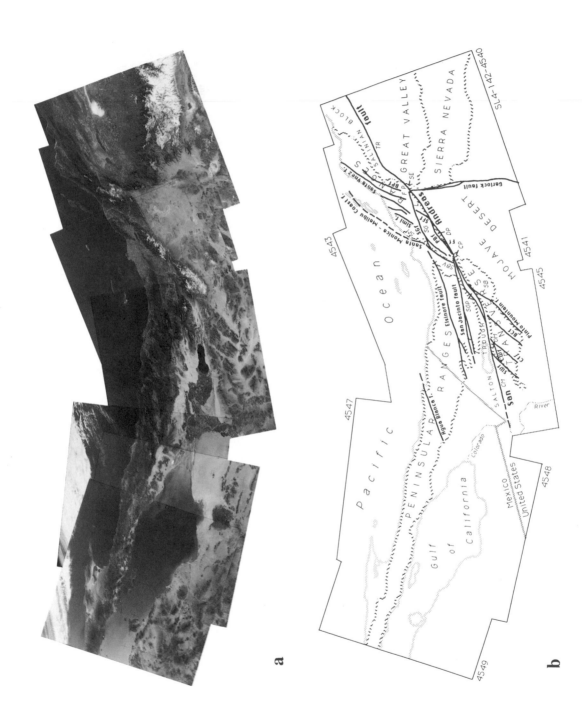

a

b

Frontispiece. San Andreas fault system, showing major faults and physiographic features in southern California and northwestern Mexico (after Silver and others, 1977, Fig. 4-3; Powell, 1981, Fig. 1-1). a, Photomosaic constructed from Skylab 4 hand-held photographs. b, Sketch map. Faults: BCf = Blue Cut fault; BPf = Big Pine fault; Cf = Chiriaco fault; CWf = Clemens Well fault; Ff = Fenner fault; PBf = Punchbowl fault; SFf = San Francisquito fault; SGf = San Gabriel fault; SWf = Salton Wash fault. Physiographic features; CM = Chocolate Mountains; CP = Cajon Pass; DP = Devils Punchbowl; F = Frazier Mountain; OM = Orocopia Mountains; P = Mount Pinos; SB = San Bernardino Mountains; SBV = San Bernardino Valley; SE = San Emigdio Mountains; SG = San Gabriel Mountains; SGP = San Gorgonio Pass; SP = Sierra Pelona; SM = Santa Monica Mountains; TR = Temblor Range.

Geological Society of America
Memoir 178

The San Andreas Fault System:
Displacement, Palinspastic Reconstruction,
and Geologic Evolution

Edited by

Robert E. Powell
U.S. Geological Survey
Room 117, U.S. Post Office
Spokane, Washington 99201-1087

R. J. Weldon, II
Department of Geological Sciences
University of Oregon
Eugene, Oregon 97403-1272

and

Jonathan C. Matti
U.S. Geological Survey
Department of Geological Sciences
University of Arizona
Tucson, Arizona 85721

1993

Published by The Geological Society of America, Inc.
3300 Penrose Place, P.O. Box 9140, Boulder, Colorado 80301

GSA Books Science Editor Richard A. Hoppin

Printed in U.S.A.

Library of Congress Cataloging-in-Publication Data
The San Andreas fault system : displacement, palinspastic
   reconstruction, and geologic evolution / edited by Robert E. Powell,
   R. J. Weldon, II, Jonathan C. Matti.
      p.   cm. — (Memoir / Geological Society of America ; 178)
      Includes bibliographical references and index.
      ISBN 0-8137-1178-9
      1. San Andreas Fault (Calif.)   I. Powell, Robert E.   II. Weldon,
   R. J., 1955–   .   III. Matti, Jonathan C.   IV. Series: Memoir
   (Geological Society of America) ; 178.
   QE606.5.U6S27   1993
   551.8'7'09794--dc20   92-39410
                                                   CIP

Cover: Composite of hand-held orbital photographs showing the
San Andreas fault system in southern California.

10   9   8   7   6   5   4   3   2   1

Contents

Plates
(in slipcase)

Description of Map Units (Plates I and II) and Correlation of Map Units

Plate I.   Geologic Map of Southern and Central California

Plate IIA. Paleogeologic Map ca. 5 or 6 Ma

Plate IIB. Paleogeologic Map ca. 10 to 13 Ma

Plate IIC. Paleogeologic Map ca. 20 Ma $\pm$ 2 Ma

Plate IIIA. Generalized Geologic Map of Southern California Showing Rock Units, Faults, and Geographic Locations Discussed in Chapter 2

Plate IIIB. Palinspastic Reconstruction for the San Andreas and San Gabriel Fault Zones Based on Restoration of the Triassic Liebre Mountain and Mill Creek Megaporphyry Bodies

Plate IV. Charts Showing Age and Correlation of Early Miocene Through Pleistocene Sedimentary Rocks Displaced by Various Strands of the San Andreas Fault System

Plate V.   Distribution and Geologic Relations of Fault Systems in the Vicinity of the Central Transverse Ranges, Southern California

Foreword

In its broadest definition, the San Andreas fault system is a complex array of late Cenozoic right- and left-lateral strike-slip faults that has deformed the western margin of North America between Cape Mendocino to the north and the Gulf of California to the south (Fig. 1). Associated with the strike-slip faults are myriad folds and reverse and normal faults that formed as the strike-slip fault blocks shifted within a broad zone of dextral shear along the continental margin (see Frontispiece, for example). The most prominent member of this fault system is the San Andreas fault proper that forms a continuous 1,100-km-long intracontinental link between the oceanic Mendocino fracture zone and the

system of spreading axes and transform faults in the Gulf of California and Salton trough (Fig. 2).

The San Andreas fault system is the most well studied of the world's great strike-slip fault systems; despite its complexity, many geologists consider it to be a satisfactorily understood geologic feature within the conceptual framework of plate tectonics. It is generally accepted that the San Andreas fault developed as a transform fault between the North American and Pacific plates that somehow has encroached into the continent. The continuity of the San Andreas fault proper is deceptively simple, however, and its evolution as an intracontinental transform cannot be fully

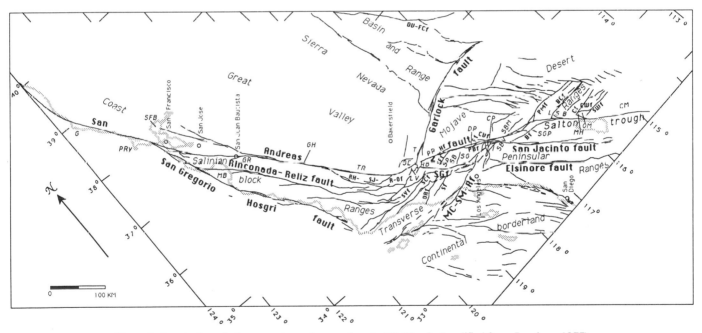

Figure 1. San Andreas fault system in southern and central California (modified from Jennings, 1977). Faults: Bf = Banning fault; BCf = Blue Cut fault; Cf = Chiriaco fault; CVf = Cajon Valley fault; CWf = Clemens Well fault; DV-FCf = Death Valley–Furnace Creek fault; Ff = Fenner fault; Hf = Hitchbrook fault; MC-SM-Rf = Malibu Coast–Santa Monica–Raymond fault; ORf = Oak Ridge fault; PBf = Punchbowl fault; PMf = Pinto Mountain fault; RH-SJ-R-Of = Red Hills–San Juan–Russell–Ozena fault; Sf = Simi fault; SCf = San Cayetano fault; SFf = San Francisquito fault; SGf = San Gabriel fault; SWf = Salton Wash fault; SYf = Santa Ynez fault. Physiographic features: CM = Chocolate Mountains; CP = Cajon Pass; DP = Devils Punchbowl; G = Gualala; GH = Gold Hill; GR = Gabilan Range; L = Liebre Mountain; LSB = Little San Bernardino Mountains; LV = Lockwood Valley; MB = Monterey Bay; MH = Mecca Hills; OM = Orocopia Mountains; PR = Portal Ridge; PRy = Point Reyes; RB = Ridge basin; SB = Soledad basin; SBM = San Bernardino Mountains; SBV = San Bernardino Valley; SE = San Emigdio Mountains; SFB = San Francisco Bay; SG = San Gabriel Mountains; SGP = San Gorgonio Pass; SP = Sierra Pelona; T = Tehachapi Mountains; TR = Temblor Range.

vii

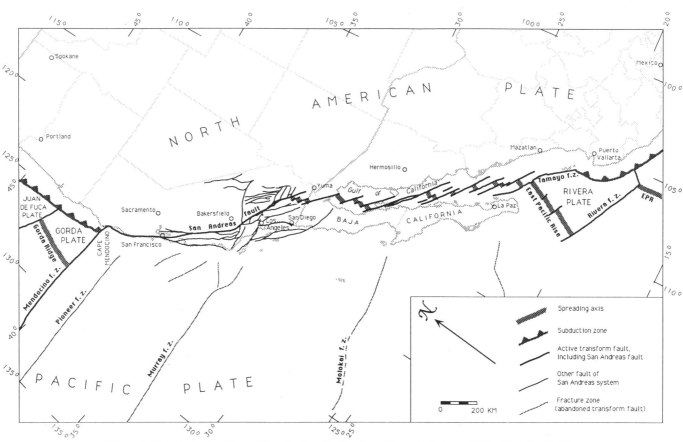

Figure 2. Plate tectonic setting of San Andreas fault system. Tectonic features: EPR = East Pacific Rise; f.z. = fracture zone. Sources include Krause (1965), Jennings (1977), Atwater (1989), and Lonsdale (1989).

understood without distinguishing the timing and magnitude of incremental displacements that have accumulated along various reaches of the fault and without comprehending the roles of other right-lateral, left-lateral, reverse, and normal faults in the San Andreas system.

The authors of the chapters in this volume critically examine the geologic evidence that constrains timing and magnitude of movement on various faults of the San Andreas system, and they develop and discuss paleogeologic reconstructions based on these constraints. The volume is presented to offer new insight into the evolution of the San Andreas fault system, to show that there are still areas of significant geologic controversy in understanding that evolution, and to stimulate further study of the geologic evolution of the fault system.

TECTONIC CONCEPTS AND MODELS: PROBLEMS AND PREDICAMENTS

In the early 1950s, Hill and Dibblee (1953) and Crowell (1952, 1954) revolutionized thinking about strike-slip faults in California when they presented convincing evidence for hundreds of kilometers of dextral displacement on the San Andreas fault and several tens of kilometers of dextral displacement on the San Gabriel fault. These landmark studies engendered similar studies further documenting large overall displacement on the major strike-slip faults in California, including not only the San Andreas and San Gabriel faults (for example, Bazeley, 1961; Crowell, 1962, 1975b, c, 1982; Dibblee, 1966; Addicott, 1968; Huffman, 1972; Clarke and Nilsen, 1973; Huffman and others, 1973; Bohannon, 1975; Ehlig and others, 1975; Matthews, 1976; Ehlig, 1981, 1982; Powell, 1981; Ehlig and Crowell, 1982; Nilsen, 1984; Ross, 1984; Frizzell and others, 1986; Stanley, 1987), but also other dextral faults such as the San Jacinto fault (Sharp, 1967; Bartholomew, 1970), the Red Hills–San Juan–Russell–Ozena fault (Schwade and others, 1958; Smith, 1977; Yeats and others, 1989), the Rinconada-Reliz fault (Durham, 1974; Dibblee, 1976; Graham, 1978; Ross, 1984), and the San Gregorio–Hosgri fault (Hall, 1975; Graham and Dickinson, 1978a, b; Nagel and Mullins, 1983; Clark and others, 1984), and sinistral faults such as the Garlock fault (Smith, 1962; Smith and Ketner, 1970; Davis and Burchfiel, 1973), the Malibu Coast–Santa Monica–Raymond fault (Barbat, 1958; Lamar, 1961; Yeats, 1968; Campbell and Yerkes, 1976), and the east-trending faults of the eastern Transverse Ranges (Dibblee, 1967, 1968, 1975; Hope, 1969; Powell, 1975, 1981, 1982; Bacheller, 1978).

Although most studies conducted subsequent to the early

1950s have confirmed that the San Andreas fault has accumulated hundreds of kilometers of dextral displacement and the San Gabriel fault has accumulated tens of kilometers, there are significant discrepancies among various measurements of displacement on these two faults. As studies on the San Andreas fault system have proliferated, persistent discrepancies have emerged not only among measurements of displacement north of the Transverse Ranges and among measurements in and south of the Transverse Ranges, but also between the two groups of measurements. Various measurements north of the Transverse Ranges between the southern tip of the Sierra Nevada and Cape Mendocino have indicated an overall displacement on Cretaceous and older rocks in the range of about 435 to 730 km (Hill and Dibblee, 1953; Dibblee, 1966; Hill and Hobson, 1968; Wentworth, 1968; Ross, 1970; Ross and others, 1973), whereas measurements in and south of the Transverse Ranges indicate displacement of no more than about 250 to 350 km on the San Andreas, San Gabriel, and other right-lateral faults (Crowell, 1962, 1975b, 1981; Bohannon, 1975; Ehlig and others, 1975; Ehlig, 1981, 1982; Powell, 1981).

Since the introduction of the transform-fault paradigm in 1965 (see below), measurements on the San Andreas fault proper between the Transverse Ranges and San Francisco Bay have converged into a range of about 295 to 325 km (Addicott, 1968; Huffman, 1972; Clarke and Nilsen, 1973; Huffman and others, 1973; Matthews, 1976; Nilsen, 1984; Ross, 1984; Stanley, 1987) for Eocene through early Miocene rocks, whereas most measurements on the San Andreas proper in and south of the Transverse Ranges have clustered in three distinct groups: one around 210 km (Crowell, 1962; Bohannon, 1975; Powell, 1981), another in the range of about 240 to 270 km (Crowell, 1975b, 1981; Ehlig and others, 1975; Ehlig, 1981, 1982), and a third in the range of about 150 to 165 km (Matti and others, 1985, p. 9; Frizzell and others, 1986). Other studies include measurements of about 100 km (Barrows and others, 1985) and 180 ± 20 km (Dillon, 1976). Measurements of large displacement on the San Gabriel fault range between 32 and 70 km, wherein widely discrepant measurements not uncommonly are based on the same rocks (Crowell, 1962, 1975c, 1982; Carman, 1964; Bohannon, 1975; Ehlig and others, 1975; Ehlig, 1981, 1982; Ehlig and Crowell, 1982; Ehlert, 1982).

The concept of a latest Cretaceous, Paleocene, and/or earliest Eocene proto–San Andreas fault was developed to explain the differences both between overall displacement of Cretaceous and older rocks (at least about 435 km to as much as 730 km) and Eocene through early Miocene rocks (about 300 km) north of the Transverse Ranges and between overall displacements of 435 to 730 km north of the Transverse Ranges and about 300 km in and south of the Transverse Ranges (Suppe, 1970; Dickinson and others, 1972; Gastil and others, 1972; Kistler and others, 1973; Howell, 1975; Nilsen and Clarke, 1975; Nilsen, 1978). In these studies, the proto–San Andreas fault was inferred to have diverged southward from the present trace of the San Andreas fault and transected the Salinian block, western Transverse Ranges, and southern California continental borderland.

Alternatively, other investigators (Gastil and others, 1972; Gastil and Jensky, 1973; Yeats, 1976; Dickinson, 1983) have postulated that the proto–San Andreas fault cut inland through the Transverse Ranges southward along or east of the future length of the Salton trough and Gulf of California. No throughgoing fault has been identified, however, that has the required dextral displacement in the range of about 100 to 400 km. Eventually, it was recognized and documented that the Salinian block was distended by late Cenozoic right-lateral faulting, chiefly by up to 150 km on the San Gregorio–Hosgri fault (Johnson and Normark, 1974; Graham, 1978; Graham and Dickinson, 1978a, b; Clark and others, 1984). This displacement on the San Gregorio–Hosgri fault, when fed onto the San Andreas fault north of San Francisco, largely accounts for the evidence for larger offset on that fault that had led to hypothesizing the existence of a proto–San Andreas fault.

With the advent of plate tectonics theory in the early 1960s, the concept that displacement on the San Andreas fault is related to rifting of Baja California from the North American continent (Wegener, 1924; Carey, 1958; Hamilton, 1961; Rusnak and others, 1964; Rusnak and Fisher, 1964; Hamilton and Myers, 1966) flourished. The new theory led to an attractively simple paradigm of the San Andreas fault as a transform fault that has sliced into the continent between segments of seafloor spreading ridges in the Gulf of California and in the Pacific Ocean northwest of Cape Mendocino (Wilson, 1965a, b). This new paradigm was embraced and enlarged on by investigators who marshalled supporting evidence from the ocean floor (Atwater, 1970; Larson and others, 1968; Moore and Buffington, 1968; Larson, 1972; Curray and Moore, 1984; Stock and Molnar, 1988; Lonsdale, 1989). The expanded paradigm incorporated the San Andreas fault and the transforms in the Gulf of California as a transform link between plate-boundary triple junctions near Cape Mendocino and the mouth of the gulf (McKenzie and Morgan, 1969; Atwater, 1970, 1989; Dickinson and Snyder, 1979).

In the transform-fault paradigm, the San Andreas fault system can have been active only since the time at which the Pacific and North American plates made contact and during those intervals in which relative motion occurred between the two plates. In most plate motion studies, the Pacific and North American plates are inferred to have made contact ca. 28 to 30 Ma and to have moved continuously, but at varying rates relative to one another since then (Atwater, 1970, 1989; Atwater and Molnar, 1973; Engebretson and others, 1985, p. 29–31; Stock and Molnar, 1988). Estimates of the relative velocity between the North American and Pacific plates over the last 4 to 5 m.y. range between 48 and 66 mm/yr based on seafloor magnetic anomalies (Minster and Jordan, 1978; DeMets and others, 1987).

In terms of rate and magnitude of displacement, the transform-fault paradigm encompasses two groups of models for the evolution of the San Andreas fault system. One group consists of relatively simple models in which the major right-lateral faults of the San Andreas fault system accommodate all the transform plate motion. For such models, the predicted slip rate for the San

Andreas is the relative velocity between the North American and Pacific plates, and the predicted displacement for the San Andreas is the product of the slip rate and the duration of fault movement. Models in this group are not consistent, however, with what is known about rates and magnitude of displacement on the San Andreas fault system. Not only are modern slip rates along the San Andreas fault (24 mm/yr southeast of its intersection with the San Jacinto fault and 34 to 37 mm/yr northwest of the intersection) unequal to the rate of relative movement along the transform boundary between the Pacific and North American plates (Minster and Jordan, 1984; Sieh and Jahns, 1984; Weldon and Sieh, 1985; Weldon, 1986; Humphreys and Weldon, 1991a), but also the overall displacement of about 300 to 450 km on the San Andreas fault system is only a fraction of the approximately 1,100 to 1,500 km of relative plate motion inferred from the magnetic signature of the ocean floor to have occurred since about 30 Ma (Stock and Molnar, 1988; Atwater, 1989, Fig. 23; reduced from earlier calculations of about 2,000 km by McKenzie and Morgan, 1969; Atwater, 1970).

In perhaps the simplest such model derived from the transform-fault paradigm, the San Andreas fault is inferred to have been an active transform fault that accompanied the opening of the Gulf of California. This model, which equates transform motion associated only with seafloor spreading in the Gulf of California to movement on the San Andreas fault (Wilson, 1965a, b; Atwater, 1970), requires that the fault system has accumulated its full 300-km displacement—equivalent to the 260 to 300 km by which Baja California has been rifted from mainland Mexico (Rusnak and others, 1964; Larson and others, 1968; Gastil and others, 1981; Curray and Moore, 1984; Lonsdale, 1989)—during the last 4 to 5 m.y. at a long-term average rate of about 60 mm/yr or more (Larson and others, 1968; Moore and Buffington, 1968; Larson, 1972; Curray and Moore, 1984). The San Andreas fault system, however, has clearly accumulated a significant fraction of its 300-km displacement within the continental plate between 5 Ma and at least about 17 to 18 Ma (Hill and Dibblee, 1953; Crowell, 1962, 1973, 1975b, c, 1981; Dibblee, 1966; Addicott, 1968; Dickinson and others, 1972; Huffman, 1972; Matthews, 1976; Powell, 1981, 1986; Stanley, 1987; Graham and others, 1989), thereby necessitating a lesser long-term average slip rate. Moreover, the plate motion rate measured in the Gulf of California and on the San Andreas fault system as a whole is only about 48 mm/yr (Weldon and Humphreys, 1986; Humphreys and Weldon, 1991a). These discrepancies in timing and rate between opening of the Gulf of California and overall movement on the San Andreas fault system renders a one-to-one link between the two events untenable and requires that inception of the San Andreas fault system predates the opening of the gulf.

The second group of transform models consists of more complicated scenarios, wherein the San Andreas fault system accommodates only part of the relative plate motion, with the remainder being accommodated by a broad zone of shear east of the San Andreas system, by transform faulting along the plate boundary, or by both mechanisms (Atwater, 1970, 1989; Atwater and Molnar, 1973; Garfunkel, 1973; Howell, 1976; Beck, 1986). In such models, the slip rate for the San Andreas is the relative velocity between the North American and Pacific plates minus the slip rate accommodated at the plate boundary or east of the San Andreas system.

Models in this group equate movement along dextral faults of the San Andreas fault system to seafloor spreading and transform motion over the last 30 m.y. (Atwater, 1970; Anderson, 1971; Atwater and Molnar, 1973; Blake and others, 1978; Dickinson, 1981; Irwin, 1990; Sedlock and Hamilton, 1991)—in the Gulf of California for the last 4 to 5 m.y., in the Pacific Ocean prior to about 12 Ma, and in one or the other or both between 12 and 5 Ma. Any such model requires the existence of faults with displacements larger than that of the San Andreas fault system to be concentrated offshore along the continental margin or dispersed inland. Such faults have yet to be identified.

Typically, transform models in the second group require large displacement on a throughgoing pre–10- to 12-Ma right-lateral fault that diverges southward from the present trace of the San Andreas fault in the Coast Ranges and either follows the continental margin offshore or in some way transects the southern Coast Ranges, western Transverse Ranges, and continental borderland. By about 10 to 12 Ma, this hypothetical fault would have been at least partially abandoned as transform motion stepped inland to the San Andreas fault north and south of the Transverse Ranges and the San Gabriel fault in the Transverse Ranges; at about 5 Ma, the San Gabriel fault was abandoned in favor of the San Andreas fault in the Transverse Ranges. The hypothetical pre–10- to 12-Ma fault has not, however, been identified as a throughgoing feature in the western Transverse Ranges, nor has the required displacement been documented on candidate faults in the Salinian block. Moreover, there is evidence for large pre–10-Ma displacement on dextral faults in the central and eastern Transverse Ranges and southernmost Salinian block (Smith, 1977; Powell, 1981; Joseph and others, 1982; Yeats and others, 1989).

In the most widely accepted model tailored to the transform-fault paradigm, approximately 300 km of rifting in the Gulf of California corresponds with right slip of about 300 km on the San Andreas system, which, in southern California, is divided between about 240 km on the San Andreas and San Jacinto faults and about 60 km on the San Gabriel fault (see Crowell, 1975b, 1981; Ehlig, 1981). For this model, a discrepancy exists in the timing of oceanic rifting in the gulf, which began 4 or 5 m.y. ago, and the timing of movement on the San Gabriel fault, which began at least 10 m.y. ago, although there is evidence for late Miocene extension and marine incursion in the gulf region (Moore and Buffington, 1968; Karig and Jensky, 1972; Moore, 1973; Ingle, 1974; Gastil and others, 1979; Curray and Moore, 1984) that some investigators have attributed to right-oblique extensional shear in the gulf region coeval with movement on the San Gabriel fault (Karig and Jensky, 1972; Crowell, 1979; Hausback, 1984; Stock and Hodges, 1989; Humphreys and Weldon,

1991b). Moreover, as is discussed in the following section, several chapters in this volume conclude that displacements on the San Andreas and San Gabriel faults are substantially less than the displacements of 240 and 60 km, respectively, required by the model.

The transform-fault paradigm has led to various tectonic models for the intracontinental evolution of the San Andreas fault system. Although each of these models has been constructed to satisfy significant geologic constraints, in general, the premises, requirements, and predictions of these models are not entirely consistent with all of the geologic evidence that has been garnered on land in California. As a result, important discrepancies exist between the various models and the geologic record. Resolution of these discrepancies, which is needed to advance our understanding of the San Andreas fault system, can be approached in several ways.

Resolution of some of the discrepancies between the geologic record and models generated from the transform-fault paradigm lies in part in demonstrating the most accurate displacements for the major strike-slip faults in the San Andreas system. Widely disparate measurements or model-based estimates of displacement have been published for many of the faults of the San Andreas system. Although it is evident that not all of the measurements of displacement can be valid, it is not necessarily obvious which measurements are the most accurate. Consequently, erroneous measurements of displacement are not infrequently incorporated into regional tectonic models. Documentation of overall displacement on various faults of the San Andreas system is a major emphasis in this volume.

Resolution of some discrepancies also lies in part in better understanding the roles played by other left- and right-lateral strike-slip faults, and, to a lesser extent, by associated reverse and normal faults in the evolution of the system—another important emphasis in this volume. In its simple version, the transform-fault paradigm does not include these complementary fault systems, and thus does not adequately address the kinematic complexity of the evolution of the intracontinental San Andreas fault system (Hill, 1971, 1974; Garfunkel, 1973; Powell, 1981, 1986; Minster and Jordan, 1984; Weldon and Humphreys, 1986; Humphreys and Weldon, 1991a). These faults are essential components not only in developing an overall present-day strain budget for the San Andreas as a plate boundary (Weldon and Humphreys, 1986; Humphreys and Weldon, 1991a), but also in maintaining long-term mass-balance in the palinspastic reconstruction of the San Andreas fault system. Sinistral faults include those in the Transverse Ranges and the Garlock fault; dextral faults include those in the Penisular Ranges, Mojave Desert, and Salinian block; and reverse faults include those in the western and central Transverse Ranges and southern Salinian block.

The roles played by these associated faults in the evolution of the San Andreas fault system have been approached in three ways in the literature, resulting in conflicting interpretations in some instances.

1. In attempting to understand the San Andreas and related faults, many geologists have addressed kinematic and mechanical connections between left- and right-lateral and reverse faults and bends in the San Andreas fault (for example, Eaton, 1939; Hill and Dibblee, 1953; Moody and Hill, 1956; Allen, 1957; Chinnery, 1966; Kupfer, 1968; Scholz and others, 1971; Davis and Burchfiel, 1973; Garfunkel, 1974; Dibblee, 1975, 1982; Bohannon and Howell, 1982; D. P. Hill, 1982; M. L. Hill, 1982; Bird and Rosenstock, 1984; Matti and others, 1985; Weldon and Humphreys, 1986; Meisling and Weldon, 1989). Many of these studies have attributed bending of the San Andreas fault and growth of associated left-lateral faults (including the Garlock and Pinto Mountain faults) and right-lateral faults (including the San Jacinto fault and faults in the Mojave Desert) to extension related to normal faulting in the Basin and Range province. As the bends developed, compressional stress fields led to the growth of contractional folding and reverse faulting in the Transverse Ranges and Salinian block.

2. In attempting to reconcile models for the long-term plate-tectonic role of the San Andreas fault with evidence obtained by geologic mapping along the fault, several investigators have incorporated left-lateral faults as integral coeval parts of the San Andreas fault system in southern California (Yeats, 1968; Anderson, 1971; Jahns, 1973; Campbell and Yerkes, 1976; Powell, 1986). In two of these publications (Campbell and Yerkes, 1976; Powell, 1986), left-lateral faulting along the southern margin of the Transverse Ranges west of the San Andreas fault is used to transfer transform motion on the pre–5-Ma San Andreas fault system in and north of the Transverse Ranges westward to the continental margin south of the Transverse Ranges. East of the Malibu Coast–Santa Monica–Raymond fault, the path and magnitude of displacement required by such a model has not been established unequivocally.

3. Other investigators have modeled regional tectonic relations between right-lateral faults of the San Andreas system and left-lateral faults of the Transverse Ranges by interpreting rotated paleomagnetic declination data (Luyendyk and others, 1980, 1985; Hornafius and others, 1986; Luyendyk and Hornafius, 1987). The tectonic models proposed in these chapters, however, incorporate undocumented large sinistral components of displacement on known reverse faults (including the San Cayetano, Oak Ridge, and Simi faults) in the western Transverse Ranges and displacements on the left-lateral faults (including the Pinto Mountain, Blue Cut, and Chiriaco faults) in the eastern Transverse Ranges that are much larger than the documented slip on those faults (Dibblee, 1967, 1968, 1975; Hope, 1969; Powell, 1975, 1981, 1982; Bacheller, 1978).

Further resolution of discrepancies can be gained by the study of sedimentary basins that have evolved prior to and during the evolution of the San Andreas fault system. The age (as determined from fossils, isotopic ages of intercalated volcanic rocks, or magnetostratigraphy) and provenance (as determined from clast population and paleocurrents) of sedimentary deposits along the various faults of the San Andreas system provide necessary constraints on incremental timing and magnitude of displacement. Published work includes studies in the informally named Late Cretaceous Gualala beds (Wentworth, 1968; Ross, 1970; Ross and others, 1973); Oligocene and early Miocene strata in the Transverse Ranges (Crowell, 1962; Bohannon, 1975; Woodburne, 1975; Dibblee, 1982); middle Miocene strata in the Transverse Ranges (Ehlig and others, 1975; Woodburne, 1975); and late Miocene and Pliocene strata in the Temblor Range (Fletcher, 1967; Huffman, 1972; Ryder and Thomson, 1989), Ridge basin (Crowell, 1952, 1975c, 1982; Ramirez, 1983), Devils Punchbowl area (Woodburne, 1975; Barrows and others, 1985; Barrows, 1987); and Miocene through Quaternary strata in the San Bernardino Mountains and adjacent parts of the western Mojave Desert (Gibson, 1971; Woodburne, 1975; Sadler and Demirer, 1986; Weldon, 1985, 1986; Meisling and Weldon, 1989). Additional observations and interpretations of these and other rock units are presented in several chapters in this volume.

CONTRIBUTIONS AND PERSPECTIVES IN THIS VOLUME

In the transform-fault paradigm, the San Andreas fault system is typically represented as one or two throughgoing right-lateral faults. As discussed above, however, a significant disparity exists between the relatively simple geometric relations predicted from on-land projection of seafloor tectonic features and the complexity of geologic relations actually observed along the San Andreas fault. This complexity is evident in the pattern of coeval left- and right-lateral and reverse faulting associated with the major right-lateral faults in southern and central California (see Frontispiece). Palinspastic reconstruction of pre–late and late Cenozoic paleogeologic patterns along the strike-slip faults from this region, together with evidence from Cenozoic sedimentary and volcanic rocks for the timing of these offsets, imposes kinematic constraints that contradict or are not satisfied completely by published models for the evolution of the San Andreas fault system. The chapters in this volume on the San Andreas fault system develop some of these constraints and elaborate on existing models or propose new models for the late Cenozoic tectonic evolution of California. These chapters provide significant new insight into the intracontinental evolution of a transform fault.

The 10 chapters in this volume deal with the problems of reassembling paleogeologic patterns disrupted by the evolving San Andreas system. As a group, the chapters examine the entire history of the San Andreas fault system from its inception in the early Miocene to the record of its Holocene activity.

Two chapters deal with reconstruction and evolution of the San Andreas fault north of the Transverse Ranges. These chapters address the reassembly of rocks in the San Juan Bautista area and Gabilan Range against rocks in the westernmost Mojave Desert and San Emigdio Mountains. In the published literature on displacement on the San Andreas fault, probably the most convincingly established links of paleogeologic features have been made between rocks of the San Juan Bautista area and the Gabilan Range west of the San Andreas fault and those of the San Emig-

dio and Tehachapi Mountains and Portal Ridge area east of the fault (Hill and Dibblee, 1953; Addicott, 1968; Huffman, 1972; Clarke and Nilsen, 1973; Huffman and others, 1973; Matthews, 1976; Nilsen, 1984; Ross, 1984). These linked features constitute a group of cross-fault ties that has withstood the scrutiny of continuing investigation. Although the magnitude of displacement measured by various components of the domain ranges from 275 to 320 km, few if any investigators dispute the validity of this cross-fault tie.

1. James, Kimbrough, and Mattinson present geochronologic evidence that supports reconstruction of rocks of the San Juan Bautista area and the San Emigdio Mountains, but refute reconstruction of these rocks with rocks in the Gualala area as proposed by Wentworth (1968) and by Ross (1970; see also Ross and others, 1973) and used by those authors to estimate displacement on the San Andreas fault of between about 435 and 570 km.

2. Sims elaborates on the palinspastic reconstruction of the San Juan Bautista–Gabilan Range rocks, the San Emigdio–Tehachapi–Portal Ridge rocks, and similar rocks found in slivers in the San Andreas fault zone such as that which includes Gold Hill; he infers the sequence of displacement on the strands that bound the slivers to elucidate the evolution of the San Andreas fault north of the Transverse Ranges.

The rest of the chapters in the volume deal with reconstruction of the San Andreas fault system in and south of the Transverse Ranges. Three of these chapters present studies of Miocene strata and their implications for reconstruction and evolution of the San Andreas fault. These chapters deal with several sedimentary basins that are especially important in understanding the evolution of the San Andreas fault system in southern California. One of the key cross-fault ties used to establish overall displacement in southern California has been provided by restoring the Diligencia Formation of Crowell (1975a), the Vasquez Formation (Noble, 1953), and the Plush Ranch Formation of Carman (1964) into proximity (Crowell, 1962, 1975b; Bohannon, 1975; Dibblee, 1982). Similarly, the Mill Creek Formation of Gibson (1971), situated within the San Andreas fault zone in the southern San Bernardino Mountains, has been used to constrain the magnitude and timing of offset along the San Andreas fault (Gibson, 1971; Matti and others, 1985; Sadler and Demirer, 1986). In addition, the Santa Ana Sandstone of Vaughan (1922) in the central San Bernardino Mountains has recently yielded sedimentologic data that illuminates the evolution of the San Andreas system (Sadler and Demirer, 1986).

1. Frizzell and Weigand present new geochemical and geochronologic evidence in support of correlating the early Miocene volcanic rocks of the Plush Ranch Formation in Lockwood Valley west of the San Gabriel and San Andreas faults, the Vasquez Formation in Soledad basin between the San Gabriel and San Andreas faults, and the Diligencia Formation in the Orocopia Mountains east of the San Andreas fault.

2. Sadler, Demirer, West, and Hillenbrand present a stratigraphic, sedimentologic, and provenance interpretation of the

Mill Creek Formation and infer that the Mill Creek strata were deposited in a middle or late Miocene pull-apart basin situated at a right step between dextral faults of the San Andreas system, probably along the Clemens Well fault zone.

3. Sadler presents a stratigraphic, sedimentologic, and provenance interpretation of the Santa Ana Sandstone, for which he documents a middle to late Miocene age, and discusses implications for the evolution of the San Andreas fault, across which some of the Santa Ana clasts were derived.

Three of the southern California chapters focus on the evidence and interpretation for overall displacement on the Salton trough and Mojave Desert segments of the San Andreas fault. In different ways, these chapters address controversial discrepancies between the transform-fault paradigm and the geologic record.

1. Dillon and Ehlig present evidence and interpretation for overall reconstruction and evolution of the Salton trough and Mojave Desert segments of the San Andreas fault, the Banning fault, and the San Gabriel fault.

2. Matti and Morton present evidence and interpretation for overall reconstruction and evolution of the Mojave Desert and Salton trough segments of the San Andreas fault, the various strands of the San Andreas fault in the southern San Bernardino Mountains, the Punchbowl fault, the San Gabriel fault, and left-lateral faults along the southern margin of the Transverse Ranges west of the San Andreas fault.

3. Weldon, Meisling, and Alexander document interrelations between terrestrial sedimentation and strike-slip faulting in the Transverse Ranges, and they interpret the Miocene through Holocene evolution of the San Andreas fault, including thorough discussion of Pliocene, Pleistocene, and Holocene slip rates on the San Andreas fault in the Transverse Ranges.

The role of the San Jacinto fault in the evolution of the San Andreas fault system is also controversial. In a chapter that interprets the evolution of the San Jacinto fault and its interaction with the San Andreas fault, Morton and Matti discuss Quaternary slip rates on the San Jacinto and San Andreas faults and propose that the San Jacinto fault formed and evolved in the Pleistocene.

A significant test of the validity of any measurement of fault displacement lies in assessing whether the restored distribution of displaced rock masses forms a coherent paleogeologic domain devoid of unexplained gaps and overlaps. In a chapter that deals with the entire San Andreas fault system, Powell presents evidence and interpretation for a balanced overall palinspastic reconstruction of all right- and left-lateral strike-slip faults of the San Andreas system that disrupt the granitic crust of southern and central California.

The contributions in this volume address aspects of the geology of California that do not fit the simple plate-tectonic paradigm for the evolution of the San Andreas fault system. These aspects include evidence for magnitude and timing of displacements that are inconsistent with those required by the plate-tectonic paradigm. The chapters in this collection present various interpretations of parts or all of the San Andreas fault system. To

varying degrees, many of the chapters challenge commonly accepted interpretations of the San Andreas system. As seems appropriate for such a complex feature, interpretations on some topics are widely diverse in this volume, yet, promisingly, interpretation of other topics appears to be converging in a new direction. Each of these topics is controversial and offers promising areas for further research:

1. Dillon and Ehlig present a version of the commonly accepted interpretation of displacement on the San Andreas fault system in southern California, in which a total slip of about 300 km is partitioned between about 240 km on the San Andreas fault zone and about 60 km on the San Gabriel fault.

2. The authors of the four chapters that deal with reconstruction of the San Andreas fault system in and south of the Transverse Ranges present evidence for overall displacement on the San Andreas fault proper in southern California that is significantly less than the widely published figure of about 240 km. Dillon and Ehlig present evidence for an overall displacement of 185 ± 20 km on the Salton trough segment of the San Andreas fault, although they advocate an overall displacement of 240 km on the San Andreas fault zone including the Banning fault. Matti and Morton present evidence for an overall displacement of about 160 km on the Mojave Desert segment of the San Andreas fault since the latest Miocene and infer a similar displacement for the Salton trough segment. Weldon, Meisling, and Alexander present evidence for a displacement of about 150 km on the Mojave Desert segment of the San Andreas. Powell incorporates a displacement of about 160 km on both the Salton trough and Mojave Desert segments of the San Andreas fault. All of these authors agree that this displacement has accumulated since the late Miocene, that is, since about 5 Ma.

3. Powell and Sims present considerable evidence that about 100 km of early and middle Miocene strike-slip movement occurred on the Clemens Well–Fenner–San Francisquito fault in and south of the Transverse Ranges and on the San Andreas fault north of the Transverse Ranges. Powell shows that about one-third of the displacement on the San Andreas fault system in southern California occurred on the pre–13-Ma Clemens Well–Fenner–San Francisquito fault.

4. The chapters of Powell, of Sims, and of James, Kimbrough, and Mattinson all corroborate that early Miocene units and crystalline basement rocks in southern and central California have been displaced by no more than about 300 km on the San Andreas fault and about 150 km on the San Gregorio–Hosgri fault. The coherence of the paleogeologic patterns restored by reversing slip on the faults of the late Cenozoic San Andreas system disallows the existence of any hypothetical Mesozoic or early Cenozoic proto–San Andreas fault that transects the granitic crust of California.

5. In different ways, three of the four chapters that address displacement on the San Gabriel fault challenge currently accepted interpretations of that fault. Dillon and Ehlig adopt the commonly held view that about 60 km of right slip passes from the north and south branches of the San Gabriel fault onto the

San Andreas fault between Cajon Pass and the north end of the Salton trough. Matti and Morton conclude that only 44 km of displacement passes from the north and south branches of the San Gabriel fault southeastward onto the Banning fault. Matti and Morton and Weldon, Meisling, and Alexander postulate that the Cajon Valley and Hitchbrook faults of the western Mojave Desert represent the offset northern extension of the San Gabriel fault. Weldon, Meisling, and Alexander and also Powell discuss evidence that seems to preclude much more than about 22 to 23 km of displacement on the San Gabriel fault rejoining the San Andreas fault to the southeast; they independently develop (but do not fully accept) the concept that displacement on the San Gabriel fault diminishes southeastward and steps over to a hypothetical fault farther southeast approximately along the paths of the Punchbowl and modern San Andreas faults. Powell concludes that the San Gabriel fault system has a maximum of only 42 to 45 km of displacement, that only 22 to 23 km of the right slip of the San Gabriel fault system reaches to and possibly through San Gorgonio Pass, and that most, if not all, of the displacement is absorbed by left-oblique extension between the Transverse Ranges and the Peninsular Ranges.

6. Four chapters discuss the intersections of the San Andreas, San Jacinto, and Punchbowl faults, and interpret their interaction from different perspectives. Two of them, by Morton and Matti and by Matti and Morton, emphasize that the San Jacinto fault cannot be mapped at the surface as continuous with either the Punchbowl or San Andreas fault; they propose that right slip on the San Jacinto fault is accommodated both by a right-stepping slip-transfer mechanism across the San Bernardino Valley between the San Jacinto and San Andreas faults, accompanied by extension in the intervening basin, and by accommodation on reverse and left-oblique faulting in the southeastern San Gabriel Mountains. Weldon, Meisling, and Alexander transfer right slip on the San Jacinto fault to the San Andreas fault both across the San Bernardino Valley and through the eastern San Gabriel Mountains via the Punchbowl fault; they explain uplift of and reverse faulting in the eastern San Gabriel Mountains by the restraining geometry of the junction of the San Jacinto and San Andreas faults. Matti and Morton, in their reconstruction, infer that about 20 km of right slip has been transferred from the San Jacinto to the San Andreas fault across the San Bernardino Valley; they further infer that about 5 km has been absorbed in compressional tectonics in the southeastern San Gabriel Mountains and in extensional tectonics in the San Bernardino Valley. Powell, in his reconstruction, transfers about half the overall right slip on the San Jacinto fault to the San Andreas fault and half to left-lateral faults in the southeastern San Gabriel Mountains.

7. Three chapters that focus on the significance of the middle(?) and late Miocene Mill Creek Formation and the late Miocene Punchbowl Formation (Noble, 1954) in the evolving San Andreas fault system present partially conflicting interpretations of the original position and tectonic significance of the basins in the San Andreas system. Sadler, Demirer, West, and Hillenbrand infer that the Mill Creek strata were deposited in a pull-apart basin situated at a right step between dextral faults of the Clemens Well–Fenner–San Francisquito fault zone. Matti and Morton reassemble the Mill Creek and Punchbowl Formations together between the Little San Bernardino Mountains to the east and the Liebre Mountains–Sierra Pelona–Soledad basin block to the west; they conclude that these, together with late Miocene sedimentary rocks of San Gorgonio Pass and the Mecca Hills, constitute a northwest-trending belt of late Miocene basins. These authors also suggest that these basins define a Salton trough-like feature that is contemporaneous with movement on the San Gabriel fault, but is aligned along the future trace of the San Andreas fault. Weldon, Meisling, and Alexander conclude that the evidence is not sufficient to distinguish between associating the Mill Creek and Punchbowl basins with waning phases of the Clemens Well–Fenner–San Francisquito fault or with an early southern segment of the San Andreas fault, perhaps including an early phase of the Punchbowl fault.

8. Five of the chapters provide a consensus basis for a new interpretation of the evolution of the San Andreas fault system. The conclusions of Powell, of Matti and Morton, of Weldon, Meisling, and Alexander, and of Sims are consistent with an evolutionary model in which about 100 km of displacement occurred on the Clemens Well–Fenner–San Francisquito–San Andreas fault prior to about 13 Ma, about 42 to 45 km of displacement occurred on the San Gabriel–San Andreas fault between about 12 to 13 and 5 Ma, and about 150 to 165 km of displacement occurred on the San Andreas fault since about 5 Ma. Powell concludes that, unlike the modern San Andreas fault, the Clemens Well–Fenner–San Francisquito–San Andreas fault did not connect with an extensional domain in the region now occupied by the Salton trough and Gulf of California. Matti and Morton conclude that no more than about 22 to 23 km of displacement on the San Gabriel fault entered the nascent Salton trough and Gulf of California, whereas Powell suggests that the San Gabriel fault may not have connected at all with the Salton trough or Gulf of California. Thus, the San Andreas fault was not unequivocally a throughgoing intracontinental plate-boundary transform fault prior to the emergence of the modern San Andreas fault at about 5 Ma, and even since then the slip rate on the San Andreas fault itself has not equalled the relative rate of movement between the North American and Pacific plates.

The editors of and the contributors to this volume believe that the time has come for this new look at the San Andreas fault system from the continental perspective of the geology of southern and central California. We hope that we have shown that the current understanding of the San Andreas fault system is far from complete either in detail or in grand scheme; we further hope that the evidence and ideas presented in this volume provide an impetus for additional study.

Robert E. Powell

REFERENCES CITED

Addicott, W. O., 1968, Mid-Tertiary zoogeographic and paleogeographic discontinuities across the San Andreas fault, California, *in* Dickinson, W. R., and Grantz, A., eds., Proceedings of conference on geologic problems of San Andreas fault system: Stanford, California, Stanford University Publications in the Geological Sciences, v. 11, p. 144–165.

Allen, C. R., 1957, San Andreas fault zone in San Gorgonio Pass, southern California: Geological Society of America Bulletin, v. 68, p. 315–349.

Anderson, D. L., 1971, The San Andreas fault: Scientific American, v. 225, no. 5, p. 52–68.

Atwater, T., 1970, Implications of plate tectonics for the Cenozoic tectonic evolution of western North America: Geological Society of America Bulletin, v. 81, p. 3513–3535.

—— , 1989, Plate tectonic history of the northeast Pacific and western North America, *in* Winterer, E. L., Hussong, D. M., and Decker, R. W., eds., The eastern Pacific Ocean and Hawaii: Boulder, Colorado, Geological Society of America, The Geology of North America, v. N, p. 21–72.

Atwater, T., and Molnar, P., 1973, Relative motion of the Pacific and North American plates deduced from sea-floor spreading in the Atlantic, Indian, and South Pacific oceans, *in* Kovach, R. L., and Nur, A., eds., Proceedings of the conference on tectonic problems of the San Andreas fault system: Stanford, California, Stanford University Publications in the Geological Sciences, v. 13, p. 136–148.

Bacheller, J., III, 1978, Quaternary geology of the Mojave Desert–eastern Transverse Ranges boundary in the vicinity of Twentynine Palms, California [M.S. thesis]: Los Angeles, University of California, 157 p.

Barbat, W. F., 1958, The Los Angeles Basin area, California, *in* Weeks, L. G., ed., Habitat of oil; A symposium: Tulsa, Oklahoma, American Association of Petroleum Geologists, p. 62–77.

Barrows, A. G., 1987, Geology of the San Andreas fault zone and adjoining terrane, Juniper Hills and vicinity, Los Angeles County, California, *in* Hester, R. L., and Hallinger, D. F., eds., San Andreas fault—Cajon Pass to Palmdale: Pacific Section, American Association of Petroleum Geologists Volume and Guidebook 59, p. 93–157.

Barrows, A. G., Kahle, J. E., and Beeby, D. J., 1985, Earthquake hazards and tectonic history of the San Andreas fault zone, Los Angeles County, California: U.S. Geological Survey Final Technical Report, Contract No. 14-08-0001-19193, 126 p., map scale 1:12,000. Also published in: 1985, California Division of Mines and Geology Open-File Report 85-10LA, 236 p., scale 1:12,000; and 1987, Pacific Section, American Association of Petroleum Geologists Volume and Guidebook 59, p. 1–92.

Bartholomew, M. J., 1970, San Jacinto fault zone in the northern Imperial Valley, California: Geological Society of America Bulletin, v. 81, p. 3161–3166.

Bazeley, W.J.M., 1961, 175 miles of lateral movement along the San Andreas fault since lower Miocene?: Pacific Petroleum Geologists News Letter, v. 15, no. 5, p. 2–3.

Beck, M. E., Jr., 1986, Model for late Mesozoic–early Tertiary tectonics of coastal California and western Mexico and speculations on the origin of the San Andreas fault: Tectonics, v. 5, p. 49–64.

Bird, P., and Rosenstock, R. W., 1984, Kinematics of present crust and mantle flow in southern California: Geological Society of America Bulletin, v. 95, p. 946–957.

Blake, M. C., Jr., Campbell, R. H., Dibblee, T. W., Jr., Howell, D. G., Nilsen, T. H., Normark, W. R., Vedder, J. C., and Silver, E. A., 1978, Neogene basin formation in relation to plate-tectonic evolution of San Andreas fault system, California: American Association of Petroleum Geologists Bulletin, v. 62, p. 344–372.

Bohannon, R. G., 1975, Mid-Tertiary conglomerates and their bearing on Transverse Range tectonics, southern California, *in* Crowell, J. C., ed., San Andreas fault in southern California: California Division of Mines and Geology Special Report 118, p. 75–82.

Bohannon, R. G., and Howell, D. G., 1982, Kinematic evolution of the junction of the San Andreas, Garlock, and Big Pine faults, California: Geology, v. 10,

no. 7, p. 358–363.

Campbell, R. H., and Yerkes, R. F., 1976, Cenozoic evolution of the Los Angeles basin area-relation to plate tectonics, *in* Howell, D. G., ed., Aspects of the geologic history of the California continental borderland: Pacific Section, American Association of Petroleum Geologists Miscellaneous Publication 24, p. 541–558.

Carey, S. W., 1958, A tectonic approach to continental drift, *in* Carey, S. W., ed., Continental drift; A symposium: Hobart, Australia, University of Tasmania Geology Department, p. 177–355.

Carman, M. F., Jr., 1964, Geology of the Lockwood Valley area, Kern and Ventura Counties, California: California Division of Mines and Geology Special Report 81, 62 p.

Chinnery, M. A., 1966, Secondary faulting II. Geological aspects: Canadian Journal of Earth Sciences, v. 3, p. 175–190.

Clark, J. C., Brabb, E. E., Greene, H. G., and Ross, D. C., 1984, Geology of Point Reyes peninsula and implications for San Gregorio fault history, *in* Crouch, J. K., and Bachman, S. B., eds., Tectonics and sedimentation along the California margin: Pacific Section, Society of Economic Paleontologists and Mineralogists, v. 38, p. 67–85.

Clarke, S. H., Jr., and Nilsen, T. H., 1973, Displacement of Eocene strata and implications for the history of offset along the San Andreas fault, central and northern California, *in* Kovach, R. L., and Nur, A., eds., Proceedings of the conference on tectonic problems of the San Andreas fault system: Stanford, California, Stanford University Publications in the Geological Sciences, v. 13, p. 358–367.

Crowell, J. C., 1952, Probable large lateral displacement on San Gabriel fault, southern California: American Association of Petroleum Geologists Bulletin, v. 36, p. 2026–2035.

—— , 1954, Strike-slip displacement on the San Gabriel fault, southern California, *in* Jahns, R. H., ed., Geology of southern California: California Division of Mines Bulletin 170, ch. 4, p. 49–52.

—— , 1962, Displacement along the San Andreas fault, California: Geological Society of America Special Paper 71, 61 p.

—— , 1973, Problems concerning the San Andreas fault system in southern California, *in* Kovach, R. L., and Nur, A., eds., Proceedings of the conference on tectonic problems of the San Andreas fault system: Stanford, California, Stanford University Publications in the Geological Sciences, v. 13, p. 125–135.

—— , 1975a, Geologic sketch of the Orocopia Mountains, southeastern California, *in* Crowell, J. C., ed., San Andreas fault in southern California: California Division of Mines and Geology Special Report 118, p. 99–110.

—— , 1975b, The San Andreas fault in southern California, *in* Crowell, J. C., ed., San Andreas fault in southern California: California Division of Mines and Geology Special Report 118, p. 7–27.

—— , 1975c, The San Gabriel fault and Ridge basin, southern California, *in* Crowell, J. C., ed., San Andreas fault in southern California: California Division of Mines and Geology Special Report 118, p. 208–219.

—— , 1979, The San Andreas fault system through time: Journal of the Geological Society of London, v. 136, p. 293–302.

—— , 1981, An outline of the tectonic history of southeastern California, *in* Ernst, W. G., ed., The geotectonic development of California; Rubey Volume 1: Englewood Cliffs, New Jersey, Prentice-Hall, p. 583–600.

—— , 1982, The tectonics of Ridge basin, southern California, *in* Crowell, J. C., and Link, M. H., eds., Geologic history of Ridge basin, southern California: Pacific Section, Society of Economic Paleontologists and Mineralogists, p. 25–42.

Curray, J. R., and Moore, D. G., 1984, Geologic history of the mouth of the Gulf of California, *in* Crouch, J. K., and Bachman, S. B., eds., Tectonics and sedimentation along the California margin: Pacific Section, Society of Economic Paleontologists and Mineralogists, v. 38, p. 17–35.

Davis, G. A., and Burchfiel, B. C., 1973, Garlock fault: An intracontinental transform structure, southern California: Geological Society of America Bulletin, v. 84, p. 1407–1422.

DeMets, C., Gordon, R. G., Stein, S., and Argus, D. F., 1987, A revised estimate of Pacific–North America motion and implications for western North America plate boundary zone tectonics: Geophysical Research Letters, v. 14, p. 911–914.

Dibblee, T. W., Jr., 1966, Evidence for cumulative offset on the San Andreas fault in central and northern California, *in* Bailey, E. H., ed., Geology of northern California: California Division of Mines and Geology Bulletin 190, p. 375–384.

——, 1967, Evidence of major lateral displacement on the Pinto Mountain fault, southeastern California: Geological Society of America Abstracts for 1967, Special Paper 115, p. 322.

——, 1968, Displacements on the San Andreas fault system in the San Gabriel, San Bernardino, and San Jacinto Mountains, southern California, *in* Dickinson, W. R., and Grantz, A., eds., Proceedings of conference on geologic problems of San Andreas fault system: Stanford, California, Stanford University Publications in the Geological Sciences, v. 11, p. 260–278.

——, 1975, Late Quaternary uplift of the San Bernardino Mountains on the San Andreas and related faults, *in* Crowell, J. C., ed., San Andreas fault in southern California: California Division of Mines and Geology Special Report 118, p. 127–135.

——, 1976, The Rinconada and related faults in the southern Coast Ranges, California, and their tectonic significance: U.S. Geological Survey Professional Paper 981, 55 p.

——, 1982, Regional geology of the Transverse Ranges province of southern California, *in* Fife, D. L., and Minch, J. A., eds., Geology and mineral wealth of the California Transverse Ranges; Mason Hill Volume: Santa Ana, California, South Coast Geological Society, Annual Symposium and Guidebook 10, p. 7–26.

Dickinson, W. R., 1981, Plate tectonics and the continental margin of California, *in* Ernst, W. G., ed., The geotectonic development of California; Rubey Volume 1: Englewood Cliffs, New Jersey, Prentice-Hall, p. 1–28.

——, 1983, Cretaceous sinistral strike slip along Nacimiento fault in coastal California: American Association of Petroleum Geologists Bulletin, v. 67, p. 624–645.

Dickinson, W. R., and Snyder, W. S., 1979, Geometry of triple junctions related to the San Andreas transform: Journal of Geophysical Research, v. 84, no. B2, p. 561–572.

Dickinson, W. R., Cowan, D. S., and Schweikert, R. A., 1972, Discussion of "Test of new global tectonics": American Association of Petroleum Geologists Bulletin, v. 56, p. 375–384.

Dillon, J. T., 1975, Geology of the Chocolate and Cargo Muchacho Mountains, southeasternmost California [Ph.D. thesis]: Santa Barbara, University of California, 405 p.

Durham, D. L., 1974, Geology of the southern Salinas Valley area, California: U.S. Geological Survey Professional Paper 819, 111 p.

Eaton, J. E., 1939, Ridge basin, California: American Association of Petroleum Geologists Bulletin, v. 23, p. 517–558.

Ehlert, K. W., 1982, Basin analysis of the Miocene Mint Canyon Formation, southern California, *in* Ingersoll, R. V., and Woodburne, M. O., eds., Cenozoic nonmarine deposits of California and Arizona: Pacific Section, Society of Economic Paleontologists and Mineralogists, p. 51–64.

Ehlig, P. L., 1981, Origin and tectonic history of the basement terrane of the San Gabriel Mountains, central Transverse Ranges, *in* Ernst, W. G., ed., The geotectonic development of California; Rubey Volume 1: Englewood Cliffs, New Jersey, Prentice-Hall, p. 253–283.

——, 1982, The Vincent thrust; Its nature, paleogeographic reconstruction across the San Andreas fault and bearing on the evolution of the Transverse Ranges, *in* Fife, D. L., and Minch, J. A., eds., Geology and mineral wealth of the California Transverse Ranges; Mason Hill Volume: Santa Ana, California, South Coast Geological Society Annual Symposium and Guidebook 10, p. 370–379.

Ehlig, P. L., and Crowell, J. C., 1982, Mendenhall Gneiss and anorthosite-related rocks bordering Ridge basin, southern California, *in* Crowell, J. C., and

Link, M. H., eds., Geologic history of Ridge basin, southern California: Pacific Section, Society of Economic Paleontologists and Mineralogists, p. 199–202.

Ehlig, P. L., Ehlert, K. W., and Crowe, B. M., 1975, Offset of the upper Miocene Caliente and Mint Canyon Formations along the San Gabriel and San Andreas faults, *in* Crowell, J. C., ed., San Andreas fault in southern California: California Division of Mines and Geology Special Report 118, p. 83–92.

Engebretson, D. C., Cox, A., and Gordon, R. G., 1985, Relative motions between oceanic and continental plates in the Pacific basin: Geological Society of America Special Paper 206, 59 p.

Fletcher, G. L., 1967, Post late Miocene displacement along the San Andreas fault zone, central California, *in* Gabilan Range and adjacent San Andreas fault: Pacific Sections, American Association of Petroleum Geologists and Society of Economic Paleontologists and Mineralogists Guidebook, p. 74–80.

Frizzell, V. A., Jr., Mattinson, J. M., and Matti, J. C., 1986, Distinctive Triassic megaporphyritic monzogranite; Evidence for only 160 km offset along the San Andreas fault, southern California: Journal of Geophysical Research, v. 91, no. B14, p. 14080–14088.

Garfunkel, Z., 1973, History of the San Andreas fault as a plate boundary: Geological Society of America Bulletin, v. 84, p. 2035–2042.

——, 1974, Model for the late Cenozoic tectonic history of the Mojave Desert, California, and for its relation to adjacent regions: Geological Society of America Bulletin, v. 85, p. 1931–1944.

Gastil, R. G., and Jensky, W., 1973, Evidence for strike-slip displacement beneath the trans-Mexican volcanic belt, *in* Kovach, R. L., and Nur, A., eds., Proceedings of the conference on tectonic problems of the San Andreas fault system: Stanford, California, Stanford University Publications in the Geological Sciences, v. 13, p. 171–180.

Gastil, G., Phillips, R. P., and Rodriguez-Torres, R., 1972, The reconstruction of Mesozoic California: Montreal, Canada, International Geological Congress, Section 3, 24th Session, p. 217–229.

Gastil, G., Krummenacher, D., and Minch, J., 1979, The record of Cenozoic volcanism around the Gulf of California: Geological Society of America Bulletin, Part I, v. 90, p. 839–857.

Gastil, G., Morgan, G., and Krummenacher, D., 1981, The tectonic history of peninsular California and adjacent Mexico, *in* Ernst, W. G., ed., The geotectonic development of California; Rubey Volume 1: Englewood Cliffs, New Jersey, Prentice-Hall, p. 284–306.

Gibson, R. C., 1971, Non-marine turbidites and the San Andreas fault, San Bernardino Mountains, California, *in* Elders, W. A., ed., Geological excursions in southern California; Geological Society of America Cordilleran Section Annual Meeting Guidebook: Riverside, University of California Campus Museum Contributions 1, p. 167–181.

Graham, S. A., 1978, Role of Salinian block in evolution of San Andreas fault system, California: American Association of Petroleum Geologists Bulletin, v. 62, p. 2214–2231.

Graham, S. A., and Dickinson, W. R., 1978a, Apparent offsets of on-land geologic features across the San Gregorio–Hosgri fault trend, *in* Silver, E. A., and Normark, W. R., eds., San Gregorio–Hosgri fault zone, California: California Division of Mines and Geology Special Report 137, p. 13–23.

——, 1978b, Evidence for 115 kilometers of right slip on the San Gregorio–Hosgri fault trend: Science, v. 199, no. 4325, p. 179–181.

Graham, S. A., Stanley, R. G., Bent, J. V., and Carter, J. B., 1989, Oligocene and Miocene paleogeography of central California and displacement along the San Andreas fault: Geological Society of America Bulletin, v. 101, p. 711–730.

Hall, C. A., Jr., 1975, San Simeon–Hosgri fault system, coastal California; Economic and environmental implications: Science, v. 190, p. 1291–1294.

Hamilton, W., 1961, Origin of the Gulf of California: Geological Society of America Bulletin, v. 72, p. 1307–1318.

Hamilton, W., and Myers, W. B., 1966, Cenozoic tectonics of the western United States: Reviews of Geophysics, v. 4, no. 4, p. 509–549.

Hausback, B. P., 1984, Cenozoic volcanic and tectonic evolution of Baja California Sur, Mexico, *in* Frizzell, V. A., Jr., ed., Geology of the Baja California Peninsula: Pacific Section, Society of Economic Paleontologists and Mineralogists, v. 39, p. 219–236.

Hill, D. P., 1982, Contemporary block tectonics: California and Nevada: Journal of Geophysical Research, v. 87, no. B7, p. 5433–5450.

Hill, M. L., 1971, A test of new global tectonics; Comparison of northeast Pacific and California structures: American Association of Petroleum Geologists Bulletin, v. 55, p. 3–9.

——, 1974, Is the San Andreas a transform fault?: Geology, v. 2, no. 11, p. 535–536.

——, 1982, Anomalous trends of the San Andreas fault in the Transverse Ranges, California, *in* Fife, D. L., and Minch, J. A., eds., Geology and mineral wealth of the California Transverse Ranges; Mason Hill Volume: Santa Ana, California, South Coast Geological Society Annual Symposium and Guidebook 10, p. 367–369.

Hill, M. L., and Dibblee, T. W., Jr., 1953, San Andreas, Garlock, and Big Pine faults; A study of the character, history, and tectonic significance of their displacements: Geological Society of America Bulletin, v. 64, p. 443–458.

Hill, M. L., and Hobson, H. D., 1968, Possible post-Cretaceous slip on the San Andreas fault zone, *in* Dickinson, W. R., and Grantz, A., eds., Proceedings of conference on geologic problems of San Andreas fault system: Stanford, California, Stanford University Publications in the Geological Sciences, v. 11, p. 123–129.

Hope, R. A., 1969, The Blue Cut fault, southeastern California: U.S. Geological Survey Professional Paper 650-D, p. D116–D121.

Hornafius, J. S., Luyendyk, B. P., Terres, R. R., and Kamerling, M. J., 1986, Timing and extent of Neogene tectonic rotation in the western Transverse Ranges, California: Geological Society of America Bulletin, v. 97, p. 1476–1487.

Howell, D. G., 1975, Hypothesis suggesting 700 km of right slip in California along northwest-oriented faults: Geology, v. 3, no. 2, p. 81–83.

——, 1976, A model to accommodate 1000 kilometres of right-slip, Neogene, displacement in the southern California area, *in* Howell, D. G., ed., Aspects of the geologic history of the California continental borderland: Pacific Section, American Association of Petroleum Geologists Miscellaneous Publication 24, p. 530–540.

Huffman, O. F., 1972, Lateral displacement of upper Miocene rocks and the Neogene history of offset along the San Andreas fault in central California: Geological Society of America Bulletin, v. 83, p. 2913–2946.

Huffman, O. F., Turner, D. L., and Jack, R. N., 1973, Offset of late Oligocene-early Miocene volcanic rocks along the San Andreas fault in central California, *in* Kovach, R. L., and Nur, A., eds., Proceedings of the conference on tectonic problems of the San Andreas fault system: Stanford, California, Stanford University Publications in the Geological Sciences, v. 13, p. 368–373.

Humphreys, E. D., and Weldon, R. J., II, 1991a, Deformation across southern California; A local determination of Pacific–North America relative plate motion: Journal of Geophysical Research (in press).

——, 1991b, Kinematic constraints on the rifting of Baja California, *in* Dauphin, J. P., and Simoneit, B.R.T., eds., The gulf and peninsular province of the Californias: American Association of Petroleum Geologists Memoir 47 (in press).

Ingle, J. C., Jr., 1974, Paleobathymetric history of Neogene marine sediments, northern Gulf of California, *in* Gastil, G., and Lillegraven, J., eds., Geology of peninsular California: Pacific Sections, American Association of Petroleum Geologists, Society of Economic Paleontologists and Mineralogists, and Society of Exploration Geophysicists Guidebook, p. 121–138.

Irwin, W. P., 1990, Geology and plate-tectonic development, *in* Wallace, R. E., ed., The San Andreas fault system, California: U.S. Geological Survey Professional Paper 1515, p. 60–80.

Jahns, R. H., 1973, Tectonic evolution of the Transverse Ranges province as related to the San Andreas fault system, *in* Kovach, R. L., and Nur, A., eds., Proceedings of the conference on tectonic problems of the San Andreas fault system: Stanford, California, Stanford University Publications in the Geolog-

ical Sciences, v. 13, p. 149–170f.

Jennings, C. W., compiler, 1977, Geologic map of California, scale 1:750,000: California Division of Mines and Geology.

Johnson, J. D., and Normark, W. R., 1974, Neogene tectonic evolution of the Salinian Block, west-central California: Geology, v. 2, no. 1, p. 11–14.

Joseph, S. E., Davis, T. E., and Ehlig, P. L., 1982, Strontium isotopic correlation of the La Panza Range granitic rocks with similar rocks in the central and eastern Transverse Ranges, *in* Fife, D. L., and Minch, J. A., eds., Geology and mineral wealth of the California Transverse Ranges; Mason Hill Volume: Santa Ana, California, South Coast Geological Society Annual Symposium and Guidebook 10, p. 310–320.

Karig, D. E., and Jensky, W., 1972, The proto–Gulf of California: Earth and Planetary Science Letters, v. 17, p. 169–174.

Kistler, R. W., Peterman, Z. E., Ross, D. C., and Gottfried, D., 1973, Strontium isotopes and the San Andreas fault, *in* Kovach, R. L., and Nur, A., eds., Proceedings of the conference on tectonic problems of the San Andreas fault system: Stanford, California, Stanford University Publications in the Geological Sciences, v. 13, p. 339–347.

Krause, D. C., 1965, Tectonics, bathymetry, and geomagnetism of the southern continental borderland west of Baja California, Mexico: Geological Society of America Bulletin, v. 76, p. 617–649.

Kupfer, D. H., 1968, A proposed deformation diagram for the analysis of fractures and folds in orogenic belts: Prague, Czechoslovakia, International Geological Congress, Section 13, 23rd Session, p. 219–232.

Lamar, D. L., 1961, Structural evolution of the northern margin of the Los Angeles basin [Ph.D. thesis]: Los Angeles, University of California, 142 p.

Larson, R. L., 1972, Bathymetry, magnetic anomalies, and plate tectonic history of the mouth of the Gulf of California: Geological Society of America Bulletin, v. 83, p. 3345–3359.

Larson, R. L., Menard, H. W., and Smith, S. M., 1968, Gulf of California; A result of ocean-floor spreading and transform faulting: Science, v. 161, p. 781–784.

Lonsdale, P., 1989, Geology and tectonic history of the Gulf of California, *in* Winterer, E. L., Hussong, D. M., and Decker, R. W., eds., The eastern Pacific Ocean and Hawaii: Boulder, Colorado, Geological Society of America, The Geology of North America, v. N, p. 499–521.

Luyendyk, B. P., and Hornafius, J. S., 1987, Neogene crustal rotations, fault slip, and basin development in southern California, *in* Ingersoll, R. V., and Ernst, W. G., eds., Cenozoic basin development of coastal California; Rubey Volume 6: Englewood Cliffs, New Jersey, Prentice-Hall, p. 259–283.

Luyendyk, B. P., Kamerling, M. J., and Terres, R., 1980, Geometric model for Neogene crustal rotations in southern California: Geological Society of America Bulletin, v. 91, p. 211–217.

Luyendyk, B. P., Kamerling, M. J., Terres, R. R., and Hornafius, J. S., 1985, Simple shear of southern California during Neogene time suggested by paleomagnetic declinations: Journal of Geophysical Research, v. 90, no. B14, p. 12454–12466.

Matthews, V., III, 1976, Correlation of Pinnacles and Neenach Volcanic Formations and their bearing on San Andreas fault problem: American Association of Petroleum Geologists Bulletin, v. 60, p. 2128–2141.

Matti, J. C., Morton, D. M., and Cox, B. F., 1985, Distribution and geologic relations of fault systems in the vicinity of the central Transverse Ranges, southern California: U.S. Geological Survey Open-File Report 85-365, 27 p., scale 1:250,000.

McKenzie, D. P., and Morgan, W. J., 1969, Evolution of triple junctions: Nature, v. 224, p. 125–133.

Meisling, K. E., and Weldon, R. J., 1989, Late Cenozoic tectonics of the northwestern San Bernardino Mountains, southern California: Geological Society of America Bulletin, v. 101, p. 106–128.

Minster, J. B., and Jordan, T. H., 1978, Present-day plate motions: Journal of Geophysical Research, v. 83, no. B11, p. 5331–5354.

——, 1984, Vector constraints on Quaternary deformation of the western United States east and west of the San Andreas fault, *in* Crouch, J. K., and Bachman, S. B., eds., Tectonics and sedimentation along the California margin: Pacific Section, Society of Economic Paleontologists and Mineralogists,

v. 38, p. 1–16.

Moody, J. D., and Hill, M. J., 1956, Wrench-fault tectonics: Geological Society of America Bulletin, v. 67, p. 1207–1246.

Moore, D. G., 1973, Plate-edge deformation and crustal growth, Gulf of California structural province: Geological Society of America Bulletin, v. 84, p. 1883–1905.

Moore, D. G., and Buffington, E. C., 1968, Transform faulting and growth of the Gulf of California since the late Pliocene: Science, v. 161, p. 1238–1241.

Nagel, D. K., and Mullins, H. T., 1983, Late Cenozoic offset and uplift along the San Gregorio fault zone; Central California continental margin, *in* Andersen, D. W., and Rymer, M. J., eds., Tectonics and sedimentation along faults of the San Andreas system: Pacific Section, Society of Economic Paleontologists and Mineralogists, p. 91–103.

Nilsen, T. H., 1978, Late Cretaceous geology of California and the problem of the proto–San Andreas fault, *in* Howell, D. G., and McDougall, K. A., eds., Mesozoic paleogeography of the western United States; Pacific Coast Paleogeography Symposium 2: Pacific Section, Society of Economic Paleontologists and Mineralogists, p. 559–573.

—— , 1984, Offset along the San Andreas fault of Eocene strata from the San Juan Bautista area and western San Emigdio Mountains, California: Geological Society of America Bulletin, v. 95, p. 599–609.

Nilsen, T. H., and Clarke, S. H., Jr., 1975, Sedimentation and tectonics in the early Tertiary continental borderland of central California: U.S. Geological Survey Professional Paper 925, 64 p.

Noble, L. F., 1953, Geology of the Pearland Quadrangle, California: U.S. Geological Survey Geologic Quadrangle Map GQ-24, scale 1:24,000.

—— , 1954, Geology of the Valyermo Quadrangle and vicinity, California: U.S. Geological Survey Geologic Quadrangle Map GQ-50, scale 1:24,000.

Powell, R. E., 1975, The Chiriaco fault; A left-lateral strike-slip fault in the eastern Transverse Ranges, Riverside County, California: Geological Society of America Abstracts with Programs, v. 7, no. 3, p. 362.

—— , 1981, Geology of the crystalline basement complex, eastern Transverse Ranges, southern California: Constraints on regional tectonic interpretation [Ph.D. thesis]: Pasadena, California Institute of Technology, 441 p.

—— , 1982, Crystalline basement terranes in the southern eastern Transverse Ranges, California, *in* Coooper, J. D., compiler, Geologic excursions in the Transverse Ranges, southern California: Geological Society of America, Cordilleran Section Annual Meeting Volume and Guidebook, Field Trip 11, p. 107–151.

—— , 1986, Palinspastic reconstruction of crystalline-rock assemblages in southern California: San Andreas fault as part of an evolving system of late Cenozoic conjugate strike-slip faults: Geological Society of America Abstracts with Programs, v. 18, no. 2, p. 172.

Ramirez, V. R., 1983, Hungry Valley Formation: Evidence for 220 kilometers of post Miocene offset on the San Andreas fault, *in* Andersen, D. W., and Rymer, M. J., eds., Tectonics and sedimentation along faults of the San Andreas system: Pacific Section, Society of Economic Paleontologists and Mineralogists, p. 33–44.

Ross, D. C., 1970, Quartz gabbro and anorthositic gabbro: Markers of offset along the San Andreas fault in the California Coast Ranges: Geological Society of America Bulletin, v. 81, p. 3647–3661.

—— , 1984, Possible correlations of basement rocks across the San Andreas, San Gregorio–Hosgri, and Rinconada–Reliz–King City faults, California: U.S. Geological Survey Professional Paper 1317, 37 p.

Ross, D. C., Wentworth, C. M., and McKee, E. H., 1973, Cretaceous mafic conglomerate near Gualala offset 350 miles by San Andreas fault from oceanic crustal source near Eagle Rest Peak, California: U.S. Geological Survey Journal of Research, v. 1, no. 1, p. 45–52.

Rusnak, G. A., and Fisher, R. L., 1964, Structural history and evolution of Gulf of California, *in* Van Andel, T. H., and Shor, G. G., Jr., eds., Marine geology of the Gulf of California: American Association of Petroleum Geologists Memoir 3, p. 144–156.

Rusnak, G. A., Fisher, R. L., and Shepard, F. P., 1964, Bathymetry and faults of Gulf of California, *in* Van Andel, T. H., and Shor, G. G., Jr., eds., Marine

geology of the Gulf of California: American Association of Petroleum Geologists Memoir 3, p. 59–75.

Ryder, R. T., and Thomson, A., 1989, Tectonically controlled fan delta and submarine fan sedimentation of late Miocene age, southern Temblor Range, California: U.S. Geological Survey Professional Paper 1442, 59 p.

Sadler, P. M., and Demirer, A., 1986, Pelona Schist clasts in the Cenozoic of the San Bernardino Mountains, southern California, *in* Ehlig, P. L., compiler, Neotectonics and faulting in southern California: Geological Society of America Cordilleran Section Annual Meeting Guidebook and Volume, p. 129–140.

Scholz, C. H., Barazangi, M., and Sbar, M. L., 1971, Late Cenozoic evolution of the Great Basin, western United States, as an ensialic interarc basin: Geological Society of America Bulletin, v. 82, p. 2979–2990.

Schwade, I. T., Carlson, S. A., and O'Flynn, J. B., 1958, Geologic environment of Cuyama Valley oil fields, California, *in* Weeks, L. G., ed., Habitat of oil; A symposium: Tulsa, Oklahoma, American Association of Petroleum Geologists, p. 78–98.

Sedlock, R. L., and Hamilton, D. H., 1991, Late Cenozoic tectonic evolution of southwestern California: Journal of Geophysical Research, v. 96, no. B2, p. 2325–2351.

Sharp, R. V., 1967, San Jacinto fault zone in the Peninsular Ranges of southern California: Geological Society of America Bulletin, v. 78, p. 705–729.

Sieh, K. E., and Jahns, R. H., 1984, Holocene activity of the San Andreas fault at Wallace Creek, California: Geological Society of America Bulletin, v. 95, p. 883–896.

Silver, L. T., Anderson, T. H., Conway, C. M., Murray, J. D., and Powell, R. E., 1977, Geologic features of southwestern North America, *in* Skylab explores the Earth: Washington, D.C., National Aeronautics and Space Administration, NASA SP-380, p. 89–135.

Smith, D. P., 1977, San Juan–St. Francis fault—Hypothesized major middle Tertiary right-lateral fault in central and southern California: California Division of Mines and Geology Special Report 129, p. 41–50.

Smith, G. I., 1962, Large lateral displacement on the Garlock fault, California, as measured from offset dike swarm: American Association of Petroleum Geologists Bulletin, v. 46, p. 85–104.

Smith, G. I., and Ketner, K. B., 1970, Lateral displacement on Garlock fault, southeastern California, suggested by offset sections of similar metasedimentary rocks: U.S. Geological Survey Professional Paper 700-D, p. D1–D9.

Stanley, R. G., 1987, New estimates of displacement along the San Andreas fault in central California based on paleobathymetry and paleogeography: Geology, v. 15, no. 2, p. 171–174.

Stock, J. M., and Hodges, K. V., 1989, Pre-Pliocene extension around the Gulf of California and the transfer of Baja California to the Pacific plate: Tectonics, v. 8, p. 99–115.

Stock, J., and Molnar, P., 1988, Uncertainties and implications of the Late Cretaceous and Tertiary position of North America relative to the Farallon, Kula, and Pacific plates: Tectonics, v. 7, p. 1339–1384.

Suppe, J., 1970, Offset of late Mesozoic basement terrains by the San Andreas fault system: Geological Society of America Bulletin, v. 81, p. 3253–3257.

Vaughan, F. E., 1922, Geology of the San Bernardino Mountains north of San Gorgonio Pass: Berkeley, University of California Publications, Department of Geological Sciences Bulletin, v. 13, p. 319–411.

Wegener, A., 1924, The origin of continents and oceans, 3rd ed., translated by Skerl, J.G.A.: New York, Dutton, 212 p.

Weldon, R., 1985, Implications of the age and distribution of the late Cenozoic stratigraphy in Cajon Pass, southern California, *in* Reynolds, R. E., ed., Geologic investigations along Interstate 15, Cajon Pass to Manix Lake: Redlands, California, San Bernardino County Museum Guidebook, p. 59–68.

Weldon, R. J., II, 1986, The late Cenozoic geology of Cajon Pass; Implications for tectonics and sedimentation along the San Andreas fault [Ph.D. thesis]: Pasadena, California Institute of Technology, 400 p.

Weldon, R., and Humphreys, E., 1986, A kinematic model of southern California: Tectonics, v. 5, p. 33–48.

Weldon, R. J., II, and Sieh, K. E., 1985, Holocene rate of slip and tentative recurrence interval for large earthquakes on the San Andreas fault, Cajon Pass, southern California: Geological Society of America Bulletin, v. 96, p. 793–812.

Wentworth, C. M., 1968, Upper Cretaceous and lower Tertiary strata near Gualala, California, and inferred large right slip on the San Andreas fault, *in* Dickinson, W. R., and Grantz, A., eds., Proceedings of conference on geologic problems of San Andreas fault system: Stanford, California, Stanford University Publications in the Geological Sciences, v. 11, p. 130–143.

Wilson, J. T., 1965a, A new class of faults and their bearing on continental drift: Nature, v. 207, p. 343–347.

—— , 1965b, Transform faults, oceanic ridges, and magnetic anomalies southwest of Vancouver Island: Science, v. 150, p. 482–485.

Woodburne, M. O., 1975, Cenozoic stratigraphy of the Transverse Ranges and adjacent areas, southern California: Geological Society of America Special Paper 162, 91 p.

Yeats, R. S., 1968, Rifting and rafting in the southern California borderland, *in* Dickinson, W. R., and Grantz, A., eds., Proceedings of conference on geologic problems of San Andreas fault system: Stanford, California, Stanford University Publications in the Geological Sciences, v. 11, p. 307–322.

—— , 1976, Extension versus strike-slip origin of the southern California borderland, *in* Howell, D. G., ed., Aspects of the geologic history of the California continental borderland: Pacific Section, American Association of Petroleum Geologists Miscellaneous Publication 24, p. 455–485.

Yeats, R. S., Calhoun, J. A., Nevins, B. B., Schwing, H. F., and Spitz, H. M., 1989, Russell fault; An early strike-slip fault of California Coast Ranges: American Association of Petroleum Geologists Bulletin, v. 73, p. 1089–1102.

MANUSCRIPT ACCEPTED BY THE SOCIETY APRIL 2, 1991

Printed in U.S.A.

Geological Society of America
Memoir 178
1993

Chapter 1

Balanced palinspastic reconstruction of pre–late Cenozoic paleogeology, southern California: Geologic and kinematic constraints on evolution of the San Andreas fault system

Robert E. Powell
U.S. Geological Survey, Room 117, U.S. Post Office, Spokane, Washington 99201-1087

ABSTRACT

The San Andreas fault system comprises an interactive network of right- and left-lateral strike-slip faults and related reverse and normal faults. In southern and central California, right- and left-lateral faults of the San Andreas system transect a crystalline terrane of Proterozoic through Mesozoic igneous and metamorphic rocks overlapped by Upper Cretaceous through Eocene marine sedimentary strata and by Oligocene and lower Miocene terrestrial volcanic and sedimentary strata. Paleogeologic patterns in these rocks define regional terranes and local reference domains, the reassembly of which permits determination of overall displacement on the strike-slip faults that disrupt them. Timing of fault movement is recorded by incremental displacements of upper Cenozoic sedimentary deposits and by the sequence of fault movements required to effect reassembly of the reference domains. Reassembling the pre–late Cenozoic regional paleogeologic framework of southern and central California leads to a balanced palinspastic reconstruction of the San Andreas system that differs from previously published reconstructions, both conceptually and in terms of magnitude of displacement restored on many of the principal strike-slip faults.

Four reference domains that constrain the balanced reconstruction consist of paleogeologic patterns reassembled from crystalline and sedimentary rocks now found: (1) in the Transverse Ranges in the Frazier Mountain–Mount Pinos area, the eastern Orocopia Mountains and vicinity, and the Sierra Pelona–northern San Gabriel Mountains area; (2) in the Salinian block in the La Panza Range and in the Transverse Ranges in the Liebre Mountain block and western San Bernardino Mountains; (3) in the Salinian block in the Gabilan Range, in the southern tail of the Sierra Nevada in the San Emigdio and Techachapi Mountains, and in the Portal Ridge area of the northwesternmost Mojave Desert; and (4) in the southern part of the Transverse Ranges west of the San Andreas fault, in the northern Peninsular Ranges, and in the southern Chocolate Mountains.

Simultaneous reconstruction of the four paleogeologic reference domains specifies the magnitude and sequence of displacement on the major right- and left-lateral faults of the San Andreas system. The Clemens Well–Fenner–San Francisquito fault is the earliest (and now abandoned) strand of the San Andreas fault system in southern California. It formed a continuous structure with the early San Andreas fault zone of central California, is today cut by the San Andreas fault in the Transverse Ranges, and diverges southeastward from the San Andreas fault east of the Salton trough at least as far as the

Powell, R. E., 1993, Balanced palinspastic reconstruction of pre–late Cenozoic paleogeology, southern California: Geologic and kinematic constraints on evolution of the San Andreas fault system, *in* Powell, R. E., Weldon, R. J., II, and Matti, J. C., eds., The San Andreas Fault System: Displacement, Palinspastic Reconstruction, and Geologic Evolution: Boulder, Colorado, Geological Society of America Memoir 178.

Little Chuckwalla Mountains. The Clemens Well–Fenner–San Francisquito–early San Andreas fault accumulated a displacement of about 100 or 110 km during the interval between 22 and 13 Ma and probably during the more restricted interval between 18 to 17 and 13 Ma. The disposition of displacement southeastward from the Transverse Ranges is problematic because the Clemens Well–Fenner–San Francisquito fault neither rejoins the modern San Andreas fault in the Salton trough nor extends indefinitely into the continent. Hypothetically, the southeastward extension of the fault is absorbed by coeval sinistral kinking along the southern margins of the Transverse Ranges and Chocolate Mountains and/or by synchronous detachment faulting in southeasternmost California and southwesternmost Arizona.

A zone of sinistral deformation trends roughly east-west along the southern boundary of the Transverse Ranges province. The earliest expression of this deformation is sinistral kinking that began to develop after 22 to 20 Ma and prior to 17 Ma. During the interval from 17 to 13 Ma, the zone of sinistral deformation was characterized by widespread volcanism and perhaps by faulting associated with left-oblique extension. Left-lateral displacement of about 40 to 45 km across this zone during this interval is attributed here to sinistral kinking. The zone of sinistral kinking and faulting apparently initiated in a right step between the southeastern terminus of the Clemens Well–Fenner–San Francisquito fault and the northwestern terminus of dextral faults such as the East Santa Cruz basin fault that were active coevally in the continental borderland. Subsequently, the zone served as the southern terminus of right-lateral strands of the San Gabriel fault system. Still later, it served as the northern terminus of the Elsinore and San Jacinto fault zones.

In the Transverse Ranges, the San Gabriel fault system, including, from oldest to youngest, the Canton, San Gabriel, and Vasquez Creek faults, began to splay southward from the older Clemens Well–Fenner–San Francisquito–early San Andreas fault as early as 12 to 13 Ma. The San Gabriel fault system has since accumulated a displacement of 42 km, deforming the older Clemens Well–Fenner–San Francisquito–early San Andreas fault in the process. The Canton fault, active between 13 and 10 Ma, accumulated a displacement of 15 to 17 km; the San Gabriel fault, active between 10 and 5 Ma, accumulated a displacement of 22 to 23 km; and the Vasquez Creek fault, active between 6 Ma and the present, accumulated a displacement of no more than about 5 km. Displacement on these faults merged northwestward with that of the early San Andreas fault north of the Transverse Ranges, whereas movement ceased on the Clemens Well–Fenner–San Francisquito fault.

In the Salinian block, the San Gregorio–Hosgri, Rinconada–Reliz, and Red Hills–Ozena faults developed coevally with the San Gabriel fault and also merged northwestward with the early San Andreas fault. Displacements of 45 km on the Rinconada–Reliz fault and 105 km on the San Gregorio–Hosgri fault south of Monterey Bay merge on the San Gregorio fault north of the bay for total of 150 km. Hypothetically, displacement on the San Gregorio fault is split between 70 km (after 6 to 6.5 Ma) west of the Montara Mountain block and 80 km (prior to 6 to 6.5 Ma) east of that block, so as not to leave the Montara Mountain block dangling north of the restored Salinian block. Displacement on the San Gregorio fault is transferred northward onto the San Andreas fault, thereby increasing its overall displacement to about 440 km north of the junction of the two faults.

None of the displacements on the Salinian block faults and no more than 22 or 23 km of slip on the San Gabriel fault have been shown to remerge southeastward with the Clemens–Well–Fenner–San Francisquito–early San Andreas fault. Hypothetically, these displacements stepped west to the continental margin across the western Transverse Ranges, where left-oblique extensional faulting continued through the late Miocene into the early Pliocene. Along the southern boundary of the Transverse Ranges west of the San Gabriel Mountains, the ancestral Malibu Coast–Santa Monica fault system accumulated a sinistral component of displacement of as much as 35 km in addition to the earlier sinistral kinking, whereas to the east, the ancestral Raymond-Cucamonga-Banning fault system accumulated no more than about 10 to 20 km.

The modern San Andreas fault emerged about 5 Ma. In central California, it coincides with the pre–5-Ma San Andreas fault, whereas in southern California it diverged from the older Clemens Well–Fenner–San Francisquito fault and actually crosscuts the older fault to merge southeastward with the Salton trough at the north end of the Gulf of California. Displacement on the post–5-Ma San Andreas fault varies along the fault because the crustal blocks adjoining the fault are deformed by coeval strike-slip faults, including right-lateral faults such as the San Jacinto, Calaveras–Hayward–Rodgers Creek–Maacama–Garberville, and San Gregorio–Hosgri faults, and left-lateral faults such as the Garlock fault and the east- to northeast-trending faults of the Transverse Ranges. Displacement restored on the modern San Andreas fault as measured along the present trace ranges from about 160 to 185 km.

Simultaneous palinspastic reconstruction of the four reference domains is possible only in conjunction with restoration of slip along a zig-zag system of secondary strike-slip faults of the San Andreas system that distort the crustal blocks adjoining the modern San Andreas fault in southern California. This system includes right-lateral faults in the Peninsular Ranges, Mojave Desert, and Death Valley area, and left-lateral faults in the Transverse Ranges and between the Sierra Nevada and Mojave Desert. Overall displacements restored on these secondary faults are generally well constrained by offsets of crystalline rocks and overlying Cenozoic strata. In the Peninsular Ranges, dextral displacement restored on the San Jacinto fault is 28 km, and that restored on the Elsinore fault is 5 km. About 10 km of Pliocene and Quaternary left slip is restored on faults along the southern boundary of the Transverse Ranges, where earlier left-oblique extensional faulting was overprinted by reverse and left-oblique reverse faulting by late Pliocene and Quaternary time. In the eastern Transverse Ranges, sinistral displacement restored on the major east-trending faults includes 16 km on the Pinto Mountain fault, 5 km on the Blue Cut fault, 11 km on the Chiriaco fault, and 8 km on the Salton Creek fault. In the Mojave Desert, dextral displacements restored on northwest-trending faults include 3 km on the Helendale fault, 3 km on the Lockhart-Lenwood fault, 4 km on the Harper–Harper Lake–Camp Rock–Emerson fault, 9 km on the Blackwater–Calico–Mesquite Lake fault, and 16 km on the Pisgah-Bullion fault. Although displacement on the central and eastern parts of the Garlock fault is well documented to be about 60 km, displacement at the western end, as limited by reassembly of the Gabilan Range–San Emigdio Mountains–Portal Ridge reference domain, can be no greater than 12 km.

INTRODUCTION

The San Andreas and related faults developed during the late Cenozoic in a dextral shear regime that accompanied motion between the North American and Pacific plates. The relative motion of these plates resulted in the growth of a transform fault between the northward migrating Mendocino triple junction and the southward migrating Rivera triple junction (Fig. 1) (Atwater, 1970, 1989; Dickinson, 1981; Irwin, 1990). It is generally accepted that the modern San Andreas has been the principal transform fault for the last 5 m.y., but that other faults in the San Andreas system also accommodate relative motion along the plate boundary (Bird and Rosenstock, 1984; Weldon and Humphreys, 1986; Humphreys and Weldon, 1991). Most of these and other investigators have concluded that the principal transform fault of the plate boundary prior to 5 Ma was located offshore along the continental margin. While the transform fault

paradigm satisfactorily explains the existence of the San Andreas fault system, the paradigm provides little insight into the complex intracontinental evolution of that system (Fig. 2). Only by deciphering the geologic record along the fault system, as well as the tectonics of the seafloor, can we hope to elucidate the evolving role of the San Andreas system in the plate boundary.

Our ability to unravel the deformation of the San Andreas system depends in large part on our understanding of pre–San Andreas paleogeology. By reassembling pre–late Cenozoic paleogeologic patterns in southern California, one can, in theory, specify both the paths and overall displacements of all the late Cenozoic faults that have disrupted those patterns. Because these patterns are unrelated to the younger faults, they provide a basis for defining preexisting stratigraphic or structural lines; in turn, the intersections of such lines with subsequent fault planes provide piercing points with which to measure displacement on the faults (Crowell, 1962).

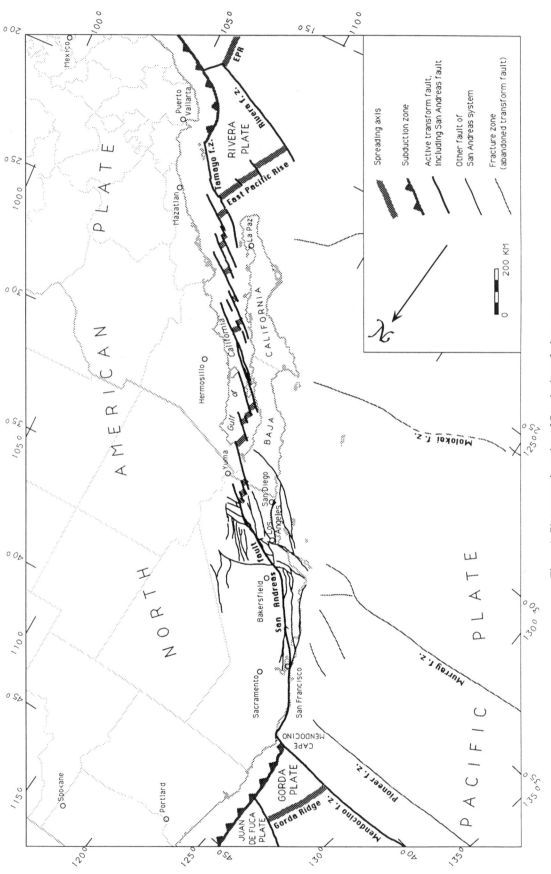

Figure 1. Plate tectonic setting of San Andreas fault system.

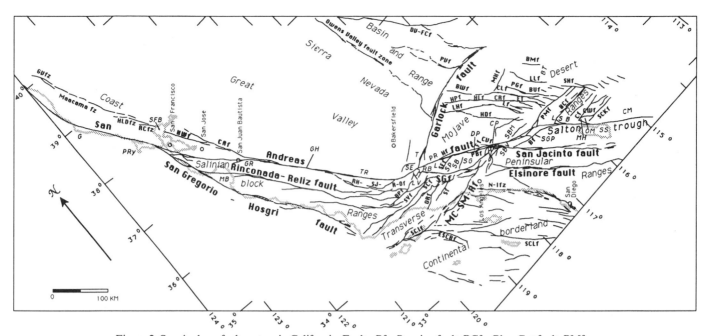

Figure 2. San Andreas fault system in California. Faults: Bf = Banning fault; BCf = Blue Cut fault; BMf = Bristol Mountains fault; BPf = Big Pine fault; BUf = Bullion fault; BWf = Blackwater fault; Cf = Chiriaco fault; CAf = Calaveras fault; CLf = Calico fault; CRf = Camp Rock fault; CVf = Cajon Valley fault; CWf = Clemens Well fault; DV-FCf = Death Valley–Furnace Creek fault; Ef = Emerson fault; ESCBf = East Santa Cruz basin fault; Ff = Fenner fault; GVfz = Garberville fault zone; Hf = Hitchbrook fault; HDf = Helendale fault; HLf = Harper Lake fault; HLDfz = Healdsburg fault zone; HPf = Harper fault; HWf = Hayward fault; Lf = Lenwood fault; LHf = Lockhart fault; LLf = Ludlow fault; MC-SM-Rf = Malibu Coast–Santa Monica–Raymond fault; MXf = Manix fault; N-Ifz = Newport-Inglewood fault zone; ORf = Oak Ridge fault; PBf = Punchbowl fault; PGf = Pisgah fault; PMf = Pinto Mountain fault; PVf = Panamint Valley fault; RCfz = Rodgers Creek fault zone; RH-SJ-R-Of = Red Hills–San Juan–Russell–Ozena fault; Sf = Simi fault; SCf = San Cayetano fault; SCIf = Santa Cruz Island fault; SCKf = Salton Creek fault; SCLf = San Clemente fault; SFf = San Francisquito fault; SGf = San Gabriel fault; SHf = Sheep Hole fault; SYf = Santa Ynez fault. Physiographic features: BT = Bristol trough; CM = Chocolate Mountains; CP = Cajon Pass; DP = Devils Punchbowl; G = Gualala; GH = Gold Hill; GR = Gabilan Range; L = Liebre Mountain; LSB = Little San Bernardino Mountains; LV = Lockwood Valley; MB = Monterey Bay; MH = Mecca Hills; OM = Orocopia Mountains; PR = Portal Ridge; PRy = Point Reyes; RB = Ridge basin; SB = Soledad basin; SBM = San Bernardino Mountains; SBV = San Bernardino Valley; SE = San Emigdio Mountains; SFB = San Francisco Bay; SG = San Gabriel Mountains; SGP = San Gorgonio Pass; SP = Sierra Pelona; SS = Salton Sea; T = Tehachapi Mountains; TR = Temblor Range.

In practice, however, reconstruction of the pre–late Cenozoic paleogeologic framework of southern California is rarely unambiguous, and indisputable piercing points are rare. Typically, paleogeologic features are incompletely preserved, which may impair the precision with which disrupted parts of any given feature can be realigned. The reconstruction of pre–late Cenozoic rock units is further complicated by the fact that these rocks exhibit a northwest-trending stratigraphic and structural grain that is transected obliquely by the right-lateral fault system. A review of investigations of the San Andreas and related faults in southern California discloses numerous instances in which matching of similar or even identical rock types in outcrop or in clast-to-source links has led to erroneous conclusions about displacement on the fault (for example, Noble, 1926; Woodford, 1960;

Baird and others, 1974; Ehlert and Ehlig, 1977; see M. L. Hill, 1981 for additional examples). In most of these instances, a match-up that seemed reasonable at the time it was made is rendered unacceptable or nonunique by subsequent recognition of a wider distribution of the rock units used to make the match, leading in turn to an alternate match-up. In this context, it is important to appraise the uniqueness of any reassembly in an effort to distinguish between a reasonable match-up and one that is well constrained. This distinction is rarely straightforward. In general, however, realigned patterns in reassembled paleogeologic domains that are composed of more than one rock unit— thereby constituting an array of probable piercing points (for example, Sharp, 1967; Davis and Burchfiel, 1973; Ehlig, 1975a; Graham and Dickinson, 1978a, b; Powell, 1981, 1982a; Clark and others, 1984; Ross, 1984)—are more reliable than match-ups based on the distribution of single rock types or paleogeologic features.

My approach has been to develop a palinspastic reconstruction of *all* late Cenozoic strike-slip faults that transect the batholithic terrane along the San Andreas fault in southern and central California (Figs. 3 through 6; Plate I). This reconstruction is based on reassembly of groups of paleogeologic features within the crystalline and sedimentary terranes along the faults (Plate II). In the first section of this chapter, these groups of features are used to define four paleogeographic reference domains, all of which must be reassembled simultaneously in the complete palinspastic reconstruction. The pre-late Cenozoic rock units and assemblages incorporated in these reference domains are described in the Appendix to this chapter and in the explanation to Plates I and II.

The reconstruction, discussed in the second section of this chapter, is balanced in plan by maintaining constant area for blocks of pre-late Cenozoic rocks as they are restored, by not allowing overlaps of the reassembled blocks, and by minimizing gaps between them. In order to accomplish a completely balanced reconstruction, it would be necessary not only to restore displacement on all of the strike-slip faults of the San Andreas fault system, but also to restore displacements on coeval reverse and normal faults and to unbend coeval folds. Restoration of just the strike-slip faults (Plate II), however, is the first step and one that closely approximates a completely balanced reconstruction. The process of assembling a balanced reconstruction emphasizes that the displacement restored on each individual strike-slip fault has consequences for the system as a whole and that not all displacements reported in the literature are viable.

The balanced palinspastic reconstruction represents a significant departure from generally accepted published reconstructions and contains elements that are certain to be controversial. The reconstruction corroborates some published measurements of displacement for faults of the San Andreas system and contradicts others. In the third and fourth sections of this chapter, therefore, each strike-slip fault is described systematically by: (1) summarizing the published evidence for fault path and displacement; (2) discussing the reckoning of fault path and displacement required to accomplish the balanced reconstruction shown in Plate II; (3) examining and, where possible, reconciling the differences between displacements restored in Plate II and previously published estimates of displacement; (4) summarizing the evidence for the timing and, in some cases, rate of fault movement; and (5) considering the kinematic role of individual faults or provincial groups of faults in the evolution of the San Andreas fault system. The final section of the chapter is a synthesis of displacement on the San Andreas fault and of the evolution of the San Andreas fault system as a whole.

PRE–LATE CENOZOIC PALEOGEOLOGY

Regional paleogeologic framework

The crust of California consists of an intricate framework of rock types constructed along an evolving continental margin. These rocks, which range in age from Early Proterozoic to late Cenozoic, are distributed among several distinct geologic terranes that record events from different eras and different crustal environments. In southern and central California, these events reveal a five-phase evolution of the regional crustal architecture: (1) growth of the Proterozoic craton, (2) accumulation of the latest Proterozoic and Paleozoic geocline, (3) accretion of the Mesozoic magmatic arc, (4) late Mesozoic and early to middle Cenozoic diastrophism and sedimentation along a convergent plate boundary, and (5) late Cenozoic volcanism and sedimentation in a dextral shear regime along a translational plate boundary. Events of the pre-late Cenozoic architectonic phases produced several regional paleogeologic terranes that were subsequently disrupted by the San Andreas system. These general features provide a regional framework for the specific palegeologic features that are grouped in the four reference domains described in the next section.

The oldest regional paleogeologic terrane within the regional framework comprises a northwest-trending belt of uplifted, metamorphosed Proterozoic and Paleozoic rocks that includes most of the crystalline Transverse Ranges and parts of the western Mojave Desert, eastern Peninsular Ranges, and several ranges in southeasternmost California and southwesternmost Arizona (see Appendix for additional discussion and references). In this belt, Early and Middle Proterozoic igneous and high-grade metamorphic rocks were unconformably overlain by upper Proterozoic and Paleozoic cratonic and miogeoclinal marine strata that were subsequently deformed and metamorphosed. This belt, which is flanked by belts of metamorphosed Triassic through Cretaceous sedimentary and volcanic strata to the east in the Mojave Desert and to the west in the Peninsular Ranges, constitutes a northwest-trending inlier of the North American craton that was isolated from the craton during accretion of the Mesozoic magmatic arc.

Other regional paleogeologic terranes formed as material was accreted to the North American craton during the petrotectonic evolution of the Mesozoic magmatic arc. A composite batholith consists of coalesced plutons that exhibit a northwest-trending lithologic grain in which different suites of plutons intruded the Proterozoic and Paleozoic rocks of the cratonic inlier and the two flanking Mesozoic metasedimentary and metavolcanic sequences (see Appendix for details). The Jurassic Coast Range ophiolite was accreted outboard of the cratonic-batholithic terrane, and was depositionally overlain by the Upper Jurassic through Upper Cretaceous Great Valley sequence.

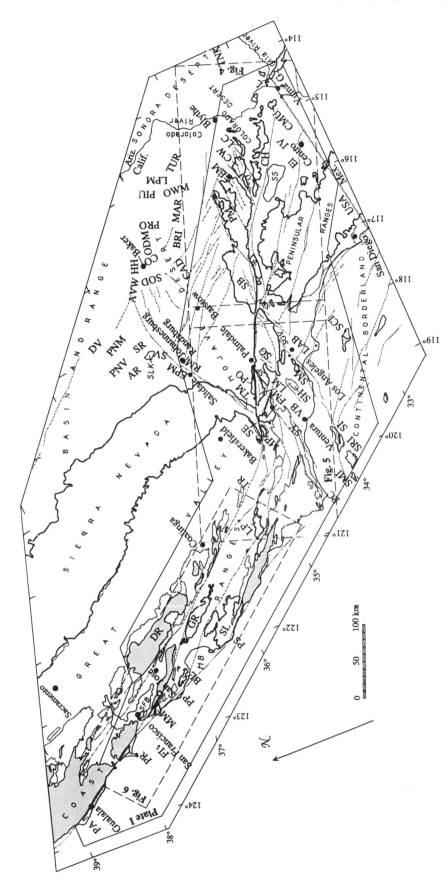

Figure 3. Geographic location map for southern and central California and southwesternmost Arizona. AR = Argus Range; AVW = Avawatz Mountains; B = Banning block; BL = Ben Lomond block; BRI = Bristol Mountains; CAD = Cady Mountains; CH = Chocolate Mountains; CMU = Cargo Muchacho Mountains; CO = Cowhole Mountains; DR = Diablo Range; DV = Death Valley; EM = Eagle Mountains; EPM = El Paso Mountains; FI = Farallon Mountains; FM = Frazier Mountain block; G = Gila Mountains; GR = Gabilan Range; HH = Hollow Hills; IV = Imperial Valley; L = Laguna Mountains; LAB = Los Angeles basin; LC = Little Chuckwalla Mountains; LM = Liebre Mountain; LP = La Panza Range; LPM = Little Piute Mountains; MAR = Marble Mountains; MB = Monterey Bay; MM = Montara Mountain; MP = Mount Pinos block; O = Orocopia Mountains; ODM = Old Dad Mountains; OWM = Old Woman Mountains; PA = Point Arena; PC = Point Conception; PIU = Piute Mountains; PM = Pinto Mountains; PNM = Panamint Mountains; PNV = Panamint Valley; PO = Portal Ridge; PP = Pigeon Point; PR = Point Reyes; PRO = Providence Mountains; PS = Point Sur; RM = Rand Mountains; SB = San Bernardino Mountains; SBV = San Bernardino Valley; SCI = Santa Catalina Island; SE = San Emigdio Mountains; SFB = San Francisco Bay; SG = San Gabriel Mountains; SGV = San Gabriel Valley; SH = Simi Hills; SI = Santa Cruz Island; SL = Santa Lucia Range; SLK = Searles Lake; SM = Santa Monica Mountains; SMI = San Miguel Island; SOD = Soda Mountains; SP = Sierra Pelona; SR = Slate Range; SRI = Santa Rosa Island; SS = Salton Sea; SV = Searles Valley; SY = Santa Ynez Range; TM = Tehachapi Mountains; TNK = Tank Mountains; TR = Temblor Range; TUR = Turtle Mountains; VB = Ventura basin. Cities and towns are shown by dots.

a

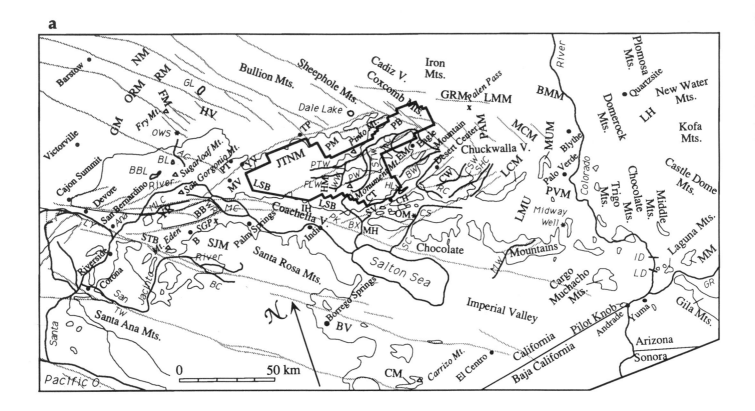

b

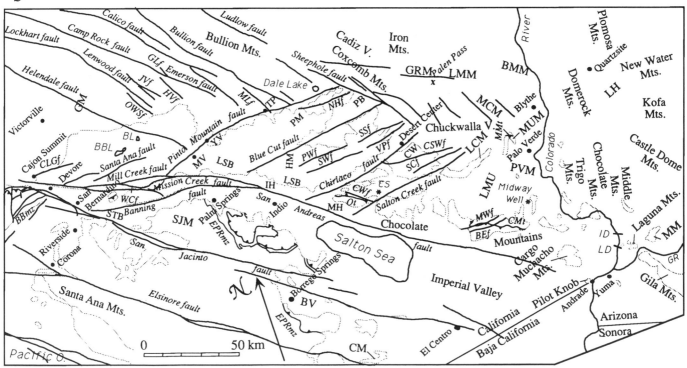

Late Mesozoic and early to middle Cenozoic diastrophism is recorded in both compressional and extensional structures associated with the Pelona Schist and related rocks, the Catalina Schist, and the Franciscan Complex. Fault-bounded terranes were delimited as the Pelona, Orocopia, and Rand Schists were underthrust beneath the cratonic-batholithic terrane; as the Franciscan Complex and the Catalina Schist were faulted beneath the Coast Range ophiolite–Great Valley terrane; and as both of these latter two terranes were faulted against the cratonic-batholithic terrane. Extensional faulting has resulted in subsequent uplift of the underthrust rocks and culminated in their exhumation in antiformal folds in the middle to late Cenozoic.

A generally northwest-trending belt of Upper Cretaceous through Eocene marine strata crops out along the boundary between the cratonic-batholithic and Coast Range ophiolite–Great Valley terranes. These strata include both those that constitute the upper part of the Great Valley sequence and overlying units, which depositionally overlie the ophiolite terrane, and those that depositionally overlap the western margin of the cratonic-

---

Figure 4. Eastern Transverse Ranges and vicinity. a, Geographic locations. Key: AC = Arrastre Creek; B = Banning; BB = Banning block; BBL = Big Bear Lake; BC = Bautista Canyon; BL = Baldwin Lake; BMM = Big Maria Mountains; BV = Borrego Valley; BW = Big Wash; BX = Box Canyon; CH = Chiriaco Summit; CM = Coyote Mountains; CS = Canyon Spring; CSW = Corn Springs Wash; CT = Cottonwood Mountains; CW = Chuckwalla Mountains; EM = Eagle Mountains; FLW = Fried Liver Wash; FM – Fry Mountains; GL – Galway Lake; GM – Granite Mountains; GR = Gila River; GRM = Granite Mountains; HL = Hayfield Lake; HM = Hexie Mountains; HV = Homestead Valley; ID = Imperial Dam; IH = Indio Hills; JTNM = Joshua Tree National Monument; LCM – Little Chuckwalla Mountains; LD = Laguna Dam; LH = Livingston Hills; LMM – Little Maria Mountains; LMU – Little Mule Mountains; LSB = Little San Bernardino Mountains; LY = Lytle Creek; MC = Mission Creek; MCM = McCoy Mountains; MH = Mecca Hills; MLC = Mill Creek; MM = Muggins Mountains; MUM = Mule Mountains; MV = Morongo Valley; MW = Mammoth Wash; NM = Newberry Mountains; OM = Orocopia Mountains; ORM = Ord Mountains; OWS = Old Woman Springs; PAM = Palen Mountains; PB = Pinto Basin; PC = Placer Canyon; PK = Pinkham Wash; PM = Pinto Mountains; PT = Pioneertown; PTW = Pinto Wash; PVM = Palo Verde Mountains; PW = Porcupine Wash; RC = Red Cloud Canyon; RM = Rodman Mountains; SC = Salt Creek; SGP = San Gorgonio Pass; SHC = Ship Creek; SJM = San Jacinto Mountains; STB = San Timoteo Badlands; STW = Smoke Tree Wash; SV = Shavers Valley; TP = Twentynine Palms; TW = Temescal Wash; WR = Whitewater River; WW = Washington Wash; YR = Yucaipa Ridge; YV = Yucca Valley. Cities and towns are shown by dots; mountain peaks by triangles; mountain passes by x's; springs and wells by asterisks. b, Faults. Key: BBmz = Black Belt mylonite zone; BEf = Black Eagle fault; CLGf = Cleghorn fault; CMt = Chocolate Mountains thrust; CSWf = Corn Springs Wash fault; CWf = Clemens well fault; EPRmz = Eastern Peninsular Ranges mylonite zone; GLf = Galway Lake fault; HVf = Homestead Valley fault; JVf = Johnson Valley fault; MLf = Mesquite Lake fault; MMt = Mule Mountains thrust; MWf = Mammoth Wash fault; NHf = Niggerhead fault; Ot = Orocopia thrust; OWSf = Old Woman Springs fault; PWf = Porcupine Wash fault; SCf = Ship Creek fault; SSf = Substation fault; SWf = Smoke Tree Wash fault; VPf = Victory Pass fault; WCf = Wilson Creek fault.

---

batholithic terrane. The Upper Cretaceous through Eocene strata contain abundant debris shed from the unroofing batholith. Along with the ophiolitic rocks and the lower part of the Great Valley sequence, these younger strata are faulted against the Franciscan-Catalina terrane.

Late Cenozoic marine and terrestrial volcanic and sedimentary rocks accumulated in basins that formed in the dextral shear regime that developed as the North American and Pacific plates came into contact. During the late Oligocene and early Miocene, such basins formed in southern California in west-northwest-trending belts that pre-dated the growth of throughgoing strike-slip faults. Since the early Miocene, additional basins have developed along the faults of the San Andreas system.

Paleogeologic reference domains

Because disrupted paleogeologic features provide the only basis for measuring prehistoric fault displacement, the validity of any palinspastic reconstruction of the San Andreas fault system depends on the validity of the paleogeologic features on which it is based. The purpose of this section is to summarize the regional paleogeologic framework and to describe the evidence for four groups of specific paleogeologic features that define reference domains within the regional paleogeologic framework. These reference domains provide a basis for palinspastic reconstruction of the San Andreas fault system in southern and central California (Fig. 7).

The four reference domains consist of paleogeologic patterns revealed by realigning the following: (1) rocks on Frazier Mountain and Mount Pinos area with rocks east of the Clemens Well fault in the Orocopia Mountains and vicinity (Fig. 8); (2) rocks in the La Panza Range, on Barrett Ridge, in the Liebre Mountain block, and in the southwestern San Bernardino Mountains (Fig. 9); (3) rocks in the Gabilan Range with rocks in the Tehachapi and San Emigdio Mountains and Portal Ridge (Fig. 10); and (4) rocks along the southern margin of the Transverse ranges west of the San Andreas with rocks in the northern Peninsular Ranges and southern Chocolate Mountains (Fig. 11). In Figures 8 through 11, these reference domains are constructed by reassembling their disrupted rock units into the closest possible fits. Although one can systematically describe the rocks of each fault block in the San Andreas fault system before reassembling them into reference domains, for brevity the rocks of all the fault blocks that reassemble into each reference domain are described together. This abbreviated approach builds on nearly 30 years of previous work in matching the rocks of these reference domains across the faults of the San Andreas system.

Frazier Mountain–Mount Pinos–eastern Orocopia Mountains reference domain. The Frazier Mountain–Mount Pinos–eastern Orocopia Mountains reference domain includes Proterozoic igneous and metamorphic rocks, Mesozoic granitic rocks, Eocene marine sedimentary strata, and upper Oligocene and lower Miocene terrestrial sedimentary and volcanic strata (Fig. 8). The Proterozoic and Mesozoic rocks comprise a crystal-

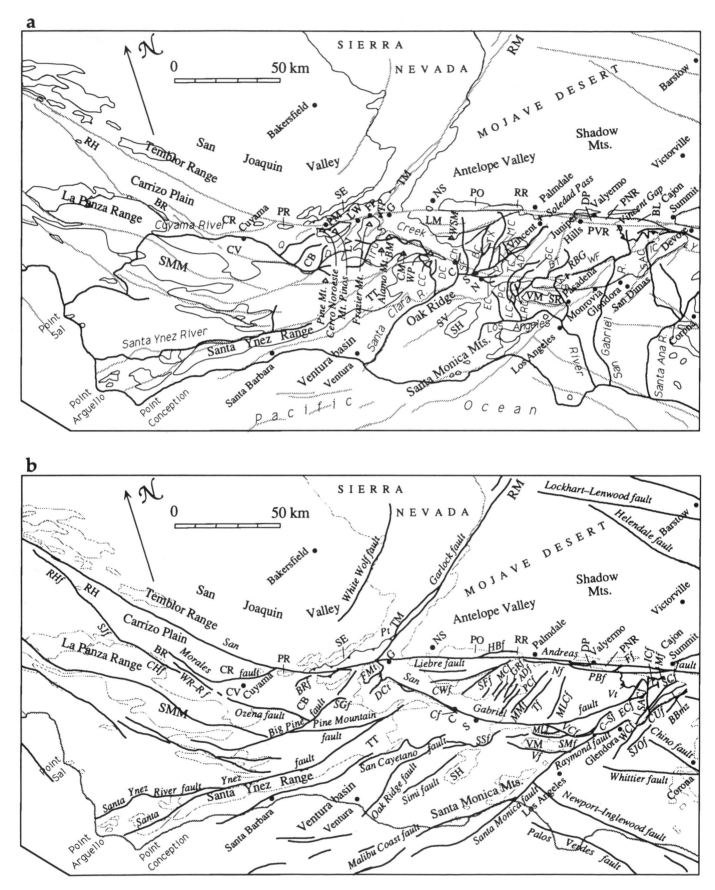

line basement that was unconformably overlapped by the Eocene and upper Oligocene to lower Miocene stratigraphic sequences. Subsequent folding has resulted in moderately to steeply dipping unconformities and basal strata. The domain is reconstructed from disrupted parts now exposed in the Frazier Mountain–Lockwood Valley–Mount Pinos area west of the San Andreas and San Gabriel faults and in the Orocopia Mountains northeast of the Clemens Well fault. Also included in this domain are rocks in the ranges north and east of the Orocopia Mountains and rocks in the Sierra Pelona, Soledad basin, and northern San Gabriel Mountains between the San Gabriel and San Andreas faults and south of the San Francisquito and Fenner faults.

Proterozoic rocks of the domain comprise the Hexie Mountains and Eagle Mountains assemblages and the anorthosite-syenite complex of the San Gabriel Mountains (see Appendix). The rocks of the Hexie Mountains assemblage, including the distinctive augen gneiss of Monument Mountain and the metasedimentary suite of Pinkham Canyon, crop out chiefly in an N-shaped belt through the reference domain. The belt trends

◄ —————————————————————

Figure 5. Western Transverse Ranges and vicinity. a, Geographic locations: AD = Agua Dulce Canyon; AS = Arroyo Seco; BC = Bouquet Canyon; BL = Blue Ridge; BM = Bear Mountain; BR = Barrett Ridge; BT = Big Tujunga Canyon; C = Castaic; CB = Cuyama Badlands; CC = Canton Canyon; Ch = Charlie Canyon; CM = Cobblestone Mountain; CR = Caliente Range; CV = Cuyama Valley; D = Dry Canyon; DC = Devil Canyon; DP = Devil's Punchbowl; EC = Elsemere Canyon; FP = Frazier Park; G = Gorman; GC = Gold Canyon; IC = Icehouse Canyon; LC = Lopez Canyon; LM = Liebre Mountain; LRC = Limerock Canyon; LV = Lockwood Valley; LW = Lake-of-the-Woods; LY = Lytle Creek; MC = Mint Canyon; N = Newhall; NS = Neenach School site; P = Placerita Canyon; PC = Pacoima Canyon; PM = Pine Mountain; PNR = Pinyon Ridge; PO = Portal Ridge; PR = Pattiway Ridge; PVR = Pleasant View Ridge; Q = Quatal Canyon; RBG = Red Box Gap; RH = Red Hills; RM = Rand Mountains; RR = Ritter Ridge; S = Saugus; SAC = San Antonio Canyon; SE = San Emigdio Mountains; SF = San Francisquito Canyon; SH = Simi Hills; SMM = Sierra Madre Mountains; SV = Simi Valley; TM = Tehachapi Mountains; SR = San Rafael Hills; TC = Tick Canyon; TP = Tejon Pass; TT = Topatopa Mountains; TX = Texas Canyon; VM = Verdugo Mountains; WF = West Fork of the San Gabriel River; WP = Whitaker Peak; WSM = Warm Springs Mountain. Cities and towns are shown by dots; mountain peaks by triangles; mountain passes by x's. b, Faults: ADf = Agua Dulce fault; BBmz = Black Belt mylonite zone; BRf = Blue Rock fault; Cf = Canton fault; C-Sf = Clamshell-Sawpit fault; CHf = Chimeneas fault; CUf = Cucamonga fault; CWf = Clearwater fault; DCt = Dry Creek thrust; ECf = Evey Canyon fault; Ff = Fenner fault; FMt = Frazier Mountain thrust; GRf = Green Ranch fault; HBf = Hitchbrook fault; ICf = Icehouse Canyon fault; MCf = Mint Canyon fault; Mf = Miller fault; MLf = Mount Lukens fault; MLCf = Mill Creek fault; MMf = Magic Mountain fault; Nf = Nadeau fault; Pt = Pastoria thrust; PBf = Punchbowl fault; PCf = Pole Canyon fault; RHf = Red Hills fault; SACf = San Antonio Canyon fault SCf = Stoddard Canyon fault; SFf = San Francisquito fault; SGf = San Guillermo fault; SJf = San Juan fault; SJO = San Jose fault; SMf = Sierra Madre fault; SSf = Santa Susana fault; T = Transmission fault; VCf = Vaszuez Creek fault; Vf = Verdugo fault; Vt = Vincent thrust; WCf = Walnut Creek fault; WR-Rf = Whiterock thrust–Russell fault.

northwest through the western San Gabriel Mountains and Frazier Mountain–southern Orocopia Mountains area, east-southeast through Soledad basin, and northwest through the Chuckwalla, Eagle, Hexie, and Pinto Mountains. In the San Gabriel, Orocopia, Little Chuckwalla, Chuckwalla, and Eagle Mountains, rocks of the anorthosite-syenite complex crop out in a parallel N-shaped belt to the east and south of the Hexie Mountains assemblage. Additional remnants of the Hexie Mountains assemblage occur east of the anorthosite-syenite belt as pendants in Mesozoic plutons in the eastern San Gabriel and Little Chuckwalla Mountains. In the Chuckwalla, Eagle, and Pinto Mountains, rocks of the Eagle Mountains assemblage, including the augen gneiss of Joshua Tree and the overlying quartzite and dolomite of the metasedimentary suite of Placer Canyon, crop out east of the Hexie Mountains assemblage and are intruded by and faulted against rocks of the anorthosite-syenite complex.

In the Mount Pinos–Frazier Mountain–Alamo Mountain block and the western San Gabriel Mountains, metamorphosed Proterozoic or Paleozoic strata of the Limerock Canyon assemblage crop out in an N-shaped belt to the west and north of the Hexie Mountains assemblage.

The Proterozoic rocks and the Proterozoic or Paleozoic rocks are intruded by Mesozoic plutonic rocks, including Triassic(?) hornblende gabbro and diorite; the Triassic Mount Lowe Granodiorite of Miller (1926), renamed the Mount Lowe intrusion by Barth and Ehlig (1988); and distinctive Jurassic and/or Cretaceous units of very coarse-grained granite, muscovite ± biotite ± garnet granite, and "polka-dot" granite (see Appendix) that intrude the Hexie Mountains assemblage, Eagle Mountains assemblage, and the anorthosite-syenite complex. Of particular significance for reconstructing the faults of the San Andreas system are the belt of gabbro and Mount Lowe intrusive bodies that intrude the eastern margins of the anorthosite-syenite complex and the Eagle Mountains assemblage in the San Gabriel, Little Chuckwalla, Chuckwalla, Eagle, and Pinto Mountains, and the association of very coarse grained granite, polka-dot granite, and muscovite granite in the Frazier Mountain–Mount Pinos–Orocopia Mountains area. Subordinate parts of the reference domain consist of Jurassic quartz-poor monzogranite and monzodiorite that intrude the eastern margin of the Proterozoic terrane in the Pinto, Eagle, and San Gabriel Mountains and of Cretaceous quartz diorite that intrudes the Limerock Canyon assemblage and western margin of the Proterozoic terrane.

On Alamo Mountain, Frazier Mountain, Mount Pinos, and in the central Orocopia Mountains, Eocene marine strata unconformably overlap the Mesozoic granitic rocks. A thick wedge of Eocene marine strata crops out west of Alamo Mountain and vicinity, where unnamed lower and middle Eocene strata overlap the crystalline basement (Carman, 1964; Nagle and Parker, 1971). Erosional remnants of unnamed lower Eocene marine strata crop out along the northern margin of Lockwood Valley where they rest on the granitic basement of Mount Pinos (Carman, 1964; Dibblee, 1982b). East of the San Andreas fault in Maniobra Valley of the Orocopia Mountains, lower and upper

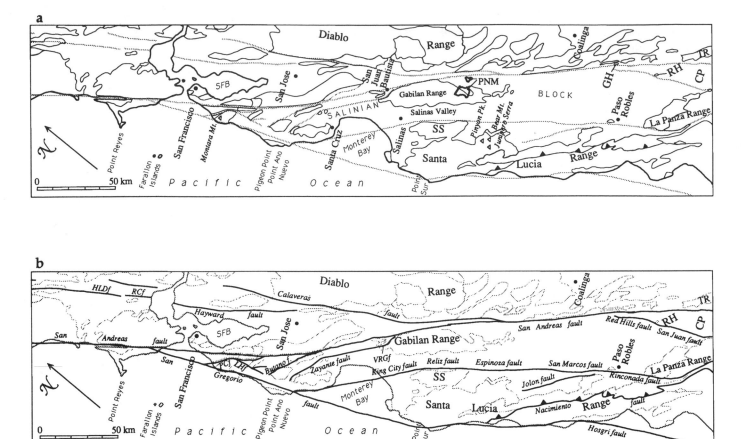

Figure 6. Southern Coast Ranges. a, Geographic locations. CP = Carrizo Plain; GH = Gold Hill; PNM = Pinnacles National Monument; RH = Red Hills; SFB = San Francisco Bay; SS = Sierra de Salinas; TR = Temblor Range. Cities and towns are shown by dots; mountain peaks by triangles. b, Faults: HLDf = Healdsburg fault zone; LHf = La Honda fault; PCf = Pilarcitos fault; RCf = Rodgers Creek fault zone; VRGf = Vergeles fault.

Eocene marine strata of the Maniobra Formation of Crowell (1975a) (see also Crowell and Susuki, 1959) also rest on granitic basement. Both on Mount Pinos and in the Orocopia Mountains, the Eocene marine strata are deposited on the very coarse-grained Mesozoic granite. In both areas, the Eocene strata consist of shale, sandstone, conglomerate, and thin limestone beds.

Upper Oligocene and lower Miocene strata in the reference domain consist of terrestrial red beds and volcanic rocks that accumulated in several extensional, fault-bounded basins subsequently truncated by the San Andreas, San Gabriel, and Clemens Well–Fenner–San Francisquito faults. East of the San Andreas fault, these strata comprise the Diligencia Formation of Crowell (1975a) in the Orocopia Mountains and unnamed volcanic and sedimentary strata in the southern Chuckwalla and Little Chuckwalla Mountains (Crowe and others, 1979; Powell, 1981). Between the San Andreas and San Gabriel faults, these strata consist of the Vasquez Formation (Noble, 1953) (see also Muehlberger, 1958; Oakeshott, 1958; Bohannon, 1975; Hendrix and

Figure 7. Palinspastic reconstruction of the San Andreas fault system (ca. 20 Ma) showing relative rotations of restored blocks. Ruled lines parallel present-day north-south on each block. Mojave Desert, Sierra Nevada, and San Bernardino Mountains are essentially unrotated relative to their present-day orientations. During San Andreas deformation subsequent to about 20 Ma, the Liebre Mountain block has undergone counterclockwise rotation; all other major blocks shown have undergone clockwise rotation, but not to the degree indicated by various paleomagnetic studies (see text for discussion). B = Banning block; BL = Ben Lomond block; CH = Chocolate Mountains; CI = Channel Islands; CM = Cargo Muchacho Mountains; CW = Chuckwalla Mountains; EM = Eagle Mountains block; FM = Frazier Mountain block; G = Gila Mountains; GR = Gabilan Range; LM = Liebre Mountain block; LP = La Panza Range; MM = Montara Mountain; MP = Mount Pinos; O = Orocopia Mountains; P = Portal Ridge; PA = Point Arena; PM = Pinto Mountains block; PP = Pigeon Point; PR = Point Reyes; PS = Point Sur; SB = San Bernardino Mountains; SC = Santa Catalina Island; SE = San Emigdio Mountains; SG = San Gabriel Mountains; SL = Santa Lucia Range; SM = Santa Monica Mountains; SY = Santa Ynez Mountains; T = Tehachapi Mountains; TR = Temblor Range. Cf = future trace of Canton fault; PMf = future trace of Pinto Mountain fault; SGf = future trace of San Gabriel fault.

Ingersoll, 1987) and the Tick Canyon Formation of Jahns (1940) (see also Muehlberger, 1958; Oakeshott, 1958; Ehlert, 1982) in the Soledad basin area between the San Gabriel Mountains and Sierra Pelona. West of both the San Andreas and San Gabriel faults, these strata include the Plush Ranch Formation of Carman (1964) in Lockwood Valley between Frazier Mountain and Mount Pinos. All of these units are complex deposomes with interfingering lithofacies of megaconglomerate, conglomerate, sandstone, shale, evaporite, mafic to intermediate flows and sills,

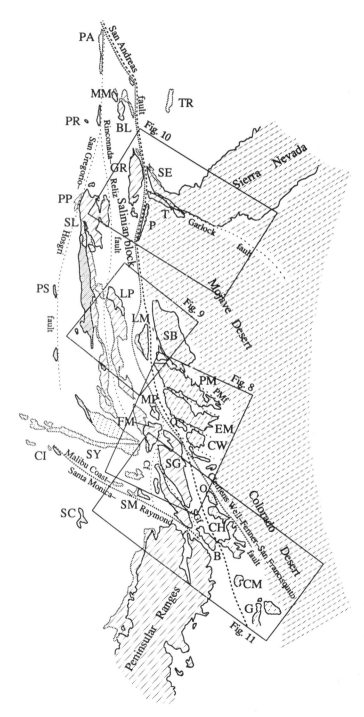

and volcanic tuff units. These deposomes accumulated in alluvial fan, fluvial, deltaic, and lacustrine environments in a tectonically active region characterized by changing source terranes within a fault-controlled paleogeography (Muehlberger, 1958; Carman, 1964; Jahns, 1973; Bohannon, 1975; Hendrix and Ingersoll, 1987).

Although specific piercing patterns have not been established within the lower Miocene strata, several similarities suggest that such patterns may exist. First, the Diligencia and Plush Ranch Formations and the Tick Canyon–Agua Dulce Canyon exposures of the Vasquez Formation all contain a section of calcite-veined and zeolitized hypersthene-augite-olivine basalt and basaltic andesite; lacustrine clay, shale, limestone, and evaporite; and volcanic tuff. Basalt flows in the Plush Ranch Formation have yielded K-Ar ages of 18 to 26 Ma, basalt flows in the Diligencia Formation have yielded K-Ar ages of about 19 to 24 Ma, and basalt and andesite flows in the Vasquez Formation have yielded K-Ar ages of 14 to 26 Ma (Crowell, 1973; Spittler and Arthur, 1982; Frizzell and Weigand, this volume). Frizzell and Weigand (this volume) argue that all ages younger than about 21 Ma in the three formations and those younger than about 24 Ma in the Vasquez Formation are altered ages. In the Agua Dulce Canyon area of the Soledad basin, the Tick Canyon Formation, which contains late Arikareean or early Hemingfordian vertebrate fossils (Durham and others, 1954; Whistler, 1967) unconformably overlies a wedge of Vasquez conglomerate and sandstone that in turn overlie Vasquez volcanic rocks. The wedge of sedimentary rocks thins northward from greater than 1,500 m thick at the Soledad fault to less than 750 m thick at the Green Ranch fault. Across the latter fault in the Tick Canyon area, the Arikareean Tick Canyon Formation rests directly on lacustrine strata. In the southern exposures of the Diligencia Formation, fluvial-deltaic sandstone strata interfingering with basalt and lacustrine strata in the Canyon Spring area also contain late Arikareean or possibly early Hemingfordian vertebrate fossils (Woodburne and Whistler, 1973; Squires and Advocate, 1982). In its restored position, the Diligencia Formation lies roughly "north" of the Tick Canyon area.

Second, the Diligencia and Plush Ranch Formations consist of similar strata deposited on similar basement rocks in similar fault-bounded, synclinally folded basins. Along the southwestern flank of the Diligencia basin and the southern flank of Plush Ranch basin, the strata were deposited on Proterozoic gneiss, whereas along the northeastern flank of the Diligencia basin and the northern flank of the Plush Ranch basin they overlap Eocene marine strata and Mesozoic granitic rocks including the distinctive very coarse grained unit. In contrast, the Vasquez Formation south of the Sierra Pelona and the unnamed strata in the southern Chuckwalla and Little Chuckwalla Mountains were deposited chiefly on Proterozoic gneiss, anorthosite, and syenite, and on the Triassic Mount Lowe intrusion.

Third, both the Diligencia and Vasquez Formations give way east-southeastward from chiefly sedimentary sections in the Orocopia Mountains and in the western Soledad basin to roughly

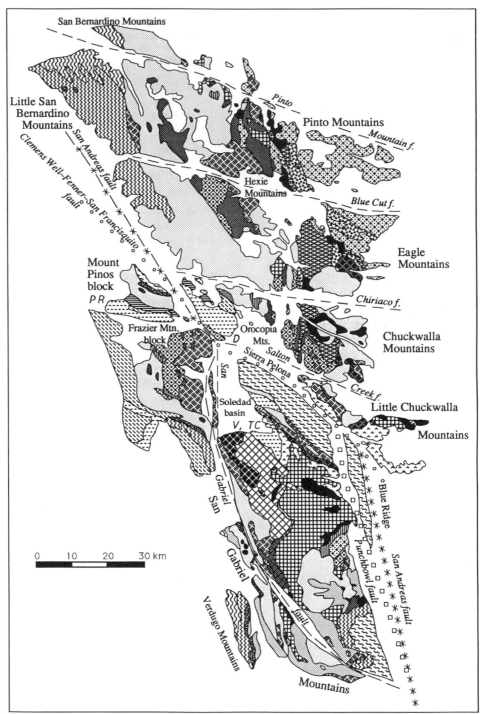

Figure 8. Paleogeologic map (ca. 20 Ma) showing reassembly of the Frazier Mountain–Mount Pinos–eastern Orocopia Mountains reference domain prior to inception of the San Andreas fault system (after Powell, 1981). Reference domain reassembled by sequentially restoring left slip of 8, 11, 5, and 6 km on the Salton Creek, Chiriaco, Blue Cut, and Pinto Mountain faults, respectively, and right slip of 162 km on the San Andreas fault proper, 42 km on the San Gabriel–San Andreas fault, and 110 km on the Clemens Well–Fenner–San Francisquito–San Andreas fault. Restoration of 100 km of right slip on the Clemens Well–Fenner–San Francisquito–San Andreas fault yields a reasonable alternate reconstruction of the Mount Pinos–Frazier Mountain blocks against the eastern Orocopia Mountains. See explanation to Plate 1 for sources of geology shown. D = Diligencia Formation of Crowell (1975); PR = Plush Ranch Formation of Carman (1964); TC = Tick Canyon Formation of Jahns (1940); V = Vasquez Formation. Blank areas are now covered by Quaternary surficial deposits, by Miocene or Pliocene sedimentary and volcanic rocks, or are minor gaps between reassembled blocks.

EXPLANATION

SEDIMENTARY AND VOLCANIC ROCKS

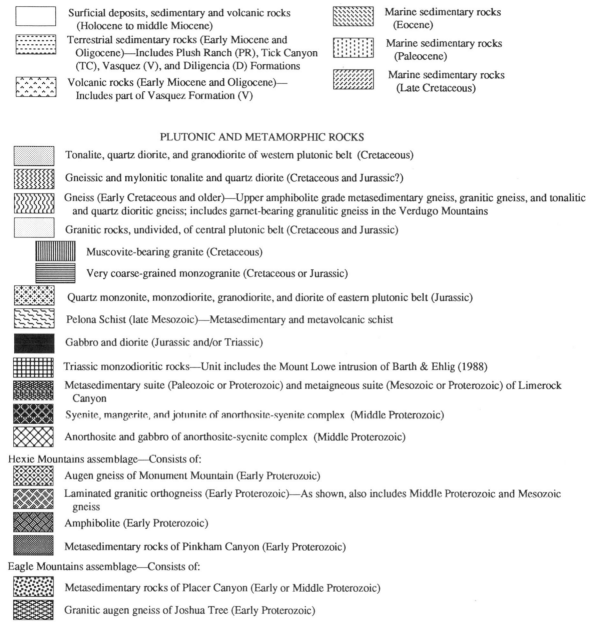

Surficial deposits, sedimentary and volcanic rocks (Holocene to middle Miocene)

Terrestrial sedimentary rocks (Early Miocene and Oligocene)—Includes Plush Ranch (PR), Tick Canyon (TC), Vasquez (V), and Diligencia (D) Formations

Volcanic rocks (Early Miocene and Oligocene)— Includes part of Vasquez Formation (V)

Marine sedimentary rocks (Eocene)

Marine sedimentary rocks (Paleocene)

Marine sedimentary rocks (Late Cretaceous)

PLUTONIC AND METAMORPHIC ROCKS

Tonalite, quartz diorite, and granodiorite of western plutonic belt (Cretaceous)

Gneissic and mylonitic tonalite and quartz diorite (Cretaceous and Jurassic?)

Gneiss (Early Cretaceous and older)—Upper amphibolite grade metasedimentary gneiss, granitic gneiss, and tonalitic and quartz dioritic gneiss; includes garnet-bearing granulitic gneiss in the Verdugo Mountains

Granitic rocks, undivided, of central plutonic belt (Cretaceous and Jurassic)

Muscovite-bearing granite (Cretaceous)

Very coarse-grained monzogranite (Cretaceous or Jurassic)

Quartz monzonite, monzodiorite, granodiorite, and diorite of eastern plutonic belt (Jurassic)

Pelona Schist (late Mesozoic)—Metasedimentary and metavolcanic schist

Gabbro and diorite (Jurassic and/or Triassic)

Triassic monzodioritic rocks—Unit includes the Mount Lowe intrusion of Barth & Ehlig (1988)

Metasedimentary suite (Paleozoic or Proterozoic) and metaigneous suite (Mesozoic or Proterozoic) of Limerock Canyon

Syenite, mangerite, and jotunite of anorthosite-syenite complex (Middle Proterozoic)

Anorthosite and gabbro of anorthosite-syenite complex (Middle Proterozoic)

Hexie Mountains assemblage—Consists of:

Augen gneiss of Monument Mountain (Early Proterozoic)

Laminated granitic orthogneiss (Early Proterozoic)—As shown, also includes Middle Proterozoic and Mesozoic gneiss

Amphibolite (Early Proterozoic)

Metasedimentary rocks of Pinkham Canyon (Early Proterozoic)

Eagle Mountains assemblage—Consists of:

Metasedimentary rocks of Placer Canyon (Early or Middle Proterozoic)

Granitic augen gneiss of Joshua Tree (Early Proterozoic)

coeval thick volcanic sections in the Chuckwalla and Little Chuckwalla Mountains and in the eastern Soledad basin. These eastern volcanic sections consist chiefly of andesite and dacite with subordinate basalt, conglomerate, and sandstone. The transitions are realigned in the reconstructed reference domain.

Fourth, similarities in clast provenance within various parts of the three formations suggest paleogeographic links. The Diligencia Formation and Texas Canyon and Tick Canyon–Aqua Dulce Canyon exposures of the Vasquez Formation contain debris flows with abundant large clasts of Proterozoic anorthosite and the Mount Lowe intrusion derived from the south (Muehlberger, 1958; L. T. Silver, oral communication, 1973, and in a 1984 lecture at the U.S. Geological Survey, Menlo Park; Spittler and Arthur, 1973; Bohannon, 1975; Squires and Advocate, 1982; Hendrix and Ingersoll, 1987). Silver's inference that the clasts, including those in the Diligencia Formation, were derived from exposures in the northwestern San Gabriel Mountains is accommodated by the restored position of the Diligencia Formation in the reassembled Frazier Mountain–Mount Pinos–eastern Orocopia Mountains reference domain. This clast-to-source match-up contrasts with the local source proposed in the Orocopia Mountains (Crowell, 1962, p. 28, 1975a; Spittler and Arthur, 1973, 1982; Bohannon, 1975), where the Mount Lowe intrusion has not been recognized. In addition, the Plush Ranch Formation contains Proterozoic norite and granulite derived from the southeast (Carman, 1964; Woodburne, 1975); the Plush Ranch and Diligencia Formations and the Texas Canyon exposures of the Vasquez Formation contain clasts of the distinctive very coarse grained Mesozoic granite derived from the north and northeast from exposures on Mount Pinos and in the eastern Orocopia Mountains (Muehlberger, 1958; Bohannon, 1975).

The palinspastic realignment attained by full restoration of their underlying crystalline basement yields a paleogeologic distribution for the lower Miocene strata (Fig. 8) that differs from previously proposed realignments (Crowell, 1962; Spittler and Arthur, 1973; Bohannon, 1975). In the reconstruction based on the crystalline rocks, the Diligencia Formation is restored to an

original position between the Plush Ranch and Vasquez sections rather than to a position "east" of those sections. The potential links suggested by the shared stratigraphic features listed above are accommodated within this new reconstruction.

La Panza Range–Liebre Mountain–San Bernardino Mountains reference domain. The La Panza Range–Liebre Mountain–San Bernardino Mountains reference domain includes probable Proterozoic metamorphic rocks, metamorphosed Proterozoic or Paleozoic strata, Mesozoic granitic and quartz dioritic rocks, Upper Cretaceous and Paleocene marine sedimentary strata, and Oligocene terrestrial sedimentary strata (Fig. 9). The Proterozoic and Mesozoic rocks comprise a crystalline basement that was overlapped successively by the Upper Cretaceous, Paleocene, and Oligocene strata. Subsequent folding and faulting has exposed steeply dipping unconformities at the bases of the sedimentary sequences. The domain is reconstructed from disrupted parts now exposed in the area of the La Panza–Caliente Ranges west of the San Andreas fault, in the Liebre Mountain block between the San Andreas and San Gabriel faults and north

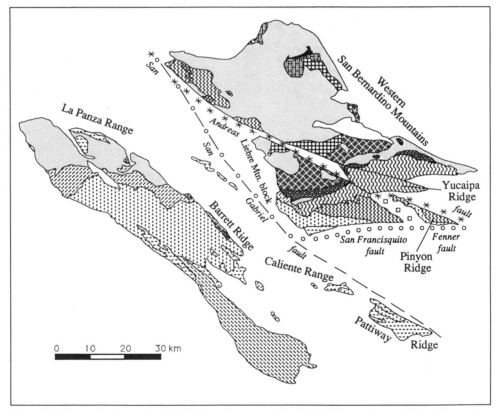

Figure 9. Paleogeologic map (ca. 20 Ma) showing reassembly of the La Panza Range–Liebre Mountain–San Bernardino Mountains reference domain prior to inception of the San Andreas fault system (Liebre Mountain–San Bernardino Mountains tie in part after Matti and others, 1985, 1986; Frizzell and others, 1986; Matti, this volume). Reference domain reassembled by restoring right slip of 162 km on the San Andreas fault proper, including 42 km on the Punchbowl fault strand; 42 km on the San Gabriel–San Andreas fault; and 110 km on the Clemens Well–Fenner–San Francisquito–San Andreas fault. See explanation to Plate I for sources of geology shown. Blank areas are now covered by Quaternary surficial deposits, by Miocene or Pliocene sedimentary and volcanic rocks, or are minor gaps between reassembled blocks.

of the San Francisquito fault, in the Pinyon Ridge area between the San Andreas and Punchbowl faults and north of the Fenner fault, and in the western San Bernardino Mountains east of the San Andreas fault. The reference domain is based on the realignment, by Matti and colleagues (Matti and others, 1985, 1986; Frizzell and others, 1986; Matti and Morton, this volume), of Triassic porphyritic monzogranite and associated crystalline and sedimentary rocks at Liebre Mountain and in the Mill Creek area of the San Bernardino Mountains; on Saul's (1983) identification of a unique assemblage of distinctive turritellas in the Upper Cretaceous strata of both the La Panza Range and Liebre Mountain block (see also Joseph and others, 1982b); on the distribution of rocks of the Limerock Canyon assemblage in the three crystalline blocks; and the trace of the quartz diorite line in the three blocks (Fig. 9; Plates I and II).

Probable Proterozoic rocks (see Appendix) of the reassembled domain include biotite-rich gneiss and small bodies of augen gneiss that may be remnants of the Proterozoic Hexie Mountains assemblage and that occur in the western San Bernardino Mountains (Cox and others, 1983a), in unnamed gneiss in the central Liebre Mountain block (R. J. Weldon, 1986, oral communication), and along Barrett Ridge (Ross, 1972 [p. 32–33], 1977, 1978, 1984 [p. 20–21]; Mattinson, 1983; Mattinson and James, 1985). South of these rocks, a belt of metamorphosed Proterozoic or Paleozoic strata of the Limerock Canyon assemblage, which includes a distinctive suite of graphite schist, vitreous white quartzite, pelitic schist, and marble, trends east-west across the southwestern San Bernardino Mountains, southern Liebre Mountain block, and Barrett Ridge.

Mesozoic plutonic rocks in the reference domain include granitic rocks and strongly deformed quartz dioritic rocks. The oldest recognized Mesozoic pluton is the megaporphyritic Triassic monzogranite identified by Matti and colleagues (1985, 1986; Frizzell and others, 1986) on opposite sides of the San Andreas fault in the northern part of the Liebre Mountain block and in the Mill Creek–San Gorgonio Mountain–Sugarloaf Mountain area of the San Bernardino Mountains. In the western San Bernardino Mountains, Liebre Mountain block, and La Panza Range the

EXPLANATION

SEDIMENTARY AND VOLCANIC ROCKS

Surficial deposits, sedimentary and volcanic rocks (Holocene to middle Miocene)

Terrestrial sedimentary rocks (Early Miocene and Oligocene)—Includes the Simmler Formation and the Charlie Canyon section of the Vasquez Formation

Marine sedimentary rocks (Eocene)

Marine sedimentary rocks (Paleocene)

Marine sedimentary rocks (Late Cretaceous)

PLUTONIC AND METAMORPHIC ROCKS

Tonalite, quartz diorite, and granodiorite of western plutonic belt (Cretaceous)

Gneissic and mylonitic tonalite and quartz diorite (Cretaceous)

Gneiss (Early Cretaceous and older)—Granitic gneiss, upper amphibolite grade metasedimentary gneiss, and quartz dioritic and tonalitic gneiss, includes garnet-bearing granulitic gneiss in the San Bernardino Mountains

Granitic rocks, undivided, of central plutonic belt (Cretaceous and Jurassic?)

Gabbro and diorite (Jurassic or Triassic)

Carbonate rocks (Triassic, Paleozoic, or Proterozoic)

Triassic monzodioritic rocks—Unit includes the megaporphyritic monzonite of Frizzell and others (1986)

Metasedimentary rocks (Paleozoic)—Miogeoclinal strata

Metasedimentary suite (Paleozoic and Proterozoic) and metaigneous suite (Mesozoic or Proterozoic) of Limerock Canyon

Hexie Mountains assemblage—Consists of:

Laminated granitic orthogneiss (Early Proterozoic)—As mapped, also includes Mesozoic gneiss

Eagle Mountains or Limerock Canyon assemblage—Consists of:

Metasedimentary rocks of Placer Canyon (Early or Middle Proterozoic) or Limerock Canyon (Paleozoic or Proterozoic)

 Punchbowl fault

older rocks were intruded by Cretaceous porphyritic monzogranite in turn intruded by Cretaceous polka-dot granite (cf. Joseph and others, 1982b). To the southwest these granitic rocks intruded foliate Cretaceous quartz dioritic rocks that themselves intruded metamorphic rocks of the Limerock Canyon assemblage in the San Bernardino Mountains, southern Liebre Mountain block, and beneath the Upper Cretaceous and Paleocene strata south of the La Panza Range.

Upper Cretaceous and Paleocene sedimentary rocks in the reference domain onlap the crystalline rocks in a south-dipping homocline of submarine fan deposits that constitute unnamed strata in the southern La Panza Range (Chipping, 1972; Nilsen and Clarke, 1975; Howell and Vedder, 1978); the Pattiway Formation (Hill and others, 1958; Dibblee, 1973) in the southernmost Caliente Range; and the San Francisquito Formation (Dibblee, 1967a; Sage, 1975; Kooser, 1982) in the Warm Springs Mountain area of the southern part of the Liebre Mountain block, the Valyermo area between the Punchbowl and San Andreas faults, and the Cajon Pass area northeast of the San Andreas fault. The Cretaceous parts of the San Francisquito Formation of the Warm Springs Mountain area and the unnamed strata of the La Panza Range share a unique assemblage of Maestrichtian turritellas (Saul, 1983, p. 2, 23, 30–33, 70, 85). In the Warm Springs Mountain area, the uppermost Cretaceous strata are pinched out by strata bearing early Paleocene turritellas, which are in turn pinched out by strata bearing late Paleocene turritellas (Saul, 1983, p. 30–33). In the Valyermo area, the San Francisquito Formation contains late Paleocene turritellas, and in the Cajon Pass area it contains a Late Cretaceous plesiosaur (Saul, 1983; Kooser, 1985).

The Oligocene strata consist of the Simmler Formation (part) (Hill and others, 1958; Bartow, 1978) in the La Panza and Caliente Ranges and the Charlie Canyon section of the Vasquez Formation in the Liebre Mountain block just north of the San Francisquito fault. Clasts from conglomerate beds in this section of the Vasquez Formation do not include any derived from the anorthosite-syenite complex, the Mount Lowe intrusion, or the augen gneiss of Monument Mountain, all of which are abundant in the Vasquez sections in Soledad basin south of the Sierra Pelona and do contain granitic rocks, sandstone, and recycled volcanic rocks that are common in the Simmler Formation of the Caliente Range (see Bohannon, 1975; Bartow, 1978; Hendrix and Ingersoll, 1987), against which the Charlie Canyon section is reconstructed.

Gabilan Range–San Emigdio Mountains–Portal Ridge reference domain. The Gabilan Range–San Emigdio Mountains–Portal Ridge reference domain (Fig. 10) includes Paleozoic(?) and Mesozoic metasedimentary rocks, Mesozoic plutonic rocks, Eocene and Oligocene marine strata, and lower Miocene volcanic rocks. The domain is reconstructed from disrupted parts now exposed west of the San Andreas fault in the Gabilan Range and vicinity, and east of the San Andreas fault in the San Emigdio and Tehachapi Mountains, the southernmost Sierra Nevada, and the westernmost Mojave Desert, including Portal Ridge and vicinity. Reassembly of this reference domain is based on the results of many previous studies (see, especially, Ross, 1984, Fig. 14; Nilsen, 1984).

Mesozoic granitic and gabbroic terranes are juxtaposed along a reconstructed fault that includes the Vergeles fault on the north flank of the Gabilan Range and the Pastoria fault in the San Emigdio and Tehachapi Mountains (Ross and Brabb, 1973; Ross, 1984, Fig. 14). Within Cretaceous quartz dioritic and granitic rocks of the batholithic terrane south of this fault, Paleozoic(?) and Mesozoic marble and pelitic schist crop out in a train of pendants, screens, and inclusions that extends through the Gabilan Range, San Emigdio and Tehachapi Mountains, and southern Sierra Nevada (Ross, 1984, Fig. 14). North of the fault, in the northern San Emigdio Mountains and in the San Juan Bautista area just north of the Gabilan Range, Jurassic gabbroic rocks that may be part of the Coast Range ophiolite are unconformably overlain by Eocene and Oligocene marine strata (Ross, 1970; Clarke and Nilsen, 1973; Nilsen, 1984; James and others, this volume).

Mesozoic schist of Sierra de Salinas that crops out in the southern part of the Gabilan Range and vicinity has been correlated tentatively with Pelona Schist that crops out on Portal Ridge and in the San Emigdio and Tehachapi Mountains (Ross, 1976, 1984 [Fig. 14]).

Distinct sequences of lower Miocene strata crop out across the future trace of the San Andreas fault in the northern and southern parts of the reference domain (Hill and Dibblee, 1953; Dibblee, 1966a; Huffman, 1972; Huffman and others, 1973; Matthews, 1976; Sims, this volume). To the north, the Eocene and Oligocene marine strata in the San Juan Bautista area and San Emigdio Mountains are overlain by lower Miocene strata in which interfingering marine strata and conglomerate red beds record a shoreline and contain interbedded 21- to 23-m.y.-old intermediate volcanic rocks. To the south, granitic and quartz dioritic rocks of the southern Gabilan Range and westernmost Mojave Desert are overlain by the 22- to 24-m.y.-old Pinnacles and Neenach Volcanics.

Southern Transverse Ranges–Chocolate Mountains reference domain. The southern Transverse Ranges–Chocolate Mountains reference domain (Fig. 11) includes Proterozoic metasedimentary and metaigneous gneiss, Paleozoic and Mesozoic metasedimentary rocks, Mesozoic plutonic rocks, and early and middle Miocene volcanic rocks. Parts of the domain west of the San Andreas fault include the north end of the Peninsular Ranges and the southern margin of the Transverse Ranges, which consists of the Santa Monica Mountains, southern San Gabriel Mountains, and Banning block of the southern San Bernardino Mountains. Parts of the domain east of the San Andreas fault include the Chocolate and Cargo Muchacho Mountains and vicinity in southeasternmost California and the Laguna, Muggins, and northern Gila Mountains and vicinity in southwesternmost Arizona. The southern Transverse Ranges–Chocolate Mountains paleogeologic reference domain incorporates Ehlig's (1966, 1975a, 1981) realignment of crystalline rock paleogeology along the San Gabriel fault in the southern San Gabriel Mountains, Dillon's (1976; Dillon and Ehlig, this volume) realignment of the San Gorgonio Pass area and Chocolate Mountains, a modified version of Campbell and Yerkes' (1976) realignment of the Santa Monica and Santa Ana Mountains, and the distribution of rocks of the metasedimentary suite of the Limerock Canyon assemblage.

Metamorphic rocks in the reference domain comprise several assemblages. Remnants of the Proterozoic and Paleozoic cratonic terrane crop out in an arc across the northern and eastern parts of the reference domain. As parts of this terrane, rocks of the Proterozoic Hexie Mountains assemblage and anorthosite-syenite complex crop out in the southern San Gabriel Mountains, Banning block, Chocolate Mountains, Laguna Mountains, and Muggins Mountains, and metamorphosed Paleozoic cratonal strata crop out in the Cargo Muchacho, northern Gila, and Muggins Mountains. In a belt to the west and south of the cratonic terrane, the Proterozoic or Paleozoic metasedimentary suite of Limerock Canyon crops out in the northern Peninsula Ranges, the southern San Gabriel Mountains, and the Verdugo Mountains. Metamorphosed upper Paleozoic(?) and Triassic rocks crop out west and south of the Limerock Canyon suite in the northern Peninsular Ranges; metamorphosed Jurassic sedimentary and volcanic rocks crop out still farther west in the northern Peninsular Ranges and Santa Monica Mountains.

Different suites of Mesozoic plutonic rocks intrude the various metamorphic rocks of the reference domain. The Hexie Mountains assemblage is intruded by Triassic(?) gabbro and diorite, the Triassic Mount Lowe intrusion, and Jurassic and Cretaceous granitic rocks in the San Gabriel Mountains, Banning block, and Chocolate Mountains. The metamorphosed Mesozoic rocks of the Peninsular Ranges are intruded by Cretaceous quartz diorite and granodiorite. The Limerock Canyon assemblage, which occurs as inclusions, pendants, and screens along the quartz diorite line in the Cretaceous batholith, is intruded by both quartz dioritic and granitic plutonic rocks.

The quartz diorite line in the northeasternmost Peninsular Ranges and southern San Gabriel Mountains was an approximate locus for synplutonic mylonitization in the Late Cretaceous and subsequent postplutonic brittle faulting in both the southern San Gabriel Mountains (Alf, 1948; Hsu, 1955; Morton, 1975b; May, 1986; Morton and Matti, 1987; May and Walker, 1989) and the eastern Peninsular Ranges (Sharp, 1966, 1967, 1979; Simpson, 1984; Erskine and Wenk, 1985; Erskine, 1986a, b). Whereas early investigators noted similarities and differences between the deformed rocks of these two areas (Miller, 1946; Allen, 1957; Woodford, 1960; Sharp, 1967), Baird and others (1974) suggested that they are correlative, and May (1986, 1989; May and Walker, 1989) has proposed a thrust-and-tear fault model that reconciles some of the differences. If the belts of deformed rocks are related, they constitute a paleogeologic feature that constrains fault displacements.

Pelona and Orocopia Schists crop out in antiforms in the southeastern San Gabriel Mountains, Banning block, and Chocolate Mountains and vicinity, where they occur beneath the Vincent and Chocolate Mountains thrust faults. Rocks of the Hexie Mountains assemblage and deformed Triassic, Jurassic, and Cretaceous plutonic rocks are superposed above the Pelona and Orocopia Schists along the thrust.

Oligocene and lower and middle Miocene volcanic and sedimentary rocks crop out throughout much of the reference domain. The lower Miocene Vaqueros Formation crops out in the northwestern Peninsular Ranges and Santa Monica Mountains, where it contains marine and nonmarine strata that interfingered aong a northwest trending shoreline (Yeats, 1968; Campbell and Yerkes, 1976; Truex, 1976). Oligocene and lower Miocene volcanic and sedimentary rocks crop out in southeasternmost California and southwesternmost Arizona, where they were deposited on Proterozoic and Mesozoic igneous and metamorphic rocks (Olmsted, 1972; Olmsted and others, 1973; Dillon, 1976; Crowe, 1978; Crowe and others, 1979; Smith and others, 1984, 1987; Sherrod and others, 1989). Middle Miocene sedimentary and volcanic rocks that crop out along the southern margin of the Transverse Ranges and in southeastern California include the Glendora Volcanics (Shelton, 1955) in the San Gabriel Valley, the Coachella Fanglomerate of Vaughan (1922) in the Banning block, and the fanglomerate of Bear Canyon of Dillon (1976) in the Chocolate Mountains and vicinity.

The terrane characterized by the Limerock Canyon metasedimentary suite, quartz diorite of the western Mesozoic plutonic belt, and synplutonic mylonitization defines a north-northeast-trending belt in the eastern Peninsular Ranges that is deflected westward in the southeastern San Gabriel Mountains (Baird and others, 1974; May, 1986, 1989). Terranes of Proterozoic rocks that crop out east and north of this belt and metamorphosed Mesozoic strata that crop out to the west and south also are deflected westward. Similarly, the western depositional(?) limit of Oligocene and lower Miocene volcanic rocks in southern Arizona may be deflected westward through the Laguna and southern Chocolate Mountains to the western limit of volcanic rocks in the northern Chocolate Mountains. Finally, the lower Miocene shoreline, inferred by Campbell and Yerkes (1976) to have been displaced sinistrally along the Malibu Coast–Santa Monica–Raymond fault, alternatively may be deflected westward beneath the younger strata of the Los Angeles basin.

BALANCED RECONSTRUCTION OF ROCKS ALONG THE SAN ANDREAS FAULT SYSTEM

San Andreas fault system

Displacement on the San Andreas and related faults is a passive response to thermal and gravitational driving mechanisms along the plate boundary. In southern and central California, the San Andreas fault system comprises an interactive plexus of northwest-trending right-lateral strike-slip and reverse faults, east-to northeast-trending left-lateral strike-slip and left-oblique reverse faults, and related folds (Plate I).

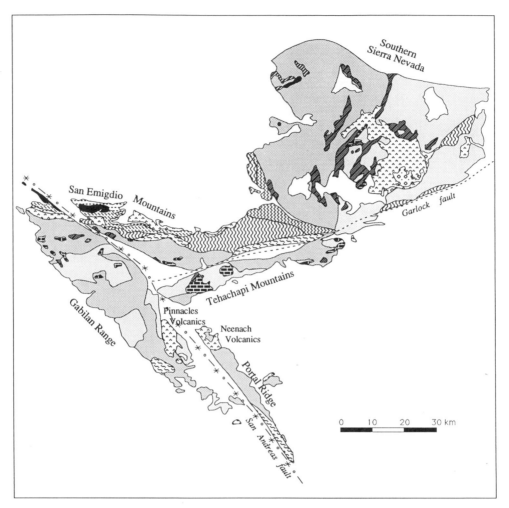

Figure 10. Paleogeologic map (ca. 20 Ma) showing reassembly of the Gabilan Range–San Emigdio Mountains–Portal Ridge reference domain prior to inception of the San Andreas fault system (after Ross, 1984). Reference domain reassembled by restoring right slip of 295 km on the San Andreas fault proper and left slip of 12 km on the western segment of the Garlock fault. See explanation to Plate I for sources of geology shown. Blank areas are now covered by Quaternary surficial deposits, by Miocene or Pliocene sedimentary and volcanic rocks, or are minor gaps between reassembled blocks.

In pattern and sequence, the dextral faults fall into six groups: (1) throughgoing faults that transect the Transverse Ranges (San Andreas, Punchbowl, San Gabriel, and Clemens Well–Fenner–San Francisquito faults); (2) faults in the Salinian block that terminate to the southeast against the western Transverse Ranges and merge northwestward with the San Andreas fault (Red Hills–Ozena, Rinconada-Reliz, and San Gregorio–Hosgri faults); (3) faults in the Peninsular Ranges that terminate to the northwest against the central and western Transverse Ranges and to the southeast in the Salton trough (San Jacinto and Elsinore faults); (4) faults in the Mojave Desert that occur between the eastern Transverse Ranges to the southeast and the left-lateral Garlock fault to the northwest; (5) faults in the Death Valley area; and (6) the Calaveras–Hayward–Rodgers Creek–Maacama–Garberville fault zone in the northern Coast Ranges that merges southward with the San Andreas fault.

The sinistral faults comprise four groups. First, the southern boundary of the Transverse Ranges province west of the San Andreas fault is characterized by a system of left-oblique reverse faults. This system includes three sets of aligned faults: one incorporating the Malibu Coast, Santa Monica, Raymond, and Clamshell–Sawpit Canyon faults; a second incorporating the Evey Canyon, San Antonio Canyon, Stoddard Canyon, and San Jose faults; and a third incorporating the Santa Susana, Sierra Madre, and Cucamonga faults. These reverse faults have been inferred to mask a throughgoing ancestral left-lateral strike-slip fault, which, if it exists, ought to have been displaced along the San Andreas fault. A second group of left-lateral faults consists of the Big Pine and Santa Ynez faults that transect the Cenozoic sedimentary rocks of the western Transverse Ranges west of the San Gabriel fault. A third group consists of northeast-trending faults in the San Gabriel Mountains that terminate westward at the San

EXPLANATION

SEDIMENTARY AND VOLCANIC ROCKS

Surficial deposits, sedimentary and volcanic rocks (Holocene to middle Miocene)

Volcanic rocks (Middle and early Miocene)

Terrestrial conglomerate (Miocene and Eocene? or Paleocene?)

Marine sedimentary rocks (Oligocene)

Marine sedimentary rocks (Eocene)

PLUTONIC AND METAMORPHIC ROCKS

Tonalite, quartz diorite, and granodiorite of western plutonic belt (Cretaceous)

Gneissic and mylonitic tonalite and quartz diorite (Cretaceous)

Gneiss (Early Cretaceous and older)—Quartz dioritic and tonalitic gneiss, granitic gneiss, and upper amphibolite grade metasedimentary gneiss; includes hypersthene- and/or garnet-bearing granulitic gneiss

Granitic rocks, undivided, of central plutonic belt (Cretaceous)

Pelona Schist and schist of Sierra de Salinas, undivided (late Mesozoic)—Metasedimentary and metavolcanic rocks

Gabbro and diorite (Jurassic)

Metasedimentary and metavolcanic rocks of western metamorphic belt (Cretaceous to Triassic)

Carbonate rocks; schist and quartzite (Jurassic to Triassic and late Paleozoic?)

* * * * * San Andreas fault

—— —— —— San Gabriel fault

o o o o o Clemens Well–Fenner–San Francisquito fault

Gabriel fault. A fourth group consists of east-trending left-lateral faults east of the San Andreas fault that terminate against that fault, including the Garlock fault between the Mojave Desert province to the south and the Sierra Nevada and Great Basin to the north, the Pinto Mountain, Blue Cut, Chiriaco, and Salton Creek faults in the eastern Transverse Ranges, and the Mammoth Wash and Black Eagle faults in the Chocolate Mountains.

Northwest-trending reverse faults include the Pine Mountain fault in the western Transverse Ranges and the Morales and Whiterock faults in the southern Salinian block.

Assembling the reconstruction

The present-day geologic base map (Plate I) for the reconstruction was modified from the central and southern parts of the 1:750,000 Geologic Map of California (Jennings, 1977) by distinguishing regionally extensive Proterozoic, Paleozoic, and Mesozoic igneous and metamorphic units. The palinspastic reconstruction (Plate II) was accomplished in three stages by cutting the base map along the strike-slip faults, then sliding the fault blocks along the cut faults to reassemble the four paleogeologic reference domains. Plate II shows successive paleogeologic reconstructions at about 5, 13, and 18 or 22 Ma.

In Figures 8 through 11, the four paleogeologic reference domains were reconstructed to achieve the closest possible fit of their disrupted rock units. The overall palinspastic reconstruction shown in Plate II was reassembled to reproduce all four of these close-fit reference domains as faithfully as possible, while simultaneously restoring known displacements on all strike-slip faults. The overall reconstruction approximately reconciles all four reference domains, but because of the difficulty of precise restoration of displacement on reverse faults and across extensional basins, it does not reproduce them exactly. Moreover, to reconstruct paleogeologic patterns as shown in Plate II, it was necessary to restore specific displacements along the various strke-slip faults of the San Andreas system. Typically, the specific displacement restored on a given fault was selected from a range of measurements (see sections on right- and left-lateral faults). In some cases, alternate choices could have been made that would change details of the reconstruction.

Reassembly of disrupted paleogeologic patterns of pre-Cenozoic plutonic and metamorphic rocks and Cenozoic sedimentary and volcanic strata not only constrains overall restoration of slip on the right- and left-lateral faults that transect the batholithic terrane of southern and central California, but also provides kinematic constraints on the sequence of faulting. The

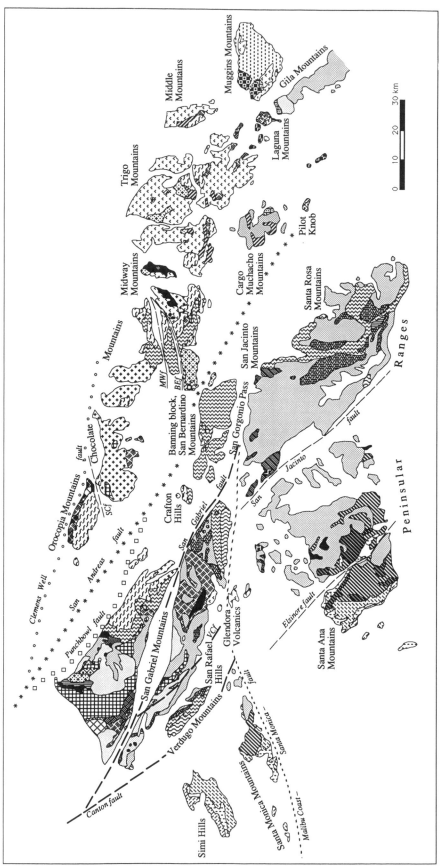

Figure 11. Paleogeologic map (ca. 20 Ma) showing reassembly of the southern boundary of Transverse Ranges province. Reconstruction constrained by cross-fault ties proposed by Ehlig (1966) for the San Gabriel fault, Dillon (1976) for the southern San Andreas fault, and Campbell and Yerkes (1976) for the Malibu Coast–Santa Monica–Raymond fault and by the distribution of the metasedimentary suite of the Limerock Canyon assemblage (see text). See explanation to Plate I for sources of geology shown. Blank areas are now covered by Quaternary surficial deposits, by Miocene or Pliocene sedimentary and volcanic rocks, or are minor gaps between reassembled blocks. BEf = Black Eagle fault; MWf = Mammoth Wash fault; SCf = Salton Creek fault; VCf = Vasquez Creek fault.

paleogeologic reference domains in Figures 8 through 11 are particularly revealing in restoring slip on the right-lateral faults that transect the Transverse Ranges because the domains have been dislocated along spatially separated faults of the San Andreas system.

In order to accomplish the reconstruction, dextral displacements were restored on the Clemens Well–Fenner–San Francis-quito fault (110 km), San Gabriel fault system (42 km), Rinconada-Reliz and San Gregorio–Hosgri fault system (150 km), San Andreas fault (162 km), San Jacinto fault (28 km), Elsinore fault (5 km), Helendale fault (3 km), Lockhart-Lenwood fault (3 km), Harper–Harper Lake–Camp Rock–Emerson fault (4 km), Blackwater–Calico–Mesquite Lake fault (9 km), Pisgah-Bullion fault (16 km), an unnamed buried fault between the

EXPLANATION

SEDIMENTARY AND VOLCANIC ROCKS

Surficial deposits, sedimentary and volcanic rocks (Holocene to middle Miocene)

Terrestrial sedimentary rocks (Early Miocene and Oligocene)

Volcanic rocks (Early Miocene and Oligocene)

Marine sedimentary rocks (Eocene)

Marine sedimentary rocks (Paleocene)

Marine sedimentary rocks (Late Cretaceous)

PLUTONIC AND METAMORPHIC ROCKS

Plutonic rocks (early Miocene and Oligocene)—Monzogranite, diorite, and felsic dike swarms; contains inclusions of Proterozoic and Mesozoic rocks

Tonalite, quartz diorite, and granodiorite of western plutonic belt (Cretaceous)

Gneissic and mylonitic tonalite and quartz diorite (Cretaceous and Jurassic?)

Gneiss (Early Cretaceous and older)—Upper amphibolite grade metasedimentary gneiss, granitic gneiss, and tonalitic and quartz dioritic gneiss; includes hypersthene- and/or garnet-bearing granulitic gneiss in the southeastern San Gabriel and Verdugo Mountains

Granitic rocks, undivided, of central plutonic belt (Cretaceous and Jurassic)

Muscovite bearing granite (Cretaceous)

Quartz monzonite, monzodiorite, granodiorite, and diorite of eastern plutonic belt (Jurassic)

Pelona and Orocopia Schists, undivided (late Mesozoic)—Metasedimentary and metavolcanic rocks

Gabbro and diorite (Cretaceous and Jurassic and/or Triassic)—Cretaceous in the Peninsular Ranges; Triassic and/or Jurassic in the Transverse Ranges and Chocolate Mountains. As shown, includes Proterozoic and/or Mesozoic amphibolite in the San Gabriel, Chocolate, and Midway Mountains

Metasedimentary and metavolcanic rocks of eastern metamorphic belt (Jurassic? and Triassic?)—Unit includes the Winterhaven Formation

Metasedimentary and metavolcanic rocks of western metamorphic belt (Cretaceous to Triassic)

Carbonate rocks; schist and quartzite (Jurassic to Triassic and late Paleozoic?)

Mount Lowe intrusion of Barth & Ehlig (1988) and related rocks (Triassic)—Predominantly monzodiorite and monzonite
Metasedimentary rocks (Paleozoic)—Unit consists of cratonal strata

Metasedimentary suite (Paleozoic or Proterozoic) and metaigneous suite (Mesozoic or Proterozoic) of Limerock Canyon

Syenite, mangerite, and jotunite of anorthosite-syenite complex (Middle Proterozoic)

Anorthosite and gabbro of anorthosite-syenite complex (Middle Proterozoic)

Porphyritic monzogranite (Middle Proterozoic)

Augen gneiss of Monument Mountain (Early Proterozoic)

Hexie Mountains assemblage—Consists of:

Laminated granitic orthogneiss (Early Proterozoic)—As shown, also includes Middle Proterozoic and Mesozoic gneiss

Amphibolite (Early Proterozoic)—As shown, also includes Mesozoic gabbroic rocks

Metasedimentary rocks of Pinkham Canyon (Early Proterozoic)

Sheep Hole and Coxcomb Mountains (3.5 km), an unnamed buried fault between the Coxcomb and Palen Mountains (4.5 km), and an unnamed buried fault between the Palen and McCoy Mountains (4 km). Displacement measured on the San Andreas fault, which includes the Pilarcitos, Punchbowl, Mill Creek, Mission Creek, and Wilson Creek faults as anastomosed strands, increases with contributions from the left-lateral faults of the eastern Transverse Ranges and the San Jacinto fault. No displacement was restored on the Red Hills–Ozena fault.

Also in order to accomplish the reconstruction, sinistral displacements were restored on the westernmost Garlock fault (12 km), Pinto Mountain fault (16 km), Niggerhead fault (1.5 to 2 km), Blue Cut fault (5 km), Chiriaco fault (11 km), Corn Spring Wash fault (3 km), Ship Creek fault (1.5 to 2 km), Salton Creek fault (8 km), a hypothesized fault between the Little Chuckwalla and Mule Mountains (5.5 km), Mammoth Wash fault (5 km), Black Eagle fault (8 km), and the frontal fault system of the western Transverse Ranges (about 10 km). No displacement was restored on the Big Pine fault.

Reconstruction of the Frazier Mountain–Mount Pinos–eastern Orocopia Mountains paleogeologic reference domain constrains displacements on several of the right-lateral faults of the San Andreas system in southern California and on the left-lateral faults of the eastern Transverse Ranges. The configuration of the reference domain differs significantly from those in the reconstructions of previous investigators and thus requires a different disposition of magnitude and timing of faulting on various strands of the San Andreas system. It does not, however, alter the generally accepted concept that about 300+ km of displacement has accumulated on the San Andreas fault system in southern California and that the Diligencia, Vasquez, and Plush Ranch Formations are displaced by an amount roughly equal to the displacement of the crystalline basement on which they were deposited. Palinspastic juxtaposition of the Frazier Mountain–Mount Pinos block against the Orocopia Mountains northeast of the Clemens Well fault (Fig. 8) provides a well-constrained cross-fault tie that dictates the total cumulative right-lateral displacement along the San Andreas fault system—including not only the San Andreas fault, but also the Clemens Well–Fenner–San Francisquito, San Gabriel, Punchbowl, and San Jacinto faults. Reconstruction of the disrupted pattern requires that the Frazier Mountain–Mount Pinos and San Gabriel Mountains blocks moved away from the eastern Transverse Ranges block first on the Clemens Well–Fenner–San Francisquito fault, then along the San Gabriel fault, and finally on the Punchbowl, San Andreas, and San Jacinto faults. Parts of the reference domain in the Frazier Mountain area southwest of the fault system and in the south-central Orocopia Mountains northeast of the fault system are separated by 326 to 336 km (measured in a straight line, which approximates the trace of the modern San Andreas fault, by comparing Fig. 8 or Plate IIC to Plate I). Because the sum of the displacements on the individual faults must equal the overall displacement, any displacement attributed to one fault affects the amount that can be attributed to the other faults in the sequence.

Restoration of dextral displacements of 162 km on the San Andreas fault, 42 km on the San Gabriel fault, and 28 km on the San Jacinto fault and of sinistral displacements on east- and northeast-trending faults in the Transverse Ranges requires a displacement of 100 to 110 km on the Clemens Well–Fenner–San Francisquito fault to reassemble the reference domain. Although a 110-km displacement is incorporated into Figures 8 and 9 and Plate IIC, a dextral displacement of 100 km would yield an equally compelling reconstruction.

Reconstruction of the La Panza Range–Liebre Mountain–San Bernardino Mountains paleogeologic reference domain constrains the overall cumulative displacement on the Clemens Well–Fenner–San Francisquito, San Gabriel, and Red Hills–Ozena faults. Palinspastic juxtaposition of patterns in crystalline and sedimentary rocks of the La Panza Range and Barrett Ridge block, the Liebre Mountain block, and the southwestern San Bernardino Mountains block (Fig. 9) requires that: (1) the La Panza Range and Liebre Mountain blocks moved apart first on the northwestward extension of the Clemens Well–Fenner–San Francisquito fault and then on the San Gabriel and Red Hills–Ozena faults, and (2) the Liebre Mountain block, accompanied by the already displaced La Panza Range, subsequently moved northwestward from the San Bernardino Mountains on the San Andreas fault. Total displacement restored on the San Andreas fault system is 295 km (measured by comparing Fig. 9 or Plate IIC to Plate I). In Plate II, the total displacement is matched by restoring 100 km on the Clemens Well–Fenner–San Francisquito fault, 42 km on the San Gabriel fault, and 160 km on the modern San Andreas fault since about 5 Ma.

Reconstruction of the Gabilan Range–San Emigdio Mountains–Portal Ridge paleogeologic reference domain (Fig. 10) constrains the overall cumulative displacement on the San Andreas fault northwest of the Transverse Ranges. Palinspastic juxtaposition of a variety of rock units in the vicinity of the Gabilan Range with correlative units in the San Emigdio and Tehachapi Mountains has provided several similar measurements of overall displacement in the range of 290 to 320 km on the San Andreas fault north of its intersection with the San Gabriel and Garlock faults (Huffman, 1972; Huffman and others, 1973; Matthews, 1976; Clarke and Nilsen, 1973; Nilsen, 1984; Ross, 1984). Overall separation measured along the modern San Andreas fault is 295 km (measured by comparing Fig. 10 or Plate IIC to Plate I). Restoration of this reference domain requires that the displacement on the San Andreas fault north of the Transverse Ranges incorporates the displacements measured separately on the Clemens Well–Fenner–San Francisquito, San Gabriel, and San Andreas faults in the Transverse Ranges. It also limits sinistral displacement at the west end of the Garlock fault to no more than about 12 km.

Reconstruction of the southern Transverse Ranges–Chocolate Mountains reference domain (Fig. 11) constrains offset on the San Gabriel fault, Salton trough and Mojave Desert segments of the San Andreas fault, San Jacinto fault, and left-lateral faults in the southern San Gabriel and Chocolate Mountains. A displacement of about 22 to 23 km is restored on the San Gabriel fault in the San Gabriel Mountains (Ehlig, 1966, 1975a, 1981), but only 5 km is restored on the Vasquez Creek fault (south branch of the San Gabriel fault). The remaining 17 km of displacement on the San Gabriel fault system is restored along a buried fault strand hypothesized to exist between the Santa Monica and Verdugo Mountains. Displacement on the San Gabriel fault system is absorbed by closing oblique left slip in the Los Angeles basin and the San Gabriel and San Bernardino Valleys and by restoring clockwise fault block rotation in the Transverse Ranges west of the San Gabriel fault. In addition, about 10 km of younger left-lateral displacement is restored along the southern margin of the Transverse Ranges. Of the 28 km restored on the San Jacinto fault in the Peninsular Ranges, 14 km is absorbed by complementary left-lateral faulting in the southeastern San Gabriel Mountains and 14 km is merged onto the San Andreas fault. A displacement of about 160 km is restored on the Salton trough segment of the San Andreas fault (see also Dillon, 1976; Dillon and Ehlig, this volume).

Integrity of the reconstruction: Gaps, overlaps, and unresolved problems

The palinspastic reconstruction presented in Plate II was constructed by restoring displacement on strike-slip faults only. In the process of reassembling the fault blocks, overlaps of blocks were avoided and gaps between them were minimized. The displacements restored were those indicated by the evidence documented here (see sections on right- and left-lateral faults), which is distilled both from original and previously published work. In two cases, it proved necessary to hypothesize faults to dispose of large displacements. In the first case, I propose the existence of a buried fault strand of the San Gabriel fault system that projects southward across the east end of the Ventura basin, then between the Santa Monica and Verdugo Mountains to the Los Angeles basin (see section on San Gabriel fault). In the second case, I have adopted previous suggestions that displacement on the Rinconada-Reliz fault merges northward with the San Gregorio–Hosgri fault along an extension of the Rinconada-Reliz fault beneath Monterey Bay. I further propose that overall displacement generally attributed to the San Gregorio–Hosgri fault north of Monterey Bay is split roughly equally between that fault and a fault that passes east of the Montara Mountain block.

Within the constraints of the documented strike-slip displacements, the reconstruction was constructed to avoid overlaps. The success of this effort is in itself a measure of the viability of the overal reconstruction. The only significant failure in this regard lies in the overlap of Oligocene and lower Miocene volcanic rocks of southeasternmost California and southwesternmost Arizona that results from reconstruction of the Clemens Well–Fenner–San Francisquito fault. Furthermore, if extended indefinitely southeastward, the restoration of displacement on the Clemens Well–Fenner–San Francisquito fault would result in overlap of crystalline basement rocks in southern Arizona east of Plate II. As discussed below, the validity of this reconstruction depends on being able to resolve this overlap by absorbing the strike-slip displacement on other early Miocene structures in southeasternmost California and southwesternmost Arizona. The obvious candidate structures are the low-angle detachment faults recognized and inferred in the region (see, for example, Tosdal and Sherrod, 1985; Drobeck and others, 1986; Sherrod and others, 1989) and the zone of sinistral deformation along the southern margin of the Transverse Ranges and the southwest margin of the Chocolate Mountains (Powell, 1986).

The only significant gaps in the overall reconstruction occur in the Transverse Ranges west of the San Andreas fault, where three major gaps occur. Given the abundant evidence for reverse faulting in this part of the Transverse Ranges, such gaps are to be expected until slip is restored on the reverse faults. The gap along the restored San Gabriel fault, as much as 10 km wide, can be closed by restoring reverse slip on the Sierra Madre fault system along the range front. Crook and others (1987) have mapped the Sierra Madre fault system as dipping 15° to 45° northward. Although rates of convergence have not been calculated for the Sierra Madre fault system, Morton and Matti (1987) have calculated a convergence rate of 5 to 10 mm/yr for the Cucamonga fault system to the east, which typically dips 35° to 45° northward (Cramer and Harrington, 1987; Morton and Matti, 1987). At a rate of 10 mm/yr on a 45°-dipping Sierra Madre fault system, 10 km of horizontal displacement would have been generated in about 1.5 m.y. (10 mm/yr × 1.5 m.y. × cos 45 − 10.6 km). At a rate of 5 mm/yr, 10 km of horizontal displacement would have been generated in about 3 m.y., or less if the dip of the fault is more shallow than 45°. This timing is consistent with the evidence that the reverse faults of the southern San Gabriel Mountains were active throughout the Quaternary (Crook and others, 1987; Morton and Matti, 1987).

The 20- to 25-km-wide gap along the restored Pine Mountain reverse fault and southern Rinconada fault (southern Nacimiento fault of some workers) can be closed, hypothetically, by restoring reverse slip on those faults and roughly parallel reverse faults including the Frazier Mountain and Dry Creek thrust fault system of the Frazier Mountain block, the San Guillermo, Ozena, and South Cuyama reverse fault system southwest of the Cuyama basin, and perhaps the Morales and Whiterock reverse fault system in the Caliente Range northeast of the Cuyama basin. Except for the southern Rinconada fault, all of these reverse faults are Pliocene and/or Quaternary in age (Crowell, 1968 [p. 357], 1982b; Bartow, 1974, p. 140–141; Dibblee, 1982b; Schwing, 1982; Davis and others, 1988). To close the gap using the Pine Mountain thrust alone, which dips 30° north (Dibblee, 1982b), would require a displacement of about 23 to 29 km on the thrust. This magnitude of displacement is consistent with interpreting the Pine Mountain fault as a major break that juxtaposes significantly different geologic terranes (Vedder and Brown, 1968; Namson and Davis, 1988). Dibblee (1982b), however, measured a displacement of 5 km and Nagle and Parker (1971, p. 282) showed a reverse separation of only about 2 km. The other reverse faults also typically have displacements of less than 10 km each (Schwade and others, 1958; Crowell, 1982b; Dibblee, 1982b; Schwing, 1982), although Davis and others (1988) measured a displacement of 5 to 14 km on the Morales and Whiterock reverse fault system. As a group, these reverse faults and associated folding accommodate 20 to 30 km of northeast-southwest shortening. Restoring this shortening would eliminate the gap in the reconstruction of strike-slip faults.

Two other gaps may indicate the existence of additional uncompensated contraction in the western Transverse Ranges. The gap along the restored Santa Ynez fault, increasing westward up to 20 km wide, requires a significant component of reverse slip on that and/or nearby faults, perhaps including the San Cayetano and other south-vergent reverse faults along the north margin of the Ventura basin. The remaining gap of about 30 km between the Santa Monica and Santa Ana Mountains may be partly due to unrestored contraction in the Transverse Ranges west of the San Gabriel fault, depending on what the basement is beneath the Los Angeles basin.

The failure of the reconstruction to accommodate restoration of a left-slip of about 15 km on the Big Pine fault west of Lockwood Valley is probably a function of not restoring reverse slip on such faults as the Pine Mountain and Ozena faults.

As discussed in the next section, the counterclockwise back-rotation of the Peninsular Ranges block in the reconstruction may be incorrect because slip was not restored on reverse faults in San Gorgonio Pass or on extensional faults in the Los Angeles basin area.

The palinspastic unslipping of strike-slip faulting in Plate II leaves unresolved the manner in whch the 12-km displacement on the western Garlock fault increases eastward to at least 48 and possibly as much as 60 to 64 km. A crucial test of the validity of the reconstruction lies in documenting the disposition of displacement along the western Garlock fault.

Relative rotations of crustal blocks

In reconstructing the paleogeology of pre–late Cenozoic rocks of southern California as shown in Plate II, A–C, it proved necessary to rotate some crustal blocks relative to others about vertical axes in order to achieve the optimal paleogeologic fit. These reconstructive back-rotations restore rotations that developed during strike-slip movement on faults of the San Andreas system (Fig. 7). In accomplishing the reconstruction, it was not necessary to back-rotate the Sierra Nevada, the Mojave Desert, or the San Bernardino Mountains. Thus, for the purposes of this discussion, these blocks are considered to be unrotated during strike-slip faulting. This interpretation is consistent with paleomagnetic studies that indicate little or no rotation in the Mojave Desert since about 18 to 19 Ma (Wells and Hillhouse, 1989; MacFadden and others, 1990) or in the Sierra Nevada since about 16 to 20 Ma (Kanter and McWilliams, 1982; Frei and others, 1984; McWilliams and Li, 1985; Plescia and Calderone, 1986). The paleomagnetic data for the Sierra Nevada are consistent with as much as 5° to 10° of middle Miocene or younger clockwise rotation, an amount that could be related to extension in the Death Valley region.

Relative to the Sierra Nevada and Mojave Desert blocks, the Peninsular Ranges are back-rotated about 17° counterclockwise in order to keep the San Jacinto range close to the Banning block of the southern San Bernardino Mountains and to close the northward-narrowing Salton trough uniformly. However, the 17° back-rotation of the Peninsular Ranges block may be an artifact of the reconstructive process in which only strike-slip displacements were restored: if the Peninsular Ranges had been held unrotated, the reconstruction would have opened a gap between the San Jacinto and Banning blocks that would in turn allow for the restoration of thrust faulting in San Gorgonio Pass. Holding the Peninsular Ranges unrotated in the reconstruction also would have narrowed the gap between the Santa Ana and Santa Monica Mountains. Paleomagnetic declinations measured on Miocene volcanic rocks from the Peninsular Ranges block in California and Baja California indicate little or no rotation of that block since the early Miocene (Marshall and others, 1979; Hagstrum and others, 1987).

Relative to the Mojave Desert, the Liebre Mountain block is back-rotated about 7° to 8° clockwise in order to keep the block close to the San Bernardino Mountains.

All other major crustal blocks in southern California are back-rotated counterclockwise in the reconstruction, indicating that they have undergone clockwise rotation, but not to the degree indicated by paleomagnetic studies in the region. Relative to the unrotated ranges of the Mojave Desert and the San Bernardino Mountains, the Transverse Ranges east of the San Andreas fault and south of the Pinto Mountain fault are back-rotated counterclockwise 20° to 25°. This amount of rotation yields the best fit of the eastern Transverse Ranges block with surrounding blocks, including the San Bernardino Mountains, the Mojave Desert ranges, and restored blocks west of the San Andreas fault system. Some variation exists in the amount of rotation of individual fault blocks due to closing of extensional components along the left-lateral faults. The restored rotations are significantly less than the 41° ± 8° inferred from paleomagnetic measurements on basalt flows in the eastern Transverse Ranges (Carter and others, 1987).

The Transverse Ranges west of the San Andreas fault have also experienced rotations. Relative to the unrotated blocks, the San Gabriel Mountains between the San Andreas and San Gabriel faults are back-rotated about 8° counterclockwise. This rotation is in the opposite sense to the post–15-Ma counterclockwise rotations of 16° ± 30° inferred from paleomagnetic measurements on samples from the Mint Canyon Formation (Terres and Luyendyk, 1985) and 23° ±4.5° inferred from measurements on samples from the Punchbowl Formation (Liu, 1990; 11° ± 9° reported by Liu and others, 1988). Relative to the unrotated blocks, the Transverse Ranges west of the San Gabriel fault are back-rotated about 50° counterclockwise in order to reconstruct the Santa Monica and Santa Ana Mountains while maintaining contact between the eastern ends of the western ranges and the San Gabriel Mountains–Alamo Mountain–Frazier Mountain crystalline blocks. As in the case of the eastern Transverse Ranges, the restored rotations for the ranges west of the San Gabriel fault are significantly less than the 70° to 92° inferred from paleomagnetic measurements (Hornafius, 1985; Kamerling and Luyendyk, 1985; Liddicoat, 1990).

Relative to the unrotated Sierra Nevada and Mojave Desert blocks, the Gabilan and La Panza Ranges are back-rotated about 12° counterclockwise in order to have the Gabilan Range about both the Tehachapi Mountains and the northwestern Mojave Desert block and the La Panza Range abut the Liebre Mountain block (see also Ryder and Thomson, 1989, p. 48, Fig. 51). Other ranges of the Salinian block have been back-rotated along with these ranges.

The unconstrained positioning of crustal blocks in the continental borderland and Channel Islands through the intermediate phases of the reconstruction is unresolved. Paleomagnetic and paleocurrent data from the northern Channel Islands have been interpreted to indicate clockwise rotation of 70° to 80° during these phases, chiefly during the middle Miocene (Kamerling and Luyendyk, 1985).

The reconstruction presented in Plate II, A-C refutes the models for the tectonic evolution of the Transverse Ranges and San Andreas fault system in southern California that have been proposed on the basis of large paleomagnetic rotations (Luyendyk and others, 1980, 1985; Hornafius and others, 1986). The magnitude and timing of fault displacements and the degree of rotation required to reconstruct the paleogeology of the region contradict some of the corresponding displacements and rotations that are key elements of those models. Perhaps it is significant that in the Transverse Ranges the rotations required for the palinspastic reconstruction are consistently between about 50 and 70 percent of the rotations interpreted from paleomagnetic data.

RIGHT-LATERAL FAULTS OF THE SAN ANDREAS SYSTEM

Transverse Ranges

Right-lateral faults of the San Andreas system that transect the Transverse Ranges province include the now-fragmented Clemens Well–Fenner–San Francisquito fault, the San Gabriel fault, and various strands of the San Andreas fault zone. The San Andreas fault zone includes the Punchbowl fault along the northeastern flank of the San Gabriel Mountains and in San Gorgonio Pass consists of a complicated network of anastomosing fault strands, including the Banning, Wilson Creek, Mission Creek, and Mill Creek faults (Allen, 1957; Matti and others, 1985; Matti and Morton, this volume).

Clemens Well–Fenner–San Francisquito fault. *Path.* The San Francisquito fault (Simpson, 1934; Clements, 1937;

Dibblee, 1961b) trends east-northeast along the north edge of the Sierra Pelona; the Fenner fault (Noble, 1954a, b) trends east in a fault sliver between the Punchbowl and San Andreas faults; and the Clemens Well fault (Crowell, 1962) trends southeast through the Orocopia Mountains. Dibblee (1967a [Fig. 72], 1968a, 1982c; see also Woodburne, 1975, Fig. 5) proposed that the San Francisquito and Fenner faults were offset segments of a once-continuous vertical fault along which the Pelona Schist to the south was brought up against Upper Cretaceous and Paleocene marine sedimentary rocks and their gneissic and granitic basement to the north. Subsequently, Bohannon (1975; see also Dibblee, 1982b, h; Goodmacher and others, 1989; Richard and Haxel, 1991) extended the Fenner–San Francisquito fault to incorporate the Clemens Well and Blue Rock faults as segments offset across the San Andreas and San Gabriel faults, respectively, and interpreted the reconstructed Clemens Well–Fenner–San Francisquito–Blue Rock fault as a normal fault.

Meanwhile, investigators involved in palinspastic reconstruction of crystalline rocks along strike-slip faults in the Salinian block and the Transverse Ranges required that some or all of these fault segments were once parts of a throughgoing right-lateral fault (Smith, 1977a; Powell, 1981, 1986; Joseph and others, 1982b). Smith used the Morales and Red Hills–San Juan–Chimeneas faults as the northwestward extension of his Clemens Well–St. Francis fault and proposed that the Clemens Well fault bends southwestward along Salt Creek (Salton Wash) between the Orocopia and Chocolate Mountains to rejoin the San Andreas fault. In his reconstruction, Smith did not incorporate the Fenner fault and argued that the San Francisquito fault was a younger left-lateral fault that offset his San Juan-St. Francis fault. Smith advocated that displacement on the older throughgoing right-lateral San Juan–Chimeneas–Morales–St. Francis–Clemens Well fault merged with the San Andreas fault at the northern end of the San Juan fault and at the southern end of the Clemens Well fault via the Salton Creek fault. Alternatively, Joseph and others linked the Red Hills–San Juan–Chimeneas fault, the Morales and Blue Rock faults, and the Clemens Well–San Francisquito fault, and extended the Clemens Well fault eastward through the Chuckwalla Mountains along the eastward extension of the Salton Creek fault.

Previous estimates of displacement. Early investigators of both the Fenner and Clemens Well faults proposed that they are right-lateral faults within the San Andreas fault system. Shortly after Hill and Dibblee (1953) documented hundreds of kilometers of displacement on the San Andreas fault northwest of Tejon Pass, Noble (1954b) suggested that the Fenner fault might have borne some of that displacement in southern California. Crowell (1962, p. 28) later suggested that the Clemens Well fault constituted a part of the San Andreas fault system with "relatively small" right-lateral displacement. This suggestion was based on his inference that anorthosite and syenite clasts found in the Diligencia Formation northeast of the fault were derived and are not far-traveled from outcrops in the crystalline rock sliver between the Clemens Well fault and Orocopia thrust (see also Bohannon,

1975; Spittler and Arthur, 1982). Still later, Crowell (1975a; see also Spittler and Arthur, 1982) concluded that displacement on the Clemens Well fault was perhaps as much as "several tens of kilometers." However, anorthosite-bearing strata in the Diligencia Formation also contain clasts of the Mount Lowe intrusion (L. T. Silver, 1973, oral communication; Bohannon, 1975; Squires and Advocate, 1982), which is not exposed in that sliver, although small bodies of the Mount Lowe intrusion are exposed in the northern Chocolate Mountains.

Three estimates of still larger right-lateral displacement have been proposed for all or parts of the Clemens Well–Fenner–San Francisquito fault. In the earliest of these, Smith (1977a) argued for a 175-km dextral displacement based on repositioning Mesozoic monzogranitic rocks that are crosscut by polka-dot–bearing granitic dikes in the La Panza Range of the southern Salinian block opposite similar monzogranite without polka-dots in the southern Little San Bernardino Mountains and opposite the only other bedrock occurrences of polka-dot granite known at the time, which lay "north of the Mecca Hills" (Ehlig, 1976) and in the Orocopia Mountains east of the Clemens Well fault (Ehlert and Ehlig, 1977). To reconstruct a paleogeologic terrane encompassing these plutonic rocks, Smith restored right-lateral displacements on successively older faults consisting of the San Andreas fault (225 km), the San Gabriel fault (55 km), and his reconstructed San Juan–Chimeneas–Morales–St. Francis–Clemens Well fault (170 to 175 km). Smith postulated that this right-lateral fault was displaced left-laterally at least 14 and probably about 20 km on the San Francisquito fault.

Discovery of additional occurrences of polka-dot granite in the Warm Springs Mountain area of the Liebre Mountain block rendered Smith's reconstruction nonunique. Based on the presence of turritella-bearing Upper Cretaceous sedimentary rocks (Saul, 1983) nonconformably above polka-dot granite in both the La Panza Range and Warm Springs Mountain area of the Liebre Mountain block, Joseph and others (1982b) proposed that the rocks in the two areas are parts of a once-continuous paleogeologic terrane. In order to effect their reconstruction of that terrane, Joseph and others sequentially restored 240 km on the San Andreas fault, 60 km on the San Gabriel fault, and 120 km on their reconstructed San Juan–Chimeneas–Morales–Blue Rock–San Francisquito–Fenner–Clemens Well fault.

Independently, I presented a reconstruction of pre-Mesozoic metamorphic rocks in the Transverse Ranges that restored 210 to 220 km on the San Andreas fault (including the Punchbowl fault as an anastomosed strand with 40 km of displacement), 45 to 50 km on the San Gabriel fault, and 85 to 90 km on the Clemens Well–Fenner–San Francisquito fault (Powell, 1981, Plate VI, 1986: 85 km is a more accurate measurement on the Plate VI map than is the 90 km cited in the caption, text, and 1986 abstract). That reconstruction fostered additional study that has led to the refined and elaborated version put forth in this chapter.

Reckoning of path. The path of the Clemens Well–Fenner–San Francisquito fault and its northwestward and southeastward extensions is constrained by reconstruction of disrupted paleogeo-logic patterns in the four reference domains. In order to reassemble the Frazier Mountain–Mount Pinos–eastern Orocopia Mountains and La Panza Range–Liebre Mountain reference domains, the path of the Clemens Well–Fenner–San Francisquito fault to the northwest of the San Francisquito fault segment must follow or be offset along the San Gabriel and San Andreas faults at least as far as the northwest end of the Mount Pinos block. In order to restore the Gabilan Range–San Emigdio Mountains–Portal Ridge reference domain, the path of the Clemens Well–Fenner–San Francisquito fault must follow the San Andreas fault between the westernmost tip of the Mojave Desert and the northwest end of the Mount Pinos block and also must follow the San Andreas fault east of the Gabilan Range. Between the Mount Pinos block and the Gabilan Range, crystalline rock exposures require only that the path of the Clemens Well–Fenner–San Francisquito fault lie between the San Andreas fault zone on the east and the Red Hills–San Juan–Chimeneas fault zone on the west, or within one of those fault zones. Outcrops of crystalline rocks on Mount Pinos and Barrett Ridge and in the La Panza and Gabilan Ranges lie west of the northwestward extension of the Clemens Well–Fenner–San Francisquito fault, and crystalline rock outcrops in the San Emigdio Mountains lie east of the fault. Well data that bear on the subsurface distribution of crystalline rocks (Ross, 1974, 1984) are not sufficiently distinctive to further constrain the location of the fault.

The distribution of lower Cenozoic sedimentary rocks, however, provides evidence for resolving whether the Clemens Well–Fenner–San Francisquito fault extends northwestward approximately along the Blue Rock–Morales–Chimeneas–San Juan–Red Hills fault (Bohannon, 1975; Joseph and others, 1982b), the Morales–Chimeneas–San Juan–Red Hills fault (Smith, 1977a), or the San Andreas fault. Blanketed by the middle Miocene Caliente Formation, the Blue Rock fault juxtaposes the Transverse Ranges crystalline complex including the Pelona Schist and Proterozoic metamorphic rocks to the south against Upper Cretaceous through lowermost Miocene marine sedimentary rocks to the north (Bohannon, 1975; Smith, 1977a; Dibblee, 1982b). The Blue Rock fault was active in the Oligocene during deposition of the Simmler Formation (Bohannon, 1975; Bartow, 1978), and the presence of coarse debris from Mount Pinos crystalline rocks in the Oligocene Simmler Formation of Pattiway Ridge (Hill and others, 1958; Bohannon, 1975; Smith, 1977a; Bartow, 1978) evidently precludes large post-Simmler dextral displacement along the Blue Rock fault. Thus, the Blue Rock fault is too old and has insufficient lateral displacement to be the northwestward extension of the Clemens Well–Fenner–San Francisquito fault (see below). Rather, it is more likely to be correlative with the Oligocene to early Miocene faults that are syndepositional with the Plush Ranch Formation in Lockwood Valley and the Vasquez Formation in Soledad basin (see later discussion of left-lateral faults).

Nor does the Morales fault yield evidence for strike-slip displacement. Between the Caliente Range and Pattiway Ridge, the Morales fault separates differing facies and thicknesses of both the Paleocene Pattiway Formation (Hill and others, 1958; Sage, 1973a; Nilsen and Clarke, 1975; Smith, 1977a) and the Oligocene Simmler Formation (Hill and others, 1958; Bartow, 1978; Smith, 1977a), indicating reverse slip rather than large dextral slip. Conversely, in either of these cases, the Paleogene and Oligocene strata of the Caliente Range and Pattiway Ridge are more reasonably situated near their present position vis-a-vis correlative rocks in the La Panza Range and Sierra Madre Mountains to the west of the Red Hills–Ozena fault than if restored northward by 120 or 175 km. The distribution of Miocene strata is also consistent with reverse rather than lateral displacement on the Morales fault (Davis and others, 1988).

For the case in which the Clemens Well–Fenner–San Francisquito fault is coincident or nearly so with the San Andreas fault, the Pattiway, Simmler, and Vaqueros Formations lie to the west of virtually all the right-lateral displacement on the evolving Clemens Well–Fenner–San Francisquito and San Andreas fault system. Thus these sedimentary rocks moved little relative to the La Panza Range–Sierra Madre Mountains block to the west, but have been displaced from an initial position opposite the Little San Bernardino and San Bernardino Mountains in the eastern Transverse Ranges to the east. This palinspastic reconstruction (Plate II) yields a reasonable paleogeologic pattern for the distribution and overlap relations of Upper Cretaceous through lower Miocene marine sedimentary units in the Salinian block and Transverse Ranges. Moreover, it permits reassembly of the Charlie Canyon section of the Vasquez Formation and the Simmler Formation of the Caliente Range in reconstructing the La Panza Range–Liebre Mountain reference domain.

The path of the Clemens Well–Fenner–San Francisquito fault southeast of the Transverse Ranges province is more problematic. The completely reconstructed southern Transverse Ranges–Chocolate Mountains reference domain shows no evidence of having been transected by the Clemens Well–Fenner–San Francisquito fault. On the other hand, offset bedrock patterns in the Chuckwalla and San Gabriel Mountains require that the fault extended southeastward between the Chuckwalla and northern Chocolate Mountains at least as far as the southernmost Chuckwalla Mountains. These constraints refute both the path proposed by Smith (1977a) for rejoining the Clemens Well–Fenner–San Francisquito fault with the San Andreas fault to the west along the Salton Creek fault, and that proposed by Joseph and others (1982b) through the Chuckwalla Mountains to the east along the Salton Creek fault. Moreover, rather than yielding evidence for right slip, detailed mapping demonstrates 8 km of left-lateral displacement on the Chuckwalla Mountains segment of the Salton Creek fault (Fig. 8; Powell, 1981, 1982a) and as much as 14 km on the Salt Creek segment of the fault (Powell, 1991).

No offset rock patterns have been recognized that specify the disposition of right-lateral displacement south or southeast from the Chuckwalla Mountains. Kinematic scenarios wherein the Clemens Well–Fenner–San Francisquito fault, after passing between the Chuckwalla and northern Chocolate Mountains, cuts southward through the Chocolate Mountains to the San Andreas fault (Fig. 12a) apparently are ruled out by the continuity of Pelona-type schist through the southern and central parts of that range (Crowell, 1973; Haxel and Dillon, 1973; Dillon, 1976). Hypothetically, this constraint could be circumvented if the Clemens Well–Fenner–San Francisquito fault were confined to the upper plate of the Chocolate Mountains thrust, perhaps as a tear fault associated with late Cenozoic detachment reactivation of the older thrust fault. In this scenario, schist would have been folded up into the subsequently eroded strike-slip fault. Alternatively, the reactivated Chocolate Mountains thrust is itself in part an exhumed low-angle part of the Clemens Well–Fenner–San Francisquito–San Andreas fault, perhaps in accord with models that represent the San Andreas fault as a decollement at depth (Lachenbruch and Sass, 1973, 1980; Eaton and Rymer, 1990). Although the symmetry of these possibilities (Fig. 12a) is appealing, any path that leads from the southern Chuckwalla Mountains southward to the San Andreas fault crosses an area where detailed mapping (Dillon, 1976; Haxel, 1977) provides little or no supporting evidence for such a fault.

By these scenarios, analogous to that shown in Figure 14c for the San Gabriel fault (see next section), one might hypothetically link right slip on the Clemens Well–Fenner–San Francisquito fault southward through the Chocolate Mountains into the region now occupied by the Gulf of California. In general, however, there is little evidence for extensional tectonics in the Gulf of California region prior to 12 to 14 Ma (Neuhaus and others, 1988; Lonsdale, 1989; Stock and Hodges, 1989; Humphreys and Weldon, 1991), although in some areas the existing evidence does not rule out earlier inception of extensional faulting (Henry, 1989).

Kinematic scenarios in which the Clemens Well–Fenner–San Francisquito fault diverges from the San Andreas fault require that the former traverse or pass beneath an extensive field of Oligocene and early and middle Miocene volcanic rocks in the Chocolate, southern Chuckwalla, Little Chuckwalla, Mule, Little Mule, and Palo Verde Mountains of southeastern California and adjacent parts of southwestern Arizona. Existing mapping in these volcanic rocks (Crowe, 1978; Crowe and others, 1979; Tosdal and Sherrod, 1985; Sherrod and others, 1989; Sherrod and Tosdal, 1991), which range in age chiefly between about 32 and 20 Ma, does not provide evidence that the volcanic field has been transected by a major strike-slip fault. Thus, the surface trace of the Clemens Well–Fenner–San Francisquito fault cannot extend undiminished to the southeast, although rejuxtapositioning of the Plush Ranch and Vasquez Formations against the Diligencia Formation and the volcanic and sedimentary strata of the Chuckwalla Mountains (Plate IIC) does require that coeval volcanic rocks in the northern Chocolate Mountains be repositioned southeast of the Chuckwalla Mountains. The fault could continue in the subsurface if it were older enough along its southeastward extension that it is coeval with and covered by younger parts of the Miocene volcanic field (ca. 22 to 20 Ma), or if it passes structurally beneath the volcanic field.

In any event, it is unreasonable to infer that the Clemens

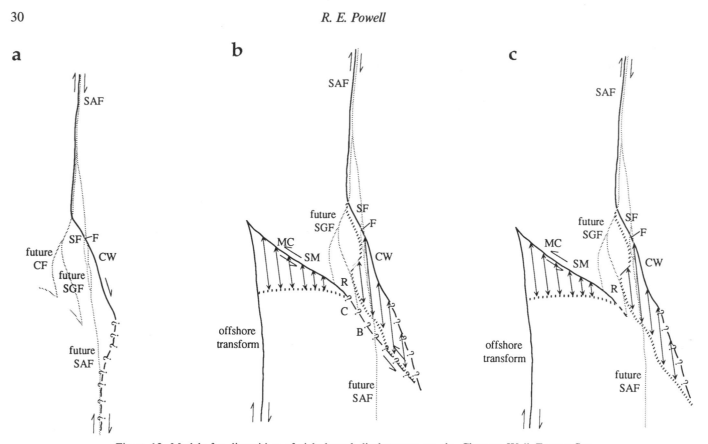

Figure 12. Models for disposition of right-lateral displacement on the Clemens Well–Fenner–San Francisquito fault as the earliest strand of the San Andreas fault system in southern California (ca. 20-13 Ma). a, Dextral displacement on the Clemens Well–Fenner–San Francisquito fault as the earliest strand of an anastomosed San Andreas fault system in southern California emanating from the Gulf of California. b, Complementary dextral displacement on the Clemens Well–Fenner–San Francisquito fault and the hypothesized East Santa Cruz basin fault and sinistral displacement on the Malibu Coast–Santa Monica–Raymond–Cucamonga fault. Strike-slip displacement is accompanied by extension and exhumation of the Catalina Schist in the Los Angeles basin area and of the Pelona and Orocopia Schists in the restored Sierra Pelona–Chocolate Mountains area. c, Dextral displacement on the Clemens Well–Fenner–San Francisquito fault accompanied by extension and exhumation of schist in the Los Angeles basin area and restored Sierra Pelona–Chocolate Mountains area, and accompanied by sinistral kinking along the southern margin of the Transverse Ranges province and the Chocolate Mountains. B = Banning fault; C = Cucamonga fault; CW = Clemens Well fault; F = Fenner fault; MC = Malibu Coast fault; R = Raymond fault; SF = San Francisquito fault; SM = Santa Monica fault. Double-tipped arrows show extensional domains required by each model. North toward top of figure.

Well–Fenner–San Francisquito fault extends indefinitely into the continent as represented in Plate IIC. It seems more likely that displacement on the fault is absorbed by deformation associated with differential extension in the Colorado Desert between the southeastward projection of the dextral Clemens Well–Fenner–San Francisquito fault and the southern Transverse Ranges–Chocolate Mountains zone of sinistral deflection (see discussion in later section on left-lateral faulting). Substantial extension is indicated by the early and middle Miocene high-angle normal faulting and low-angle detachment faulting observed and inferred in this region (Frost and others, 1982; Garner and others, 1982; Haxel and others, 1982; Silver, 1982; Tosdal and Sherrod, 1985; Drobeck and others, 1986; Spencer and Reynolds, 1986; Sherrod

and others, 1987, 1989; Powell, 1991; Richard and Haxel, 1991; Sherrod and Tosdal, 1991).

Possible models for relating strike-slip faulting and extension are shown in Figure 12, b, c, although field relations between strike-slip and extensional deformation have yet to be deciphered. If it were possible to document large left slip along an eastward continuation of the ancestral Malibu Coast–Santa Monica–Raymond fault through San Gorgonio Pass and along the southwest flank of the Chocolate Mountains, then the scenario in Figure 12b would apply. If, however, there is no major throughgoing left-lateral strike-slip fault but rather sinistral kinking (as advocated in a later section), then the scenario shown in Figure 12c applies. Because late Oligocene and early Miocene volcanic and sedimentary strata in southeasternmost California are not transected by throughgoing strike-slip faults, each of these scenarios requires large extension in and east of the Chocolate Mountains in order to avoid overlap in the reconstruction.

The metamorphic and plutonic rocks in the northern Chocolate, western Orocopia, and eastern San Gabriel Mountains—which were displaced dextrally along the Clemens Well-Fenner-San francisquito fault and sinistrally along the south flanks of the Transverse Ranges and Chocolate Mountains—either slipped from beneath the Tertiary strata and/or moved northwestward as the domain of Tertiary strata was extended by listric normal faulting and/or vertical-axis fault block rotation. Extensional deformation that was, or may have been, coeval with the strike-slip deformation includes reactivation of the Orocopia thrust as a steeply to shallowly east-dipping normal fault (Frost and others, 1982; Haxel and others, 1988; R. E. Powell and J. C. Matti, 1982, 1991–1992, unpublished mapping), moderate- and low-angle normal faults observed between volcanic strata and crystalline basement rocks along the northeast flank of the Chocolate Mountains (Powell, 1991; Richard and Haxel, 1991; R. E. Powell, 1991–1992, unpublished mapping), moderate- to high-angle normal faulting in the Colorado Desert (Tosdal and Sherrod, 1985; Sherrod and Tosdal, 1991), and possibly; domains of left- and right-lateral faulting (Silver and others, 1977; Fig. 4-4; Sherrod and Tosdal, 1991). Hypothetically, restoration of slip on these extensional faults would lead to a more realistic reconstruction in the southeast corner of Plate IIC, with the crystalline rocks of the western Orocopia, southern Chocolate, Cargo Muchacho, Gila, and Laguna Mountains reassembled in closer proximity to those of the San Gabriel, Chuckwalla, and eastern Orocopia Mountains.

Reckoning of displacement. The magnitude of right-lateral displacement restored on the Clemens Well-Fenner-San Francisquito fault is constrained by overall reconstruction of the reference domains, but also depends on the interpretation of events subsequent to accumulation of that displacement. These events include specifically the displacements that accumulated on the younger San Gabriel, Red Hills-Ozena, and San Andreas faults as they evolved from the Clemens Well-Fenner-San Francisquito fault. The sum of these displacements, however, cannot exceed the total displacement required to realign the paleogeology of the reference domains. In a straight line, the Frazier Mountain-Mount Pinos-eastern Orocopia Mountains reference domain is displaced a total of as much as 326 to 336 km, the Gabilan Range-San Emigdio Mountains-Portal Ridge reference area is displaced by a total of 295 km, and the La Panza Range-Liebre Mountain block reference area is displaced by 141 km. Smith (1977a) combined a displacement of 175 km on his San Juan-Chimeneas-Morales-St. Francis-Clemens Well fault with 55 km on the San Gabriel fault and 225 or 240 km on the San Andreas fault for a total of about 455 or 470 km, which overshoots that based on the Frazier Mountain-Mount Pinos-eastern Orocopia Mountains and Gabilan Range-San Emigdio Mountains-Portal Ridge reference domains by about 150 km. Joseph

and others (1982b) combined a displacement of 120 km on their San Juan-Chimeneas-Blue Rock-San Francisquito-Fenner-Clemens Well fault with 240 km of displacement on the San Andreas fault and 60 km of displacement on the San Gabriel fault for a total of 420 km, which again overshoots by about 120 km that based on the Frazier Mountain-Mount Pinos-eastern Orocopia Mountains and Gabilan Range-San Emigdio Mountains-Portal Ridge reference domains. A displacement of 175 or 120 km on the Clemens Well-Fenner-San Francisquito fault is possible only if considerably less displacement has accumulated on the San Gabriel and San Andreas faults than 55 to 60 and 225 to 240 km, respectively.

In previous attempts at reassembling the Frazier Mountain-Mount Pinos-Eastern Orocopia Mountains reference domain (Powell, 1981, 1986), I have restored displacements of 209 km on the San Andreas fault (including 42 km on the Punchbowl fault as an anastomosed strand of the San Andreas fault), 42 on the San Gabriel fault, and 85 km on the Clemens Well-Fenner-San Francisquito fault, for a total of 336 km. Although this disposition of displacement in the evolving San Andreas fault system satisfactorily reassembles the Frazier Mountain-Mount Pinos-eastern Orocopia Mountains reference domain, it does not satisfactorily restore the Liebre Mountains-San Bernardino Mountains reference domain (Fig. 9), which is overreached by about 25 to 30 km, even after the left-lateral displacements on the faults of the eastern Transverse Ranges are restored. In order to satisfactorily reconstruct all the reference domains, it is necessary either to decrease by about 25 to 30 km the displacement on the Punchbowl-San Andreas fault system that disrupts the Liebre Mountains-San Bernardino Mountains domain, and correspondingly to increase the amount that follows a path to the south and west of the Liebre Mountain block, namely the Clemens Well-Fenner-San Francisquito fault.

In constructing Plate II, A–C, therefore, I have restored 162 km on the San Andreas fault (including 42 km on the Punchbowl fault as an anastomosed strand of the San Andreas), 42 km on the San Gabriel fault, and about 110 km on the Clemens Well-Fenner-San Francisquito fault, for a total of 314 km. Displacement on the left-lateral faults in the eastern Transverse Ranges adds 22 km to the offset of the Frazier Mountain-Mount Pinos-eastern Orocopia Mountains reference domain (see later sections on left-lateral faults and on reconciliation and synthesis), bringing the total to 336 km. With the same displacements restored on the other faults, a displacement of as little as 100 km on the Clemens Well-Fenner-San Francisquito fault, for a total of 326 km, would yield an alternate but equally reasonable reconstruction of the Frazier Mountain-Mount Pinos-eastern Orocopia Mountains reference domain as that shown in Figure 8. Further discussion of the magnitude of displacement on the Clemens Well-Fenner-San Francisquito fault is deferred pending assessments of displacement on these younger faults.

Timing and rate of movement. Upper Oligocene and lower Miocene strata are crosscut by segments of the Clemens Well-Fenner-San Francisquito fault and its extension along the San

Andreas fault northwest from about Tejon Pass. Strata of this age are displaced by amounts equal to displacement of the Proterozoic and Mesozoic basement on which they are deposited. The Plush Ranch, Diligencia, and Vasquez Formations contain flows of basalt and basaltic andesite that have been dated in the range of about 20 to 26 Ma. Above the volcanic rocks, the Diligencia and Tick Canyon Formations contain Arikareean vertebrate fossils. On a geochronometric time scale (Fig. 13), the Arikareean ends at 21 (Turner, 1970; Berggren and others, 1985) or 20 Ma (Tedford and others, 1987; Woodburne, 1987; Bartow, 1990).

Middle Miocene strata overlap two segments of the fault. In the western Sierra Pelona, the west end of the San Francisquito fault is overlapped by the San Francisquito Canyon breccia unit of Sams (1964) and the overlying Mint Canyon Formation (Clements, 1937; Oakeshott, 1958; Stitt and Yeats, 1982), although only the uppermost part of the Mint Canyon Formation overlaps the fault (see Eaton, 1939, p. 534; Jahns, 1940, p. 167). The San Francisquito Canyon breccia unit, which underlies the Mint Canyon Formation either unconformably (Szatai, 1961) or in an interfingering relation (Sams, 1964), reportedly has yielded Barstovian vertebrate fossil fragments (Szatai, 1961), whereas the Mint Canyon Formation in Soledad basin contains late Barstovian and possibly early Clarendonian vertebrate faunal assemblages in its lower part and late Clarendonian assemblages in its

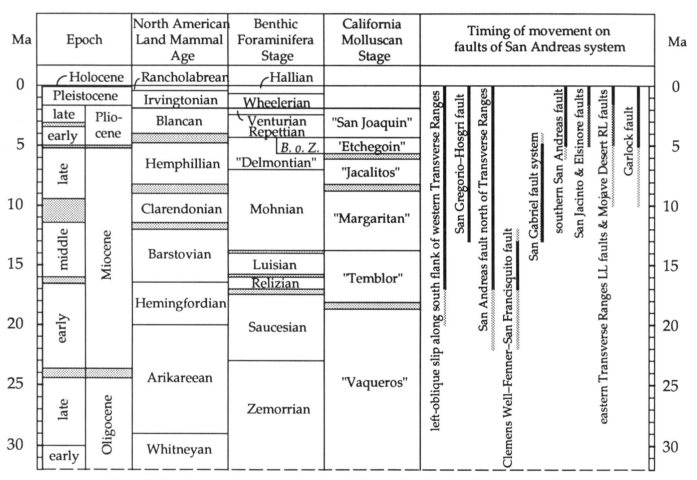

Figure 13. Late Cenozoic time scales and timing of slip on faults of the San Andreas system. Shaded intervals on time scales indicate geochronometric ranges of recently published epoch, age, and stage boundaries. Shading of fault-slip bars: black indicates intervals for which there is well-documented direct evidence of slip; dark gray indicates intervals for which there is plausible direct evidence of slip; light gray indicates intervals for which there is evidence for tectonism that may or may not be accompanied by fault slip. Abbreviations: *B. o.* Z. = *Bolivina obliqua* Zone; LL = left-lateral; RL = right-lateral. Sources for epoch boundaries: Poore and others (1981), Palmer (1983), Barron and others (1985), Berggren and others (1985), Woodburne (1987), Bartow (1990), Blake (1991). Sources for North American land mammal age boundaries: Lindsay and others (1984), Lundelius and others (1987), Repenning (1987), Tedford and others (1987), Woodburne (1987), Bartow (1990). Sources for benthic foraminifera stage boundaries: Armentrout and Echols (1981), Poore and others (1981), Blake (1991), Bartow (1990). Sources for California molluscan stage boundaries. Poore and others (1981), Ensley and Verosub (1982), Bartow (1990).

upper part (Durham and others, 1954, p. 66–67; Whistler, 19 1967, p. 1; for additional references, see Oakeshott, 1958; Winterer and Durham, 1962). Clasts in the upper Barstovian lower part of the Mint Canyon Formation that were derived from the north include Pelona Schist from south of the San Francisquito fault and Paleogene marine sandstone from north of the fault (Ehlert, 1982), indicating that most of the movement on the fault was over by the late Barstovian. On a geochronometric time scale (Fig. 13), the Barstovian spans an interval from 17 or 16 to 12 or 11 Ma and the Clarendonian spans an interval from 12 or 11 to 10 or 9 or 8 Ma (Turner, 1970; Berggren and Van Couvering, 1974; Berggren and others, 1985; Tedford and others, 1987; Woodburne, 1987). Fission track ages on zircons from tuff beds in the upper part of the Mint Canyon Formation are 10.1 ± 0.8 and 11.6 ± 1.2 Ma (J. D. Obradovich and T. H. McCulloh, 1982, personal communication cited in Terres and Luyendyk, 1985).

At the south end of the Devil's Punchbowl area, the west end of the Fenner fault is overlapped by the lowermost part of the Punchbowl Formation in Devils's Punchbowl (Noble, 1954a, b; Dibblee, 1967a, 1968a), which contains vertebrate fossils that are no older than early Clarendonian in age (Woodburne and Golz, 1972; Woodburne, 1975). Magnetostratigraphic correlation yields an age of 12.5 to 12.7 Ma for the base of the Punchbowl Formation (Liu, 1990).

Reconstruction of the paleogeologic reference domains requires that the Clemens Well–Fenner–San Francisquito fault was continuous with the San Andreas fault north of the Transverse Ranges. If the combined fault began to move along its entire length simultaneously, then movement on both the Clemens Well–Fenner–San Francisquito fault in southern California and the San Andreas fault in central California may have begun as late as 18 or 17 Ma (see section on San Andreas fault).

In summary, movement on the Clemens Well–Fenner–San Francisquito fault is bracketed between about 22 to 20 and 13 Ma, and perhaps in the more restricted interval between 18 to 17 and 13 Ma (Fig. 13). In displacing crystalline rocks and the lower Miocene strata by equal amounts, the Clemens Well–Fenner–San Francisquito fault is constrained to have begun to move at or after about 22 to 20 Ma. The overlap of the west end of the San Francisquito fault by the uppermost part of the Mint Canyon Formation, and the overlap of the west end of the Fenner fault by the lowermost part of the Punchbowl Formation in Devil's Punchbowl constrain principal movement on the Fenner and San Francisquito segments of the fault to have ceased by about 13 Ma. No middle Miocene units are known in the vicinity of the Clemens Well segment of the fault, and it has not been shown to be overlapped by any units older than Quaternary or, possibly, Pliocene. Without elaboration, Crowell and Ramirez (1979) stated that Quaternary gravels are uplifted along the Clemens Well fault and infer that the fault was active into the Quaternary. The northwest projection of the fault, however, is buried beneath the Pleistocene Ocotillo Conglomerate and perhaps beneath upper Pliocene strata in the Mecca Hills. Moreover, I have observed that the southern part of the Clemens Well fault is capped by unbroken, deeply incised fanglomerate that is probably Pliocene or Pleistocene in age.

The long-term average slip rate on the Clemens Well–Fenner–San Francisquito–early San Andreas fault can be calculated for several cases. For the preferred case developed above, a displacement of 110 km (Plate II) accumulated in the 4-m.y. interval between 17 and 13 Ma, yielding an average slip rate of 28 mm/yr. If the fault began to move as early as 18 Ma and continued as late as 12 Ma, then the average slip rate over 6 m.y. would have been 18 mm/yr. For a displacement of 100 km, the average slip rate would have been 25 or 17 mm/yr, respectively. Alternatively, if fault movement occurred over the 7- to 10-m.y. interval between 20 to 22 and 13 to 12 Ma, the the slip rate would have been in the range of 16 to 11 mm/yr for a displacement of 110 km, or 14 to 10 mm/yr for a displacement of 100 km. For comparison, the slip rates calculated for a displacement of 85 km (Powell, 1981), as a minimum possible displacement on the Clemens Well–Fenner–San Francisquito–early San Andreas fault, are 21 to 14 mm/yr for the 4- to 6-m.y. interval, or 12 to 9 mm/yr for the 7- to 10-m.y. interval.

Kinematic role in the evolving San Andreas fault system. In the palinspastic reconstruction presented in this chapter, the Clemens Well–Fenner–San Francisquito fault was continuous with the early San Andreas fault northwest of the Transverse Ranges, and probably was linked via that fault to the Mendocino triple junction at the northern end of the transform system. To the south, dextral movement on the Clemens Well–Fenner–San Francisquito fault probably stepped westward to the continental margin across a zone of sinistral deformation along the southern boundary of the Transverse Ranges. As the early San Andreas fault in southern California, the Clemens Well–Fenner–San Francisquito fault bore about a third of the overall displacement of the San Andreas fault system in the Transverse Ranges.

San Gabriel fault. Path. The San Gabriel fault transects the southwestern San Gabriel Mountains from Placerita Creek southeastward to Big Tujunga Canyon, where it splits into two branches (Kew, 1924, p. 99). An east-southeast–trending branch curves gradually eastward within the mountain range along the upper part of the Arroyo Seco and along the east and west forks of the San Gabriel River to San Antonio Canyon where it is truncated by the northeast-trending San Antonio Canyon fault and evidently offset sinistrally to a fault in Icehouse Canyon that continues east to the San Jacinto fault system in Lytle Creek (Hsu, 1955, p. 307–308; Dibblee, 1968a; Ehlig, 1975a; Morton, 1975b; Crowell, 1973, 1975c, d, 1982a; Powell and Silver, 1979, Fig. 13; Evans, 1982; Weber, 1982a; Matti and Morton, this volume). The other branch continues southeastward across the Arroyo Seco to the range front where it intersects and either merges with or is truncated by the Sierra Madre fault of the frontal thrust system (Kew, 1924; Hill, 1928; Miller, 1928, 1934; Eckis, 1934, Plate A; Dibblee, 1968a; Ehlig, 1973, 1975a, 1986; Crowell, 1975d, 1982b; Crook and others, 1987). The former branch is the San Gabriel fault (Kew, 1924; Miller, 1928; Ehlig, 1968), whereas the latter branch has been referred to variously as the Vasquez or Vasquez Creek fault (Miller, 1928; Jahns and Proctor, 1975; Crook and others, 1987), the Sierra Madre fault (Eckis, 1934; Ehlig, 1968), and the south or southeast branch of the San Gabriel fault (Crowell, 1962, 1981; Ehlig, 1973, 1975a, 1981, 1982; Weber, 1982a). In this chapter, the latter branch is referred to as the Vasquez Creek fault.

Northwest of the San Gabriel Mountains, Kew (1924, p. 99, Plate 10) mapped the San Gabriel fault through the sedimentary rocks of the easternmost Ventura basin, tracing the fault just north of Saugus west-northwest to the Topatopa anticlinorium, where he concluded that displacement on the fault was absorbed by folding. Eaton (1939) modified Kew's interpretation by extending the San Gabriel fault northwest from Saugus along the west margin of the Ridge basin to the San Andreas fault, following the Palomas Canyon fault of Clements (1937). Still farther northwest, the intersection of the San Gabriel and San Andreas faults is obscured beneath the Dry Creek and Frazier Mountain thrust faults that superpose crystalline rocks above Pliocene sedimentary strata of the Ridge basin (Eaton, 1939; Crowell, 1950, 1964, 1975d, 1982b).

Previous estimates of displacement. Early investigators interpreted the San Gabriel fault in the San Gabriel Mountains as a reverse fault that resulted in uplift of the San Gabriel Mountains (Kew, 1924; R. T. Hill, 1928; M. L. Hill, 1930). Eaton (1939) concluded that the fault scissored at Castaic, north of which it was a normal fault with as much as 7,000 m of dip slip, and that the San Gabriel fault developed as a consequence of the interaction of the right-lateral San Andreas fault and the left-lateral Garlock fault. Crowell (1950) estimated 1,800 to 4,270 m of pre-Pleistocene normal dip slip on the San Gabriel fault and 550 m of Pleistocene reverse dip slip on the fault.

Although Eaton and Crowell each had suggested the possibility of a strike-slip component of displacement on the San Gabriel fault, dextral displacement was first identified by Oakeshott (1950, p. 64), who presented evidence for 4.4-km offset of streams and marine Pliocene strata. Early measurements of large (20+ km) right-lateral displacement on the fault, however, were based on relating conglomerate clasts to crystalline bedrock sources. The initial measurements ranged from 24 to 40 km (Crowell, 1952, 1954b, 1960, 1962; Carman, 1964), although later investigators working with the same sedimentary units in conjunction with displaced paleogeographic and paleostructural features have argued for offsets ranging from 50 to 70 km (Ehlig, 1973, 1982; Bohannon, 1975; Ehlig and others, 1975; Ehlert, 1982). From the first, however, proposals of large lateral displacement have been disputed by a few workers (Paschall and Off, 1961; Weber, 1982a, 1986a, b) who have cited evidence for a large dip-slip component of movement and have maintained that the lateral component is smaller by an order of magnitude than that favored by those advocating large lateral slip.

Six distinct paleogeologic features, when reassembled, have provided variously well-constrained measures of large lateral displacement on the San Gabriel fault. The reassembled features consist of the paleogeologic distribution of distinctive Proterozoic and Mesozoic crystalline rocks; the distribution of marine Paleocene and Eocene strata; two Miocene depositional basins (eastern Ventura and Ridge basins) in which coarse, angular clasts in proximal breccia and conglomerate units are restored to the vicinity of their crystalline source terranes; a throughgoing Miocene paleodrainage; and a suite of Oligocene and early Miocene paleostructures comprising an antiformal uplift of crystalline rocks flanked by faults and syntectonic depositional basins. In contrast, there is no direct evidence for large displacement on the Vasquez Creek fault (Powell and Silver, 1979; Powell and others, 1983a; Crook and others, 1987).

Reassembly of kindred igneous and metamorphic rocks in the southern and western San Gabriel Mountains and in the Alamo Mountain–Frazier Mountain area has provided measures of overall displacement on the San Gabriel fault. In the crystalline rocks of the southern San Gabriel Mountains, Ehlig (1968, 1975a, 1981) has convincingly documented a total displacement of 21 to 23 km along the San Gabriel fault. The offset of about 22 km is tightly constrained by a piercing pattern of crystalline rocks that includes three subunits within the Mount Lowe intrusion, a mylonitic western contact between the granodiorite and Proterozoic gneiss, and intrusive contacts between the granodiorite and younger Mesozoic plutonic rocks (Fig. 11, Plate II). Northwest of the junction between the San Gabriel and Vasquez Creek faults, crystalline rocks displaced along the San Gabriel fault include the Pelona Schist and several Proterozoic units, including augen gneiss, retrograded granulite, and anorthosite. Exposures of the displaced correlative crystalline rock units in the Mount Pinos–Frazier Mountain–Alamo Mountain–Whitaker Peak block and in the Sierra Pelona and western San Gabriel Mountains have been variously reassembled to support greatly differing measurements of offset along the San Gabriel fault, including 32 km (Carman, 1964, p. 36, 56), 45 km (Powell, 1981; Plate VI: 45 km measured on map is erroneously cited as 60 km in caption and text), 48 km (Crowell, 1960; Woodburne, 1975, p. 6–10, Figs. 3-4), 32 to 48 km (Crowell, 1962, p. 40), and 60 km (Bohannon, 1975; Ehlig and Crowell, 1982).

Marine Upper Cretaceous, Paleocene, and Eocene rocks that occur in four areas along the San Gabriel fault zone have also yielded measures of overall displacement. Northeast of the fault, the Upper Cretaceous and Paleocene San Francisquito Formation (Dibblee, 1967a; Kooser, 1982; Saul, 1983) crops out between the crystalline rocks of the Liebre Mountain block and the San Francisquito fault. Sage (1973a) argued that the San Francisquito Formation has been displaced about 60 km from Paleocene strata in the Caliente Range. A sliver of Paleocene rocks occurs within the San Gabriel fault zone between Pacoima and Gold Canyons in the western San Gabriel Mountains (Clements and Oakeshott, 1935; Oakeshott, 1950, 1958, p. 58). Southwest of the fault, marine Paleocene and Eocene strata crop out in Elsmere Canyon in the western San Gabriel Mountains, occur in the subsurface in the Newhall area, and crop out in Piru Creek northwest of Castaic (Oakeshott, 1950, 1958; Nagle and Parker, 1971; Howell, 1975a; Nilsen and Clarke, 1975). The Pacoima

Canyon sliver can be rejuxtaposed against the Paleocene rocks at Newhall by restoring 22 km (Yeats, 1981b) or 10 to 25 km (Weber, 1982a, p. 85) along the San Gabriel fault. Yeats (1981b) observed that rejuxtaposing Paleogene rocks of the Piru Creek and Newhall area would require the existence of an ancestral San Gabriel fault west of the present fault, although some investigators have argued that the two marine sequences are unrelated (Stitt, 1986).

Southwest of the San Gabriel fault, upper Oligocene and lower to middle Miocene strata of the Sespe and lower part of the Modelo Formations in Canton and Devils Canyons south to southeast of Whitaker Peak contain conglomerate beds with coarse, angular, west-transported clasts of anorthosite, syenite, the Mendenhall Gneiss, and the Mount Lowe intrusion that were derived from an originally nearby crystalline source terrane northeast of the fault (Crowell, 1952, 1954a, b, 1975d, 1982b; Paschall and Off, 1961; Bohannon, 1975; Stitt, 1986). Today, the nearest known exposures of anorthosite, syenite, and the Mendenhall Gneiss are located along the San Gabriel fault in the San Gabriel Mountains, between 30 and 60 km southeast of Canton Canyon. The Mount Lowe intrusion crops out east of the anorthosite and related rocks and occurs in proximity to the San Gabriel fault from 50 to 80 km southeast of Canton Canyon and along the Vasquez Creek fault from 50 to 70 km southeast. Crowell, who first identified clasts of anorthosite and related rocks, has variously matched the clasts to their source region by restoring dextral displacements on the San Gabriel fault of 24 to 40 km (1952, 1954b), 32 km (1960), 35 km (1975d), and 35 to 56 km (1982a). Bohannon (1975), who identified clasts of the Mount Lowe intrusion in the Sespe Formation, argued that the Sespe strata are displaced 60 km from their source.

Published observations on the distribution of clasts in the marine Towsley Formation in the eastern Ventura basin and along the west flank of the San Gabriel Mountains place tentative limits on syn- and post-Towsley displacement on the San Gabriel fault. Anorthosite and related rocks that occur as clasts in the Miocene part of the upper Miocene and lower Pliocene Towsley Formation of the eastern Ventura basin are also derived from the western San Gabriel Mountains, although some of the clasts were probably cycled through the Sespe and Modelo Formations (Paschall and Off, 1961; Weber, 1982a; Stitt, 1986). Although the magnitude of slip is difficult to establish, most investigators have concluded that fans derived from the San Gabriel Mountains have been beheaded by strike-slip movement along the San Gabriel fault (Winterer and Durham, 1962; Stitt, 1986). Ehlig (1975b) mentioned the presence of clasts of the Mount Lowe intrusion and cataclastic gneiss and an absence of anorthosite clasts in lower Pliocene marine sandstone and conglomerate in Gold Canyon along the San Gabriel fault and west of the mouth of Big Tujunga Canyon in the western San Gabriel Mountains. Although not so identified by Ehlig, these strata are evidently part of the Elsmere Member of the Repetto Formation of Oakeshott (1950, 1958) that are now included in the Towsley Formation by Winterer and Durham (1962) and Stitt (1986). In seeking paleo-

drainages that would tap a source area including the Mount Lowe intrusion but excluding anorthosite, Ehlig inferred a minimum displacement of 10 to 13 km on the San Gabriel fault if the paleodrainages transported sediment from east to west and a maximum displacement of 19 to 32 km if paleodrainages transported sediment from north to south. Weber (1982a, Fig. 21a) mentioned clasts of anorthosite at the base of the Towsley Formation in Lopez and Elsmere Canyons. The locations of these occurrences in relation to cross-fault outcrops of anorthosite permit a displacement within Ehlig's minimum-maximum range, but are also consistent with a lesser displacement if the paths of the paleodrainages had a northerly component, perhaps as a result of deflection along the San Gabriel fault.

The Plush Ranch and Caliente Formations in Lockwood Valley and the Cuyama badlands and the Vasquez, Tick Canyon, and Mint Canyon Formations in Soledad basin are correlative Miocene terrestrial units disrupted by the San Gabriel fault (Carman, 1964; see also Crowell, 1960, 1962, 1975d). Based on the presence in both the lower part of the Mint Canyon Formation and the Caliente Formation of rounded clasts of anorthosite derived from the San Gabriel Mountains and volcanic rocks thought to be derived from the Mojave Desert, Carman (1964, p. 36, 42) inferred that the Mint Canyon and Caliente Formations were deposited in a once-continuous late Miocene paleodrainage system, part of which is buried beneath sedimentary strata in the southern Ridge basin. In reconstructing the paleogeography, Carman measured a displacement of 32 km along the San Gabriel fault. Ehlig and colleagues (Ehlig and others, 1975; Ehlert, 1982) recognized the presence of clasts of the Mount Lowe intrusion that were also derived from the San Gabriel Mountains in the fluvial debris and identified a source for the volcanic rocks in the northern Chocolate Mountains. By directly rejuxtaposing present-day exposures of the Mint Canyon and Caliente Formations across the San Gabriel fault, Ehlig and colleagues have measured displacements of 60 km (Ehlig and others, 1975; Ehlig, 1981, 1982) and 70 km (Ehlert, 1982).

Northwest of Castaic, movement along the San Gabriel fault has exposed a block of crystalline rocks between Whitaker Peak and Frazier Mountain that created a western source (Eaton, 1939; Crowell, 1952, 1952a, 1975d, 1982b, c) for the sedimentary Violin Breccia of Crowell (1954a, 1982c) (see also Link, 1982, 1983), which is exposed along the fault for a distance of 30 to 35 km and found in the subsurface for an additional 1 to 2 km (Stitt, 1986). By matching the distribution of clasts of augen gneiss, retrograded granulite gneiss, amphibolite, and anorthosite and related rocks in the Violin Breccia east of the fault with the distribution of these rock types on Frazier and Alamo Mountain in the crystalline source terrane west of the fault, Crowell has measured dextral offset of 24 km (1952, 1954a, b), 32 km (1960), 30 km (1975d), and at least 35 but perhaps as much as 60 km (1982b).

The distribution of anorthosite and related rocks derived from a point source along Piru Creek east of Alamo Mountain has also provided a measure of incremental displacement on the

San Gabriel fault. In contrast to the larger offsets of lower parts of the Violin Breccia, anorthosite slide blocks and clasts in the upper part of the Violin Breccia—where it interfingers with the uppermost part of the Peace Valley Formation of Crowell (1964) (equivalent to the Peace Valley beds of Crowell [1950]) and lowermost part of the Hungry Valley Formation (Shepard, 1962)—are separated about 9 km from bedrock exposures of anorthosite just north of Piru Creek on the south flank of Bear Mountain (J. C. Crowell, cited in Woodburne, 1975; Ehlig and Crowell, 1982).

In a paleostructural reconstruction, Bohannon (1975) inferred that, prior to disruption by the San Gabriel fault, the San Francisquito fault was once continuous with the subsurface Blue Rock fault (which is inferred from stratigraphic discontinuities beneath the Caliente Formation), and the antiformal uplifts of the Pelona Schist on Abel Mountain and in the Sierra Pelona were once parts of a continuous fold. By reassembling these two structures, Bohannon measured a displacement of 60 km (see also Ehlig, 1981; Dibblee, 1982b, g; Ehlig and Crowell, 1982; Joseph and others, 1982b). In one alternate interpretation, Smith (1977a) reconstructed the San Francisquito and Morales faults by restoring a displacement of 55 km along the San Gabriel fault, but in fact that realignment requires a displacement of about 90 km on the San Gabriel fault. A second alternate interpretation, in which the San Francisquito fault is realigned with the segment of the San Andreas fault that flanks Frazier Mountain and Mount Pinos by restoring a displacement of about 45 km along the San Gabriel fault (Powell, 1981; Plate VI), is elaborated below in a discussion of kinematic constraints.

In addition to the paleogeologic features that indicate large lateral displacement, there are four younger features that indicate relatively small lateral displacement. First, from the northwestward deflection of Big Tujunga and Pacoima Canyons as they cross the San Gabriel fault zone respectively at and northeast of its junction with the Vasquez Creek fault, Oakeshott (1950, p. 64, 1958, p. 91) has inferred as much as 6 km of right-lateral displacement on the fault since Big Tujunga Canyon was incised and as much as 2.4 km since Pacoima Canyon was incised. Second, Ehlig (1975b) mentioned clasts of anorthosite and the Mendenhall Gneiss in "middle-to-upper Pliocene" marine conglomerate east of Newhall, from which he inferred between 6 and 19 km of displacement on the San Gabriel fault. Although not so identified by Ehlig, these strata are in the lower member of the Pico Formation of Oakeshott (1950, 1958, restricted to the upper part of the Pico Formation of Kew, 1924; see also Winterer and Durham, 1962; Stitt, 1986). Third, in the eastern Ventura basin, strata in the Pico Formation are displaced 1.8 km (Stitt, 1986; Stitt and Yeats, 1982) and in the overlying Saugus Formation about 0.5 km (Weber, 1982a, b). Fourth, in the vicinity of the northwest terminus of the San Gabriel fault, Weber (1982a, b, 1986a, b) measured 1.2 km of right separation of contacts between subunits of the Hungry Valley Formation.

Reckoning of path and displacement northwest of Castaic-Saugus area. The kinematics of reassembling the igneous and

metamorphic terrane of the Transverse Ranges and vicinity (Plate II) further constrains the magnitude of displacement on the San Gabriel and related faults. Hypothetically, the intersection of the San Gabriel fault and the older Clemens Well–Fenner–San Francisquito fault could have evolved in one of two ways: either the San Gabriel fault disrupted and offset the abandoned Clemens Well–Fenner–San Francisquito fault or it merged northward with the older fault. In the latter case, the older Clemens Well–Fenner–San Francisquito fault would have been abandoned in favor of the younger San Gabriel fault to the southeast of their intersection, whereas to the northwest of the intersection the "two" faults would have constituted older and younger episodes of activity on a single fault. In the first case, overall displacement on the San Gabriel fault would be 65 or 90 km, depending on whether the San Francisquito fault was matched with the Blue Rock fault (Bohannon, 1975; Ehlert, 1982; Joseph and others, 1982b) or the Morales fault (Smith, 1977a). As discussed in the section on the Clemens Well–Fenner–San Francisquito fault, the available evidence does not indicate large lateral displacement on either the Blue Rock or the Morales fault. In the second case, overall displacement on the San Gabriel fault would be the amount by which the abandoned segment of the older fault (the San Francisquito fault segment) had become separated from the segment that remained active (the early San Andreas fault northwest of its junction with the San Gabriel fault) as it absorbed displacement along the younger San Gabriel fault.

Reassembly of the Frazier Mountain–Mount Pinos–eastern Orocopia Mountains reference domain (Fig. 8) requires that the sequence outlined in this second case is the correct one. In the first case, the crystalline rocks of Frazier Mountain and Mount Pinos presently exposed along the San Andreas fault between Gorman and Pine Mountain would have been offset initially from the western San Gabriel Mountains along the San Gabriel fault, indicating that the Frazier Mountain–Mount Pinos stretch of the San Andreas fault is also part of the San Gabriel fault, but not part of the Clemens Well–Fenner–San Francisquito fault, which would not allow the reference domain to be restored. In the second case, the Frazier Mountain–Mount Pinos stretch of the San Andreas fault is realigned with the San Francisquito fault, indicating that it is also part of the Clemens Well–Fenner–San Francisquito fault as well as part of the San Gabriel fault, thereby allowing the reference domain to be restored. The San Francisquito fault is displaced from the Frazier Mountain–Mount Pinos stretch of the San Andreas fault by 40 to 45 km, depending on the location of the San Francisquito fault beneath the Ridge Basin strata. Within the context of this reckoning, it is probably not coincidental that the southern limit of the Violin Breccia occurs 42 km south of the intersection between the San Gabriel fault and the Frazier Mountain-Mount Pinos stretch of the San Andreas fault. A displacement of 40 to 45 km falls within the range of measurements limited by previous work, and 42 km is restored in Figure 8 and Plate IIA, B.

Reckoning of path and displacement southeast of Castaic-Saugus area. Between Big Tujunga and San Antonio Canyons,

Ehlig's (1968, 1981) reconstruction convincingly demonstrates 22 to 23 km of displacement on the San Gabriel fault and clearly defines its path as far east as San Antonio Canyon. East of that canyon, the Icehouse Canyon and Miller faults juxtapose unrelated rock assemblages, which is consistent with but does not prove large lateral offset. Farther east, however, Matti and others (1985, p. 4) have documented 16 to 25 km of dextral displacement on the ancestral Banning fault between 7.5 and 4 or 5 Ma, the magnitude and timing of which approximates that along the San Gabriel fault between Big Tujunga and San Antonio Canyons. On the basis of this circumstantial evidence, these authors hypothesized an unspecified link between the two faults through the southeastern San Gabriel Mountains. The Icehouse Canyon and Miller faults provide the most likely path for eastward disposition of the displacement on the San Gabriel fault and provide a likely link with the Banning fault (see also May, 1986, p. 122–123; Matti and Morton, this volume).

A dextral displacement of about 5 km on the Vasquez Creek fault is consistent with the distribution of Cretaceous quartz diorite and monzogranite (Powell and others, 1983a; Crook and others, 1987; R. E. Powell, 1981–1987, unpublished mapping) and realigns the Mount Lukens and Clamshell-Sawpit faults. Conversely, the distribution of crystalline rock units along the Vasquez Creek fault (Fig. 11 and Plates I, II; Powell and Silver, 1979; Powell and others, 1983a; Crook and others, 1987) seems to preclude large lateral displacement on that fault. Restoration of any displacement greater than a few kilometers results in a misalignment and apparent repetition of the quartz diorite line in the southern San Gabriel Mountains. Moreover, south or east from Pasadena, no path has been compellingly or consistently specified in the literature for a fault with large dextral displacement that would merge eastward with either the San Gabriel fault or with the San Andreas fault, despite the fact that the unspecified fault is generally assigned nearly 40 km of right-lateral displacement (compare, for example, Crowell, 1960, 1975c, d, 1982b; Dibblee, 1968a; Ehlig, 1981). Furthermore, the 15-Ma Glendora Volcanics, which straddle the eastward projection of the Vasquez Creek fault, are not disrupted by large lateral displacement. Finally, rather than right-lateral displacement, most of the known faults between the San Gabriel River and the Peninsular Ranges, including the Raymond, San Antonio Canyon, Stoddard Canyon, and Cucamonga faults, exhibit components of left-lateral displacement (Morton, 1973, 1975b; Matti and others, 1985).

Reconciliation. The paradigm of the San Gabriel fault as an early, anastomosed (as opposed to splayed) strand of the San Andreas fault, and, similarly, of the Vasquez Creek fault as an anastomosed strand of the San Gabriel fault (Crowell, 1960, 1962, 1975c, d, 1981, 1982b; Ehlig, 1981) can be challenged for a lack of evidence supporting the offsets required by the paradigm. While it is widely accepted that the San Gabriel and Vasquez Creek faults remerge eastward with each other and with the San Andreas fault, the eastward convergence of the faults has not been demonstrated compellingly. Moreover, whereas the San Gabriel fault with a displacement of 22 km probably connects

with the Banning fault via the Icehouse Canyon and Miller faults in the southeastern San Gabriel Mountains (Matti and others, 1985; Matti and Morton, this volume), neither the Vasquez Creek fault nor any path proposed for extending it south or east from Pasadena has been shown to bear large dextral displacement. In fact, with the exception of the San Gabriel fault and splays of the San Jacinto fault, all known faults in the southern San Gabriel Mountains exhibit reverse or left-oblique displacement. It has been suggested ad hoc that movement on the postulated southeastward extension of the Vasquez Creek fault displaces the eastward extension of the Malibu Coast–Santa Monica–Raymond fault to the position of the modern Cucamonga fault (Woodburne, 1975, p. 7, Fig. 4) or to the Evey Canyon–San Antonio Canyon fault (May, 1986, 1989; Matti and Morton, this volume). Because all of the modern reverse faults are much younger than the interval of time during which most displacement accumulated on the San Gabriel fault, this suggestion depends on the hypothesized existence of a throughgoing ancestral strike-slip(?) fault that was reactivated by the reverse faults subsequent to its postulated disruption by the Vasquez Creek fault.

The 42-km displacement restored along the San Gabriel fault northwest of the Castaic-Saugus area in Plate II is roughly equivalent to the 48-km displacement estimated by Crowell (1960, 1968) and Woodburne (1975) and is compatible both with the 32- to 48-km range of displacement permitted by Crowell's (1962) reconstruction of the Soledad and Tejon areas and with the 35- to 60-km range permitted by Crowell's (1982b) interpretations of clast provenance in the Violin Breccia and the Modelo Formation. Because it is, however, significantly more than the 22-km displacement on the fault in the southern San Gabriel Mountains and significantly less than the widely accepted 60 km of slip (Bohannon, 1975; Ehlig and others, 1975; Ehlig, 1981, 1982; Ehlig and Crowell, 1982), an attempt is made here to reconcile the differences. To set the stage for such a reconciliation, my assessment of the large body of literature published on the displacement along the San Gabriel fault supports three general statements. First, overall displacement on the San Gabriel fault east of its intersection with the Vasquez Creek fault has been convincingly demonstrated to be about 22 km. Second, overall displacement on the San Gabriel fault northwest of the Castaic-Saugus area is substantially greater than displacement on the San Gabriel fault east of its intersection with the Vasquez Creek fault and lies in the range of 30 to 60 km. Third, no direct evidence has been recognized that establishes either large displacement on the Vasquez Creek fault or any alternate disposition of the difference in displacement between the two reaches of the San Gabriel fault.

In evaluating the arguments for displacement on the San Gabriel fault northwest of its intersection with the Vasquez Creek fault, it is important to recognize that the paleogeographic and paleostructural features usually interpreted as piercing points supporting the largest estimates of displacement (60 to 70 km) are reasonable but not compelling paleogeologic reconstructions. For example, the 10- to 13-Ma paleodrainage of Ehlig and colleagues (Ehlig and others, 1975; Ehlig, 1981; Ehlert, 1982) is constrained only where its deposits are exposed in the Mint Canyon and Caliente Formations, and although these authors reconstructed the shortest possible drainage by directly juxtaposing these remnants, their data do not rule out the possibility that a part of the sedimentary record is missing and that the drainage once flowed through intervening deposits now buried under younger Ridge Basin deposits or eroded from above the subsequently uplifted Frazier Mountain–Whitaker Peak crystalline block. By minimizing the length of the drainage, Ehlig and others maximized the displacement on the San Gabriel fault, but if a longer drainage existed, postdrainage displacement required on the San Gabriel fault is less. Carman (1964), who initially proposed the fluvial connection between the lower part of the Mint Canyon and Caliente Formations, suggested a minimum displacement of 32 km. the 42-km displacement indicated in Figure 8 and Plate II lies within the range bracketed by these end-member estimates.

Similarly, the offset indicated by the inferred paleostructural features depends on the uniqueness of the way in which their composite elements have been reassembled. Thus, the displacement on the San Gabriel fault is 90 km if the San Francisquito fault is matched with the Morales fault, 60 km if it is matched with the Blue Rock fault, or 42 km if it is aligned with the San Andreas fault northwest of Frazier Park. Because the Pelona Schist occurs beneath a regional thrust fault, the Abel Mountain–Sierra Pelona match-up is a unique piercing point only if it can be demonstrated that the two bodies of schist were exhumed along a single continuous antiform prior to disruption by the San Gabriel fault. Alternate interpretations of the paleotectonic setting, such as one in which those bodies were uplifted in right-stepping en echelon antiforms (Powell, 1981, p. 371–372, Plate VII), permit a lesser displacement than the single-antiform interpretation.

No unambiguous piercing point has been established in either the crystalline terrane or the marine Paleogene strata transected by the San Gabriel fault. The paleogeologic reconstruction of Ehlig and Crowell (1982), showing 60 km of displacement, looks reasonable, but their generalized geologic units also reassemble into reasonable looking paleogeologic maps showing lesser displacements within a 42- to 60-km range. For example, Dibblee (1982b, h), in two articles in a single volume, showed paleogeologic maps with 45 and 60 km restored, respectively.

Although the disparity in overall displacement measured along the San Gabriel fault is reduced by using a magnitude of 42 km rather than 60 km for the fault segment northwest of the Castaic-Saugus area, the lesser magnitude is still about twice as large as the 22- to 23-km displacement on the fault east of its intersection with the Vasquez Creek fault. The disposition of the unaccounted 22 or 23 km of displacement is a major unresolved problem regarding the San Gabriel fault. It is widely accepted that the excess displacement is accommodated by the Vasquez Creek fault, but given the conspicuous lack of evidence for large displacement on the Vasquez Creek fault and the evidence presented herein for about 5 km of displacement on the fault, one is justified and perhaps compelled to entertain the notion that the

displacement is not absorbed by that fault, but rather is accommodated by some other structure or combination of structures.

In view of the ambiguities involving the generally accepted hypothesis, it seems prudent to consider alternate hypotheses for accommodating southward and southeastward disposition of displacement along the San Gabriel fault. Alternatives in which the San Gabriel fault is offset left-laterally from an early strand of the San Andreas fault in San Gorgonio Pass (see Jahns, 1973) are untenable both in terms of timing of movement (refer below to section on ancestral Malibu Coast-Santa Monica fault; Woodburne, 1975, p. 5–17) and within the context of the reassembly of the metasedimentary suite of Limerock Canyon and the Mesozoic plutonic rocks (Fig. 11, Plate II). Alternatives in which right-lateral faults of the San Gabriel fault system are absorbed by a zone of sinistral kinking and/or strike slip faulting along the southern margin of Transverse Ranges (see later section on left-lateral faults) are inconsistent with evidence that indicates incompatible timing for movement on the two fault systems.

There are at least three alternate hypotheses that could have operated singly or in combination to accommodate the missing 22 or 23 km of displacement to the southeast. First, the missing displacement may have been accommodated by as much as 22 km of late Miocene north-south, left-oblique extension in the Los Angeles or Ventura basin or both (Fig. 14b). The Los Angeles basin subsided rapidly and accumulated as much as 10 km of sediment during the middle and late Miocene and the Pliocene (Yerkes and others, 1965; Crowell, 1987; Mayer, 1987). During the late Miocene and early Pliocene, the Ventura basin was a deep-water marine turbidite trough that shallowed abruptly against the San Gabriel Mountains at its eastern end, where turbidite gave way to conglomerate (Eaton, 1939, Fig. 1, p. 522; Winterer and Durham, 1962, p. 339; Weber, 1982a, p. 63, 65). Moreover, in the Santa Monica Mountains, Simi Hills, and Ventura basin region, there is abundant structural and stratigraphic evidence for middle and late Miocene north-south crustal extension (Yeats, 1968, 1983; Campbell and Yerkes, 1976).

Second, some or all of the missing displacement may have

Figure 14. Models for disposition of right-lateral displacement on the San Gabriel fault system. North toward top of figure. a, San Gabriel fault as an anastomosed strand of San Andreas fault system (ca. 5 Ma). b, Left-oblique extension in the Los Angeles basin associated with slip on the Canton fault, in the San Bernardino Valley associated with slip on the San Gabriel fault, and in the San Gabriel Valley associated with slip on the Vasquez Creek fault. c, Early left-oblique extension in the Los Angeles basin associated with slip on the Canton fault, followed by later right-oblique extension in the Salton trough area associated with slip on the San Gabriel fault. d, Complementary dextral displacement on the San Gabriel and Punchbowl faults and sinistral displacement on currently northeast-trending faults of western San Gabriel Mountains and Soledad basin. CW = Clearwater fault; LM = Liebre Mountain fault; PB = Punchbowl fault; VC = Vasquez Creek fault. Double-tipped arrows show extensional domains required by each model.

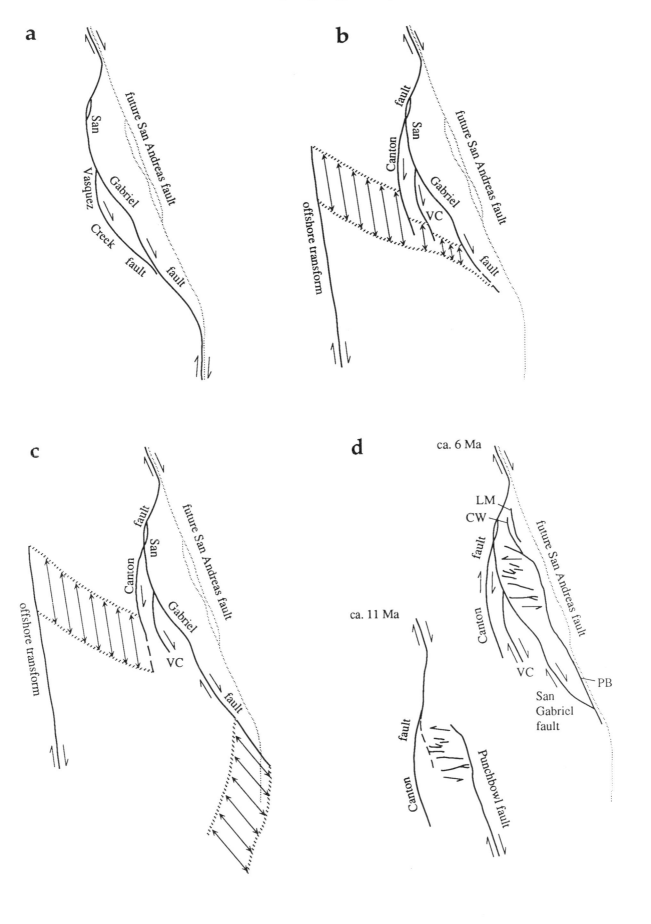

splayed south onto a right-lateral fault other than the Vasquez Creek fault: because it is not exposed anywhere else, the only possible path for such a fault diverges from the San Gabriel fault in the Castaic-Saugus area beneath young strata in the eastern Ventura basin (Fig. 14b, c). From about Castaic, the hypothetical order fault, now buried, extends south between the Santa Susana and San Gabriel Mountains and then between the Santa Monica and Verdugo Mountains and carries as much as 23 km of displacement. There is some evidence for the existence of this hypothetical fault in the pre-Mohnian subsurface map and cross sections shown by Stitt (1986, Figs. 3, 10–12, 16–19) for the eastern Ventura basin. Specifically, along the eastern edge of the Ventura basin the fault may form the western limit of a buried basement ridge of igneous and metamorphic rocks, and may at the same time constitute the northwest-trending linear contact between an eastern outlier of marine Luisian strata and coeval nonmarine strata to the west. The Canton fault between granitic and gneissic rocks southeast of Whitaker Peak (Crowell, 1954a, b, 1975d) is considered to be an early strand of the San Gabriel fault (Stitt and Yeats, 1982; Stitt, 1986; see also Crowell, 1982b) and may be an exposed part of this hypothetical buried fault.

Third, the greater measured displacement to the northwest along the San Gabriel fault may have been in part accommodated by components contributed by coeval left-lateral faults in the western San Gabriel Mountains and Soledad basin (Fig. 14d; see later discussion; cf. Weldon and others, this volume). The sinistral faults of the San Gabriel Mountains and Soledad basin have a cumulative displacement of at least 18 km, as much as 32 km if the San Francisquito fault was reactivated as a left-lateral fault of this set with a displacement of about 14 km (see Smith, 1977a). Movement on these faults occurred chiefly in late Barstovian to Clarendonian time during deposition of the Mint Canyon Formation, or about 15 or 13 to 10 Ma, and they may have absorbed displacement on the Canton fault strand of the San Gabriel system. Although this model is appealing conceptually, I was unable to make it work in constructing Plate II, A through C.

The southward disposition of displacement on the strands of the San Gabriel fault system is represented in Figures 11 and 14b and Plate II. I assign 22 km to the San Gabriel fault southeast of its junction with the Vasquez Creek fault and 5 km to the Vasquez Creek fault. The 22-km displacement on the San Gabriel fault hypothetically has been linked southeastward with the Banning fault (see Matti and Morton, this volume), and in Plate II, A, B that displacement and the 5-km displacement on the Vasquez Creek fault are absorbed by closing hypothetical left-oblique extension in the San Gabriel and San Bernardino Valleys. An alternate interpretation would be to carry the 22-km displacement southeastward along the Banning fault beyond San Gorgonio Pass (see Matti and Morton, this volume) and absorb it by closing early right-oblique extension in the Coachella Valley that may have accompanied deposition of the marine Imperial Formation (Fig. 14c). I assign the remaining 15 km of displacement measured on the San Gabriel fault northwest of its intersection with the Vasquez Creek fault to the Canton fault and its hypothetical

extension to the south. That 15-km displacement is absorbed southeastward by closing left-oblique extension in the Los Angeles basin. In assembling the reconstruction, I closed the Los Angeles basin rather than the Ventura basin in part because that choice helps reassemble the rocks of the Santa Monica and Santa Ana Mountains. Geometric analysis of the relation between the San Gabriel fault and the sinistral faults of the Transverse Ranges west of the San Gabriel fault (see section on left-lateral faults), however, is more consistent with closing the Ventura basin.

Early displacement of about 10 to 15 km on the Canton-San Gabriel fault prior to deposition of the Violin Breccia offers a way to partially reconcile some of the observations and interpretations of previous investigators with the reconstruction shown in Plate II. If one accepts Crowell's interpretation that the coarse Violin Breccia east of the San Gabriel fault is progressively younger from southeast to northwest and was shed continuously from a northwestward-traveling crystalline block west of the fault, then one might expect that the length of the breccia deposit (about 30 to 35 km) would approximate the offset accumulated during that interval of deposition. Moreover, beds at the northwest end (top) of the breccia should exhibit no displacement from their source during that interval, whereas beds at the southeast end (bottom) of the breccia should have recorded the full offset of the fault during that interval. In the model proposed here, movement on the Canton–San Gabriel fault displaced the Frazier–Alamo Mountains block by about 13 km from the Soledad basin basement and the gneiss block now located between the Canton and San Gabriel faults while that block was still attached to the Mendenhall Gneiss in the Soledad basin–San Gabriel Mountains crystalline block. Subsequently, the gneiss block now located between the Canton and San Gabriel faults was sliced from the Soledad–San Gabriel block, and the Frazier Mountain–Alamo Mountain–Whitaker Peak block was displaced another 32 km along the San Gabriel fault, the amount recorded in the Violin Breccia "conveyor belt." If, as it appears, uplift of the Frazier Mountain–Alamo Mountain–Whitaker Peak block accompanied movement on the younger San Gabriel fdault but did not accompany movement on the older Canton fault, then the throughgoing Mint Canyon–Caliente fluvial system could have carried debris along and across a coevally active Canton fault, and then been displaced 32 km along the San Gabriel fault as measured by Carman (1964).

North of the intersection between the San Gabriel fault and the Clemens Well–Fenner–San Francisquito–early San Andreas fault, displacement on the San Gabriel fault is constrained to have followed the early San Andreas fault. If the angle of intersection between the two faults has remained constant (see discussion below in section on kinematic role of Salinian block faults), the amount of dextral displacement transferred northward to the early San Andreas fault is given by the relation

$$D_{SA} = D_{SG} \cos \alpha, \tag{1}$$

where D_{SA} is the increase in displacement on the San Andreas fault, D_{SG} is the displacement on the San Gabriel fault, and α is the angle of intersection between the two faults (Fig. 15). In plates I and II, the measured angle of intersection (α) between the two faults lies in the range of 30° to 40° and displacement (D_{SG}) restored on the San Gabriel fault is 42 km. Substituting these values into equation (1), displacement (D_{SA}) contributed to the San Andreas fault is 32 to 36 km. By this analysis, one should expect the remaining 6 to 10 km to have been accommodated by contractional structures along the San Andreas fault north of its intersection with the San Gabriel fault. The acceleration in rate of fold growth between about 12.5 and 7.4 Ma along the west side of the southern San Joaquin basin (Harding, 1976) may represent part of the contractional component of movement on the San Andreas fault that is related to movement on the San Gabriel fault.

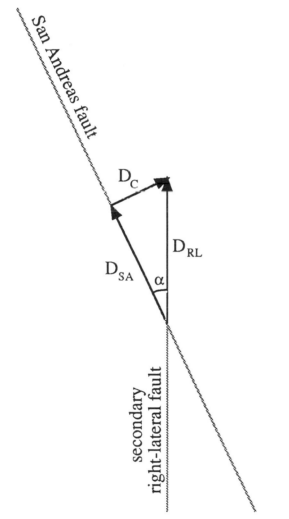

Figure 15. Vector diagram for interaction between San Andreas fault and coeval right-lateral faults that join it from the south, including the San Gabriel, San Gregorio–Hosgri, and San Jacinto faults. Symbols: α, angle of intersection between the dextral faults; D_{SA} displacement on San Andreas fault; D_{RL} displacement on secondary right-lateral fault; D_c displacement perpendicular to San Andreas fault on contractional structures, including folds and reverse faults.

Timing and rate of movement. Overall displacement along the Canton, San Gabriel, and Vasquez Creek faults has accumulated since sometime after deposition of the Sespe Formation in the Canton Canyon area and probably after deposition of the Vaqueros Formation and Rincon Shale that overlie the Sespe. Proterozoic and Mesozoic crystalline basement rocks on the Mount Pinos–Frazier Mountain–Alamo Mountain block and the anorthosite- and Mount Lowe intrusion-bearing conglomerate in the Oligocene Sespe Formation in Canton Canyon have been translated northwestward from the western San Gabriel Mountains by the total displacement (42 to 45 km) on the three faults. The Sespe Formation is Refugian and Zemorrian in age, the Vaqueros Formation in the Canton Canyon area is Zemorrian in age, and the Rincon is Zemorrian and Saucesian in age, and the Rincon is Zemorrian and Saucesian in age, according to Paschall and Off (1961) and Stitt (1986). Inasmuch as there is no evidence for movement prior to the end of the Saucesian, it is likely that all movement on the San Gabriel fault is Relizian or younger. On a geochronometric time scale (Fig. 13), the Saucesian-Relizian boundary has been placed variously at 15.3 Ma (Turner, 1970; Berggren and Van Couvering, 1974), 16.7 Ma (Armentrout and Echols, 1981), about 17 Ma (Kleinpell and others, 1980; Bartow, 1990), 17.5 Ma (Boellstorff and Steinbeck, 1975), and between about 19 and 17.5 Ma (Poore and others, 1981). Thus, movement on the Canton–San Gabriel fault began sometime after about 17 to 15 Ma.

The lower part of the Mint Canyon Formation is late Barstovian and possibly early Clarendon in age, according to Durham and others (1954, p. 66–67) and Whistler (1967, p. 1). It and the lower two members of the Caliente Formation in Lockwood Valley and the Dry Canyon area of the Cuyama badlands—which range in age from Hemingfordian through early Clarendonian, according to James (1963), Carman (1964, including D. E. Savage, personal communication, 1957), and Kelly and Lander (1988)—are offset at least 32 km, and perhaps by the full displacement of 42 to 45 km on the San Gabriel fault. In the former case, movement on the fault started prior to or during deposition of these strata, whereas in the latter case movement began after their deposition. The upper part of the Mint Canyon, which is late Clarendonian according to Durham and others (1954) and contains tuff beds dated at 10.1 and 11.6 Ma (J. D. Obradovich and T. H. McCulloh, 1982, cited in Terres and Luyendyk, 1985), has been interpreted to have been deposited during movement on the San Gabriel fault (Ehlert, 1982; Stitt, 1986). On a geochronometric time scale (Fig. 13), the Barstovian spans an interval from 17 or 16 to 12 or 11 Ma and the Clarendonian spans an interval from 12 or 11 to 10 or 9 or 8 Ma (Turner, 1970; Berggren and Van Couvering, 1974; Berggren and others, 1985; Tedford and others, 1987; Woodburne, 1987; Bartow, 1990). If early movement on the Canton–San Gabriel fault postdates the lower part of the Mint Canyon Formation and is coeval with the upper part of the formation, then displacement associated with that early movement accumulatd between 13 or 12 Ma and 10 to 8 Ma. On the other hand, if early movement is coeval with both the lower and upper parts of the Mint Canyon, then displacement accumulated between 15 or 14 and 10 to 8 Ma.

The west-transported, anorthosite-bearing conglomerate beds in the Modelo Formation in Canton Canyon that are offset from a probable source in the San Gabriel Mountains were originally thought to be Mohnian and younger than the Mint Canyon Formation (Crowell, 1952, 1954b). On the basis of that age, Crowell (1960, 1975d) suggested that the offset recorded by those beds is slightly less than the total. The Modelo Formation is Relizian, Luisian, and Mohnian in age, and the anorthosite-bearing conglomerate beds are now known to contain or are correlated with rocks that contain Luisian and early Mohnian microfaunas (Paschall and Off, 1961; Crowell, 1982b, p. 31; Stitt, 1986), roughly coeval with the late Barstovian to early Clarendonian parts of the Mint Canyon and Caliente Formations. On a geochronometric time scale, the Relizian is variously shown as ranging from 15.3 to 14.5 or 13.7 (Turner, 1970), 15.3 to 14 (Berggren and Van Couvering, 1974), 16.7 to 15.8 (Armentrout and Echols, 1981), 17.5 to 16 (Boellstorff and Steinbeck, 1975), 17 to 15 (Kleinpell and others, 1980), 19 to 16 (Poore and others, 1981), and 17 to 15.8 Ma (Bartow, 1990); and the Luisian is variously shown as ranging from 14.5 or 13.7 to about 12 Ma (Turner, 1970), 14 to 13 Ma (Berggren and Van Couvering, 1974), 16 to 13 Ma (Boellstorff and Steinbeck, 1975), 15 to 14.5 or 14 Ma (Kleinpell and others, 1980), 16 to 14 Ma (Poore and others, 1981), and 15.8 to 13.8 Ma (Bartow, 1990). These strata of the Modelo Formation have not been shown conclusively to be offset by the total displacement along the San Gabriel fault. If they are fully offset, then movement on the Canton–San Gabriel fault began no earlier than about 13 or 12 Ma. On the other hand, if the anorthosite-bearing Modelo conglomerate beds record tectonism associated with strike-slip faulting, movement on the Canton–San Gabriel fault may have begun as early as about 16 to 14 Ma. This early movement accumulated on the Canton fault, which juxtaposes Mesozoic granitic rocks onlapped by a section of the Zemorrian and Saucesian Rincon Shale and the Relizian, Luisian, and the early Mohnian part of the Modelo Formation against Proterozoic(?) gneiss overlain by early Mohnian Modelo Formation. The fault is overlapped by early Mohnian strata of the Modelo Formation (Crowell, 1952, 1954b, 1975d, 1982b; Stitt and Yeats, 1982; Stitt, 1986), indicating that its movement had ceased by about the middle of the Mohnian, or by 10 to 9 Ma.

Northwest of Castaic, movement along the San Gabriel fault exposed a block of crystalline rocks between Whitaker Peak and Alamo Mountain that provided the source for the Violin Breccia. In Crowell's interpretation, the Violin Breccia east of the fault gets progressively younger from southeast to northwest and the breccia was shed continuously from a northwestward-traveling crystalline block west of the fault, indicating that the San Gabriel fault moved at least during the interval of time spanned by the range in age of the breccia. Although the breccia itself is unfossiliferous, its age has been established through its stratigraphic

relations with finer grained, fossiliferous upper Miocene strata of the Ridge basin. Successively, from southeast to northwest, the Violin Breccia interfingers eastward with the upper part of the Castaic Formation of Crowell (1954a, b), the Peace Valley Formation of Crowell (1964) (see also Crowell, 1950; Dibblee, 1967a; Link, 1982, 1983), and the lower part of the Hungry Valley Formation of Crowell (1950, 1952, 1982a) (see also Dibblee, 1967a; Link, 1982, 1983; Ramirez, 1983). The Peace Valley and lower part of the Hungry Valley Formations in turn interfinger farther eastward with the conglomeratic Ridge Route Formation of Clements (1937) (see also Link, 1982, 1983). The lowermost part of the Peace Valley Formation interfingers with and overlies the upper part of the Castaic Formation and the uppermost part of the Peace Valley Formation interfingers with and underlies the lower part of the Hungry Valley Formation. The marine Castaic Formation, in turn, overlaps the Mint Canyon Formation with angular unconformity (Kew, 1924; Jahns, 1940; Oakeshott, 1958; Winterer and Durham, 1962; Ehlert, 1982; Stitt, 1986).

Fossils and the interfingering relations establish that the age of the Violin Breccia, corresponding to the period of movement on the San Gabriel fault that resulted in deposition of the breccia, ranges from the latest Clarendonian, latest "Margaritan," and middle(?) Mohnian to the latest Hemphillian, latest "Jacalitos," and latest(?) Mohnian. Marine microfossils in the Castaic Formation are late Mohnian and Delmontian in age (Kleinpell, 1938, p. 71; P. P. Goudkoff, 1951, written communication cited in Paschall and Off, 1961; Skolnick and Arnal, 1959; Kleinpell and others, 1980, p. 34–35), and megafossils are late "Margaritan" and "Jacalitos" (Neroly and Cierbo) in age (Jahns, 1940, p. 167; Adegoke, 1969; Ensley and Verosub, 1982; for additional references, see Link, 1982). The Ridge Route and Peace Valley Formations contain Clarendonian and Hemphillian vertebrate fossils (Miller and Downs, 1974; Welton and Link, 1982; for additional references, see Link, 1982), and the uppermost part of the Ridge Route and Peace Valley Formations and lowermost part of the Hungry Valley Formation contain late Hemphillian terrestrial vertebrate fauna (Chester Stock, cited in Crowell, 1950; Miller and Downs, 1974; Welton and Link, 1982).

On the basis of the fossil evidence for the age of the Violin Breccia, the San Gabriel fault accumulated displacement between about 10 and 6 or 5 Ma. Turner (1970) stated that the lower boundary of the Mohnian is younger than about 13 Ma and published ranges include 13 to 6 Ma (Berggren and Van Couvering, 1974), 14.5 or 14.0 to 11 Ma (Kleinpell and others, 1980), 13.5 to 6.0 Ma (Poore and others, 1981), and 13.8 to 7.0 Ma (Bartow, 1990). The "Margaritan"-"Jacalitos" boundary lies in the range of 8.5 to 9.0 Ma (Poore and others, 1981; Ensley and Verosub, 1982); the Hemphillian has been shown to span intervals from 10 to 3.8 Ma (Berggren and Van Couvering, 1974), from 9 to 5 Ma (Berggren and others, 1985), from 9[+] to 5 Ma (Ensley and Verosub, 1982), and from 8.2 to 5 or 4.5 Ma (Woodburne, 1987); and the "Jacalitos" to span an interval from about 9 or 8.5 to 6 or 5.5 Ma (Poore and others, 1981; Ensley and Verosub, 1982). Because deposition of Hemphillian strata of the Hungry Valley Formation continued after deposition of the Violin Breccia had ceased, the interval of movement indicated by the breccia came to a halt somewhat prior to the end of the Hemphillian (Crowell, 1950, 1982b).

Magnetostratigraphic correlation (on the magnetic polarity time scale of Ness and others, 1980) of strata in the Ridge Route and Peace Valley Formations yield an age range from about 8.3 to 5.5 Ma (Ensley and Verosub, 1982), providing corroborating evidence for the age of the Violin Breccia and the corresponding interval of movement on the San Gabriel fault. Because the Violin Breccia also interfingers with strata above and below the Peace Valley Formation, the interval of movement began somewhat earlier than 8.3 Ma and continued somewhat later than 5.5 Ma. Thus, on the basis of magnetostratigraphic evidence, fault movement associated with deposition of the breccia began about 10 or 9 Ma and ended between about 5 to 5.5 Ma.

Movement along the San Gabriel fault in the eastern Ventura basin continued during deposition of the Towsley Formation (Crowell, 1952; Winterer and Durham, 1962; Ehlig, 1975b; Stitt, 1986; Weber, 1982a). Marine microfauna indicate that the Towsley Formation is Mohnian and Delmontian in age (Winterer and Durham, 1962). In the westernmost San Gabriel Mountains, the Towsley Formation includes strata mapped as the Elsmere Member of the Repetto Formation by Oakeshott (1950, 1958) and Winterer and Durham (1962), which has yielded "Jacalitos" marine megafossils (Grant and Gale, 1931, p. 29–32). On a geochronometric time scale (Fig. 13), the end of the Mohnian-Delmontian interval (equivalent to the top of the *Bolivina obliqua* Zone) is variously placed at 2.7 Ma (Berggren and Van Couvering, 1974) and at about 4 to 5 Ma (Kleinpell and others, 1980; Armentrout and Echols, 1981; Poore and others, 1981; Bartow, 1990). These ages are consistent with movement on the San Gabriel fault between about 10 and 5 Ma.

Anorthosite slide blocks and clasts in the Violin Breccia where it interfingers with the Hemphillian-age uppermost part of the Peace Valley and lowermost part of the Hungry Valley Formations (Shepard, 1962) are separated about 9 km from bedrock exposures of anorthosite just north of Piru Creek on the south flank of Bear Mountain (J. C. Crowell, cited in Woodburne, 1975; Ehlig and Crowell, 1982). Thus, about 9 km of offset has occurred since about 6 Ma.

About 1 to 2 km of displacement has occurred since about 3 or 4 Ma. In the eastern Ventura basin, Pliocene strata of the Pico Formation that are "Repettian" in age (2.9 to 2.1 Ma, Berggren and Van Couvering, 1974; 4.3 to 2.5 Ma, Bartow, 1990) and Venturian (2.1 to 1.8 Ma, Berggren and Van Couvering, 1974; 2.5 to 1.9 Ma, Bartow, 1990) on the basis of microfaunal assemblages (Winterer and Durham, 1962; Stitt, 1986) are displaced 1.8 km. Although Ehlig (1975b) used the presence of clasts of anorthosite and the Mendenhall Gneiss in marine conglomerate beds in the Pico Formation east of Newhall to estimate between 6 and 19 km of displacement on the San Gabriel fault, his argument was not unique and the distribution of anorthosite clasts is also

consistent with a displacement of about 1 or 2 km. In the same area, the Pleistocene Saugus Formation is offset 0.5 km and consists of basal marine strata that contain a Wheelerian (1.8 to 0.7 Ma, Berggren and Van Couvering, 1974; 1.9 to 0.7 Ma, Bartow, 1990) microfaunal assemblage (Winterer and Durham, 1962; Stitt, 1986) and overlying terrestrial strata that contain a Pleistocene vertebrate fauna (Winterer and Durham, 1962; Stitt, 1986) and the 0.7-Ma Bishop ash bed (Levi and others, 1986). The magnetostratigraphic record of the terrestrial strata contains the Matuyama Reversed-Polarity Chron–Brunhes Normal-Polarity Chron boundary (0.73 Ma, magnetic polarity time scale of Mankinen and Dalrymple, 1979), and possibly the Jaramillo Normal-Polarity Subchron (of the Matuyama) (0.9 to 0.97 Ma) and the Olduvai Normal-Polarity Subchron (of the Matuyama) (1.67 to 1.87 Ma) (Levi and others, 1986). In the vicinity of the northwest terminus of the San Gabriel fault, strata of the late Hemphillian Hungry Valley Formation that are displaced 1.2 km (Weber, 1982b, 1986a, b) were deposited in the interval between about 5.0 and 3.5 Ma. The apparent 2.4- and 6-km offsets of Pacoima and Big Tujunga Canyons, respectively, noted by Oakeshott (1950, 1958) indicate either that the streams were incised prior to 3 or 4 Ma or that the true offset is less than the apparent offset.

In summary, in the context of the displacement model developed above, movement occurred on the San Gabriel and related faults prior to, during, and after deposition of the Violin Breccia (Fig. 13) (cf. Crowell, 1950, 1982b; Woodburne, 1975). The oldest strand of the San Gabriel fault, following the Canton fault south of Piru Creek and its hypothetical buried extension south of about Castaic, accumulated as much as 13 km of displacement in the interval between 15 to 12 and 10 to 9 Ma. Displacement during this interval disrupted Proterozoic and Mesozoic crystalline rocks, marine Upper Cretaceous and Paleogene strata, and Oligocene and lower Miocene marine and terrestrial strata. This interval of movement is inferred to have been coeval with the evolving middle Miocene Mint Canyon–Caliente paleodrainage, the changing course of which kept pace with fault movement through the Barstovian (through about 12 Ma) but ultimately was diverted elsewhere. The loss of a throughgoing drainage led to differing Clarendonian stratigraphic records in the upper part of the Mint Canyon Formation of Soledad basin and the upper part of the Caliente Formation of Lockwood Valley (Carman, 1964; Woodburne, 1975; Ehlert, 1982).

Following movement on the Canton–San Gabriel fault, the San Gabriel fault splayed from it near Piru Creek and ran subparallel to the older fault between Piru Creek and about Castaic, from where the younger fault diverged southeastward. Most of the displacement on the San Gabriel fault accumulated during the deposition of the Violin Breccia. Movement during this phase began about 10 or 9 Ma (see Crowell, 1981; Ensley and Verosub, 1982) and by about 5 Ma the fault had acquired a displacement of about 30 km which merged with the 13-km displacement on the Canton fault near the intersection of the two faults near Piru Creek. Of the younger 30-km displacement, 21 to 23 km was acquired between 10 or 9 and 6 Ma, and 7 to 8 km was acquired between 6 and 5 or 4 Ma. The San Gabriel fault between its intersections with the Vasquez Creek and San Andreas faults has acquired up to another 1 or 2 km from about 3 or 4 Ma to the present, for a total displacement of 42 to 46 km. The most recent 5-km displacement probably followed the Vasquez Creek fault southeast of its intersection with the San Gabriel fault.

The long-term average slip rate for the San Gabriel fault system can be calculated for several cases. For the preferred case—in which the fault system accumulated displacements of 13 km between 13 and 10 Ma, 22 km between 10 and 6 Ma, and 7 to 8 km between 6 and 4 Ma—the average slip rate is 4.3, 5.5, and 3.5 to 4.0 mm/yr for successive intervals. The overall long-term average slip rate for a displacement of 42 km between 13 and 5 Ma is 5.3 mm/yr. For the same displacement to accumulate between 13 and 4 Ma would require a rate of 4.7 mm/yr. These rates are significantly less than the rate of 12 mm/yr calculated for a displacement of 60 km accumulating in the 5 m.y. between about 10 and 5 Ma.

Kinematic role in the evolving San Andreas fault system. In the prevailing and almost universally accepted regional tectonic model, the San Gabriel fault is interpreted as an early strand of the San Andreas fault system that was abandoned in favor of the modern trace of the San Andreas fault (Fig. 14a). In this model, the San Gabriel and San Andreas faults merge northwestward in the vicinity of Tejon Pass and southeastward in the vicinity of San Gorgonio Pass, and overall displacement of about 300 km on the San Andreas fault system typically is divided between these two principal anastomosed strands at about 60 and 240 km, respectively. The existence of the Clemens Well–Fenner–San Francisquito fault as a component of the San Andreas system with a displacement of 100 or 110 km, however, calls into question the validity of this model.

After the Clemens Well–Fenner–San Francisquito fault was abandoned at about 113 Ma, the San Gabriel fault system developed as the active strand of the San Andreas system in the Transverse Ranges until about 5 Ma. The San Gabriel fault connected with the early San Andreas fault north of the Transverse Ranges, and probably was linked via that fault to the Mendocino triple junction. Reconstruction of the southern Transverse Ranges–Chocolate Mountains reference domain requires that at least half of the displacement on the San Gabriel fault system be accommodated by extensional and sinistral deformation along the southern margin of the Transverse Ranges. For at least the time it took to accumulate half of the overall slip on the San Gabriel fault system, dextral movement continued to step west to the continental margin, as it had during the time that the Clemens Well–Fenner–San Francisquito fault was active. During the time it took to accumulate the latter half of its displacement, the San Gabriel fault system extended to San Gorgonio Pass. During this time, displacement either continued to step westward to the continental margin (Fig. 14b) or connected with incipient extension in the Salton trough and Gulf of California (Fig. 14c) (see Matti and Morton, this volume).

Clearwater and Liebre faults. The Clearwater and Liebre faults between the San Francisquito and San Andreas faults are two reverse-dextral oblique faults that were active in the interval between about 10 and 5 Ma (Link, 1983). Based on their disruption of magnetostratigraphically dated upper Miocene strata in the Ridge basin, the Clearwater fault is inferred to have accumulated a right separation of 1.6 to 4 km (Crowell, 1982a; Link, 1983) between about 10 to 9 and 7.8 Ma (Ensley and Verosub, 1982), and the Liebre fault is inferred to have accumulated a right separation of about 8 km by Crowell (1982a) and 6.5 km by Link (1983) between about 7.3 and 6 Ma (Ensley and Verosub, 1982), for a total between about 8 and 12 km. Hypothetically, these two faults developed as sequential transitional structures between the Clemens Well–Fenner–San Francisquito and San Andreas faults in both space and time.

Punchbowl fault. Path. The Punchbowl fault is exposed along the northeast flank of the San Gabriel Mountains. Initially mapped as a northwestern extension of the San Jacinto fault zone (Johnson and Warren, 1927; Dutcher and Garrett, 1963; Ehlig, 1968), the Punchbowl fault was given a separate name by Noble (1954a, b), although he too considered the fault a part of the San Jacinto fault zone. More recently, continuity between the Punchbowl and San Jacinto faults in the southeasternmost San Gabriel Mountains has been disputed by several investigators who instead have mapped the Punchbowl fault southeastward onto the San Andreas fault (Dibblee, 1967a, 1968a, 1975b, 1982e; Morton, 1975a, b; Morton and Matti, this volume; Ehlig, 1981, 1982).

Previous estimates of displacement. Measurements of about 40 to 50 km of right-lateral offset on the Punchbowl fault have been based on matching the San Francisquito and Fenner faults, both of which juxtapose the Pelona Schist to the south against Mesozoic batholithic rocks overlain by the Upper Cretaceous and Paleocene San Francisquito Formation to the north (Dibblee, 1967a [Fig. 72, p. 114], 1968a, 1975b; Ehlig, 1968, 1981, 1982; Sage, 1975; Woodburne, 1975).

The upper Miocene Punchbowl Formation crops out along and east of the Punchbowl fault between Valyermo and Vincent Gap. The unit has not been recognized to the west of the fault. Several investigators (Farley and Ehlig, 1977; Ehlert and Ehlig, 1977; Barrows and others, 1985, p. 44–45, 58) have noted the presence of clasts of polka-dot granite, foliated quartz diorite, and sandstone and recycled volcanic clasts derived from the San Francisquito Formation; a few (Barrows and others, 1985, p. 58) have noted the absence of clasts of the anorthosite-syenite complex and Mount Lowe intrusion in the Punchbowl Formation. Based on their observations of the distribution of clast types, these authors inferred that the source terrane for the Punchbowl Formation was not the San Gabriel Mountains but was likely to have been the southern Liebre Mountain block. From these provenance inferences, they concluded that, along with the basement on which it rests, the Punchbowl Formation has been offset 44 to 45 km along the Punchbowl fault from its place of deposition adjacent to the southeastern Liebre Mountain block.

The Pliocene(?) formation of Juniper Hills of Barrows (1980) crops out west of the Punchbowl fault in the area west of Palmdale and between the Punchbowl and Nadeau faults southeast of there. From the distribution in the Juniper Hills formation of boulders derived from the Pelona Schist and San Francisquito Formation and of volcanic and granitic clasts recycled from the Punchbowl Formation into the formation of Juniper Hills, Barrows and others (1985), proposing a conveyor belt model similar to that of the Violin Breccia of Crowell (1954a), measured 13 km of displacement during deposition of the formation of Juniper Hills west of Palmdale and greater amounts along the fault to the southeast.

For the reckoning of its displacement from the palinspastic reconstruction in Plate II, the Punchbowl fault is discussed as part of the San Andreas fault in the following section.

Timing of movement. The timing of displacement on the Punchbowl fault, although seemingly well understood, is not unequivocal. On the basis of terrestrial vertebrate fossils, the Punchbowl Formation, which is disrupted by the Punchbowl fault, ranges from no older than early Clarendonian in age at the base to at least Hemphillian (Woodburne and Golz, 1972, p. 34; Woodburne, 1975, p. 65; Barrows and others, 1985). On the basis of magnetostratigraphy, the base of the formation is 12.5 to 12.7 Ma (Liu, 1990). Barrows and others (1985) argued that displacement on the Punchbowl fault occurred after deposition of the Clarendonian and Hemphillian Punchbowl Formation and during deposition of the formation of Juniper Hills that is probably Pliocene and Pleistocene (Blancan) in age. They noted that the northern part of the fault is covered by Quaternary alluvial fan deposits. Sharp and Silver (1971) inferred 17 to 32 km of Quaternary displacement along the Punchbowl fault. Woodburne (1975), on the other hand, proposed a two-stage movement history, in which he advocated 20 km of pre-Barstovian displacement and 6 to 20 km of Quaternary displacement. Weldon and others (this volume) also advocate two periods of movement on the Punchbowl fault, with Clarendonian as well as Pliocene and Pleistocene movement.

San Andreas fault. Path. Segmented by big bends at Mount Pinos and at San Gorgonio Pass, the trace of the San Andreas fault transects several physiographic provinces, trending southeastward obliquely through the Coast Ranges province from Cape Mendocino to Mount Pinos, east-southeastward obliquely along the edge of the Mojave Desert and through the Transverse Ranges province from Mount Pinos to San Gorgonio Pass, and southeastward along the east margin of the Salton trough from San Gorgonio Pass to the Salton Sea. Although the concept of a San Andreas fault system is commonly used in a broad sense to refer to the entire array of Cenozoic right-lateral faults in California (Crowell, 1962, 1975c; Dibblee, 1968a), the use of San Andreas fault or fault zone typically is restricted to the modern throughgoing fault zone that is no more than about 10 km wide. In this chapter, the three segments of the fault are referred to as the Coast Ranges, Mojave Desert, and Salton trough segments. In each of these segments, displacement along the San Andreas fault

zone commonly is divided among several anastomosed fault strands that for the most part are not distinguished herein (see, for example, Allen, 1957; Barrows and others, 1985; Matti and others, 1985; Sims, this volume). In three exceptions, I refer separately to the Punchbowl fault along the Mojave Desert segment; the Wilson Creek, Mission Creek, Mill Creek, and Banning strands of the San Andreas fault zone just north of San Gorgonio Pass; and the Pilarcitos fault along the Coast Ranges segment west of San Francisco Bay. The literature that reports on investigations of the San Andreas fault is voluminous and for relevant articles and references, the reader is referred to volumes edited by Jahns (1954), Bailey (1966), Dickinson and Grantz (1968), Kovach and Nur (1973), Crowell (1975b), Ernst (1981), Sylvester (1984), Sylvester and Crowell (1989), and Wallace (1990), to a paper by Hill (1981), and to other chapters in this volume.

Previous estimates of displacement northwest of Tejon Pass. The modern era in the history of study of the San Andreas fault began with the landmark paper in which Hill and Dibblee (1953) first documented hundreds of kilometers of right-lateral displacement on the San Andreas fault in central California (see M. L. Hill, 1981, for a synopsis of the history of study of the San Andreas fault up to that time). In matching successively older paleogeologic features across the fault, Hill and Dibblee measured increasingly large displacements, from which they inferred that displacement was cumulative since the Jurassic.

One of their measurements was constrained by reconstruction of a disrupted lower Miocene shoreline along which marine strata and terrestrial conglomeratic redbeds were overlapped by intermediate volcanic rocks. Remnants of this shoreline assemblage crop out in the San Emigdio Mountains east of the San Andreas fault and at the north end of the Gabilan Range west of the fault, indicating a displacement variously measured at about 275 ka (Addicott, 1968), 284 km (Hill and Dibblee, 1953; Bazeley, 1961; Dibblee, 1966a; Turner, 1969; Huffman, 1970), 295 km (Huffman and others, 1973), 280 to 305 km (Dickinson and others, 1972), and 300 to 308 km (Huffman, 1972).

Other measurements of displacement are based on the reconstruction of marine basins north of the shoreline. Paleobathymetric gradients in upper Oligocene (Zemorrian) and lower Miocene (Saucesian) marine strata in the La Honda basin and San Juan Bautista area have been matched across the San Andreas fault with gradients in the southern San Joaquin basin. Measurements include 275 km of displacement subsequent to the Zemorrian and Saucesian (Addicott, 1968), and 325 to 330 km of post-Zemorrian displacement and 320 to 325 km of post-Saucesian displacement (Stanley, 1987). Reconstruction of parts of a Zemorrian deep-sea fan in the La Honda basin and southern San Joaquin basin (Temblor Range) has yielded a measurement of 315 to 320 km of post-Zemorrian displacement (Graham and others, 1989).

Lower Miocene strata at the north end of the Gabilan Range and in the San Emigdio Mountains unconformably overlie Eocene and Oligocene marine strata that in turn overlie Jurassic gabbroic rocks. The Eocene and Oligocene strata have been reassembled in various paleogeographic configurations that have yielded displacement measurements of about 284 km (Bazeley, 1961), 325 km (Addicott, 1968), 305 to 330 km (Clarke and Nilsen, 1973; Nilsen and Link, 1975; Nilsen, 1984). Reassembly of the Jurassic gabbroic rocks has yielded a measurement of about 325 km (Ross, 1970).

Correlative early Miocene intermediate and felsic volcanic rocks that unconformably overlie Cretaceous tonalite, granodiorite, and granite crop out in the Pinnacles area of the southern Gabilan Range and the Neenach area of the northwesternmost Mojave Desert. Palinspastic restoration of these units has yielded displacement measurements of about 284 km (Turner and others, 1970), 300 to 316 km (Huffman, 1972; Huffman and others, 1973), 305 km (Dickinson and others, 1972), 308 km (Huffman, 1970), 314 to 315 km (Matthews, 1976). The Cretaceous granitoid rocks on which the volcanic units are deposited in the two areas are similar petrographically and chemically (Kistler and others, 1973; Ross, 1984, p. 31–37).

Debris from the Mesozoic metamorphic and plutonic rocks and the Pinnacles Volcanics of the Gabilan Range that was incorporated into the upper Miocene Santa Margarita Formation of the southern Temblor Range provides a measure of incremental displacement on the San Andreas fault zone in central California. The distribution of these clasts in the Santa Margarita Formation, which indicates the magnitude of displacement since they were shed eastward across the fault, has been measured variously as 240 to 245 km (Huffman, 1970, 1972; Turner and others, 1970) or 260 km (Fletcher, 1967). Of this late Miocene and younger slip, various amounts have been attributed to late Miocene movement, including 8 to 34 km (Huffman, 1972), 24 km (Thomson and Ryder, 1976), and 45 km (Sims, this volume); the remainder has been attributed to post-Miocene movement, including estimates of 234 ± 8 km (Huffman, 1972), 200 to 205 km (Berry and others, 1966; Thomson and Ryder, 1976), and 165 km (Sims, this volume). Moreover, prior to shedding debris into the Santa Margarita depositional basin, the Pinnacles Volcanics had been displaced by as little as about 25 to 50 km (Huffman, 1972; Ryder and Thomson, 1989), or as much as about 90 km from the Neenach area to the area of the southern Temblor Range (Dickinson and others, 1972; Matthews, 1976; Sims, this volume).

Several investigators have postulated still greater, pre–early Miocene displacement on the San Andreas fault in central and northern California (Hill and Dibblee, 1953; Dibblee, 1966a; Hill and Hobson, 1968; Wentworth, 1968; Ross, 1970; Hill, 1971; Ross and others, 1973). These estimates of larger displacement are now known to be largely accommodated by the northwestward addition of displacement on the Rinconada-Reliz and San Gregorio–Hosgri faults in the Salinian block (see below and Johnson and Normark, 1974; Graham and Dickinson, 1978a, b).

The Pilarcitos fault is generally mapped as merging to the southeast with the San Andreas fault and northwest with the San Gregorio and San Andreas fault (Jennings, 1977; Brabb and Pampeyan, 1983). Because displacement on the intervening segment of the San Andreas fault appears to be limited to 26 to 40 km by distinctive limestone beds in the Franciscan Complex (Wakabayashi and Moores, 1988; Griscom and Jachens, 1989), the Pilarcitos fault is commonly interpreted as an ancestral anastomosed strand of the San Andreas fault that bears all but 26 to 40 km of the overall displacement on that fault (see Champion and others, 1984; Griscom and Jachens, 1989). In this interpretation, the Pilarcitos fault was the active strand of the San Andreas fault until 1 or 2 Ma (Griscom and Jachens, 1989). As such, it is the strand of the San Andreas fault with which the San Gregorio fault merged throughout most of its history. (In a paper in 1988, Wakabayashi and Moores, however, interpreted the Pilarcitos fault as a Paleogene thrust fault along which rocks of the Franciscan Complex have been thrust southwestward over the granitic rocks of the Salinian block.)

Previous estimates of displacement southeast of Tejon Pass. In contrast to central California, no consensus has emerged for displacement on the San Andreas fault in southern California. Since 1953, when Hill and Dibblee tentatively suggested that the San Andreas fault in southern California had displaced the Pelona and Orocopia Schists about 260 km between the Sierra Pelona and Orocopia Mountains, measurements of displacement on the San Andreas fault zone inclusive of the Punchbowl fault have clustered in four groups.

One group of measurements ranges between about 210 and 225 km. In the early 1960s, Crowell (1960, 1962) and Crowell and Walker (1962) measured a displacement of about 210 km based on a proximal repositioning of exposures of the Proterozoic anorthosite-syenite complex and of Oligocene and lower Miocene sedimentary and volcanic strata in the Vasquez and Diligencia Formations as well as the Pelona and Orocopia Schists. Although subsequent investigators of the Vasquez and Diligencia Formations have concluded that these formations accumulated in separate basins and thus do not constitute a piercing point (Spittler and Arthur, 1973; Bohannon, 1975), additional paleostructural evidence for a displacement in the range of about 210 to 225 km on the San Andreas fault zone is provided by realignment of antiformal axes in the Sierra Pelona and Orocopia Mountains exposures of the Pelona and Orocopia Schists, and by realignment of the San Francisquito, Fenner, and Clemens Well faults (Dibblee, 1968a, 1982h; Bohannon, 1975; Smith, 1977a; Powell, 1981). Ramirez (1983) concluded that clasts of marble, quartzite, and olivine basalt in the Pliocene Hungry Valley Formation were derived from a source in the southern San Bernardino Mountains near Morongo Valley, which requires that the Hungry Valley Formation of Crowell (1950) has been displaced about 220 km from its source, although other investigators have built a more convincing case for a source in the San Bernardino Mountains between Cajon Pass and the Santa Ana River, which requires a displacement in the range of 150 to 170 km (Crowell, 1982a, b; Matti and others, 1985, 1986; Frizzell and others, 1986; Meisling and Weldon, 1989; Matti and Morton, this volume;

Weldon and others, this volume). In their initial report on the Chocolate–Orocopia Mountains source terrane for clasts of rapakivi-textured volcanic rocks in the Mint Canyon and Caliente Formations, Ehlig and Ehlert (1972) measured a displacement of 226 km along the San Andreas fault.

Other investigators, however, have advocated the larger and now generally accepted displacement of 240 km or more on the San Andreas fault in southern California, again inclusive of the Punchbowl fault (Crowell, 1975c, 1981; Ehlig and others, 1975; Ehlig, 1976, 1981, 1982; Ehlert and Ehlig, 1977; Buesch and Ehlig, 1982). Displacement in the range of 240 to 270 km is based chiefly on measurements derived from realignment of the Mint Canyon–Caliente drainage with a source area in the Orocopia and northern Chocolate Mountains (Ehlig and others, 1975) and on reassembly of a paleogeologic domain now exposed in the Soledad Pass area just north of the San Gabriel Mountains and in the Salt Creek (Salton Wash) area between the Orocopia and Chocolate Mountains (Ehlig, 1981, 1982; Buesch and Ehlig, 1982; Dillon and Ehlig, this volume). In each of these latter two areas, Proterozoic syenite is juxtaposed across a fault against parts of the Triassic Mount Lowe intrusion and late Oligocene and early Miocene volcanic rocks.

In contrast, still other well-documented studies along the fault have yielded displacement measurements of 160 km on the Mojave Desert segment (Matti and others, 1985, 1986; Frizzell and others, 1986; Matti and Morton, this volume) and of 180 to 185 ± 20 km on the Salton trough segment (Dillon, 1976; Farley, 1979; Dillon and Ehlig, this volume). From exposures in the Mill Creek area of the San Bernardino Mountains and in the Liebre Mountain area, Matti and others (1985, 1986; Frizzell and others, 1986; Matti and Morton, this volume) defined a disrupted paleogeologic domain comprising a pluton of Triassic porphyritic monzonite that is intruded by mafic dikes and overlapped by middle and upper Miocene conglomerate. By reassembling this domain, these authors measured a displacement of about 160 km for the Mojave Desert segment. Similarly, from exposures in the Whitewater River area of the San Bernardino Mountains and in the central Chocolate Mountains, Dillon (1976; see also Farley, 1979) defined a disrupted paleogeologic domain comprising Proterozoic gneiss intruded by the Triassic Mount Lowe intrusion and Jurassic plutonic rocks, all of which are thrust over the Pelona and Orocopia Schists. These crystalline rocks are overlapped by middle and upper Miocene fanglomerate and basalt (the Coachella Fanglomerate of Vaughan [1922] and the fanglomerate of Bear Canyon of Dillon [1976]) and by upper Miocene and Pliocene marine and estuarine shale, siltstone, and sandstone (the Imperial Formation of Vaughan [1922] and the Bouse Formation). In reassembling this domain, Dillon measured a displacement of about 180 ± 20 km for the Salton trough segment. Alternatively, Peterson (1975), in restoring the Coachella Fanglomerate alone, measured a displacement of 215 km for the Salton trough segment.

Occasionally since 1953, arguments have been advanced for far less overall displacement on the San Andreas fault zone inclusive of the Punchbowl fault, including about 48 to 80 km (Noble, 1954b; Woodford, 1960; Baird and others, 1971, 1974; Woodburne, 1975), and 102 km (Barrows and others, 1985). However, most of the observations that have led to these arguments also can be interpreted in ways that are consistent with larger measurements of displacement. For example, the presence of rocks of the Limerock Canyon assemblage associated with the quartz dioritic and monzogranitic suites of Mesozoic plutonic rocks in the southeasternmost San Gabriel Mountains west of the San Andreas fault and in the southwestern San Bernardino Mountains east of the fault yields an apparent offset of 40 to 50 km. The apparent offset indicated by this cross-fault match-up was a factor in the thinking of some of the investigators who have advocated overall displacements in that range (Woodford, 1960; Baird and others, 1974). Overall distribution of these rocks, however, allows an alternate realignment that provides a greater measure of displacement on the San Andreas fault. Barrows and others (1985) admitted that their estimate of 102 km was a minimum measurement of overall displacement.

Reckoning of displacement on San Andreas fault northwest of Tejon Pass. Reassembly of the paleogeologic pattern of the Gabilan Range–San Emigdio Mountains–Portal Ridge reference domain is accomplished by restoring displacements along the various strands of the San Andreas fault zone in central California and the left-lateral Garlock fault. In reconstructing that domain, I have reassembled the Jurassic gabbro, Eocene and Oligocene marine strata, and early Miocene shoreline and volcanic rocks of the San Juan Bautista area at the north end of the Gabilan Range and the Eagle Rest Peak area of the San Emigdio Mountains by restoring a displacement of 295 km on the San Andreas fault. The displacement of 295 km is measured by direct comparison of the reassembled reference domain (Fig. 10, Plate IIC) to Plate I, which is modified from the 1:750,000-scale Geologic Map of California (Jennings, 1977). Previous investigators who have measured the offset of various components of the reference domain used herein have reported displacements in the range of 275 to 330 km (Hill and Dibblee, 1953; Bazeley, 1961; Dibblee, 1966a; Addicott, 1968; Huffman, 1970, 1972; Ross, 1970, 1984; Turner and others, 1970; Dickinson and others, 1972; Clarke and Nilsen, 1973; Huffman and others, 1973; Nilsen and Link, 1975; Matthews, 1976; Nilsen, 1984). Reassembly of the reference domain also restores about 12 km of left slip along the Garlock fault at its west end. Restoring both the right- and left-lateral displacements requires counterclockwise back-rotations of 38° for the Gabilan Range and 22° for the San Emigdio and Tehachapi Mountains, all relative to Portal Ridge held fixed. One effect of these back-rotations is to reposition the Pinnacles and Neenach Volcanics side by side, thus reassembling them by restoring an overall displacement of 290 km on the San Andreas fault.

Reckoning of displacements on San Andreas and Punchbowl faults southeast of Tejon Pass. Reassembly of the La Panza Range–Liebre Mountain–San Bernardino Mountains reference domain (Fig. 9) necessitates restoring a displacement of 159 km on the San Andreas fault, including any displacement contributed by the Punchbowl fault. Relative to the San Bernardino Mountains, the Liebre Mountain block is back-rotated 10° counterclockwise.

In order to reconstruct the paleogeologic pattern of the Frazier Mountain–Mount Pinos–eastern Orocopia Mountains reference domain, it is necessary to realign the Clemens Well and Fenner faults by restoration along the San Andreas fault and the Fenner and San Francisquito faults by restoration along the Punchbowl fault (Fig. 8, Plate II, A, B). To realign the Fenner and San Francisquito faults requires restoring a displacement of 44 to 45 km on the Punchbowl fault. Because the northwest end of the Clemens Well fault is covered by Pliocene and younger rocks in the vicinity of the Mecca and Indio Hills, the magnitude of displacement required to realign the Clemens Well and Fenner faults is not as clear. If the Clemens Well fault curves from its northwesterly trend to the westerly trend of the Fenner fault to intersect the San Andreas fault in the northern Mecca Hills, then realignment of the Clemens Well and Fenner faults requires restoring a displacement in the range of 160 to 165 km on the San Andreas fault. A lesser displacement of about 115 to 120 km is required, however, if the Clemens Well fault projects on-strike to the Chiriaco fault, is offset about 11 km left-laterally (see later section on left-lateral faults), then continues on-strike beneath the Indio Hills before curving westward to intersect the San Andreas fault. In the former case, a fault segment buried beneath the Mecca Hills intervenes between the Clemens Well and Fenner faults in the reconstruction; in the latter case, a fault segment now buried beneath both the Mecca and Indio Hills intervenes between the Clemens Well and Fenner faults. If the Punchbowl fault is considered to be an anastomosed strand of the San Andreas fault, and a displacement of 43 to 45 km on the Punchbowl fault is included as part of the offset along the San Andreas fault, the overall displacement on the San Andreas–Punchbowl fault system is 204 to 210 km in the case that links the Fenner and Clemens Well faults directly (cf. Dibblee, 1968a, 1982h; Bohannon, 1975; Smith, 1977a; Powell, 1981), or 159 to 165 km in the case that links the Fenner and Clemens Well faults via a fault segment buried beneath the Indio and Mecca Hills (cf. Matti and Morton, this volume).

The apparent disparity between measurements of displacement based on restoring the Liebre Mountain–San Bernardino Mountains reference domain and the San Francisquito–Fenner–Clemens Well fault along the San Andreas fault requires reconciliation. In part, the disparity between these measurements can be resolved by components of displacement on the San Andreas fault contributed by coeval left-lateral faults in the Transverse Ranges east of the San Andreas (see later section on left-lateral faults). In attempting to resolve the remainder of the disparity, one can consider four hypotheses. First, if one argues that the pre–5-Ma strand of the San Andreas fault was located along the Mojave Desert segment of the post–5-Ma San Andreas fault rather than along the Clemens Well–Fenner–San Francis-

quito fault, then some of the crystalline rocks of the San Bernardino Mountains could have been thrust over the pre–5-Ma San Andreas fault onto the Liebre Mountain block along the Squaw Peak–Liebre Mountain thrust fault (Meisling and Weldon, 1986). The crystalline rocks in the upper plate of the Liebre Mountain thrust would than have been displaced by only about 160 km along the post–5-Ma San Andreas fault, whereas the lower plate rocks would have been displaced by as much as about 300 km on both the pre– and post–5-Ma San Andreas fault. Although this hypothesis would account for the 160-km offset of the Triassic porphyritic monzogranite and other rocks on the upper plate, it does not account for the equivalent offset of metasedimentary rocks of the Limerock Canyon assemblage in the lower plate.

Second, if some of the displacement on the Punchbowl or San Andreas fault was pre–5-Ma in age and was shunted westward along some combination of the San Francisquito, Clearwater, and Liebre faults, then only the post–5-Ma displacement merges with the San Andreas fault northwest of its intersection with the San Francisquito fault (see Fig. 14b). Although this hypothesis is conceptually appealing, the displacements measured on these more westerly faults do not appear to accommodate an additional 20 to 40 km.

Third, the segment of the San Andreas fault between the north end of the Punchbowl fault and the east end of the San Francisquito fault could have also been part of the Clemens Well–Fenner–San Francisquito fault—analogous to the San Andreas fault north of its junction with the San Gabriel fault. In this case, displacement on the Punchbowl fault would be only about 15 to 20 km, and overall displacement on the San Andreas–Punchbowl fault would be 175 to 180 km. To further reduce displacement on the San Andreas–Punchbowl fault to 160 km would require shunting 15 to 20 km to the west as discussed in the first case. As in the first case, this redistribution of displacement does not seem to permit reconstruction of the paleogeologic reference domains.

Fourth, the Fenner and Clemens Well faults can be connected via a segment buried beneath the Indio and Mecca Hills, requiring less displacement on the San Andreas fault than a direct connection (see also Matti and Morton, this volume). This alternative, which is incorporated in the reconstruction shown in Plate II, A through C, provides the best fit of measured displacements on all faults of the San Andreas fault system (see later synthesis section).

The 160-km displacement restored on the San Andreas fault southeast of Tejon Pass in Plate II is considerably less than widely cited measurements of displacement in the range of 240 to 270 km. As is discussed in the final section, the difference is resolved by displacement attributed herein to the Clemens Well–Fenner–San Francisquito fault.

Timing and rates of movement northwest of Tejon Pass. Movement on the San Andreas fault in southern and central California postdates eruption of the Pinnacles and Neenach Volcanics and the intermediate volcanic rocks that crop out in the San Juan Bautista area just north of the Gabilan Range and in the San Emigdio Mountains. These early Miocene volcanic units are displaced along with the crystalline basement by about 295 km. Samples from Pinnacles-Neenach volcanic field have yielded K-Ar ages between about 22 and 24 Ma (Turner, 1968, p. 67, 74; Turner and others, 1970; see also Fox and others, 1985; Sims, this volume) and samples from the San Juan Bautista–San Emigdio field have yielded K-Ar ages between 21 and 25 Ma (Turner, 1968, p. 13, 15, 53, 1970, p. 101; Thomas, 1986; see also Sims, this volume). Movement also postdates the Zemorrian and Saucesian paleobathymetric gradients in the La Honda and southern San Joaquin basins (Stanley, 1987), indicating that full displacement on the San Andreas fault northwest of Tejon Pass accumulated since the Saucesian, which ended at about 17 or 18 Ma (Kleinpell and others, 1980; Poore and others, 1981). In studying the provenance of Miocene sandstone in the Temblor Range, however, O'Day and Sims (1986) noted an abrupt influx of abundant volcanic debris during the late Saucesian from a northward-migrating source to the west: they concluded that the volcanic debris was shed from a source on the Salinian block as it moved northward on the early San Andreas fault. By this interpretation, movement on the San Andreas was underway by 19 or 18 Ma. Right-stepping en echelon folds that developed in the Temblor Range and vicinity in the late Saucesian are probably coincident with beginning of strike slip on the San Andreas fault (Harding, 1976). By the beginning of the Mohnian, cumulative displacement on the San Andreas fault was as little as about 25 (Ryder and Thomson, 1989, p. 48, Fig. 51) to 50 km (Huffman, 1972) or as much as 90 km (Matthews, 1976; Sims, this volume). The latter displacement is consistent with that restored in Plate II.

Younger strata are displaced by less than 295 km on the San Andreas fault. The upper part of the Monterey Formation in the southern Temblor Range that contains clastic debris shed from the Gabilan Range also contains marine microfossils that span the Mohnian (Bandy and Arnal, 1969; Huffman, 1972; Thomson and Ryder, 1976; Webb, 1981; Graham and Williams, 1985 Ryder and Thomson, 1989). The Santa Margarita Formation of the Temblor Range, which interfingers with and overlies the Monterey Formation (Dibblee, 1973; Graham and Williams, 1985), contains coarse debris shed from the Gabilan Range and contains a "Margaritan" molluscan fauna (Addicott, 1968; Dibblee, 1973). In the southern Temblor Range, the Santa Margarita Formation grades eastward into upper Mohnian sandstone in the upper part of the Monterey Formation (Webb, 1981). In the northern Temblor Range, Santa Margarita-like strata extend into the Delmontian (Graham and Williams, 1985). On a geochronometric time scale (Fig. 13), the Mohnian has been estimated to range from 13.5 or 13 to 6 Ma (Berggren and Van Couvering, 1974; Poore and others, 1981), 14 to 11 Ma (Kleinpell and others, 1980), 14.2 to 7.2 Ma (Armentrout and Echols, 1981), and 13.8 to 7 Ma (Bartow, 1990), and the "Margaritan" from about 14 to 9 Ma (Poore and others, 1981). Thus, movement on the San Andreas fault recorded by the clastic debris in the Monterey and Santa Margarita Formations of the Temblor Range occurred from 14 or 13 Ma through at least 7 or 6 Ma and possibly 5 Ma, during which time displacement of as little as 8 to 34 km

(Huffman, 1972), about 45 km (Sims, this volume), or as much as 110 to 115 km (Ryder and Thomson, 1989, Fig. 51) accumulated on the San Andreas fault. Estimates of the magnitude of displacement since 5 Ma range from 160 to 255 km (Berry and others, 1966; Fletcher, 1967; Huffman, 1972; Thomson and Ryder, 1976; Ryder and Thomson, 1989; Sims, this volume). The lower limit of this range is consistent with the displacement restored in Plate II.

In summary, movement on the San Andreas fault northwest of Tejon Pass occurred in three stages between about 18 Ma and the present. Rate of movement for the San Andreas fault varied for the three stages. Between 18 to 17 and 14 to 13, the long-term average slip rate was in the range between 18 and 30 mm/yr if displacement was 90 km or between 5 and 8.3 mm/yr if displacement was 25 km. The former case is consistent with linking this part of the San Andreas fault to the coeval Clemens Well–Fenner–San Francisquito fault, with its displacement of 100 to 110 km and long-term average slip rate of 17 and 28 mm/yr. Between 14 to 13 and 6 to 5 Ma, the long-term average slip rate on the San Andreas fault northwest of Tejon Pass was in the range between 2.6 and 3.4 mm/yr if displacement was 24 km or between 5 and 6.4 mm/yr if displacement was 45 km. The latter scenario is consistent with the timing and rates calculated above for the San Gabriel fault (see also Sims, this volume).

Timing of movement southeast of Tejon Pass. The distribution of middle and upper Miocene strata is consistent with displacement of at least 170 km on the San Andreas fault. The Mint Canyon Formation in Soledad basin is separated from a source area in the Orocopia and northernmost Chocolate Mountains. Similarly, the Coachella Fanglomerate just east of San Gorgonio Pass is displaced from its crystalline source terrane and the correlative fanglomerate of Bear Canyon in the southern Chocolate Mountains. The latter two sedimentary units contain interbedded olivine basalt flows that have yielded K-Ar ages of 10 and 13 Ma, respectively (Peterson, 1975; Dillon, 1976, p. 297–315; Crowe, 1978). The upper Miocene Punchbowl Formation, which in the Devils Punchbowl area ranges in age from 13 to 8 Ma on the basis of magnetostratigraphy (Liu, 1990), is displaced from its crystalline- and volcanic-rock source terrane in the southern Little San Bernardino and Orocopia Mountains (Ehlig, 1976, 1981; Ehlert and Ehlig, 1977; Matti and others, 1985).

The youngest units that are inferred to have been displaced about 160 km or more on the San Andreas fault include the upper Miocene and Pliocene Imperial and Bouse Formations in San Gorgonio Pass and the southern Chocolate Mountains (Dillon, 1976, p. 339–341, p. 359–360), and the Pliocene Hungry Valley Formation of the Ridge basin from a crystalline rock source terrane in the San Bernardino Mountains (Crowell, 1982a, c; Ramirez, 1983; Matti and others, 1985, 1986; Frizzell and others, 1986; Weldon and others, this volume). All of these formations include rocks that are about 5 m.y. old, indicating that movement on the Mojave Desert and Salton trough segments of the San Andreas fault, and possibly on the Punchbowl fault, began no earlier than 5 Ma (Fig. 13). This post–5-Ma displacement of the Mojave Desert and Salton trough segments of the San Andreas fault in southern California continues northwest of Tejon Pass, where its path is spatially coincident with the earlier San Andreas fault along the Coast Ranges segment of the fault.

Numerous studies have yielded abundant evidence that various strands of the San Andreas fault have been active throughout the Pliocene and Quaternary along the Coast Ranges segment (see, for example, Hill and Dibblee, 1953; Dibblee, 1966a; Wallace, 1975; Sieh and Jahns, 1984; Perkins and others, 1989; Prentice, 1989; Brown, 1990; Prentice and others, 1991; Sims, this volume), the Mojave Desert segment (see, for example, Morton and Miller, 1975; Barrows and others, 1985; Weldon and Sieh, 1985; Harden and Matti, 1989; Weldon and others, this volume), in San Gorgonio Pass (Allen, 1957; Matti and others, 1985), and along the Salton trough segment (Allen and others, 1972; Sylvester and Smith, 1976; Keller and others, 1982; Sieh, 1986). Currently, the San Andreas fault proper is actively seismic along the Coast Ranges segment between about Cholame and San Jose, but elsewhere is largely inactive, or locked (Allen, 1968, 1981; Hill and others, 1990).

Kinematic role in the evolving San Andreas fault system. The San Andreas fault proper in southern California began to develop at about 5 Ma. At that time, it became the first major throughgoing strike-slip fault to demonstrably connect with the Salton trough and Gulf of California. It is this latest part of the San Andreas fault system that forms a transform fault linking the spreading axis in the Gulf of California with the Mendocino fracture zone on the Pacific floor.

Salinian block

The Salinian block is transected by three major north-northwest–trending right-lateral composite faults. From east to west, these faults are the Red Hills–Ozena, Rinconada-Reliz, and San Gregorio–Hosgri faults.

Red Hills–Ozena fault. Path. From north to south, the Red Hills–Ozena composite fault consists of the Red Hills, San Juan, Chimeneas, Russell, and Ozena faults. Ross (1972, p. 22, Plate 1, 1984, Fig. 12, p. 20) has mapped the San Juan fault of Hill and Dibblee (1953; Hill, 1954; Schwade and others, 1958) in three segments—the Red Hills, San Juan, and Chimeneas faults—that reach successively southward from the San Andreas fault along the west margin of crystalline rocks exposed in the Red Hills, along the east flank of crystalline rocks exposed in the La Panza Range, and along the west margin of crystalline rocks exposed on Barrett Ridge (see also Bartow, 1974, 1978). Most investigators have extended the Red Hills–San Juan–Chimeneas fault southeastward via the Russell fault, a northwest-trending, high-angle subsurface fault beneath the central Cuyama Valley between the Caliente Range and the Sierra Madre, and via the Ozena fault to intersect the left-lateral Big Pine fault (Schwade and others, 1958; Ross, 1972, p. 33, Fig. 16, 1984, Fig. 12; Bartow, 1978; Howell and Vedder, 1978). The link between the Red Hills–San Juan–Chimeneas and Russell faults is overridden

and obscured by the southwestward vergent Whiterock thrust fault, and the link between the Russell and Ozena faults is overridden and obscured by the northeastward vergent South Cuyama thrust fault, but the subsurface geology requires a throughgoing fault (Schwade and others, 1958; Bartow, 1978; Schwing, 1982; Yeats and others, 1988, 1989). Alternatively, Smith (1977a) and Joseph and others (1982b) have advocated connecting the Red Hills–San Juan–Chimeneas fault southeastward with the Morales thrust fault, although their proposed paths for large strike-slip displacement are inconsistent with detailed surface and subsurface mapping that shows no connection between the Red Hills–San Juan–Chimeneas and Morales faults (Schwade and others, 1958; Bartow, 1978; Schwing, 1982). Yeats and others (1988, 1989), however, mapped a branch of the Russell fault in the subsurface east of Barrett Ridge. Hill and Dibblee (1953) suggested that the San Guillermo fault is the offset continuation of the Ozena fault across the Big Pine fault (see also Poynor, 1960, p. 84; Dibblee, 1976, p. 44; Smith, 1977a), but the San Guillermo fault is a low-angle reverse fault along which much right slip seems unlikely.

Previous estimates of displacement. Measurements of offset on the Red Hills–Ozena fault are based on restoration of inferred paleogeologic patterns involving crystalline basement rocks, Upper Cretaceous and Paleogene marine strata, and Oligocene and lower Miocene continental and marine strata. Measured displacements include 8 to 19 km (Hill, 1954), 8 km (Dibblee, 1976, Fig. 17), 13 to 14.5 km (Bartow, 1974, 1978), 26 to 29 km (Yeats and others, 1988, 1989), and 37 km (Schwade and others, 1958). In restoring 13 to 14.5 km, Bartow (1974) matched a paleogeologic pattern on Barrett Ridge and in the La Panza Range that includes plutonic and metamorphic basement rocks, a southward-dipping contact between Upper Cretaceous strata and the crystalline basement, and a northward-dipping contact between lower Miocene strata and the crystalline basement. In studying Oligocene through Pliocene strata along the Russell fault in the subsurface of Cuyama Valley, Schwing (1982), although he postulated recurrent right slip on the fault, emphasized that the only piercing point he recognized consisted of an elongate dome in the Miocene and Pliocene Morales(?) Formation that is offset about 1.4 km. Yeats and others (1988, 1989), in measuring a displacement of 26 to 29 km on the Russell fault on the basis of proposed piercing points in the Simmler Formation and older rocks, concluded that displacement on the Russell fault is equally distributed northward between the Chimeneas fault and a buried fault to the east of Barrett Ridge.

Many recent investigators have emphasized a pronounced contrast in the metamorphic host rocks of the Mesozoic batholith across the Red Hills–Ozena fault zone and some have suspected much larger displacement than the estimates listed in the previous paragraph (Ross, 1972, p. 32, Plate 1, 1984, p. 20–21; Kistler and others, 1973; Howell and Vedder, 1978), although others have concluded that the discontinuity in metamorphic rock types represents a pre–Late Cretaceous structural feature (Champion and others, 1984). Right-lateral offsets of 175 (Smith, 1977a) and 120 km (Joseph and others, 1982b) have been proposed for the Red Hills–San Juan–Chimeneas fault as an early strand of the San Andreas fault (see section on Clemens Well–Fenner–San Francisquito fault).

Reckoning of displacement. In combination with displacements on the Clemens Well–Fenner–San Francisquito, San Gabriel, and San Andreas faults, reassembly of the Gabilan Range–San Emigdio Mountains–Portal Ridge and La Panza Range–Liebre Mountain–San Bernardino Mountains reference domains limits displacement on the Red Hills–Ozena fault to 0 to 15 km (Figs. 9, 10). No displacement is restored in Figures 9 or 10 or in Plate II; however, if a displacement of 100 rather than 110 km were restored on the Clemens Well–Fenner–San Francisquito–San Andreas fault, the paleogeologic reconstruction would permit restoration of about 10 km on the Red Hills–Ozena fault.

The Red Hills–Ozena fault illustrates the need to assess critically the evidence for displacement and to distinguish between a displacement that is based on seemingly reasonable conjectural interpretation of inconclusive evidence and one that is solidly established by the evidence. In this regard, most of the evidence cited in measuring moderate to large displacement along the Red Hills–San Juan–Chimeneas(–Russell) fault is inconclusive. For example, the depositional contact between Upper Cretaceous and Paleocene marine strata and underlying crystalline basement rocks has been used as evidence for dextral displacements of about 8 km (Dibblee, 1976), 26 to 29 km (Yeats and others, 1988, 1989), 13 to 14.5 km (Bartow, 1974), 37 km (Schwade and others, 1958), and 120 km (Joseph and others, 1982b). Moreover, both Smith (1977a) and Joseph and others (1982b), in proposing 175- or 120-km displacements, did not demonstrate that the Barrett Ridge basement rocks and Paleocene marine strata are incompatible with the cross-fault basement rocks and Upper Cretaceous and Paleocene strata in the La Panza Range, nor did they show compatibility for the Barrett Ridge rocks with rocks in their restored position 120 or 175 km to the northwest.

The larger estimates of 120 (Joseph and others, 1982b) and 175 km (Smith, 1977a) that have been proposed for offset along the Red Hills–San Juan–Chimeneas fault on the basis of offset Mesozoic batholithic rocks—including the polka-dot granite—and overlapping Upper Cretaceous and Paleocene sedimentary units include offset that is attributed to the northwestward extension of the Clemens Well–Fenner–San Francisquito fault in the palinspastic reconstruction of this study (Figs. 9, 10; Plate II). The Clemens Well–Fenner–San Francisquito fault cannot, however, have followed any part of the Red Hills–Ozena fault and probably followed a path closer to that of the modern San Andreas fault (see discussion in earlier section on the Clemens Well–Fenner–San Francisquito fault).

The estimates with the most specific constraints are the 13- to 14.5-km measurement of Bartow (1974, p. 139; 1978) and the 26- to 29-km measurement of Yeats and others (1988, 1989). Yeats and others (1988, 1989), however, were unable to specify a path along which a displacement of 26 to 29 km on the Russell fault could be accommodated to the south or southeast, and I was unable to accommodate such displacement in the reconstruction in this study. Bartow's (1974) displacement measurement of 13 to 14.5 km in reassembling a paleogeologic pattern of crystalline rocks and sedimentary strata on Barrett Ridge and in the La Panza Range can be considered an upper limit to displacement on the Chimeneas-Russell segment of the Red Hills–Ozena fault. Two observations, however, permit restoring a lesser displacement on the fault: the chiefly metamorphic basement rocks of Barrett Ridge are different than the chiefly monzogranitic plutonic rocks of the La Panza Range (Plate I; Ross, 1972, p. 27–29, Plate I); and the northwest trend of the contacts between the basement rocks and both the Upper Cretaceous and lower Miocene strata intersect the Chimeneas fault at a highly oblique angle (Jennings, 1977), such that the trend of the contacts could explain the distribution of the strata across the fault without much fault displacement. Although I chose to restore little or no right-lateral displacement on the Red Hills–Ozena fault zone, Plate II could be modified to accommodate a displacement of 10 to 15 km, as measured by Bartow (1974, 1978) and by Yeats and others (1988, 1989) on the Chimeneas fault segment, if compensating adjustments were made elsewhere in the reconstruction—for example, by restoring 10 km less on the Clemens Well–Fenner–San Francisquito–San Andreas fault.

Timing of movement. The evidence for timing of movement on the various segments of the Red Hills–Ozena fault may not be consistent with linking these segments into a throughgoing fault zone. Hill's (1954) evidence for offset of stratigraphic features in the lower Miocene Vaqueros Formation indicates that movement on both the San Juan and Russell faults is post–early Miocene in age. Schwade and others (1958) concluded that Miocene and Pliocene movement on the San Juan–Russell fault segment recurred on an older pre-Miocene, post-Cretaceous fault, but they did not specify the amounts of pre-, syn-, and post-Miocene slip. On the basis of the distribution of coarse conglomerate in the Simmler Formation, however, Bartow (1974, 1978) concluded that the Ozena and Blue Rock faults were active in the late Oligocene, whereas the Chimeneas-Russell fault was not; he further concluded that the Red Hills–Chimeneas-Russell fault accumulated most of its displacement during the early Pliocene. Dibblee (1976, Fig. 16) showed that the upper Miocene Santa Margarita Formation is disrupted by the San Juan fault. Bartow (1974, p. 140) stated that the Red Hills fault disrupts the Pliocene and Pleistocene Paso Robles Formation with minor displacement; Smith (1977a), on the other hand, stated that the Red Hills fault is buried beneath the Paso Robles Formation. The Ozena fault truncates the Pliocene and early Pleistocene Morales Formation (Jennings, 1977).

Of the various segments of the Red Hills–Ozena fault, a detailed history of movement has been established only for the Russell fault. Beneath Cuyama Valley, the Russell fault crosscuts the Vaqueros Formation, Monterey Formation, Branch Canyon Sandstone, and Santa Margarita Formation (Schwing, 1982;

Yeats and others, 1988, 1989) and is overlain unconformably by the Morales Formation (Yeats and others, 1988, 1989). In the Salinian block, the Vaqueros Formation is late Oligocene and early Miocene (late Zemorrian and early Saucesian) in age, the Monterey Formation and Branch Canyon Sandstone are late Saucesian, Relizian, and Luisian in age, the Santa Margarita Formation is Mohnian in age, and the Morales Formation is Pliocene and early Pleistocene in age (Hill and others, 1958; Durham, 1974; Dibblee, 1973, 1976; Lagoe, 1981, 1984, 1987, 1988; Schwing, 1982; Yeats and others, 1988, 1989). Yeats and others (1988, 1989) inferred that the Russell fault is overlapped by the Morales Formation, which contains a vertebrate fauna that is possibly Blancan in age (James, 1963), although Schwing (1982) concluded that an elongate dome in the Morales Formation has been displaced 1.4 km along the Russell fault. In the former case, movement on the Russell fault would have ended no later than about 5 Ma; in the latter case, movement would have ended between 5 and 2 Ma.

Yeats and others (1988, 1989) concluded that movement on the Russell fault postdates deposition of the Simmler Formation, is coeval with deposition of the Soda Lake Shale Member and overlying Saucesian Painted Rock Member of the Vaqueros Formation, and predates deposition of the Morales Formation. Because the Simmler Formation contains interbedded basalt flows that have yielded whole-rock K-Ar ages of 22.9 ± 0.7 and 23.4 ± 0.8 Ma (Ballance and others, 1983), movement is younger than about 23 Ma. The basal Quail Canyon Sandstone Member of the Vaqueros Formation was deposited unconformably on the Simmler Formation and underlying Eocene strata (Bartow, 1974; Yeats and others, 1988, 1989); the Soda Lake Shale and Painted Rock Sandstone Members of the Vaqueros, which overlie the Quail Canyon Sandstone, were deposited syntectonically in a syncline that Yeats and others (1988, 1989) inferred developed in response to dextral slip on the Russell fault. The syncline is displaced 3.7 km right-laterally along the Russell fault (Yeats and others, 1989). The Soda Lake Shale is Zemorrian and Saucesian, containing nannoplankton correlated with zones CN1b and CN1c of Bukry (1981) that range in age from 21.6 to 23.7 Ma on the time scale of Berggren and others (1985) (Lagoe, 1988; Yeats and others, 1989) and from 18.2 to 22.7 Ma on the time scale of Bartow (1990). The Painted Rock Sandstone is Saucesian (Lagoe, 1987). thus, most of the movement on the Russell fault occurred during the interval between about 23 or 22 and 20 or 19 Ma. Only 3.7 km of displacement has accumulated during and since the growth of the Saucesian syncline. Of this amount, 3.0 to 3.4 km has accumulated since deposition of the Monterey Formation and Branch Canyon Sandstone, which continued through the end of the Luisian, or about 14 or 13 Ma on the time scales of Berggren and Van Couvering (1974), Kleinpell and others (1980), Armentrout and Echols (1981), Poore and others (1981), and Bartow (1990).

San Gregorio-Hosgri and Rinconada-Reliz fault system. *Paths.* The Rinconada-Reliz fault trends north-northwest on the west flank of the La Panza Range, and the Reliz fault trends

north-northwest between the Gabilan and Santa Lucia Ranges. For more complete discussions of nomenclature and the history of study of this fault zone, see Dibblee (1976) and Ross (1972, p. 34, Plate 1; 1984, p. 16–18). The Rinconada fault is usually connected northwestward with the San Marcos, Espinosa, and Reliz faults (Hill and Dibblee, 1953; Schwade and others, 1958; Dibblee, 1972, 1976; Ross, 1972, 1984; Kistler and others, 1973; Hart, 1976; Graham, 1978), although some investigators have advocated an alternate connection with the Jolon fault via the southern part of the San Marcos fault (Durham, 1965, 1974; Smith, 1977b; Joseph and others, 1982b). The preponderance of evidence, however, indicates that the Jolon fault splays off the main path of the Rinconada–San Marcos–Espinosa–Reliz fault. The path of the continuation of the Rinconada-Reliz fault still farther to the northwest is controversial. Some investigators have distributed its displacement on faults in the Santa Lucia Range (Dibblee, 1976; Champion and Kistler, 1991), and Champion and Kistler (1991) asserted that displacement on the Rinconada-Reliz fault resulted in oroclinal bending of the Vergeles-Zayante fault. Most investigators have concluded, however, that the Rinconada-Reliz fault projects beneath Monterey Bay to join the San Gregorio fault (Martin and Emery, 1967; Greene and others, 1973; Ross and Brabb, 1973; Greene, 1977, p. 112–116; Graham, 1978; Ross, 1984; Champion, 1989), although there is disagreement as to the exact location of the connecting fault.

The San Gregorio–Hosgri fault zone, which trends along the coast of California between Bolinas and Point Sal, comprises several aligned faults, including the Seal Cove, San Gregorio, and Sur faults between Point Reyes and Point Sur, and the Sur, San Simeon, and Hosgri faults between Point Sur and Point Arguello (Hall, 1975; Silver, 1978; Greene and Clark, 1979; Payne and others, 1979; Leslie, 1981). For more complete discussions of nomenclature and the history of study of this fault zone, see Silver and Normark (1978) and Leslie (1981).

Previous estimates of displacement. Measurements of displacement on the Rinconada-Reliz fault vary depending on the age of the offset feature. The upper Miocene strata of the Santa Margarita and Pancho Rico Formations are separated by 18 km (Durham, 1965, 1974, p. 65–67; Dibblee, 1976, p. 36; Addicott, 1978), although no compelling piercing point has been established to show that the separation is due to strike-slip movement. Very large granitic boulders in the middle Miocene Tierra Redonda Formation (Durham, 1974) are displaced 24 km from a possible source either in granite exposures near Paso Robles (Durham, 1974, p. 65–66; Dibblee, 1976, p. 37–38) or 56 km from a possible source in the granitic rocks of the La Panza Range (Dibblee, 1976, p. 38). Hart (1976, p. 35) matched these boulder beds in the Tierra Redonda Formation with similar strata (referred to the Vaqueros Formation by Hart, 1976, and by Dibblee, 1976) in the La Panza Range east of the fault indicating 53 km of displacement. In matching buried basement highs, basement troughs filled with Miocene sediments, and Miocene shorelines, two investigators have measured displacements in the range of 35 to 48 km (Dibblee, 1976, p. 40; Graham, 1978), whereas in matching Upper Cretaceous and lower Tertiary strata overlying granitic basement, other investigators have measured displacements in the range of 60 to 72 km (Schwade and others, 1958; Dibblee, 1976, p. 38–40). A few investigators have suggested as much as about 170 to 200 km displacement on the Rinconada-Reliz fault (Kistler and others, 1973; Howell, 1975b; Smith, 1977b; Joseph and others, 1982b), whereas Ross (1972 [p. 87], 1978, 1984) has argued on the basis of detailed mapping and petrography of exposed and subsurface crystalline rock patterns that there has been little or no dextral slip on the fault, although his evidence permits as much as, but not more than, about 45 km (Ross, 1984, p. 18). A displacement of about 45 km is consistent with the measurements of Hart (1976), Graham (1978), and Ross (1984), and some of the measurements of Dibblee (1976).

For the various segments of the San Gregorio–Hosgri fault zone south of Pigeon Point, several investigators initially measured large overall dextral displacement ranging from 80 to 100 km (Silver, 1974; Hall, 1975, 1978; Mullins and Nagel, 1981). Their measurements are based on the offset of explicit geologic features in Upper Cretaceous through lower Miocene rocks and the igneous and metamorphic rocks on which they were deposited. Subsequently, reassembly of these and other geologic features along the segment of the fault between Point Reyes and Point Sur were found to be consistent with an offset of 110 to 115 km (Greene, 1977, including the Ascension fault; Graham and Dickinson, 1978a, b; Hall, 1981; Griscom and Jachens, 1989). Most recently, for the segments of the fault north of Point Sur, Clark and others (1984), Ross (1984), and Champion and Kistler (1991) have established an overall displacement of 150 km, whereas for segments of the fault between about Point Ano Nuevo and Point Sur, Nagel and Mullins (1983) have established an overall displacement of 105 km. James and Mattinson (1985) concluded that tonalite bodies of Bodega Head and the northern part of the Ben Lomond block are disrupted parts of an originally continuous pluton that have been displaced by at least 110, and probably about 160 km, along the San Gregorio fault. A few investigators have suggested somewhat less displacement (Howell and Vedder, 1978; Seiders, 1978), but only Hamilton and Willingham (1977, 1978) disputed the evidence for large offset on the San Gregorio–Hosgri fault, arguing instead for a maximum dextral displacement of 10 to 20 km. Hamilton (1982) and Sedlock and Hamilton (1991) postulated that large displacement accumulated on the San Gregorio–Hosgri and Rinconada-Reliz faults during the Paleocene, but that not more than 5 km has accumulated during the late Cenozoic.

Reckoning of displacement. Displacement on the Rinconada-Reliz and San Gregorio–Hosgri fault system is not constrained by the reference domains shown in Figures 8 through 11. In Plate II, A, B, I have restored right-lateral displacements of 45 km along the Rinconada-Reliz fault (see Graham, 1978), 150 km along the San Gregorio–Hosgri fault north of Point Ano Nuevo (see Clark and others, 1984), and 105 km along the fault zone south of Point Ano Nuevo (see Nagel and Mullins, 1983). Because the difference between the measurements of displacement

on the San Gregorio–Hosgri fault north and south of Point Ano Nuevo is precisely the displacement of the Rinconada-Reliz fault, the northward disposition of displacement on the latter fault in the reconstruction follows the King City fault as mapped by Ross and Brabb (1973) beneath Monterey Bay to merge with the displacement on the San Gregorio fault at Point Ano Nuevo. This interpretation is not inconsistent with the ranges of offset measured by Graham and Dickinson (1978a, b) for various features displaced along the San Gregorio–Hosgri fault.

In order to achieve a best-fit reconstruction of the Montara Mountain granitic block, an early part (about 80 km) of the 150-km displacement is passed east of the Montara Mountain block along the La Honda fault to the Pilarcitos fault strand of the San Andreas zone. Champion (1989; Champion and Kistler, 1991) independently proposed about 40 km of pre–5-Ma right slip on this fault in order to realign a steep gradient in Sr initial values located in both the Ben Lomond and Montara Mountain blocks. This model implies that the Pilarcitos fault is part of the San Gregorio fault as well as the San Andreas fault (see earlier section on San Andreas fault). If 80 km is restored on the Pilarcitos fault as an early strand of the San Gregorio fault, the Pilarcitos fault could then have been displaced by 70 km or less on the later strand of the San Gregorio fault (Plate II). Alternatively, if 40 km is restored on the Pilarcitos strand, then it could have been displaced by 110 km or less (Champion, 1989; Champion and Kistler, 1991). As the principal strand of the San Andreas fault until 1 or 2 Ma, however, the Pilarcitos fault cannot have been offset by the San Gregorio fault as proposed by Graham and Dickinson (1978a, b) and Griscom and Jachens (1989). Because the San Andreas–Pilarcitos and San Gregorio faults have been active contemporaneously since 12 or 13 Ma, the San Gregorio fault has merged with rather than offset the Pilarcitos fault. In this model, the San Andreas–Pilarcitos–San Gregorio fault may initially have followed the inferred offshore fault west of Point Arena, then switched to the present path of the San Andreas fault onshore at Point Arena.

To the south, the Rinconada-Reliz and San Gregorio–Hosgri faults do not transect the western Transverse Ranges. Their displacements probably are accommodated by a combination of left slip, clockwise rotation, and extension on structures in the western Transverse Ranges.

Timing of movement. The timing of movement on the Rinconada-Reliz fault is not well established. Crystalline rocks, Upper Cretaceous and lower Tertiary marine strata, and lower and middle Miocene strata of the Vaqueros and Tierra Redonda Formations evidently are displaced by equal amounts along the fault. Because the Tierra Redonda Formation, which overlies the Vaqueros Formation and underlies and interfingers with the lowest member of the Monterey Formation, ranges in age from late Saucesian to Luisian (Durham, 1974), movement on the fault is middle Miocene or younger in age (Hart, 1976; Graham, 1978). Graham suggested that the Relizian and Luisian Monterey Formation was deposited in en echelon basins, the development of which preceded and accompanied movement on the Rinconada-

Reliz fault. The upper Miocene Santa Margarita Formation and lower Pliocene Pancho Rico Formation display about 18 km of right separation. Durham's (1974) estimate of 18 km of post–early Pliocene displacement, based on an early Pliocene age for the Pancho Rico Formation, must now be considered post–late Miocene displacement, based on a late Miocene age for the formation (Addicott, 1978).

The youngest unit that is fully displaced along the San Gregorio–Sur segment of the fault zone is the porcellaneous and cherty member of the middle and upper Miocene (Relizian, Luisian, and Mohnian) Monterey Formation (Clark and others, 1984). Fission track counts on zircons from volcanic ash units in the Monterey Formation indicate that the formation ranges in age from about 15 to 8 Ma, and that an ash near the base of the diatomite member that overlies the porcellaneous and cherty member is 11.3 ± 0.9 Ma (Obradovich and Naeser, 1981). Uppermost Miocene and Pliocene (Mohnian and Delmontian) strata of the Santa Margarita Sandstone, *Bolivina obliqua*–bearing Santa Cruz Mudstone, and Purisima Formation (Clark and Brabb, 1978; Clark, 1981) are offset only about 70 km along the San Gregorio fault (Clark and others, 1984), which is a little less than half the overall displacement on the San Gregorio–Sur segment of the fault. The *Bolivina obliqua* zone spans a time interval of 6.5 to 4 Ma (Obradovich and Naeser, 1981) or 5.7 to 4.4 Ma (Bartow, 1990), although glauconite from the basal part of the Purisima Formation above the Santa Cruz Mudstone has yielded a K-Ar age of 6.7 ± 0.6 Ma (Clark, 1981). The San Gregorio–Sur–San Simeon–Hosgri fault zone has been active in the Quaternary and is seismically active (Greene and others, 1973; Coppersmith and Griggs, 1978; Gawthrop, 1978; Payne and others, 1979; Weber and others, 1979; Leslie, 1981; Weber and Cotton, 1981; Hanson and Lettis, 1990).

In summary, the San Gregorio–Hosgri and Rinconada-Reliz fault system accumulated slip during the middle and late Miocene and the Pliocene (Fig. 13). The San Gregorio–Sur segment of the San Gregorio–Hosgri fault zone began moving at some time after about 13 to 12 Ma and prior to about 11 Ma (Clark and others, 1984; cf. Hall, 1975; Graham and Dickinson, 1978a, b; Greene and Clark, 1979), had accumulated 80 km by about 6 or 7 Ma, and has moved an additional 70 km since then (Clark and others, 1984). According to Hall (1981), 80 to 115 km of right slip has occurred along the Hosgri–San Simeon fault between about 13 and 8 Ma.

Kinematic role of Salinian block faults in the evolving San Andreas fault system. The Red Hills–Ozena, Rinconada-Reliz, and San Gregorio–Hosgri faults comprise a group of en echelon, northwest-trending right-lateral faults that have extended the Salinian block northwestward (Schwade and others, 1958; Johnson and Normark, 1974; Graham, 1978, 1979; Page, 1982; Hornafius and others, 1986). This faulting accompanied the opening and deforming of en echelon middle and late Tertiary sedimentary basins (Schwade and others, 1958; Graham, 1978). The faults in this group terminate northwestward against the San Andreas fault but diverge southeastward from it; a component of

their overall displacement is additive to the San Andreas fault northwest of their intersections with that fault. Most of the evidence for timing of movement on the Red Hills–Ozena, Rinconada-Reliz, and San Gregorio–Hosgri faults indicates middle and late Miocene and Pliocene activity. Although precursive basins developed during the Relizian, most of the displacement on these faults accumulated since the Relizian, or since about 13 Ma, and more than half had accumulated by the end of the Miocene, or by about 5 Ma.

Although one might infer from these observations that the dextral faults of the Salinian block, along with the San Gabriel fault, define a domain of fault blocks that have rotated counterclockwise relative to the San Andreas fault, palinspastic reconstruction of the paleogeologic reference domains requires that the Salinian fault blocks and the San Andreas fault rotated clockwise as they evolved (Fig. 7, Plate II) (see also Ryder and Thomson, 1989, Fig. 51). Thus, the values measured from Plates I and II for fault displacement, fault length, and angles of fault intersection and rotation do not satisfy the equations derived from the geometry of the rotational fault block model (Freund, 1974; Garfunkel, 1974; Garfunkel and Ron, 1985; see also later section on left-lateral faults). During the evolution of the fault system, the angle between the San Andreas and the Salinian block and San Gabriel faults appears to have remained constant at about 15°, and the length of the San Andreas fault between its intersections with the San Gabriel and San Gregorio–Hosgri faults has remained constant at about 500 km. Meanwhile, the Salinian fault blocks has rotated clockwise by about 12° and the San Andreas fault north of the San Emigdio Mountains has rotated clockwise by about 15° to 20°. Because the San Gregorio–Hosgri fault intersects the San Andreas fault at an angle of about 15°, the amount of displacement transferred northward onto the San Andreas fault calculated from equation (1) is (150 km) cos 15° = 145 km (Fig. 15).

By this analysis, one should expect the postulated small clockwise rotation of the San Andreas fault and Salinian block to have been accommodated by contractional structures in the Coast Ranges along the east side of the San Andreas fault north of its intersection with the San Gabriel fault. Contractional deformation that is coeval with the modern San Andreas fault since about 5 or 6 Ma is well documented in the growth of anticlines along the eastern margin of the Coast Ranges (Reed and Hollister, 1936; Harding, 1976; Rymer and Ellsworth, 1991).

The Salinian block faults have been interpreted speculatively in other ways that are contradicted by the reconstruction presented herein. For example, the Red Hills–San Juan–Chimeneas fault also has been interpreted as an anastomosed strand of the San Andreas fault (Ross, 1972, p. 32–33; Graham, 1979) or an older throughgoing San Andreas–like right-lateral fault with a 100- to 200-km displacement (Smith, 1977a; Joseph and others, 1982b). In this role, the fault has been proposed to curve back to the San Andreas fault via a fault concealed beneath Cuyama Valley (Ross, 1972, p. 32), presumably either the Morales or Blue Rock faults (Smith, 1977a; Joseph and others, 1982b). In another example, the Rinconada-Reliz and/or San Gregorio–Hosgri faults also have been interpreted as parts of a pre-Neogene proto–San Andreas fault (Kistler and others, 1973; Hamilton, 1982). The evidence for magnitude and timing of displacement presented above disputes these interpretations.

The Russell fault, together with the Blue Rock and the ancestral Ozena faults were probably right-oblique normal faults that developed in the tectonic regime that immediately preceded the development of throughgoing strike-slip faults.

Peninsular Ranges

San Jacinto fault. Path. The San Jacinto fault zone transects the northeastern Peninsular Ranges between Borrego Valley in the Salton trough and the San Bernardino Valley, north of which it splays as it enters the southeastern San Gabriel Mountains. (For a history of study of the San Jacinto fault zone, see Sharp, 1967). Although some investigators have concluded that the San Jacinto fault merges northward with the San Andreas fault between the San Gabriel and San Bernardino Mountains (Noble, 1954a, b; Ehlig, 1975a, 1981; R. J. Weldon, 1989, written communication), others have demonstrated that the recognized strands of the San Jacinto fault in the southeastern San Gabriel Mountains do not join the San Andreas fault at the surface (Dibblee, 1968a; Morton, 1975a, b; Morton and Matti, this volume).

Previous estimates of displacement. Measurements of right-lateral displacement of Mesozoic crystalline rocks along the San Jacinto fault zone in the Peninsular Ranges province range between about 24 and 30 km (Sharp, 1967; Baird and others, 1970; Bartholomew, 1970; R. I. Hill, 1981). Eckis (1930) postulated that the distribution of Miocene and Pliocene strata along the San Jacinto fault in the vicinity of Borrego Valley indicates a dextral displacement of 29 to 32 km. Fluvial and lacustrine units near the base of the Pliocene and Pleistocene San Timoteo badlands section, initially estimated to be offset 18 to 19 km (English, 1953; Dutcher and Garrett, 1963, p. 38; see also Sharp, 1967; Bartholomew, 1970) along the San Jacinto fault, have more recently been demonstrated to be offset 23 to 26 km (Matti and Morton, 1975). The Pleistocene Bautista Beds of Frick (1921) are distributed along the San Jacinto fault and evidently accumulated during movement (Sharp, 1967, 1981). Restoration of 18 to 24 km of dextral slip aligns the Cucamonga and Banning faults, and some investigators have inferred that these faults are in part an old feature that was disrupted by the San Jacinto fault (Allen, 1957; Sharp, 1967; Dibblee, 1968a; Baird and others, 1971; Jahns, 1973), although recent reverse movement on the Cucamonga and Banning faults is coeval with movement on the San Jacinto fault (Woodford, 1960; Jahns, 1973; Matti and others, 1985). At its northern end where it enters the San Gabriel Mountains, however, the main splays of the San Jacinto fault offset contacts in the crystalline rocks only about 11 to 13 km (Arnett, 1949; Dibblee, 1968a; Morton, 1975a, b), although subordinate fault splays account for another 2 or 3 km (Morton, 1975a, b).

Reckoning of displacement. In the reconstruction shown in Figure 11 and Plate II, a displacement of about 28 km is restored along the San Jacinto fault. Of the overall dextral displacement on the San Jacinto fault, 13 km has been documented to enter the southeastern San Gabriel Mountains, where it is absorbed by left-lateral faulting on the San Antonio Canyon and Stoddard Canyon faults (see Reconciliation and Synthesis section). The disposition of the remaining 15 km is not resolved but seems too large to be absorbed in compression associated with uplift of the San Gabriel Mountains. In Figures 8 through 11 and in Plate II, the remaining 15 km is passed onto the San Andreas fault east of the splays recognized in the southeastern San Gabriel Mountains by means of a stepping transfer mechanism (Matti and others, 1985; Morton and Matti, this volume), perhaps in combination with movement on a connecting fault buried beneath the San Bernardino basin.

Timing of movement. Movement on the San Jacinto fault zone began no earlier than about 5 or 6 Ma and probably no later than about 1.5 Ma (Fig. 13), although Morton and Matti (this volume) argue that movement on the fault began at about 1 Ma. The strata near the base of the San Timoteo badlands section contain the late Hemphillian Mount Eden mammalian fauna (Frick, 1921; May and Repenning, 1982) that is about 5 m.y. old on the basis of magnetostratigraphic correlation of these vertebrate faunal assemblages in fossil occurrences in Texas and Kansas (Lindsay and others, 1975). The estimate of 23 to 26 km of displacement of the strata containing the Mount Eden fauna is either slightly less than or the same as the overall displacement of older units. If this difference is real, then movement on the San Jacinto fault had begun by about 5 Ma; otherwise, it began later. The Bautista Beds of Frick (1921; see also Sharp, 1967), which accumulated during movement on the fault, include the 0.7-Ma Bishop ash bed (Merriam and Bischoff, 1975; Sarna-Wojcicki and others, 1984) and locally contain an Irvingtonian mammalian fauna (Hibbard and others, 1965) that is less than 2 m.y. old on the basis of magnetostratigraphic correlation of these vertebrate faunal assemblages in fossil occurrences in Texas and Kansas (Lindsay and others, 1975; Lundelius and others, 1987).

There is abundant geologic and seismic evidence that the San Jacinto fault is currently active (see, for example, U.S. Geological Survey, 1972; Sharp, 1981; Sanders and Kanamori, 1984; Hill and others, 1990; Wesnousky and others, 1991).

Elsinore fault. Path. The Elsinore fault transects the Peninsular Ranges from about the United States–Mexico border on the west side of the Salton trough to about Corona at the north end of the Peninsular Ranges (for discussions of nomenclature and histories of study of the fault zone, see Kennedy, 1977; Weber, 1977; Lamar and Rockwell, 1986). Just south of Corona, the Elsinore fault zone bifurcates northward into the Whittier and Chino faults.

Previous estimates of displacement. Measurements of right-lateral displacement of crystalline rocks along the Elsinore fault zone range from less than 1 (Mann, 1955; Gray, 1961) to about 40 km (Lamar, 1961; Sage, 1973a, b; Campbell and Yerkes,

1976). The most convincing measurements of displacement, however, lie in the range from about 5 to 11 km (Woodford, 1960; Sharp, 1966; Baird and others, 1970; Woodford and others, 1971; Kennedy, 1977; Weber, 1977; Johnson and others, 1983). Crystalline rocks interpreted by some investigators to limit strike-slip displacement along the southern part of the Elsinore fault to about 2 km (Todd, 1978; Todd and Hoggatt, 1979; Lowman, 1980) have been interpreted by others to comprise a landslide that partially covers the fault (Pinault, 1984; Rockwell and others, 1986).

Reckoning of displacement. In the reconstruction shown in Figure 11 and Plate II, a displacement of about 5 km is restored along the Elsinore fault.

Timing of movement. Movement on the Elsinore fault zone apparently began no earlier than about 5 or 6 and no later than about 1.5 to 2 Ma (Fig. 13). An abrupt facies boundary in an unnamed Pleistocene formation of sandstone and conglomerate is offset more than 5 km along one strand of the Elsinore fault (Kennedy, 1977). The unit contains cobbles of basalt derived from flows that were extruded at about 7 Ma (Morton and Morton, 1979) and an intrastratified ash that is correlated with the 0.7-Ma Bishop ash bed (Merriam and Bischoff, 1975; Sarna-Wojcicki and others, 1984). The unnamed unit has been correlated in part with the Temecula Arkose of Mann (1955) (see also Kennedy, 1977), which contains a late Blancan mammalian fauna (Golz and others, 1977) that is about 2 to 3 m.y. old on the basis of magnetostratigraphic correlation of a similar fauna found in fossil occurrences in Texas and Kansas (Lindsay and others, 1975; Lundelius and others, 1987). If the offset of the unnamed unit is the full displacement on the Elsinore fault, then movement began after 2 or 3 Ma.

There is abundant geologic and seismic evidence that the Elsinore fault is currently active (see, for example, Kennedy, 1977; Weber, 1977; Rockwell and Lamar, 1986; Hill and others, 1990).

Kinematic role of Peninsular Ranges faults in the evolving San Andreas fault system. Along with the San Andreas fault zone, the San Jacinto and Elsinore faults were associated with the opening of the Salton trough at the north end of the Gulf of California (Hamilton, 1961; Biehler and others, 1964; Hamilton and Myers, 1966; Lomnitz and others, 1970; Elders and others, 1972; Fuis and others, 1982; Fuis and Kohler, 1984; Sharp, 1982). Evidently, the San Jacinto fault and presumably the Elsinore fault as well evolved in conjunction with the left-lateral faults in the eastern Transverse Ranges and southern San Gabriel Mountains and with the growth of the deflection in the San Andreas fault through San Gorgonio Pass (Allen, 1957; Garfunkel, 1974; Crowell, 1981; Matti and others, 1985; Meisling and Weldon, 1989). The kinematic model incorporated into the reconstruction in Plate II is one in which sinistral faulting along the southern margin of the San Gabriel Mountains and along the Pinto Mountain fault and dextral faulting along the San Andreas fault mutually deflected one another. Coeval dextral faulting developed along the San Jacinto and Elsinore faults in the Penin-

sular Ranges block south of the sinistral fault zone in the San Gabriel Mountains. Displacement on the sinistral fault system kinematically absorbed the complementary right-lateral displacement on the Elsinore fault and about the first half of that on the San Jacinto fault. Ultimately, the San Andreas fault, as it was deflected, intersected the San Jacinto and the second half of displacement on the San Jacinto fault was transferred northward onto the San Andreas. As discussed above in the section on relative rotations of crustal blocks, it seems likely from the reconstruction that the domain of dextral faulting in the Peninsular Ranges remained fixed in orientation. If the current angle of intersection of about 25° between the San Jacinto and San Andreas faults has also remained constant since they were linked, then, from equation (1), displacement on the San Andreas fault increases by an amount equal to (14 km)cos 25°, or by 13 km, north of the intersection (see Fig. 15). The timing of movement on the various faults permits this scenario to have evolved in as little as the last 1.5 m.y. or as much as the last 5 m.y.

From their alignment, one might be tempted to infer a link between the Elsinore fault and the Vasquez Creek fault of the San Gabriel system, but such an inference is inconsistent with what is known about the timing of movement on the two faults.

Mojave Desert

Paths. The Mojave Desert crustal block has been sliced by numerous northwest- to north-northwest–trending, steeply dipping faults arrayed en echelon in a belt that reaches from near the Garlock fault in the northwest at least to the eastern end of the Pinto Mountains (Dibblee, 1961a; Garfunkel, 1974; Powell, 1981; Dokka, 1983) and probably from there through the southeastern Mojave Desert (Hope, 1966; Pelka, 1973; Rotstein and others, 1976; Powell, 1981; Stone and Pelka, 1989) and possibly across the Colorado River into Arizona (Miller and McKee, 1971; Sumner and Thompson, 1974; Powell, 1981). In the north-central Mojave Desert, these faults comprise zones of aligned faults, including the Helendale, Lockhart-Lenwood, Harper–Harper Lake–Camp Rock–Emerson, Blackwater–Calico–Mesquite Lake, and Pisgah-Bullion faults. Smaller faults, including the Old Woman Springs, Johnson Valley, Homestead Valley, and Galway Lake faults, splay off these major zones. The Ludlow and Bristol Mountains faults are two additional north-northwest–trending faults located in the east-central Mojave Desert. Identified and inferred northwest-trending faults in the southeastern Mojave Desert include the Sheep Hole fault and several unnamed faults within and bounding the Coxcomb, Palen, McCoy, Little Maria, and Maria Mountains. In southwestern Arizona, unnamed northwest-trending faults have been identified in the Plomosa Mountains and the area southeast of Yuma.

Previous estimates of displacement. Right-lateral displacement has occurred along the north-northwest–trending faults in the northwestern Mojave Desert (Dibblee, 1961a, 1967a, 1980; Hawkins, 1975; Miller and Morton, 1980; Miller, 1980; Powell, 1981, p. 334–336; Dokka and Glazner, 1982a; Dokka, 1983;

Dokka and Travis, 1990a), southeastern Mojave Desert (Hope, 1966, p. 102–104; Silver and others, 1977; Pelka, 1973; Powell, 1981, p. 336–338; Miller and others, 1982), and in southwestern Arizona (Miller, 1970; Miller and McKee, 1971; Sumner and Thompson, 1974). Most of the individual faults have accumulated displacements of less than 5 km, although a few have displacements of as much as 10 km. Garfunkel (1974) liberally estimated a cumulative dextral displacement in the range of 65 to 105 km on the faults in the northwestern Mojave Desert, whereas subsequent investigators have provided more conservative estimates of about 14 km (Miller, 1980, not including estimates for the Helendale and Bristol Mountains faults), 25 to 65 km (Powell, 1981, not including an estimate for the Bristol Mountains fault), and 26.7 to 38.4 km (Dokka, 1983). Dokka and Travis (1990a) tabulated observed displacements on dextral faults in the central Mojave Desert that summed to 28 to 40 km, and predicted an additional 39 km on the Bristol Mountains and Granite Mountains faults of the northeastern Mojave Desert based on their tectonic model, for total of about 65 km. These latter two faults lie east of the reconstruction in Plate II. Estimated dextral separation on the north-northwest–trending faults of the southeastern Mojave Desert sum to something in the range of 25 to 35 km (Powell, 1981), and cumulative right separation on the faults in southwesternmost Arizona includes 5 to 10 km on faults in the Plomosa Mountains (Miller and McKee, 1971) and 16 to 23 km on faults southeast of Yuma (Sumner and Thompson, 1974).

Reckoning of displacements. In Plate II, dextral displacements restored on northwest-trending active faults in the Mojave Desert include 3 km on the Helendale fault, 3 km on the Lockhart-Lenwood fault, 4 km on the Harper–Harper Lake–Camp Rock–Emerson fault, 9 km on the Blackwater–Calico–Mesquite Lake fault, and 16 km on the Pisgah-Bullion fault. These displacements are consistent with previously published measurements (Miller, 1980; Miller and Morton, 1980; Dokka, 1983; Dokka and Travis, 1990a). In addition, dextral displacements are restored on burial faults between the Sheep Hole and Coxcomb Mountains (3.5 km), the Coxcomb and Palen Mountains (4.5 km), and the Palen and McCoy Mountains (4 km).

Timing of movement. Right-lateral faults of the Mojave Desert northwest of the Pinto Mountains show evidence of more recent movement than those farther southeast. Various of the faults northwest of the Pinto Mountains have displaced 2.5-Ma basalt (Dibblee, 1967a, 1980; Burke and others, 1982) and Pleistocene and Holocene alluvial fans (Hawkins, 1975; Dokka and Glazner, 1982a; Dokka, 1983; Meisling, 1984; Meisling and Weldon, 1989; Dokka and Travis, 1990a). Many of these faults are seismically active (see, for example, Fuis and others, 1977, Plate 10; Hill and others, 1990, Fig. 5.9), and recent earthquakes at Galway Lake and in Homestead Valley were accompanied by right-lateral ground-breakage and have yielded right-lateral fault plane solutions (see, for example, Beeby and Hill, 1975; Hill and Beeby, 1977; Hawkins and McNey, 1979; Hill and others, 1980). In contrast, right-lateral faults in the Plomosa Mountains, Arizona and northwest-trending faults in California southeast of the Pinto Mountains are covered by Quaternary and Holocene alluvium (Miller and McKee, 1971; Rotstein and others, 1976).

The time of inception of dextral faults in the Mojave Desert is not well constrained (Fig. 13). There is abundant evidence that right-lateral strike-slip faults of the Mojave and Sonora Deserts developed later than about 20 to 14 Ma. In the central Mojave Desert, right-lateral faults displace volcanic and sedimentary strata as young as about 20 Ma by the same amount as Mesozoic crystalline rocks (Dokka, 1983; Dokka and Travis, 1990a). The lower and middle parts of the Barstow Formation, which range in age from 19 to 14 Ma on the basis of K-Ar and ^{40}Ar/^{39}Ar ages on biotite and sanidine (MacFadden and others, 1990), are fully offset along the Calico-Blackwater fault (Dokka and Travis, 1990a). The right-lateral faults in the Plomosa Mountains crosscut 20-Ma rhyodacite flows (Miller and McKee, 1971). Movement on the dextral faults probably began no earlier than about 10 Ma (Carter, 1987; Dokka and Travis, 1990a). It seems unlikely, however, that much, if any, displacement accumulated on the Mojave Desert faults any earlier than the inception of the San Andreas fault proper in southern California at about 5 Ma; Dokka and Glazner (1982b) suggested, on the basis of field observations, that right-lateral faults in the central Mojave Desert initiated during the Pliocene. Some authors have inferred that movement along the currently active right-lateral faults of the Mojave Desert commenced no earlier than about 1.5 Ma (Meisling and Weldon, 1989; Dokka and Travis, 1990a).

Kinematic role of Mojave Desert faults in the evolving San Andreas fault system. The northwest-trending right-lateral faults of the central Mojave Desert form a conjugate set with the coeval east-trending left-lateral faults of the eastern Transverse Ranges to the southwest and terminate near the Garlock fault to the north (Hope, 1966; Garfunkel, 1974; Powell, 1981, p. 339; Dokka and Travis, 1990a). All of these faults have developed in conjunction with the growth of deflections in trend of the post–5-Ma San Andreas fault at Tejon Pass and San Gorgonio Pass (see, for example, Garfunkel, 1974; Powell, 1981; Matti and others, 1985). It is evident that the crustal blocks adjacent to the San Andreas fault in southern California have not behaved rigidly and the Mojave Desert faults are one set of faults by which the crust northeast of the San Andreas fault has been distorted. As such, they take up plate motion not accommodated by the San Andreas fault (Bird and Rosenstock, 1984; Weldon and Humphreys, 1986; Dokka and Travis, 1990b). In this context, it seems reasonable to postulate that movement on the right-lateral faults in the western Mojave Desert began no earlier than about 5 Ma and no later than about 1.5 to 2 Ma.

Garfunkel (1974) modeled the Mojave Desert faulting as a set of fault blocks that rotated clockwise relative to the left-lateral Garlock fault, which he held fixed as the northern boundary of the domain of right-lateral faulting. After subsequent paleomagnetic studies showed that the Mojave Desert fault blocks are unrotated, however, Garfunkel and Ron (1985) revised the earlier model by requiring the Garlock fault to rotate while the

Mojave Desert faults maintained a constant orientation. The dextral displacements utilized in Garfunkel's model, however, are much larger than the displacements measured by other investigators.

Dokka and Travis (1990a) reported that the dextral faults of the central Mojave Desert terminate south of the Garlock fault and that their displacements diminish to zero northward. They concluded that the Mojave Desert dextral faults were not linked kinematically to the Garlock, but, rather, to a domain of clockwise-rotating sinistral faults that trends northeast across the east-central Mojave Desert to the north of the domain of dextral faulting. Dokka and Travis (1990a) postulated that right slip on the dextral faults of the central and east-central Mojave Desert accompanied slight counterclockwise rotation of the fault blocks against the domain of sinistral faulting, and that the dextral faults of the east-central Mojave Desert developed earlier than those of the central Mojave. They further postulated that strike-slip faulting in the Mojave Desert is linked to the hypothetical southward extension of the dextral Death Valley fault zone.

The reconstruction in Plate II, assembled prior to publication of the model of Dokka and Travis (1990a), links the Mojave Desert faults kinematically to the Garlock fault. The reconstruction would require very little change to accommodate the disposition of displacement in Dokka and Travis's model by routing that displacement northeastward off the plate rather than northwestward to the Garlock fault.

In interpreting the seismic structure of the Transverse Ranges and Mojave Desert, Hadley and Kanamori (1977) postulated that the current boundary between the North American and Pacific plates follows the San Andreas fault at the surface, whereas at a depth of 30 to 35 km the plate boundary has stepped eastward across a horizontal zone of shear and forms a broad zone of shear in the mantle beneath the northwest-trending belt of en echelon right-lateral faults in the Mojave Desert.

LEFT-LATERAL FAULTS

Transverse Ranges east of the San Andreas fault

Paths. R. T. Hill (1928, p. 144–146, Plate I/II) first depicted the easterly physiographic and structural trend in the eastern Transverse Ranges province, recognizing three prominent fault-controlled physiographic lineaments. Subsequent investigators have shown that the east-west physiographic grain of the province is controlled by several east-trending left-lateral strike-slip faults, the largest of which are 60 to 85 km long (Hope, 1969; Merifield and Lamar, 1975; Silver and others, 1977, Figs. 4-4, 4-5; Powell, 1981, p. 315–333, Plate I, 1982a).

From north to south, the principal throughgoing left-lateral faults in the eastern Transverse Ranges include the three recognized by Hill, now known as the Pinto Mountain, Blue Cut, and Chiriaco faults, and the Salton Creek fault (Plate I). The Pinto Mountain fault (Hill, 1928, p. 146; Allen, 1957; Bacheller, 1978) extends eastward from the Mission Creek fault of the San An-

dreas zone, through the southernmost San Bernardino Mountains, then along the northern margin of the Little San Bernardino and Pinto Mountains. The trace of the Pinto Mountain fault is marked by abundant evidence for the breakage of surficial deposits between its intersection with the Mission Creek fault and its junction with the Mesquite fault just east of Twentynine Palms (Allen, 1957; Dibblee, 1967e, f, 1968b; Bacheller, 1978). The lack of such evidence east of that junction led Dibblee (1975a, 1980, Fig. 8, 1982d) to conclude that the Pinto Mountain fault terminates at the Mesquite fault and Hatheway and West (1975) to conclude that it is deflected into the Pinto Mountains east of the junction. However, the continued linear escarpment along the northern front of the Pinto Mountains is seemingly controlled by a major fault, probably the buried eastward extension of the Pinto Mountain fault (Bassett and Kupfer, 1964, p. 40; Bacheller, 1978, p. 127–128; Powell, 1981, p. 317), and the relatively small displacement of crystalline rock patterns across east-trending faults within the range indicate that they are at most only splays of the Pinto Mountain fault.

The trace of the Blue Cut fault (Pruss and others, 1959; Hope, 1966, 1969) extends eastward from the central Little San Bernardino Mountains through the Hexie Mountains and Pinto Basin. The fault, exposed only in the vicinity of the Blue Cut in the Little San Bernardino Mountains, forms the central break in an array of faults responsible for Hill's (1928, p. 147) Eagle Mountain lineament.

The Chiriaco fault (Powell, 1975, 1981, 1982a) extends eastward from the northern Mecca Hills through Chiriaco Summit, Hayfield Lake, and Desert Center. The fault underlies alluvium in Shavers and Chuckwalla Valleys (Brown, 1923, p. 53, 238; Miller, 1944, p. 72), but may be responsible for linear shoreline segments of Hayfield Lake. The fault has also been referred to as the Orocopia lineament (Hill, 1928, p. 147–148; Merifield and Lamar, 1975) and the Hayfield fault (Dibblee, 1982h).

The Salton Creek fault (Crowell, 1962, Fig. 3; Crowe and others, 1979; Crowell and Ramirez, 1979; Powell, 1981, 1982a) extends eastward from the southern Mecca Hills up Salt Creek (Salton Wash) between the Orocopia and Chocolate Mountain, then transects the Chuckwalla Mountains along an unnamed wash between the wide northern part of the range and the narrow southern part. The fault is exposed along the north flank of the Chocolate Mountains along Salt Creek.

Shorter, less prominent faults are present between the major faults, some as splays. The Corn Springs Wash fault splays southeastward from the Chiriaco fault between the Orocopia and Chuckwalla Mountains and transects the latter range along Corn Springs Wash; similarly, the Niggerhead fault probably splays southeastward from the Pinto Mountain fault before curving eastward through the eastern Pinto Mountains.

Other subordinate faults do not connect with the throughgoing major faults. The Porcupine Wash and Substation faults and the Smoke Tree Wash and Victory Pass faults occur as two aligned pairs separated by about 3 to 5 km between the Blue Cut and Chiriaco faults, the Ship Creek fault occurs between the Corn

Springs Wash and Salton Creek faults, and the Graham Pass fault is located south of the Salton Creek fault. Although the Porcupine Wash and Substation faults and the Victory Pass and Smoke Tree Wash faults are aligned and their displacements are nearly equal, the two faults that make up each pair are separated by gaps of 12 to 13 km across which geologic units and contacts are not broken. The Porcupine Wash fault extends from the lip of the Little San Bernardino escarpment through the Hexie Mountains to the Pinto Basin Road, where it disappears within the alluviated granodiorite of the southernmost Pinto Basin; along strike, the Substation fault transects the east-central Eagle Mountains before disappearing beneath alluvium in Chuckwalla Valley. The Smoke Tree Wash fault, which underlies the east-west reaches of Smoke Tree Wash and Pinkham Canyon, dies out westward as it nears the Little San Bernardino Mountains and disappears eastward beneath alluvium; along strike, the Victory Pass fault transects the southeastern Eagle Mountains, where it controls much of the course of Big Wash before passing eastward beneath the alluvium of Chuckwalla Valley. The Ship Creek fault transects the Chuckwalla Mountains along Ship Creek, dying out in the western part of the range and passing east beneath alluvium in Chuckwalla Valley.

Small left-lateral displacements also have been proposed to accompany or follow reverse displacement on the Santa Ana and Cleghorn faults in the San Bernardino Mountains (Meisling and Weldon, 1982; Powell and others, 1983b). Oblique displacement with components of left and reverse slip makes the style of these faults more akin to coeval east- to northeast-trending faults in the Transverse Ranges west of the San Andreas fault than to the other left-slip faults east of the fault.

Displacements. Overall displacements on the left-lateral faults are well constrained by disrupted paleogeologic patterns in the Proterozoic and Mesozoic crystalline rocks of the eastern Transverse Ranges (Fig. 8, Plate II). Disrupted bedrock patterns in the eastern Transverse Ranges include northwest-trending belts of Proterozoic plutonic and metamorphic units of the Eagle and Hexie Mountains assemblages, Mesozoic plutonic units of the central and eastern plutonic belts, and northwest- and northeast-trending dike swarms, and the Orocopia Schist. For the major sinistral faults, reassembly of these features into their pre–late Cenozoic paleogeologic pattern indicate displacements of 16 km on the Pinto Mountain fault (within the range of 16 to 20 km measured by Dibblee, 1967b, 1975a, 1980, 1982d; Bacheller, 1978), 5 km on the Blue Cut fault (Hope, 1966, 1969; Powell, 1981, 1982a), 11 km on the Chiriaco fault (Powell, 1975, 1981, 1982a), 8 km on the Chuckwalla Mountains segment of the Salton Creek fault (Powell, 1982a), and as much as 14 km on the Salt Creek segment of the Salton Creek fault (Powell, 1991).

For the major faults, incremental displacement has been established only on the Pinto Mountain fault. By matching clasts of Jurassic porphyritic monzogranite in lower Quaternary fanglomerate in the Twentynine Palms area north of the Pinto Mountain fault to the nearest exposures of such monzogranite that could have been a source for the clasts in the Pinto Moun-

tains south of the fault, Bacheller (1978) measured about 9 km of displacement on the Pinto Mountain fault since the deposition of the fanglomerate. About 7 km of displacement occurred prior to deposition of the fanglomerate.

For the subordinate sinistral faults, reassembly of the crystalline terrane into its pre–late Cenozoic paleogeologic pattern indicates overall displacements of about 2 km on the Niggerhead fault (Hope, 1966, 1969), 3 km on the Porcupine Wash and Substation faults (Hope, 1966; Powell, 1981, 1982a), 1 to 1.5 km on the Smoke Tree Wash and Victory Pass faults (Hope, 1966; Powell, 1981, 1982a), 2.5 to 3 km on the Corn Springs Wash fault (Powell, 1981, 1982a), and about 2 km on the Ship Creek fault (Powell, 1981, 1982a). In the San Bernardino Mountains, the Santa Ana fault shows 2.4 km of left separation of Early Proterozoic gneiss, upper Proterozoic quartzite, and Mesozoic quartz diorite (Powell and others, 1983b), and the Cleghorn fault exhibits 3 to 4 km of displacement, both of an antiformal axis in a crystalline terrane of foliated Mesozoic plutonic rocks and marble pendants and of a steeply dipping contact between the crystalline rocks and the lower and middle Miocene Punchbowl Formation of Cajon Pass (Meisling and Weldon, 1982, 1989; Weldon, 1986; Weldon and Springer, 1988).

In addition to the left-lateral displacement of crystalline bedrock patterns across the Blue Cut and Chiriaco fault zones, vertical displacement of late Cenozoic basalt flows and broad alluviated valleys provide evidence for young graben-like faulting associated with the strike-slip faults (Hope, 1966, p. 61, 67, 91, 92; Powell, 1981, Plate I) and for extension contemporaneous with shear in the province. Similarly, the Morongo Valley fault, branching south from the Pinto Mountain fault near its west end, appears to be a normal fault with the north side down. The Cleghorn fault, which formed as a reverse fault in the late Miocene, also may have experienced normal displacement in association with left slip in the Quaternary (Weldon and others, 1981; Weldon and Springer, 1988; Meisling and Weldon, 1989). This association of extension with sinistral shear in the eastern Transverse Ranges contrasts with a combination of compression and sinistral shear indicated by oblique reverse and left-lateral slip on coeval east-trending faults in the Transverse Ranges west of the San Andreas fault.

Timing and rates of movement. Although the magnitude of overall displacement on the left-lateral faults is well constrained by matching crystalline rock patterns, much of the evidence for the timing of that movement is less clear. The time of inception of faulting is especially difficult to ascertain. Because the Chiriaco and Salton Creek faults appear to truncate the Eocene Maniobra and the upper Oligocene and lower Miocene Diligencia Formations in the Orocopia Mountains and unnamed upper Oligocene and lower Miocene volcanic rocks in the Chuckwalla Mountains, movement on the left-lateral faults in the province is probably no older than middle Miocene in age. Although the Salton Creek fault has been inferred to have been a locus of intrusion for 30-Ma volcanic domes and to have been overlapped by 17-Ma basalt (Crowell, 1975a; Crowe and others, 1979; Crowell and

R. E. Powell

Ramirez, 1979; Buesch and Ehlig, 1982), the inferred domes are surrounded by alluvium and the relation between the volcanic rocks and the fault is conjectural. In my view, these isolated exposures consist of rhyolite, pumiceous tuff, and minor conglomerate that have been displaced by as much as 14 km from similar, better exposed rocks that comprise a rhyolitic dome field in the Tabaseca Tank area of the northeastern Chocolate Mountains.

Some evidence is consistent with an interpretation that movement on the left-lateral faults is no older than middle or perhaps late Miocene. First, both the Chiriaco and Salton Creek faults appear to offset or deflect the Clemens Well fault, along which movement inferentially had ceased by about 13 Ma (see earlier discussion). Second, the left-lateral fault system disrupts an erosion surface that existed during the Miocene. Remnants of a paleosol on that surface are capped by basalt flows at Pioneer Town in the eastern San Bernardino Mountains and in the Mojave Desert at Old Woman Springs and Fry Mountain that have yielded K-Ar ages that range in age from about 6 to 11 Ma (Oberlander, 1972; Petersen, 1976, p. 25, 99; Neville and Chambers, 1982; J. L. Morton, 1980, written communication cited in Neville, 1983). Basalt flows in the central San Bernardino Mountains that are broken by the Santa Ana fault have yielded K-Ar ages in the range of about 6 Ma (F. K. Miller, 1974, written communication cited in Woodburne, 1975, p. 83). Basalt flows that rest on a warped erosion surface in the Pinto and Eagle Mountains are interbedded with lacustrine and alluvial clay, siltstone, sandstone, and conglomerate and are displaced vertically on the normal faults associated with Pinto Basin and the Blue Cut fault (Hope, 1966; Powell, 1981). A tilted basalt flow in the Cottonwood Mountains is sinistrally offset by the Smoke Tree Wash fault. These and other basalts in the region have yielded K-Ar ages in the range of about 8 to 15 Ma (Carter and others, 1987). If extrusion of the basalt flows is related to extension that has accompanied the strike-slip faulting, then the faulting could be at least as old as about 15 to 11 Ma; however, if the volcanism largely preceded the strike-slip faulting, then the inception of faulting could be as young or younger than about 8 to 6 Ma. The latter interpretation is favored by the abundance of evidence that the basalt flows are broken by the faults and the lack of evidence that basalt has intruded any of the faults, whereas the former interpretation is perhaps favored by the observation that the left-lateral faults appear to be loci for the occurrence of basalt flows in the eastern Transverse Ranges.

The documented age range of sedimentary units that have accumulated in basins along the left-lateral faults is entirely Quaternary. Along the Pinto Mountain fault, Quaternary deposits consist of alluvial fanglomerate and sand and lacustrine silt and clay, and contain an ash bed that has been correlated chemically with the 0.7-Ma Bishop ash bed (Bacheller, 1978, p. 36–43) and a silty clay bed with a vertebrate fauna identified as Rancholabrean in age (R. E. Reynolds, 1978, oral communication cited in Bacheller, 1978, p. 70; see also Bassett and Kupfer, 1964, p. 31). A Pleistocene vertebrate fauna has been collected from an un-

specified part of a section of lacustrine and alluvial clay, siltstone, sandstone, and conglomerate with interbedded basalt in the northeasternmost Eagle Mountains (Scharf, 1935; Bassett and Kupfer, 1964; Hope, 1966), but presumably came from above the basalt, which has yielded a K-Ar age of 7.8 Ma (Carter and others, 1987). At its west end, the Chiriaco fault appears to be covered by the Pleistocene Ocotillo Conglomerate of Dibblee (1954), and the westward projection of the buried fault lies approximately along the northern limit of pre-Ocotillo uppermost Pliocene or lowermost Pleistocene (Blancan) *Equus*-bearing strata in the Mecca Hills (Downs, 1957; D. E. Savage, 1953, personal communication, and T. E. Downs, 1957, written communication cited in Hays, 1957, p. 82–84, 184–185).

The age of the youngest alluvial units that are broken by the sinistral faults is younger along those faults in the northern half of the eastern Transverse Ranges province than along those in the southern half, which in turn may be younger than those broken by left-lateral faults in the Chocolate Mountains to the south of the province. In the northern part of the eastern Transverse Ranges, the Pinto Mountain, Blue Cut, Porcupine Wash, and Smoke Tree Wash faults are characterized by scarps in all but the youngest Holocene alluvium (Allen, 1957; Dibblee, 1967e, f, 1968b; Hope, 1966; Bacheller, 1978). The Pinto Mountain fault shows evidence for ground water damming; and the Blue Cut fault, the aligned Porcupine Wash and Substation faults, and the aligned Smoke Tree Wash and Victory Pass faults are seismically active. The Pinto Mountain fault accumulated as much as 7 km of displacement prior to and 9 km during the deposition of Quaternary strata in the Twentynine Palms area. The syndepositional displacement is recorded by the distribution of clasts of quartzite and Jurassic porphyritic monzogranite in fanglomerate that underlies, interfingers with, and overlies a section of sandstone, siltstone, and clay that contains the Bishop ash bed and the Rancholabrean silty clay (Bacheller, 1978, p. 29–70). The age of the Bishop ash bed is 0.7 Ma and the Rancholabrean begins after the Matuyama-Brunhes boundary (0.73 Ma) or at about 0.6 Ma (Lindsay and others, 1975). Left-lateral movement on the Cleghorn fault is entirely Quaternary in age and about one-third of the overall displacement has accumulated in the last 0.5 m.y. (Meisling and Weldon, 1982).

In the southern part of the eastern Transverse Ranges, modern alluvium is unbroken along the Chiriaco fault, and the fault is covered at its west end by the Pleistocene Ocotillo Conglomerate (see Dibblee, 1954; Hays, 1957; Ware, 1958). No surface expression of the Chiriaco fault has been found, except perhaps the rather straight southern margin of the west half of Hayfield Lake aligned with the roughly linear northern margin of the eastern part of the lake (Powell, 1981, p. 323). Hayfield Lake may be either a sag feature related to movement on the Chiriaco fault or a depression due to aggradation of alluvial fans that block or dam exterior drainage. Of possible significance is the observation that the drainage divide in the Hexie and Cottonwood Mountains between the Colorado River and the Salton trough steps about 14 km eastward along the Chiriaco fault to a position between the Orocopia and Chuckwalla Mountains. The interior drainage that empties into Hayfield Lake lies along the fault between its intersections with the drainage divide. The formation of the drainage divide is probably no older than the inception of the Salton trough at about 4 or 5 Ma. If the step in the divide along the Chiriaco fault is an offset, then the overall displacement on that fault has occurred since about 5 Ma and some of the displacement may have been young enough to have truncated the uppermost Pliocene or lowermost Pleistocene *Equus*-bearing strata of the Mecca Hills. According to Lindsay and others (1975), *Equus* appeared in North America at about 3.2 Ma.

Pre-Holocene alluvial deposits of unestablished age are broken along the Corn Springs Wash and Salton Creek faults. At two exposures along the Corn Springs Wash fault in the Chuckwalla Mountains, Pleistocene(?) or Pliocene(?) conglomeratic sandstone is faulted against shattered Mesozoic batholithic rocks (Powell, 1981, p. 326) and is capped by unbroken, deeply incised, indurated Pleistocene(?) fanglomerate. At a third locality along the fault, steeply tilted Pleistocene(?) or Pliocene(?) conglomeratic sandstone is capped by deeply incised, indurated Pleistocene(?) fanglomerate. Crystalline rocks in the Chocolate Mountains south of the Salton Creek fault are faulted against Pleistocene(?) or Pliocene(?) fanglomerate to the north of the fault (see also Jennings, 1977; Crowe and others, 1979), whereas younger alluvium has buried the fault. In the Chuckwalla Mountains, however, no surface break has been recognized along the Salton Creek fault in a deeply dissected unit of pre-Holocene fanglomerate that appears to be one of the oldest such units in the region.

Although little documentation exists for the timing of movement on the left-lateral faults in the eastern Transverse Ranges province, the few data that are available can be combined with some general observations to postulate limiting cases. Data for the Pinto Mountain fault indicate that about 9 km of its displacement has accumulated in about the last 0.7 m.y. (Bacheller, 1978). For at least the accumulation of the latter half of its displacement, then, the Pinto Mountain fault had an average slip rate of about 12 to 13 mm/yr. Similarly, data for the Cleghorn fault indicate that its overall left slip of 3.5 to 4 km has accumulated since about 4 Ma and that about a third of its overall displacement has accumulated in the last 0.5 m.y. (Meisling and Weldon, 1982, 1989). The average slip rate for the Cleghorn fault is roughly 2 to 3 mm/yr (Weldon and others, 1981; Meisling and Weldon, 1982, 1989). If a constant slip rate is assumed, the overall displacements indicate that the age of inception of faulting was no older than about 1.5 Ma for the northern left-lateral faults. Thus, one limiting case is that sinistral faulting in the eastern Transverse Ranges began about 1.5 Ma. This timing is consistent with the idea that Quaternary uplift of the San Bernardino Mountains, beginning between about 2.0 and 1.5 Ma, occurred as a result of coeval movement on the right-lateral San Andreas fault and the left-lateral Pinto Mountain fault (Dibblee, 1968a, 1975a; Farley, 1979, p. 115–120; Matti and others, 1985, p. 12–13; Meisling and Weldon, 1989). Other cases can be postulated,

however, by assuming lesser slip rates prior to about 0.7 Ma or by hypothesizing that movement on the faults was episodic with intervals of activity and intervals of quiescence. The most likely other limiting case is that sinistral faulting began no earlier than the end of late Miocene basaltic volcanism around 6 Ma, although the available data do not seem to rule out the possibility that left-lateral faulting began with the inception of that volcanism as early as 15 Ma.

Kinematic role in the evolving San Andreas fault system. The left-lateral faults of the eastern Transverse Ranges comprise a kinematically related group of faults across which the western Mojave Desert block has moved more than 50 km westward relative to the Chocolate Mountains block. All of the faults in the set terminate to the east at right-lateral faults of the Mojave Desert province and to the west at either right-lateral faults of the San Andreas fault system or within the fractured crystalline terrane of the Little San Bernardino Mountains escarpment, which, on the basis of earthquake focal mechanisms, is characterized by left-oblique extension (Williams and others, 1990). What evidence there are for the timing of movement of the faults indicates that they have interacted coevally with right-lateral faults in the Mojave Desert province and along the evolving San Andreas fault.

Relation to right-lateral faults in the Mojave Desert. The boundary between the domains of left-lateral faulting in the eastern Transverse Ranges province and right-lateral faulting in the Mojave Desert province is marked by wedge-shaped alluvial valleys, including, from northwest to southeast, Dale Lake, Pinto Basin, and Chuckwalla Valley. No crosscutting relation has been observed between right- and left-lateral faults (Allen, 1957; Hope, 1966; Dibblee, 1968b, 1975a, 1980; Powell, 1981, 1982a), and geophysical evidence in Chuckwalla Valley indicates the presence of a thick alluvial wedge bounded in the subsurface by northwest-trending faults to the northeast and by east-trending faults to the south (Rotstein and others, 1976). The pattern, timing, and magnitudes of displacement on the two sets of faults permit the interpretation that they are a conjugate shear system superimposed on the crystalline rocks of the region (cf. Hope, 1966, p. 111; Powell, 1981, p. 339). From paleogeologic patterns in the crystalline rocks of the eastern Transverse Ranges and Mojave Desert (Plates I and II), I infer that the conjugate domains have maintained contact along their mutual boundary, with extensional deformation limited to the wedge-shaped valleys and with little or no lateral displacement along the boundary.

Because both the right- and left-lateral faults exhibit more recent activity to the northwest vs. inactivity to the southeast, in terms both of faulted young alluvium and of seismicity, the conjugate faults may have developed sequentially, as each sinistral fault was in alignment with the southern bend (San Gorgonio Pass) in the San Andreas fault, as the eastern Transverse Ranges migrated southeastward relative to the southern bend (Crowell and Ramirez, 1979; Dibblee, 1982h; cf. Hadley and Kanamori, 1977). An alternate interpretation of the pattern and timing of the left- and right-lateral faults, however, involves a penecontemporaneous inception of the faulting with southeast to northwest younging of the cessation of activity (Powell, 1981).

At present, the conjugate strike-slip faults of the eastern Transverse Ranges and Mojave Desert provinces intersect with acute angles of 40° to 60° bisected by west-northwest azimuths. Assuming a nonrotating stress field and symmetrical development of the left- and right-lateral fault sets, the regional stress field inferred from the conjugate fault system has a principal stress axis oriented N30°E, which bisects the obtuse angle between the shear planes. By the Mohr-Coulomb theory of brittle failure, however, the principal stress axis should bisect the acute angle between conjugate shear planes, which ideally is 60°. By analogy with progressive deformational patterns in clay-cake experiments (see, for example, Cloos, 1955; Badgely, 1965, p. 108–112; Hoeppner and others, 1969), one can infer models for the evolution of the present geometry in which the Mojave Desert and eastern Transverse Ranges faults initiated with an acute angle of 60° bisected by the principal stress axis. If the principal stress axis is assumed to have maintained a constant northeast orientation throughout the evolution of the faulting, then the initial shear planes of the conjugate system had roughly north-south and east-northeast orientations, respectively, then rotated with increasing deformation into their present orientations (Powell, 1981, p. 339). This model is consistent with the magnitude and sense of displacement on the two fault sets and predicts about 30° of clockwise rotation of the left-lateral faults and 30° of counterclockwise rotation of the right-lateral faults (cf. Garfunkel, 1974; Carter and others, 1987). These predicted rotations were apparently supported by the results of some paleomagnetic studies in the Mojave and Sonora Deserts that suggested counterclockwise rotations of about 15° to 30° (Burke and others, 1982; Calderone and Butler, 1984; Valentine and others, 1987, 1988b) and a more recent study in the eastern Transverse Ranges that indicated clockwise rotations of about 41° ± 8° (Carter and others, 1987).

Balanced palinspastic reconstruction of pre–late Cenozoic paleogeologic patterns (Fig. 8, Plate II), however, does not permit restoration of 30° to 40° of clockwise rotation in the eastern Transverse Ranges or of 30° counterclockwise rotation in the Mojave Desert. Rather, the tightest reconstruction of movement along the Pinto Mountain and Blue Cut faults involves clockwise back-rotation that restores about 15° of counterclockwise rotation of the Pinto Mountains block relative to the San Bernardino Mountains and Mojave Desert blocks and about 5° of counterclockwise rotation of the Eagle Mountains and Chuckwalla Mountains–Orocopia Mountains blocks relative to the Pinto Mountains block, or about 20° relative to the San Bernardino Mountains and Mojave Desert blocks. Little or no back-rotation is required for reconstruction of faults in the Mojave Desert, a conclusion that is consistent with the results of the most recent paleomagnetic studies in the Mojave and Sonora Deserts. These studies show little or no rotation for samples from 18- to 19-Ma and younger rocks (Valentine and others, 1988a; Wells and Hillhouse, 1989; MacFadden and others, 1990; Calderone and others, 1990) and clockwise rotation of about 20° to 50° for samples

from 19- to 20-Ma and older rocks (Golombek and Brown, 1988; Ross and others, 1989).

Tectonic models that were first proposed for the Transverse Ranges province on the basis of paleomagnetic data (Luyendyk and others, 1980, 1985, Fig. 7; Luyendyk and Hornafius, 1987) are inconsistent with critical geologic data available on the displacement of left-lateral faults of the eastern Transverse Ranges. The proposed clockwise rotation of about 90° requires displacements of as much as 30 km on individual left-lateral faults in the eastern Transverse Ranges, a result that is unequivocally contradicted by the documented displacements (Hope, 1966, 1969; Dibblee, 1967b, 1975a, 1980, 1982d; Powell, 1975, 1981, 1982a; Bacheller, 1978). Later models (Hornafius and others, 1986; Carter and others, 1987), which incorporate a clockwise rotation of about 40° for the eastern Transverse Ranges, require displacements of as much as 23 km on individual sinistral faults, much closer to the measured displacements but still too large.

Relation to right-lateral faults of the San Andreas system. The evidence for timing of movement on the left-lateral faults of the eastern Transverse Ranges requires that their displacement accumulated coevally with that on the San Andreas fault. In accord with this interpretation, there is no evidence for a crosscutting relation between right-lateral strands of the San Andreas fault zone and any of the left-lateral faults in the eastern Transverse Ranges. In particular, at the intersection of the Pinto Mountain and Mission Creek faults, the two faults mutually deflect one another (Allen, 1957; Matti and others, 1985), indicating that they are coeval and interact with one another. Moreover, restoration of crystalline rock patterns does not permit those models in which the left-lateral faults in the eastern and western Transverse Ranges are displaced along the San Andreas fault (see, for example, Hope, 1966, p. 140–143; Jahns, 1973).

Rather than behave as a rigid block along the northeast wall of the San Andreas fault, the eastern Transverse Ranges block has deformed internally by rotation of subblocks along the left-lateral faults, and in changing shape, the block has been extended along the San Andreas fault (Fig. 16). Rotation of the sinistral fault blocks must be accompanied by lateral displacement along either the San Andreas fault or the boundary with the Mojave Desert, or both. In Plate II, the eastern boundary is held fixed and all the required displacement is accommodated along the San Andreas fault. In effect, points south of the block have been displaced farther along the San Andreas fault than points north of the block, or conversely, points north of the block have lagged behind points to the south in moving along the San Andreas fault. The same kinematic argument applies to each left-lateral fault involved in the distortion of the eastern Transverse Ranges block. Consequently, the Salton trough segment of the San Andreas fault is longer than it would have been if the left-lateral faults had not formed. One might expect that displacement along the Salton trough segment of the San Andreas fault would be greater than along the Mojave Desert segment, but the lengthening of the San Andreas fault was accompanied by a corresponding distortion in the Peninsular Ranges and southernmost San Gabriel Mountains west of the fault. This distortion has occurred principally on the dextral San Jacinto fault, which contributes additional displacement to the Mojave Desert segment of the San Andreas. The process of assembling the reconstruction shown in Figures 8 through 11 and Plate II, A through C graphically balances displacements on the various faults.

The geometry of the rotational fault block model (Freund, 1974; Garfunkel, 1974; Garfunkel and Ron, 1985) yields the following trigonometric identities (Fig. 16):

$$\sin(\alpha - \rho) = \frac{W}{l},$$

$$\sin \alpha = \frac{W}{l_o},$$

and, by the law of sines,

$$\frac{\sin \rho}{D_{LL}} = \frac{\sin(\alpha - \rho)}{l_o}.$$

Equations for displacement on the left-lateral faults of the eastern Transverse Ranges and on the San Andreas fault can be derived from these identities. Displacement on the left-lateral faults is given by

$$D_{LL} = W\left(\frac{\sin \rho}{\sin \alpha \sin(\alpha - \rho)}\right), \qquad (2)$$

where D_{LL} is the overall displacement on the left-lateral faults, W is the overall width of the rotated fault blocks measured perpendicular to fault strike, α is the initial angle between the sinistral faults and the San Andreas fault, and ρ is the angle of rotation of the sinistral fault blocks.

The San Andreas fault serves as the west boundary against which the fault blocks of the eastern Transverse Ranges are rotated. If one requires that the northwest corner of the rotated domain remain in contact with the unrotated San Bernardino Mountains, then the rotation of the eastern Transverse Ranges fault blocks has contributed southeastwardly increasing right-lateral displacement to the San Andreas fault. The overall increase in displacement along the segment of the San Andreas fault that intersects the left-lateral faults is equal to the amount by

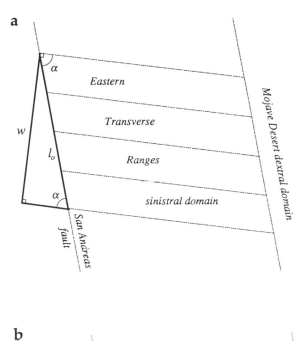

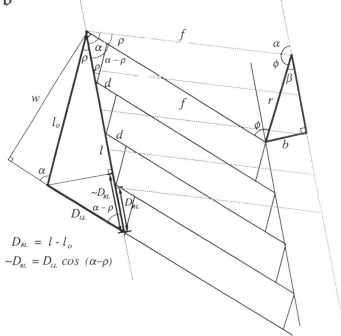

Figure 16. Fault-block rotation model for domain of clockwise rotation and sinistral faults of the eastern Transverse Ranges. For San Andreas system, this figure represents the interaction between the sinistral faults of the eastern Transverse Ranges and the San Andreas fault to the west and the Mojave Desert domain of dextral shear to the east. a, Initial orientation of faults. Symbols: α = initial angle of intersection between sinistral faults and left boundary of domain of clockwise rotation; w = width of domain of sinistral faulting; l_o = initial length of domain of clockwise rotation along its left boundary. b, Orientation of faults after clockwise rotation, superimposed on initial orientation. Symbols: ρ = angle of clockwise rotation; $\alpha - \rho$ = angle of intersection between sinistral faults and left boundary of domain of clockwise rotation after rotation; l = length of domain of clockwise rotation along its left boundary after rotation; D_{LL} = overall displacement on sinistral faults; D_{RL} = dextral displacement along left boundary of rotated domain, where upper left corner of rotated domain is held fixed; $\sim D_{RL}$ = approximation of D_{RL}; b = amount by which the eastern Transverse Ranges block has decreased in width as a consequence of clockwise fault block rotation. D_{RL} corresponds to additional displacement that occurs along the San Andreas fault as a consequence of clockwise rotation and sinistral faulting in the eastern Transverse Ranges; b corresponds with the magnitude of the bend in the San Andreas fault at the north end of the sinistral domain. See text for discussion.

which the San Andreas has been lengthened by the rotation associated with slip between the fault blocks that abut it. This amount is given by

$$D_{SA} = l - l_o = W \left(\frac{1}{\sin (\alpha - \rho)} - \frac{1}{\sin \alpha} \right),$$ (3)

where D_{SA} is the increase in displacement on the San Andreas fault along the segment bounding the rotated domain of the eastern Transverse Ranges block between the Pinto Mountains and Salton Creek faults, l_o is the length of the San Andreas fault bounding the eastern Transverse Ranges block prior to movement on and rotation of the left-lateral faults, and l is the present length of the San Andreas bounding the rotated domain of the eastern Transverse Ranges block. As an upper limit, a useful approximation of the increase in displacement on the San Andreas fault is given by

$$D_{SA} < D_{LL} \cos (\alpha - \rho).$$ (4)

Values measured from the palinspastic reconstruction in Plates I and II, when substituted for parameters in the equations for the rotational fault block model, provide a test of the applicability of that model for the sinistral fault blocks of the eastern Transverse Ranges. In Plate I, the width (W) of the sinistrally faulted domain measured perpendicular to the strike of the faults is about 70 km, the present-day angle of intersection ($\alpha - \rho$) between the Transverse Ranges sinistral faults and the San Andreas fault zone is about 45°, and l measured along the San Andreas fault between the Pinto Mountain and Salton Creek faults is about 100 km. In Plate IIC, l_o is about 85 to 90 km. Progressing from Plate IIC to Plate I, the angle of clockwise rotation (ρ) of the sinistral faults is 15° for the Pinto Mountain block and 20° for the Eagle Mountains and Chuckwalla Mountains–Orocopia Mountains blocks.

When the measured values of $\alpha - \rho = 45°$, $\rho = 20°$, and $W = 70$ km are substituted into equation (2), they yield a displacement (D_{LL}) of 37 km, which corresponds reasonably well with the 40-km sum of the measured displacements on the throughgoing Pinto Mountain, Blue Cut, Chiriaco, and Salton Creek faults. On the other hand, if the paleomagnetically determined clockwise rotation of 41° ± 8° (Carter and others, 1987) is substituted for ρ, equation (2) yields a displacement (D_{LL}) of 65 ± 10 km. This result is much greater than the sum of the measured displacements, even if as much as 10 km is added for the subordinate left-lateral faults that were not restored in Plate II because they do not completely transect the sinistral domain. Of uncertain significance is the paleomagnetic data from 14-Ma basalt flows in the Palen Mountains that indicate 31° of clockwise rotation in the Mojave Desert domain of right-lateral faulting east of the eastern Transverse Ranges (Carter and others, 1987).

The southeastward increase in displacement (D_{SA}) contributed to the San Andreas fault by the sinistral faults can be calculated in three ways. First, when the measured values of $\alpha - \rho = 45°$, $\rho = 20°$, and $W = 70$ km are substituted into equation (3),

they yield a displacement increase of 22 km. Second, when measured values of $l = 100$ km and $l_o = 85$ km are substituted into equation (3), they yield a displacement increase of only 15 km. In assembling the reconstruction, however, l_o could have been made shorter by as much as 5 km not allowing a gap to open along the Morongo Valley fault, which would increase the calculated value for D_{SA} to 20 km. Finally, when measured values of $\alpha - \rho = 45°$ and D_{LL} of 40 km are substituted into equation (4), they yield an upper limit of 28 km for the displacement increase.

The rotational fault block model for the eastern Transverse Ranges sinistral domain has another important consequence for the evolution of the southern San Andreas fault system. The model requires that the width of the eastern Transverse Ranges domain—as measured between the San Andreas fault and the boundary with the Mojave Desert dextral domain—has diminished with increasing fault block rotation (Fig. 16). If contact is maintained to the east along the boundary between the eastern Transverse Ranges and Mojave Desert domains, then the western boundary of the sinistrally rotated domain—i.e., the San Andreas fault—is dragged east. The amount of drag is given by

$$b = r \sin \beta,$$ (5)

where

$$\beta = 180 - \alpha - \phi,$$

$$r = f \frac{\sin \rho}{\sin \phi},$$

$$\phi = \frac{180 - \rho}{2},$$

and f is the length of the throughgoing left-lateral faults of the eastern Transverse Ranges. For $\alpha - \rho = 45°$, $\rho = 20°$, and an average fault length $f = 75$ km for the sinistral faults, equation (5) shows that the San Andreas fault has been pulled east by about 15 km, which is roughly the magnitude of the step in the fault across the southern "big bend" through San Gorgonio Pass.

Chocolate Mountains

Left-lateral faults also have been recognized in the Chocolate Mountains to the south of the eastern Transverse Ranges, including the Black Eagle and Mammoth Wash faults (Dillon, 1976). The Black Eagle and Mammoth Wash faults displace the Chocolate Mountains thrust by 8 and 5 km, respectively (Dillon, 1976).

Tentative correlation of synfaulting fanglomerate in Mammoth Wash with the upper Miocene fanglomerate of Bear Canyon, and of syn- to postfaulting fanglomerate in Mammoth Wash with fanglomerate that to the south overlies lower Pliocene Bouse Formation, led Dillon (1976, p. 330–331) to conclude that displacement along the Mammoth Wash and Black Eagle faults occurred from the late Miocene to the late Pliocene. The timing of movement on these faults is thus older than at least the latest movement on the faults of the eastern Transverse Ranges.

Transverse Ranges west of the San Andreas fault

Faults along southern boundary of Transverse Ranges.
The southern boundary of the Transverse Ranges west of the San Andreas fault has been an intermittently or continuously active zone of left-oblique extensional and left-oblique reverse faulting since the middle Miocene. The Quaternary tectonics of this boundary is characterized by north-south contraction, characterized by reverse faulting that has uplifted the Transverse Ranges (Barbat, 1958; Jahns, 1973; Morton and Yerkes, 1987). West of the Los Angeles River, the zone of faulting now forms the boundary between the Santa Monica Mountains and the Los Angeles basin. East of the river, the zone of deformation not only forms the boundary between the San Gabriel Mountains and the San Gabriel and San Bernardino Valleys, it also encompasses a complex terrane of sinistrally kinked and ductilely faulted and brittlely fractured Proterozoic and Mesozoic igneous and metamorphic rocks along the southern flank of the range. Middle and late Miocene sinistral strike slip has been inferred along the southern boundary of the Transverse Ranges. The displacement on various fault strands within the system is only partly understood, and the overall magnitude of displacement and its disposition across this fault system is difficult to assess.

Previous assessments of path and displacement on modern fault system. Within this zone of faulting, components of left slip occur on three sets of Quaternary reverse faults. The range-bounding set of reverse faults of the southern San Gabriel Mountains, consisting of the Sierra Madre fault (Crook and others, 1987) to the west and the Cucamonga fault (Morton and Matti, 1987) to the east, at least locally exhibits a component of left slip. West-northwest of the San Gabriel Mountains, the Santa Susana and San Cayetano reverse faults lie on trend with these faults (Yeats, 1979, 1983, 1986, 1987; Ehlig, 1986). Uplift on this set of faults is Quaternary in age; its inception is probably not much older than about 3 Ma and may be as young as 1.5 Ma (Eckis, 1928; Treiman and Saul, 1986; Morton and Matti, 1987; Yerkes and Lee, 1987).

The range-bounding fault set is transected by and appears to step across two sets of east-northeast–trending left-oblique reverse faults. One of these sets consists of the Malibu Coast, Santa Monica, Raymond or Raymond Hill, and Clamshell-Sawpit faults (Eldridge and Arnold, 1907; Johnson and Warren, 1927; Miller, 1928; Eckis, 1934; Barbat, 1958; Jahns, 1973; Morton, 1973; Campbell and Yerkes, 1976; Crook and others, 1987). The Malibu Coast–Santa Monica fault forms the boundary between the Santa Monica Mountains and the Los Angeles basin. Along trend to the east-northeast, the Raymond fault traces south of the Verdugo Mountains and converges on the southern flank of the San Gabriel Mountains at Monrovia. In turn, the Clamshell-Sawpit fault lies within the San Gabriel Mountains on trend with the Raymond fault. Although these faults are aligned on an east-northeast trend, the Raymond fault evidently does not connect at the surface with either the Clamshell-Sawpit fault (Eckis, 1934; Morton, 1973; Crook and others, 1987) or the Malibu Coast–

Santa Monica fault (Lamar, 1970). All of these faults are steeply north-dipping Quaternary reverse faults with a few thousand feet of displacement. The Malibu Coast–Santa Monica and Raymond faults are seismically active and have documented components of left slip (Barbat, 1958; Jahns, 1973; Crook and others, 1987; Real, 1987; Jones and others, 1990).

In the southeastern San Gabriel Mountains, another east-northeast–trending set of left-lateral strike-slip faults enters the range front at San Dimas and San Antonio Canyon (Johnson and Warren, 1927; Hsu, 1955, p. 307–308; Morton, 1975b; Morton and others, 1983; May, 1986; Cramer and Harrington, 1987; Morton and Matti, 1987). This set includes the Evey Canyon fault, the San Antonio or San Antonio Canyon fault, and the Stoddard or Stoddard Canyon fault in the San Gabriel Mountains, and the Walnut Creek and San Jose faults south of the range. Earthquakes on the San Jose and Walnut Creek faults have yielded left-lateral focal mechanisms (Cramer and Harrington, 1987; Real, 1987; Hauksson and Jones, 1991).

Evidence for the timing of movement on the San Antonio Canyon fault is conflicting. The fault is seismically active and its sense of movement is interpreted as oblique-slip with components of east-side-up reverse and left slip (Cramer and Harrington, 1987). Moreover, the San Antonio Canyon fault evidently offsets the San Gabriel fault, probably from the Icehouse Canyon fault. On the other hand, May (1986, p. 120) stated that unbroken Quaternary terraces in San Antonio Canyon overlap the San Antonio Canyon fault, and two investigators (Hsu, 1955, p. 310; Ehlig, 1981) observed that dikes of andesite, diabase, olivine basalt, and quartz latite intrude gouge in the San Antonio Canyon fault and are also broken by the fault. These dikes also intrude Miocene monzogranite (K-Ar ages 14 to 19 Ma, Miller and Morton, 1977; U-Pb zircon age 25.6 ± 1 Ma, May and Walker, 1989) and are considered by Ehlig (1981) to be related to the Glendora Volcanics (Shelton, 1955), which interfinger with middle Miocene (Luisian) sandstone and conglomerate (Woodford and others, 1946; Shelton, 1955).

The magnitudes of lateral displacement on the San Antonio Canyon–Evey Canyon and Stoddard Canyon faults appear to be small. The Glendora Volcanoes do not appear to be offset along the westward projection of the San Antonio Canyon and Stoddard Canyon faults, which precludes large post-Luisian displacement. The Icehouse Canyon fault has been interpreted as the eastward extension of the San Gabriel fault, in which case the component of left slip on the San Antonio Canyon fault is about 3 km (Dibblee, 1968a, p. 264–266; Crowell, 1975d; Morton, 1975b, Fig. 2; Powell and Silver, 1979, Figs. 8, 13; Evans, 1982; Cramer and Harrington, 1987). The 5-km magnitude of left separation of the Vincent thrust across the fault is consistent with 3 km of left slip. Similarly, the Miller fault may represent the eastward continuation of the Icehouse Canyon fault offset by about 2 km of left slip along the Stoddard Canyon fault. The total displacement of about 5 km on these two faults, plus the left slip distributed on other east- to northeast-trending faults in the southeastern part of the San Gabriel Mountains (Morton, 1975b;

Morton and others, 1983; Morton and Matti, 1987), corresponds roughly with the post-Mohnian left slip on the Malibu Coast–Santa Monica–Raymond fault (see below). From cross-fault stratigraphic and sedimentologic correlations in the Mohnian Puente and Modelo Formations, Lamar (1961) estimated a minimum of 11 km of post-Mohnian left slip on the Malibu Coast–Santa Monica fault, whereas from restoration of a Luisian fault, Truex (1976) estimated 6.4 km of post-Mohnian left slip. The apparently unbroken Glendora Volcanics, however, indicate that displacement has stepped between the en echelon San Antonio Canyon–San Jose and Malibu Coast–Santa Monica–Raymond fault sets rather than following a fault that links the two.

Alternatively, the Evey Canyon, Icehouse Canyon, and Miller faults have been interpreted as parts of an eastward extension of the left-lateral ancestral Malibu Coast–Santa Monica fault (see next section). In this interpretation, the older fault is displaced sinistrally about 9 km along the San Antonio Canyon fault (Matti and others, 1985; May, 1986, p. 123; Matti and Morton, this volume), which is inconsistent with interpreting the San Gabriel and Icehouse Canyon faults as parts of a once-continuous fault that has been displaced 3 km on the San Antonio Canyon fault.

Previous assessments of path and displacement on ancestral Malibu Coast–Santa Monica fault. The Malibu Coast–Santa Monica fault and the Raymond fault have been inferred to mask an ancestral strike-slip fault along which large displacement occurred during the middle and late Miocene (Barbat, 1958; Jahns, 1973; Campbell and Yerkes, 1976). Estimates of large left-lateral displacement on the Malibu Coast–Santa Monica fault include 15 km (Barbat, 1958, Fig. 2), 21 to 22 km (Lamar, 1961; Dibblee, 1982f), 0 to 40 km (Platt and Stuart, 1974), 60 km (Sage, 1973b; Truex, 1976), and 90 km (Yeats, 1968, 1976; Yeats and others, 1974; Campbell and Yerkes, 1976). The 60- and 90-km estimates are anchored on the offset of a northwest-trending early Miocene (Saucesian?) shoreline in the Vaqueros Formation, remnants of which crop out in the Santa Ana and Santa Monica Mountains. Given that the match requires extrapolation across the intervening Los Angeles basin, the actual offset probably lies somewhere between a minimum of 40 km and a maximum of 90 km (see also Campbell and Yerkes, 1976). The 6.4- to 11-km range of Pliocene or younger displacement indicates that between about 30 and 80 km of slip occurred on the ancestral fault during the late Saucesian, Relizian, and Luisian.

The eastward disposition of 30 to 80 km of displacement on the ancestral Malibu Coast–Santa Monica fault is controversial and not uniquely constrained. Hypothetically, the displacement passes eastward to the San Andreas fault and possibly beyond (Barbat, 1958; Jahns, 1973; Campbell and Yerkes, 1976; Powell, 1986). The disposition of left slip to the east generally has been postulated to follow one of two paths: either east along the range front, then into the San Gabriel Mountains on the Evey Canyon and San Antonio Canyon faults, then east along the Icehouse Canyon and Miller faults (Matti and others, 1985; May and

Walker, 1989), or east along the range front on the Sierra Madre and Cucamonga faults (Jahns, 1973; Campbell and Yerkes, 1976). If it once extended east of the modern trace of the San Andreas fault, the continuation of the ancestral Malibu Coast–Santa Monica-Raymond fault has been offset southeastward to the Chocolate Mountains (Garfunkel, 1973; Jahns, 1973; Woodburne, 1975, p. 42).

Reckoning from reassembly of crystalline rock paleogeologic patterns. The late Cenozoic east- and east-northeast–trending fault sets within the southern San Gabriel Mountains are subparallel to lithologic trends in the Mesozoic and older plutonic and metamorphic rocks, and mimic older syn- and postbatholithic deformation in attitude and sense of displacement. This situation renders it difficult to establish piercing points and to distinguish the relative contributions of pre–late Cenozoic and late Cenozoic faults to observed discontinuities in bedrock patterns. Without convincing piercing points, it is not possible to reconstruct the left-lateral fault system uniquely. Despite the lack of compelling piercing points, however, the known distribution of crystalline rocks and faults (Fig. 11; Plates I, II) nonetheless limits possible reconstructions.

After restoration of about 5 km of left slip on the Quaternary faults of the southeastern San Gabriel Mountains and 11 km of post-Mohnian slip on the Malibu Coast–Santa Monica–Raymond fault, further reassembly of the southern Transverse Ranges–Chocolate Mountains reference domain (Fig. 11) along sinistral faults in the Transverse Ranges is problematic. Belts of the Limerock Canyon assemblage, Mesozoic plutonic rocks, and synplutonic mylonitization can be reconstructed to be consistent with little or no left slip on the eastward extension of the ancestral Malibu Coast–Santa Monica fault along the southern boundary of the Transverse Ranges, or with as much as about 40 km of left slip, depending on the trend of the belts where they are concealed between the Peninsular Ranges and Transverse Ranges.

One distinction between these two end-member possibilities lies in the position in which one reassembles the crystalline rocks of the Banning block vis-a-vis the northern Peninsular Ranges and Cargo Muchacho Mountains. The Banning block contains two assemblages of metamorphic rocks that are caught up within younger gneissic quartz diorite of Jurassic or Cretaceous age. Along the northeastern half of the block are inclusions of both gneissic rocks that are probably part of the Proterozoic Hexie Mountains assemblage, including the augen gneiss of Monument Mountain, and the deformed Mount Lowe intrusion. These rocks are displaced along the San Jacinto fault zone from their lithic equivalents in the southeastern San Gabriel Mountains north of the San Gabriel, Icehouse Canyon, and Miller faults (Matti and others, 1985; Matti and Morton, this volume), and along the San Andreas fault zone from lithic equivalents in the central Chocolate Mountains (Dillon, 1976; Dillon and Ehlig, this volume). To the southwest of these rocks, inclusions of marble in the Banning block (J. C. Matti, 1984, oral communication) are likely to belong to one of three metasedimentary suites: the metasedimentary suite of the Limerock Canyon assemblage that crops out in the southeastern San Gabriel Mountains south of the San Gabriel, Icehouse Canyon, and Miller faults and high in the San Jacinto and Santa Rosa Mountains of the northeasternmost Peninsular Ranges; the easternmost suite in the western intrabatholithic belt of Mesozoic metasedimentary strata low in the San Jacinto Mountains; or the Paleozoic strata that crop out in the Cargo Muchacho and Gila Mountains west of the Salton trough and on Coyote Mountain and possibly in the southern Santa Rosa Mountains west of the Salton trough. In either of the first two cases, little or no left slip is required on any eastward continuation of the ancestral Malibu Coast–Santa Monica–Raymond fault through San Gorgonio Pass; whereas in the third case, several tens of kilometers of left slip seem to be required on the eastward continuation of that fault both in San Gorgonio Pass and east of the San Andreas fault between the Chocolate Mountains to the northeast and the Cargo Muchacho and Gila Mountains to the southwest. If the fault exists in pre-Cenozoic crystalline rocks east of the San Andreas fault, it is concealed by late Oligocene and early Miocene volcanic strata along the southwestern margin of the southern Chocolate Mountains (Haxel and Dillon, 1973; Crowe, 1978), which is inconsistent with the timing of displacement on the Malibu Canyon–Santa Monica fault (see below). If, however, large dextral displacement on the Clemens Well–Fenner–San Francisquito fault was somehow transformed into an extensional regime in and beneath these volcanic rocks, then it is not inconceivable that large sinistral displacement on the eastward projection of the Malibu Coast–Santa Monica–Raymond fault could have been similarly transformed.

Large left slip on the postulated eastward continuation of the ancestral Malibu Coast–Santa Monica–Raymond fault through or along the southeastern San Gabriel Mountains, through San Gorgonio Pass, and along the Chocolate Mountains would explain the apparent geologic differences between the Peninsular and Transverse Ranges (Campbell and Yerkes, 1976; Matti and others, 1985; Powell, 1986; May and Walker, 1989; Matti and Morton, this volume). Much of the profound difference, however, occurs across a pre- or synplutonic Mesozoic structure or structures along which postplutonic late Mesozoic and early Cenozoic movement also has occurred, and does not clearly require additional large displacement during the late Cenozoic. The reconstruction in this chapter incorporates no displacement on the eastern projection of the ancestral Malibu Coast–Santa Monica–Raymond fault because the case for such slip is not convincingly unique.

The westerly to west-northwesterly lithologic and structural trends in the crystalline rocks along the southern flank of the Transverse Ranges suggest a resolution for the seemingly contradictory evidence for the magnitude of displacement. From northwest-trending outcrops north and south of the southern boundary of the Transverse Ranges province, the outcrops of rocks of the Hexie Mountains assemblage and the Limerock Canyon assemblage, as well as various belts of Mesozoic plutonic rocks, are deflected to more westerly trends (Plate I). Rocks north of the kink are offset westward by tens of kilometers from rocks south of the kink, but are connected by west-trending belts of outcrop in the kink rather than displaced along a fault. Similarly, the western depositional(?) limit of Oligocene and lower Miocene volcanic rocks in the northern Chocolate Mountains is several tens of kilometers farther west than the western limit of coeval volcanic rocks in southern Arizona, but these western limits are connected more or less continuously by the southern depositional(?) limit of volcanic rocks in the southern Chocolate Mountains of southeasternmost California and the Laguna and Muggins Mountains of southwesternmost Arizona. Thus, if one postulates that a kink with about a 40-km step formed in the middle Miocene, then that much less displacement is required on the inferred ancestral Malibu Coast–Santa Monica fault and no major sinistral strike-slip fault is required on any eastward projection of that fault. Evidently, rotation of the kink is counterclockwise.

This kink is hypothesized here to have served as a locus for subsequent middle and late Miocene and perhaps early Pliocene left-oblique extensional faulting in the Los Angeles basin that accompanied right slip on the San Gabriel fault system. The left-oblique extension restored in Figure 11 and Plate II accounts for about 35 km of left separation between the Santa Ana and Santa Monica Mountains. Middle and upper Miocene volcanic rocks and conglomerate in the Los Angeles basin, Puente Hills, and vicinity were deposited in basins that may have resulted from the extension. The kink demonstrably served as a locus for late Pliocene and Quaternary left-oblique reverse faulting.

Timing of movement. Movement on the ancestral Malibu Coast–Santa Monica fault began prior to 16 Ma and most of the offset had accumulated by 13 Ma. The youngest feature fully offset by the ancestral Malibu Coast–Santa Monica fault is an early Miocene (Saucesian?) shoreline in the Vaqueros Formation exposed in the Santa Monica and Santa Ana Mountains (Yeats, 1968, 1976; Yeats and others, 1974; Campbell and Yerkes, 1976; Truex, 1976). Middle Miocene basaltic volcanism between about 16 and 13 Ma (Turner, 1970) and deposition of the late Saucesian to early Luisian San Onofre Breccia and related rocks (Yeats and others, 1974; Stuart, 1979) are interpreted to have been coeval with diastrophism during which most of the sinistral offset accumulated (Yeats, 1968, 1976; Yeats and others, 1974; Campbell and Yerkes, 1976; Truex, 1976). The inferred eastward continuation of the ancestral fault is also a locus of middle Miocene volcanism, including the Luisian Glendora Volcanics (Shelton, 1955) and the 13-Ma Black Mountain Basalt in the Cargo Muchacho and southern Chocolate Mountains (Haxel and Dillon, 1973; Dillon, 1976; Haxel, 1977; Crowe, 1978).

There is circumstantial evidence that faulting continued to develop through the middle and late Miocene associated with uplift of the Transverse Ranges. In the Los Angeles basin area, Luisian and Mohnian rocks that overlie older rocks with angular unconformity and the Mohnian strata are offset sinistrally at least 11 km, but much less than the older rocks (Lamar, 1961; Yeats and others, 1974; Truex, 1976). Along the eastward projection of the ancestral fault, coarse fanglomerate derived from the north accumulated on the Banning block (Coachella Fanglomerate) and fanglomerate derived from the west accumulated in the southern Chocolate Mountains (fanglomerate of Bear Canyon).

Kinematic role in the evolving San Andreas fault system. The role of the zone of left-lateral faulting along the southern boundary of the Transverse Ranges in the evolution of the San Andreas fault system is enigmatic and controversial. Menard's (1955, 1960) suggestion that this zone of faulting is the landward continuation of the Murray fracture zone has persisted (Barbat, 1958; Jahns, 1973; Campbell and Yerkes, 1976; Hadley and Kanamori, 1977), despite a lack of evidence for continuity across the continental margin (Von Huene, 1969). In an analogy with their model for the Garlock fault, Davis and Burchfiel (1973) suggested that the Malibu Coast–Santa Monica–Raymond–Sierra Madre fault was an intracontinental transform fault that separates a region of extension to the south from an unextended region to the north. Campbell and Yerkes (1976) proposed that the Malibu Coast–Santa Monica–Raymond fault continues east to the San Andreas fault and that prior to about 5 Ma it functioned as part of a zone of left slip and extension along which right slip on the San Andreas fault in and north of the Transverse Ranges stepped to right slip along the continental margin south of the Transverse Ranges. In their model, the plate boundary prior to about 5 Ma followed the San Andreas fault southeastward from the Mendocino fracture zone to its intersection with the Malibu Coast–Santa Monica–Raymond fault, where it stepped westward along the eastward projection of the Murray fracture zone to the continental margin. With the opening of the Gulf of California after about 5 Ma, the modern San Andreas fault developed as the plate boundary and the southern boundary of the Transverse Ranges developed into a zone of convergence and contraction. Luyendyk and colleagues (Luyendyk and others, 1980, 1985; Hornafius, 1985; Kamerling and Luyendyk, 1985; Hornafius and others, 1986) postulated that left slip on the Malibu Coast–Santa Monica fault accompanied middle and late Miocene clockwise rotation in the western Transverse Ranges.

All of these concepts have merit and are incorporated to various degrees within the role assigned to the zone of sinistral kinking and faulting in this section. I consider the zone of kinking and faulting, accompanied by extension and rotation, to have initiated as a link between the southeastern terminus of the Clemens Well–Fenner–San Francisquito fault and the northwestern terminus of a fault or faults that were active coevally in the southern California continental borderland and connected with the southward migrating triple junction at the southern end of the transform system. One such fault may have been the East Santa Cruz basin fault, which has been postulated to merge northward with the west end of the left-lateral Santa Cruz Island fault and to have accumulated 120 to 180 km of right-lateral displacement between 20 and 13 Ma (Howell and others, 1974, 1976; Howell, 1976), although nowhere near that amount is accommodated in the palinspastic reconstruction shown in Plate

II, A through C. The intersection and interaction between the zone of sinistral deformation and the dextral Clemens Well–Fenner–San Francisquito fault is cryptic, but the two systems appear to mutually terminate one another.

The earliest expression of sinistral deformation was kinking and/or strike-slip faulting that began to develop after about 22 to 20 Ma and probably by about 18 to 17 Ma. During and subsequent to sinistral kinking, during the interval from about 18 or 17 to 13 or 12 Ma, the zone of deformation was characterized by widespread volcanism—including intrusion of diabase dikes and sills in the Santa Monica Mountains (Turner, 1970; Yerkes and Campbell, 1980; Weigand, 1982) and mafic, intermediate, and felsic dikes in the San Gabriel Mountains (Ehlig, 1981; May, 1986)— and by faulting associated with left-oblique(?) extension in the western Transverse Ranges and Chocolate Mountains (Fig. 12c). In the western Transverse Ranges and Channel Islands, this deformation was accompanied by clockwise rotation of sedimentary and volcanic strata (Hornafius and others, 1986; Liddicoat, 1990), possibly as they detached from a substrate composed of the Catalina Schist (Yeats, 1968). Detachment also occurred above the Orocopia Schist in the Chocolate Mountains between 18 and 13 Ma, but rotation has not been documented there.

Subsequently, the zone served as the southern terminus of right-lateral strands of the San Gabriel fault system. It is hypothesized here that left-oblique extensional faulting in the Los Angeles basin and San Gabriel and San Bernardino Valleys continued through at least the late Miocene (about 5 Ma) and perhaps into the early Pliocene, providing a mechanism by which at least half of the right-lateral displacement on the faults of the San Gabriel fault system stepped westward to the continental margin where offshore dextral faults probably connected to the southern triple junction.

By late Pliocene and Quaternary time, extensional faulting was overprinted by left-oblique reverse faulting along the southern margin of the Transverse Ranges. Reverse and sinistral displacement on the eastern half of this fault system is kinematically related to complementary right-lateral displacement on the Elsinore and San Jacinto faults (cf. Morton, 1975; Cramer and Harrington, 1987; Hauksson and Jones, 1991) and a late phase of sinistral kinking involving the San Andreas fault in San Gorgonio Pass and vicinity. In the reconstruction in Plate II, left-lateral displacement on this fault system transforms all the right-lateral displacement on the Elsinore fault and about half that on the San Jacinto fault.

Faults in the western San Gabriel Mountains and Soledad basin. *Paths.* Left-lateral strike-slip faults in the western San Gabriel Mountains and Soledad basin comprise a set of northeast-trending faults that include the Pole Canyon, Magic Mountain, Transmission Line, Mill Creek, and several smaller faults (Oakeshott, 1958; Carter and Silver, 1972; Powell and Silver, 1979; Carter, 1980b [p. 154–165], 1982a) and the Mint Canyon, Green Ranch, and Agua Dulce faults in the Soledad basin (Jahns and Muehlberger, 1954; Muehlberger, 1958; Oakeshott, 1958; Jahns, 1973). The east-northeast–trending San Francisquito fault may have been reactivated as a left-lateral fault of this set (Smith, 1977a). These faults are subparallel to and in some cases partly coincident with older east- to northeast-trending extensional faults, movement on which was syndepositional with the lower Miocene Vasquez Formation (Bailey and Jahns, 1954; Jahns and Muehlberger, 1954; Muehlberger, 1958; Oakeshott, 1958; Jahns, 1973; Bohannon, 1975; Carter, 1980b, 1982a; Hendrix and Ingersoll, 1987).

Estimates of displacement. Sinistral separation on individual faults in the San Gabriel Mountains, as established by offset contacts within the anorthosite-syenite complex and related rocks, ranges from 0.5 to 5.5 km (Oakeshott, 1958; Carter, 1980b, 1982a) and totals about 11 km (Carter, 1980b, p. 154). Sinistral separation on the Soledad basin faults, as established by offset contacts in the lower and middle Miocene strata, ranges from about 0.5 to 3 km each and totals about 6.5 to 7.5 km (Jahns and Muehlberger, 1954; Muehlberger, 1958; Oakeshott, 1958; Jahns, 1973). Smith (1977a) estimated a left-lateral displacement of about 14 to 20 km on part of the San Francisquito fault. That displacement, however, has not been verified on the basis of offset rock units. Because these faults are not throughgoing between the San Gabriel and San Andreas faults, their displacements are not restored in the palinspastic reconstruction in Plate II.

Timing of movement. Faults of the older extensional set in the Soledad basin are coeval with the upper Oligocene and lower Miocene Vasquez Formation and are overlapped by the Arikareean Tick Canyon and Clarendonian Mint Canyon Formations, whereas faults of the younger left-lateral set displace strata of both the Tick Canyon and Mint Canyon Formations (Jahns and Muehlberger, 1954; Muehlberger, 1958; Oakeshott, 1958; Carter, 1980b, 1982a). Thus, movement on the left-lateral faults began after 21 Ma. Offset on most faults of the younger set dies out within the Mint Canyon Formation indicating that movement had ceased by the end of deposition of the Mint Canyon Formation at about 10 Ma, whereas movement on other faults of the set continued after deposition of the Mint Canyon Formation but had ended prior to deposition of the Saugus Formation (Jahns and Muehlberger, 1954; Muehlberger, 1958; Oakeshott, 1958; Carter, 1980b, 1982a).

Kinematic role in the evolving San Andreas fault system. Oakeshott (1958) concluded that the right-lateral San Gabriel fault and the left-lateral faults of the western San Gabriel Mountains and Soledad basin constitute conjugate faults in a regional strain-pattern associated with north-south shortening, implying that the sinistral faults were coeval with the San Gabriel fault. The left-lateral faults can also be interpreted, however, as having been obliquely truncated by the San Gabriel fault (Powell and Silver, 1979). The timing of movement on these left-lateral faults indicates that they were coeval with the Canton–San Gabriel fault and perhaps with the Clemens Well–Fenner–San Francisquito fault, and largely older than the strands of the San Gabriel fault that transect the San Gabriel Mountains. The left-lateral faults are thus too old to have formed as a result of clockwise

rotation of basement blocks between the right-lateral San Gabriel and southern San Andreas faults as depicted by Dibblee (1982h, Fig. 11). Rather, this rotation was coeval with the growth of Canton–San Gabriel fault (Fig. 14d).

On the basis of paleomagnetic data from late Oligocene and early Miocene volcanic flows in the Vasquez Formation of Soledad basin, Terres and Luyendyk (1985) inferred that these sinistral faults accompanied middle Miocene clockwise block rotation of about 50°. In a kinematic and mechanical analysis of these faults, Nur and others (1986, 1989) concluded that an initial set of sinistral faults (now east-trending) was replaced by the more favorably oriented northeast-trending second set as clockwise rotation progressed beyond ~45°.

The slight clockwise rotation of the San Gabriel Mountains indicated on Plate II, however, is consistent with the sense, but not the magnitude, of rotation inferred for the San Gabriel Mountains–Soledad basin block from the paleomagnetic data. Moreover, the small magnitudes of slip on the sinistral faults, the wide spacing between the faults relative to their slip, and the relatively short strike-length of individual faults, as currently mapped, suggest a rotation for the San Gabriel Mountains–Soledad basin block as a whole that is much less than 50°.

Faults west of the San Gabriel fault. Paths. The Transverse Ranges province west of the San Gabriel fault is characterized by several throughgoing east-west faults. In addition to the Malibu Coast–Santa Monica reverse fault and the Santa Susana–San Cayetano thrust discussed in an earlier section, these faults include the Big Pine, Santa Ynez–Santa Ynez River, Oak Ridge, and Simi faults. The Big Pine fault (Hill and Dibblee, 1953) is generally considered to extend westward from the San Andreas fault at Lake of the Woods through Lockwood Valley and along the southern margin of the Cuyama badlands.

From Cobblestone Mountain just west of the San Gabriel fault, the Santa Ynez–Santa Ynez River fault extends westward along the north flank of the Topatopa and Santa Ynez Mountains to Point Conception. At its eastern end, the Santa Ynez fault abuts the Frazier Mountain crystalline block, although it does not crosscut the granitic and metamorphic rocks of that block.

The Oak Ridge and San Cayetano reverse faults, respectively, make up the structural southern and northern margins of the axial part of the Ventura basin synclinal trough (Bailey and Jahns, 1954; Jennings and Troxel, 1954, p. 27–29; Nagle and Parker, 1971; Yerkes and Lee, 1987), and the trace of the Simi reverse fault runs along the northern margin of Simi Valley (Hanson, 1983; Seedorf, 1983; Yeats, 1983).

The Malibu Coast–Santa Monica(-Raymond?) fault discussed above is also part of this group of faults.

Estimates of displacement. A sinistral offset of 13 or 16 km has been measured along the Big Pine fault (Hill and Dibblee, 1953; Carman, 1964; Poynor, 1960; Crowell, 1962; Vedder and Brown, 1968; Dibblee, 1982b). Most of the evidence for lateral displacement lies within Cenozoic sedimentary rocks along the southern margin of the Cuyama badlands and to the west. The most convincing evidence consists of facies relations in a synclinal fold axis in late Eocene strata, all of which are displaced 16 km. Corroborating evidence for this magnitude of displacement is provided by cross-fault ties between the Ozena and San Guillermo faults and the Nacimiento and Pine Mountain faults. To the east, in Lockwood Valley, the chief evidence for large lateral offset on the Big Pine fault is a coarse, monolithologic fanglomerate member of the Plush Ranch Formation north of the fault that is separated from the nearest exposed bedrock source of augen gneiss of the San Gabriel assemblage south of the fault by 13 km (Carman, 1964).

While many investigators have advocated an unspecified but apparently minor amount of left slip on the Santa Ynez, Santa Ynez River, and related faults (Dibblee, 1966b, 1982a, g; Gordon, 1979; Sylvester and Darrow, 1979), other investigators have presented evidence for greater than 15 km (Page and others, 1951), 20 to 35 km (McCracken, 1969, 1971), and 37 km (McCulloh, 1981). The lack of evidence for significant displacement in the crystalline rocks of the Frazier Mountain block indicates that if there is large displacement on the Santa Ynez fault, it diminishes eastward (see also Gordon, 1979; Dibblee, 1982a). The central reach of the Santa Ynez fault exhibits a maximum vertical displacement of at least a few kilometers, but that displacement dies out at each end in folded Miocene strata (Dibblee, 1966c, 1982a, g; Sylvester and Darrow, 1979).

As discussed earlier, sinistral displacement on the Malibu Coast–Santa Monica fault is as little as 40 km (restored in Plate II) or as much as 90 km.

Other faults along which large left-lateral displacement has been hypothesized include the San Cayetano, Oak Ridge, nad Simi faults between the Santa Monica and Santa Ynez Mountains (Yeats, 1968, Hornafius and others, 1986). (Truex [1976] postulated about 60 km of *right* slip on the Simi fault.) However, demonstrated slip on the Oak Ridge and San Cayetano faults is reverse (Bailey and Jahns, 1954; Jennings and Troxel, 1954, p. 27–29; Nagle and Parker, 1971; Morton and Yerkes, 1987; Yerkes and others, 1987), and the hypothesized ancestral left slip on either of these faults, although neither proved nor disproved, seems unlikely. Demonstrated slip on the Simi fault is also reverse, and detailed subsurface cross-fault correlations in Eocene and Oligocene strata preclude large lateral slip (Hanson, 1983; Seedorf, 1983).

Reckoning from reassembly of crystalline rock paleogeologic patterns. The evidence for magnitude and timing of displacement on and the tectonic role of the Big Pine fault is equivocal. Some of the evidence is consistent with but insufficient to prove the case for a throughgoing left-lateral fault, whereas other evidence appears to contradict such an interpretation. From the evidence available, I suggest that the so-called Big Pine fault is a composite of aligned features. The Big Pine fault along the southern margin of the Cuyama badlands and west of the badlands probably has a left-lateral component of slip (as much as 16 km) conjugate to the right-lateral Rinconada and Red Hills–

Ozena faults, and a reverse component of slip continuous with the San Guillermo fault. I infer that the Cuyama badlands segment of the Big Pine fault, as a reverse fault connecting the San Guillermo and Ozena reverse faults, was instrumental in absorbing contractional shortening that accompanied rotation of this area associated with the northern "big bend" in the San Andreas fault. Undoing such shortening in the region is necessary to arrive at simultaneous restoration of all three reference domains along the San Andreas fault system (Plate II, A, B). From the lack of evidence for left slip in members of the Caliente Formation and in the Proterozoic rocks of Mount Pinos and Frazier Mountain, I further infer that the faulting in Lockwood Valley is not part of a throughgoing left-lateral Big Pine fault. Some of the faulting in Lockwood Valley is extensional and coeval with deposition of the Plush Ranch Formation and some of it is Pliocene(?) and Pleistocene in age (Carman, 1964; Crowell, 1968). The younger faulting, which may have reactivated the older extensional faults, may also be left-lateral, but overall displacement is minor. The older set of faults in Lockwood Valley is coeval with the faults in Soledad basin and the Blue Rock fault in the Cuyama badlands that are syndepositional with the Vasquez and Simmler Formations, respectively.

Displacement on the Santa Ynez fault is constrained at its eastern end by the observation that the fault, where it enters the Frazier Mountain crystalline block, does not significantly laterally displace the granitic and metamorphic rocks of that block. In making the reconstruction shown in Plate II, A, B, left-oblique reverse slip is removed to back-rotate the rocks along the Santa Ynez fault counterclockwise about a pivot point at the eastern terminus of the fault. Such a model is consistent with the evidence for westward-increasing, but small, lateral displacement on the Santa Ynez fault. The sinistral component of slip restored in Plate II is 5 to 10 km.

Similarly, in order to make the reconstruction shown in Plate II, A, B, left-oblique reverse movement is restored on the east-trending Oak Ridge and San Cayetano reverse faults by back-rotating the rocks along the faults counterclockwise about pivot points at the east end of the Ventura basin. For these faults, there simply is no evidence for large left-lateral displacement. Fault plane solutions for earthquakes in the vicinity of these faults show predominantly reverse slip with only minor components of left-lateral slip (Jackson and Molnar, 1990).

Timing of movement. The Big Pine fault shows evidence for two intervals of movement (Carman, 1964, p. 52; Poynor, 1960; Crowell, 1968; Jahns, 1973; Bohannon, 1975). The earlier interval was coeval with deposition of megabreccia in the upper Oligocene and lower Miocene Plush Ranch Formation, whereas the later interval is Pliocene(?) and Pleistocene in age. Movement on the Santa Ynez fault is post–middle Miocene in age (Gordon, 1979) and at least in part Quaternary (Dibblee, 1966c, 1982a, g). The Oak Ridge and San Cayetano faults are Pliocene and Pleistocene in age (Jennings and Troxel, 1954, p. 27–29; Nagle and

Parker, 1971; Yerkes and Lee, 1987; Yerkes and others, 1987). The Simi fault is pre-middle Miocene and Pleistocene in age (Truex, 1976; Hanson, 1983; Seedorf, 1983; Yeats, 1983).

Kinematic role in the evolving San Andreas fault system. In Plate II, the back-rotation accommodated graphically in reconstructing the crystalline rocks and Upper Cretaceous and Paleogene strata (Plate II, A, B) is about 50°. This back-rotation resulted from restoring a sinistral displacement of 35 to 40 km on the Malibu Coast–Santa Monica–Raymond fault, westward-increasing left-oblique reverse slip on the Santa Ynez and San Cayetano–Oak Ridge faults, and dextral displacement on the San Gabriel fault. A component of 5 to 10 km of sinistral slip was restored on the Santa Ynez fault. Rotation of the western Transverse Ranges blocks is limited by reconstruction of paleogeologic patterns in the plutonic and metamorphic rocks in the Transverse and Peninsular Ranges, by keeping the east ends of the blocks attached to the crystalline rocks of the Frazier Mountain–Alamo Mountain block and the San Gabriel Mountains, and by requiring realignment of the Saucesian shoreline in the Santa Monica and Santa Ana Mountains.

The rotating western Transverse Ranges blocks, pivoted from the east end, helps accommodate the large fraction of San Gabriel fault displacement that has not been accounted for to the east. The Malibu Coast–Santa Monica–Raymond fault acquired some of its lateral displacement coevally with movement on the Canton–San Gabriel fault between about 14 or 13 and 10 or 9 Ma, but also in part during movement on the Vasquez Creek–San Gabriel fault between about 6 and 4 Ma. Paleomagnetic studies (Hornafius and others, 1986) and the reconstruction in Plate II both indicate that most, but not all, of the clockwise rotation in the Transverse Ranges west of the San Gabriel fault occurred between about 15 and 10 Ma.

In fault block rotation models based primarily on paleomagnetic evidence for rotation, large left slip has been suggested for the Big Pine, Santa Ynez, San Cayetano, Oak Ridge, Simi, and Malibu Coast–Santa Monica faults to accommodate 75° to 95° of clockwise rotation indicated for middle Miocene (ca. 16 to 10 Ma) and older volcanic and sedimentary strata in the Transverse Ranges west of the San Gabriel fault (Luyendyk and others, 1980, 1985; Hornafius and others, 1986; Liddicoat, 1990). However, the lack of geologic evidence for large left slip on the San Cayetano and Oak Ridge faults and the evidence against left slip on the Simi fault contradict modeling these faults as left-lateral faults. Moreover, these block rotation models require slip on the San Jacinto and Elsinore faults beginning about 16 Ma, and require large slip on the Elsinore fault. Existing evidence, however, shows that these dextral faults have developed since 5 Ma and that the Elsinore fault has an overall displacement of not more than about 10 km (see section on Peninsular Ranges dextral faults).

The above arguments lead to the conclusion that about half of the rotation indicated by the paleomagnetic data is related to structures other than sinistral faults of the western Transverse Ranges. In the western Transverse Ranges, clockwise rotation is plausible in the upper plates of detachment faults or within a dextral orocline.

Garlock fault

Path. The Garlock fault zone trends northeast from the San Andreas fault to the vicinity of Koehn Lake, from where it curves eastward to the Avawatz Mountains (for discussions of nomenclature and the history of study of the Garlock fault zone, see Smith, 1962, and Davis and Burchfiel, 1973). The fault zone separates the Sierra Nevada and Basin and Range provinces from the Mojave Desert.

Displacement. Large overall left-lateral displacement along the central and eastern reaches of the Garlock fault east of about Koehn Lake is well documented. Measurements based on restoring offset lithologic and structural features include 64 km to realign the Jurassic Independence dike swarm (Smith, 1962), 74 km to realign an inferred northwest-trending boundary in structural style (Michael, 1966), 48 to 64 km to realign exposures of a section of Paleozoic eugeoclinal metasedimentary and metavolcanic rocks (Smith and Ketner, 1970), 65 km to realign upper Proterozoic and lower Paleozoic miogeoclinal strata (R. H. Jahns, B. W. Troxel, and L. A. Wright, 1968, 1971, and oral communication cited in Davis and Burchfiel, 1973), and 56 to 64 km to realign a moderately west-dipping intracrystalline thrust fault (Davis and Burchfiel, 1973). The most convincing of these estimates lie in the range of 48 to 64 km.

Displacements measured by offset late Miocene through Holocene features along the central reach of the Garlock are less than the overall displacement by magnitudes commensurate with the various ages of the features. Specifically, the Mesquite Schist of former usage (abandoned by Carr and others, 1984), which crops out as a point source in the central El Paso Mountains north of the Garlock fault, has shed clasts southward across the fault into Pleistocene terrace and alluvial fan deposits now located up to 19 km east of the source, into conglomerate interlayered with the Pliocene Almond Mountain Volcanics about 32 km east of the source, and into the upper Miocene Bedrock Spring Formation about 47 km east of the source (Carter, 1980a, 1987).

Displacement along the western reach of the Garlock fault is not as well documented as that along the central and eastern reaches. Large overall displacement along the western reach of the Garlock west of about Koehn Lake has been suggested, but is less convincingly demonstrated than displacement along the central and eastern reaches. Displacement in the range of 48 to 100 km has been measured by realigning exposures of the Pelona and Rand Schists (Dibblee, 1967a [p. 115], 1980; Davis and Burchfiel, 1973; Ross, 1989, p. 90), although the published evidence does not prove that the bodies of schist along the Garlock fault were originally juxtaposed. Ross (1989, p. 89–91) concluded that the distribution of igneous and metamorphic rocks in the southern Sierra Nevada, San Emigdio, and Tehachapi Mountains is consistent with a sinistral displacement of about 50 km at the west end of the Garlock fault, but he admitted that the paleogeologic reconstruction on which his measurement is based is reasonable but not compelling.

Geologic mapping at the east terminus of the Garlock fault also has revealed a lack of evidence for large lateral displacement (Troxel and others, 1979; Brady and others, 1980; Brady, 1982).

Reckoning. Reassembly of the Gabilan Range–San Emigdio Mountains–Portal Ridge reference domain (Fig. 10) juxtaposes rocks of the northern Gabilan Range against rocks of the San Emigdio Mountains north of the Garlock fault and simultaneously juxtaposes rocks of the southern Gabilan Range against rocks of the Tehachapi Mountains and Portal Ridge in the northwesternmost Mojave Desert south of the Garlock fault. The continuity of the Gabilan Range massif with crystalline blocks both north and south of the Garlock fault at its western terminus limits displacement on the western part of the Garlock fault to no more than about 12 km. Although Ross (1989, p. 89–91) from his paleogeologic reconstruction inferred a displacement of about 50 km along the western part of the Garlock fault, a displacement in the range of 8 to 12 km provides an equally reasonable restoration of igneous and metamorphic geologic units and contacts shown in Figure 10 and Plates I and II. Moreover, restoration of 8 to 12 km does not disrupt the present-day proximity of Tertiary terrestrial volcanic and sedimentary rocks north and south of the Garlock fault in the vicinity of Tehachapi and Mojave (Fig. 10).

The difference between the generally accepted displacement of 48 to 64 km and 12 km is not easily explained. Assuming that the 12-km displacement is correct, then either displacement on the Garlock fault diminishes westward toward the San Andreas fault (Troxel and others, 1972; LaViolette and others, 1980) or overall displacement is considerably less (by 36 to 52 km) than generally thought.

In the latter case, one might postulate ad hoc that the difference can be accounted for by a sinistral oroclinal kink analogous to, but narrower and probably younger than, the kink inferred along the southern boundary of the Transverse Ranges province. Although Proterozoic strata and Mesozoic metavolcanic and plutonic rocks at the east end of the Garlock fault (see Jennings, 1977; Brady and others, 1980; Brady, 1982) seem to be deflected sinistrally by as much as about 50 km, the mapped distribution of the offset features along the central reach of the fault does not appear to permit this interpretation. There is, however, paleomagnetic evidence for as much as 15° to 20° counterclockwise rotation of terrestrial strata of the Ricardo Group of Loomis and Burbank (1988) between about 10 and 7 Ma (Burbank and Whistler, 1987; Loomis and Burbank, 1988).

In the former case, the reconstruction in Plate II requires that overall slip on the Garlock fault decreases westward from a maximum of at least 48 km along the central and eastern reaches to not more than 12 km at the west end of the fault. The most convincing paleogeologic features used to constrain displacement on the central reach of the Garlock fault are the Independence dike swarm, the intracrystalline thrust fault, and the Paleozoic eugeoclinal section, all of which are reasonably consistent with an overall displacement of as little as 48 km. Depending on how one projects the dike swarm from its exposure in the Argus Range southward across Searles Valley to the Garlock fault, the offset of the swarm could be as much as the 64 km measured by Smith (1962, Fig. 3) or as little as about 48 km. Furthermore, north of the Garlock fault the trend of the dike swarm as emplaced is characterized by lateral steps (Smith, 1962, Fig. 1); if the Garlock fault happened to coincide with a primary sinistral step in the dike swarm, then displacement could be even less than 48 km. As noted by Davis and Burchfiel (1973), lateral offset of the intracrystalline thrust fault, which dips west at 30° to 40°, may be less than 56 km if the dip slip on the Garlock fault has lifted the north wall relative to the south wall. Westward diminution from at least 48 to 12 km of left slip on the Garlock fault would require extension in the westernmost Mojave Desert to accommodate the difference in displacement.

Timing of movement. The Garlock fault has been continuously or intermittently active during the Quaternary, Pliocene, and perhaps the late Miocene (Dibblee, 1967a, p. 112–115, 1980; Clark, 1973; Carter, 1980a, 1987; Clark and LaJoie, 1974; LaViolette and others, 1980). By matching distinctive bedrock clasts in gravel units south of the fault with point sources north of the fault, Carter (1980a) concluded that at least 18 to 19 km of displacement occurred during the Quaternary, that about 32 km has occurred since extrusion of the Pliocene Almond Mountain Volcanics, and that 47 km has occurred since deposition of the Bedrock Spring Formation. This formation contains a mammalian fauna formerly considered middle or possibly late early Pliocene in age (G. E. Lewis in Smith, 1964, p. 42), which on current time scales translates to a late Miocene Hemphillian or possibly late Clarendonian age.

The time of inception of strike slip and the overall average slip rate on the Garlock fault are controversial. Several investigators have suggested that the Garlock fault or an ancestral fault zone near the modern Garlock fault was probably active between the Paleocene and early Miocene (Hewett, 1954a, b; Nilsen and Clarke, 1975; Dibblee, 1967a, Figs. 74–75; Sharry, 1982; Cox, 1982; Cox and Diggles, 1986). It also has been suggested that early movement on the fault was linked to the inception of extension in the Basin and Range province (Davis and Burchfiel, 1973; Wright, 1976) thought to be coeval with basaltic volcanism between about 17 and 14 Ma (McKee, 1971; Christiansen and McKee, 1978). By extrapolating a late Miocene through Quaternary average slip rate of 7 mm/yr farther back into time, Carter (1987) calculated that at a constant slip rate full displacement of 64 km on the central and eastern reaches of the Garlock fault would have accumulated since about 9 Ma; alternatively, by using the Holocene average slip rate of 7 mm/yr (Carter, 1971, 1987; Clark and Lajoie, 1974; Astiz and Allen, 1983) and by inferring a faster slip rate of 11 to 12 mm/yr during the Pliocene and Pleistocene, Carter (1980, 1982b) calculated that displacement on the Garlock fault could have accumulated since about 5.5 Ma. Burbank and Whistler (1987) and Loomis and Burbank (1988) inferred that sinistral slip on the fault commenced with the start of counterclockwise rotation and a sharp increase in the apparent rate of sediment accumulation in middle Miocene strata of their Ricardo Group at 10 or 9 Ma. If overall displacement on the Garlock fault is as little as 48 km, the lower limit of the displacement range generally ascribed to it, then the 47-km offset of the Bedrock Spring Formation from its source is equivalent to the overall displacement. In this case, overall displacement would have accumulated since about 8 to 10 Ma if the Bedrock Spring Formation is late Clarendonian or early Hemphillian in age, or since about 4 to 6 Ma if the formation is late Hemphillian in age. Calculated overall average slip rates are 5 to 6 mm/yr and 8 to 12 mm/yr, respectively, for these two possibilities.

Slip rates measured and inferred on the Garlock fault vary along the fault, perhaps supporting the hypothesis that overall slip decreases from a maximum of 48 to 64 km along the central and eastern reaches of the fault to as little as 12 km at the west end. The Holocene slip rate is well established at 7 to 8 mm/yr for the central reach of the fault (Carter, 1971, 1987; Clark and Lajoie, 1974), and Carter (1980, 1982b, 1987) inferred a slip rate of 7 to 8 or 11 to 12 mm/yr for the Pleistocene, but slip rates are less well known for the eastern and western reaches of the fault. On the western segment, geodetic measurements indicate a modern sinistral strain rate in the range of 1 to 8 mm/yr (Keller and others, 1978; Rodgers, 1979), whereas geomorphic evidence indicates a Holocene slip rate in the range of 1.6 to 3.3 mm/yr (Rodgers, 1979; LaViolette and others, 1980). If these rates persisted throughout the evolution of the Garlock fault, then the central reach would have accumulated its displacement in 6 m.y., during which time the western reach would have accumulated as little as 6 to 18 km, which is consistent with the reconstruction in Plate II.

Kinematic role in the evolving San Andreas fault system. Some early investigators interpreted the left-lateral Garlock fault and the right-lateral San Andreas fault as coevally interacting conjugate faults associated with a north-south shortening and east-west extension (Hill and Dibblee, 1953; Hill, 1954). Subsequent investigators suggested that the Garlock fault is a secondary fault that developed as a consequence of movement on the San Andreas fault (McKinstry, 1953; Moody and Hill, 1956; Chinnery, 1966) or in conjunction with the bend in the San Andreas fault near its intersection with the Garlock (Stuart, 1991). Other investigators (Eaton, 1932, 1939; Hamilton and Myers, 1966; Scholz and others, 1971; Davis and Burchfiel, 1973; Wernicke and others, 1982; Weldon and Humphreys, 1986), on the other hand, interpreted the Garlock fault as a boundary structure along which east-west extension in the Basin and Range province was transformed into strike-slip displacement at the southern limit of the province, and concluded that although the Garlock fault deflected the San Andreas fault, the two were unrelated structures. Still other geologists interpreted the Garlock fault as a northern boundary structure against which

right-lateral faulting in the Mojave Desert terminated (Garfunkel, 1974; Cummings, 1976).

In the latter case, some of the difference in displacement between the central and western reaches of the Garlock may have been accommodated by right-lateral displacement on the Mojave Desert faults (Garfunkel, 1974; Garfunkel and Ron, 1985), but probably not much. The sum of observed overall displacements on the Mojave Desert faults northwest of the Pinto Mountains probably lies in the range of 26 to 40 km (Dokka, 1983; Dokka and Travis, 1990a) and is unlikely to be greater than about 65 km (Powell, 1981, p. 334–335). Dokka and Travis (1990a), moreover, concluded that cumulative displacement on the right-lateral faults north of Barstow is not more than 10 to 12 km, that the Mojave Desert faults do not reach the Garlock fault at the surface, and that displacements on the Mojave Desert faults and the Garlock fault are not related kinematically.

In its interaction with the San Andreas fault, the Garlock fault does not displace the San Andreas fault and separates the Mojave Desert and Sierra Nevada blocks, both of which exhibit little or no paleomagnetic evidence for rotation during the fault's existence. It seems likely that the Garlock fault developed its curving trend and the San Andreas fault developed a sinistral deflection in order to maintain contact among the unrotating blocks north and south of the Garlock fault and the block west of the San Andreas fault.

At its eastern end, the Garlock fault has interacted with the coeval right-lateral Death Valley fault zone and either the two conjugate faults mutually terminate one another (Jahns and Wright, 1960; Brady and others, 1980; Brady and Troxel, 1981; Brady, 1982), or mutually deflect one another allowing a buried projection of the Garlock fault to extend eastward beyond the intersection (Davis and Burchfiel, 1973) and the Death Valley fault zone to extend southward into the eastern Mojave Desert (Noble, 1926; Grose, 1959; Hewett, 1954a, Plate 1; Brady and Dokka, 1989).

In Plate II, strike-slip displacement on the Garlock fault is modeled as having occurred between about 5 Ma and the present.

SYNTHESIS

Disposition of displacement along the San Andreas fault

Palinspastic reconstruction of pre–middle Miocene paleogeologic patterns in southern and central California yields measurements of overall displacements on late Cenozoic strike-slip faults in California and in some instances constrains their sequence. Evidence for timing and sequence of movement is provided by upper Cenozoic rocks that are incrementally offset along various faults, or, alternatively, cover the faults. In combination, these two bodies of evidence reveal the progressive late Cenozoic crustal deformation that accompanied the evolution of the San Andreas fault system. The San Andreas fault proper is only one manifestation of this regional late Cenozoic crustal strain. Throughout much of its evolution, the San Andreas fault has interacted with other right- and left-lateral faults of the San Andreas system. As a consequence, displacement varies along the San Andreas fault, and it is necessary to reconcile measurements of displacement on right- and left-lateral faults of the San Andreas fault system in order to understand its evolution.

The disposition of displacement on various faults, which is balanced graphically in the reconstruction (Figs. 8 through 11, Plate II), can also be balanced analytically. The displacements used in the following analysis are derived from the reconstruction in Plate II, the assembly of which required making specific choices from within ranges of possble displacement. Despite the uncertainties discussed above for each fault, limits of certainty are not included in the analysis because it is offered to illustrate that a reasonable balance can be achieved using the displacements restored in the reconstruction. The absence of explicit uncertainties, however, is not intended to imply that the displacements are precise or accurate to the nearest kilometer. This exercise is a first cut at achieving an overall balance; its results will need to be revised and refined as we learn more about fault displacements.

Reconciliation of displacements measured on the post-5 Ma San Andreas fault. Reconciliation of seemingly different measurements of displacement on the various segments of the modern San Andreas fault zone is achieved by accounting for deformation in blocks adjacent to the fault that have not behaved rigidly. This deformation includes sinistral and dextral faulting both east and west of the San Andreas fault proper.

Reconstruction of the Frazier Mountain–Mount Pinos–eastern Orocopia Mountains reference domain is accomplished by restoring displacements sequentially on the Salton trough and Mojave Desert segments of the San Andreas fault zone, the San Gabriel fault, and the Clemens Well–Fenner–San Francisquito fault. As measured in Figures 8 through 11 and Plates I and II, this reassembly restores about 185 km on the various strands of the San Andreas fault zone, including the Punchbowl fault, 42 km on the San Gabriel fault, and 110 km on the Clemens Well–Fenner–San Francisquito fault, for a total of 337 km, which is comparable to the measurement of 336 km made by comparing Figure 8 to the present-day geologic map (Plate I). (An alternate reconstruction could have been assembled by aligning paleogeologic features in Figure 8 to indicate an overall measurement of 326 km and displacement on the Clemens Well–Fenner–San Francisquito fault of 100 km.) The 185-km displacement sums all strands of the San Andreas fault zone. This sum includes a component of at least 16 km contributed by coeval clockwise block rotation of the sinistral domain of the eastern Transverse Ranges, as calculated using equation (3) for a domain width of about 50 km between the Chiriaco and Pinto Mountain faults. It also includes a component of 13 km contributed by the San Jacinto fault as it merges northward with the San Andreas fault zone, as calculated using equation (1). By removing the components contributed by the eastern Transverse Ranges and San Jacinto faults, one can calculate a base displacement of 185 km – 16 – 13 = 156 km that would have accumulated on the San Andreas fault proper in southern California if the crustal blocks adjacent to it

has not been distorted. This displacement is not measurable, however, because of the contributions of the coeval right- and left-lateral faults.

Reconstruction of the La Panza Range–Liebre Mountain–San Bernardino Mountains reference domain is accomplished by restoring displacements of 159 km on the Mojave Desert segment of the San Andreas fault zone to reassemble the Liebre Mountains–San Bernardino Mountains part of the domain (see also Matti and Morton, this volume), and 0 km on the Red Hills–Ozena fault, 32 km on the San Gabriel and related faults, calculated using equation (1), and 110 km on the Clemens Well–Fenner–San Francisquito fault to reassemble the La Panza Range–Liebre Mountain part. The latter two faults merge with the San Andreas fault at Tejon Pass, and total displacement is 301 km, which compares with the 295 km measured by comparing Figure 9 to the present-day geologic map (Plate I). The 159-km displacement on the Mojave Desert segment of the San Andreas fault zone includes a component of 13 km contributed by the San Jacinto fault, but does not include any significant contribution from the left-lateral faults in the eastern Transverse Ranges. By removing the component contributed by the San Jacinto fault, one can calculate a nonmeasurable base displacement on the Mojave Desert–Coast Ranges segment of the San Andreas fault of 159 km – 13 = 146 km.

Reconstruction of similar geologic domains in the Whitewater River area of the Banning block and the southern Chocolate Mountains is accomplished by restoring a displacement of 162 km on the Salton trough segment of the San Andreas fault zone (Fig. 11, Plates I and II; see also Dillon, 1976; Dillon and Ehlig, this volume). The 162-km displacement includes a component of about 17 km contributed by coeval clockwise block rotation of the sinistral domain of the eastern Transverse Ranges, as calculated using equation (3) for a domain width of about 55 km between the Salton Creek and a line midway between the Blue Cut and Pinto Mountain faults. It does not include contributions from the right-lateral Punchbowl or San Jacinto faults. This displacement is the overall displacement on the San Andreas fault south of its intersection with the zone of left-lateral deformation along the southern boundary of the Transverse Ranges province west of the San Andreas. When the component contributed by the left-lateral faults of the eastern Transverse Ranges is removed, the calculated base displacement on the San Andreas fault is 145 km.

Reconstruction of the Gabilan Range–San Emigdio Mountains–Portal Ridge reference domain (Fig. 10, Plate II) is accomplished by restoring a displacement of 295 km on the San Andreas fault zone. The 295-km measurement includes components of 13 km contributed by the San Jacinto fault, 0 km contributed by the Red Hills–Ozena fault, 32 km contributed by the San Gabriel fault, and 110 km contributed by the Clemens Well–Fenner–San Francisquito fault as they merge northward with the San Andreas fault zone. Subtracting the combined displacement contributed by the Clemens Well–Fenner–San Francisquito and San Gabriel faults (110 km + 32 = 142 km) yields a post–5-Ma displacement of 153 km on the Coast Ranges segment of the San Andreas fault. By also subtracting the contribution of the San Jacinto fault, one can calculate a base displacement of 140 km on the Coast Ranges segment of the San Andreas fault. Alternately, if only 100 km is contributed by the Clemens Well–Fenner–San Francisquito fault, post–5-Ma displacement would be increased to 163 km, yielding a base displacement of 150 km on the Coast Ranges segment. The base displacements calculated for the various segments of the post–5-Ma San Andreas fault indicate that the fault slip for the San Andreas fault system can be balanced spatially and temporally. The calculated base displacement for the San Andreas fault proper as a whole in southern and central California is 150 ± 10 km since 4.5 to 5 Ma. These figures yield a long-term average slip rate in the range of 28 to 35 mm/yr, which is equal to or greater than Holocene slip rates measured along the San Andreas fault proper (Keller and others, 1982; Sieh and Jahns, 1984; Weldon and Sieh, 1985; Sieh, 1986; Weldon, 1986; Harden and Matti, 1989; Perkins and others, 1989; Prentice, 1989; Prentice and others, 1991).

Reconstruction of displacement measured on the pre–5-Ma San Andreas fault. The San Andreas fault has yielded very different measurements of overall displacement north and south of its junction with the San Gabriel fault. North of that junction, its displacement is about 300 km; south of the junction, its displacement is about 160 km. The difference of 140 km equals the sum of displacement components contributed by the Clemens Well–Fenner–San Francisquito fault (110 km) and the San Gabriel fault (~30 km).

Another large difference is evident between the total displacement on the San Andreas and related faults (including the Clemens Well–Fenner–San Francisquito, San Gabriel, San Jacinto, and Elsinore faults) north and south of the zone of sinistral deformation that extends along the southern boundary of the Transverse Ranges province and the southwestern flank of the Chocolate Mountains. North of that zone, the San Andreas, San Gabriel, and Clemens Well–Fenner–San Francisquito faults have an overall displacement of 336 km, including displacements contributed by the San Jacinto fault and by clockwise block rotation of the sinistral domain of the eastern Transverse Ranges. South of the zone, the San Andreas, San Jacinto, and Elsinore faults have a total displacement of 162 km + 28 + 5 = 195 km. (The 162-km displacement on the San Andreas fault includes displacement contributed by clockwise block rotation of the sinistral domain of the eastern Transverse Ranges.) The difference of 141 km in total displacement nearly equals the sum of displacements on the Clemens Well–Fenner–San Francisquito fault (100 or 110 km) and the San Gabriel fault system (42 km). This accounting requires that displacement on the Clemens Well–Fenner–San Francisquito fault and most of the displacement on the San Gabriel fault be disposed of along paths that do not rejoin the San Andreas fault proper to the south (see previous discussion of kinematic roles of the Clemens Well–Fenner–San Francisquito and San Gabriel faults). In the reconstruction in Plate II, A, B, the 42-km dextral displacement of the San Gabriel fault system is absorbed by clockwise fault block rotation on sinistral faults west of the San Gabriel fault and by left-oblique extension in the Los Angeles basin and San Gabriel–San Bernardino Valley. The southeastward disposition of displacement on the Clemens Well–Fenner–San Francisquito fault is not portrayed satisfactorily in Plate II, and ultimately needs to be resolved within the context of one of the scenarios shown in Figure 12.

Disposition of displacement on the Punchbowl fault. One can, for the purpose of analysis, take what now appears to be a false reconstructive step and realign the Fenner–San Francisquito fault directly with the Clemens Well fault by restoring a displacement of 209 km on the San Andreas fault (Powell, 1981). It is a false step because restoration of 42 to 44 km of post–5- to 6-Ma displacement on the Punchbowl fault and 159 km on the Mojave Desert segment of the San Andreas fault does not allow one to reassemble both the Frazier Mountain–Mount Pinos–eastern Orocopia Mountains reference domain and the Triassic porphyritic monzonite bodies of Liebre Mountain and the Mill Creek area of the San Bernardino Mountains. If the Fenner and Clemens Well faults are realigned en route to reassembling the Frazier Mountain–Mount Pinos–eastern Orocopia Mountains reference domain by restoring 44 km on the Punchbowl fault in combination with about 160 km on the San Andreas fault, then the Liebre Mountain body of porphyritic monzonite overshoots the Mill Creek body by about 25 to 30 km.

If one takes this false step, the apparent 209-km measurement on the Salton trough and Mojave Desert segments can be only partially reconciled with the 145- to 146-km calculated base displacement on the Mojave Desert and Salton trough segments by subtracting the contributions of the San Jacinto fault and the left-lateral Pinto Mountain, Blue Cut, and Chiriaco faults of the eastern Transverse Ranges from the larger measurement (209 km – 13 – 16 = 180 km). The excess displacement is 34 to 35 km.

For the purposes of analysis, at least three distinct kinematic scenarios can be invoked to balance the displacements: (1) overall displacement on the Punchbowl fault is less than 44 km by the amount of the 34- to 35-km discrepancy, or 9 to 10 km; (2) 44 km of displacement on the Punchbowl fault is distributed to the north of the Sierra Pelona between the San Andreas fault (9 to 10 km) to the northeast of the Liebre Mountain block and other faults (34 to 35 km) to the south and west of the Liebre Mountain block; or (3) the 159-km measurement of displacement of the Liebre Mountain–San Bernardino Mountains reference domain on the San Andreas fault system includes the 44-km displacement on the Punchbowl fault, thus requiring that there is only about 115 km of displacement on other strands of the San Andreas system including the San Andreas and San Jacinto faults. Each scenario has kinematic consequences that permit an assessment of its validity.

In the first scenario, one can consider the possibility that displacement on the Punchbowl fault is less than 44 km by the amount of the discrepancy, or about 10 km. If displacement on the Punchbowl fault were only about 10 km, then restoration of displacement on the San Andreas and Punchbowl faults would satisfy the requirement that it reconstruct the Liebre Mountain–San Bernardino reference domain. This scenario would not, however, achieve a realignment of the Fenner and San Francisquito faults, requiring that a segment of the San Andreas fault between the San Francisquito fault and Valyermo was also a segment of the Clemens Well–Fenner–San Francisquito fault. This configuration, which is analogous to that by which the San Andreas fault northwest of about Tejon Pass is also part of the Clemens Well–Fenner–San Francisquito fault, requires that the 34- to 35-km-displacement subtracted from the Punchbowl fault be added to displacement on the Clemens Well–Fenner–San Francisquito fault in order to reassemble the Frazier Mountain–Mount Pinos–eastern Orocopia Mountains reference domain. Although this scenario is easy to conceive, it is difficult or impossible to work into a balanced reconstruction. Such an attempt opens a 35-km gap between the rocks north and south of the San Francisquito fault.

In the second scenario, one can consider the possibility that the Punchbowl fault has a displacement of 44 km that does not entirely pass along the San Andreas fault northwest of its intersection with the San Francisquito fault, but rather is split between 9 or 10 km on the San Andreas fault and 34 or 35 km on other faults to the south and west of the Liebre Mountain block. In this scenario, the base displacement calculated from the Liebre Mountain–San Bernardino Mountains measurement (159 km – 13 = 146 km) and from the false measurement based on reconnecting the Fenner fault directly to the Clemens Well fault along the San Andreas fault proper (209 km – 13 – 16 – 35 = 145 km) seemingly are reconciled. If the excess 35-km displacement is shunted west and north along the San Francisquito fault, however, it requires that part of the Punchbowl fault displacement to be older than about 13 Ma, which is older than the time of movement generally recognized for the Punchbowl fault. Moreover, if one postulates that the Punchbowl fault has a two-stage history (Woodburne, 1975; Weldon and others, this volume) and that the earlier stage is coeval with the Clemens Well–Fenner–San Francisquito fault, the resulting reconstruction overshoots the Gabilan Range–San Emigdio Mountains–Portal Ridge reference domain by about 35 km.

Alternately, one can postulate that the Punchbowl fault was active coevally with the San Gabriel fault system, and that 34 or 35 km of dextral displacement stepped left from the Punchbowl fault to the San Gabriel fault system (Fig. 14d). By this interpretation, which is independently proposed by Weldon and others (this volume), the Punchbowl fault would have had a multi-stage movement history. The model requires that, analogous with the San Gabriel fault system, the Punchbowl fault had late Barstovian to Clarendonian (ca. 15 or 13 to 10 Ma) movement that was absorbed in part by the coeval east-northeast–trending sinistral faults in the San Gabriel Mountains–Soledad basin block between the two dextral faults (cf. Dibblee, 1982h). The slight clockwise rotation (8°) shown in Figure 7, and the observations that the sinistral faults do not completely transect the San Gabriel Mountains block between the San Gabriel and Punchbowl faults

and that the sinistral faults are widely spaced relative to their displacements, indicate that the amount of dextral displacement transformed is probably less than 5 km. During the next phase of the model, hypothetical Hemphillian (ca. 10 to 6 Ma) movement on the Punchbowl fault might have been distributed onto the Clearwater and Liebre faults, providing a link to carry as much as 12 km displacement northward from the Punchbowl fault to the San Gabriel fault. Analogously, the San Gabriel fault, in extending southeastward through the southern San Gabriel Mountains, might have provided a link southward between the two faults. Pliocene and younger movement on the Punchbowl fault—corresponding with the 13-km offset of Blancan(?) strata (Barrows and others, 1985)—passed northwestward along the San Andreas fault. The overall displacement on the left- and right-lateral faults that developed between the Punchbowl and San Gabriel faults in the first two stages leaves much of the left-stepping 35-km displacement unaccounted for at the surface. Moreover, the process of drafting a reconstruction again leads to an artificial gap between the rocks north and south of the San Francisquito fault. Although one might invoke ad hoc large clockwise back-rotation to close the gap, such a mechanism severely disrupts other parts of the reconstruction in southern California (Plate II).

In the third scenario, incorporated in Plate II, restoration of the Clemens Well and Fenner faults is modified. Rather than linking the east end of the Fenner fault to the northwest end of the Clemens Well fault (Powell, 1981), the Fenner fault is restored along the San Andreas fault to a position north of the Clemens Well fault and the gap is filled by a fault-segment buried beneath the Indio Hills (Matti and Morton, this volume), although this scenario is ad hoc in that there is no direct evidence for the proposed fault segment.

The disposition of Punchbowl displacement to the southeast is constrained by timing and kinematics: it does not follow the Fenner–Clemens Well fault, which is older than the Punchbowl fault, nor does it follow the San Jacinto fault or the youngest fault strands of the San Andreas system in San Gorgonio Pass, including the San Bernardino and Mill Creek strands, all of which are younger than the Punchbowl fault. It is unlikely that displacement on the Punchbowl fault follows the Banning fault because the crystalline rock patterns in the Banning block and the southeastern San Gabriel Mountains are adequately realigned by restoration of displacement on the San Jacinto and San Gabriel–Banning faults. Any additional displacement contributed by the Punchbowl fault leads to a mismatch. By this process of elimination, displacement on the Punchbowl fault heads toward San Gorgonio Pass where it must follow one of the older strands of the San Andreas system, including the Wilson Creek and Mission Creek faults (Matti and Morton, this volume).

Evolving San Andreas fault system

The palinspastic reconstruction in Plate II specifies the sequence of faulting and the interactions among various faults of the San Andreas system. The development of faulting required by the reconstruction provides new insight into the evolution of the San Andreas fault system as an intracontinental transform system.

Dextral shear accompanied by right-oblique extension probably occurred in southern California prior to movement on any throughgoing strike-slip fault (Powell, 1986; see also Weldon and others, this volume). The Pelona and Orocopia Schists have been uplifted in a train of right-stepping en echelon antiforms that prefigure part of the later course of the Clemens Well–Fenner–San Francisquito fault (Powell, 1981, p. 365, 369–372, Plates VI, VII). The antiforms, which developed along a tectonic belt that is delineated by en echelon(?) sedimentary and volcanic basins of Oligocene and early Miocene age, began growing in the late Oligocene or early Miocene, but also were active in the middle, and, perhaps, late Miocene. This tectonic belt shows evidence for dextral oroclinal bending of crystalline rocks units (Plate IIC) that occurred prior to movement on the Clemens Well–Fenner–San Francisquito fault and for pre–12 Ma clockwise rotation of late Oligocene and early Miocene volcanic rocks (Luyendyk and others, 1985; Terres and Luyendyk, 1985) that occurred prior to and/or during movement on the Clemens Well–Fenner–San Francisquito fault. The basins—including those in which the Simmler, Plush Ranch, Vasquez, Tick Canyon, and Diligencia Formations and the volcanic rocks of southeasternmost California accumulated—were bounded by normal faults and characterized by syntectonic deposition of coarse sedimentary breccia in terrestrial sections. The Tick Canyon and Diligencia Formations have yielded late Arikareean vertebrate fossils. Most of the volcanic rocks have yielded K-Ar ages that range from 26 or 25 to 23 or 22 Ma, although some ages from southeasternmost California fall in the range of 32 to 26 Ma and some in the range of 22 to 18 or 17 Ma. Thus, early dextral shear accompanied by extensional tectonism was distributed in a northwest-trending belt in southern California between 26 and 22 Ma and probably also between 32 and 26 Ma. This belt of dextral shear included the future path of the throughgoing, right-lateral Clemens Well–Fenner–San Francisquito fault.

The interval between 22 and 17 Ma was transitional between the early stage of distributed dextral shear and later stages characterized by the discrete, throughgoing strike-slip faults of the San Andreas system. The Clemens Well–Fenner–San Francisquito fault formed either during or at the end of this interval.

The Clemens Well–Fenner–San Francisquito fault, as a throughgoing strike-slip fault was the earliest manifestation of the San Andreas fault system in southern California. Prior to the formation of the San Gabriel fault, the Clemens Well–Fenner–San Francisquito–San Andreas fault accumulated a displacement of 100 or 110 km during some or all of the interval between about 22 and 13 Ma (Plate IIC to IIB). Either the Clemens Well–Fenner–San Francisquito fault formed at 22 Ma and moved at a relatively slow rate (~10 to 15 mm/yr) until 13 Ma, or it formed at 17 or 18 Ma and moved at a relatively fast rate (~20 to 30 mm/yr). The latter interpretation is favored by the evidence that the Diligencia Formation, which contains an Arikareean (-Hemingfordian?) vertebrate fauna and volcanic rocks as young

as about 20 Ma, is fully displaced along the Clemens Well–Fenner–San Francisquito, San Gabriel, and San Andreas faults, and by Stanley's (1987) evidence that Saucesian strata in the Coast Ranges are offset by the full displacement of the San Andreas fault.

The Clemens Well–Fenner–San Francisquito fault was the active strand of the San Andreas fault system in southern California during the late early and early middle Miocene, and it formed a continuous structure with the early San Andreas fault zone northwest of Tejon Pass. This fault probably connected to the northward-migrating Mendocino triple junction at the north end of the San Andreas transform system. The Clemens Well–Fenner–San Francisquito fault is crosscut by the San Andreas fault proper in the Transverse Ranges, and diverges southeastward from that fault east of the Salton trough. The available evidence indicates that, unlike displacement on the younger San Andreas fault proper in southern California, displacement on the Clemens Well–Fenner–San Francisquito fault was not associated with the right-stepping system of transforms and spreading axes that, beginning around 4 or 5 Ma, accompanied the rifting of the Salton trough and the Gulf of California. The timing of movement on the Clemens Well–Fenner–San Francisquito fault is older than that of the seafloor spreading and most of the extensional tectonics in the gulf, and the southwestern end of the strike-slip fault does not merge into either the Salton trough or the gulf. Extensional faulting that occurred after 15 and before 13 Ma on Isla Tiburon (Neuhaus and others, 1988) and perhaps elsewhere in the southern part of the gulf (Henry, 1989) may have been coeval with movement on the Clemens Well–Fenner–San Francisquito fault, but no structural connection has been established.

Displacement on the Clemens Well–Fenner–San Francisquito fault is inferred to have stepped westward to the continental margin, where a dextral fault is postulated to have connected with the southward-migrating triple junction at the south end of the San Andreas transform system (cf. Campbell and Yerkes, 1976). The East Santa Cruz basin fault, which is inferred to have accumulated a right-lateral displacement of 180 km in the southern California continental borderland during the interval between 20 and 13 Ma (Howell and others, 1974; Howell, 1976), may be that fault. Movement on the Clemens Well–Fenner–San Francisquito fault probably accompanied both extension in southeasternmost California and southwesternmost Arizona, and sinistral kinking and faulting along the southern margin of the Transverse Ranges margin (Fig. 12c).

Although extensional tectonics began in the late Oligocene or early Miocene, prior to movement on the Clemens Well–Fenner–San Francisquito–San Andreas fault, as indicated both by syndepositional normal faulting and by widespread volcanism, detachment faulting continued through the early Miocene in both southeasternmost California and southwestern Arizona. Detachment led to truncation of the lower Miocene formations and eventually to exhumation of the Pelona and Orocopia Schists in the middle Miocene. The detachment faults are overlapped by middle Miocene units such as the Caliente and Mint Canyon

Formations in the Transverse Ranges and the fanglomerate of Bear Canyon of Dillon (1976) in the Chocolate Mountains. This timing is coeval with slip on the Clemens Well–Fenner–San Francisquito fault, and, hence, consistent with the suggestion that the detachment and strike-slip faulting are somehow related.

Between 22 to 20 and 18 or 17 Ma, a zone of sinistral deformation began to develop along the southern boundary of the incipient Transverse Ranges province and in southeasternmost California. During the interval from 17 to 13 Ma, left-lateral kinking and faulting were accompanied by extension and widespread volcanism in the western Transverse Ranges and vicinity (Yeats, 1968; Weigand, 1982) and in southeasternmost California and southwesternmost Arizona (Crowe and others, 1979; Shafiqullah and others, 1980; Frost and others, 1982; Murray, 1982; Tosdal and Sherrod, 1985; Sherrod and Tosdal, 1991). During the middle Miocene, this zone of deformation accumulated 50 to 80 km of sinistral displacement west of the present-day Los Angeles River and about 40 km east of the river. As much as about 40 km of this displacement may have been accommodated by kinking between 22 to 20 and 13 Ma and the rest by later faulting. This zone of sinistral deformation is postulated here to have formed between the southeastern terminus of the Clemens Well–Fenner–San Francisquito–San Andreas fault and the northwestern terminus of the dextral East Santa Cruz basin fault offshore in the continental borderland. Evidence for timing of deformation permits one to infer either that the left- and right-lateral faults and the zone of sinistral deformation developed synchronously or that the right-lateral faults developed slightly earlier.

In the Transverse Ranges, the San Gabriel fault system, including from oldest to youngest, the Canton, San Gabriel, and Vasquez Creek faults began to develop as a splay off the Clemens Well–Fenner–San Francisquito–San Andreas fault perhaps as early as 13 Ma (Plate IIB to IIA). The San Gabriel fault system has accumulated a displacement of 42 to 45 km since then, moving at an overall average rate of about 5 mm/yr and deforming the older fault in the process. Between about 13 and 10 Ma, a displacement of as much as 15 km accumulated on the Canton fault and its now-buried extension southward between the Santa Monica and Verdugo Mountains. Subsequently, between 10 and 5 Ma, the San Gabriel fault propagated southeastward through the southern San Gabriel Mountains, accumulating a displacement of 22 to 23 km, followed by 5 km on the Vasquez Creek fault. The San Gabriel fault system merged northwestward with the early San Andreas segment of the Clemens Well–Fenner–San Francisquito–early San Andreas fault, while movement ceased to the southeast on the Clemens Well–Fenner–San Francisquito segment of the fault.

To the south, at least about half of the displacement on the faults of the San Gabriel system is hypothesized to have been absorbed by clockwise fault block rotation in the sinistral domain of the Transverse Ranges west of the San Gabriel system and by left-oblique extension along the zone of sinistral deformation between the Transverse and Peninsular Ranges. The magnitude of the left-lateral faulting component of the middle Miocene deformation is difficult to assess, but if the hypothesized 40 km of kinking is real, then a displacement of only 10 to 40 km is required during the middle Miocene on the hypothetical ancestral Malibu Coast–Santa Monica sinistral fault, and no displacement is required on any eastward projection of that fault. The remaining 10 to 40 km of displacement (35 km in Plate II) inferred for the middle Miocene Malibu Coast–Santa Monica fault is a reasonable magnitude to be associated with a clockwise rotation of 50° in the sinistral domain of the western Transverse Ranges west of the San Gabriel system. This displacement, together with left-oblique extension associated with opening of the Ventura and Los Angeles basin, is enough to absorb a displacement of 20 km on the Canton and Vasquez Creek faults of the San Gabriel system. Presumably, right slip on the San Gabriel fault system stepped west via the domain of sinistral deformation to a dextral fault on the continental margin that connected to the southward-migrating triple junction at the south end of the San Andreas transform system.

The displacement of 22 to 23 km that follows the San Gabriel–Icehouse Canyon–Miller–Banning fault path of the San Gabriel system can be disposed of in two ways. Either it is transformed by left-oblique extension along the zone of sinistral deformation between the Transverse and Peninsular Ranges (Fig. 14b, Plate II), or it is linked with early rifting indicated by the deposition of upper Miocene marine strata at the north end of the "proto-Gulf" in the Salton trough–Gulf of California region (Fig. 14c). In the latter case, the San Gabriel–Icehouse Canyon–Miller–Banning fault is the first strike-slip fault to link the Mendocino triple junction with extension in the Gulf of California region that began as early as 15 to 13 Ma (Neuhaus and others, 1988; Henry, 1989), but most of which has occurred since about 12 or 13 Ma (Lonsdale, 1989; Stock and Hodges, 1989; Humphreys and Weldon, 1991).

In the Salinian block, the San Gregorio–Hosgri and Rinconada–Reliz faults developed coevally with the San Gabriel fault and also merged northwestward with the early San Andreas fault (=Clemens Well–Fenner–San Francisquito fault) (Plate IIB to IIA), but unlike the San Gabriel fault have continued to accumulate displacement since the end of the Miocene. To the southeast, displacement on the Salinian block faults terminates against the western Transverse Ranges and presumably are absorbed by sinitral deformation there. South of their junction, the San Gregorio–Hosgri and Rinconada–Reliz faults have displacements of 105 and 45 km, respectively. The combined displacement on the San Gregorio–Hosgri and Rinconada–Reliz faults north of Monterey Bay is about 150 km (Clark and others, 1984; Ross, 1984), which increases the overall displacement on the San Andreas fault proper to nearly 450 km between San Francisco and Point Arena. The San Gregorio–Hosgri and Rinconada–Reliz faults are largely responsible for extending the Salinian block along the San Andreas fault (Johnson and Normark, 1974).

The modern San Andreas fault developed at about 5 Ma (Plate IIA to Plate I). In southern California, the modern San

Andreas fault diverged eastward from the San Gabriel fault and crosscut the Clemens Well–Fenner–San Francisquito fault to merge southeastward with the Salton trough at the north end of the Gulf of California. The Punchbowl fault primarily served as an early fault strand in the anastomosing San Andreas fault zone along its Mojave Desert segment (Plate IIA to Plate I) (Dibblee, 1967a, 1968a, 1975b; Morton, 1975a, b; Barrows and others, 1985; Morton and Matti, this volume). In conjunction with the system of spreading axes and transform faults in the Gulf of California, the modern San Andreas fault provided a new transform link between the Mendocino triple junction to the north and the southern triple junction that had migrated south of the mouth of the gulf.

Despite its role as a transform fault, the long-term average slip rate on the post–5-Ma San Andreas fault proper is only about 20 to 35 mm/yr (Sieh and Jahns, 1984; Weldon and Sieh, 1985; Harden and Matti, 1989; Perkins and others, 1989; Prentice, 1989; Prentice and others, 1991; Weldon and others, this volume), which is roughly ½ to ¾ of the rate of plate motion between Baja California and the North American plate, most recently calculated at about 48 mm/yr (DeMets and others, 1987, 1990; Kroger and others, 1987; Ward, 1990). The deficit in slip rate and the resulting shortfall in overall displacement on the San Andreas fault are compensated by the growth of an auxiliary system of right- and left-lateral strike-slip faults since the inception of the modern San Andreas (see also Bird and Rosenstock, 1984; Weldon and Humphreys, 1986).

In southern California, faults that developed coevally with the San Andreas include the right-lateral San Jacinto and Elsinore faults west of the San Andreas fault in the Peninsular Ranges, right-lateral faults east of the San Andreas in the Mojave Desert and the Death Valley area, left-lateral faults in the Transverse Ranges both east and west of the San Andreas, and the left-lateral Garlock fault (Plate IIA to Plate I). The existence of these faults indicates that crustal blocks adjoining the San Andreas fault have not behaved rigidly, and the faults serve both to increase the magnitude of displacement measured along the San Andreas fault and to provide paths for displacement in addition to that measured on the San Andreas. The left-lateral faults east of the San Andreas fault have interacted both with conjugate right-lateral faults in the Mojave Desert and Death Valley area and with the San Andreas fault. Clockwise fault block rotation that accompanied left-lateral faulting in the eastern Transverse Ranges has contributed a component of displacement to the San Andreas fault and opened triangular holes along the San Andreas that have filled with Pliocene and Pleistocene strata, including those in the Mecca and Indio Hills. Interaction between the San Andreas and both the Pinto Mountain and Garlock faults has resulted in sinistral deflections in the trend of the San Andreas near San Gorgonio and Tejon Passes, respectively, where these deflections have been accompanied by contractional deformation. The San Jacinto and Elsinore faults interacted with the zone of sinistral deformation along the southern boundary of the Transverse Ranges; the San Jacinto fault also has interacted with

the San Andreas. All of the displacement on the Elsinore fault and about the first half of that on the San Jacinto fault were absorbed by compensating left slip or left-oblique slip of as much as 10 km along the zone of sinistral deformation during the Pliocene and Quaternary. The remaining half of the displacement on the San Jacinto fault has been transferred to the San Andreas fault and contractional deformation in the southeastern San Gabriel Mountains. Since about 1.5 to 3 Ma, about 10 km left-oblique contraction has occurred along the southern flank of the San Gabriel Mountains and as much as a few tens of kilometers of contraction has occurred in the Transverse Ranges west of the San Gabriel fault.

In central and northern California, the dextral Calaveras-Hayward fault zone and its northward continuation developed coevally with the modern San Andreas, and movement continued coevally on the San Gregorio–Hosgri fault (see, for example, Brown, 1990). The reconstruction in Plate II requires that the Salinian block and the Coast Ranges segment of the San Andreas fault north of the San Emigdio Mountains rotated clockwise by as much as 20° as the San Gabriel and Salinian block faults developed, about half of which has occurred since the inception of the modern San Andreas. This rotation has been accompanied by contractional deformation in the Coast Ranges east of the San Andreas fault, which by the reconstruction should increase northward to tens of kilometers of shortening.

The coherence of the palinspastic reconstruction (Plate II) provides a perspective for considering regional thrust and detachment fault systems that are known or have been hypothesized in the vicinity of the San Andreas fault system. Some such fault systems are older than the San Andreas system and are better understood in the pre-San Andreas paleogeologic setting of Plate IIC. These thrust and detachment systems include well documented late Mesozoic and early and middle Cenozoic faults associated with underthrusting, uplift, and exhumation of the rocks of the metamorphic infrastructure (Pelona, Orocopia, and Rand Schists, Catalina Schist, Franciscan Complex) and with movement on the Eastern Peninsular Ranges mylonite zone and correlative structures in the San Gabriel Mountains and Alamo Mountain block. These faults are older than the San Andreas fault system and reconstruction of the San Andreas system restores their paleogeologic setting.

Except in the area from the Chocolate Mountains southeast into Arizona, the restored paleogeologic patterns in the reconstruction in Plate II yield little or no evidence or need for large (more than ~10 to 15 km) disruption of the surface distribution of Proterozoic and Mesozoic crystalline rocks on any late Cenozoic regional thrust or detachment fault system that may have evolved coevally with the San Andreas system. The paleogeologic coherence of the reconstruction does not, however, preclude the existence of a subhorizontal decollement that does not break the surface. Many investigators have postulated the existence of a low-angle decollement either at the base of the crust or near the base of the seismogenic zone, above which brittle deformation has produced the strike-slip faults of the San Andreas system and below which ductile deformation has occurred, with a structural discontinuity between the two domains (Lachenbruch and Sass, 1973, 1980; Hadley and Kanamori, 1977, 1978; Powell, 1981, p. 328–332; Yeats, 1981a; Webb and Kanamori, 1985; Namson and Davis, 1988; Eaton and Rymer, 1990).

Balanced palinspastic restoration of slip along the faults of the San Andreas system is a necessary step toward improving our understanding of the regional paleogeologic framework and of the events that led to the evolution of the San Andreas system as an intracontinental transform.

ACKNOWLEDGMENTS

This chapter is a culmination of 15 years mapping in and around the Transverse Ranges of southern California, both in pursuit of a Ph.D. degree at the California Institute of Technology and as a geologist with the U.S. Geological Survey. In developing the ideas presented here, I have benefited particularly from lengthy discussions with L. T. Silver, J. C. Matti, R. J. Weldon, and D. M. Morton, and generally from more casual interactions with many others. R. O. Castle, P. Stone, and R. J. Weldon provided thorough and thoughtful technical reviews; their comments and questions led to substantial improvements to the content and organization of this chapter. Carol Ann Miller prepared the plates for publication.

APPENDIX 1

SELECTED PRE-LATE CENOZOIC ROCK UNITS AND ASSEMBLAGES

The paleogeologic patterns restored in Plate II depend largely on an unpublished regional synthesis of Proterozoic through Paleogene rock units and assemblages in southern California. For more detailed lithologic descriptions of the units, refer to Plate I.

EARLY AND MIDDLE PROTEROZOIC ROCKS

Proterozoic rocks in the igneous and metamorphic complex of the Transverse Ranges province and vicinity (Plate I) constitute a cratonic outlier caught up in the Mesozoic magmatic arc along the Pacific margin of North America. Early and Middle Proterozoic plutonic and metaplutonic rocks comprise units of amphibolite and granitic orthogneiss, at least four units of porphyritic granite, typically deformed to augen gneiss, and an intrusive complex of anorthosite, syenite, and related rocks. Early and Middle(?) Proterozoic metasedimentary rocks comprise two high-grade suites. These Proterozoic rock units and suites are grouped into three assemblages.

Eagle Mountains assemblage

In the Pinto, Eagle, and Chuckwalla Mountains, Proterozoic granitic augen gneiss is overlain nonconformably by a section of metamorphosed Proterozoic strata. The granitic augen gneiss, which crops out extensively in the Chuckwalla, central Eagle, and south-central Pinto Mountains, is referred to here as the augen gneiss of Joshua Tree after its exposures along the eastern boundary of Joshua Tree National Monument in the Eagle and Pinto Mountains. The augen gneiss was deformed from a protolith of porphyritic granite, some of which crops out in the central Pinto Mountains.

The augen gneiss of Joshua Tree is capped by a metamorphosed

paleosol that is overlain by quartzite, pelitic schist, pelitic granofels, and dolomite. These rocks were metamorphosed along an andalusite-sillimanite P-T trajectory, reaching peak conditions of about P = 4 kbar and T = 600°C. This metasedimentary section, which is well exposed on Pinto Mountain and vicinity in the Pinto Mountains and in the northeastern Eagle Mountains between Placer Canyon and the town of Eagle Mountain south to Big Wash, is referred to here as the metasedimentary suite of Placer Canyon. Identical strata crop out in the northeasternmost Mojave Desert about 15 km north of Baker, providing a tentative stratigraphic link between Proterozoic rocks of the Transverse Ranges cratonic outlier and the North American craton.

Hexie Mountains assemblage

Metasedimentary and metaigneous rocks of this assemblage are widespread in the Transverse Ranges and vicinity. East of the San Andreas fault, the assemblage crops out extensively in the Pinto, Hexie, Cottonwood, Eagle, Orocopia, Chuckwalla, and Little Chuckawalla Mountains, and also on Pilot Knob and in the southern Chocolate Mountains of southeasternmost California and in the Laguna Mountains of southwesternmost Arizona. West of the San Andreas fault, the assemblage is exposed in the Alamo Mountain–Frazier Mountain–Mount Pinos block, in the San Gabriel Mountains, and in Soledad basin (Plate I). The assemblage consists of metasedimentary rocks, amphibolite, granitic orthogneiss, and augen gneiss. These units commonly are intermingled as ductilely deformed migmatite, within which intrusive and structural contacts are overprinted by intense folding and penetrative foliation that have pervasively deformed all the units. Sequencing relations are, however, evident in less deformed domains.

The oldest unit in the assemblage is the metasedimentary suite of Pinkham Canyon, which consists of amphibolite-grade pelitic gneiss, pelitic granofels, quartzose granofels, and minor vitreous quartzite. These rocks were regionally metamorphosed along an andalusite-sillimanite facies-series P-T trajectory. Comparison of the pelitic assemblages with published results of experimental studies suggests peak conditions of P_t = 3.5 to 4 kbar, T = 525° to 625°C (Powell, 1981, 1982a). The Pinkham Canyon suite was metamorphosed prior to and/or during the intrusive events described in the following paragraph. Although the metasedimentary suite of Pinkham Canyon was metamorphosed from a largely different stratigraphic sequence than that of the Placer Canyon suite, the predominant lithosomes in each suite are present in minor amounts in the other suite, suggesting a stratigraphic link between the Eagle and Hexie Mountains assemblages. Moreover, a single set of metamorphic isograds has been mapped across the metasedimentary rocks of both assemblages, and the two assemblages both were pervasively deformed prior to being intruded by bodies of the Proterozoic anorthosite-syenite complex.

The metasedimentary rocks were successively intruded by the protoliths of foliated metamorphosed hornblendic gabbro and diorite and amphibolite, mesocratic laminated biotite-quartz-feldspar granodioritic to monzogranitic orthogneiss, and various units of leucocratic granitic orthogneiss that are commonly garnetiferous. During and/or after the intrusive episodes, rocks of the assemblage underwent west-vergent, locally isoclinal folding (Powell, 1981, 1982a; Postlethwaite, 1988) that resulted in interlayering of the various rock types. These rocks constitute much of the Mendenhall Gneiss of Oakeshott (1958) in the western San Gabriel Mountains, the Pinto Gneiss of Miller (1938) in the central Pinto Mountains, and the Augustine gneiss unit of Powell (1981), although in all of these areas, the named units also contain domains of the metasedimentary suite of Pinkham Canyon and the augen gneiss of Monument Mountain. In addition, the Mendenhall Gneiss and the Augustine gneiss unit contain retrograded granulite gneiss associated with intrusion of the anorthosite-syenite complex.

The augen gneiss of Monument Mountain is a distinctive, dark colored deformed porphyritic granodiorite-monzogranite. Dikes and inclusions clearly indicate that the porphyritic granodiorite intruded the metasedimentary suite of the Pinkham Canyon, the amphibolite, and most of the granitic orthogneiss units. During the late stages of or after its intrusion, the porphyritic granite and its host rocks were affected by a metamorphic event during which abundant garnets grew both in inclusions in the porphyritic granite and in the granite that surrounds the inclusions. In the Hexie Mountains, these relations are evident in both the augen gneiss and its porphyritic granodiorite protolith. The porphyritic granodiorite and its host rocks were also intruded by dikes of garnetiferous biotite-quartz-feldspar monzogranitic rocks and leucocratic quartz-feldspar granitic rocks. Deformation of the augen gneiss of Monument Mountain occurred during and/or after the garnet-producing metamorphism. Identical relations are observed in the Piute Mountains of the east-central Mojave Desert in Early Proterozoic augen gneiss of the Fenner Granite Gneiss of Hazzard and Dosch (1937), providing another tentative link between Proterozoic rocks of the Transverse Ranges cratonic outlier and the North American craton. This potential link is strengthened by geochemical and isotopic studies relating these and other augen gneiss units (Bennett and DePaolo, 1987; Bender, 1989; Wooden and others, 1991).

In proximity to bodies of the younger anorthosite-syenite complex, granoblastic granodioritic orthogneiss is characterized by rutilated(?) blue to lavendar quartz, plagioclase, mesoperthite to antiperthite, and mafic clots and coronas of uralite and biotite after hypersthene and augite (Oakeshott, 1958; Silver and others, 1963; Ehlig, 1975, 1981; Carter, 1980b; Powell, 1981, 1982a). These minerals and textures indicate that the rock underwent granulite-facies metamorphism and were subsequently retrograded. Relict domains of the augen gneiss of Monument Mountain and its host rocks occur within the retrograded granulite and biotite augen gneiss can be traced into the granulitic gneiss where the augen gneiss gradually loses its protolithic texture and composition and assumes the texture and mineralogy of the granulitic orthogneiss (see also Ehlig, 1981, p 262; Powell, 1981, p. 192). These rocks constitute parts of the Mendenhall Gneiss and the Augustine gneiss unit that in Plates I and II have not been distinguished from the granitic orthogneiss unit that predates the augen gneiss of Monument Mountain.

In the Laguna Mountains, strongly deformed rocks of the Hexie Mountains assemblage are intruded by mildly deformed light colored granitic augen gneiss that grades into undeformed porphyritic granite (Olmsted, 1972; R. E. Powell, 1986, unpublished mapping). This younger granite also crops out at Laguna Dam, in the Muggins Mountains, and in small outcrops in Yuma (Olmsted and others, 1973; Smith and others, 1984; R. E. Powell, 1986, unpublished mapping).

Anorthosite-syenite complex

The anorthosite-syenite complex is exposed west of the San Andreas fault in the San Gabriel Mountains and Soledad basin and east of the fault in the Orocopia, Little Chuckwalla, Chuckwalla, and Eagle Mountains. The anorthosite and related rocks constitute a massif of the least deformed Proterozoic plutonic rocks in the Transverse Ranges. Whereas most other Proterozoic units typically are pervasively deformed, the units of the anorthosite-syenite complex are deformed only in discrete fault zones. The massif consists of progressively younger suites that, respectively, comprise anorthosite and gabbro; syenite, mangerite, and jotunite; and jotunite and jotunitic gabbro. The greatest thickness of these units is exposed in the western San Gabriel Mountains, where at least 7 km of alternately layered andesine anorthosite and gabbro are overlain and intruded by as much as 5 km of layered syenite, mangerite, and jotunite (Carter and Silver, 1972; Carter, 1980b, 1982a).

The anorthosite-syenite complex intruded units of the Hexie and Eagle Mountains assemblages subsequent to most of the deformation of the older rocks. In the San Gabriel Mountains, syenite and jotunite contain inclusions of anorthosite, laminated granitic gneiss that exhibits retrograded granulite texture and mineralogy, and the augen gneiss of Monument Mountain, and undeformed dikes of jotunite intrude granitic gneiss and amphibolite of the Mendenhall Gneiss (Carter, 1980b, 1982a, oral communication, 1987). Granitic orthogneiss and amphibolite of the Mendenhall Gneiss contain inclusions of siliceous and garnetiferous rocks that are probably part of the metasedimentary suite of Pinkham Canyon, but neither amphibolite nor metasedimentary rocks have been recognized unequivocally as inclusions in the anorthosite-syenite complex. In the Mill Creek window, amphibolite and laminated granitic gneiss are tectonically intermingled with anorthosite along an intensely deformed contact (Carter, 1980b, 1982a). In the Orocopia, Chuckwalla, and Eagle Mountains, the undeformed syenite-mangerite-jotunite unit intruded strongly deformed metasedimentary rocks, laminated granitic gneiss, and augen gneiss of the Hexie Mountains assemblage. In Ship Creek of the Chuckwalla Mountains, the undeformed syenite-mangerite-jotunite unit intruded the strongly deformed augen gneiss of Joshua Tree (C. E. Postlethwaite, 1986, oral communication; Davis, 1989; the observations of these two investigators agree with those recorded on my original field maps and notes, but those observations were erroneously overlooked in my writings [Powell, 1981, 1982a; Powell and others, 1984a]). Rusty-weathering mangerite and jotunite in the Chuckwalla and Eagle Mountains are characterized by mesoperthite and hornblende and mafic clots of uralite and biotite after pyroxene. Inclusions of the biotite granite gneiss of Joshua Tree within the mangerite-jotunite commonly are hybridized and contain the mafic minerals found in mangerite-jotunite.

Ages of Proterozoic units

The isotopic geochronology of the Proterozoic units is reasonably consistent with the field relations. From the Eagle Mountains assemblage, the porphyritic granite-augen gneiss of Joshua Tree has yielded slightly discordant zircon U-Pb ages in a range of 1.65 to 1.70 Ga (L. T. Silver, 1978–1980, oral communication). Amphibolite from the Hexie Mountains assemblage has yielded an Nd-Sm model age of 1.71 Ga (Bennett and DePaolo, 1987), and a hornblende ^{40}Ar-^{39}Ar age of greater than 1.4 Ga (C. E. Jacobson, cited in C. E. Postlethwaite, 1988, written communication). Leucocratic granitic orthogneiss that intrudes amphibolite in the central Pinto Mountains has yielded a zircon U-Pb minimum age of 1.68 Ga (Wooden and others, 1991). The augen gneiss of Monument Mountain, which intruded amphibolite and granitic orthogneiss, has yielded zircon U-Pb ages of 1.65 to 1.68 Ga (Silver, 1971). The anorthosite-syenite complex, which intruded the augen gneiss units of Joshua Tree and Monument Mountain, has yielded zircon U-Pb ages of about 1.2 Ga (Silver and others, 1963; Silver, 1971; L. T. Silver, 1978, oral communication). Granitic orthogneiss with relict granulite-facies mineral assemblages in the Mendenhall Gneiss, where it is intruded by 1.2-Ga the anorthosite-syenite complex in the Pacoima Canyon area of the western San Gabriel Mountains, has yielded zircon U-Pb ages of about 1.42 to 1.44 Ga (Silver and others, 1963; Silver, 1971). Silver and others (1963, p. 211) concluded that these "apparent ages are the result of a combination of two or more growth and/or recrystallization events," at least one of which "must be distinctly older than" 1.44 Ga and the youngest of which is 1.2 Ga. These authors further concluded (p. 211–212) "that it is most likely that the principal metamorphic stages of the gneiss including much of the retrogression were intimately related to the conditions attending emplacement and crystallization of the" 1.2-Ga anorthosite-syenite complex and that the "original rock was probably older than" 1.44 Ga. Plausible older events include zircon growth associated with intrusion of the 1.68-Ga leucogranitic gneiss and the somewhat younger augen gneiss of Monument Mountain. The younger porphyritic granite that crops out in the Laguna Mountains, at Laguna Dam, in the Muggins Mountains, and in small outcrops in Yuma has yielded U-Pb zircon ages of about 1.44 to 1.45 Ga (L. T. Silver, 1968, written communication, cited in Olmsted and others, 1973, p. 32; L. T. Silver, 1975, oral communication cited in Dillon, 1976, p. 56).

PROTEROZOIC OR PALEOZOIC ROCKS

Limerock Canyon assemblage

In the Limerock Canyon assemblage, metasedimentary strata are typically intruded by and intimately intermingled with foliated, hornblendic gabbro-diorite and amphibolite and laminated, garnetiferous granitic orthogneiss. These rocks form a metasedimentary and metaigneous complex that crops out west of the San Andreas fault along the eastern flank of the Peninsular Ranges in the San Jacinto and Santa Rosa Mountains, in Bautista Canyon, and to the west of Borrego Springs, in the San Gabriel Mountains south and west of the San Gabriel fault, in the San Rafael Hills and Verdugo Mountains; on the Frazier Mountain–Mount Pinos block on Cobblestone Mountain and on the north flank of Mount Pinos, and in the southern half of the Liebre Mountain block (Plate I). The metasedimentary rocks of the southwestern San Gabriel Mountains, San Rafael Hills, and Verdugo Mountains were named the Placerita Formation by Miller (1934) (see also Oakeshott, 1958), who later (Miller, 1946) also identified the metasedimentary rocks of the southeastern San Gabriel Mountains as the Placerita Formation (see also Ehlig, 1975a, 1981). Subsequently, the rocks in the southeastern San Gabriel Mountains were named the San Antonio Canyon and Rainbow Flat Groups by Hsu (1955) (see also Morton, 1975b; Evans, 1982; Morton and others, 1983; May, 1986; May and Walker, 1989). The metasedimentary rocks along the crest of the San Jacinto Mountains were named the Desert Divide Group by Brown (1980, 1981) (see also Matti and others, 1983b; Calzia and others, 1988). Metasedimentary and metaigneous rocks mapped by Ross (1972, p. 28–29, Plate 1) on Barrett Ridge are probably part of the Limerock Canyon assemblage, and those mapped by Compton (1966) and Seiders and others (1983) on the Pinyon Peak–Bear Mountain ridge in the Santa Lucia Range may also include parts of it. East of the San Andreas fault, the metasedimentary suite crops out in the San Bernardino Mountains in a folded section reaching from the south flank of Sugarloaf Mountain around the east and south flanks of San Gorgonio Mountain to Mill Creek and Yucaipa Ridge (see also Dibblee, 1964d, 1967f; Farley, 1979; Cox and others, 1983a; Powell and others, 1983b). Because the stratigraphic sequence of the metasedimentary rocks and the intrusive relations of the metaigneous rocks are particularly well exposed in Limerock Canyon of the southwestern San Gabriel Mountains, the assemblage is referred to here as the Limerock Canyon assemblage. Except for the exposures in the southern San Bernardino Mountains, the known outcrop belt of the Limerock Canyon assemblage lies to the west of the San Andreas fault.

The metasedimentary suite of the Limerock Canyon assemblage consists of several regionally mappable rock types in an upper amphibolite-grade metamorphic terrane. The suite comprises a section of graphite schist, vitreous quartzite, pelitic schist, marble, and calc-silicate rocks typically intruded by and intermingled with granitic orthogneiss and amphibolite. The granitic orthogneiss ranges from leucocratic to biotite-rich and is commonly garnetiferous; the amphibolite contains domains of little deformed gabbro and diorite. The amphibolite and granitic gneiss are similar to corresponding lithodemes in the Hexie Mountains assemblage. The metamorphic grade of the metasedimentary suite of Limerock Canyon is characterized by sillimanite-garnet-cordierite- and cordierite-orthoamphibole-bearing pelitic rocks, plagioclase-diopside-garnet±scapolite- and plagioclase-wollastonite-scapolite-bearing calc-silicate rocks, and diopside-forsterite–, diopside-wollastonite–, and scapolite-bearing calcite marble (Hsu, 1955; Shay, 1975; Brown, 1981; May, 1986; May and Walker, 1989). The mixed package of metasedimentary and metaigneous rocks has undergone pervasive deformation prior to and during the intrusion of Cretaceous quartz diorite and monzogranite, in which it occurs as inclusions and relatively small pendants and screens.

The ages of the protolith and metamorphism of the metasedimentary suite of Limerock Canyon have not been established unequivocally.

The age of the protolith is demonstrably older than the Late Cretaceous plutons that intruded it between about 75 and 90 Ma; it is also older than the Early Cretaceous deformational and metamorphic event that produced quartz dioritic and tonalitic gneiss and granulite gneiss in the Peninsular Ranges, Transverse Ranges, Salinian block, and southern Sierra Nevada. Relations with Triassic and Proterozoic plutonic rocks in the San Gabriel and San Bernardino Mountains are obscured by the intrusion of Cretaceous plutonic rocks and the accompanying deformation and by late Cenozoic strike-slip faulting. No fossils have been recovered from the high-grade metasedimentary strata. Mineral paragenesis and microfabrics of the Limerock Canyon metamorphic rocks appear to have formed in a single metamorphic event, which is reasonably inferred to be related to Cretaceous plutonism and tectonism (Ehlig, 1975a, 1981; May and Walker, 1989), but could be older.

It seems most likely that the protolith of the Limerock Canyon metasedimentary suite is Paleozoic or Proterozoic in age (see also Miller, 1946; Oakeshott, 1958; Evans, 1982), and there is evidence for either interpretation. Circumstantial evidence that supports a Paleozoic age includes the abundance of marble in the suite and its position between known Proterozoic and Paleozoic rocks to the east and north and Mesozoic strata to the west and south. Carbonate rocks, quartzite, and graphite schist in the Limerock Canyon suite generally have been correlated lithologically with the Paleozoic miogeoclinal sections in the central San Bernardino Mountains and southeastern Peninsular Ranges (Miller, 1946; Woodford, 1960; Ehlig, 1975a, 1981; Matti and others, 1983b; Powell and others, 1983b; Gastil, 1985; Todd and others, 1988).

Other circumstantial evidence, however, supports an Early or Middle Proterozoic age for the Limerock Canyon assemblage. First, the distribution of the Limerock Canyon assemblage coincides with the western margin of known Proterozoic rocks of the Hexie Mountains assemblage in the Transverse Ranges and vicinity. Second, metamorphic mineral assemblages indicate that these metasedimentary rocks have been metamorphosed to the same grade as the metasedimentary suites of Placer Canyon and Pinkham Canyon, which were metamorphosed in the Middle Proterozoic. Third, at one locality in Tick Canyon in the Soledad basin basement, marble and calc-silicate rocks that are probably part of the metasedimetary suite of the Limerock Canyon assemblage are included within metaplutonic rocks of the Middle(?) and Early Proterozoic Mendahall Gneiss of the Hexie Mountains assemblage. Fourth, in the central San Bernardino Mountains, the degree of deformation and grade of metamorphism of the metasedimentary rocks of the Limerock Canyon assemblage and of the Early Proterozoic Baldwin Gneiss are equivalent, whereas the Upper Proterozoic and Paleozoic strata that unconformably overlie the Baldwin Gneiss are less pervasively deformed and metamorphosed at a lower grade than the metasedimentary rocks of the Limerock Canyon assemblage.

LATE PROTEROZOIC AND PALEOZOIC ROCKS

Metamorphosed Late Proterozoic and Paleozoic strata crop out along the San Andreas fault in association with the Early and Middle Proterozoic rocks of the cratonic outlier. Miogeoclinal strata are exposed in the San Bernardino Mountains and western Mojave Desert east of the San Andreas and in the southeastern Peninsular Ranges west of the fault. Metamorphosed cratonal strata are exposed east of the fault in the Gila and Muggins Mountains of southwesternmost Arizona and in the Cargo Muchacho Mountains of southeasternmost California.

Metamorphosed strata in the central San Bernardino Mountains. The Early Proterozoic Baldwin Gneiss of Guillou (1953) (see also Dibblee, 1964a, d, 1967f, g; Silver, 1971; Cameron, 1981, 1982; Powell and others, 1983b) in the central San Bernardino Mountains is depositionally overlain by metamorphosed upper Proterozoic and Paleozoic strata in the vicinity of Sugarloaf Mountain. The Baldwin Gneiss lies at the northwest end of the Transverse Ranges belt of Proterozoic rocks and consists of rocks assigned to the Hexie Mountains and Limerock Canyon assemblages.

In and around Wildhorse Canyon, quartz-pebble conglomerate at the base of the upper Proterozoic and Paleozoic miogeoclinal section is deposited in channels cut into the Baldwin Gneiss (Cameron, 1981, 1982; Powell and others, 1983b). Elsewhere in the central San Bernardino Mountains, the contact is obscured by younger mylonite zones and low-angle faulting, Mesozoic plutons, and Quaternary landslides and talus. Above a basal uppermost Proterozoic and Lower Cambrian section that is chiefly siliciclastic, the rest of the miogeoclinal assemblage is virtually all carbonate rocks. The dominantly siliciclastic lower part of the section consists of strata correlative with the Late Proterozoic Johnnie Formation and Stirling Quartzite, Late Proterozoic and Cambrian Wood Canyon Formation, and Cambrian Zabriskie Quartzite and Carrara Formation. The dominantly carbonate upper part of the section consists of strata correlative with the Cambrian Bonanza King and Nopah Formations, Devonian Sultan and Mississippian Limestone, Mississippian Monte Cristo Limestone, and Mississippian to Lower Permian Bird Spring Formation. The *Skolithus*-bearing supermature Zabriskie Quartzite, banded dolomite in the Bonanza King Formation, and the stromatoporoid(?)-bearing Sultan Limestone have proven particularly useful markers in correlating the metamorphosed sections of the San Bernardino Mountains and western Mojave Desert with the miogeoclinal section of the southernmost Great Basin (Burchfiel and Davis, 1972, 1981; Stewart and Poole, 1975; Miller and Cameron, 1982; Powell and others, 1983b; Stewart and others, 1984).

Metamorphosed strata in the Yuma area. Early and Middle Proterozoic rocks in the Laguna Mountains, Muggins Mountains, Yuma, and vicinity in southwesternmost Arizona that are included in the Hexie Mountains assemblage constitute the southeast end of the belt of Proterozoic rocks in the Transverse Ranges and vicinity. Abutting these Proterozoic rocks, metamorphosed strata that are lithologically equivalent to parts of the Paleozoic cratonal section are found in the Cargo Muchacho Mountains of southeastern California and in the Gila and Muggins Mountains of southwestern Arizona, although these strata typically have been assigned a Precambrian age (Wilson, 1933; Olmsted and others, 1973; Dillon, 1976; Smith and others, 1984). Marble, quartzite, and calc-silicate rocks in the Gila Mountains and southeastern Cargo Muchacho Mountains have strong lithologic affinity to metamorphosed upper Paleozoic cratonal section (the Supai Group in the Cargo Muchacho Mountains and the Kaibab Limestone, Coconino Sandstone, and possibly the Hermit Shale in the Gila Mountains) and are overlain by probable Mesozoic pelitic schist, calc-silicate rocks, and fine-grained, micaceous cross-bedded quartzite. In the Muggins Mountains, the probable Paleozoic strata are caught up in a mylonite zone that forms the southern limit of Proterozoic granite in that range (see Smith and others, 1984). These rocks are lithologically equivalent to those found in cratonic sections in northeastern Sonora (Leveille and Frost, 1984), southwestern Arizona (Miller, 1970), and southeastern California (Stone and others, 1983).

Metamorphosed strata in the southeastern Peninsular Ranges. As part of the belt of Proterozoic and Paleozoic rocks, rocks of the Paleozoic miogeoclinal assemblage are found in the easternmost Peninsular Ranges of California and Baja California (Miller and Dockum, 1983; Gastil, 1985). On Carrizo Mountain in the Coyote Mountains of the southeastern Peninsular Ranges, the discovery of Ordovician conodonts in a section of predominantly carbonate rocks with subordinate quartzite and schist led to correlation of that section with the Pogonip Group of the Great Basin (Miller and Dockum, 1983). Although the Ordovician limestone is the only unit from which conodonts were recovered, the structurally complicated section on Carrizo Mountain includes quartzite, pelitic schist, and banded dolomite that are clearly equivalent to the metamorphosed Cambrian Tapeats Sandstone (or Late Proterozoic and Cambrian Wood Canyon Formation), Bright Angel Shale (or Carrara Formation), and Bonanza King Formation, respectively, and stromatoporoid(?)-bearing carbonate rocks that are lithologically equivalent to the Devonian and Mississippian Sultan Lime-

stone. The contact between these strata and those of the metasedimentary suite of the Limerock Canyon assemblage that crop out farther west has been obscured by younger events.

MESOZOIC ROCKS

Together, the Proterozoic and Paleozoic rocks of the Transverse Ranges and vicinity define an elongate, uplifted terrane that is flanked to the east and west by thick sections of metamorphosed Mesozoic strata. To the east, a north-northwest–trending belt of greenschist-grade metamorphosed Triassic, Jurassic, and Cretaceous continental strata and Jurassic and Cretaceous plutonic rocks in the central Mojave Desert and Colorado Desert isolates the Proterozoic and Paleozoic domain of the Transverse Ranges from the main mass of the North American craton, which stretches east uninterrupted from the eastern Mojave Desert. In the western Mojave Desert, the base of the Mesozoic section was deposited unconformably on Paleozoic miogeoclinal strata (Dibblee, 1967a; E. L. Miller, 1978; Miller and Carr, 1978), whereas it was deposited conformably on Paleozoic cratonal strata in Palen Pass (Hamilton, 1987; Stone and Pelka, 1989) and apparently also in the Gila Mountains.

To the west, the Mesozoic paleogeology is more complicated than to the east. Proterozoic and Paleozoic rocks are juxtaposed against a composite assemblage of distinct suites of amphibolite- and greenschist-grade Mesozoic marine strata in the Peninsular Ranges and Sierra Nevada, including suites of upper Paleozoic(?) and lower Mesozoic schist-quartzite-marble; sequences of Triassic and Jurassic flysch with subordinate conglomerate, limestone, and mafic to intermediate volcanic rocks; and metamorphosed Upper Jurassic to Lower Cretaceous(?) intermediate volcanic rocks and marine argillite and sandstone. Between the Peninsular Ranges and Sierra Nevada, however, the Proterozoic rocks of the Transverse Ranges and southern Salinian block and amphibolite-grade schist, quartzite, and marble of late Paleozoic(?) and early Mesozoic age in the central and northern Salinian block are juxtaposed against the rocks of the Franciscan Complex; the Triassic and Jurassic flysch and volcanic sections that intervene to the north in the western Sierra Nevada and to the south in the Peninsular Ranges are largely absent.

The metamorphosed Proterozoic, Paleozoic, and Mesozoic rocks of southern California were intruded by at least four Mesozoic plutonic supersuites (Plate I) that comprise part of the Mesozoic magmatic arc in southern California. During the late Mesozoic and early Cenozoic, this composite cratonic crystalline terrane was emplaced tectonically above Mesozoic strata that had accumulated along the continental margin and were metamorphosed prior to and during overthrusting.

Metamorphic infrastructure

Pelona, Orocopia, and Rand Schists and schist of Sierra de Salinas. A widespread stratigraphic assemblage of metamorphosed Mesozoic strata that is discontinuously exposed in southern California and southwesternmost Arizona has been variously named the Pelona, Orocopia, and Rand Schists. These exposures share distinctive lithologic and structural features that have prompted most investigators to correlate them and to incorporate them into a single tectonostratigraphic package (see reviews in Ehlig, 1968; Haxel and Dillon, 1978; Jacobson and others, 1988). In addition, Ross (1976) has suggested that the schist of Sierra de Salinas in the Santa Lucia and Gabilan Ranges of the Salinian block is correlative with the Pelona Schist on Portal and Ritter Ridges in the westernmost Mojave Desert. Nd-Sm and Pb isotopic studies indicate that much of the protolith of the Pelona, Orocopia, and Rand Schists and the schist of Sierra de Salinas was derived from a Proterozoic continental source (Bennett and DePaolo, 1982; Silver and others, 1984; Mattinson and James, 1986; Haxel and Tosdal, 1986).

No stratigraphic or structural base has been observed for any exposure of Pelona-type schist, and no direct paleontologic or isotopic evidence has been found to indicate the age of the protolith. By inference, however, the age of the Orocopia Schist protolith is older than a deformed Jurassic diorite that intruded the schist in the Chocolate Mountains (zircon U-Pb date of about 163 Ma, Mukasa and others, 1984; Haxel and Tosdal, 1986). Similarly, the age of the Rand Schist protolith is older than Cretaceous granitic rocks intrusive into the schist in the Rand Mountains (discordant K-Ar age of 74.2 Ma on hornblende and 18.2 Ma on coexisting biotite, Kistler and Peterman, 1978; see also Miller and Morton, 1977; zircon U-Pb age of 79 ± 1 Ma, Silver and Nourse, 1986) and older than Late Jurassic or Early Cretaceous tonalite in the San Emigdio Mountains (E. W. James, 1986). The schist of Sierra de Salinas also was intruded by Cretaceous plutonic rocks (Ross, 1976; zircon U-Pb ages of about 80 to 85 Ma, Mattinson and James, 1985, 1986).

The Pelona, Orocopia, and Rand Schists and the schist of Sierra de Salinas are exposed in antiformal domes exposed as windows beneath layers of crystalline rocks that consist of Mesozoic plutonic rocks and the metamorphosed Proterozoic, Paleozoic, and Mesozoic rocks intruded by the plutons. The tectonic growth of these antiforms is complex, and, as a group, the antiforms reveal evidence for several deformational events during and prior to disruption along the San Andreas fault system. In the youngest event, sedimentation in Miocene basins recorded the erosion of the overlying crystalline rocks, culminating in exhumation of the Pelona and Orocopia Schists in the middle Miocene during or just prior to movement on the Clemens Well–Fenner–San Francisquito fault. An older event is recorded by fission track and K-Ar data that indicate Late Cretaceous and Paleogene uplift of the schists, thrust mylonite, and overlying crystalline rocks from 75 or 70 to 52 Ma (Miller and Morton, 1980; Ehlig, 1981, 1982; Dillon, 1986; Mahaffie and Dokka, 1986). Complex intrusive and structural contact relations in the Rand Schist in the Rand Mountains (Silver and others, 1984; Silver and Nourse, 1986) and San Emigdio Mountains (A. I. James, 1986), and in the Orocopia Schist in the Chocolate Mountains and vicinity (Haxel and others, 1985; Haxel and Tosdal, 1986) reveal evidence for Late Cretaceous extensional faulting and uplift prior to the intrusion of a 79-Ma pluton in the Rand Mountains and a 60-Ma pluton in the Chocolate Mountains; these relations also provide evidence for the earlier, subduction-related thrusting event that superposed rocks of the North American craton above the schist units (see also Jacobson and others, 1988). Prior to displacement on the faults of the San Andreas system, Pelona and Orocopia Schists were folded up in a set of right-stepping en echelon antiforms aligned along the future trace of the Clemens Well–Fenner–San Francisquito fault, and Rand and Pelona Schists were folded up in a set of left-stepping en echelon antiforms aligned along the future trace of Garlock fault.

Mesozoic plutonic arc

As part of the Cordilleran system of Mesozoic batholiths, voluminous plutonic rocks intruded Proterozoic and Phanerozoic host rocks in the Mojave and Sonoran Deserts and the eastern two-thirds of the Transverse Ranges, and Phanerozoic host rocks in the Sierra Nevada and most of the Peninsular Ranges and Salinian block. Mesozoic plutonic rocks define three broad northwest-trending belts in the area shown in Plates I and II. Plutons of the central belt roughly coincide in distribution with the Proterozoic and Paleozoic rocks; plutons of the eastern belt intrude the eastern margin of the domain of Proterozoic and Paleozoic rocks and the domain of metamorphosed Mesozoic strata that lies to the east; and plutons of the western belt intrude the western margin of the domain of Proterozoic and Paleozoic rocks and the domain of metamorphosed Mesozoic strata that lies to the west. Intrusive relations exist between plutons of adjacent belts.

Eastern plutonic belt. The eastern plutonic belt consists for the most part of a calc-alkaline to alkaline suite of porphyritic hornblende-biotite quartz monzonite, monzodiorite, and quartz diorite. These rocks are characterized by a paucity of quartz and by phenocrysts of alkali feldspar with a distinct lavendar cast. Chlorite and epidote are ubiquitous and abundant alteration products of hornblende and biotite. These plutonic rocks have yielded Middle and Late Jurassic K-Ar and U-Pb ages (about 165 to 150 Ma) in the Transverse Ranges (Bishop, 1963; L. T. Silver, 1978, oral communication) and the Mojave and Sonoran Deserts (Miller and Morton, 1980; Anderson and Silver, 1986; Tosdal, 1986; Karish and others, 1987; Tosdal and others, 1989), where they form part of a belt of Jurassic magmatic rocks along the Pacific margin of the North American continent (Kistler, 1974; Anderson and others, 1979; Burchfiel and Davis, 1981; Gastil, 1985; Tosdal and others, 1989).

The Jurassic plutons are intruded by Cretaceous plutons of leucocratic, biotite ± hornblende ± muscovite–bearing monzogranite and granodiorite. These plutonic rocks range from equigranular to porphyritic with phenocrysts of alkali feldspar, and are locally foliated to mylonitic (Miller and others, 1982; Howard and others, 1982). Granodiorite in the Coxcomb Mountains has yielded a zircon U-Pb age of 70 Ma (Calzia and others, 1986).

Central plutonic belt. The central plutonic belt consists of Triassic, Jurassic, and Cretaceous plutonic rocks. The oldest part of the central belt consists of an undated suite or suites of hornblende gabbro and diorite. This suite consists of medium- to extremely coarse grained hornblende diorite, gabbro, gabbroic pegmatite, anorthositic gabbro, and hornblendite. Rocks of this suite intrude and metamorphose Paleozoic carbonate strata in the San Bernardino Mountains and are found as inclusions in the Triassic Mount Lowe intrusion of Barth and Ehlig (1988) see below) and younger Mesozoic plutonic rocks of both the eastern and central belts.

The gabbro-diorite unit is intruded by a suite of alkalic Triassic plutonic rocks that intrude the Proterozoic and Paleozoic domain in the Transverse Ranges and vicinity (Plate I). These rocks have yielded zircon U-Pb ages of 207 to 241 Ma (Silver, 1971; L. T. Silver, 1975, oral communication cited in Dillon, 1976, p. 74; C. F. Miller, 1977, 1978; Frizzell and others, 1986; J. E. Wright, cited in Karish and others, 1987; Barth and others, 1990), an Rb-Sr whole-rock isochron age of 208 Ma (Joseph and others, 1982a), and hornblende ^{40}Ar-^{39}Ar ages of 214 and 233 Ma (Cameron, 1981; Miller and Sutter, 1981). As a suite, they are characterized by felsic rocks with clinopyroxene and hornblende, a paucity of quartz, and in places garnet, primary epidote, or biotite. Several of the Triassic plutons are also characterized by abundant large alkali feldspar phenocrysts and are locally strongly foliated. The largest of these is the zoned plutonic body referred to as the Mount Lowe Granodiorite by Miller (1926), the Lowe Granodiorite and Parker Quartz Diorite by Miller (1934, 1946), the Lowe Granodiorite by Ehlig (1975a, 1981), the Lowe igneous pluton by Joseph and others (1982a), and the Mount Lowe intrusion by Barth and Ehlig (1988), which exhibits mylonitic contacts against older rock units, including laminated granitic orthogneiss, amphibolite, anorthosite, and syenite-mangerite-jotunite (Carter and Silver, 1972; Carter, 1980b, 1982a; Powell and others, 1983a). The largest body of the Mount Lowe intrusion crops out in the San Gabriel Mountains, and smaller displaced bodies crop out in the southernmost San Bernardino, Chocolate, and Little Chuckwalla Mountains (Silver, 1971; Crowell, 1975a; Ehlig, 1975a, 1977, 1981; Dillon, 1976, p. 74–75; Farley, 1979; Dillon and Ehlig, this volume). Bodies of another suite of Triassic plutonic rocks crop out in the Mule, Palo Verde, and Trigo Mountains (Tosdal, 1986; Barth and others, 1990).

Two disrupted plutons of biotite-hornblende porphyritic monzonite are characterized by abundant 4- to 8-cm equant alkali-feldspar phenocrysts. Displaced bodies of one of these plutons in the Mill Creek area of the San Bernardino Mountains and in the northern Liebre Mountain block (Matti and others, 1985) have yielded zircon U-Pb ages of about 215 Ma (Frizzell and others, 1986). Displaced bodies of the second pluton crop out in the Pinto Mountains near Twentynine Palms and the adjacent part of the Mojave Desert to the northwest (Miller, 1938; Rogers, 1961; Dibblee, 1967e, 1968b; K. A. Howard, T. T. Fitzgibbon, and R. E. Powell, 1984–1985, unpublished mapping).

The alkalic suite also includes an undated pluton of clinopyroxene-bearing, porphyritic monzodiorite in the western Eagle Mountains (Powell, 1981) and nonporphyritic clinopyroxene-hornblende alkalic plutons in the San Bernardino Mountains and the adjacent western Mojave Desert (C. F. Miller, 1977; Cameron, 1981; Miller and Cameron, 1982).

The youngest part of the central belt consists of quartz-rich, calc-alkaline, biotite and hornblende-biotite monzogranite and granodiorite. In the Transverse Ranges east of the San Andreas fault, generally the intrusive sequence, from oldest to youngest, is sphene-biotite-hornblende-bearing granodiorite, coarse-grained biotite-bearing porphyritic monzogranite with zoned, pink phenocrysts of alkali feldspar, coarse- to very coarse grained biotite-bearing nonporphyritic monzogranite that grades into muscovite ± garnet-bearing monzogranite, and fine- to medium-grained biotite-bearing porphyritic monzogranite or granodiorite that grades into nonporphyritic rocks. The fine-grained nonporphyritic granitoid locally contains 0.5- to 4-cm long dark gray quartz-biotite-muscovite ± garnet spherules with white to light gray quartzofeldspathic rims that impart a distinctive polka-dot texture to the rocks (Pelka, 1971; see also Smith, 1977a; Joseph and others, 1982b). The coarse- to very coarse grained nonporphyritic monzogranite is locally muscovitic, and at one locality in the Pinto Mountains also contains polka-dots. Each of these rock types occurs in separate plutons that are aligned subparallel to the northwest-trending axis of a composite batholith (Powell, 1981). Lithologically equivalent plutons with the same intrusive sequence crop out in the western Mojave Desert, the Transverse Ranges west of the San Andreas fault, and the Salinian block. In the Eagle and Chuckwalla Mountains, all of these plutonic units are cut by a swarm of mafic to felsic dikes, which also intrude rocks of the eastern plutonic belt in the Eagle Mountains.

Plutonic rocks of this youngest sequence have yielded zircon U-Pb ages of about 75 to 85 Ma in the Salinian block and San Gabriel Mountains (Carter and Silver, 1971; Silver, 1971; L. T. Silver, 1976, oral communication; Mattinson, 1978; Mattinson and James, 1985; May, 1986; Silver and Nourse, 1986; May and Walker, 1989) and have yielded both Jurassic and Cretaceous zircon U-Pb ages in the Chuckwalla Mountains (L. T. Silver, 1976, oral communication; Wooden and others, 1991). Plutons of polka-dot granite have yielded Late Cretaceous Rb-Sr ages (ranging from about 72 to 104 Ma) from the La Panza Range of the Salinian block, from Warm Springs Mountain in the Liebre Mountain block, and from the Orocopia, Little San Bernardino, and Pinto Mountains of the eastern Transverse Ranges (Joseph, 1981; Joseph and others, 1982b). A dike of the swarm in the Eagle Mountains has yielded a zircon U-Pb age of about 145 Ma where it intruded the Middle Proterozoic syenite-mangerite-jotunite unit of the anorthosite-syenite complex (James, 1987, 1989). In the Chuckwalla Mountains, however, dikes that are similar in lithology to those of the Eagle Mountains swarm have intruded all of the plutonic rocks, including Cretaceous granodiorite and bodies of polka-dot granite. Either more than one plutonic sequence is present or the zircons from each phase of the sequence have yielded different ages at different localities. The discrepancies between the sequence derived from field observations and the zircon U-Pb ages have not yet been resolved. Biotite and hornblende K-Ar ages in the Transverse Ranges and Mojave Desert yield uplift ages of 60 to 80 Ma (Armstrong and Suppe, 1973; Miller and Morton, 1980).

Western plutonic belt. The western plutonic belt consists of Cretaceous biotite-hornblende quartz diorite, tonalite, and granodiorite. The boundary between the central and western belts is the quartz diorite line of Moore (1959; see also Yeats, 1968), along which quartz dioritic rocks of the western belt were intruded by granitic rocks of the central belt. In detail, quartz dioritic and granitic rocks are intermingled across the line. Rocks of the western plutonic belt have yielded zircon U-Pb ages of about 85 to 120 Ma in the Salinian block (Mattinson and James, 1985), about 85 to 90 in the Transverse Ranges and Mojave Desert (Silver, 1971; May, 1986; Silver and Nourse, 1986; May and Walker, 1989),

and about 90 to 120 Ma in the Peninsular Ranges (Hill and Silver, 1979; Silver and others, 1979).

The quartz dioritic plutonic rocks of the western plutonic belt near the boundary with the granitic rocks of the central plutonic belt are commonly but not invariably intensely foliated, whereas the granitic rocks along the boundary typically are less commonly and less intensely deformed. Deformed quartz dioritic rocks of this ilk are widespread in southern California (James and Mattinson, 1988; Todd and others, 1988; May, 1989), both west and east of the San Andreas fault (Plate I). West of the San Andreas fault, they crop out in the eastern Peninsular Ranges (Sharp, 1967; Dibblee, 1981), the southern San Gabriel Mountains (Hsu, 1955; Morton, 1973; 1975b; Evans, 1982; Cox and others, 1983b; Morton and others, 1983; May, 1986; Morton and Matti, 1987; May and Walker, 1989), the southern part of the Frazier Mountain–Alamo Mountain block, the southern part of the Liebre Mountain block (Simpson, 1934; Dibblee, 1967a; Ross, 1972, p. 48), and in the Salinian block including widespread exposures in the Santa Lucia Range (Compton, 1966; Ross, 1972). Within the San Andreas fault zone, similar rocks constitute much of the block bounded by the San Andreas, Punchbowl, and Fenner faults (Dibblee, 1967a; Ross, 1972, p. 39–42; Barrows and others, 1985) and perhaps some of the Banning block in San Gorgonio Pass (Allen, 1957; Farley, 1979; Matti and others, 1982a, 1983a). East of the San Andreas fault, they crop out in the Chocolate Mountains (Dillon, 1976), the Little San Bernardino Mountains (Dibblee, 1967f; Proctor, 1968; Powell, 1981, 1982a), the southern and western San Bernardino Mountains (Allen, 1957; Dibblee, 1964d, 1967f; Weldon and others, 1981; Cox and others, 1983a), the western Mojave Desert (Dibblee, 1967a), and the southern Sierra Nevada and San Emigdio Mountains (Ross, 1984, 1985, 1989; Sams and Saleeby, 1988). In the granitic rocks of the western Mojave Desert, inliers of quartz dioritic gneiss intrusive into and interlayered with metasedimentary rocks crop out in the Rand Mountains and the Barstow area of the Mojave Desert (Dibblee, 1967a), where they constitute the Johannesburg Gneiss of Hulin (1925) and the Waterman Gneiss of Bowen (1954), respectively. The distribution of deformed quartz diorite along the boundary between the two belts in the Salinian block, Transverse Ranges, and Mojave Desert suggests that the boundary is folded.

Much of this pervasive deformation is Early Cretaceous in age. Gneissose plutonic rocks have yielded U-Pb zircon ages of 105 to 120 Ma in the Peninsular Ranges (Silver and others, 1979; Todd and Shaw, 1979; Todd and others, 1988), 108 to 130 Ma in the Salinian block (Mattinson and James, 1985; James and Mattinson, 1988), and 110 to 120 Ma in the southern Sierra Nevada (Sams and Saleeby, 1988). In the southeastern San Gabriel Mountains, a late-stage granulite pegmatite in a granulite domain engulfed in deformed quartz dioritic and tonalitic rocks has yielded a U-Pb zircon age of about 108 Ma (May, 1986; May and Walker, 1989). Undeformed and little-deformed tonalitic plutons have yielded U-Pb zircon ages of 90 to 100 Ma in the Peninsular Ranges (Silver and others, 1979; Todd and Shaw, 1979; Todd and others, 1988), 76 to 100 Ma in the Salinian block (Mattinson and James, 1985; James and Mattinson, 1988), and 100 Ma in the southern Sierra Nevada (Sams and Saleeby, 1988).

In some areas, however, deformation persisted through much of the Late Cretaceous. In the San Gabriel Mountains, the quartz dioritic plutonic suite becomes increasingly deformed eastward along the southern flank of the range until, in the southeastern San Gabriel Mountains and northeastern Peninsular Ranges, it includes highly deformed quartz diorite of Late Cretaceous age (Alf, 1948; Hsu, 1955; Morton, 1975b; May, 1986, 1989; Morton and Matti, 1987 and this volume; May and Walker, 1989). Plutonic rocks from the southeastern San Gabriel Mountains have yielded U-Pb zircon ages of about 88 Ma for highly deformed tonalite, about 85 Ma for slightly deformed tonalite, and about 78 Ma for undeformed granodiorite (May and Walker, 1989). In the northeastern Peninsular Ranges and southeastern San Gabriel Mountains, the eastern margin of the quartz dioritic and tonalitic rocks was evidently the approximate locus of synplutonic mylonitization in the Late Cretaceous and subsequent postplutonic brittle faulting (see for example, Alf, 1948; Hsu, 1955; Sharp, 1966, 1967, 1979; Matti and others, 1983b; Simpson, 1984; Erskine, 1986a, b; May, 1986, 1989; Morton and Matti, 1987; Todd and others, 1988; May and Walker, 1989). The mylonite transects tonalite plutons in the San Jacinto Mountains that have yielded U-Pb zircon ages of 97 Ma (Hill and Silver, 1979).

Plutons of the western plutonic belt intruded metamorphosed Mesozoic rocks of the western metamorphic belt, Proterozoic or Paleozoic rocks of the Limerock Canyon assemblage, and Proterozoic rocks of the Hexie Mountain assemblage. Cretaceous quartz dioritic rocks that intruded each of these assemblages are variously deformed.

UPPER CRETACEOUS TO EOCENE MARINE OVERLAP SEQUENCE

The plutonic crystalline terrane of southern California, including the Peninsular Ranges, Transverse Ranges, Salinian block, and southern Sierra Nevada, was overlapped unconformably from the west by Upper Cretaceous through Eocene marine strata. Clast populations indicate that these strata were deposited prior to exhumation of the Pelona Schist and probably prior to exhumation of most of the Proterozoic rocks in the Transverse Ranges and vicinity. The abrupt eastern limit of these overlap strata, with its overlap and pinch-out relations, provides a paleogeologic marker with which to measure displacement on late Cenozoic strike-slip faults. These strata are characterized by lithosomes and sedimentary structures indicative of a submarine-fan environment of deposition, and conglomerate clasts indicate a provenance of Mesozoic plutonic and Paleozoic and Mesozoic metasedimentary and metavolcanic rocks (see, for example, Chipping, 1972; Nilsen and Clarke, 1975; Kooser, 1982; Vedder and others, 1983). In the Peninsular Ranges and Great Valley–Sierra Nevada region, these strata overlap metamorphosed Mesozoic volcanic and sedimentary strata intruded by quartz dioritic plutonic rocks, whereas in the Transverse Ranges and Salinian block they rest on high-grade metamorphic rocks of the western assemblage of Transverse Ranges igneous and metamorphic complex and on quartz-rich Jurassic and Cretaceous plutonic rocks of the central batholithic belt as well as scraps of more westerly rocks. Pre-Eocene strata west of the present trace of the San Andreas fault and the Late Cretaceous plutonic rocks that underlie them have yielded anomalously low paleomagnetic inclinations that have been interpreted as indicative of deposition at low paleolatitudes and subsequent northward transport of 2,000 to 2,500 km (McWilliams and Howell, 1982; Vedder and others, 1983; Champion and others, 1984, 1986; Fry and others, 1985; Hagstrum and others, 1985; Kanter and Debiche, 1985; Beck, 1986; Morris and others, 1986). The coherence of paleogeologic patterns in the reconstruction in Plate II refutes this interpretation and requires another explanation for the low inclinations (for example, see Butler and others, 1991).

REFERENCES CITED

Addicott, W. O., 1968, Mid-Tertiary zoogeographic and paleogeographic discontinuities across the San Andreas fault, California, in Dickinson, W. R., and Grantz, A., eds., Proceedings of Conference on Geologic Problems of San Andreas Fault System: Stanford, California, Stanford University Publications in the Geological Sciences, v. 11, p. 144–165.

—— , 1978, Revision of the age of the Pancho Rico Formation, central Coast Ranges, California: U.S. Geological Survey Bulletin 1457-A, p. A88–A89.

Adegoke, O. S., 1969, Stratigraphy and paleontology of the marine Neogene formations of the Coalinga region, California: Berkeley, University of California Publications in Geological Sciences, v. 80, 269 p.

Alf, R. M., 1948, A mylonite belt in the southeastern San Gabriel Mountains, California: Geological Society of America Bulletin, v. 59, p. 1101–1120.

Allen, C. R., 1957, San Andreas fault zone in San Gorgonio Pass, southern California: Geological Society of America Bulletin, v. 68, p. 315–349.

—— , 1968, The tectonic environments of seismically active and inactive areas along the San Andreas fault system, in Dickinson, W. R., and Grantz, A., eds., Proceedings of Conference on Geologic Problems of San Andreas Fault System: Stanford, California, Stanford University Publications in the Geological Sciences, v. 11, p. 70–82.

—— , 1981, The modern San Andreas fault, in Ernst, W. G., ed., The geotectonic development of California; Rubey Volume 1: Englewood Cliffs, New Jersey, Prentice-Hall, p. 511–534.

Allen, C. R., Wyss, M., Brune, J. N., Grantz, A., and Wallace, R. E., 1972, Displacements on the Imperial, Superstition Hills, and San Andreas faults triggered by the Borrego Mountain earthquake, in The Borrego Mountain earthquake of April 9, 1968: U.S. Geological Survey Professional Paper 787, p. 87–104.

Anderson, T. H., and Silver, L. T., 1986, The border connection; Geological correlations and contrasts between Arizona and Sonora, in Beatty, B., and Wilkinson, P.A.K., eds., Frontiers in geology and ore deposits of Arizona and the Southwest: Arizona Geological Society Digest, v. 16, p. 72–73.

Anderson, T. H., Eells, J. L., and Silver, L. T., 1979, Precambrian and Paleozoic rocks of the Caborca region, Sonora, Mexico, in Anderson, T. H., and Roldán-Quintana, J., eds., Geology of northern Sonora; Geological Society of America Annual Meeting Guidebook for Field Trip 27: Hermosillo, Sonora, Mexico, Institute of Geology, U.N.A.M., and Pittsburgh, Pennsylvania, University of Pittsburgh, p. 1–22.

Armentrout, J. M., and Echols, R. J., 1981, Biostratigraphic-chronostratigraphic scale of the northeastern Pacific Neogene, in Proceedings of International Geological Correlation Programme Project 114 International Workshop on Pacific Neogene Biostratigraphy, 6th International Working Group Meeting: Osaka, Japan, Osaka Museum of Natural History, p. 7–27.

Armstrong, R. L., and Suppe, J., 1973, Potassium-argon geochronometry of Mesozoic igneous rocks in Nevada, Utah, and southern California: Geological Society of America Bulletin, v. 84, p. 1375–1391.

Arnett, G. R., 1949, Geology of the Lytle Creek area, California: Compass, v. 26, p. 294–304.

Astiz, L., and Allen, C. R., 1983, Seismicity of the Garlock fault, California: Seismological Society of America Bulletin, v. 73, p. 1721–1734.

Atwater, T., 1970, Implications of plate tectonics for the Cenozoic tectonic evolution of western North America: Geological Society of America Bulletin, v. 81, p. 3513–3535.

—— , 1989, Plate tectonic history of the northeast Pacific and western North America, in Winterer, E. L., Hussong, D. M., and Decker, R. W., eds., The eastern Pacific Ocean and Hawaii: Boulder, Colorado, Geological Society of America, The Geology of North America, v. N, p. 21–72.

Bacheller, J., III, 1978, Quaternary geology of the Mojave Desert–eastern Transverse Ranges boundary in the vicinity of Twentynine Palms, California [M.S. thesis]: Los Angeles, University of California, 157 p.

Badgley, P. C., 1965, Structural and tectonic principles: New York, Harper and Row, 521 p.

Bailey, E. H., ed., 1966, Geology of northern California: California Division of Mines and Geology Bulletin 190, 508 p.

Bailey, E. H., Irwin, W. P., and Jones, D. L., 1964, Franciscan and related rocks, and their significance in the geology of western California: California Division of Mines and Geology Bulletin 183, 177 p.

Bailey, T. L., and Jahns, R. H., 1954, Geology of the Transverse Range province, southern California, in Jahns, R. H., ed., Geology of southern California: California Division of Mines Bulletin 170, chapter 2, p. 83–106.

Baird, A. K., Welday, E. E., and Baird, K. W., 1970, Chemical variations in batholithic rocks of southern California: Geological Society of America Abstracts with Programs, v. 2, p. 69.

Baird, A. K., Baird, K. W., Woodford, A. O., and Morton, D. M., 1971, The Transverse Ranges; A unique structural-petrochemical belt across the San Andreas fault system: Geological Society of America Abstracts with Programs, v. 3, p. 77–78.

Baird, A. K., Morton, D. M., Woodford, A. O., and Baird, K. W., 1974, Transverse Ranges province: A unique structural-petrochemical belt across the San Andreas fault system: Geological Society of America Bulletin, v. 85, p. 163–174.

Ballance, P. F., Howell, D. G., and Ort, K., 1983, Late Cenozoic wrench tectonics along the Nacimiento, South Cuyama, and La Panza faults, California, indicated by depositional history of the Simmler Formation, *in* Andersen, D. W., and Rymer, M. J., eds., Tectonics and sedimentation along faults of the San Andreas system: Pacific Section, Society of Economic Paleontologists and Mineralogists, p. 1–9.

Bandy, O. L., and Arnal, R. E., 1969, Middle Tertiary basin development, San Joaquin Valley, California: Geological Society of America Bulletin, v. 80, p. 783–819.

Barbat, W. F., 1958, The Los Angeles basin area, California, *in* Weeks, L. G., ed., Habitat of oil; A symposium: Tulsa, Oklahoma, American Association of Petroleum Geologists, p. 62–77.

Barron, J. A., Keller, G., and Dunn, D. A., 1985, A multiple microfossil biochronology for the Miocene, *in* Kennett, J. P., ed., The Miocene ocean; Paleoceanography and biogeography: Geological Society of America Memoir 163, p. 21–36.

Barrows, A. G., 1980, Geologic map of the San Andreas fault zone and adjoining terrane, Juniper Hills and vicinity, Los Angeles County, California: California Division of Mines and Geology Open-File Report 80-2 LA, scale 1:9,600.

Barrows, A. G., Kahle, J. E., and Beeby, D. J., 1985, Earthquake hazards and tectonic history of the San Andreas fault zone, Los Angeles County, California: U.S. Geological Survey Final Technical Report, Contract 14-08-0001-19193, 126 p., scale 1:12,000.

Barth, A. P., and Ehlig, P. L., 1988, Geochemistry and petrogenesis of the marginal zone of the Mount Lowe intrusion, central San Gabriel Mountains, California: Contributions to Mineralogy and Petrology, v. 100, p. 192–204.

Barth, A. P., Tosdal, R. M., and Wooden, J. L., 1990, A petrologic comparison of Triassic plutonism in the San Gabriel and Mule Mountains, southern California: Journal of Geophysical Research, v. 95, no. B12, p. 20075–20096.

Bartholomew, M. J., 1970, San Jacinto fault zone in the northern Imperial Valley, California: Geological Society of America Bulletin, v. 81, p. 3161–3166.

Bartow, J. A., 1974, Sedimentology of the Simmler and Vaqueros Formations in the Caliente Range–Carrizo Plain area, California: U.S. Geological Survey Open-File Report 74-338, 163 p.

—— , 1978, Oligocene continental sedimentation in the Caliente Range area, California: Journal of Sedimentary Petrology, v. 48, p. 75–98.

—— , compiler, 1990, Neogene time scale: U.S. Geological Survey Open-File Report 90-636A.

Bassett, A. M., and Kupfer, D. H., 1964, A geologic reconnaissance in the southeastern Mojave Desert, California: California Division of Mines and Geology Special Report 83, 43 p.

Bazeley, W.J.M., 1961, 175 miles of lateral movement along the San Andreas fault since lower Miocene?: Pacific Petroleum Geologist News Letter, v. 15, no. 5, p. 2–3.

Beck, M. E., Jr., 1986, Model for late Mesozoic-early Tertiary tectonics of coastal California and western Mexico and speculations on the origin of the San Andreas fault: Tectonics, v. 5, p. 49–64.

Beeby, D. L., and Hill, R. L., 1975, Galway Lake fault; A previously unmapped active fault in the Mojave Desert, San Bernardino County, California: California Division of Mines and Geology, California Geology, v. 28, p. 219–221.

Bender, E. E., 1989, Petrology of Early Proterozoic granitic gneisses from the eastern Transverse Ranges; Implications for the tectonics of southeastern

California: Geological Society of America Abstracts with Programs, v. 21, p. 56.

Bennett, V., and DePaolo, D. J., 1982, Tectonic implications of Nd isotopes in the Pelona, Rand, Orocopia and Catalina Schists, southern California: Geological Society of America Abstracts with Programs, v. 14, p. 442.

Bennett, V. C., and DePaolo, D. J., 1987, Proterozoic crustal history of the western United States as determined by neodymium isotopic mapping: Geological Society of America Bulletin, v. 99, p. 674–685.

Berggren, W. A., and van Couvering, J. A., 1974, The late Neogene; Biostratigraphy, geochronology, and paleoclimatology of the last 15 million years in marine and continental sequences: Paleogeography, Paleoclimatology, Paleoecology, v. 16, p. 1–216.

Berggren, W. A., Kent, D. V., Flynn, J. J., and van Couvering, J. A., 1985, Cenozoic geochronology: Geological Society of America Bulletin, v. 96, p. 1407–1418.

Berry, F.A.F., Huffman, O. F., and Turner, D. L., 1966, Post-Miocene movement along the San Andreas fault, California, *in* Abstracts for 1966: Geological Society of America Special Paper 101, p. 15–16.

Biehler, S., Kovach, R. L., and Allen, C. R., 1964, Geophysical framework of northern end of Gulf of California structural province, *in* Van Andel, T. H., and Shor, G. G., Jr., eds., Marine geology of the Gulf of California; A symposium: American Association of Petroleum Geologists Memoir 3, p. 126–143.

Bird, P., and Rosenstock, R. W., 1984, Kinematics of present crust and mantle flow in southern California: Geological Society of America Bulletin, v. 95, p. 946–957.

Bishop, C. C., compiler, 1963, Geologic map of California; Needles sheet: California Division of Mines and Geology, scale 1:250,000.

Blake, G. H., 1991, Review of the Neogene biostratigraphy and stratigraphy of the Los Angeles basin and implications for basin evolution, *in* Biddle, K. T., ed., Active margin basins: American Association of Petroleum Geologists Memoir 52, p. 135–184.

Blake, M. C., Jr., Jayko, A. S., McLaughlin, R. J., and Underwood, M. B., 1988, Metamorphic and tectonic evolution of the Franciscan Complex, northern California, *in* Ernst, W. G., ed., Metamorphism and crustal evolution of the western United States; Rubey Volume 7: Englewood Cliffs, New Jersey, Prentice-Hall, p. 1035–1060.

Boellstorff, J. D., and Steineck, P. L., 1975, The stratigraphic significance of fission-track ages on volcanic ashes in the marine late Cenozoic of southern California: Earth and Planetary Science Letters, v. 27, p. 143–154.

Bohannon, R. G., 1975, Mid-Tertiary conglomerates and their bearing on Transverse Range tectonics, southern California, *in* Crowell, J. C., ed., San Andreas fault in southern California: California Division of Mines and Geology Special Report 118, p. 75–82.

Bortugno, E. J., and Spittler, T. E., compilers, 1986, Geologic map of the San Bernardino Quadrangle: California Division of Mines and Geology Regional Geologic Map Series Map 3A, scale 1:250,000.

Bowen, O. E., Jr., 1954, Geology and mineral deposits of Barstow Quadrangle, San Bernardino County, California: California Division of Mines Bulletin 165, p. 7–185.

Brabb, E. E., and Pampeyan, E. H., compilers, 1983, Geologic map of San Mateo County, California: U.S. Geological Survey Miscellaneous Investigations Series Map I-1257-A, scale 1:62,500.

Brady, R. H., III, 1982, Geology at the intersection of the Garlock and Death Valley zones, northeastern Avawatz Mountains, California; A field guide, *in* Cooper, J. D., Troxel, B. W., and Wright, L. A., eds., Geology of selected areas in the San Bernardino Mountains, western Mojave Desert, and southern Great Basin, California; Geological Society of America Cordilleran Section Annual Meeting Volume and Guidebook: Shoshone, California, The Death Valley Publishing Company, p. 53–59.

Brady, R. H., III, and Dokka, R. K., 1989, The eastern Mojave shear zone; A major tectonic boundary in the southwestern Cordillera: Geological Society of America Abstracts with Programs, v. 21, p. 59.

Brady, R. H., III, and Troxel, B. W., 1981, Eastern termination of the Garlock fault in the Avawatz Mountains, San Bernardino County, California: Geological Society of America Abstracts with Programs, v. 13, p. 46–47.

Brady, R. H., III, Troxel, B. W., and Butler, P. R., 1980, Tectonic and stratigraphic elements of the northern Avawatz Mountains, San Bernardino County, California, *in* Fife, D. L, and Brown, A. R., eds., Geology and mineral wealth of the California desert; Dibblee Volume: Santa Ana, California, South Coast Geological Society, p. 224–234.

Brown, A. R., 1980, Limestone deposits of the Desert Divide, San Jacinto Mountains, California, *in* Fife, D. L., and Brown, A. R., eds., Geology and mineral wealth of the California desert; Dibblee Volume: Santa Ana, California, South Coast Geological Society, p. 284–293.

Brown, A. R., 1980, Limestone deposits of the Desert Divide, San Jacinto Mountains, California, *in* Fife, D. L., and Brown, A. R., eds., Geology and mineral wealth of the California desert; Dibblee Volume: Santa Ana, California, South Coast Geological Society, p. 284–293.

—— , 1981, Structural history of the metamorphic, granitic and cataclastic rocks in the southeastern San Jacinto Mountains, *in* Brown, A. R., and Ruff, R. W., eds., Geology of the San Jacinto Mountains: Santa Ana, California, South Coast Geological Society Annual Field Trip Guidebook 9, p. 100–138.

Brown, H. J., 1982, Possible Cambrian miogeoclinal strata, northern Shadow Mountains, western Mojave Desert, California, *in* Fife, D. L., and Minch, J. A., eds., Geology and mineral wealth of the California Transverse Ranges; Mason Hill Volume: Santa Ana, California, South Coast Geological Society Annual Symposium and Guidebook 10, p. 355–365.

Brown, J. S., 1923, The Salton Sea region, California; A geographic, geologic, and hydrologic reconnaissance with a guide to desert watering places: U.S. Geological Survey Water-Supply Paper 497, 292 p.

Brown, R. D., Jr., 1990, Quaternary deformation, *in* Wallace, R. E., ed., The San Andreas fault system, California: U.S. Geological Survey Professional Paper 1515, p. 82–113.

Buesch, D. C., and Ehlig, P. L., 1982, Structural and lower Miocene volcanic rock correlation between Soledad Pass and Salton Wash along the San Andreas fault: Geological Society of America Abstracts with Programs, v. 14, p. 153.

Bukry, D., 1981, Cenozoic coccoliths from the Deep Sea Drilling Project, *in* Warme, J. E., Douglas, R. G., and Winterer, E. L., eds., The Deep Sea Drilling Project; A decade of progress: Society of Economic Paleontologists and Mineralogists Special Publication 32, p. 335–353.

Burbank, D. W., and Whistler, D. P., 1987, Temporally constrained tectonic rotations derived from magnetostratigraphic data; Implications for the initiation of the Garlock fault, California: Geology, v. 15, p. 1172–1175.

Burchfiel, B. C., and Davis, G. A., 1972, Structural framework and evolution of the southern part of the Cordilleran orogen, western United States: American Journal of Science, v. 272, p. 97–118.

—— , 1981, Mojave Desert and environs, *in* Ernst, W. G., ed., The geotectonic development of California; Rubey Volume 1: Englewood Cliffs, New Jersey, Prentice-Hall, p. 217–252.

Burke, D. B., Hillhouse, J. W., McKee, E. H., Miller, S. T., and Morton, J. L., 1982, Cenozoic rocks in the Barstow basin area of southern California; Stratigraphic relations, radiometric ages, and paleomagnetism: U.S. Geological Survey Bulletin 1529-E, 16 p.

Butler, R. F., Dickinson, W. R., and Gehrels, G. E., 1991, Paleomagnetism of coastal California and Baja California; Alternatives to large-scale northward transport: Tectonics, v. 10, p. 561–576.

Calderone, G., and Butler, R. F., 1984, Paleomagnetism of Miocene volcanic rocks from southwestern Arizona; Tectonic implications: Geology, v. 12, p. 627–630.

Calderone, G. J., Butler, R. F., and Acton, G. D., 1990, Paleomagnetism of middle Miocene volcanic rocks in the Mojave-Sonora Desert region of western Arizona and southeastern California: Journal of Geophysical Research, v. 95, no. B1, p. 625–647.

Calzia, J. P., DeWitt, E., and Nakata, J. K., 1986, U-Th-Pb age and initial strontium isotopic ratios of the Coxcomb granodiorite, and a K-Ar date of olivine basalt from the Coxcomb Mountains, southern California: Isochron/West, v. 47, p. 3–8.

Calzia, J. P., Madden-McGuire, D. J., Oliver, H. W., and Schreiner, R. A., 1988, Mineral resources of the Santa Rosa Mountains Wilderness Study Area, Riverside County, California: U.S. Geological Survey Bulletin 1710-D, 14 p.

Cameron, C. S., 1981, Geology of the Sugarloaf and Delamar Mountain areas, San Bernardino Mountains, California [Ph.D. thesis]: Cambridge, Massachusetts Institute of Technology, 399 p.

—— , 1982, Stratigraphy and significance of the Upper Precambrian Big Bear Group, *in* Cooper, J. D., Troxel, B. W., and Wright, L. A., eds., Geology of selected areas in the San Bernardino Mountains, western Mojave Desert, and southern Great Basin, California; Field trip number 9; Geological Society of America, Cordilleran Section Annual Meeting Volume and Guidebook: Shoshone, California, Death Valley Publishing Company, p. 5–20.

Campbell, R. H., and Yerkes, R. F., 1976, Cenozoic evolution of the Los Angeles basin area; Relation to plate tectonics, *in* Howell, D. G., ed., Aspects of the geologic history of the California continental borderland: Pacific Section, American Association of Petroleum Geologists Miscellaneous Publication 24, p. 541–558.

Carman, M. F., Jr., 1964, Geology of the Lockwood Valley area, Kern and Ventura Counties, California: California Division of Mines and Geology Special Report 81, 62 p.

Carr, M. D., Poole, F. G., Harris, A. G., and Christiansen, R. L., 1981, Western facies Paleozoic rocks in the Mojave Desert, California, *in* Howard, K. A., Carr, M. D., and Miller, D. M., eds., Tectonic framework of the Mojave and Sonoran Deserts, California and Arizona: U.S. Geological Survey Open-File Report 81-503, p. 15–17.

Carr, M. D., Christiansen, R. L., and Poole, F. G., 1984, Pre-Cenozoic geology of the El Paso Mountains, southwestern Great Basin, California; A summary, *in* Lintz, Joseph, Jr., ed., Western geological excursions, Volume 4; Geological Society of America Annual Meeting Guidebook: Reno, University of Nevada MacKay School of Mines Department of Geological Sciences, p. 84–93.

Carter, B. A., 1971, Quaternary displacement on the Garlock fault, California: EOS Transactions of the American Geophysical Union, v. 52, no. 4, p. 350.

—— , 1980a, Quaternary displacement on the Garlock fault, California, *in* Fife, D. L., and Brown, A. R., eds., Geology and mineral wealth of the California desert; Dibblee Volume: Santa Ana, California, South Coast Geological Society, p. 457–466.

—— 1980b, Structure and petrology of the San Gabriel anorthosite-syenite body, Los Angeles, County, California [Ph.D. thesis]: Pasadena, California Institute of Technology, 393 p.

—— , 1982a, Geology and structural setting of the San Gabriel anorthosite-syenite body and adjacent rocks of the western San Gabriel Mountains, Los Angeles County, California; Field trip number 5, *in* Cooper, J. D., compiler, Geologic excursions in the Transverse Ranges, southern California; Geological Society of America Cordilleran Section Annual Meeting Volume and Guidebook: Fullerton, California State University Department of Geological Sciences, p. 1–53.

—— , 1982b, Neogene displacement on the Garlock fault, California: EOS Transactions of the American Geophysical Union, v. 63, p. 1124.

—— , 1987, Quaternary fault-line features of the central Garlock fault, Kern County, California, *in* Hill, M. L., ed., Cordilleran Section of the Geological Society of America: Boulder, Colorado, Geological Society of America, Centennial Field Guide, v. 1, p. 133–135.

Carter, B. A., and Silver, L. T., 1971, Post-emplacement structural history of the San Gabriel anorthosite complex: Geological Society of America Abstracts with Programs, v. 3, p. 92–93.

—— , 1972, Structure and petrology of the San Gabriel anorthosite-syenite body, California: 24th International Geological Congress, Montreal, Section 2, p. 303–311.

Carter, J. N., Luyendyk, B. P., and Terres, R. R., 1987, Neogene clockwise tectonic rotation of the eastern Transverse Ranges, California, suggested by paleomagnetic vectors: Geological Society of America Bulletin, v. 98, p. 199–206.

Champion, D. E., 1989, Identification of the Rinconada fault in western San Mateo County, CA through the use of strontium isotopic studies: Geological Society of America Abstracts with Programs, v. 21, p. 64.

Champion, D. E., and Kistler, R. W., 1991, Paleogeographic reconstruction of the northern Salinian block using Sr isotopic properties: Geological Society of America Abstracts with Programs, v. 23, p. 12.

Champion, D. E., Howell, D. G., and Grommé, C. S., 1984, Paleomagnetic and geologic data indicating 2500 km of northward displacement for the Salinian and related terranes, California: Journal of Geophysical Research, v. 89, no. B9, p. 7736–7752.

Champion, D. E., Howell, D. G., and Marshall, M., 1986, Paleomagnetism of Cretaceous and Eocene strata, San Miguel Island, California borderland and the northward translation of Baja California: Journal of Geophysical Research, v. 91, no. B11, p. 11557–11570.

Chinnery, M. A., 1966, Secondary faulting; II. Geological aspects: Canadian Journal of Earth Sciences, v. 3, p. 175–190.

Chipping, D. H., 1972, Early Tertiary paleogeography of central California: American Association of Petroleum Geologists Bulletin, v. 56, p. 480–493.

Christiansen, R. L., and McKee, E. H., 1978, Late Cenozoic volcanic and tectonic evolution of the Great Basin and Columbia intermontane regions, *in* Smith, R. B., and Eaton, G. P., eds., Cenozoic tectonics and regional geophysics of the western Cordillera: Geological Society of America Memoir 152, p. 283–311.

Clark, J. C., 1981, Stratigraphy, paleontology, and geology of the central Santa Cruz Mountains, California Coast Ranges: U.S. Geological Survey Professional Paper 1168, 51 p.

Clark, J. C., and Brabb, E. E., 1978, Stratigraphic contrasts across the San Gregorio fault, Santa Cruz Mountains, west central California, *in* Silver, E. A., and Normark, W. R., eds., San Gregorio Hosgri fault zone, California: California Division of Mines and Geology Special Report 137, p. 3–12.

Clark, J. C., Brabb, E. E., Greene, H. G., and Ross, D. C., 1984, Geology of Point Reyes peninsula and implications for San Gregorio fault history, *in* Crouch, J. K., and Bachman, S. B., eds., Tectonics and sedimentation along the California margin: Pacific Section, Society of Economic Paleontologists and Mineralogists, v. 38, p. 67–85.

Clark, M. M., 1973, Map showing recently active breaks along the Garlock and associated faults, California: U.S. Geological Survey Miscellaneous Geologic Investigations Map I-741, scale 1:24,000.

Clark, M. M., and Lajoie, K. R., 1974, Holocene behavior of the Garlock fault: Geological Society of America Abstracts with Programs, v. 6, p. 156–157.

Clarke, S. H., Jr., and Nilsen, T. H., 1973, Displacement of Eocene strata and implications for the history of offset along the San Andreas fault, central and northern California, *in* Kovach, R. L., and Nur, A., eds., Proceedings of the Conference on Tectonic Problems of the San Andreas Fault System: Stanford, California, Stanford University Publications in the Geological Sciences, v. 13, p. 358–367.

Clements, T., 1937, Structure of southeastern part of Tejon Quadrangle, California: American Association of Petroleum Geologists Bulletin, v. 21, p. 212–232.

Clements, T., and Oakeshott, G. B., 1935, Lower Eocene (Martinez) of San Gabriel Mountains [abs.]: Geological Society of America Proceedings 1934, p. 310.

Cloos, E., 1955, Experimental analysis of fracture patterns: Geological Society of America Bulletin, v. 66, p. 241–256.

Cloos, M., 1982, Flow melanges; Numerical modeling and geologic constraints on their origin in the Franciscan subduction complex, California: Geological Society of America Bulletin, v. 93, p. 330–345.

Compton, R. R., 1966, Granitic and metamorphic rocks of the Salinian block, California Coast Ranges, *in* Bailey, E. H., ed., Geology of northern California: California Division of Mines and Geology Bulletin 190, p. 277–287.

Coppersmith, K. J., and Griggs, G. B., 1978, Morphology, recent activity, and seismicity of the San Gregorio fault zone, *in* Silver, E. A., and Normark, W. R., eds., San Gregorio–Hosgri fault zone, California: California Division of Mines and Geology Special Report 137, p. 33–43.

Cox, B. F., 1982, Stratigraphy, sedimentology, and structure of the Goler Formation (Paleocene), El Paso Mountains, California; Implications for Paleogene tectonism on the Garlock fault zone [Ph.D. thesis]: Riverside, University of California, 248 p.

——, 1987, Stratigraphy, depositional environments, and paleotectonics of the Paleocene and Eocene Goler Formation, El Paso Mountains, California; Geologic summary and roadlog, *in* Cox, B. F., ed., Basin analysis and paleontology of the Paleocene and Eocene Goler Formation, El Paso Mountains, California: Pacific Section, Society of Economic Paleontologists and Mineralogists, p. 1–29.

Cox, B. F., and Diggles, M. F., 1986, Geologic map of the El Paso Mountains Wilderness Study Area: U.S. Geological Survey Miscellaneous Field Studies Map MF-1827, scale 1:24,000.

Cox, B. F., Matti, J. C., Oliver, H. W., and Zilka, N. T., 1983a, Mineral resource potential map of the San Gorgonio Wilderness, San Bernardino County, California: U.S. Geological Survey Miscellaneous Field Studies Map MF-1161-C, scale 1:62,500.

Cox, B. F., Powell, R. E., Hinkle, M. E., and Lipton, D. A., 1983b, Mineral resource potential map of the Pleasant View Roadless Area, Los Angeles County, California: U.S. Geological Survey Miscellaneous Field Studies Map MF-1649-A, scale 1:62,500.

Cramer, C. H., and Harrington, J. M., 1987, Seismicity and tectonics of the Cucamonga fault and the eastern San Gabriel Mountains, San Bernardino County, *in* Recent reverse faulting in the Transverse Ranges, California: U.S. Geological Survey Professional Paper 1339, p. 7–26.

Criscione, J. J., Davis, T. E., and Ehlig, P., 1978, The age of sedimentation/diagenesis for the Bedford Canyon Formation and the Santa Monica Formation in southern California; A Rb/Sr evaluation, *in* Howell, D. G., and McDougall, K. A., eds., Mesozoic paleogeography of the western United States: Pacific Section, Society of Economic Paleontologists and Mineralogists Pacific Coast Paleogeography Symposium 2, p. 385–396.

Crook, R., Jr., Allen, C. R., Kamb, B., Payne, C. M., and Protor, R. J., 1987, Quaternary geology and seismic hazard of the Sierra Madre and associated faults, western San Gabriel Mountains, *in* Recent reverse faulting in the Transverse Ranges, California: U.S. Geological Survey Professional Paper 1339, p. 27–63.

Crowe, B. M., 1978, Cenozoic volcanic geology and probable age of inception of basin-range faulting in the southeasternmost Chocolate Mountains, California: Geological Society of America Bulletin, v. 89, p. 251–264.

Crowe, B. M., Crowell, J. C., and Krummenacher, D., 1979, Regional stratigraphy, K-Ar ages, and tectonic implications of Cenozoic volcanic rocks, southeastern California: American Journal of Science, v. 279, p. 186–216.

Crowell, J. C., 1950, Geology of the Hungry Valley area, southern California: American Association of Petroleum Geologists Bulletin, v. 34, p. 1623–1646.

——, 1952, Probable large lateral displacement on San Gabriel fault, southern California: American Association of Petroleum Geologists Bulletin, v. 36, p. 2026–2035.

——, 1954a, Geology of the Ridge basin area, Los Angeles and Ventura Counties, *in* Jahns, R. H., ed., Geology of southern California: California Division of Mines Bulletin 170, Map Sheet 7.

——, 1954b, Strike-slip displacement of the San Gabriel fault, southern California, *in* Jahns, R. H., ed., Geology of southern California: California Division of Mines Bulletin 170, chapter 4, p. 49–52.

——, 1960, The San Andreas fault in southern California: Report of the 21st International Geologic Congress, Copenhagen, part 18, p. 45–52.

——, 1962, Displacement along the San Andreas fault, California: Geological Society of America Special Paper 71, 61 p.

——, 1964, The San Andreas fault zone from the Temblor Mountains to Antelope Valley, southern California: Pacific Sections, American Association of Petroleum Geologists and Society of Economic Paleontologists and Mineralogists, and San Joaquin Geologic Society Guidebook, p. 1–39.

——, 1968, Movement histories of faults in the Transverse Ranges and speculations on the tectonic history of California, *in* Dickinson, W. R., and Grantz, A., eds., Proceedings of Conference on Geologic Problems of San Andreas Fault System: Stanford, California, Stanford University Publications in the Geological Sciences, v. 11, p. 323–341.

——, 1973, Problems concerning the San Andreas fault system in southern California, *in* Kovach, R. L., and Nur, A., eds., Proceedings of the Conference on Tectonic Problems of the San Andreas Fault System: Stanford, California, Stanford University Publications in the Geological Sciences, v. 13, p. 125–135.

——, 1975a, Geologic sketch of the Orocopia Mountains, southeastern California, *in* Crowell, J. C., ed., San Andreas fault in southern California: California Division of Mines and Geology Special Report 118, p. 99–110.

——, ed., 1975b, San Andreas fault in southern California: California Division of Mines and Geology Special Report 118, 272 p.

——, 1975c, The San Andreas fault in southern California, *in* Crowell, J. C., ed., San Andreas fault in southern California: California Division of Mines and Geology Special Report 118, p. 7–27.

——, 1975d, The San Gabriel fault and Ridge basin, southern California, *in* Crowell, J. C., ed., San Andreas fault in southern California: California Division of Mines and Geology Special Report 118, p. 208–219.

——, 1981, An outline of the tectonic history of southeastern California, *in* Ernst, W. G., ed., The geotectonic development of California; Rubey Volume 1: Englewood Cliffs, New Jersey, Prentice-Hall, p. 583–600.

——, 1982a, Pliocene Hungry Valley Formation, Ridge basin, southern California, *in* Crowell, J. C., and Link, M. H., eds., Geologic history of Ridge basin, southern California: Society of Economic Paleontologists and Mineralogists, Pacific Section, p. 143–149.

——, 1982b, The tectonics of Ridge basin, southern California, *in* Crowell, J. C., and Link, M. H., eds., Geologic history of Ridge basin, southern California: Pacific Section, Society of Economic Paleontologists and Mineralogists, p. 25–42.

——, 1982c, The Violin Breccia, Ridge basin, southern California, *in* Crowell, J. C., and Link, M. H., eds., Geologic history of Ridge basin, southern California: Pacific Section, Society of Economic Paleontologists and Mineralogists, p. 89–98.

——, 1987, Late Cenozoic basins of onshore southern California; Complexity is the hallmark of their tectonic history, *in* Ingersoll, R. V., and Ernst, W. G., eds., Cenozoic basin development of coastal California; Rubey Volume 6: Englewood Cliffs, New Jersey, Prentice-Hall, p. 207–241.

Crowell, J. C., and Ramirez, V. R., 1979, Late Cenozoic faults in southeastern California, *in* Crowell, J. C., and Sylvester, A. G., eds., Tectonics of the juncture between the San Andreas fault system and the Salton trough, southeastern California; Geological Society of America Annual Meeting Guidebook: Santa Barbara, University of California Department of Geological Sciences, p. 27–39.

Crowell, J. C., and Susuki, T., 1959, Eocene stratigraphy and paleontology, Orocopia Mountains, southeastern California: Geological Society of America Bulletin, v. 70, p. 581–592.

Crowell, J. C., and Walker, J.W.R., 1962, Anorthosite and related rocks along the San Andreas fault, southern California: Berkeley, University of California Publications in Geological Sciences, v. 40, no. 4, p. 219–288.

Cummings, D., 1976, Theory of plasticity applied to faulting, Mojave Desert, southern California: Geological Society of America Bulletin, v. 87, p. 720–724.

Davis, G. A., and Burchfiel, B. C., 1973, Garlock fault; An intracontinental transform structure, southern California: Geological Society of America Bulletin, v. 84, p. 1407–1422.

Davis, T. L., and 5 others, 1988, Structure of the Cuyama Valley, Caliente Range, and Carrizo Plain and its significance to the structural style of the southern Coast Ranges and western Transverse Ranges, *in* Bazeley, W.J.M., ed., Tertiary tectonics and sedimentation in the Cuyama basin, San Luis Obispo, Santa Barbara, and Ventura Counties, California: Society of Economic Paleontologists and Mineralogists, v. 59, p. 141–158.

DeMets, C., Gordon, R. G., Stein, S., and Argus, D. F., 1987, A revised estimate of Pacific–North America motion and implications for western North America plate boundary zone tectonics: Geophysical Research Letters, v. 14, p. 911–914.

DeMets, C., Gordon, R. G., Argus, D. F., and Stein, S., 1990, Current plate motions: Geophysical Journal International, v. 101, p. 425–478.

Dibblee, T. W., Jr., 1954, Geology of the Imperial Valley region, California, *in* Jahns, R. H., ed., Geology of southern California: California Division of Mines Bulletin 170, chapter 2, p. 21–28.

——, 1960, Geologic map of the Barstow Quadrangle, San Bernardino County, California: U.S. Geological Survey Mineral Investigations Field Studies Map MF-233, scale 1:62,500.

——, 1961a, Evidence of strike-slip movement on northwest-trending faults in Mojave Desert, California: U.S. Geological Survey Professional Paper 424-B, p. B197–B199.

——, 1961b, Geologic map of the Bouquet Reservoir Quadrangle, Los Angeles County, California: U.S. Geological Survey Mineral Investigations Field Studies Map MF-79, scale 1:62,500.

——, 1963, Geology of the Willow Springs and Rosamond Quadrangles, California: U.S. Geological Survey Bulletin 1089-C, p. 141–253.

——, 1964a, Geologic map of the Lucerne Valley Quadrangle, San Bernardino County, California: U.S. Geological Survey Miscellaneous Geologic Investigations Map I-426, scale 1:62,500.

——, 1964b, Geologic map of the Ord Mountains Quadrangle, San Bernardino County, California: U.S. Geological Survey Miscellaneous Geologic Investigations Map I-427, scale 1:62,500.

——, 1964c, Geologic map of the Rodman Mountains Quadrangle, San Bernardino County, California: U.S. Geological Survey Miscellaneous Geologic Investigations Map I-430, scale 1:62,500.

——, 1964d, Geologic map of the San Gorgonio Mountain Quadrangle, San Bernardino and Riverside Counties, California: U.S. Geological Survey Miscellaneous Geologic Investigations Map I-431, scale 1:62,500.

——, 1966a, Evidence for cumulative offset on the San Andreas fault in central and northern California, *in* Bailey, E. H., ed., Geology of northern California: California Division of Mines and Geology Bulletin 190, p. 375–384.

——, 1966b, Geologic map of the Lavic Quadrangle, San Bernardino County, California: U.S. Geological Survey Miscellaneous Geologic Investigations Map I-472, scale 1:62,500.

——, 1966c, Geology of the central Santa Ynez Mountains, Santa Barbara County, California: California Division of Mines and Geology Bulletin 186, 99 p.

——, 1967a, Areal geology of the western Mojave Desert, California: U.S. Geological Survey Professional Paper 522, 153 p.

——, 1967b, Evidence of major lateral displacement on the Pinto Mountain fault, southeastern California: Geological Society of America Abstracts for 1967, Special Paper 115, p. 322.

——, 1967c, Geologic map of the Deadman Lake Quadrangle, San Bernardino County, California: U.S. Geological Survey Miscellaneous Geologic Investigations Map I-488, scale 1:62,500.

——, 1967d, Geologic map of the Emerson Lake Quadrangle, San Bernardino County, California: U.S. Geological Survey Miscellaneous Geologic Investigations Map I-490, scale 1:62,500.

——, 1967e, Geologic map of the Joshua Tree Quadrangle, San Bernardino and Riverside Counties, California: U.S. Geological Survey Miscellaneous Geologic Investigations Map I-516, scale 1:62,500.

——, 1967f, Geologic map of the Morongo Valley Quadrangle, San Bernardino and Riverside Counties, California: U.S. Geological Survey Miscellaneous Geologic Investigations Map I-517, scale 1:62,500.

——, 1967g, Geologic map of the Old Woman Springs Quadrangle, San Bernardino County, California: U.S. Geological Survey Miscellaneous Geologic Investigations Map I-518, scale 1:62,500.

——, 1968a, Displacements on the San Andreas fault system in the San Gabriel, San Bernardino, and San Jacinto Mountains, southern California, *in* Dickinson, W. R., and Grantz, A., eds., Proceedings of Conference on Geologic Problems of San Andreas Fault System: Stanford, California, Stanford University Publications in the Geological Sciences, v. 11, p. 260–278.

——, 1968b, Geologic map of the Twentynine Palms Quadrangle, San Bernardino and Riverside Counties, California: U.S. Geological Survey Miscellaneous Geologic Investigations Map I-561, scale 1:62,500.

——, 1968c, Geology of the Fremont Peak and Opal Mountain Quadrangles, California: California Division of Mines and Geology Bulletin 188, 64 p.

——, 1970, Geologic map of the Daggett Quadrangle, San Bernardino County, California: U.S. Geological Survey Miscellaneous Geologic Investigations Map I-592, scale 1:62,500.

——, 1971, Geologic map of the Redlands Quadrangle, California: U.S. Geological Survey Open-File Map, scale 1:62,500.

——, 1972, Rinconada fault in the southern Coast Ranges, California, and its significance: Geological Society of America Abstracts with Programs, v. 4, p. 145.

——, 1973, Stratigraphy of the southern Coast Ranges near the San Andreas fault from Cholame to Maricopa, California: U.S. Geological Survey Professional Paper 764, 45 p.

——, 1975a, Late Quaternary uplift of the San Bernardino Mountains on the San Andreas and related faults, *in* Crowell, J. C., ed., San Andreas fault in southern California: California Division of Mines and Geology Special Report 118, p. 127–135.

——, 1975b, Tectonics of the western Mojave Desert near the San Andreas fault, *in* Crowell, J. C., ed., San Andreas fault in southern California: California Division of Mines and Geology Special Report 118, p. 155–161.

——, 1976, The Rinconada and related faults in the southern Coast Ranges, California, and their tectonic significance: U.S. Geological Survey Professional Paper 981, 55 p.

——, 1980, Geologic structure of the Mojave Desert, *in* Fife, D. L., and Brown, A. R., eds., Geology and mineral wealth of the California desert; Dibblee Volume: Santa Ana, California, South Coast Geological Society, p. 69–100.

——, 1981, Geology of the San Jacinto Mountains and adjacent areas, *in* Brown, A. R., and Ruff, R. W., eds., Geology of the San Jacinto Mountains: Santa Ana, California, South Coast Geological Society Annual Field Trip Guidebook 9, p. 1–47.

——, 1982a, Geologic structure and tectonics along the eastern Santa Ynez fault, western Transverse Ranges, California: Geological Society of America Abstracts with Programs, v. 14, p. 159.

——, 1982b, Geology of the Alamo Mountain, Frazier Mountain, Lockwood Valley, Mount Pinos, and Cuyama badlands areas, southern California, *in* Fife, D. L., and Minch, J. A., eds., Geology and mineral wealth of the California Transverse Ranges; Mason Hill Volume: Santa Ana, California, South Coast Geological Society Annual Symposium and Guidebook 10, p. 57–77.

——, 1982c, Geology of the Castaic block, the mountains and hills northwest of the San Gabriel Mountains, southern California, *in* Fife, D. L., and Minch, J. A., eds., Geology and mineral wealth of the California Transverse Ranges; Mason Hill Volume: Santa Ana, California, South Coast Geological Society Annual Symposium and Guidebook 10, p. 78–93.

——, 1982d, Geology of the San Bernardino Mountains, southern California, *in* Fife, D. L., and Minch, J. A., eds., Geology and mineral wealth of the California Transverse Ranges; Mason Hill Volume: Santa Ana, California, South Coast Geological Society Annual Symposium and Guidebook 10, p. 149–169.

——, 1982e, Geology of the San Gabriel Mountains, southern California, *in* Fife, D. L., and Minch, J. A., eds., Geology and mineral wealth of the California Transverse Ranges; Mason Hill Volume: Santa Ana, California, South Coast Geological Society Annual Symposium and Guidebook 10, p. 131–147.

——, 1982f, Geology of the Santa Monica Mountains and Simi Hills, southern California, *in* Fife, D. L., and Minch, J. A., eds., Geology and mineral wealth of the California Transverse Ranges; Mason Hill Volume: Santa Ana, California, South Coast Geological Society Annual Symposium and Guidebook 10, p. 94–130.

——, 1982g, Geology of the Santa Ynez–Topatopa Mountains, southern Cali-

fornia, *in* Fife, D. L., and Minch, J. A., eds., Geology and mineral wealth of the California Transverse Ranges; Mason Hill Volume: Santa Ana, California, South Coast Geological Society Annual Symposium and Guidebook 10, p. 41–56.

——, 1982h, Regional geology of the Transverse Ranges province of southern California, *in* Fife, D. L., and Minch, J. A., eds., Geology and mineral wealth of the California Transverse Ranges; Mason Hill Volume: Santa Ana, California, South Coast Geological Society, Annual Symposium and Guidebook 10, p. 7–26.

Dibblee, T. W., Jr., and Bassett, A. M., 1966a, Geological map of the Cady Mountains Quadrangle, San Bernardino County, California: U.S. Geological Survey Miscellaneous Geologic Investigations Map I-467, scale 1:62,500.

——, 1966b, Geologic map of the Newberry Quadrangle, San Bernardino County, California: U.S. Geological Survey Miscellaneous Geologic Investigations Map I-461, scale 1:62,500.

Dibblee, T. W., Jr., and Chesterman, C. W., 1953, Geology of the Breckenridge Mountain Quadrangle, California: California Division of Mines Bulletin 168, 56 p.

Dibblee, T. W., Jr., and Louke, G. P., 1970, Geologic map of the Tehachapi Quadrangle, Kern County, California: U.S. Geological Survey Miscellaneous Geologic Investigations Map I-607, scale 1:62,500.

Dibblee, T. W., Jr., and Warne, A. H., 1970, Geologic map of the Cummings Mountain Quadrangle, Kern County, California: U.S. Geological Survey Miscellaneous Geologic Investigations Map I-611, scale 1:62,500.

Dickinson, W. R., 1981, Plate tectonics and the continental margin of California, *in* Ernst, W. G., ed., The geotectonic development of California; Rubey Volume 1: Englewood Cliffs, New Jersey, Prentice-Hall, p. 1–28.

Dickinson, W. R., and Grantz, A., eds., 1968, Proceedings of Conference on Geologic Problems of San Andreas Fault System: Stanford, California, Stanford University Publications in the Geological Sciences, v. 11, 374 p.

Dickinson, W. R., Cowan, D. S., and Schweikert, R. A., 1972, Test of new global tectonics; Discussion: American Association of Petroleum Geologists Bulletin, v. 56, p. 375–384.

Dillon, J. T., 1976, Geology of the Chocolate and Cargo Muchacho Mountains, southeasternmost California [Ph.D. thesis]: Santa Barbara, University of California, 405 p.

——, 1986, Timing of thrusting and metamorphism along the Vincent–Chocolate Mountain thrust system, southern California: Geological Society of America Abstracts with Programs, v. 18, p. 101.

Dokka, R. K., 1983, Displacements on late Cenozoic strike-slip faults of the central Mojave Desert, California: Geology, v. 11, p. 305–308.

Dokka, R. K., and Glazner, A. F., 1982a, Aspects of early Miocene extension of the central Mojave Desert, *in* Cooper, J. D., compiler, Geologic excursions in the California desert; Geological Society of America Cordilleran Section Annual Meeting Volume and Guidebook: Fullerton, California State University Department of Geological Sciences, p. 31–45.

——, 1982b, Field trip roadlog; Late Cenozoic tectonic and magmatic evolution of the central Mojave Desert, California, *in* Cooper, J. D., compiler, Geologic excursions in the California desert; Geological Society of America Cordilleran Section Annual Meeting Volume and Guidebook: Fullerton, California State University Department of Geological Sciences, p. 1–30.

Dokka, R. K., and Travis, C. J, 1990a, Late Cenozoic strike-slip faulting in the Mojave Desert, California: Tectonics, v. 9, p. 311–340.

——, 1990b, Role of the eastern California shear zone in accommodating Pacific-North American plate motion: Geophysical Research Letters, v. 17, p. 1323–1326.

Downs, T., 1957, Late Cenozoic vertebrates from the Imperial Valley region, California [abs.]: Geological Society of America Bulletin, v. 68, p. 1822–1823.

Drobeck, P. A., Hillemeyer, F. L., Frost, E. G., and Liebler, G. S., 1986, The Picacho mine; A gold mineralized detachment in southeastern California, *in* Beatty, B., and Wilkinson, P.A.K., eds., Frontiers in geology and ore deposits of Arizona and the Southwest: Arizona Geological Society Digest, v. 16, p. 187–221.

Dunne, G. C., Moore, J. N., Anderson, D., and Galbraith, G., 1975, The Bean Canyon Formation of the Tehachapi Mountains, California; An early Mesozoic arc-trench gap deposit?: Geological Society of America Abstracts with Programs, v. 7, p. 314.

Durham, D. L., 1965, Evidence of large strike-slip displacement along a fault in the southern Salinas Valley, California, *in* Geological Survey research 1965: U.S. Geological Survey Professional Paper 525-D, p. D106–D111.

—— , 1974, Geology of the southern Salinas Valley area, California: U.S. Geological Survey Professional Paper 819, 111 p.

Durham, J. W., Jahns, R. H., and Savage, D. E., 1954, Marine-nonmarine relationships in the Cenozoic section of California, *in* Jahns, R. H., ed., Geology of southern California: California Division of Mines Bulletin 170, chapter 3, p. 59–71.

Dutcher, L. C., and Garrett, A. A., 1963, Geologic and hydrologic features of the San Bernardino area, California: U.S. Geological Survey Water-Supply Paper 1419, 114 p.

Eaton, J. E., 1932, Decline of Great Basin, southwestern United States: American Association of Petroleum Geologists Bulletin, v. 16, p. 1–49.

—— , 1939, Ridge basin, California: American Association of Petroleum Geologists Bulletin, v. 23, p. 517–558.

Eaton, J. P., and Rymer, M. J., 1990, Regional seismotectonic model for the southern Coast Ranges, *in* Rymer, M. J., and Ellsworth, W. L., eds., The Coalinga, California, earthquake of May 2, 1983: U.S. Geological Survey Professional Paper 1487, p. 97–111.

Eckis, R., 1928, Alluvial fans of the Cucamonga district, southern California: Journal of Geology, v. 36, p. 224–247.

—— , 1930, Geology of a portion of the Indio Quadrangle [M.S. thesis]: Pasadena, California Institute of Technology.

—— , 1934, South coastal basin investigation; Geology and groundwater storage capacity of valley fill: California Division of Water Resources Bulletin 45, 279 p.

Ehlert, K. W., 1982, Basin analysis of the Miocene Mint Canyon Formation, southern California, *in* Ingersoll, R. V., and Woodburne, M. O., eds., Cenozoic nonmarine deposits of California and Arizona: Pacific Section, Society of Economic Paleontologists and Mineralogists, p. 51–64.

Ehlert, K. W., and Ehlig, P. L., 1977, The "polka-dot" granite and rate of displacement on the San Andreas fault in southern California: Geological Society of America Abstracts with Programs, v. 9, p. 415–416.

Ehlig, P. L., 1966, Displacement along the San Gabriel fault, San Gabriel Mountains, southern California, *in* Abstracts for 1966: Geological Society of America Special Paper 101, p. 60.

—— , 1968, Causes of distribution of Pelona, Rand, and Orocopia Schists along the San Andreas and Garlock faults, *in* Dickinson, W. R., and Grantz, A., eds., Proceedings of Conference on Geologic Problems of San Andreas Fault System: Stanford, California, Stanford University Publications in the Geological Sciences, v. 11, p. 294–306.

—— , 1973, History, seismicity and engineering geology of the San Gabriel fault, *in* Moran, D. E., Slosson, J. E., Stone, R. O., and Yelverton, C. A., eds., Geology, seismicity, and environmental impact: Association of Engineering Geologists Special Publication October 1973, p. 247–251.

—— , 1975a, Basement rocks of the San Gabriel Mountains, south of the San Andreas fault, southern California, *in* Crowell, J. C., ed., San Andreas fault in southern California: California Division of Mines and Geology Special Report 118, p. 177–186.

—— , 1975b, Geologic framework of the San Gabriel Mountains, *in* Oakeshott, G. B., ed., San Fernando, California, earthquake of 9 February 1971: California Division of Mines and Geology Bulletin 196, p. 7–18.

—— , 1976, Magnitude and timing of displacement on the San Andreas fault in southern California and its palinspastic implications [abs.]: American Association of Petroleum Geologist Bulletin, v. 60, p. 668.

—— , 1977, Structure of the San Andreas fault zone in San Gorgonio Pass, southern California: Geological Society of America Abstracts with Programs, v. 9, p. 416.

—— , 1981, Origin and tectonic history of the basement terrane of the San Gabriel Mountains, central Transverse Ranges, *in* Ernst, W. G., ed., The geotectonic development of California; Rubey Volume 1: Englewood Cliffs, New Jersey, Prentice-Hall, p. 253–283.

—— , 1982, The Vincent thrust; Its nature, paleogeographic reconstruction across the San Andreas fault and bearing on the evolution of the Transverse Ranges, *in* Fife, D. L., and Minch, J. A., eds., Geology and mineral wealth of the California Transverse Ranges; Mason Hill Volume: Santa Ana, California, South Coast Geological Society Annual Symposium and Guidebook 10, p. 370–379.

—— , 1986, Neotectonics in the area between the central and western Transverse Ranges; An introduction, *in* Ehlig, P. L., compiler, Neotectonics and faulting in southern California; Geological Society of America Cordilleran Section Annual Meeting Guidebook and Volume: Los Angeles, California State University Department of Geology, p. 1–4.

Ehlig, P. L., and Crowell, J. C., 1982, Mendenhall Gneiss and anorthosite-related rocks bordering Ridge basin, southern California, *in* Crowell, J. C., and Link, M. H., eds., Geologic history of Ridge basin, southern California, *in* Crowell, J. C., and Link, M. H., eds., Geologic history of Ridge basin, southern California: Pacific Section, Society of Economic Paleontologists and Mineralogists, p. 199–202.

Ehlig, P. L., and Ehlert, K. W., 1972, Offset of Miocene Mint Canyon Formation from volcanic source along San Andreas fault, southern California: Geological Society of America Abstracts with Programs, v. 4, p. 154.

Ehlig, P. L., Ehlert, K. W., and Crowe, B. M., 1975, Offset of the upper Miocene Caliente and Mint Canyon Formations along the San Gabriel and San Andreas faults, *in* Crowell, J. C., ed., San Andreas fault in southern California: California Division of Mines and Geology Special Report 118, p. 83–92.

Elders, W. A., Rex, R. W., Meidav, T., Robinson, P. T., and Biehler, S., 1972, Crustal spreading in southern California: Science, v. 178, no. 4056, p. 15–24.

Eldridge, G. H., and Arnold, R., 1907, The Santa Clara Valley, Puente Hills and Los Angeles oil districts, southern California: U.S. Geological Survey Bulletin 309, 266 p.

English, H. D., 1953, The geology of the San Timoteo badlands, Riverside County, California [M.A. thesis]: Pomona, California, Pomona College.

Ensley, R. A., and Verosub, K. L., 1982, Biostratigraphy and magnetostratigraphy of southern Ridge basin, central Transverse Ranges, California, *in* Crowell, J. C., and Link, M. H., eds., Geologic history of Ridge basin, southern California: Pacific Section, Society of Economic Paleontologists and Mineralogists, p. 13–24.

Ernst, W. G., ed., 1981, The geotectonic development of California; Rubey Volume 1: Englewood Cliffs, New Jersey, Prentice-Hall, 706 p.

Erskine, B. G., 1986a, Mylonitic deformation and associated low-angle faulting in the Santa Rosa mylonite zone, southern California [Ph.D. thesis]: Berkeley, University of California, 247 p.

—— , 1986b, Syn-tectonic granitic intrusion and mylonitic deformation along the eastern margin of the northern Peninsular Ranges batholith, southern California: Geological Society of America Abstracts with Programs, v. 18, p. 105.

Erskine, B. G., and Wenk, H.-R., 1985, Evidence for Late Cretaceous crustal thinning in the Santa Rosa mylonite zone, southern California: Geology, v. 13, p. 274–277.

Evans, J. G., 1982, Geology of the Sheep Mountain Wilderness Study Area and the Cucamonga Wilderness and additions, Los Angeles and San Bernardino Counties, California: U.S. Geological Survey Bulletin 1506-A, p. 5–28.

Farley, T., 1979, Geology of a part of northern San Gorgonio Pass, California [M.S thesis]: Los Angeles, California State University, 159 p.

Farley, T., and Ehlig, P. L., 1977, Displacement on the Punchbowl fault based on occurrence of "polka-dot" granite clasts: Geological Society of America Abstracts with Programs, v. 9, p. 419.

Fletcher, G. L., 1967, Post late Miocene displacement along the San Andreas fault zone, central California, *in* Gabilan Range and adjacent San Andreas fault: Pacific Sections, American Association of Petroleum Geologists and Society of Economic Paleontologists and Mineralogists Guidebook, p. 74–80.

Fox, K. F., Jr., 1976, Melanges in the Franciscan Complex, a product of triple-junction tectonics: Geology, v. 4, p. 737–740.

Fox, K. F., Jr., Fleck, R. J., Curtis, G. H., and Meyer, C. E., 1985, Implications of the northwestwardly younger age of the volcanic rocks of west-central California: Geological Society of America Bulletin, v. 96, p. 647–654.

Frei, L. S., Magill, J. R., and Cox, A., 1984, Paleomagnetic results from the central Sierra Nevada; Constraints on reconstructions of the western United States: Tectonics, v. 3, p. 157–177.

Freund, R., 1974, Kinematics of transform and transcurrent faults: Tectonophysics, v. 21, p. 93–134.

Frick, C., 1921, Extinct vertebrate faunas of the badlands of Bautista Creek and San Timoteo Cañon, southern California: Berkeley, University of California Publications in Geology, v. 12, p. 277–424.

Frizzell, V. A., Jr., Mattinson, J. M., and Matti, J. C., 1986, Distinctive Triassic megoporphyritic monzogranite; Evidence for only 160 km offset along the San Andreas fault, southern California: Journal of Geophysical Research, v. 91, no. B14, p. 14080–14088.

Frost, E. G., Martin, D. L., and Krummenacher, D., 1982, Mid-Tertiary detachment faulting in southwestern Arizona and southeastern California and its overprint on the Vincent thrust system: Geological Society of America Abstracts with Programs, v. 14, p. 164.

Fry, J. G., Bottjer, D. J., and Lund, S. P., 1985, Magnetostratigraphy of displaced Upper Cretaceous strata in southern California: Geology, v. 13, p. 648–651.

Fuis, G. S., and Kohler, W. M., 1984, Crustal structure and tectonics of the Imperial Valley region, California, *in* Rigsby, C. A., ed., The Imperial basin; Tectonics, sedimentation, and thermal aspects: Pacific Section, Society of Economic Paleontologists and Mineralogists, p. 1–13.

Fuis, G. S., Friedman, M. E., and Hileman, J. A., 1977, Preliminary catalog of earthquakes in southern California, July 1974–September, 1976: Pasadena, California Institute of Technology Seismological Laboratory, U.S. Geological Survey Open-File Report 77-181, 107 p.

Fuis, G. S., Mooney, W. D., Healey, J. H., McMechan, G. A., and Lutter, W. J., 1982, Crustal structure of the Imperial Valley region, *in* The Imperial Valley, California, earthquake of October 15, 1979: U.S. Geological Survey Professional Paper 1254, p. 25–49.

Garfunkel, Z., 1973, History of the San Andreas fault as a plate boundary: Geological Society of America Bulletin, v. 84, p. 2035–2042.

—— , 1974, Model for the late Cenozoic tectonic history of the Mojave Desert, California, and for its relation to adjacent regions: Geological Society of America Bulletin, v. 85, p. 1931–1944.

Garfunkel, Z., and Ron, H., 1985, Block rotation and deformation by strike-slip faults; 2. The properties of a type of macroscopic discontinuous deformation: Journal of Geophysical Research, v. 90, no. B10, p. 8589–8602.

Garner, W. E., Frost, E. G., Tanges, S. E., and Germinario, M. P., 1982, Mid-Tertiary detachment faulting and mineralization in the Trigo Mountains, Yuma County, Arizona, *in* Frost, E. G., and Martin, D. L., eds., Mesozoic-Cenozoic tectonic evolution of the Colorado River region, California, Arizona, and Nevada; Anderson-Hamilton Volume; Geological Society of America Cordilleran Section Annual Meeting Symposium and Field Trip Volume: San Diego, California, Cordilleran Publishers, p. 159–171.

Gastil, G., 1985, Terranes of peninsular California and adjacent Sonora, *in* Howell, D. G., ed., Tectonostratigraphic terranes of the circum-Pacific region: Circum-Pacific Council for Energy and Mineral Resources Earth Science Series, no. 1, p. 273–283.

Gastil, G., Morgan, G., and Krummenacher, D., 1981, The tectonic history of Peninsular California and adjacent Mexico, *in* Ernst, W. G., ed., The geotectonic development of California; Rubey Volume 1: Englewood Cliffs, New Jersey, Prentice-Hall, p. 284–306.

Gawthrop, W. H., 1978, Seismicity and tectonics of the central California coastal region, *in* Silver, E. A., and Normark, W. R., eds., San Gregorio–Hosgri fault zone, California: California Division of Mines and Geology Special Report 137, p. 45–56.

Gibson, R. C., 1971, Non-marine turbidites and the San Andreas fault, San Bernardino Mountains, California, *in* Elders, W. A., ed., Geological excur-

sions in southern California; Geological Society of America Cordilleran Section Annual Meeting Guidebook: Riverside, University of California Campus Museum Contributions, no. 1, p. 167–181.

Golombek, M. P., and Brown, L. L., 1988, Clockwise rotation of the western Mojave Desert: Geology, v. 16, p. 126–130.

Golz, D. J., Jefferson, G. T., and Kennedy, M. P., 1977, Late Pliocene vertebrate fossils from the Elsinore fault zone, California: Journal of Paleontology, v. 51, p. 864–866.

Goodmacher, J., Barnett, L., Buckner, G., Ouachrif, L., Vidigal, A., and Frost, E., 1989, The Clemens Well fault in the Orocopia Mountains of southern California; A strike-slip or normal fault structure?: Geological Society of America Abstracts with Programs, v. 21, p. 85.

Gordon, S. A., 1979, Relations between the Santa Ynez fault zone and the Pine Mountain thrust fault system, Piru Mountains, California: Geological Society of America Abstracts with Programs, v. 11, p. 80.

Graham, S. A., 1978, Role of Salinian block in evolution of San Andreas fault system, California: American Association of Petroleum Geologists Bulletin, v. 62, p. 2214–2231.

—— , 1979, Tertiary paleotectonics and paleogeography of the Salinian block, *in* Armentrout, J. M., Cole, M. R., and TerBest, H., Jr., eds., Cenozoic paleogeography of the western United States: Pacific Section, Society of Economic Paleontologist and Mineralogists Pacific Coast Paleogeography Symposium 3, p. 45–51.

Graham, S. A., and Dickinson, W. R., 1978a, Apparent offsets of on-land geologic features across the San Gregorio–Hosgri fault trend, *in* Silver, E. A., and Normark, W. R., eds., San Gregorio–Hosgri fault zone, California: California Division of Mines and Geology Special Report 137, p. 13–23.

—— , 1978b, Evidence for 115 kilometers of right slip on the San Gregorio–Hosgri fault trend: Science, v. 199, no. 4325, p. 179–181.

Graham, S. A., and Williams, L. A., 1985, Tectonic, depositional, and diagenetic history of Monterey Formation (Miocene), central San Joaquin basin, California: American Association of Petroleum Geologists Bulletin, v. 69, p. 385–411.

Graham, S. A., Stanley, R. G., Bent, J. V., and Carter, J. B., 1989, Oligocene and Miocene paleogeography of central California and displacement along the San Andreas fault: Geological Society of America Bulletin, v. 101, p. 711–730.

Grant, U. S., IV, and Gale, H. R., 1931, Catalogue of the marine Pliocene and Pleistocene Mollusca of California and adjacent regions: San Diego Society of Natural History Memoir, v. 1, 1036 p.

Gray, C. H., Jr., 1961, Geology and mineral resources of the Corona South Quadrangle and the Santa Ana Narrows area, Riverside, Orange, and San Bernardino Counties, California: California Division of Mines Bulletin 178, 120 p.

Greene, H. G., 1977, Geology of the Monterey Bay region: U.S. Geological Survey Open-File Report 77-718, 347 p.

Greene, H. G., and Clark, J. C., 1979, Neogene paleogeography of the Monterey Bay area, California, *in* Armentrout, J. M., Cole, M. R., and TerBest, H., Jr., eds., Cenozoic paleogeography of the western United States: Pacific Section, Society of Economic Paleontologists and Mineralogists Pacific Coast Paleogeography Symposium 3, p. 277–296.

Greene, H. G., Lee, W.H.K., McCulloch, D. S., and Brabb, E. E., 1973, Faults and earthquakes in the Monterey Bay region, California: U.S. Geological Survey Miscellaneous Field Studies Map MF-518, scale 1:200,000.

Griscom, A., and Jachens, R. C., 1989, Tectonic history of the north portion of the San Andreas fault system, California, inferred from gravity and magnetic anomalies: Journal of Geophysical Research, v. 94, no. B3, p. 3089–3099.

Grose, L. T., 1959, Structure and petrology of the northeast part of the Soda Mountains, San Bernardino County, California: Geological Society of America Bulletin, v. 70, p. 1509–1547.

Guillou, R. B., 1953, Geology of the Johnston Grade area, San Bernardino County, California: California Division of Mines Special Report 31, 18 p.

Hadley, D., and Kanamori, H., 1977, Seismic structure of the Transverse Ranges, California: Geological Society of America Bulletin, v. 88, p. 1469–1478.

Hadley, D. M., and Kanamori, H., 1978, Recent seismicity in the San Fernando region and tectonics in the west-central Transverse Ranges, California: Seismological Society of America Bulletin, v. 68, p. 1449–1457.

Hagstrum, J. T., McWilliams, M., Howell, D. G., and Grommé, S., 1985, Mesozoic paleomagnetism and northward translation of the Baja California peninsula: Geological Society of America Bulletin, v. 96, p. 1077–1090.

Hagstrum, J. T., Sawlan, M. G., Hausback, B. P., Smith, J. G., and Grommé, C. S., 1987, Miocene paleomagnetism and tectonic setting of the Baja California Peninsula, Mexico: Journal of Geophysical Research, v. 92, no. B3, p. 2627–2639.

Hall, C. A., Jr., 1975, San Simeon–Hosgri fault system, coastal California; Economic and environmental implications: Science, v. 190, no. 4221, p. 1291–1294.

—— , 1978, Origin and development of the Lompoc–Santa Maria pull-apart basin and its relation to the San Simeon–Hosgri strike-slip fault, western California, in Silver, E. A., and Normark, W. R., eds., San Gregorio–Hosgri fault zone, California: California Division of Mines and Geology Special Report 137, p. 25–31.

—— , 1981, Evolution of the western Transverse Ranges microplate; Late Cenozoic faulting and basinal development, in Ernst, W. G., ed., The geotectonic development of California; Rubey Volume 1: Englewood Cliffs, New Jersey, Prentice-Hall, p. 559–582.

Hamilton, D. H., 1982, The proto–San Andreas fault and the early history of the Rinconada, Nacimiento, and San Gregorio–Hosgri faults of coastal central California: EOS Transactions of the American Geophysical Union, v. 63, p. 1124.

Hamilton, D. H., and Willingham, C. R., 1977, Hosgri fault zone; Structure, amount of displacement, and relationship to structures of the western Transverse Ranges: Geological Society of America Abstracts with Programs, v. 9, p. 429.

—— , 1978, Evidence for a maximum of 20 km of Neogene right slip along the San Gregorio fault zone of central California: EOS Transactions of the American Geophysical Union, v. 59, p. 1210.

Hamilton, W., 1961, Origin of the Gulf of California: Geological Society of America Bulletin, v. 72, p. 1307–1318.

—— , 1987, Mesozoic geology and tectonics of the Big Maria Mountains region, southeastern California, in Dickinson, W. R., and Klute, M. A., Mesozoic rocks of southern Arizona and adjacent areas: Arizona Geological Society Digest, v. 18, p. 33–47.

Hamilton, W., and Myers, W. B., 1966, Cenozoic tectonics of the western United States: Reviews of Geophysics, v. 4, no. 4, p. 509–549.

Hanson, D. W., 1983, Faulting in the northern Simi Valley area, Ventura County, California, in Squires, R. L., and Filewicz, M. V., eds., Cenozoic geology of the Simi Valley area, southern California: Pacific Section, Society of Economic Paleontologists and Mineralogists, p. 225–232.

Hanson, K. L., and Lettis, W. R., 1990, Pleistocene slip rates for the San Simeon fault zone based on offset marine terraces and displaced drainages, in Lettis, W. R., Hanson, K. L., Kelson, K. I., and Wesling, J. R., eds., Neotectonics of south-central coastal California; Pacific Cell, Friends of the Pleistocene Field Trip Guidebook: San Francisco, Geomatrix, p. 191–224.

Harden, J. W., and Matti, J. C., 1989, Holocene and late Pleistocene slip rates on the San Andreas fault in Yucaipa, California, using displaced alluvial-fan deposits and soil chronology: Geological Society of America Bulletin, v. 101, p. 1107–1117.

Harding, L. E., and Coney, P. J., 1985, The geology of the McCoy Mountains Formation, southeastern California and southwestern Arizona: Geological Society of America Bulletin, v. 96, p. 755–769.

Harding, T. P., 1976, Tectonic significance and hydrocarbon trapping consequences of sequential folding synchronous with San Andreas faulting, San Joaquin Valley, California: American Association of Petroleum Geologists Bulletin, v. 60, p. 356–378.

Hart, E. W., 1976, Basic geology of the Santa Margarita area, San Luis Obispo County, California: California Division of Mines and Geology Bulletin 199, 45 p.

Hatheway, A. W., and West, L. M., 1975, Terminus of the Pinto Mountain fault, near Twentynine Palms California: Geological Society of America Abstracts with Programs, v. 7, p. 323.

Hauksson, E., and Jones, L. M., 1991, The 1988 and 1990 Upland earthquakes; Left-lateral faulting adjacent to the central Transverse Ranges: Journal of Geophysical Research, v. 96, no. B5, p. 8143–8165.

Hawkins, H. G., 1975, Strike-slip displacement along the Camp Rock–Emerson fault, central Mojave Desert, San Bernardino County, California: Geological Society of America Abstracts with Programs, v. 7, p. 324.

Hawkins, H. G., and McNey, J. L., 1979, Homestead Valley earthquake swarm, San Bernardino County, California: California Division of Mines and Geology, California Geology, v. 32, p. 222–224.

Haxel, G. B., 1977, The Orocopia Schist and the Chocolate Mountain thrust, Picacho–Peter Kane Mountain area, southeasternmost California [Ph.D. thesis]: Santa Barbara, University of California, 277 p.

Haxel, G. B., and Dillon, J., 1973, The San Andreas fault system in southeasternmost California, in Kovach, R. L., and Nur, A., eds., Proceedings of the Conference on Tectonic Problems of the San Andreas Fault System: Stanford, California, Stanford University Publications in the Geological Sciences, v. 13, p. 322–333.

—— , 1978, The Pelona-Orocopia Schist and Vincent–Chocolate Mountain thrust system, southern California, in Howell, D. G., and McDougall, K. A., eds., Mesozoic paleogeography of the western United States: Pacific Section, Society of Economic Paleontologists and Mineralogists Pacific Coast Paleogeography Symposium 2, p. 453–469.

Haxel, G. B., and Tosdal, R. M., 1986, Significance of the Orocopia Schist and Chocolate Mountains thrust in the late Mesozoic tectonic evolution of the southeastern California–southwestern Arizona region; Extended abstract, in Beatty, B., and Wilkinson, P.A.K., eds., Frontiers in geology and ore deposits of Arizona and the Southwest: Arizona Geological Society Digest, v. 16, p. 52–61.

Haxel, G. B., Mueller, K., Frost, E., and Silver, L. T., 1982, Mid-Tertiary detachment faulting in the northernmost Mohawk Mountains, southwestern Arizona: Geological Society of America Abstracts with Programs, v. 14, p. 171.

Haxel, G. B., Tosdal, R. M., and Dillon, J. T., 1985, Tectonic setting and lithology of the Winterhaven Formation; A new Mesozoic stratigraphic unit in southeasternmost California and southwestern Arizona: U.S. Geological Survey Bulletin 1599, 19 p.

Haxel, G. B., and 6 others, 1988, Mineral resources of the Orocopia Mountains Wilderness Study Area, Riverside County, California: U.S. Geological Survey Bulletin 1710-E, 22 p.

Hays, W. H., 1957, Geology of the central Mecca Hills, Riverside County, California [Ph.D. thesis]: New Haven, Connecticut, Yale University, 324 p.

Hazzard, J. C., and Dosch, E. F., 1937, Archean rocks in the Piute and Old Woman Mountains, San Bernardino County, California [abs.]: Geological Society of America Proceedings for 1936, p. 308–309.

Hendrix, E. D., and Ingersoll, R. V., 1987, Tectonics and alluvial sedimentation of the upper Oligocene/lower Miocene Vasquez Formation, Soledad basin, southern California: Geological Society of America Bulletin, v. 98, p. 647–663.

Henry, C. D., 1989, Late Cenozoic Basin and Range structure in western Mexico adjacent to the Gulf of California: Geological Society of America Bulletin, v. 101, p. 1147–1156.

Hewett, D. F., 1954a, A fault map of the Mojave Desert region, in Jahns, R. H., ed., Geology of southern California: California Division of Mines Bulletin 170, chapter 4, p. 15–18.

—— , 1954b, General geology of the Mojave Desert region, California, in Jahns, R. H., ed., Geology of southern California: California Division of Mines Bulletin 170, chapter 2, p. 5–20.

Hibbard, C. W., Ray, D. E., Savage, D. E., Taylor, D. W., and Guilday, J. E., 1965, Quaternary mammals of North America, in Wright, H. E., Jr., and Frey, D. G., eds., The Quaternary of the United States: Princeton, New Jersey, Princeton University Press, p. 509–525.

Hill, D. P., Eaton, J. P., and Jones, L. M., 1990, Seismicity, 1980–86, *in* Wallace, R. E., ed., The San Andreas fault system, California: U.S. Geological Survey Professional Paper 1515, p. 115–151.

Hill, M. L., 1930, Structure of the San Gabriel Mountains, north of Los Angeles, California, with a foreword by F. S. Hudson: Berkeley, University of California Department of Geological Sciences Bulletin, v. 19, no. 6, p. 137–170.

—— , 1954, Tectonics of faulting in southern California, *in* Jahns, R. H., ed., Geology of southern California: California Division of Mines Bulletin 170, chapter 4, p. 5–13.

—— , 1971, A test of new global tectonics; Comparison of northeast Pacific and California structures: American Association of Petroleum Geologists Bulletin, v. 55, p. 3–9.

—— , 1981, San Andreas fault; History of concepts: Geological Society of America Bulletin, v. 92, part 1, p. 112–131.

Hill, M. L., and Dibblee, T. W., Jr., 1953, San Andreas, Garlock, and Big Pine faults, California; A study of the character, history, and tectonic significance of their displacements: Geological Society of America Bulletin, v. 64, p. 443–458.

Hill, M. L., and Hobson, H. D., 1968, Possible post-Cretaceous slip on the San Andreas fault zone, *in* Dickinson, W. R., and Grantz, A., ed., Proceedings of Conference on Geologic Problems of San Andreas Fault System: Stanford, California, Stanford University Publications in the Geological Sciences, v. 11, p. 123–129.

Hill, M. L., Carlson, S. A., and Dibblee, T. W., Jr., 1958, Stratigraphy of Cuyama Valley–Caliente Range area, California: American Association of Petroleum Geologists Bulletin, v. 42, p. 2973–3000.

Hill, R. I., 1981, Geology of Garner Valley and vicinity, *in* Brown, A. R., and Ruff, R. W., eds., Geology of the San Jacinto Mountains: Santa Ana, California, South Coast Geological Society Annual Field Trip Guidebook 9, p. 90–99.

Hill, R. I., and Silver, L. T., 1979, Strontium isotopic variability in the pluton of San Jacinto Peak, southern California: EOS Transactions of the American Geophysical Union, v. 61, p. 411.

Hill, R. L., and Beeby, D. J., 1977, Surface faulting associated with the 5.2 magnitude Galway Lake earthquake of May 31, 1975; Mojave Desert, San Bernardino County, California: Geological Society of America Bulletin, v. 88, p. 1378–1384.

Hill, R. L., and 5 others, 1980, Geologic study of the Homestead Valley earthquake swarm of March 15, 1979: California Division of Mines and Geology, California Geology, 33, p. 60–67.

Hill, R. T., 1928, Southern California geology and Los Angeles earthquakes: Los Angeles, Southern California Academy of Sciences, 232 p.

Hoeppener, R., Kalthoff, E., and Schrader, P., 1969, Zur physikalischen Tektonik; Bruchbildung bei verschiedenen affinen Deformationen im Experiment: Geologische Rundschau, v. 59, p. 179–193.

Hoots, H. W., 1931, Geology of the eastern part of the Santa Monica Mountains, Los Angeles County, California, *in* Shorter contributions to general geology, 1930: U.S. Geological Survey Professional Paper 165, p. 83–134.

Hope, R. A., 1966, Geology and structural setting of the eastern Transverse Ranges, southern California [Ph.D. thesis]: Los Angeles, University of California, 158 p.

—— , 1969, The Blue Cut fault, southeastern California, *in* Geological Survey Research 1969: U.S. Geological Survey Professional Paper 650-D, p. D116–D121.

Hornafius, J. S., 1985, Neogene tectonic rotation of the Santa Ynez Range, western Transverse Ranges, California, suggested by paleomagnetic investigation of the Monterey Formation: Journal of Geophysical Research, v. 90, no. B14, p. 12503–12522.

Hornafius, J. S., Luyendyk, B. P., Terres, R. R., and Kamerling, M. J., 1986, Timing and extent of Neogene tectonic rotation in the western Transverse Ranges, California: Geological Society of America Bulletin, v. 97, p. 1476–1487.

Howard, K. A., Miller, D. M., and John, B. E., 1982, Regional character of mylonitic gneiss in the Cadiz Valley area, southeastern California, *in* Frost, E. G., and Martin, D. L., eds., Mesozoic-Cenozoic tectonic evolution of the Colorado River region, California, Arizona, and Nevada; Anderson-Hamilton Volume; Geological Society of America Cordilleran Section Annual Meeting Symposium and Field Trip Volume: San Diego, California, Cordilleran Publishers, p. 441–447.

Howell, D. G., 1975a, Early and middle Eocene shoreline offset by the San Andreas fault, southern California, *in* Crowell, J. C., ed., San Andreas fault in southern California: California Division of Mines and Geology Special Report 118, p. 69–74.

—— , 1975b, Hypothesis suggesting 700 km of right slip in California along northwest-oriented faults: Geology, v. 3, p. 81–83.

—— , 1976, A model to accommodate 1000 kilometres of right-slip, Neogene, displacement in the southern California area, *in* Howell, D. G., ed., Aspects of the geologic history of the California continental borderland: Pacific Section, American Association of Petroleum Geologists Miscellaneous Publication 24, p. 530–540.

Howell, D. G., and Vedder, J. G., 1978, Late Cretaceous paleogeography of the Salinian block, California, *in* Howell, D. G., and McDougall, K. A., eds., Mesozoic paleogeography of the western United States: Pacific Section, Society of Economic Paleontologists and Mineralogists Pacific Coast Paleogeography Symposium 2, p. 523–534.

Howell, D. G., Stuart, C. J., Platt, J. P., and Hill, D. J., 1974, Possible strike-slip faulting in the southern California borderland: Geology, v. 2, p. 93–98.

Howell, D. G., McLean, H., and Vedder, J. G., 1976, Cenozoic tectonism on Santa Cruz Island, *in* Howell, D. G., ed., Aspects of the geologic history of the California continental borderland: Pacific Section, American Association of Petroleum Geologists Miscellaneous Publication 24, p. 392–415.

Hsu, K. J., 1955, Granulites and mylonites of the region about Cucamonga and San Antonio Canyons, San Gabriel Mountains, California: Berkeley, University of California Publications in Geological Sciences, v. 30, no. 4, p. 223–352.

Hudson, F. S., 1922, Geology of the Cuyamaca region of California, with special reference to the origin of the nickeliferous pyrrhotite: Berkeley, University of California Department of Geological Sciences Bulletin, v. 13, p. 175–252.

Huffman, O. F., 1970, Miocene and post Miocene offset on the San Andreas fault in central California: Geological Society of America Abstracts with Programs, v. 2, p. 104–105.

—— , 1972, Lateral displacement of upper Miocene rocks and the Neogene history of offset along the San Andreas fault in central California: Geological Society of America Bulletin, v. 83, p. 2913–2946.

Huffman, O. F., Turner, D. L., and Jack, R. N., 1973, Offset of Late Oligocene-early Miocene volcanic rocks along the San Andreas fault in central California, *in* Kovach, R. L., and Nur, A., eds., Proceedings of the Conference on Tectonic Problems of the San Andreas Fault System: Stanford, California, Stanford University Publications in the Geological Sciences, v. 13, p. 368–373.

Hulin, C. D., 1925, Geology and ore deposits of the Randsburg Quadrangle, California: California Mining Bureau Bulletin 95, 152 p.

Humphreys, E. D., and Weldon, R. J., II, 1991, Kinematic constraints on the rifting of Baja California, *in* Dauphin, J. P., and Simoneit, B.R.T., eds., The gulf and peninsular province of the Californias: American Association of Petroleum Geologists Memoir 47, p. 217–229.

Imlay, R. W., 1963, Jurassic fossils from southern California: Journal of Paleontology, v. 37, p. 97–107.

—— , 1964, Middle and Upper Jurassic fossils from southern California: Journal of Paleontology, v. 38, p. 505–509.

Irwin, W. P., 1990, Geology and plate-tectonic development, *in* Wallace, R. E., ed., The San Andreas fault system, California: U.S. Geological Survey Professional paper 1515, p. 60–80.

Jackson, J., and Molnar, P., 1990, Active faulting and block rotations in the western Transverse Ranges, California: Journal of Geophysical Research, v. 95, no. B13, p. 22073–22087.

Jacobson, C. E., Dawson, M. R., and Postlethwaite, C. E., 1988, Structure, metamorphism, and tectonic significance of the Pelona, Orocopia, and Rand

Schists, southern California, *in* Ernst, W. G., ed., Metamorphism and crustal evolution of the western United States; Rubey Volume 7: Englewood Cliffs, New Jersey, Prentice-Hall, p. 976–997.

Jahns, R. H., 1940, Stratigraphy of the easternmost Ventura basin, California, with a description of a new lower Miocene mammalian fauna from the Tick Canyon Formation: Carnegie Institution of Washington Publication 514, p. 145–194.

——, ed., 1954, Geology of southern California: California Division of Mines Bulletin 170.

——, 1973, Tectonic evolution of the Transverse Ranges province as related to the San Andreas fault system, *in* Kovach, R. L., and Nur, A., eds., Proceedings of the Conference on Tectonic Problems of the San Andreas Fault System: Stanford, California, Stanford University Publications in the Geological Sciences, v. 13, p. 149–170f.

Jahns, R. H., and Muehlberger, W. R., 1954, Geology of the Soledad basin, Los Angeles County, *in* Jahns, R. H., ed., Geology of southern California: California Division of Mines Bulletin 170, Map Sheet 6.

Jahns, R. H., and Proctor, R. J., 1975, The San Gabriel and Santa Susana–Sierra Madre fault zones in the western and central San Gabriel Mountains, southern California: Geological Society of America Abstracts with Programs, v. 7, p. 329.

Jahns, R. H., and Wright, L. A., 1960, Garlock and Death Valley fault zones in the Avawatz Mountains, California [abs.]: Geological Society of America Bulletin, v. 71, p. 2063–2064.

James, A. I., 1986, Structural geology of the Rand thrust, northwestern Mojave Desert, California: Geological Society of America Abstracts with Programs, v. 18, p. 120–121.

James, E. W., 1986, U/Pb age of the Antimony Peak tonalite and its relation to Rand Schist in the San Emigdio Mountains, California: Geological Society of America Abstracts with Programs, v. 18, p. 121.

——, 1987, Extension of the Independence dike swarm to the western Mojave Desert and eastern Transverse Ranges of California: Geological Society of America Abstracts with Programs, v. 19, p. 715.

——, 1989, Southern extension of the Independence dike swarm of eastern California: Geology, v. 17, p. 587–590.

James, E. W., and Mattinson, J. M., 1985, Evidence for 160 km of post–mid-Cretaceous slip on the San Gregorio fault, coastal California: EOS Transactions of the American Geophysical Union, v. 66, p. 1093.

——, 1988, Metamorphic history of the Salinian block; An isotopic reconnaissance, *in* Ernst, W. G., ed., Metamorphism and crustal evolution of the western United States; Rubey Volume 7: Englewood Cliffs, New Jersey, Prentice-Hall, p. 938–952.

James, G. T., 1963, Paleontology and nonmarine stratigraphy of the Cuyama Valley badlands, California; Part 1. Geology, faunal interpretations, and systematic descriptions of Chiroptera, Insectivora, and Rodentia: California University Publications in Geological Sciences, v. 45, 171 p.

Jennings, C. W., compiler, 1977, Geologic map of California: California Division of Mines and Geology, scale 1:750,000.

Jennings, C. W., and Troxel, B. W., 1954, Ventura basin, *in* Jahns, R. H., ed., Geology of southern California: California Division of Mines Bulletin 170, Geologic Guide 2, 63 p.

Johnson, H. R., and Warren, V. C., 1927, Geological and structural conditions of the San Gabriel Valley region, *in* Conkling, Harold, San Gabriel Investigation: California Division of Water Rights Bulletin 5, p. 73–100.

Johnson, J. D., and Normark, W. R., 1974, Neogene tectonic evolution of the Salinian block, west-central California: Geology, v. 2, p. 11–14.

Johnson, N. M., and 5 others, 1983, Rates of late Cenozoic tectonism in the Vallecito–Fish Creek basin, western Imperial Valley, California: Geology, v. 11, p. 664–667.

Jones, D. L., Blake, M. C., Jr., and Rangin, C., 1976, The four Jurassic belts of northern California and their significance to the geology of the southern California borderland, *in* Howell, D. G., ed., Aspects of the geologic history of the California continental borderland: Pacific Section, American Association of Petroleum Geologists Miscellaneous Publication 24, p. 343–362.

Jones, L. M., Sieh, K. E., Hauksson, E., and Hutton, L. K., 1990, The 3 December 1988, Pasadena, California earthquake; Evidence for strike-slip motion on the Raymond fault: Seismological Society of America Bulletin, v. 80, p. 474–482.

Joseph, S. E., 1981, Isotopic correlation of the La Panza Range granitic rocks with similar rocks in the central and eastern Transverse Ranges [M.S. thesis]: Los Angeles, California State University, 77 p.

Joseph, S. E., Criscione, J. J., Davis, T. E., and Ehlig, P. L., 1982a, The Lowe igneous pluton, *in* Fife, D. L., and Minch, J. A., eds., Geology and mineral wealth of the California Transverse Ranges; Mason Hill Volume: Santa Ana, California, South Coast Geological Society Annual Symposium and Guidebook 10, p. 307–309.

Joseph, S. E., Davis, T. E., and Ehlig, P. L., 1982b, Strontium isotopic correlation of the La Panza Range granitic rocks with similar rocks in the central and eastern Transverse Ranges, *in* Fife, D. L., and Minch, J. A., eds., Geology and mineral wealth of the California Transverse Ranges; Mason Hill Volume: Santa Ana, California, South Coast Geological Society Annual Symposium and Guidebook 10, p. 310–320.

Kamerling, M. J., and Luyendyk, B. P., 1985, Paleomagnetism and Neogene tectonics of the northern Channel Islands, California: Journal of Geophysical Research, v. 90, no. B14, p. 12485–12502.

Kanter, L. R., and Debiche, M., 1985, Modeling the motion histories of the Point Arena and central Salinia terranes, *in* Howell, D. G., ed., Tectonostratigraphic terranes of the circum-Pacific region: Circum-Pacific Council for Energy and Mineral Resources Earth Science Series, no. 1, p. 227–238.

Kanter, L. R., and McWilliams, M. O., 1982, Rotation of the southernmost Sierra Nevada, California: Journal of Geophysical Research, v. 87, no. B5, p. 3819–3830.

Karish, C. R., Miller, E. L., and Sutter, J. F., 1987, Mesozoic tectonic and magmatic history of the central Mojave Desert, *in* Dickinson, W. R., and Klute, M. A., eds., Mesozoic rocks of southern Arizona and adjacent areas: Arizona Geological Society Digest, v. 18, p. 15–32.

Keller, E. A., Bonkowski, M. S., Korsch, R. J., and Shlemon, R. J., 1982, Tectonic geomorphology of the San Andreas fault zone in the southern Indio Hills, Coachella Valley, California: Geological Society of America Bulletin, v. 93, p. 46–56.

Keller, R. P., Allen, C. R., Gilman, R., Goulty, N. R., and Hileman, J. A., 1978, Monitoring slip along major faults in southern California: Seismological Society of America Bulletin, v. 68, p. 1187–1190.

Kelly, T. S., and Lander, E. B., 1988, Biostratigraphy and correlation of Hemingfordian and Barstovian land mammal assemblages, Caliente Formation, Cuyama Valley area, California, *in* Bazeley, W.J.M., ed., Tertiary tectonics and sedimentation in the Cuyama basin, San Luis Obispo, Santa Barbara, and Ventura Counties, California: Pacific Section, Society of Economic Paleontologists and Mineralogists, v. 59, p. 1–19.

Kennedy, M. P., 1977, Recency and character of faulting along the Elsinore fault zone in southern Riverside County, California: California Division of Mines and Geology Special Report 131, 12 p.

Kew, W.S.W., 1924, Geology and oil resources of a part of Los Angeles and Ventura Counties, California: U.S. Geological Survey Bulletin 753, 202 p.

Kistler, R. W., 1974, Phanerozoic batholiths in western North America; A summary of some recent work on variations in time, space, chemistry, and isotopic compositions: Annual Review of Earth and Planetary Sciences, v. 2, p. 403–418.

Kistler, R. W., and Peterman, Z. E., 1978, Reconstruction of crustal blocks of California on the basis of initial strontium isotopic compositions of Mesozoic granitic rocks: U.S. Geological Survey Professional Paper 1071, 17 p.

Kistler, R. W., Petersman, Z. E., Ross, D. C., and Gottfried, D., 1973, Strontium isotopes and the San Andreas fault, *in* Kovach, R. L., and Nur, A., eds., Proceedings of the Conference on Tectonic Problems of the San Andreas Fault System: Stanford, California, Stanford University Publications in the Geological Sciences, v. 13, p. 339–347.

Kleinpell, R. M., 1938, Miocene stratigraphy of California: Tulsa, Oklahoma, American Association of Petroleum Geologists, 450 p.

Kleinpell, R. M., Hornaday, G. R., Warren, A. D., and Donnelly, A. T., 1980, The Miocene stratigraphy of California revisited: American Association of Petroleum Geologists Studies in Geology, no. 11, p. 1–182.

Kooser, M. A., 1982, Stratigraphy and sedimentology of the type San Francisquito Formation, southern California, *in* Crowell, J. C., and Link, M. H., eds., Geologic history of Ridge basin, southern California: Pacific Section, Society of Economic Palentologists and Mineralogists, p. 53–61.

—— , 1985, Paleocene plesiosaur(?), *in* Reynolds, R. E., compiler, Geologic investigations along Interstate 15, Cajon Pass to Manix Lake, California: Redlands, California, San Bernardino County Museum Publications, p. 43–48.

Kovach, R. L., and Nur, A., 1973, Proceedings of the Conference on Tectonic Problems of the San Andreas Fault System: Stanford, California, Stanford University Publications in the Geological Sciences, v. 13, 494 p.

Kroger, P. M., Lyzenga, G. A., Wallace, K. S., and Davidson, J. M., 1987, Tectonic motion in the western United States inferred from very long baseline interferometry measurements, 1980–1986: Journal of Geophysical Research, v. 92, no. B13, p. 14151–14163.

Lachenbruch, A. H., and Sass, J. H., 1973, Thermo-mechanical aspects of the San Andreas fault system, *in* Kovach, R. L. and Nur, A., eds., Proceedings of the Conference on Tectonic Problems of the San Andreas Fault System: Stanford, California, Stanford University Publications in the Geological Sciences, v. 13, p. 192–205.

—— , 1980, Heat flow and energetics of the San Andreas fault zone: Journal of Geophysical Research, v. 85, no. B11, p. 6185–6223.

Lagoe, M. B., 1981, Subsurface facies analysis of the Saltos Shale Member, Monterey Formation (Miocene) and associated rocks, Cuyama Valley, California, *in* Garrison, R. E., and Douglas, R. G., eds., The Monterey Formation and related siliceous rocks of California: Pacific Section, Society of Economic Paleontologists and Mineralogists Special Publication, p. 199–211.

—— , 1984, Paleogeography of Monterey Formation, Cuyama basin, California: American Association of Petroleum Geologists Bulletin, v. 68, p. 610–627.

—— , 1987, Middle Cenozoic basin development, Cuyama basin, California, *in* Ingersoll, R. V., and Ernst, W. G., eds., Cenozoic basin development of coastal California; Rubey Volume 6: Englewood Cliffs, New Jersey, Prentice-Hall, p. 172–206.

—— , 1988, An outline of foraminiferal biofacies in the Soda Lake Shale Member, Vaqueros Formation, Cuyama basin, California, *in* Bazeley, W.J.M., ed., Tertiary tectonics and sedimentation in the Cuyama basin, San Luis Obispo, Santa Barbara, and Ventura Counties, California: Society of Economic Paleontologists and Mineralogists, v. 59, p. 21–27.

Lamar, D. L., 1961, Structural evolution of the northern margin of the Los Angeles Basin [Ph.D. thesis]: Los Angeles, University of California, 142 p.

—— , 1970, Geology of the Elysian Park–Repetto Hills area, Los Angeles County, California: California Division of Mines and Geology Special Report 101, 45 p., scale 1:24,000.

Lamar, D. L., and Rockwell, T. K., 1986, An overview of the tectonics of the Elsinore fault zone, *in* Ehlig, P. L., compiler, Neotectonics and faulting in southern California; Geological Society of America Cordilleran Section Annual Meeting Guidebook and Volume: Los Angeles, California State University Department of Geology, p. 149–158.

Lamb, T. N., 1970, Fossiliferous Triassic(?) meta-sedimentary rocks near Sun City, Riverside County, California: Geological Society of America Abstracts with Programs, v. 2, p. 110–111.

Larsen, E. S., Jr., 1948, Batholith and associated rocks of Corona, Elsinore, and San Luis Rey Quadrangles, southern California: Geological Society of America Memoir 29, 182 p.

LaViolette, J. W., Christenson, G. E., and Stepp, J. C., 1980, Quaternary displacement on the western Garlock fault, southern California, *in* Fife, D. L., and Brown, A. R., eds., Geology and mineral wealth of the California desert; Dibblee volume: Santa Ana, California, South Coast Geological Society, p. 449–456.

Leslie, R. B., 1981, Continuity and tectonic implications of the San Simeon–Hosgri fault zone, central California: U.S. Geological Survey Open-File Report 81-430, 59 p.

Leveille, G., and Frost, E. G., 1984, Deformed upper Paleozoic–lower Mesozoic cratonic strata, El Capitan, Sonora, Mexico: Geological Society of America Abstracts with Programs, v. 16, p. 575.

Levi, S., Schultz, D. L., Yeats, R. S., Stitt, L. T., and Sarna-Wojcicki, A. M., 1986, Magnetostratigraphy and paleomagnetism of the Saugus Formation near Castaic, Los Angeles County, California, *in* Ehlig, P. L., compiler, Neotectonics and faulting in southern California; Geological Society of America Cordilleran Section Annual Meeting Guidebook and Volume: Los Angeles, California State University Department of Geology, p. 103–108.

Liddicoat, J. C., 1990, Tectonic rotation of the Santa Ynez Range, California, recorded in the Sespe Formation: Geophysical Journal International, v. 102, p. 739–745.

Lindsay, E. H., Johnson, N. M., and Opdyke, N. D., 1975, Preliminary correlation of North American land mammal ages and geomagnetic chronology, *in* Smith, G. R., and Friedland, N. E., eds., Studies on Cenozoic paleontology and stratigraphy in honor of Claude W. Hibbard: University of Michigan Papers on Paleontology 12, p. 111–119.

Lindsay, E. H., Opdyke, N. D., and Johnson, N. M., 1984, Blancan-Hemphillian land mammal ages and late Cenozoic mammal dispersal events: Annual Review of Earth and Planetary Sciences, v. 12, p. 445–488.

Link, M. H., 1982, Stratigraphic nomenclature and age of Miocene strata, Ridge basin, southern California, *in* Crowell, J. C., and Link, M. H., eds., Geologic history of Ridge basin, southern California: Pacific Section, Society of Economic Paleontologists and Mineralogists, p. 5–12.

—— , 1983, Sedimentation, tectonics, and offset of Miocene-Pliocene Ridge basin, California, *in* Andersen, D. W., and Rymer, M. J., eds., Tectonics and sedimentation along faults of the San Andreas system: Pacific Section, Society of Economic Paleontologists and Mineralogists, p. 17–31.

Liu, W., 1990, Paleomagnetism of Miocene sedimentary rocks in the Transverse Ranges; The implications for tectonic history [Ph.D. thesis]: Pasadena, California Institute of Technology, 233 p.

Liu, W., Kirschvink, J. L., and Weldon, R., 1988, Paleomagnetic study of the Punchbowl Formation, Devil's Punchbowl, California: EOS Transactions of the American Geophysical Union, v. 69, p. 1164.

Lomnitz, C., Mooser, F., Allen, C. R., Brune, J. N., and Thatcher, W., 1970, Seismicity and tectonics of the northern Gulf of California region, Mexico. Preliminary results: Geofisica Internacional, v. 10, no. 2, p. 37–48.

Lonsdale, P., 1989, Geology and tectonic history of the Gulf of California, *in* Winterer, E. L., Hussong, D. M., and Decker, R. W., eds., The eastern Pacific Ocean and Hawaii: Boulder, Colorado, Geological Society of America, The Geology of North America, v. N, p. 499–521.

Loomis, D. P., and Burbank, D. W., 1988, The stratigraphic evolution of the El Paso basin, southern California; Implications for the Miocene development of the Garlock fault and uplift of the Sierra Nevada: Geological Society of America Bulletin, v. 100, p. 12–18.

Lowman, P. D., Jr., 1980, Vertical displacement on the Elsinore fault of southern California; Evidence from orbital photographs: Journal of Geology, v. 88, p. 415–432.

Lundelius, E. L., Jr., and 8 others, 1987, The North American Quaternary sequence, *in* Woodburne, M. O., ed., Cenozoic mammals of North America; Geochronology and biostratigraphy: Berkeley, University of California Press, p. 211–235.

Luyendyk, B. P., and Hornafius, J. S., 1987, Neogene crustal rotations, fault slip, and basin development in southern California, *in* Ingersoll, R. V., and Ernst, W. G., eds., Cenozoic basin development of coastal California; Rubey Volume 6: Englewood Cliffs, New Jersey, Prentice-Hall, p. 259–283.

Luyendyk, B. P., Kamerling, M. J., and Terres, Richard, 1980, Geometric model for Neogene crustal rotations in southern California: Geological Society of America Bulletin, v. 91, p. 211–217.

Luyendyk, B. P., Kamerling, M. J., Terres, R. R., and Hornafius, J. S., 1985, Simple shear of southern California during Neogene time suggested by paleomagnetic declinations: Journal of Geophysical Research, v. 90, no. B14,

p. 12454–12466.

MacFadden, B. J., Swisher, C. C., III, Opdyke, N. D., and Woodburne, M. O., 1990, Paleomagnetism, geochronology, and possible tectonic rotation of the middle Miocene Barstow Formation, Mojave Desert, southern California: Geological Society of America Bulletin, v. 102, p. 478–493.

Mahaffie, M. J., and Dokka, R. K., 1986, Thermochronologic evidence for the age and cooling history of the upper plate of the Vincent thrust, California: Geological Society of America Abstracts with Programs, v. 18, p. 153.

Mankinen, E. A., and Dalrymple, G. B., 1979, Revised geomagnetic polarity time scale for the interval 0-5 m.y.B.P.: Journal of Geophysical Research, v. 84, no. B2, p. 615–626.

Mann, J. F., Jr., 1955, Geology of a portion of the Elsinore fault zone, California: California Division of Mines Special Report 43, 22 p.

Marshall, M., Pischke, G., Mace, N., Gastil, G., and Luyendyk, B. P., 1979, Paleomagnetism and circum-Gulf of California tectonics: Geological Society of America Abstracts with Programs, v. 11, p. 472.

Martin, B. D., and Emery, K. O., 1967, Geology of Monterey Canyon, California: American Association of Petroleum Geologists Bulletin, v. 51, p. 2281–2304.

Matthews, V., III, 1976, Correlation of Pinnacles and Neenach Volcanic Formations and their bearing on San Andreas fault problem: American Association of Petroleum Geologists Bulletin, v. 60, p. 2128–2141.

Matti, J. C., and Morton, D. M., 1975, Geologic history of the San Timoteo badlands, southern California: Geological Society of America Abstracts with Programs, v. 7, p. 344.

Matti, J. C., and 6 others, 1982a, Mineral resource potential map of the Whitewater Wilderness Study Area, Riverside and San Bernardino Counties, California: U.S. Geological Survey Miscellaneous Field Studies Map MF-1478-A, scale 1:24,000.

Matti, J. C., and 8 others, 1982b, Mineral resource potential map of the Bighorn Mountains Wilderness Study Area (CDCA-217), San Bernardino County, California: U.S. Geological Survey Miscellaneous Field Studies Map MF-1493-A, scale 1:48,000.

Matti, J. C., Cox, B. F., and Iverson, S. R., 1983a, Mineral resource potential map of the Raywood Flat Roadless Area, San Bernardino and Riverside Counties, California: U.S. Geological Survey Miscellaneous Field Studies Map MF-1563-A, scale 1:62,500.

Matti, J. C., Cox, B. F., Powell, R. E., Oliver, H. W., and Kuizon, Lucia, 1983b, Mineral resource potential map of the Cactus Spring Roadless Area, Riverside County, California: U.S. Geological Survey Miscellaneous Field Studies Map MF-1650-A, scale 1:24,000.

Matti, J. C., Morton, D. M., and Cox, B. F., 1985, Distribution and geologic relations of fault systems in the vicinity of the central Transverse Ranges, southern California: U.S. Geological Survey Open-File Report 85-365, 27 p., 1:250,000.

Matti, J. C., Frizzell, V. A., and Mattinson, J. M., 1986, Distinctive Triassic megaporphyritic monzogranite displaced 160 ± 10 km by the San Andreas fault, southern California; A new constraint for palinspastic reconstructions: Geological Society of America Abstracts with Programs, v. 18, p. 154.

Mattinson, J. M., 1978, Age, origin, and thermal histories of some plutonic rocks from the Salinian block of California: Contributions to Mineralogy and Petrology, v. 67, p. 233–245.

——— , 1983, Basement rocks of the southeastern Salinian block, California; U-Pb isotopic relationships: Geological Society of America Abstracts with Programs, v. 15, p. 414.

——— , 1986, Geochronology of high-pressure–low-temperature Franciscan metabasites; A new approach using the U-Pb system, in Evans, B. W., and Brown, E. H., eds., Blueschists and eclogites: Geological Society of America Memoir 164, p. 95–105.

Mattinson, J. M., and James, E. W., 1985, Salinian block U/Pb age and isotopic variations; Implications for origin and emplacement of the Salinian terrane, in Howell, D. G., ed., Tectonostratigraphic terranes of the circum-Pacific region: Houston, Texas, Circum-Pacific Council for Energy and Mineral Resources Earth Science Series, no. 1, p. 215–226.

——— , 1986, The Sierra de Salinas Schist, Salinian block, California; Isotopic constraints on age and origin: Geological Society of America Abstracts with Programs, v. 18, p. 154.

Maxwell, J. C., 1974, Anatomy of an orogen: Geological Society of America Bulletin, v. 85, p. 1195–1204.

May, D. J., 1986, Amalgamation of metamorphic terranes in the southeastern San Gabriel Mountains, California [Ph.D. thesis]: Santa Barbara, University of California, 325 p.

——— , 1989, Late Cretaceous intra-arc thrusting in southern California: Tectonics, v. 8, p. 1159–1173.

May, D. J., and Walker, N. W., 1989, Late Cretaceous juxtaposition of metamorphic terranes in the southeastern San Gabriel Mountains, California: Geological Society of America Bulletin, v. 101, p. 1246–1267.

May, S. R., and Repenning, C. A., 1982, New evidence for the age of the Mount Eden fauna, southern California: Journal of Vertebrate Paleontology, v. 2, p. 109–113.

Mayer, L., 1987, Subsidence analysis of the Los Angeles basin, in Ingersoll, R. V., and Ernst, W. G., eds., Cenozoic basin development of coastal California; Rubey Volume 6: Englewood Cliffs, New Jersey, Prentice-Hall, p. 299–320.

McCracken, W. A., 1969, Environmental reconstruction; Sespe Formation, Ventura basin, California: Geological Society of America Abstracts with Programs for 1969, Part 7, p. 145–146.

——— , 1971, Paleogeographic and tectonic considerations, Sespe Formation, Ventura basin, California: Geological Society of America Abstracts with Programs, v. 3, p. 158–159.

McCulloh, T. H., 1981, Middle Tertiary laumontite isograd offset 37 km by left-lateral strike-slip on Santa Ynez fault, California [abs.]: American Association of Petroleum Geologists Bulletin, v. 65, p. 956.

McDowell, F. W., Lehman, D. H., Gucwa, P. R., Fritz, D., and Maxwell, J. C., 1984, Glaucophane schists and ophiolites of the northern California Coast Ranges; Isotopic ages and their tectonic implications: Geological Society of America Bulletin, v. 95, p. 1373–1382.

McKee, E. H., 1971, Tertiary igneous chronology of the Great Basin of western United States; Implications for tectonic models: Geological Society of America Bulletin, v. 82, p. 3497–3502.

McKinstry, H. E., 1953, Shears of the second order: American Journal of Science, v. 251, p. 401–414.

McWilliams, M., and Li, Y., 1985, Oroclinal bending of the southern Sierra Nevada batholith: Science, v. 230, p. 172–175.

McWilliams, M. O., and Howell, D. G., 1982, Exotic terranes of western California: Nature, v. 297, p. 215–217.

Meisling, K. E., 1984, Neotectonics of the north frontal fault system of the San Bernardino Mountains, southern California; Cajon Pass to Lucerne Valley [Ph.D. thesis]: Pasadena, California Institute of Technology, 394 p.

Meisling, K. E., and Weldon, R. J., 1982, The late-Cenozoic structure and stratigraphy of the western San Bernardino Mountains, in Cooper, J. D., compiler, Geologic excursions in the Transverse Ranges, southern California; Geological Society of America Cordilleran Section Annual Meeting Volume and Guidebook: Fullerton, California State University Department of Geological Sciences, p. 75–81.

——— , 1986, Cenozoic uplift of the San Bernardino Mountains; Possible thrusting across the San Andreas fault: Geological Society of America Abstracts with Programs, v. 18, p. 157.

——— , 1989, Late Cenozoic tectonics of the northwestern San Bernardino Mountains, southern California: Geological Society of America Bulletin, v. 101, p. 106–128.

Menard, H. W., 1955, Deformation of the northeastern Pacific basin and the west coast of North America: Geological Society of America Bulletin, v. 66, p. 1149–1198.

——— , 1960, The East Pacific Rise: Science, v. 132, no. 3441, p. 1737–1746.

Merifield, P. M., and Lamar, D. L., 1975, Faults on Skylab imagery of the Salton trough area, southern California: Houston, Texas, NASA Lyndon B. Johnson Space Center Technical Report 75-1, Contract NAS 2-7698, 23 p.

Merriam, R., and Bischoff, J. L., 1975, Bishop Ash; A widespread volcanic ash extended to southern California: Journal of Sedimentary Petrology, v. 45, p. 207–211.

Michael, E. D., 1966, Large lateral displacement on Garlock fault, California, as measured from offset fault system: Geological Society of America Bulletin, v. 77, p. 111–113.

Miller, C. F., 1977, Alkali-rich monzonites, California; Origin of near silica-saturated alkaline rocks and their significance in a calc-alkaline batholithic belt [Ph.D. thesis]: Los Angeles, University of California, 283 p.

——, 1978, An early Mesozoic alkalic magmatic belt in western North America, *in* Howell, D. G., and McDougall, K. A., eds., Mesozoic paleogeography of the western United States: Society of Economic Paleontologists and Mineralogists, Pacific Section, Pacific Coast Paleogeography Symposium 2, p. 163–173.

Miller, D. M., Howard, K. A., and John, B. E., 1982, Preliminary geology of the Bristol Lake region, Mojave Desert, California, *in* Cooper, J. D., compiler, Geologic excursions in the California desert: Geological Society of America Cordilleran Section Annual Meeting Volume and Guidebook: Fullerton, California State University Department of Geological Sciences, p. 91–100.

Miller, E. L., 1978, The Fairview Valley Formation; A Mesozoic intraorogenic deposit in the southwestern Mojave Desert, *in* Howell, D. G., and McDougall, K. A., eds., Mesozoic paleogeography of the western United States: Pacific Section, Society of Economic Paleontologists and Mineralogists Pacific Coast Paleogeography Symposium 2, p. 277–282.

——, 1981, Geology of the Victorville region, California: Geological Society of America Bulletin, v. 92, Part I, p. 160–163; Part II, p. 554–608.

Miller, E. L., and Cameron, C. S., 1982, Late Precambrian to Late Cretaceous evolution of the southwestern Mojave Desert, California, *in* Cooper, J. D., Troxel, B. W., and Wright, L. A., eds., Geology of selected areas in the San Bernardino Mountains, western Mojave Desert, and southern Great Basin, California; Field trip number 9, Geological Society of America Cordilleran Section Annual Meeting Volume and Guidebook: Shoshone, California, The Death Valley Publishing Company, p. 21–34.

Miller, E. L., and Carr, M. D., 1978, Recognition of possible Aztec-equivalent sandstones and associated Mesozoic metasedimentary deposits within the Mesozoic magmatic arc in the southwestern Mojave Desert, California, *in* Howell, D. G., and McDougall, K. A., eds., Mesozoic paleogeography of the western United States: Pacific Section, Society of Economic Paleontologists and Mineralogists Pacific Coast Paleogeography Symposium 2, p. 283–289.

Miller, E. L., and Sutter, J. F., 1981, $^{40}Ar/^{39}Ar$ age spectra for biotite and hornblende from plutonic rocks in the Victorville region, California: Geological Society of America Bulletin, v. 92, Part I, p. 164–169.

Miller, F. K., 1970, Geologic map of the Quartzsite Quadrangle, Yuma County, Arizona: U.S. Geological Survey Geologic Quadrangle Map GQ-841, scale 1:62,500.

——, 1979, Geologic map of the San Bernardino North Quadrangle, California: U.S. Geological Survey Open-File Report 79-770, scale 1:24,000.

——, 1987, Reverse-fault system bounding the north side of the San Bernardino Mountains, *in* Recent reverse faulting in the Transverse Ranges, California: U.S. Geological Survey Professional Paper 1339, p. 83–95.

Miller, F. K., and McKee, E. H., 1971, Thrust and strike-slip faulting in the Plomosa Mountains, southwestern Arizona: Geological Society of America Bulletin, v. 82, p. 717–722.

Miller, F. K., and Morton, D. M., 1977, Comparison of granitic intrusions in the Pelona and Orocopia Schists, southern California: U.S. Geological Survey Journal of Research, v. 5, no. 5, p. 643–649.

——, 1980, Potassium-argon geochronology of the eastern Transverse Ranges and southern Mojave Desert, southern California: U.S. Geological Survey Professional Paper 1152, 30 p.

Miller, R. H., and Dockum, M. S., 1983, Ordovician conodonts from metamorphosed carbonates of the Salton trough, California: Geology, v. 11, p. 410–412.

Miller, S. T., 1980, Geology and mammalian biostratigraphy of a part of the northern Cady Mountains, Mojave Desert, California: U.S. Geological Survey Open-File Report 80-878, 122 p.

Miller, W. E., and Downs, T., 1974, A Hemphillian local fauna containing a new genus of antilocaprid from southern California: Los Angeles County, Natural History Museum Contributions in Science, no. 258, 36 p.

Miller, W. J., 1926, Crystalline rocks of the middle-southern San Gabriel Mountains, California [abs.]: Geological Society of America Bulletin, v. 37, p. 149.

——, 1928, Geomorphology of the southwestern San Gabriel Mountains of California: Berkeley, University of California Department of Geological Sciences Bulletin, v. 17, p. 193–240.

——, 1934, Geology of the western San Gabriel Mountains of California: University of California at Los Angeles Publications in Mathematical and Physical Sciences, v. 1, no. 1, 114 p.

——, 1938, Pre-Cambrian and associated rocks near Twenty-nine Palms, California: Geological Society of America Bulletin, v. 49, p. 417–446.

——, 1944, Geology of Palm Springs–Blythe strip, Riverside County, California: California Journal of Mines, v. 40, p. 11–72.

——, 1946, Crystalline rocks of southern California: Geological Society of America Bulletin, v. 57, p. 457–542.

Moody, J. D., and Hill, M. J., 1956, Wrench-fault tectonics: Geological Society of America Bulletin, v. 67, p. 1207–1246.

Moore, J. G., 1959, The quartz diorite boundary line in the western United States: Journal of Geology, v. 67, p. 198–210.

Morris, L. K., Lund, S. P., and Bottjer, D. J., 1986, Paleolatitude drift history of displaced terranes in southern and Baja California: Nature, v. 321, p. 844–847.

Morton, D. M., 1973, Geology of parts of the Azusa and Mount Wilson Quadrangles, San Gabriel Mountains, Los Angeles County, California: California Division of Mines and Geology Special Report 105, 21 p.

——, 1975a, Relations between major faults, eastern San Gabriel Mountains, southern California: Geological Society of America Abstracts with Programs, v. 7, p. 352–353.

——, 1975b, Synopsis of the geology of the eastern San Gabriel Mountains, southern California, *in* Crowell, J. C., ed., San Andreas fault in southern California: California Division of Mines and Geology Special Report 118, p. 170–176.

Morton, D. M., and Matti, J. C., 1987, The Cucamonga fault zone; Geologic setting and Quaternary history, *in* Recent reverse faulting in the Transverse Ranges, California: U.S. Geological Survey Professional Paper 1339, p. 179–203.

Morton, D. M., and Miller, F. K., 1975, Geology of the San Andreas fault zone north of San Bernardino between Cajon Canyon and Santa Ana Wash, *in* Crowell, J. C., ed., San Andreas fault in southern California: California Division of Mines and Geology Special Report 118, p. 136–146.

Morton, D. M., and Yerkes, R. F., 1987, Recent reverse faulting in the Transverse Ranges, California; Introduction, *in* Recent reverse faulting in the Transverse Ranges, California: U.S. Geological Survey Professional Paper 1339, p. 1–5.

Morton, D. M., Rodriguez, E. A., Obi, C. M., Simpson, R. W., Jr., and Peters, T. J., 1983, Mineral resource potential map of the Cucamonga Roadless Areas, San Bernardino County, California: U.S. Geological Survey Miscellaneous Field Studies Map MF-1646-A, scale 1:31,680.

Morton, J. L., and Morton, D. M., 1979, K-Ar ages of Cenozoic volcanic rocks along the Elsinore fault zone, southwestern Riverside County, California: Geological Society of America Abstracts with Programs, v. 11, p. 119.

Muehlberger, W. R., 1958, Geology of northern Soledad basin, Los Angeles County, California: American Association of Petroleum Geologists Bulletin, v. 42, p. 1812–1844.

Mukasa, S. B., Dillon, J. T., and Tosdal, R. M., 1984, A late Jurassic minimum age for the Pelona-Orocopia Schist protolith, southern California: Geological Society of America Abstracts with Programs, v. 16, p. 323.

Mullins, H. T., and Nagel, D. K., 1981, Franciscan-type rocks off Monterey Bay, California; Implications for western boundary of Salinian block: Geo-Marine Letters, v. 1, no. 2, p. 135–139.

Murray, K., 1982, Tectonic implications of Cenozoic volcanism in southeastern California, *in* Frost, E. G., and Martin, D. L., eds., Mesozoic-Cenozoic tectonic evolution of the Colorado River region, California, Arizona, and

Nevada: Anderson-Hamilton Volume: San Diego, California, Cordilleran Publishers, Geological Society of America Cordilleran Section Annual Meeting Symposium and Field Trip Volume, p. 77–83.

Nagel, D. K., and Mullins, H. T., 1983, Late Cenozoic offset and uplift along the San Gregorio fault zone; Central California continental margin, *in* Andersen, D. W., and Rymer, M. J., eds., Tectonics and sedimentation along faults of the San Andreas system: Pacific Section, Society of Economic Paleontologists and Mineralogists, p. 91–103.

Nagle, H. E., and Parker, E. S., 1971, Future oil and gas potential of onshore Ventura basin, California, *in* Cram, I. H., ed., Future petroleum provinces of the United States; Their geology and potential: American Association of Petroleum Geologists Memoir 15, v. 1, p. 254–297.

Namson, J., and Davis, T., 1988, Structural transect of the western Transverse Ranges, California; Implications for lithospheric kinematics and seismic risk evaluation: Geology, v. 16, p. 675–679.

Ness, G., Levi, S., and Couch, R., 1980, Marine magnetic anomaly timescales for the Cenozoic and Late Cretaceous; A precis, critique, and synthesis: Review of Geophysics and Space Physics, v. 18, p. 753–770.

Neuhaus, J. R., Cassidy, M., Krummenacher, D., and Gastil, R. G., 1988, Timing of protogulf extension and transtensional rifting through volcanic/sedimentary stratigraphy of S.W. Isla Tiburon, Gulf of Calif., Sonora, Mexico: Geological Society of America Abstracts with Programs, v. 20, p. 218.

Neville, S. L., 1983, Late Miocene alkaline volcanism, south-central Mojave Desert and northeast San Bernardino Mountains, California [M.S. thesis]: Riverside, University of California, 156 p.

Neville, S. L., and Chambers, J. M., 1982, Late Miocene alkaline volcanism, northeastern San Bernardino Mountains and adjacent Mojave Desert, *in* Cooper, J. D., compiler, Geologic excursions in the Transverse Ranges, southern California; Geological Society of America, Cordilleran Section Annual Meeting Volume and Guidebook: Fullerton, California State University Department of Geological Sciences, p. 103–106.

Nilsen, T. H., 1984, Offset along the San Andreas fault of Eocene strata from the San Juan Bautista area and western San Emigdio Mountains, California: Geological Society of America Bulletin, v. 95, p. 599–609.

Nilsen, T. H., and Clarke, S. H., Jr., 1975, Sedimentation and tectonics in the early Tertiary continental borderland of central California: U.S. Geological Survey Professional Paper 925, 64 p.

Nilsen, T. H., and Link, M. H., 1975, Stratigraphy, sedimentology and offset along the San Andreas fault of Eocene to lower Miocene strata of the northern Santa Lucia Range and the San Emigdio Mountains, Coast Ranges, central California, *in* Weaver, D. W., Hornaday, G. R., and Tipton, Ann, eds., Paleogene symposium and selected technical papers; Conference on Future Energy Horizons of the Pacific Coast: Pacific Sections, American Association of Petroleum Geologists, Society of Economic Paleontologists and Mineralogists, and Society of Exploration Geophysicists, Annual Meeting Proceedings, p. 367–400.

Noble, L. F., 1926, The San Andreas rift and some other active faults in the desert region of southeastern California: Carnegie Institution of Washington Year Book No. 25, p. 415–428.

—— , 1953, Geology of the Pearland Quadrangle, California: U.S. Geological Survey Geologic Quadrangle Map GQ-24, scale 1:24,000.

—— , 1954a, Geology of the Valyermo Quadrangle and vicinity, California: U.S. Geological Survey Geologic Quadrangle Map GQ-50, scale 1:24,000.

—— , 1954b, The San Andreas fault zone from Soledad Pass to Cajon Pass, California, *in* Jahns, R. H., ed., Geology of southern California: California Division of Mines Bulletin 170, chapter 4, p. 37–48.

Nur, A., Ron, H., and Scotti, O., 1986, Fault mechanics and the kinematics of block rotations: Geology, v. 14, p. 746–749.

—— , 1989, Mechanics of distributed fault and block rotation, *in* Kissel, C., and Laj, C., eds., Paleomagnetic rotations and continental deformation; NATO ASI Series C, Mathematical and Physical Sciences, v. 254: Boston, Kluwer Academic Publishers, p. 209–228.

Oakeshott, G. B., 1950, Geology of the Placerita oil field, Los Angeles County, California: California Journal of Mines and Geology, v. 46, p. 43–79.

—— , 1958, Geology and mineral deposits of San Fernando Quadrangle, Los Angeles County, California: California Division of Mines Bulletin 172, 147 p.

Oberlander, T. M., 1972, Morphogenesis of granitic boulder slopes in the Mojave Desert, California: Journal of Geology, v. 80, p. 1–20.

Obradovich, J. D., and Naeser, C. W., 1981, Geochronology bearing on the age of the Monterey Formation and siliceous rocks in California, *in* Garrison, R. E., and Douglas, R. G., eds., The Monterey Formation and related siliceous rocks of California: Pacific Section, Society of Economic Paleontologists and Mineralogists Special Publication, p. 87–95.

O'Day, P. A., and Sims, J. D., 1986, Sandstone composition and paleogeography of the Temblor Formation, central California; Evidence for early to middle Miocene right-lateral displacement on the San Andreas fault system: Geological Society of America Abstracts with Programs, v. 18, p. 165.

Olmsted, F. H., 1972, Geologic map of the Laguna Dam 7.5-minute Quadrangle, Arizona and California: U.S. Geological Survey Geologic Quadrangle Map GQ-1014, scale 1:24,000.

Olmsted, F. H., Loeltz, O. J., and Irelan, B., 1973, Geohydrology of the Yuma area, Arizona and California: U.S. Geological Survey Professional Paper 486-H, 227 p.

Page, B. M., 1966, Geology of the Coast Ranges of California, *in* Bailey, E. H., ed., Geology of northern California: California Division of Mines and Geology Bulletin 190, p. 255–276.

—— , 1981, The southern Coast Ranges, *in* Ernst, W. G., ed., The geotectonic development of California; Rubey Volume 1: Englewood Cliffs, New Jersey, Prentice-Hall, p. 329–417.

—— , 1982, Migration of Salinian composite block, California, and disappearance of fragments: American Journal of Science, v. 282, p. 1694–1734.

Page, B. M., Marks, J. G., and Walker, G. W., 1951, Stratigraphy and structure of mountains northeast of Santa Barbara, California: American Association of Petroleum Geologists Bulletin, v. 35, p. 1727–1780.

Palmer, A. R., compiler, 1983, The Decade of North American Geology 1983 geologic time scale: Geology, v. 11, p. 503–504.

Paschall, R. H., and Off, T., 1961, Dip-slip versus strike-slip movement on San Gabriel fault, southern California: American Association of Petroleum Geologists, v. 45, p. 1941–1956.

Payne, C. M., Swanson, O. E., and Schell, B. A., 1979, Investigation of the Hosgri fault, offshore southern California, Point Sal to Point Conception: U.S. Geological Survey Open-File Report 79-1199, 17 p.

Pelka, G. J., 1971, Paleocurrents of the Punchbowl Formation and their interpretation: Geological Society of America Abstracts with Programs, v. 3, p. 176.

—— , 1973, Geology of the McCoy and Palen Mountains, southeastern California [Ph.D. thesis]: Santa Barbara, University of California, 162 p.

Perkins, J. A., Sims, J. D., and Sturgess, S. S., 1989, Late Holocene movement along the San Andreas fault at Melendy Ranch; Implications for the distribution of fault slip in central California: Journal of Geophysical Research, v. 94, no. B8, p. 10217–10230.

Petersen, R. M., 1976, Patterns of Quaternary denudation and deposition at Pipes Wash (Mohave Desert) California [Ph.D. thesis]: Riverside, University of California, 279 p.

Peterson, M. S., 1975, Geology of the Coachella fanglomerate, *in* Crowell, J. C., ed., San Andreas fault in southern California: California Division of Mines and Geology Special Report 118, p. 119–126.

Pinault, C. T., 1984, Structure, tectonic geomorphology and neotectonics of the Elsinore fault zone between Banner Canyon and the Coyote Mountains, southern California [M.S. thesis]: San Diego, California, San Diego State University, 231 p.

Platt, J. P., 1975, Metamorphic and deformational processes in the Franciscan Complex, California; Some insights from the Catalina Schist terrane: Geological Society of America Bulletin, v. 86, p. 1337–1347.

—— , 1976, The petrology, structure, and geologic history of the Catalina Schist terrain, southern California: Berkeley, University of California Publications in Geological Sciences, v. 112, 111 p.

—— , 1986, Dynamics of orogenic wedges and the uplift of high-pressure metamorphic rocks: Geological Society of America Bulletin, v. 97, p. 1037–1053.

Platt, J. P., and Stuart, C. J., 1974, Newport-Inglewood fault zone, Los Angeles basin, California: Discussion: American Association of Petroleum Geologists, v. 58, p. 877–883.

Plescia, J. B., and Calderone, G. J., 1986, Paleomagnetic constraints on the timing and extent of rotation of the Tehachapi Mountains, California: Geological Society of America Abstracts with Programs, v. 18, p. 171.

Poore, R. Z., Barron, J. A., and Addicott, W. O., 1981, Biochronology of the northern Pacific Miocene, *in* Proceedings of International Geological Correlation Programme Project 114 International Workshop on Pacific Neogene Biostratigraphy, 6th International Working Group Meeting: Osaka, Japan, Osaka Museum of Natural History, p. 91–97.

Postlethwaite, C. E., 1988, The structural geology of the Red Cloud thrust system, southern Eastern Transverse Ranges, California [Ph.D. thesis]: Ames, Iowa State University, 135 p.

Powell, R. E., 1975, The Chiriaco fault; A left-lateral strike-slip fault in the eastern Transverse Ranges, Riverside County, California: Geological Society of America Abstracts with Programs, v. 7, p. 362.

——, 1981, Geology of the crystalline basement complex, eastern Transverse Ranges, southern California; Constraints on regional tectonic interpretation [Ph.D. thesis]: Pasadena, California Institute of Technology, 441 p.

——, 1982a, Crystalline basement terranes in the southern eastern Transverse Ranges, California; Field trip number 11, *in* Cooper, J. D., compiler, Geologic excursions in the Transverse Ranges, southern California; Geological Society of America Cordilleran Section Annual Meeting Volume and Guidebook: Fullerton, California State University Department of Geological Sciences, p. 107–151.

——, 1982b, Prebatholithic terranes in the crystalline basement of the Transverse Ranges, southern California: Geological Society of America Abstracts with Programs, v. 14, p. 225.

——, 1986, Palinspastic reconstruction of crystalline-rock assemblages in southern California; San Andreas fault as part of an evolving system of late Cenozoic conjugate strike-slip faults: Geological Society of America Abstracts with Programs, v. 18, p. 172.

——, 1991, Eagle Mts. 30' × 60' Quadrangle, southern California; I. Geologic mapping: Geological Society of America Abstracts with Programs, v. 23, p. A478.

Powell, R. E., and Silver, L. T., 1979, Geologic analysis of ASTP photographs of parts of southern California, *in* El-Baz, F., and Warner, D. M., eds., Earth observations and photography: Apollo-Soyuz Test Project summary science report, v. 2: Washington, D.C., National Aeronautics and Space Administration NASA SP-412, p. 9–27.

Powell, R. E., Cox, B. F., Matti, J. C., and Gabby, P. N., 1983a, Mineral resource potential map of the Arroyo Seco Roadless Area, Los Angeles County, California: U.S. Geological Survey Miscellaneous Field Studies Map MF-1607-A, scale 1:24,000.

Powell, R. E., and 5 others, 1983b, Mineral resource potential map of the Sugarloaf Roadless Area, San Bernardino County, California: U.S. Geological Survey Miscellaneous Field Studies Map MF-1606-A, scale 1:24,000.

Powell, R. E., Watts, K. C., and Lane, M. E., 1984a, Mineral resource potential of the Chuckwalla Mountains Wilderness Study Area (CDCA-348), Riverside County, California: U.S. Geological Survey Open-File Report 84-674, 25 p., scale 1:62,500.

Powell, R. E., Whittington, C. L., Grauch, V.J.S., and McColly, R. A., 1984b, Mineral resource potential of the Eagle Mountains Wilderness Study Area (CDCA-334), Riverside County, California: U.S. Geological Survey Open-File Report 84-631, 25 p., scale 1:62,500.

Poynor, W. D., 1960, Geology of the San Guillermo area and its regional correlation, Ventura County, California [M.A. thesis]: Los Angeles, University of California, 119 p.

Prentice, C. S., 1989, The northern San Andreas fault; Russian River to Point Arena, *in* Sylvester, A. G., and Crowell, J. C., leaders, The San Andreas transform belt; 28th International Geologic Congress field trip guidebook T309: Washington, D.C., American Geophysical Union, p. T309:49-51.

Prentice, C. S., Niemi, T. M., and Hall, N. T., 1991, Quaternary tectonics of the northern San Andreas fault, San Francisco Peninsula, Point Reyes, and Point Arena, California, *in* Sloan, D., and Wagner, D. L., eds., Geologic excursions in northern California; San Francisco to the Sierra Nevada: California Division of Mines and Geology Special Publication 109, p. 25–34.

Proctor, R. J., 1968, Geology of the Desert Hot Springs–upper Coachella Valley area, California: California Division of Mines and Geology Special Report 94, 50 p.

Pruss, D. E., Olcott, G. W., and Oesterling, W. A., 1959, Areal geology of a portion of the Little San Bernardino Mountains, Riverside and San Bernardino Counties, California [abs.]: Geological Society of America Bulletin, v. 70, no. 12, pt. 2, p. 1741.

Ramirez, V. R., 1983, Hungry Valley Formation; Evidence for 220 kilometers of post Miocene offset on the San Andreas fault, *in* Andersen, D. W., and Rymer, M. J., eds., Tectonics and sedimentation along faults of the San Andreas system: Pacific Section, Society of Economic Paleontologists and Mineralogists, p. 33–44.

Real, C. R., 1987, Seismicity and tectonics of the Santa Monica–Hollywood–Raymond Hill fault zone and northern Los Angeles basin, *in* Recent reverse faulting in the Transverse Ranges, California: U.S. Geological Survey Professional Paper 1339, p. 113–124.

Reed, R. D., and Hollister, J. S., 1936, Structural evolution of southern California: American Association of Petroleum Geologists Bulletin, v. 20, p. 1529–1704.

Repenning, C. A., 1987, Biochronology of the microtine rodents of the United States, *in* Woodburne, M. O., ed., Cenozoic mammals of North America; Geochronology and biostratigraphy: Berkeley, University of California Press, p. 236–268.

Richard, S. M., and Haxel, G. B., 1991, Progressive exhumation of the Orocopia and Pelona Schists along a composite normal fault system, southeastern California and southwestern Arizona: Geological Society of America Abstracts with Programs, v. 23, p. 92.

Rockwell, T. K., and Lamar, D. L., trip leaders, 1986, Neotectonics of the Elsinore fault, southern California; Field trip number 18, *in* Ehlig, P. L., compiler, Neotectonics and faulting in southern California; Geological Society of America Cordilleran Section Annual Meeting Guidebook and Volume: Los Angeles, California State University Department of Geology, p. 147–208.

Rockwell, T. K., and 5 others, 1986, Roadlog; Neotectonics of the Elsinore fault fieldtrip, *in* Ehlig, P. L., compiler, Neotectonics and faulting in southern California; Geological Society of America Cordilleran Section Annual Meeting Guidebook and Volume: Los Angeles, California State University Department of Geology, p. 197–208.

Rodgers, D. A., 1979, Vertical deformation, stress accumulation, and secondary faulting in the vicinity of the Transverse Ranges of southern California: California Division of Mines and Geology Bulletin 203, 74 p.

Rogers, J.J.W., 1961, Igneous and metamorphic rocks of the western portion of Joshua Tree National Monument, Riverside and San Bernardino Counties, California: California Division of Mines Special Report 68, 26 p.

Ross, D. C., 1970, Quartz gabbro and anorthositic gabbro; Markers of offset along the San Andreas fault in the California Coast Ranges: Geological Society of America Bulletin, v. 81, p. 3647–3661.

——, 1972, Petrographic and chemical reconnaissance study of some granitic and gneissic rocks near the San Andreas fault from Bodega Head to Cajon Pass, California: U.S. Geological Survey Professional Paper 698, 92 p.

——, 1974, Map showing basement geology and location of wells drilled to basement, Salinian block, central and southern Coast Ranges, California: U.S. Geological Survey Miscellaneous Field Studies Map MF-588, scale 1:500,000.

——, 1976, Metagraywacke in the Salinian block, central Coast Ranges, California; And a possible correlative across the San Andreas fault: U.S. Geological Survey Journal of Research, v. 4, no. 6, p. 683–696.

——, 1977, Pre-intrusive metasedimentary rocks of the Salinian block, California; A paleotectonic dilemma, *in* Stewart, J. H., Stevens, C. H., Fritsche,

A. E., eds., Paleozoic paleogeography of the western United States: Pacific Section, Society of Economic Paleontologists and Mineralogists Pacific Coast Paleogeography Symposium 1, p. 371–380.

——— , 1978, The Salinian block; A Mesozoic granitic orphan in the California Coast Ranges, *in* Howell, D. G., and McDougall, K. A., eds., Mesozoic paleogeography of the western United States: Pacific Section, Society of Economic Paleontologists and Mineralogists, Pacific Coast Paleogeography Symposium 2, p. 509–522.

——— , 1984, Possible correlations of basement rocks across the San Andreas, San Gregorio–Hosgri, and Rinconada–Reliz–King City faults, California: U.S. Geological Survey Professional Paper 1317, 37 p.

——— , 1985, Mafic gneissic complex (batholithic root?) in the southernmost Sierra Nevada, California: Geology, v. 13, p. 288–291.

——— , 1989, The metamorphic and plutonic rocks of the southernmost Sierra Nevada, California, and their tectonic framework: U.S. Geological Survey Professional Paper 1381, 159 p.

Ross, D. C., and Brabb, E. E., 1973, Petrography and structural relations of granitic basement rocks in the Monterey Bay area, California: U.S. Geological Journal of Research, v. 1, no. 3, p. 273–282.

Ross, D. C., Wentworth, C. M., and McKee, E. H., 1973, Cretaceous mafic conglomerate near Gualala offset 350 miles by San Andreas fault from oceanic crustal source near Eagle Rest Peak, California: U.S. Geological Survey Journal of Research, v. 1, no. 1, p. 45–52.

Ross, T. M., Luyendyk, B. P., and Haston, R. B., 1989, Paleomagnetic evidence for Neogene clockwise tectonic rotations in the central Mojave Desert, California: Geology, v. 17, p. 470–473.

Rotstein, Y., Combs, J., and Biehler, S., 1976, Gravity investigation in the southeastern Mojave Desert, California: Geological Society of America Bulletin, v. 87, p. 981–993.

Ryder, R. T., and Thomson, A., 1989, Tectonically controlled fan delta and submarine fan sedimentation of late Miocene age, southern Temblor Range, California: U.S. Geological Survey Professional Paper 1442, 59 p.

Rymer, M. J., and Ellsworth, W. L., eds., 1990, The Coalinga, California, earthquake of May 2, 1983: U.S. Geological Survey Professional Paper 1487, 417 p.

Sage, O. G., Jr., 1973a, Paleocene geography of southern California [Ph.D. thesis]: Santa Barbara, University of California, 250 p.

——— , 1973b, Paleocene geography of the Los Angeles region, *in* Kovach, R. L., and Nur, A., eds., Proceedings of the Conference on Tectonic Problems of the San Andreas Fault System: Stanford, California, Stanford University Publications in the Geological Sciences, v. 13, p. 348–357.

——— , 1975, Sedimentological and tectonic implications of the Paleocene San Francisquito Formation, Los Angeles County, California, *in* Crowell, J. C., ed., San Andreas fault in southern California: California Division of Mines and Geology Special Report 118, p. 162–169.

Saleeby, J. B., Goodin, S. E., Sharp, W. D., and Busby, C. J., 1978, Early Mesozoic paleotectonic-paleogeographic reconstruction of the southern Sierra Nevada region, *in* Howell, D. G., and McDougall, K. A., eds., Mesozoic paleogeography of the western United States: Pacific Section, Society of Economic Paleontologists and Mineralogists Pacific Coast Paleogeography Symposium 2, p. 311–336.

Sams, D. B., and Saleeby, J. B., 1988, Geology and petrotectonic significance of crystalline rocks of the southernmost Sierra Nevada, California, *in* Ernst, W. G., ed., Metamorphism and crustal evolution of the western United States; Rubey Volume 7: Englewood Cliffs, New Jersey, Prentice-Hall, p. 865–893.

Sams, R. H., 1964, Geology of the Charlie Canyon area, northwest Los Angeles County, California [M.A. thesis]: Los Angeles, University of California, 101 p.

Sanders, C. O., and Kanamori, H., 1984, A seismotectonic analysis of the Anza seismic gap, San Jacinto fault zone, southern California: Journal of Geophysical Research, v. 89, no. B7, p. 5873–5890.

Sarna-Wojcicki, A. M., and 9 others, 1984, Chemical analyses, correlations, and ages of upper Pliocene and Pleistocene ash layers of east-central and south-ern California: U.S. Geological Survey Professional Paper 1293, 40 p.

Saul, L. R., 1983, Turritella zonation across the Cretaceous-Tertiary boundary, California: Berkeley, University of California Publications in Geological Sciences, v. 125, 165 p.

Scharf, D., 1935, The Quaternary history of the Pinto basin, *in* Campbell, E.W.C., and Campbell, W. H., The Pinto basin site: Southwest Museum Paper, no. 9, p. 11–20.

Schoellhamer, J. E., Vedder, J. G., Yerkes, R. F., and Kinney, D. M., 1981, Geology of the northern Santa Ana Mountains, California: U.S. Geological Survey Professional Paper 420-D, 109 p.

Scholz, C. H., Barazangi, M., and Sbar, M. L., 1971, Late Cenozoic evolution of the Great Basin, western United States, as an ensialic interarc basin: Geological Society of America Bulletin, v. 82, p. 2979–2990.

Schwade, I. T., Carlson, S. A., and O'Flynn, J. B., 1958, Geologic environment of Cuyama Valley oil fields, California, *in* Weeks, L. G., ed., Habitat of oil; A symposium: Tulsa, Oklahoma, American Association of Petroleum Geologists, p. 78–98.

Schwarcz, H. P., 1969, Pre-Cretaceous sedimentation and metamorphism in the Winchester area, northern Peninsular Ranges, California: Geological Society of America, Special Paper 100, 63 p.

Schweickert, R. A., 1981, Tectonic evolution of the Sierra Nevada Range, *in* Ernst, W. G., ed., The geotectonic development of California; Rubey Volume 1: Englewood Cliffs, New Jersey, Prentice-Hall, p. 87–131.

Schwing, H. F., 1982, Interaction between the Transverse Ranges and Salinian block in the south Cuyama oil field, Cuyama Valley, California, *in* Fife, D. L., and Minch, J. A., eds., Geology and mineral wealth of the California Transverse Ranges; Mason Hill Volume: Santa Ana, California, South Coast Geological Society Annual Symposium and Guidebook 10, p. 395–402.

Sedlock, R. L., and Hamilton, D. H., 1991, Late Cenozoic tectonic evolution of southwestern California: Journal of Geophysical Research, v. 96, no. B2, p. 2325–2351.

Seedorf, D. C., 1983, Upper Cretaceous through Eocene subsurface stratigraphy, Simi Valley and adjacent regions, California, *in* Squires, R. L., and Filewicz, M. V., eds., Cenozoic geology of the Simi Valley area, southern California: Pacific Section, Society of Economic Paleontologists and Mineralogists Volume and Guidebook, p. 109–128.

Seiders, V. M., 1978, Onshore stratigraphic comparisons across the San Simeon and Hosgri faults, California: Geological Society of America Abstracts with Programs, v. 10, p. 146.

Seiders, V. M., Joyce, J. M., Leverett, K. A., and McLean, H., 1983, Geologic map of part of the Ventana Wilderness and the Black Butte, Bear Mountain, and Bear Canyon Roadless Areas, Monterey County, California: U.S. Geological Survey Miscellaneous Field Studies Map MF-1559-B, scale 1:50,000.

Shafiqullah, M., and 5 others, 1980, K-Ar geochronology and geologic history of southwestern Arizona and adjacent areas, *in* Jenney, J. P., and Stone, C., eds., Studies in western Arizona: Arizona Geological Society Digest, v. 12, p. 201–260.

Sharp, R. V., 1966, Ancient mylonite zone and fault displacements in the Peninsular Ranges of southern California, *in* Abstracts for 1966: Geological Society of America Special Paper 101, p. 333.

——— , 1967, San Jacinto fault zone in the Peninsular Ranges of southern California: Geological Society of America Bulletin, v. 78, p. 705–729.

——— , 1979, Some characteristics of the eastern Peninsular Ranges mylonite zone, *in* Proceedings of conference VIII; Analysis of Actual Fault Zones in Bedrock: U.S. Geological Survey Open-File Report 79-1239, p. 258–267.

——— , 1981, Variable rates of late Quaternary strike slip on the San Jacinto fault zone, southern California: Journal of Geophysical Research, v. 86, no. B3, p. 1754–1762.

——— , 1982, Tectonic setting of the Imperial Valley region, *in* The Imperial Valley, California, earthquake of October 15, 1979: U.S. Geological Survey Professional Paper 1254, p. 5–14.

Sharp, R. V., and Silver, L. T., 1971, Quaternary displacement on the San Andreas and Punchbowl faults at the San Gabriel Mountains, southern California: Geological Society of America Abstracts with Programs, v. 3,

p. 191.

Sharry, J., 1982, Minimum age and westward continuation of the Garlock fault zone, Tehachapi Mountains, California: Geological Society of America Abstracts with Programs, v. 14, p. 233.

Shay, K., 1975, Mineralogical zoning in a scapolite-bearing skarn body on San Gorgonio Mountain, California: American Mineralogist, v. 60, p. 785–797.

Shelton, J. S., 1955, Glendora volcanic rocks, Los Angeles basin, California: Geological Society of America Bulletin, v. 66, p. 45–89.

Shepard, J. B., Jr., 1962, San Gabriel fault zone: American Association of Petroleum Geologists Bulletin, v. 46, p. 1938–1941.

Sherrod, D. R., and Tosdal, R. M., 1991, Geologic setting and Tertiary structural evolution of southwestern Arizona and southeastern California: Journal of Geophysical Research, v. 96, p. B7, p. 12407–12423.

Sherrod, D. R., Pickthorn, L. G., Tosdal, R. M., Grubensky, M. J., and Koch, R. D., 1987, Major early Miocene extensional deformation in southwestern Arizona and southeastern California: Geological Society of America Abstracts with Programs, v. 19, p. 841.

Sherrod, D. R., and 5 others, 1989, Mineral resources of the Trigo Mountains Wilderness Study Area, La Paz County, Arizona: U.S. Geological Survey Bulletin 1702-J, 16 p.

Sieh, K., 1986, Slip rate across the San Andreas fault and prehistoric earthquakes at Indio, California: EOS Transactions of the American Geophysical Union, v. 67, p. 1200.

Sieh, K. E., and Jahns, R. H., 1984, Holocene activity of the San Andreas fault at Wallace Creek, California: Geological Society of America Bulletin, v. 95, p. 883–896.

Silver, E. A., 1974, Structural interpretation from free-air gravity on the California continental margin, 35° to 40° N: Geological Society of America Abstracts with Programs, v. 6, p. 253.

——, 1978, The San Gregorio–Hosgri fault zone; An overview, *in* Silver, E. A., and Normark, W. R., eds., San Gregorio–Hosgri fault zone, California: California Division of Mines and Geology Special Report 137, p. 1–2.

Silver, E. A., and Normark, W. R., eds., 1978, San Gregorio–Hosgri fault zone, California: California Division of Mines and Geology Special Report 137, 56 p.

Silver, L. T., 1971, Problems of crystalline rocks of the Transverse Ranges: Geological Society of America Abstracts with Programs, v. 3, p. 193–194.

——, 1982, Evidence and a model for west-directed early to mid-Cenozoic basement overthrusting in southern California: Geological Society of America Abstracts with Programs, v. 14, p. 617.

Silver, L. T., and Nourse, J. A., 1986, The Rand Mountains "thrust" complex in comparison with the Vincent thrust–Pelona Schist relationship, southern California: Geological Society of America Abstracts with Programs, v. 18, p. 185.

Silver, L. T., McKinney, C. R., Deutsch, S., and Bolinger, J., 1963, Precambrian age determinations in the western San Gabriel Mountains, California: Journal of Geology, v. 71, p. 196–214.

Silver, L. T., Anderson, T. H., Conway, C. M., Murray, J. D., and Powell, R. E., 1977, Geologic features of southwestern North America, *in* Skylab explores the Earth: Washington, D.C., National Aeronautics and Space Administration NASA SP-380, p. 89–135.

Silver, L. T., Taylor, H. P., Jr., and Chappell, B., 1979, Some petrological, geochemical and geochronological observations of the Peninsular Ranges batholith near the international border of the U.S.A. and Mexico, *in* Abbott, P. L., and Todd, V. R., eds., Mesozoic crystalline rocks; Peninsular Ranges batholith and pegmatites, Point Sal ophiolite; Geological Society of America Annual Meeting Guidebook: San Diego, California, San Diego State University Department of Geological Sciences, p. 83–110.

Silver, L. T., and 6 others, 1984, Some observations of the tectonic history of the Rand Mountains, Mojave Desert, California: Geological Society of America Abstracts with Programs, v. 16, p. 333.

Simpson, C., 1984, Borrego Springs–Santa Rosa mylonite zone; A late Cretaceous west-directed thrust in southern California: Geology, v. 12, p. 8–11.

Simpson, E. C., 1934, Geology and mineral deposits of the Elizabeth Lake Quadrangle, California: California Journal of Mines and Geology, v. 30, p. 371–415.

Skolnick, H., and Arnal, R. E., 1959, Ventura basin edge environment: American Association of Petroleum Geologists Bulletin, v. 43, p. 477–483.

Smith, D. B., Tosdal, R. M., Ardian, B. M., and Vaughn, R. B., 1984, Assessment of mineral resources in the Muggins Mountains Bureau of Land Management Wilderness Study Area (AZ-050-53A), Yuma County, Arizona: U.S. Geological Survey Open-File Report 84-662, 31 p.

Smith, D. B., and 8 others, 1987, Mineral resources of the Indian Pass and Picacho Peak Wilderness Study Areas, Imperial County, California: U.S. Geological Survey Bulletin 1711-A, 21 p.

Smith, D. P., 1977a, San Juan–St. Francis fault; Hypothesized major middle Tertiary right-lateral fault in central and southern California: California Division of Mines and Geology Special Report 129, p. 41–50.

——, 1977b, The San Juan–St. Francis fault and the Rinconada-Jolon fault; Proposed major middle Tertiary right-lateral faults in southern California: Geological Society of America Abstracts with Programs, v. 9, p. 501–502.

Smith, G. I., 1962, Large lateral displacement on the Garlock fault, California, as measured from offset dike swarm: American Association of Petroleum Geologists Bulletin, v. 46, p. 85–104.

——, 1964, Geology and volcanic petrology of the Lava Mountains, San Bernardino County, California: U.S. Geological Survey Professional Paper 457, 97 p.

Smith, G. I., and Ketner, K. B., 1970, Lateral displacement on Garlock fault, southeastern California, suggested by offset sections of similar metasedimentary rocks: U.S. Geological Survey Professional Paper 700-D, p. D1–D9.

Sorensen, S., 1988, Tectonometamorphic significance of the basement rocks of the Los Angeles basin and the inner California continental borderland, *in* Ernst, W. G., ed., Metamorphism and crustal evolution of the western United States; Rubey Volume 7: Englewood Cliffs, New Jersey, Prentice-Hall, p. 998–1022.

Spencer, J. E., and Reynolds, S. J., 1986, Some aspects of the middle Tertiary tectonics of Arizona and southeastern California, *in* Beatty, B., and Wilkinson, P.A.K., eds., Frontiers in geology and ore deposits of Arizona and the Southwest: Arizona Geological Society Digest, v. 16, p. 102–107.

Spittler, T. E., and Arthur, M. A., 1973, Post early Miocene displacement along the San Andreas fault in southern California, *in* Kovach, R. L., and Nur, A., eds., Proceedings of the Conference on Tectonic Problems of the San Andreas Fault System: Stanford, California, Stanford University Publications in the Geological Sciences, v. 13, p. 374–382.

——, 1982, The lower Miocene Diligencia Formation of the Orocopia Mountains, southern California; Stratigraphy, petrology, sedimentology and structure, *in* Ingersoll, R. V., and Woodburne, M. O., eds., Cenozoic nonmarine deposits of California and Arizona: Pacific Section, Society of Economic Paleontologists and Mineralogists, p. 83–99.

Squires, R. L., and Advocate, D. M., 1982, Sedimentary facies of the nonmarine lower Miocene Diligencia Formation, Canyon Spring area, Orocopia Mountains, southern California, *in* Ingersoll, R. V., and Woodburne, M. O., eds., Cenozoic nonmarine deposits of California and Arizona: Pacific Section, Society of Economic Paleontologists and Mineralogists, p. 101–106.

Stanley, R. G., 1987, New estimates of displacement along the San Andreas fault in central California based on paleobathymetry and paleogeography: Geology, v. 15, p. 171–174.

Stewart, J. H., and Poole, F. G., 1975, Extension of the Cordilleran miogeosynclinal belt to the San Andreas fault, southern California: Geological Society of America Bulletin, v. 86, p. 205–212.

Stewart, J. H., McMenamin, M.A.S., and Morales-Ramirez, J. M., 1984, Upper Proterozoic and Cambrian rocks in the Caborca region, Sonora, Mexico; Physical stratigraphy, biostratigraphy, paleocurrent studies, and regional relations: U.S. Geological Survey Professional Paper 1309, 36 p.

Stitt, L. T., 1986, Structural history of the San Gabriel fault and other Neogene structures of the central Transverse Ranges, California, *in* Ehlig, P. L., compiler, Neotectonics and faulting in southern California: Geological Society of America Cordilleran Section Annual Meeting Guidebook and Volume: Los

Angeles, California State University Department of Geology, p. 43–102.

Stitt, L. T., and Yeats, R. S., 1982, A structural sketch of the Castaic area, northeastern Ventura basin and northern Soledad basin, California, *in* Fife, D. L., and Minch, J. A., eds., Geology and mineral wealth of the California Transverse Ranges; Mason Hill Volume: Santa Ana, California, South Coast Geological Society of Annual Symposium and Guidebook 10, p. 390–394.

Stock, J. M., and Hodges, K. V., 1989, Pre-Pliocene extension around the Gulf of California and the transfer of Baja California to the Pacific plate: Tectonics, v. 8, p. 99–115.

Stone, P., and Pelka, G. J., 1989, Geologic map of the Palen-McCoy Wilderness Study Area and vicinity, Riverside County, California: U.S. Geological Survey Miscellaneous Field Studies Map MF-2092, scale 1:62,500.

Stone, P., Howard, K. A., and Hamilton, W., 1983, Correlation of metamorphosed Paleozoic strata of the southeastern Mojave Desert region, California and Arizona: Geological Society of America Bulletin, v. 94, p. 1135–1147.

Stone, P., Page, V. M., Hamilton, W., and Howard, K. A., 1987, Cretaceous age of the upper part of the McCoy Mountains Formation, southeastern California and southwestern Arizona, and its tectonic significance; Reconciliation of paleobotanical and paleomagnetic evidence: Geology, v. 15, p. 561–564.

Stuart, C. J., 1979, Middle Miocene paleogeography of coastal southern California and the California borderland; Evidence from schist-bearing sedimentary rocks, *in* Armentrout, J. M., Cole, M. R., TerBest, H., Jr., eds., Cenozoic paleogeography of the western United States: Pacific Section, Society of Economic Paleontologists and Mineralogists Pacific Coast Paleogeography Symposium 3, p. 29–44.

Stuart, W. D., 1991, Cause of the Garlock fault: Geological Society of America Abstracts with Programs, v. 23, p. A198.

Sumner, J. R., and Thompson, G. A., 1974, Esimates of strike-slip offset in southwestern Arizona: Geological Society of America Bulletin, v. 85, p. 943–946.

Suppe, J., and Armstrong, R. L., 1972, Potassium-argon dating of Franciscan metamorphic rocks: American Journal of Science, v. 272, p. 217–233.

Sylvester, A. G., compiler, 1984, Wrench fault tectonics: Tulsa, Oklahoma, American Association of Petroleum Geologists, 374 p.

Sylvester, A. G., and Crowell, J. C., leaders, 1989, The San Andreas transform belt; 28th International Geologic Congress field trip guidebook T309: Washington, D.C., American Geophysical Union, 119 p.

Sylvester, A. G., and Darrow, A. C., 1979, Structure and neotectonics of the western Santa Ynez fault system in southern California: Tectonphysics, v. 52, p. 389–405.

Sylvester, A. G., and Smith, R. R., 1976, Tectonic transpression and basement-controlled deformation in San Andreas fault zone, Salton trough, California: American Association of Petroleum Geologists Bulletin, v. 60, p. 2081–2102.

Szatai, J. E., 1961, The geology of parts of the Redrock Mountain, Warm Spring, Violin Canyon, and Red Mountain Quadrangles, Los Angeles County, California [Ph.D. thesis]: Los Angeles, University of Southern California.

Tedford, R. H., and 8 others, 1987, Faunal succession and biochronology of the Arikareean through Hemphillian interval (late Oligocene through earliest Pliocene Epochs) in North America, *in* Woodburne, M. O., ed., Cenozoic mammals of North America; Geochronology and biostratigraphy: Berkeley, University of California Press, p. 153–210.

Terres, R. R., and Luyendyk, B. P., 1985, Neogene tectonic rotation of the San Gabriel region, California, suggested by paleomagnetic vectors: Journal of Geophysical Research, v. 90, no. B14, p. 12467–12484,

Thomas, J. M., 1986, Correlation and petrogenesis of the Miocene volcanic rocks in the San Emigdio and San Juan Bautista areas, California [M.S. thesis]: Northridge, California State University, 94 p.

Thomson, A., and Ryder, R. T., 1976, Patterns of diachronous deep-water sedimentation as evidence for an actively shifting source terrane; Santa Margarita Formation and equivalent units (upper Miocene), southern Temblor Range, California: Geological Society of America Abstracts with Programs, v. 8, p. 1139–1140.

Todd, V. R., 1978, Geologic map of Monument Peak Quadrangle, San Diego County, California: U.S. Geological Survey Open-File Report 78-697, 47 p.,

scale 1:24,000.

Todd, V. R., and Hoggatt, W. C., 1979, Vertical tectonics in the Elsinore fault zone south of 33°7'30": Geological Society of America Abstracts with Programs, v. 11, p. 528.

Todd, V. R., and Shaw, S. E., 1979, Structural, metamorphic and intrusive framework of the Peninsular Ranges batholith in southern San Diego County, California, *in* Abbott, P. L., and Todd, V. R., eds., Mesozoic crystalline rocks; Peninsular Ranges batholith and pegmatites, Point Sal ophiolite; Geological Society of America Annual Meeting Guidebook: San Diego, California, San Diego State University Department of Geological Sciences, p. 177–231.

Todd, V. R., Erskine, B. G., and Morton, D. M., 1988, Metamorphic and tectonic evolution of the northern Peninsular Ranges batholith, southern California, *in* Ernst, W. G., ed., Metamorphism and crustal evolution of the western United States; Rubey Volume 7: Englewood Cliffs, New Jersey, Prentice-Hall, p. 894–937.

Tosdal, R. M., 1982, The Mule Mountains thrust in the Mule Mountains, California and its probable extension in the southern Dome Rock Mountains, Arizona; A preliminary report, *in* Frost, E. G., and Martin, D. L., eds., Mesozoic-Cenozoic tectonic evolution of the Colorado River region, California, Arizona, and Nevada; Anderson-Hamilton Volume; Geological Society of America Cordilleran Section Annual Meeting Symposium and Field Trip Volume: San Diego, California, Cordilleran Publishers, p. 55–60.

—— , 1986, Mesozoic ductile deformations in the southern Dome Rock Mountains, northern Trigo Mountains, Trigo Peaks and Livingston Hills, southwestern Arizona, and Mule Mountains, southeastern California, *in* Beatty, B., and Wilkinson, P.A.K., eds., Frontiers in geology and ore deposits of Arizona and the Southwest: Arizona Geological Society Digest, v. 16, p. 62–71.

—— , 1988, Mesozoic rock units along the Lake Cretaceous Mule Mountains thrust system, southeastern California and southwestern Arizona [Ph.D. thesis]: Santa Barbara, University of California, 365 p.

Tosdal, R. M., and Sherrod, D. R., 1985, Geometry of Miocene extensional deformation, lower Colorado River region, southeastern California and southwestern Arizona; Evidence for the presence of a regional low-angle normal fault, *in* Papers Presented to the Conference on Heat and Detachment in Crustal Extension on Continents and Planets, Sedona, Arizona: Houston, Texas, Lunar and Planetary Institute Contribution 575, p. 147–151.

Tosdal, R. M., Haxel, G. B., and Wright, J. E., 1989, Jurassic geology of the Sonoran Desert region, southern Arizona, southeastern California, and northernmost Sonora; Construction of a continental-margin magmatic arc, *in* Jenny, J. P., and Reynolds, S. J., eds., Geologic evolution of Arizona: Arizona Geological Society Digest, v. 17, p. 397–434.

Treiman, J., and Saul, R., 1986, The mid-Pleistocene inception of the Santa Susana Mountains, *in* Ehlig, P. L., Neotectonics and faulting in southern California; Geological Society of America Cordilleran Section Annual Meeting Guidebook and Volume: Los Angeles, California State University Department of Geology, p. 7–12.

Troxel, B. W., and Gunderson, J. N., 1970, Geology of the Shadow Mountains and northern part of the Shadow Mountains Southeast Quadrangles, western San Bernardino County, California: California Division of Mines and Geology Preliminary Report 12, scale 1:24,000.

Troxel, B. W., Wright, L. A., and Jahns, R. H., 1972, Evidence for differential displacement along the Garlock fault zone, California: Geological Society of America Abstracts with Programs, v. 4, p. 250.

Troxel, B. W., Jahns, R. H., and Butler, P. R., 1979, Quaternary and Tertiary history of offsets along the easternmost segment of the Garlock fault zone: Geological Society of America Abstracts with Programs, v. 11, p. 132.

Truex, J. N., 1976, Santa Monica and Santa Ana Mountains; Relation to Oligocene Santa Barbara basin: American Association of Petroleum Geologists Bulletin, v. 60, p. 65–86.

Turner, D. L., 1968, Potassium-argon dates concerning the Tertiary foraminiferal time scale and San Andreas fault displacement [Ph.D. thesis]: Berkeley,

University of California, 99 p.

——, 1969, K-Ar ages of California Coast Range volcanics; Implications for San Andreas fault displacement: Geological Society of America Abstracts with Programs for 1969, Part 3, p. 70.

——, 1970, Potassium-argon dating of Pacific coast Miocene foraminiferal stages, *in* Bandy, O. L., ed., Radiometric dating and paleontologic zonation: Geological Society of America Special Paper 124, p. 91–129

Turner, D. L., Curtis, G. H., Berry, F.A.F., and Jack, R. N., 1970, Age relationship between the Pinnacles and Parkfield felsites and felsite clasts in the southern Temblor Range, California; Implications for San Andreas fault displacement: Geological Society of America Abstracts with Programs, v. 2, p. 154–155.

U.S. Geological Survey, 1972, The Borrego Mountain earthquake of April 9, 1968: U.S. Geological Survey Professional Paper 787, 207 p.

Valentine, M. J., Brown, L., and Golombek, M. P., 1987, Tectonic rotation of the Barstow area, central Mojave Desert, California inferred from paleomagnetic data: EOS Transactions of the American Geophysical Union, v. 68, p. 1254.

——, 1988a, Constraints on timing of block rotations in the western Mojave Desert: EOS Transactions of the American Geophysical Union, v. 69, p. 1164.

——, 1988b, Counterclockwise Cenozoic rotations of the Mojave Desert: Geological Society of America Abstracts with Programs, v. 20, p. 239.

Vaughan, F. E., 1922, Geology of the San Bernardino Mountains north of San Gorgonio Pass: Berkeley, University of California Publications in Geological Sciences, v. 13, p. 319–411.

Vedder, J. G., and Brown, R. D., Jr., 1968, Structural and stratigraphic relations along the Nacimiento fault in the southern Santa Lucia Range and San Rafael Mountains, California, *in* Dickinson, W. R., and Grantz, A., eds., Proceedings of Conference on Geologic Problems of San Andreas Fault Systems: Stanford, California, Stanford University Publications in the Geological Sciences, v. 11, p. 242–259.

Vedder, J. G., Howell, D. G., and McLean, H., 1983, Stratigraphy, sedimentation, and tectonic accretion of exotic terranes, southern Coast Ranges, California, *in* Watkins, J. S., and Drake, C. L., eds., Studies in continental margin geology: American Association of Petroleum Geologists Memoir 34, p. 471–496.

Von Huene, R., 1969, Geologic structure between the Murray fracture zone and the Transverse Ranges: Marine Geology, v. 7, p. 475–499.

Wakabayashi, J., and Moores, E. M., 1988, Evidence for the collision of the Salinian block with the Franciscan subduction zone, California: Journal of Geology, v. 96, p. 245–253.

Walker, J. D., 1987, Permian to middle Triassic rocks of the Mojave Desert, *in* Dickinson, W. R., and Klute, M. A., eds., Mesozoic rocks of southern Arizona and adjacent areas: Arizona Geological Society Digest, v. 18, p. 1–14.

Wallace, R. E., 1975, The San Andreas fault in the Carrizo Plain–Temblor Range region, California, *in* Crowell, J. C., ed., San Andreas fault in southern California: California Division of Mines and Geology Special Report 118, p. 241–250.

——, ed., 1990, The San Adreas fault system, California: U.S. Geological Survey Professional Paper 1515, 283 p.

Ward, S. N., 1990, Pacific–North America plate motions: New results from very long baseline interferometry: Journal of Geophysical Research, v. 95, no. B13, p. 21965–21981.

Ware, G. C., Jr., 1958, The geology of a portion of the Mecca Hills, Riverside County, California [M.A. thesis]: Los Angeles, University of California, 60 p.

Webb, G. W., 1981, Stevens and earlier Miocene turbidite sandstones, southern San Joaquin Valley, California: American Association of Petroleum Geologists Bulletin, v. 65, p. 438–465.

Webb, T. H., and Kanamori, H., 1985, Earthquake focal mechanisms in the eastern Transverse Ranges and San Emigdio Mountains, southern California and evidence for a regional decollement: Seismological Society of America Bulletin, v. 75, p. 737–757.

Weber, F. H., Jr., 1977, Seismic hazards related to geologic factors, Elsinore and Chino fault zones, northwestern Riverside County, California: U.S. Geological Survey Final Technical Report, Grants. 14-08-0001-G-74, -G-133, and -G-263, 96 p. (California Division of Mines and Geology Open File Report 77-4 LA).

——, 1982a, Geology and geomorphology along the San Gabriel fault zone, Los Angeles and Ventura Counties, California: U.S. Geological Survey Final Technical Report, Contract 14-08-0001-16600, Modification 1, 157 p. (California Division of Mines and Geology Open-File Report 82-2 LA).

——, 1982b, New interpretation of structural elements in the vicinity of the northwestern terminus of the trace of the San Gabriel fault zone, Ventura County, California, *in* Fife, D. L., and Minch, J. A., eds., Geology and mineral wealth of the California Transverse Ranges; Mason Hill Volume: Santa Ana, California, South Coast Geological Society Annual Symposium and Guidebook 10, p. 384–389.

——, 1986a, Geologic relationships along the San Gabriel fault between Castaic and the San Andreas fault, Kern, Los Angeles, and Ventura Counties, California, *in* Ehlig, P. L., compiler, Neotectonics and faulting in southern California; Geological Society of America Cordilleran Section Annual Meeting Guidebook and Volume: Los Angeles, California State University Department of Geology, p. 109–122.

——, 1986b, Geologic relationships between the San Gabriel and San Andreas faults: California Division of Mines and Geology, California Geology, v. 39, p. 5–14.

Weber, G. E., and Cotton, W. R., 1981, Geologic investigation of recurrence intervals and recency of faulting along the San Gregorio fault zone, San Mateo County, California: U.S. Geological Survey Open-File Report 81-263, 99 p.

Weber, G. E., Lajoie, K. R., and Wehmiller, J. F., 1979, Quaternary crustal deformation along a major branch of the San Andreas fault in central California: Tectonophysic, v. 52, p. 378–379.

Weigand, P. W., 1982, Middle Cenozoic volcanism of the western Transverse Ranges, *in* Fife, D. L., and Minch, J. A., eds., Geology and mineral wealth of the California Transverse Ranges; Mason Hill Volume: Santa Ana, California, South Coast Geological Society Annual Symposium and Guidebook 10, p. 170–188.

Weldon, R. J., II, 1986, The late Cenozoic geology of the Cajon Pass; Implications for tectonics and sedimentation along the San Andreas fault [Ph.D. thesis]: Pasadena, California Institute of Technology, 400 p.

Weldon, R. J., II, and Humphreys, E., 1986, A kinematic model of southern California: Tectonics, v. 5, p. 33–48.

Weldon, R. J., II, and Sieh, K. E., 1985, Holocene rate of slip and tentative recurrence interval for large earthquakes on the San Andreas fault, Cajon Pass, southern California: Geological Society of America Bulletin, v. 96, p. 793–812.

Weldon, R. J., II, and Springer, J. E., 1988, Active faulting near the Cajon Pass well, southern California; Implications for the stress orientation near the San Andreas fault: Geophysical Research Letters, v. 15, p. 993–996.

Weldon, R. J., II, Meisling, K. E., Sieh, K. E., and Allen, C. R., 1981, Neotectonics of the Silverwood Lake area, San Bernardino County: Report to the California Department of Water Resources, 22 p.

Wells, R. E., and Hillhouse, J. W., 1989, Paleomagnetism and tectonic rotation of the lower Miocene Peach Springs Tuff; Colorado Plateau, Arizona, to Barstow, California: Geological Society of America Bulletin, v. 101, p. 846–863.

Welton, B. J., and Link, M. H., 1982, Vertebrate paleontology of Ridge basin, southern California, *in* Crowell, J. C., and Link, M. H., eds., Geologic history of Ridge basin, southern California: Pacific Section, Society of Economic Paleontologists and Mineralogists, p. 205–210.

Wentworth, C. M., 1968, Upper Cretaceous and lower Tertiary strata near Gualala, California, and inferred large right slip on the San Andreas fault, *in* Dickinson, W. R., and Grantz, A., eds., Proceedings of Conference on Geologic Problems of San Andreas Fault System: Stanford, California, Stanford University Publications in the Geological Sciences, v. 11, p. 130–143.

Wernicke, B., Spencer, J. E., Burchfiel, B. C., Guth, P. L., 1982, Magnitude of crustal extension in the southern Great Basin: Geology, v. 10, p. 499–502.

Wesnousky, S. G., Prentice, C. S., and Sieh, K. E., 1991, An offset Holocene stream channel and the rate of slip along the northern reach of the San Jacinto fault zone, San Bernardino Valley, California: Geological Society of America Bulletin, v. 103, p. 700–709.

Whistler, D. P., 1967, Oreodonts of the Tick Canyon Formation, southern California: PaleoBios, no. 1, 14 p.

Williams, P. L., Sykes, L. R., Nicholson, Craig, and Seeber, Leonardo, 1990, Seismotectonics of the easternmost Transverse Ranges, California: Relevance for seismic potential of the southern San Andreas fault: Tectonics, v. 9, no. 1, p. 185–204.

Wilson, E. D., 1933, Geology and mineral deposits of southern Yuma County, Arizona: Arizona Bureau of Mines Bulletin 134, 236 p.

Winterer, E. L., and Durham, D. L., 1962, Geology of southeastern Ventura basin, Los Angeles County, California: U.S. Geological Survey Professional Paper 334-H, p. 275–366.

Woodburne, M. O., 1975, Cenozoic stratigraphy of the Transverse Ranges and adjacent areas, southern California: Geological Society of America Special Paper 162, 91 p.

—— , 1987, A prospectus of the North American mammal ages, *in* Woodburne, M. O., ed., Cenozoic mammals of North America; Geochronology and biostratigraphy: Berkeley, University of California Press, p. 285–290.

Woodburne, M. O., and Golz, D. J., 1972, Stratigraphy of the Punchbowl Formation, Cajon Valley, southern California: Berkeley, University of California Publications in Geological Sciences, v. 92, 73 p.

Woodburne, M. O., and Whistler, D. P., 1973, An early Miocene oreodont (Merychyinae, Mammalia) from the Orocopia Mountains, southern California: Journal of Paleontology, v. 47, p. 908–912.

Wooden, J. L., Powell, R. E., Howard, K. A., and Tosdal, R. M., 1991, Eagle Mtns. 30′ × 60′ Quadrangle, southern California; II. Isotopic and chronologic studies: Geological Society of America Abstracts with Programs, v. 23, p. A478.

Woodford, A. O., 1960, Bedrock patterns and strike-slip faulting in southwestern California: American Journal of Science, v. 258A, p. 400–417.

Woodford, A. O., Moran, T. G., and Shelton, J. S., 1946, Miocene conglomerates of Puente and San Jose Hills, California: American Association of Petroleum Geologists Bulletin, v. 30, p. 514–560.

Woodford, A. O., Shelton, J. S., Doehring, D. O., and Morton, R. K., 1971, Pliocene-Pleistocene history of the Perris block, southern California: Geological Society of America Bulletin, v. 82, p. 3421–3448.

Wright, L., 1976, Late Cenozoic fault patterns and stress fields in the Great Basin and westward displacement of the Sierra Nevada block: Geology, v. 4, p. 489–494.

Yeats, R. S., 1968, Rifting and rafting in the southern California borderland, *in* Dickinson, W. R., and Grantz, A., eds., Proceedings of Conference on Geologic Problems of San Andreas Fault System: Stanford, California, Stanford University Publications in the Geological Sciences, v. 11, p. 307–322.

—— , 1976, Extension *versus* strike-slip origin of the southern California borderland, *in* Howell, D. G., ed., Aspects of the geologic history of the California continental borderland: Pacific Section, American Association of Petroleum

Geologists Miscellaneous Publication 24, p. 455–485.

—— , 1979, Stratigraphy and paleogeography of the Santa Susana fault zone, Transverse Ranges, California, *in* Armentrout, J. M., Cole, M. R., and TerBest, H., Jr. eds., Cenozoic paleogeography of the western United States: Pacific Section, Society of Economic Palentologists and Mineralogists Pacific Coast Paleogeography Symposium 3, p. 191–204.

—— , 1981a, Quaternary flake tectonics of the California Transverse Ranges: Geology, v. 9, p. 16–20.

—— , 1981b, Subsurface geology of the San Gabriel, Holser, and Simi–Santa Rosa faults, Transverse Ranges, California: U.S. Geological Survey Final Technical Report, Contract 14-08-0001-16747, 14 p.

—— , 1983, Simi; A structural essay, *in* Squires, R. L., and Filewicz, M. V., eds., Cenozoic geology of the Simi Valley area, southern California: Pacific Section, Society of Economic Paleontologists and Mineralogists Volume and Guidebook, p. 233–239.

—— , 1986, The Santa Susana fault at Aliso Canyon oil field, *in* Ehlig, P. L., compiler, Neotectonics and faulting in southern California; Geological Society of America Cordilleran Section Annual Meeting Guidebook and Volume: Los Angeles, California State University Department of Geology, p. 13–22.

—— , 1987, Late Cenozoic structure of the Santa Susana fault zone, *in* Recent reverse faulting in the Transverse Ranges, California: U.S. Geological Survey Professional Paper 1339, p. 137–160.

Yeats, R. S., Cole, M. R., Merschat, W. R., and Parsley, R. M., 1974, Poway fan and submarine cone and rifting in the inner southern California borderland: Geological Society of America Bulletin, v. 85, p. 293–302.

Yeats, R. S., Calhoun, J. A., Nevins, B. B., Schwing, H. F., and Spitz, H. M., 1988, The Russell fault; An early strike-slip fault of the California Coast Ranges, *in* Bazeley, W.J.M., ed., Tertiary tectonics and sedimentation in the Cuyama basin, San Luis Obispo, Santa Barbara, and Ventura Counties, California: Society of Economic Paleontologists and Mineralogists, v. 59, p. 127–139.

Yeats, R. S., Calhoun, J. A., Nevins, B. B., Schwing, H. F., and Spitz, H. M., 1989, Russell fault; Early strike-slip fault of the California Coast Ranges: American Association of Petroleum Geologists Bulletin, v. 73, p. 1089–1102.

Yerkes, R. F., and Campbell, R. H., 1980, Geologic map of east-central Santa Monica Mountains, Los Angeles County, California: U.S. Geological Survey Miscellaneous Investigations Series Map I-1146, scale 1:24,000.

Yerkes, R. F., and Lee, W.H.K., 1987, Late Quaternary deformation in the western Transverse Ranges, *in* Recent reverse faulting in the Transverse Ranges, California: U.S. Geological Survey Professional Paper 1339, p. 71–82.

Yerkes, R. F., McCulloh, T. H., Schoellhamer, J. E., and Vedder, J. G., 1965, Geology of the Los Angeles basin, California; An introduction: U.S. Geological Survey Professional Paper 420-A, 57 p.

Yerkes, R. F., Sarna-Wojcicki, A. M., and Lajoie, K. R., 1987, Geology and Quaternary deformation of the Ventura area, *in* Recent reverse faulting in the Transverse Ranges, California: U.S. Geological Survey Professional Paper 1339, p. 169–178.

MANUSCRIPT ACCEPTED BY THE SOCIETY SEPTEMBER 6, 1990

Geological Society of America
Memoir 178
1993

Chapter 2

Paleogeographic evolution of the San Andreas fault in southern California: A reconstruction based on a new cross-fault correlation

Jonathan C. Matti
U.S. Geological Survey, Department of Geological Sciences, University of Arizona, Tucson, Arizona 85721
Douglas M. Morton
U.S. Geological Survey, Department of Earth Sciences, University of California, Riverside, California 92521

ABSTRACT

Distinctive porphyritic bodies of alkalic monzogranite and quartz monzonite of Triassic age that occur in the Mill Creek region of the San Bernardino Mountains and on the opposite side of the San Andreas fault at the northwest end of Liebre Mountain appear to be segments of a formerly continuous pluton that has been severed by the fault and displaced about 160 km. Reassembly of the megaporphyritic bodies by restoration of sequential right-lateral displacements on various strands of the San Andreas and San Gabriel fault zones leads to a palinspastic reconstruction for southern California that reassembles crystalline and sedimentary terranes differently from widely cited reconstructions.

Reassembled crystalline rocks establish three coherent patterns: (1) The reunited Liebre Mountain and Mill Creek Triassic megaporphyry bodies form western outliers of a province of Permian and Triassic alkalic granitoid rocks that occurs in the western Mojave Desert and San Bernardino Mountains. (2) Following sequential restorations of 160 km on the San Andreas fault and 44 km on the San Gabriel and Cajon Valley faults, the Table Mountain and Holcomb Ridge basement slices currently positioned along the west edge of the Mojave Desert near Wrightwood and Valyermo area reassembled with the Liebre Mountain and San Bernardino Mountains blocks. This restoration unites terranes of Mesozoic granitoid rock and Paleozoic(?) marble, metaquartzite, and pelitic gneiss that have strong lithologic and compositional similarities, and brings together within a single province quartz diorite and granodiorite that are hosts for three known bedrock occurrences of aluminous dike rocks ("polka-dot" granite) that previously have been used to assemble a different reconstruction for the San Andreas fault. (3) Crystalline rocks of the San Gorgonio Pass region, including Triassic monzogranite and granodiorite of the Lowe igneous pluton and Jurassic blastoporphyritic quartz monzonite, are juxtaposed adjacent to the southern Chocolate Mountains where similar rocks have been mapped.

The reconstructed crystalline rocks provide a paleogeographic framework for Pliocene and late Miocene sedimentary basins now scattered along the San Gabriel and San Andreas faults. Ridge Basin is juxtaposed adjacent to the southwestern San Bernardino Mountains in an orogenic setting compatible with stratigraphic relations, depositional fabrics, and clast compositions in the Ridge Basin fill. Synorogenic sediments of the upper Miocene Ridge Route and Peace Valley Formations accumulated in this setting at

Matti, J. C., and Morton, D. M., 1993, Paleogeographic evolution of the San Andreas fault in southern California: A reconstruction based on a new cross-fault correlation, *in* Powell, R. E., Weldon, R. J., II, and Matti, J. C., eds., The San Andreas Fault System: Displacement, Palinspastic Reconstruction, and Geologic Evolution: Boulder, Colorado, Geological Society of America Memoir 178.

a time (9 to 5 Ma) when the western San Bernardino Mountains were undergoing uplift and erosion. The Pliocene Hungry Valley Formation was deposited toward the end of this orogenic pulse (5 to 4 Ma) when early movements on the San Andreas fault began to displace Ridge Basin away from the San Bernardino Mountains. Other late Miocene basins widely dispersed today are reassembled within an early Salton Trough that included, from northwest to southeast, the Punchbowl and Mill Creek basins, the Coachella Fanglomerate and Hathaway Formation, and the fanglomerate of Bear Canyon in the southern Chocolate Mountains.

The new reconstruction allows only about 205 km of displacement on the combined San Andreas and San Gabriel fault zones, which is about 110 km short of the 315 ± 10 km post–early Miocene offsets documented for the San Andreas in central California. This shortfall requires the existence of another Miocene strand of the San Andreas fault system in southern California. The Clemens Well–Fenner–San Francisquito fault system of Powell (this volume) provides this additional fault strand. Together, the Clemens Well, San Gabriel, and San Andreas faults generated about 315 ± 10 km of right slip comparable to displacements documented for the San Andreas fault in central California. The displacements in southern California have occurred in middle Miocene through Holocene time, and extend the history of late Cenozoic right-lateral faulting farther back into the Miocene than envisioned by traditional reconstructions for the San Andreas fault system in southern California.

INTRODUCTION

In 1953, M. L. Hill and T. W. Dibblee proposed that stratigraphic and structural elements of the California continental margin have been rearranged by large right-lateral displacements on the San Andreas fault. In the 40 yr since this revolutionary proposal exploded conventional views of Earth history, numerous attempts have been made to document the distribution and geologic history of faults in the San Andreas family and to reconstruct the palinspastic configuration of sedimentary and crystalline terranes displaced by these faults. The search for ancient paleogeography has been particularly challenging in southern California, where structural and stratigraphic elements have been rearranged not only by displacements on the San Andreas proper, but also by displacements on a variety of right-lateral, left-lateral, compressional, and extensional fault complexes that contributed to and interacted with the San Andreas system.

Any attempt to reconstruct pre–San Andreas paleogeography in southern California must first determine how right-lateral displacements have been distributed among individual strands of the system. Total displacement on the San Andreas since its inception in the Miocene is widely accepted to be about 315 ± 10 km. In central California, this displacement has occurred mainly along the San Andreas proper (Hill and Dibblee, 1953; Crowell, 1981); however, in southern California the total displacement has been taken up by several discrete fault strands—including the San Andreas, San Jacinto, Punchbowl, San Gabriel, and Banning faults—as well as other less well-known structures (Fig. 1). The San Andreas fault alone accounts

for a large fraction of the total displacement in southern California, but a comparison of palinspastic reconstructions proposed in the literature (Crowell, 1960, 1962, 1975a, 1979, 1981; Jahns, 1973; Dillon, 1975; Woodburne, 1975; Smith, 1977; Ehlig, 1981, 1982; Powell, 1981, 1986; Ross, 1984; Matti and others, 1985) shows that significant disagreement exists regarding the timing and amount of displacement on the San Andreas proper and on associated right-lateral faults. This disparity of interpretation has led to conflicting palinspastic reconstructions for dismembered terranes that now are distributed across southern California between the Chocolate Mountains and the Frazier Mountain region (Plate IIIA).

In this chapter, we add to this disparity by proposing a new model for the paleogeographic evolution of the San Andreas fault in southern California. This model is based on the hypothesis that crystalline rocks of the Liebre Mountain region have been displaced by the San Andreas fault from their original position adjacent to the southwestern San Bernardino Mountains (Plate IIIA; Matti and others, 1986; Frizzell and others, 1986). From this starting point, we use new information about the movement history of various strands of the San Andreas fault system to reconstruct how crystalline and sedimentary terranes in southern California might originally have been assembled before they were dismembered and rearranged by right-slip displacements within the system. This palinspastic reconstruction requires that the San Andreas fault proper in southern California has no more than 160 km of displacement since its inception 4 to 5 m.y. ago, and that the full history of right slip within the San Andreas transform system in southern California must be extended farther back into

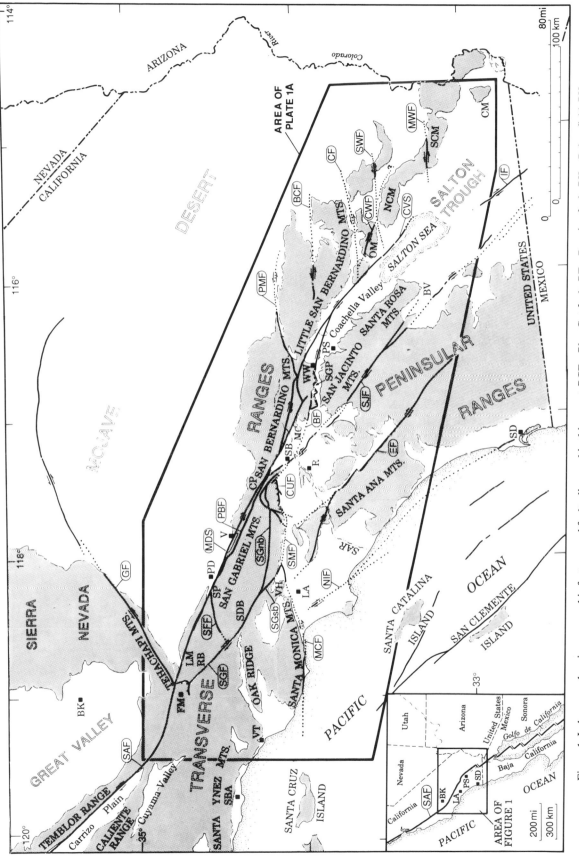

Figure 1. Index map showing geographic features and faults discussed in this report. BCF = Blue Cut fault; BF = Banning fault; BK = Bakersfield; BV = Borrego Valley region; CF = Chiriaco fault; CM = Cargo Muchacho Mountains; CP = Cajon Pass region; CVS = Coachella Valley segment, San Andreas fault; CWF = Clemens Well fault; EF = Elsinore fault; FM = Frazier Mountain region; GF = Garlock fault; IF = Imperial fault; LA = Los Angeles; LM = Liebre Mountain; MCF = Malibu Coast–Santa Monica–Raymond fault; MC = Mill Creek; MDS = Mojave Desert segment, San Andreas fault; MWF = Mammoth Wash fault; NCM = northern Chocolate Mountains; NIF = Newport-Inglewood fault; OM = Orocopia Mountains; PBF = Punchbowl fault; PD = Palmdale; PMF = Pinto Mountain fault; R = Riverside; RB = Ridge Basin; SAF = San Andreas fault; SAR = Santa Ana River; SB = San Bernardino; SBA = Santa Barbara; SBM = San Bernardino Mountains; SCM = southern Chocolate Mountains; SD = San Diego; SDB = Soledad basin; SFF = San Francisquito fault; SGF = Sar. Gabriel fault; SGnb = north branch, San Gabriel fault; SGsb = south branch, San Gabriel fault; SGP = San Gorgonio Pass; SJF = San Jacinto fault; SMF = Sierra Madre fault; SP = Sierra Pelona; SWF = Salton Wash fault; V = Valyermo; VT = Ventura; WW = Whitewater.

the late Cenozoic than predicted by most paleogeographic models and must be apportioned among other faults in the San Andreas system.

GEOLOGIC SETTING

A brief review of geologic relations in southern California provides a framework for reconstructing the displacement history of faults associated with the San Andreas system.

Basement rocks

The diverse basement terranes of southwestern California can be grouped into rocks of Peninsular Ranges type, San Gabriel Mountains type, and San Bernardino Mountains type (Fig. 2, Plate IIIA).

Basement rocks of *San Gabriel Mountains type* (Fig. 2A) consist of two crustal layers separated by a low-angle tectonic contact—the Vincent thrust (Ehlig, 1982). The upper thrust plate consists of Mesozoic plutons of various compositions, ages, and deformational styles that have intruded prebatholithic crystalline rocks. One distinctive Mesozoic granitoid unit is the Lowe igneous pluton of Triassic age (Joseph and others, 1982a). The prebatholithic rocks largely are Proterozoic, and include orthogneisses and the anorthosite-syenite complex that is so well known from the western San Gabriel Mountains (Silver, 1971; Ehlig, 1981; Carter, 1982). The lower plate of the Vincent thrust consists of Pelona Schist—late Mesozoic quartzofeldspathic sandstone and siltstone, limestone, quartzite, chert, and mafic volcanic rocks that have been metamorphosed to greenschist and lower amphibolite facies, presumably during late Mesozoic to early Tertiary emplacement of the upper plate (Ehlig, 1968b, 1981, 1982). From the Frazier Mountain region southeast to the Salton Trough, Pelona Schist occurs as windows and fault-bounded blocks (Ehlig, 1968b), including (Plate IIIA): (1) the Sierra Pelona window in the western San Gabriel Mountains; (2) the Lytle Creek window in the eastern San Gabriel Mountains; (3) the Blue Ridge slice between the Punchbowl and San Andreas faults; (4) a large window that mostly has been buried beneath sedimentary fill of the San Bernardino Valley; and (5) the Orocopia Mountains and Chocolate Mountains windows of Pelona-type schist that are referred to as Orocopia Schist and that are separated from rocks of San Gabriel Mountains type by the Orocopia–Chocolate Mountain thrust. Both lower and upper plates are intruded by high-level Miocene granitoid plutons and dikes that were emplaced after initial juxtaposition of the two plates (Miller and Morton, 1977; Crowe and others, 1979).

Basement rocks of *Peninsular Ranges type* (Fig. 2B) consist of Jurassic and Cretaceous granitoid rocks (granodiorite, quartz diorite, tonalite, gabbro) that have intruded prebatholithic metasedimentary rocks (pelitic schist, metaquartzite, marble, quartzofeldspathic gneiss, and schist). Along its north and northeast edge, the Peninsular Ranges block is bordered by a mylonitic belt of ductile deformation that separates lower plate plutonic and meta-

sedimentary rocks of typical Peninsular Ranges type from broadly similar upper plate rocks that appear to be parautochthonous equivalents of the more typical Peninsular Ranges suite; the autochthonous and parautochthonous suites have been telescoped along the mylonite zone (Sharp, 1979; Erskine, 1985). The ductile zone is referred to by different names locally, but Sharp (1979) recognized its regional significance and named it the Eastern Peninsular Ranges mylonite zone. The ductile zone probably continues westward into the southeastern San Gabriel Mountains, where mylonitic rocks have been mapped a few kilometers north of the Mountain front (Alf, 1948; Hsu, 1955; Morton and Matti, 1987); if so, then the San Gabriel mylonitic belt may separate lower and upper plate rocks of Peninsular Ranges type—a speculative proposal we adopt here (Plate IIIA). Juxtaposition of Peninsular Ranges–type and San Gabriel Mountains–type rocks in the southeastern San Gabriel Mountains represents either (1) original intrusive relations between the two suites, (2) their tectonic juxtaposition by Neogene strike-slip displacements, (3) their tectonic juxtaposition by Paleogene or latest Cretaceous thrust faulting (May and Walker, 1989), or (4) a combination of some or all of these mechanisms.

Basement rocks of *San Bernardino Mountains type* (Fig. 2C) are similar to those in the Mojave Desert, and consist of Triassic through Cretaceous granitoid rocks of various compositions that have intruded prebatholithic orthogneiss (Proterozoic) and metasedimentary rocks (Late Proterozoic and Paleozoic metaquartzite, marble, pelitic schist, and gneiss). The metasedimentary rocks are comparable with rocks of the Cordilleran miogeocline (Stewart and Poole, 1975). The Mesozoic plutonic rocks include both deformed and undeformed suites that extend southeastward into the Little San Bernardino Mountains, where they intrude rocks of San Gabriel Mountains type. Strongly deformed Mesozoic granodiorite, tonalite, and quartz diorite form a discrete belt in the southeastern San Bernardino Mountains and along the western margin of the Little San Bernardino Mountains (Plate III). Like rocks of San Gabriel Mountains type, rocks of San Bernardino Mountains type may be a layered terrane with batholithic and prebatholithic rocks in an upper plate separated from Pelona Schist in a lower plate by a low-angle fault comparable to the Vincent thrust.

Late Cenozoic sedimentary basins

During late Oligocene through early Pleistocene time, several major marine and nonmarine depositional basins developed within the evolving paleogeography of the San Andreas fault system (Plate IV). The stratigraphic sequences deposited in these basins have been dispersed throughout southern California by displacements on various strands of the San Andreas system, and analysis of their facies patterns, depositional fabrics, and clast assemblages assists in the reconstruction of late Cenozoic paleogeography. The biostratigraphy and physical stratigraphy of these sedimentary sequences were reviewed by Woodburne (1975), and our Plate IV summarizes and updates some of the information contained in Figure 2 of Woodburne's report.

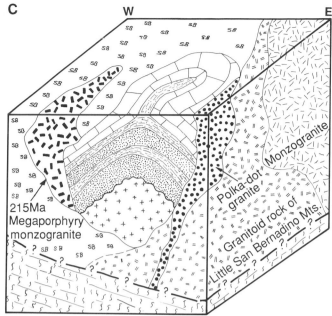

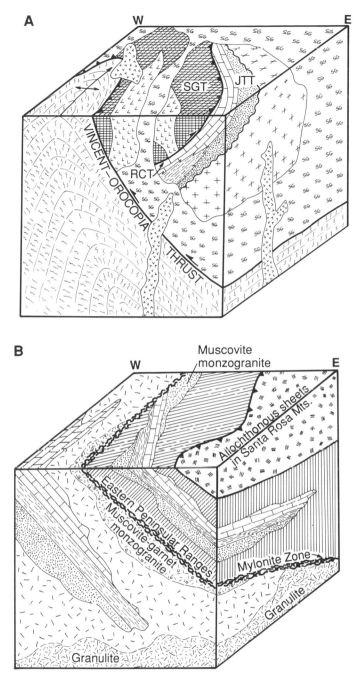

Figure 2. Block diagrams schematically illustrating relations among major lithologies and structures in rocks of San Gabriel Mountains type, Peninsular Ranges type, and San Bernardino Mountains type. Symbols same as in Plate III. W = west; E = east. A, Rocks of San Gabriel Mountains type, including rocks of Powell's (1982a, b) Joshua Tree terrane (JTT) and San Gabriel terrane (SGT) separated by the pre-Cretaceous Red Cloud thrust (RCT). The Joshua Tree terrane includes Proterozoic metaquartzite, pelitic schist and gneiss, and marble resting nonconformably on granitic gneiss; the San Gabriel terrane includes Proterozoic anorthosite and gabbro, syenite-mangerite, and orthogneiss (augen gneiss and retrograded granulite). In Plates IIIA and B, the Joshua Tree and San Gabriel terranes are lumped into the cross-hatched pattern shown here for the San Gabriel terrane. B, Rocks of Peninsular Ranges type, including lower and upper plates of the Eastern Peninsular Ranges mylonite zone of Sharp (1979). In both plates, metasedimentary rocks include metaquartzite, pelitic gneiss and schist, graphitic schist, and marble. C, Rocks of San Bernardino Mountains type. Metasedimentary rocks include metaquartzite, pelitic gneiss and schist, and marble, with younger units comparable to uppermost Proterozoic and Paleozoic rocks of the Cordilleran miogeocline (Stewart and Poole, 1975). SB indicates undifferentiated granitoid and metasedimentary rocks of San Bernardino Mountains type.

STRIKE-SLIP FAULTS OF THE SAN ANDREAS FAULT SYSTEM

Introduction

Rocks of Peninsular Ranges type, San Gabriel Mountains type, and San Bernardino Mountains type are traversed by a series of northwest-trending strike-slip faults (Fig. 1, Plate IIIA) that most workers assign to the San Andreas transform-fault system—a family of right-lateral faults that has evolved along the continental margin of western North America since middle Miocene time in response to interactions between the North American plate and various oceanic plates to the west (Atwater, 1970; Crowell, 1979; Dickinson and Snyder, 1979a, b). Youthful faults commonly viewed as modern components of the San Andreas system include the San Andreas fault proper; the San Jacinto, Whittier-Elsinor, and Newport-Inglewood faults; and various northwest-trending faults occurring in the offshore continental borderland (see Allen, 1957, p. 346; Crowell, 1975a, p. 10–11, 1981, p. 593). Older faults commonly viewed as abandoned components of the San Andreas system include the Punchbowl, San Gabriel, and Banning faults (Allen, 1957; Crowell, 1962, 1975a, 1981; Ehlig, 1976, 1981). Even older right-lateral faults that may belong to the San Andreas system include the San Juan–St. Francis and San Francisquito–Fenner–Clemens Well fault zones proposed in various forms by Smith (1977), Powell (1981, 1986), and Joseph and others (1982b).

Fault nomenclature. The complex pattern of late Cenozoic right-lateral faults in southern California has led to nomenclature for the San Andreas fault that is more complex than for central California. There, most if not all Miocene and younger displacement on the San Andreas has occurred within a narrow, singular (though complex) zone that extends the length of central California to Hill and Dibblee's (1953) "Big Bend" at the latitude of the Garlock fault. The term "San Andreas fault" has been used by most workers in central California (Hill, 1981), and little confusion has arisen with regard to which geologic structure bears the name San Andreas fault or whether multiple fault strands have generated sequential displacements. By contrast, the San Andreas fault in southern California consists of multiple strands, each representing some portion of the geologic history allocated to the more singular zone in central California (Noble, 1932; Dibblee, 1954, 1968a; Allen, 1957; Crowell, 1962; Woodburne, 1975; Matti and others, 1985). This structural complexity has presented two challenges: (1) to document the distribution of the various fault strands and to determine their sequencing and amount of right-lateral displacement, and (2) to establish nomenclature that provides a logical framework for understanding faulting history and for relating fault segments having similar and/or dissimilar movement histories.

A nomenclatural framework for the San Andreas fault in southern California has evolved through the efforts of many workers (see the historical review by Hill, 1981). Early workers recognized that the characteristic geomorphic and geologic features of the San Andreas rift zone in central California extend southeast beyond Hill and Dibblee's Big Bend region and intervene between the Mojave Desert and the massifs of Liebre Mountain and the San Gabriel Mountains. On this basis, the term San Andreas fault originally was extended into southern California. However, Nobel (1926, 1932) was among the first to observe that the San Andreas in southern California splays into several major branches—some occurring close to each other within the narrow Mojave Desert zone, others forming discrete strands that follow independent traces many kilometers apart. As Allen (1957) pointed out, this strand complexity creates a dilemma: which strand should bear the name San Andreas fault and which strand generated the large displacements proposed by Hill and Dibblee (1953) for the fault in central California?

Since the pioneering work by Noble, right-lateral strike-slip faults in the offshore and onshore of southern California have come to be viewed as elements of the San Andreas family or system of faults (Allen, 1957; Crowell, 1962, 1975a). This system embraces a hierarchical nomenclature that includes fault "strand," fault "zone," and fault "system" (Crowell, 1975a, p. 10–12): thus, the San Andreas fault (along with other fault strands) occurs within the San Andreas fault zone that (along with other fault zones) occurs within the San Andreas fault system. Within this hierarchical classification, we adopt the fault nomenclature identified in Table 1 and Plates III and V.

Whenever possible, we refer to the San Andreas fault in southern California by one of the specific strand names identified in Table 1. However, because each of these strands merged with or fed into the San Andreas fault in central California, each southern California strand during its lifetime represented the San Andreas before it was abandoned and succeeded by the next "San Andreas." This iterative pattern has culminated in the modern San Andreas fault in southern California—the genetically related set of strands that most workers believe originated onshore in response to Pliocene opening of the Gulf of California by seafloor spreading and transform faulting (Atwater, 1970). The term San Andreas fault usually is applied to this modern strand. Within this complex geologic and nomenclatural framework, we use the term San Andreas (sensu lato) for all strands of the San Andreas in southern California that have fed into the central California segment of the fault and contributed to its total history; we use the term San Andreas (sensu stricto) for those strands in southern California that have contributed to displacements on the fault only in Pliocene and Quaternary time. Thus, we can refer to the San Andreas fault generically without reference to a particular strand, but at the same time distinguish between broader vs. narrower interpretations of the name.

San Jacinto fault

The San Jacinto fault traverses rocks of Peninsular Ranges type and the alluviated San Bernardino Valley before entering the southeastern San Gabriel Mountains (Plates IIIA, V). There, the fault splays into several subparallel strands that traverse rocks of Peninsular Ranges type and San Gabriel Mountains type (Fig. 3). Early workers show these strands continuing to the northwest and joining the San Andreas fault by way of the Punchbowl fault (Noble, 1954b; Ehlig, 1975, Fig. 1, 1981, Fig. 10-2). However, Dibblee (1968a, Fig. 1, p. 266; 1975b, p. 156) and Morton (1975a, Figs. 1, 2, p. 175) concluded that the San Jacinto fault cannot be mapped into either the Punchbowl or San Andreas faults; instead the strands of the San Jacinto appear to curve westward into the San Gabriel Mountains without joining the San Andreas (Morton, 1975a; Morton and others, 1983). Morton (1975b) viewed the San Jacinto fault complex in the southeastern San Gabriel Mountains as a schuppen-like structure within which right slip on northwest-oriented strands of the San Jacinto is transformed into reverse left slip on east- and northeast-oriented faults that appear to be continuations of the San Jacinto zone. We adopt this interpretation here, but modify it by proposing that the existing geometry of faults related to the San Jacinto zone is inherited from a geometry established earlier by faults of the Malibu Coast–Banning system and the San Gabriel–Banning system (discussed below). These earlier Neogene faults established a structural template that has been modified by right slip within the San Jacinto zone and that led to the schuppen transform proposed by Morton (1975b).

The San Jacinto fault is the youngest of several Neogene faults in the southeastern San Gabriel Mountains that we believe

TABLE 1. NOMENCLATURAL USAGE FOR FAULTS OF THE SAN ANDREAS FAULT SYSTEM

Usage in this Chapter	Previous Usage
	SAN ANDREAS FAULT ZONE
Salton Trough	
Coachella Valley segment, San Andreas fault	North Branch, San Andreas fault (Dibblee, 1954, 1967b, 1975a; Dillon, 1975; Peterson, 1975)
	Mission Creek fault (Allen, 1957; Farley,1979)
Coachella Valley segment, Banning fault	South Branch, San Andreas fault (Dibblee, 1954, 1968a, 1975a; Dillon, 1975)
	Banning fault (Allen, 1957)
San Bernardino Mountains	
Wilson Creek strand, San Andreas fault	Wilson Creek fault (Dibblee, 1964, 1982b; Gibson, 1964, 1971; Sadler and Demirer, 1986)
Mission Creek strand, San Andreas fault	Mission Creek fault (Allen, 1957; Ehlig, 1977; Farley, 1979; Matti and others, 1983a, 1985)
Mill Creek strand, San Andreas fault	Mill Creek fault (Allen, 1957; Gibson, 1964, 1971; Ehlig, 1977; Farley, 1979); north branch, San Andreas fault (Dibblee, 1964, 1967b, 1968a, 1975a, 1982b)
San Bernardino strand, San Andreas fault	South Branch, San Andreas fault (Dibblee, 1964, 1968a, 1975a, 1982b)
San Gabriel Mountains–Mojave Desert	
Punchbowl fault	Noble (1953; 1954a, b); Dibblee (1967a, 1968a, 1982a); Ehlig (1975, 1981); Barrows and others (1985, 1987); Barrows (1975, 1987)
Nadeau fault (south branch)	Barrows and others (1985, 1987); Barrows (1987)
Nadeau fault (north branch)	Barrows and others (1985, 1987); Barrows (1987)
Mojave Desert strand	Dibblee (1967a, 1968a, 1982a); "main trace" of Barrows and others (1985, 1987); Barrows (1975, 1987)
Little Rock fault	Barrows and others (1985, 1987); Barrows (1975, 1987)
	SAN GABRIEL FAULT ZONE
Salton Trough through San Bernardino Valley	
Banning fault	Matti and Morton (1982); Matti and others (1985)
San Gabriel Mountains	
San Gabriel fault (south branch)	No consistent or documented usage
Stoddard Canyon fault	Interpreted as a left-lateral fault by Morton (1975a)
San Gabriel fault (north branch)	Ehlig (1973, 1975, 1981, 1982); Dibblee (1968a, 1982a); Crowell (1975a, c; 1981, 1982a)
Icehouse Canyon fault	Evans (1982); Matti and others (1985); May and Walker (1989)
Cajon Pass	
Cajon Valley fault	Dibblee (1967a, 1968a); Woodburne and Golz (1972); Matti and others (1985); Weldon (1986); Meisling and Weldon (1989)
	SAN JACINTO FAULT ZONE
San Jacinto Mountains to San Bernardino Valley	
San Jacinto fault	Sharp (1967); Matti and others (1985)
San Gabriel Mountains	
San Jacinto fault	Morton (1975a); Dibblee (1968a, 1975a, 1982a)
Glen Helen fault	Morton (1975a); Matti and others (1985); Morton and Matti (1987, this volume)
	CLEMENS WELL–FENNER–SAN FRANCISQUITO FAULT ZONE
Orocopia Mountains	
Clemens Well fault	Crowell (1975b); Powell (1981, 1986, this volume)
San Gabriel Mountains	
Fenner fault	Dibblee (1967a, 1968a, 1982a); Powell (1981, 1986, this volume); Ehlig (1975, 1981); Barrows and others (1985, 1987); Barrows (1975, 1987)
Soledad Basin region	
San Francisquito fault	Dibblee (1967a, 1968a, 1982a); Powell (1981,1986, this volume); Ehlig (1975, 1981); Barrows and others (1985, 1987); Barrows (1975, 1987)

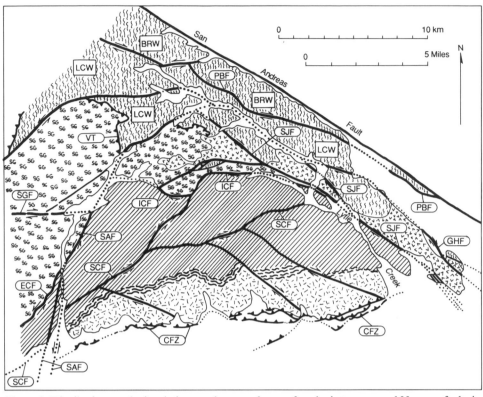

Figure 3. Distribution, geologic relations, and nomenclature of geologic terranes and Neogene faults in the southeastern San Gabriel Mountains (modified from Morton, 1975a, Figs. 1, 2). Symbols same as Plates IIIA and B. BRW = Blue Ridge window of Pelona Schist; CFZ = Cucamonga fault zone; ECF = Evey Canyon fault; GHF = Glen Helen fault; ICF = Icehouse Canyon fault; LCW = Lytle Creek window of Pelona Schist; PBF = Punchbowl fault; SAF = San Antonio fault; SCF = Stoddard Canyon fault (includes San Gabriel fault [south branch] west of the San Antonio fault); SGF = San Gabriel fault (north branch); SJF = San Jacinto fault; VT = Vincent thrust.

have juxtaposed and rejuxtaposed rocks of San Gabriel Mountains type and Peninsular Ranges type (Fig. 3). The distribution and history of the San Jacinto strands therefore can be determined only after the distribution and history of earlier Neogene faults have been determined. In our discussion of the San Gabriel and Banning faults (below), we propose that isolated fault segments like the Evey Canyon, Icehouse Canyon, and Stoddard Canyon faults in the southeastern San Gabriel Mountains (Fig. 3) are segments of the middle Miocene left-lateral Malibu Coast–Raymond–Banning fault and the late Miocene San Gabriel–Banning fault. We also propose that the main trace of the San Jacinto fault in lower Lytle Creek Canyon once was part of those two fault zones, and shared their sequential left- and right-slip histories in addition to episodes of Quaternary right slip related to the San Jacinto fault farther south in the Peninsular Ranges. This conclusion provides a foundation for our interpretation of fault patterns within the San Jacinto zone in the southeastern San Gabriel Mountains.

We reiterate here that faults attributable to or associated with the Quaternary San Jacinto fault of the Peninsular Ranges cannot be mapped beyond the southeastern San Gabriel Mountains. Instead, right-lateral strands of the San Jacinto fault curve westward and southwestward into the San Gabriel Mountains and transform into east- to northeast-trending north-dipping faults that have reverse left-slip displacements (Morton, 1975a, Figs. 1, 2). One of these has reactivated the Stoddard Canyon segment of the San Gabriel fault to yield latest movement indicators interpreted by Morton (1975a) as evidence for oblique left slip. A second oblique left-lateral fault (the San Antonio) trends down San Antonio Canyon and has displaced three older fault segments (Fig. 3): (1) the north branch of the San Gabriel fault about 3 km from the Icehouse Canyon fault, (2) the Evey Canyon fault about 3 km from the Icehouse Canyon fault, and (3) the south branch of the San Gabriel fault about 3 km from the Stoddard Canyon fault. The oblique left-lateral faults all root into the northwest-oriented San Jacinto fault zone in lower Lytle Creek Canyon, which has generated at least 8 to 13 km of Quaternary right slip that has displaced plutonic contacts between Miocene granitoid rock and Pelona Schist (Morton, 1975a).

We are not certain how the San Jacinto fault has contributed right slip to the San Andreas fault system. Within the Peninsular Ranges, 25 km of Quaternary right slip can be documented for the San Jacinto on the basis of displaced basement rocks (Sharp, 1967) and displaced Pliocene sediments of the San

Timoteo Badlands (of Frick, 1921, as used by Matti and Morton, 1975). The San Jacinto fault thus should be viewed as a strand of the San Andreas system. However, the geometric complexity in the southeastern San Gabriel Mountains makes it unclear to us exactly how right slip on the San Jacinto has been transferred to the San Andreas, if at all. Two extreme interpretations apply: (1) right slip on the San Jacinto is transferred entirely to the San Andreas by way of a right-step that takes place between San Gorgonio Pass and the southeastern San Gabriel Mountains; or (2) right slip on the San Jacinto fault is transformed entirely into oblique left slip along faults within the southeastern San Gabriel Mountains, and thereby does not contribute at all to right slip on the Mojave Desert segment of the San Andreas. We suspect that kinematic interaction between the San Jacinto and San Andreas faults has involved both right-stepping slip transfer and left-stepping slip bypass, but until the relative scale of either process is documented, the amount of slip transfer from the San Jacinto to the Mojave Desert segment of the San Andreas can only be inferred. For the reconstruction proposed in this chapter, we arbitrarily infer that 20 km of slip has been transferred between the two faults by way of the right step across the San Bernardino Valley; we infer that the remaining 5 km has been transformed into oblique left slip within the southeastern San Gabriel Mountains, where this transformation has led to convergence, uplift, block rotations, and severe shape changes of blocks bounded by brittle faults.

San Andreas fault

The main strand of the San Andreas fault in southern California consists of two segments (Fig. 1, Plate IIIA): (1) the Mojave Desert segment that mainly separates rocks of San Gabriel Mountains type from rocks of San Bernardino Mountains type, and (2) the Coachella Valley segment that separates rocks of Peninsular Ranges type from rocks of San Bernardino Mountains type and San Gabriel Mountains type. In plan view (Plate IIIA), the main strand has a left-stepping geometry: the Mojave Desert segment has been stepped left (west) about 15 km from the Coachella Valley segment, with the step occurring in the San Gorgonio Pass region of the southeastern San Bernardino Mountains (Matti and others, 1985). This left step forms a structural knot in the San Andreas that has influenced the evolution of the entire transform-fault system and has led to multiple fault strands that evolved sequentially (Matti and others, 1985). Multiple strands also have developed along the Mojave Desert segment of the San Andreas (Barrows and others, 1985). This strand complexity has made it difficult to identify the distribution and displacement history of faults in the San Andreas family, and has led to variable fault nomenclature and to conflicting fault-movement scenarios. For these reasons, we will review briefly the geologic setting of the San Andreas fault in the Coachella Valley, the San Bernardino Mountains, and the San Gabriel Mountains.

Geologic setting, Coachella Valley. The Coachella Valley strand of the San Andreas fault forms a relatively simple fault zone, although locally it is complicated by en echelon strands and lateral splays (Clark, 1984). The strand is relatively straight and generally has a uniform regional strike of about N45°W, although local departures from this trend occur (Bilham and Williams, 1985). Youthful tectonic landforms and faulted Holocene alluvial deposits indicate that the Coachella Valley segment is a modern neotectonic element (Keller and others, 1982). However, the antiquity and tectonic significance of the segment cannot be judged directly because Quaternary and late Tertiary sediments of the Salton Trough conceal older rocks bearing evidence for the fault's full history.

In the southern Indio Hills the Coachella Valley segment of the San Andreas fault is joined by another right-lateral fault (Plate V). Dibblee (1954, p. 26, Plate 3; 1968a; 1975a) referred to these two faults as north and south branches of the San Andreas, but Allen (1957, p. 336–339, p. 346) cited geologic relations that discouraged him from assigning the name "San Andreas" to any of the faults in this region. Accordingly, Allen (1957, Fig. 1) applied the names "Mission Creek fault" and "Banning fault," respectively, to Dibblee's north and south branches of the San Andreas.

We share Dibblee's view that the northern of his two branches is the main strand of the San Andreas fault in the Salton Trough, and that the term San Andreas can properly be applied to the strand. However, we are not certain that Dibblee's southern branch is a major throughgoing fault to which the term San Andreas should be applied. Our nomenclature reflects these interpretations (Plates IIIA, V). We refer to Dibblee's north branch as the Coachella Valley segment of the San Andreas fault, and apply that nomenclature to the entire extent of the fault in the Salton Trough. Like Allen (1957) we refer to Dibblee's south branch as the Banning fault, and we suggest that it is a reactivated segment of the old Banning fault (an idea shared by Dibblee, 1975a, p. 134).

The southeast and northwest terminations of the Coachella Valley strand of the San Andreas involve complex interactions with other fault zones. To the southeast, the surface expression of the modern San Andreas terminates near the southeast margin of the Salton Sea, where it interacts with the Brawley seismic zone and the Imperial fault (Sharp, 1982; Johnson and Hill, 1982). Most workers view this segment of the San Andreas as the northwesternmost of a series of right-stepping transform faults that extends from the Gulf of California onshore into the Salton Trough (Moore and Buffington, 1968, Fig. 4; Elders and others, 1972, Fig. 1; Crowell and Ramirez, 1979; Crowell, 1981, Fig. 18-4; Lonsdale and Lawver, 1980, Fig. 1; Johnson and Hill, 1982, Fig. 6; Curray and Moore, 1984, Figs. 1, 9).

At its northwest end the Coachella Valley strand is deflected westward and splays into the Mission Creek and Mill Creek strands of the southeastern San Bernardino Mountains. We cannot prove that these two strands merge to the southeast because they are buried by unfaulted late Quaternary alluvium where they exit the San Bernardino Mountains. However, their map pattern in the mountains strongly suggests that the strands coalesce to the southeast, and by the simplest interpretation they ultimately form a single fault zone—the Coachella Valley strand of the San Andreas.

Geologic setting, San Bernardino Mountains. The San Andreas fault in the San Bernardino Mountains consists of four discrete strands, including from oldest to youngest, the Wilson Creek, Mission Creek, Mill Creek, and San Bernardino strands (Plates IIIA, V; Matti and others, 1985). The San Bernardino strand is the modern neotectonic trace of the San Andreas, but the Mission Creek strand is the major trace that has juxtaposed far-traveled rocks of San Gabriel Mountains type against rocks of San Bernardino Mountains type (Plate IIIA). Southeast of the San Bernardino Mountains the Mission Creek and Mill Creek strands merge to form the Coachella Valley segment of the San Andreas fault; to the northwest, these two strands merge within the Mojave Desert segment of the San Andreas. The San Bernardino strand also merges northwestward with the Mojave Desert segment, but to the southeast the San Bernardino strand is aligned with and has interacted with the neotectonic Coachella Valley segment of the Banning fault. The Wilson Creek fault also merges northwestward within the Mojave Desert segment, but its southeastward continuation has been truncated by the Mission Creek fault and presumably has been displaced to the northwest by that fault and by younger strands of the San Andreas.

Wilson Creek strand. The Wilson Creek strand is a little-studied fault that extends for about 40 km along the southwest flank of the San Bernardino Mountains (Plates IIIA, V; Matti and others, 1985). The strand is characterized by its sinuous trace and west-dipping reverse dips. The fault juxtaposes crystalline rocks of unconfirmed provincial affinity and overlying nonmarine sedimentary rocks of the Mill Creek Formation (of Gibson, 1971; Plate IIIA, sequence 11) against crystalline rocks of San Bernardino Mountains type and overlying nonmarine sedimentary rocks that are unnamed (Plates III–V).

Mission Creek strand. The Mission Creek strand of the San Andreas fault crops out only in the southeastern San Bernardino Mountains, where the fault forms an east-trending convex-north arc (Plates IIIA, V; Matti and others, 1983a, 1985). The strand is not exposed anywhere else in the region, but it can be inferred to the northwest beneath the alluviated San Bernardino Valley and to the southeast beneath the alluviated Coachella Valley (Plates IIIA, V). Throughout its extent, the Mission Creek fault juxtaposes lower and upper plate rocks of San Gabriel Mountains type against rocks of San Bernardino Mountains type. Thus, the Mission Creek fault is a major strand of the San Andreas that is responsible for bringing together distinctly different types of crystalline rock. Along the southwest margin of the San Bernardino Mountains, a slice of crystalline rocks of uncertain provincial correlation locally is caught up between the Mission Creek and Wilson Creek faults and thus intervenes between rocks of San Gabriel Mountains type and San Bernardino Mountains type; we refer to this crystalline slice and its overlying sedimentary rocks as the Wilson Creek block (Plates IIIA, V). From its structural position between rocks of San Gabriel Mountains type and San Bernardino Mountains type, we conclude that the Wilson Creek

block was emplaced against the San Bernardino Mountains by right slip on the Wilson Creek fault before the San Gabriel Mountains block was emplaced outboard (west) of these two blocks by right slip on the Mission Creek fault (Matti and others, 1985). The Mission Creek strand thus is younger than the Wilson Creek strand, and generated a major share of the San Andreas fault's right-slip history before the strand was deformed into the left-stepping geometry it displays in the San Gorgonio Pass structural knot. In late Quaternary time this left step evolved to such a degree that the Mission Creek strand was abandoned by the San Andreas fault (Plate IV; Matti and others, 1985).

Mill Creek and San Bernardino strands. The Mill Creek and San Bernardino strands of the San Andreas are the youngest strands of the zone, and evolved in late Quaternary time as the San Andreas fault attempted to straighten out and bypass the San Gorgonio Pass knot that had developed in the Mission Creek strand (Matti and others, 1985; Morton and Matti, this volume). The Mill Creek strand forms a fairly straight trace for about 60 km along the southeast flank of the San Bernardino Mountains. The strand generated about 8 km of displacement during late Pleistocene time (Plates III–V; Matti and others, 1985).

The San Bernardino strand is the modern trace of the San Andreas fault in the vicinity of the San Bernardino Mountains, and formed by reactivating the old Mission Creek strand (Matti and others, 1985). The San Bernardino strand is aligned with the neotectonic Coachella Valley segment of the Banning fault, but the late Quaternary belt of thrust and wrench faults of the San Gorgonio Pass fault zone intervenes between the two right-lateral faults. Neotectonic right slip on the Coachella Valley segment of the Banning fault steps left around the structural knot in San Gorgonio Pass and onto the San Bernardino strand, which generated 2 to 3 km of displacement in approximately the last 125,000 yr (Matti and others, 1985).

Faulting chronology. The relative chronology of displacements on the Wilson Creek, Mission Creek, Mill Creek, and San Bernardino strands can be established by structural and geomorphic relations among them: the sinuous and deformed Wilson Creek strand was succeeded by the Mission Creek strand, which gradually developed an insurmountable left step and ultimately was abandoned and succeeded sequentially by the Mill Creek and San Bernardino strands (Matti and others, 1985).

The absolute chronology of displacements on the four strands is moderately to well constrained (Plate IV). In the Whitewater area, south of all strands of the San Andreas fault but north of the Banning fault, Miocene and Pliocene sedimentary units all record about the same amount of right-lateral displacement away from their probable depositional positions (Plate IIIA, sequences 3 and 7). These relations can be interpreted as evidence for initiation of right slip on the San Andreas (sensu stricto) after about 5 or 6 Ma (Matti and others, 1985, p. 10–11). The Mission Creek fault in the southeastern San Bernardino Mountains is buried by Pleistocene alluvium that probably is older than 0.5 Ma, and the Mill Creek strand is buried by unfaulted alluvium of latest Quaternary age (Matti and others, 1985, p. 10–12). The San Bernardino strand displaces all geologic materials except for youngest active alluvium, and is the youngest strand of the San Andreas fault in the San Bernardino Mountains region.

Geologic setting in the San Gabriel Mountains. In the San Gabriel Mountains, Barrows and others (1985, 1987; Barrows, 1987) recognize five discrete strands of the San Andreas fault, including from west to east, the Punchbowl, Nadeau south, Nadeau north, Mojave Desert, and Little Rock strands (our Mojave Desert strand is the main San Andreas trace of Barrows and others, 1985, 1987). Careful mapping by Barrows and others (1985) shows that the Punchbowl, Nadeau (north and south branches), and Little Rock faults all are truncated on their northwest and southeast ends by the Mojave Desert strand of the San Andreas. Thus, they can be viewed as anastomosed and abandoned strands of the San Andreas that probably are related to strands of the fault farther to the southeast in the San Bernardino Mountains and/or Salton Trough.

Regional correlation of the Punchbowl fault is of particular interest because the fault appears to be a significant strand of the San Andreas, and yet its isolated position outboard (west) of the main San Andreas trace makes its role in the overall history of the San Andreas difficult to evaluate. The fault extends for about 75 km along the northeast flank of the San Gabriel Mountains (Plates III, V; Barrows and others, 1987). The strand is characterized by its locally sinuous trace and west-dipping reverse dips, a geometry attributed by Barrows and others (1987) to deformation following its right-slip history. The Punchbowl fault originally was identified by Noble (1953), who interpreted it as a reverse dip-slip fault and implied that it merged with the San Jacinto fault by way of the Glen Helen fault (Noble, 1954a, b). Subsequently, Dibblee (1967a, 1968a) showed that the Punchbowl fault is a right-lateral fault that he interpreted as an old strand of the San Andreas family. Because it is truncated on the southeast and northwest by younger strands of the San Andreas zone, the Punchbowl should be viewed like any other linear structure that has been displaced by a strike-slip fault: its northwest and southeast terminations have been displaced from cross-fault counterparts that should be identifiable after right slip is restored on younger strands of the San Andreas. This interpretation motivated us to search southeastward for a possible displaced counterpart for the Punchbowl fault on the opposite side of the main trace of the San Andreas.

The multiple San Andreas strands in the San Gabriel Mountains evolved sequentially about 5 m.y. ago (Barrows and others, 1985, 1987). The Punchbowl fault appears to be the oldest strand, although Barrows and others (1987, p. 2, 84, 86) indicate that sequencing relations between the Punchbowl and Little Rock strands are not completely clear. According to Barrows (1987, p. 149–153, Figs. 6-8), stratigraphic relations between the Punchbowl fault and sediments of the Juniper Hills Formation (of Barrows, 1987) indicate that the Punchbowl fault may have generated right slip continuously between early Blancan (5 Ma) and late Blancan time (2 Ma). However, the age of the Juniper Hills Formation is poorly constrained (Barrows, 1987,

p. 129–130), and relevant beds in the unit cannot be confidently assigned to a particular part of the Blancan (A. G. Barrows, oral communication, 1990). The Nadeau faults appear to be Pliocene in age, while the Mojave Desert strand may have originated in middle Pleistocene time.

This chronology compares generally with that proposed by Matti and others (1985) for the Wilson Creek, Mission Creek, Mill Creek, and San Bernardino strands of the San Andreas in the San Bernardino Mountains. By this interpretation, the first-generation Punchbowl and Wilson Creek strands would have been coeval in the time frame 6 Ma to 3.5 Ma, followed by the Little Rock–Nadeau–Mojave Desert strands and the Mission Creek–Mill Creek–San Bernardino strands in the time frame 3.5 Ma to present.

Displacement history. Widely cited estimates for displacement on the San Andreas fault (sensu stricto) in southern California are based on cross-fault correlations between the Soledad basin–Sierra Pelona region northwest of the San Gabriel Mountains and the Orocopia Mountains–Chocolate Mountains region east of the Salton Trough (Fig. 1). Hill and Dibblee (1953, p. 450) originally suggested that the San Andreas fault has displaced Pelona Schist of Sierra Pelona about 260 km from similar rocks in the Orocopia Mountains. Later, Crowell and Walker (1962) correlated Precambrian gneiss, syenite, anorthosite, and gabbro in the Soledad region with similar rocks in the Orocopia region, and proposed that the two regions formerly were contiguous and have been separated 210 km by right-lateral displacement on the San Andreas fault. Crowell (1962, p. 25–44; 1975a) embellished this theme, and further suggested that the San Andreas fault has displaced Oligocene and Miocene sedimentary and volcanic rocks of the Soledad basin (Vasquez Formation; Plate IIIA, sequence 18) 210 to 240 km from similar rocks in the Orocopia Mountains (Diligencia Formation of Crowell, 1975b; Plate IIIA, sequence 19). Bohannon (1975) elaborated on correlation of the Oligocene and Miocene rocks, and proposed that the Diligencia and Vasquez Formations originally were deposited in contiguous nonmarine basins formed by extensional tectonics.

Volcanic and plutonic clasts contained in upper Miocene sedimentary rocks have been cited as evidence for 240 km of displacement on the San Andreas fault. Ehlig and others (1975; Ehlert, 1982b) suggested that clasts of rapakivi-textured volcanic and plutonic rocks in the Mint Canyon Formation of the Soledad region (Plate IIIA, sequence 12) were derived from bedrock sources in the northern Chocolate Mountains; these authors reconstructed the Mint Canyon Formation opposite Salton Wash between the Chocolate and Orocopia Mountains (Ehlig and others, 1975, Fig. 3). This cross-fault correlation was supported by Buesch and Ehlig (1982), who correlated volcanic rocks and faults between Soledad basin and Salton Wash. Aluminous quartz monzonite clasts ("polka-dot" granite of Ehlig and Joseph, 1977) in upper Miocene sedimentary rocks also have been used to establish cross-fault correlations. Polka-dot granite clasts in the Miocene Punchbowl Formation near Valyermo (Pelka, 1971; Plate IIIA, sequence 9) and similar clasts in beds of the Ridge

Basin Group of Crowell (1954a, b) led Farley and Ehlig (1977) to suggest that the Punchbowl basin and Ridge Basin once were part of a continuous drainage system that has been disrupted and displaced 40 km by the Punchbowl fault. Ehlert and Ehlig (1977) then used the polka-dot granite clasts to propose that the Ridge Basin–Punchbowl couplet was displaced about 240 km from the Orocopia Mountains region, where possible bedrock sources for polka-dot granite clasts have been identified (Ehlig and Joseph, 1977; Ehlert, 1982a, described additional bedrock occurrences of polka-dot granite).

These cross-fault correlations have led to a widely accepted general model in which the San Andreas fault in southern California has generated about 240 km of displacement since the opening of the Gulf of California 4 or 5 m.y. ago (Crowell, 1981). This general model has been questioned by several workers, some doubting whether large displacement has occurred on the San Andreas (Woodford, 1960; Baird and others, 1974), some pointing out discrepancies in the stratigraphic evidence supporting specific cross-fault correlations (Spittler and Arthur, 1973, 1982), some acknowledging large displacements overall but proposing limited displacements since the late Miocene (Woodburne, 1975), others reporting evidence for considerably less than 240 km of right slip on the San Andreas (Barrows and others, 1985, 1987). These alternative points of view continue to force detailed examinations of the accepted paradigm, no matter how well established that paradigm might be in the broad scientific community.

Recently, Matti and others (1985, 1986) suggested that total displacement on the San Andreas (sensu stricto) in southern California is about 160 km. This proposal is based on cross-fault correlation of Triassic megaporphyritic granitoid rocks that crop out in the Mill Creek region of the San Bernardino Mountains and in the Liebre Mountain region on the opposite side of all strands of the San Andreas fault. The rocks consist of 215-Ma hornblende monzogranite characterized by subhedral to euhedral potassium-feldspar phenocrysts as much as 8 cm long (Fig. 4A,B). Petrologic, geochemical, and geochronologic data demonstrate that the Mill Creek and Liebre Mountain rocks have similar modal and chemical compositions, intrusive ages, and thermal histories (Frizzell and others, 1986), and appear to be segments of a formerly continuous pluton that was cut by the fault and displaced 160 ± 10 km.

We accept that large right-lateral displacements have occurred on the San Andreas fault (sensu stricto) in southern California. Although we favor the 160-km displacement, both this figure and the more widely accepted 240-km displacement need to be tested in the light of specific displacement histories for specific strands of the system:

Displacements in the Coachella Valley. Although small youthful displacements on the Coachella Valley segment of the San Andreas fault have been recognized on the basis of cross-fault correlations among Quaternary alluvial materials (e.g., Keller and others, 1982; Matti and others, 1985), large older displacements recognized on the basis of cross-fault correlation between

Cenozoic and pre-Cenozoic units generally have not been recognized. In part this reflects the fact that older rocks of appropriate age containing evidence for large-scale displacements largely are buried by young Quaternary sediment that has filled the Salton Trough. In addition, the widely cited model for 240 km of Pliocene and Quaternary displacement on the San Andreas fault in effect has dampened the search for pre-Pliocene cross-fault counterparts in the Salton Trough because the 240-km model requires that pre-Pliocene rocks have been displaced completely out of the Salton Trough region.

A study by Dillon (1975) suggests that this may not be the case. In his study of the southern Chocolate Mountains, Dillon (1975, Fig. 70, p. 334–365) proposed that rocks in the southeastern San Bernardino Mountains have been displaced from the southern Chocolate Mountains by 180 ± 20 km of right slip on the Coachella Valley strand of the San Andreas (Dillon's north branch of the San Andreas). Dillon's proposal is based on three cross-fault correlations: (1) crystalline rocks of San Gabriel Mountains type northeast of San Gorgonio Pass correlated with similar rocks in the vicinity of Mammoth Wash in the Chocolate Mountains (Dillon, 1975, p. 59–60, 351–353); (2) the Miocene Coachella Fanglomerate in the Whitewater area correlated with the fanglomerate of Bear Canyon in the southern Chocolate Mountains (Dillon, 1975, p. 341–346); and (3) the inferred strandline position of the marine Imperial Formation in the Whitewater area correlated with the inferred strandline position for the marine Bouse Formation in the Chocolate Mountain region (Dillon, 1975, p. 347–350, Fig. 69). Dillon's reconstruction contrasts with that of Peterson (1975), who restored the Coachella Fanglomerate farther south in the Salton Trough and called for 215 km of displacement on the Coachella Valley strand of the San Andreas.

Displacements in the San Bernardino Mountains. Following Crowell's (1962) proposal that the San Andreas fault in southern California has 210 km of right-lateral displacement, many workers have attempted to apportion this displacement among various San Andreas strands in the southeastern San Bernardino Mountains. Gibson (1964, 1971) inferred on the basis of paleo-current and clast-provenance studies that the Mill Creek fault has displaced the Miocene Mill Creek Formation (of Gibson, 1971) about 120 km from its original position adjacent to the Orocopia Mountains. Dibblee (1968a, p. 269) concluded that if Crowell's 210 km of right slip has occurred along strands of the San Andreas in the San Bernardino Mountains, then the largest movement probably occurred along the north branch (Mill Creek fault). Later, Dibblee (1975a, p. 134) proposed that the north branch generated about 96 km of right slip and displaced crystalline rocks in the southeastern San Bernardino Mountains from presumed cross-fault counterparts in the Orocopia Mountains. Dibblee (1982b, p. 164) subsequently increased this value to 120 km—a displacement identical to Gibson's (1964, 1971) and presumably based on Gibson's palinspastic restoration of the Mill Creek strata to the Orocopia Mountains region.

Neither Gibson (1964, 1971) nor Dibblee (1968a, 1975a)

Figure 4. Outcrop photographs of Triassic megaporphyritic monzogranite and quartz monzonite; coin is about 2 cm in diameter. A, Outcrop in the Mill Creek region along State Highway 38 in the southeastern San Bernardino Mountains, north of all strands of the San Andreas fault. B, Outcrop on the Liebre Mountain block along the Old Ridge Route, south of all strands of the San Andreas fault.

accounted for the large difference between their proposed displacements on the Mill Creek fault (96 to 120 km) and Crowell's (1962) proposal for total displacement on the San Andreas (210 km). This difference presumably was made up by other faults in the region. However, Dibblee (1968a, p. 168) had ruled out his south branch of the San Andreas because the fault has only a few kilometers of right slip in the San Gorgonio Pass area, leaving only the Banning and San Jacinto faults to take up the missing 90 km. However, neither of these faults qualifies for the following reasons: (1) the Banning did not generate right-lateral displacements at the same time as the San Andreas (sensu stricto) and therefore does not figure into the 210-km reconstruction proposed by Crowell (1962); (2) the San Jacinto has no more than 25 or 30 km of displacement (Sharp, 1967); and (3) neither fault figures into palinspastic reconstruction of rocks in the southeastern San Bernardino Mountains because both faults pass outboard of that region.

Dillon's (1975) proposal that the north branch of the San Andreas has displaced rocks in the southeastern San Bernardino Mountains from counterparts in the southern Chocolate Mountains provided a major alternative to the proposals by Gibson and Dibblee. Dillon's displacement on Dibblee's north branch (180 ± 20 km) is considerably greater than the displacements proposed by Gibson and Dibblee (96 to 120 km), but both estimates seemed equally attractive based on the merits of their underlying cross-fault correlations.

This conflict was resolved by later studies in the San Bernardino Mountains that reinterpreted the bounding faults between distinctive basement terranes. Ehlig (1977), Farley (1979), and Matti and others (1983a, 1985) showed that the Mill Creek fault (Dibblee's north branch) does not form a major break between crystalline rocks as Gibson (1971) and Dibblee (1968a, 1975a, 1982a) believed. Instead, lithologic similarities between crystalline rocks on either side of the fault preclude large right-lateral displacements on this strand of the San Andreas (Matti and others, 1985, p. 9). Thus, the Mill Creek fault could not have displaced Gibson's Mill Creek strata by more than a few kilometers from their depositional position. Building on Ehlig's (1977) earlier work, Farley (1979) demonstrated that the major lithologic break between crystalline rocks is formed by the Mission Creek fault of Allen (1957), an interpretation refined by Matti and others (1983a, 1985). This fault, not the Mill Creek strand (Dibblee's north branch), was shown to be the San Andreas strand that juxtaposed rocks of San Gabriel Mountains type against rocks of San Bernardino Mountains type as proposed by Dillon (1975; see Farley, 1979, p. 120–129; Matti and others, 1985, p. 9–10). This clarified the distribution of major basement terranes and their bounding faults and set limits on displacements for individual strands, but left two factors still unresolved: (1) the palinspastic position of the Mill Creek Formation (discussed below), and (2) the 40- to 80-km discrepancy between Dillon's estimate of 180 ± 20 km of displacement on the San Andreas and the widely cited estimated of 240 km (upgraded by Crowell, 1975a, 1981, and Ehlig, 1981, from Crowell's original estimate of 210 km).

The slip discrepancy was reemphasized when Matti and others (1985, 1986) proposed that total right slip on the San Andreas fault in the vicinity of the San Bernardino Mountains may be no more than 160 ± 10 km. This estimate corresponds nicely with the lower limit of Dillon's estimate of 180 ± 20 km, and both proposals indicate that estimates of 210 to 240 km for total right slip on the San Andreas fault (sensu stricto) in southern California may be too large. We adopt this position, and use the 160-km displacement together with estimates for displacement on other strands of the San Andreas zone to reconstruct fault-movement histories for individual strands of the San Andreas in the vicinity of the San Bernardino Mountains.

Total displacement on the San Andreas fault (sensu stricto) in the vicinity of the San Bernardino Mountains is the sum of displacements on individual faults that jointly have displaced the Liebre Mountain megaporphyry body away from its cross-fault counterpart in the Mill Creek region:

$$A + B + C + D + E = 160 \text{ km},$$

where A = San Bernardino strand, B = Mill Creek Strand, C = San Jacinto fault, D = Mission Creek strand, and E = Wilson Creek strand. The equation can be partly solved using the following values:

A = about 3 km = right slip on the San Bernardino strand in the last 125,000 yr (Matti and others, 1985, p. 9, 11). This displacement has fed into the Mojave Desert segment of the San Andreas.

B = about 8 km = right slip on the Mill Creek strand in the late Pleistocene (Matti and others, 1985, p. 9, 11). This displacement fed into the Mojave Desert segment of the San Andreas.

C = <25 km = right slip on the San Jacinto fault in Quaternary time (Sharp, 1967; Matti and Morton, 1975; Morton and Matti, this volume). Most of this slip has been transferred to the San Andreas fault between San Gorgonio Pass and Cajon Pass, and thereby has contributed to total slip on that structure (Matti and others, 1985, p. 16–18; Morton and Matti, this volume). However, an unknown percentage of right slip on the San Jacinto

has been dissipated by extension within the right-stepping region of the San Bernardino Valley and absorbed by convergence within the Cucamonga fault zone and uplift of the southeastern San Gabriel Mountains (Matti and others, 1982b, 1985; Morton and Matti, 1987; Morton and Matti, this volume). In this report we arbitrarily apportion 20 km of right slip on the San Jacinto fault to the San Andreas and 5 km to contractional and extensional deformation related to strain transfer between the San Jacinto and San Andreas faults.

Using slip data for A, B, and C, the displacement equation is complete except for D and E, displacements on the Mission Creek and Wilson Creek faults:

$$3 \text{ km} + 8 \text{ km} + 20 \text{ km} + D + E = 160 \text{ km}.$$

Assuming values A, B, and C are correct and sum to 31 km, then values D and E (the combined displacements on the Mission Creek and Wilson Creek faults) sum to 129 km. In their discussion of combined slip on the Wilson Creek and Mission Creek faults, Matti and others (1985, p. 8–10) concluded that the Wilson Creek was the major strand; they proposed that the two faults together generated about 150 km of displacement, with the Wilson Creek strand generating about 110 km of slip and the Mission Creek strand generating about 40 km. This scenario accommodated both Gibson's (1971) proposal that the Mill Creek Formation overlying the Wilson Creek block has been displaced from the Orocopia Mountains region and Dillon's (1975) proposal that rocks of San Gabriel Mountains type in San Gorgonio Pass have been displaced from the southern Chocolate Mountains. However, Matti and others (1985, p. 10) pointed out that an equally reasonable palinspastic restoration of displaced blocks could be obtained if the proposed displacements on the Wilson Creek and Mission Creek faults (110 and 40 km, respectively) were exactly reversed (40 and 110 km, respectively). Matti and others (1985) could not resolve this dilemma because independent evidence for the amount and timing of right slip on the Wilson Creek and Mission Creek faults does not yet exist in the San Bernardino Mountains.

To arrive at values for D and E in the displacement equation, we turn to strike-slip faults in the San Gabriel Mountains that might be counterparts of either the Mission Creek fault or the Wilson Creek fault. Given our conclusion that the Mission Creek strand is middle Pliocene and younger in age and is a major fault that juxtaposed rocks of San Gabriel Mountains type and San Bernardino Mountains type, its counterpart in the San Gabriel Mountains should include faults of the same age that also juxtapose the two crystalline suites. By these criteria the Mission Creek strand is correlative with the Mojave Desert strand, which must be viewed as a major strand of the San Andreas that has a right-slip history extending from the present back to at least middle Pleistocene (Barrows and others, 1985, 1987). Although Barrows and others (1987, p. 2, 83, 85) indicated that the Mojave Desert strand (their "main trace") may be no older than 1 to 1.4 Ma, they (1987, p. 82) also pointed out that ". . . all rock units older than late Pleistocene that are juxtaposed along the main trace are dissimilar. . . ." We take this to mean that the Mojave Desert strand is the major trace of the San Andreas in the vicinity of the San Gabriel Mountains, and is responsible for juxtaposing rocks of San Gabriel Mountains type against rocks of San Bernardino Mountains type. Like other workers, we suspect that the Mojave Desert strand has a longer lived history than proposed by Barrows and others (1985, 1987). Thus, we correlate the Mission Creek strand (and the younger Mill Creek and San Bernardino strands) with the Mojave Desert strand (and potentially with the Pliocene Nadeau faults).

If the Mission Creek and Mojave Desert strands are the major throughgoing trace of the San Andreas, then restoration of significant slip on that strand, together with restored slip on the Mill Creek and San Bernardino strands, would bring the northwest termination of the Punchbowl fault near Palmdale closer to the southeast termination of the Wilson Creek fault in the southeastern San Bernardino Mountains (Plate IIIA). We propose that the Wilson Creek and Punchbowl faults originally formed a single throughgoing structure that was severed by younger strands of the San Andreas fault; the displaced segments of the once-continuous structure now are situated in the San Bernardino and San Gabriel Mountains. Comparison of the Wilson Creek and Punchbowl faults is supported by three lines of evidence: (1) they generated right-lateral displacements during the same time period (5 or 6 Ma to about 3.5 Ma, although Barrows, 1987, would have activity on the Punchbowl fault be as young as 2 Ma); (2) they both have sinuous traces and west-dipping reverse dips; and (3) they both bound crystalline rock slices (the Wilson Creek block and crystalline rocks of Pinyon Ridge in the San Gabriel Mountains) that have no documented affinity with particular crystalline terranes in southern California but are broadly similar to each other. We propose that the Punchbowl and Wilson Creek strands of the San Andreas once formed a continuous throughgoing right-lateral fault that had about 40 to 45 km of displacement based on estimates for the Punchbowl fault in the San Gabriel Mountains. In this report, we use the 40-km figure and insert this value for E in the displacement equation.

Using slip data for A, B, C, and E, the displacement equation

$$3 \text{ km} + 8 \text{ km} + 20 \text{ km} + D + 40 \text{ km} = 160 \text{ km}$$

can be solved for D, yielding a displacement of 89 km for the Mission Creek strand of the San Andreas fault.

Displacements in the San Gabriel Mountains. Barrows and others (1985, Table 4, 1987, p. 86) were able to document no more than 102 km of right slip on all strands of the San Andreas fault in the eastern San Gabriel Mountains: 21 km on the main or Mojave Desert strand of the San Andreas; 16 km on the Nadeau fault (north and south branches); 21+ km on the Little Rock fault; and 44 km on the Punchbowl fault. Their estimates for the Punchbowl fault are comparable with those proposed by other workers. Dibblee (1967a, Fig. 72; 1968a, p. 263–264, Fig. 1)

recognized between 32 and 48 km of displacement on the Punchbowl based on cross-fault correlations between the San Francisquito and Fenner faults, between marine rocks of the San Francisquito Formation, and between the Sierra Pelona and Blue Ridge windows of Pelona Schist. Farley and Ehlig (1977) and Ehlig (1981, Fig. 10-4) proposed about 40 km of displacement based on their suggestion that the Punchbowl fault has displaced strata in Ridge Basin that contain polka-dot granite clasts from strata in the Punchbowl Formation that contain similar clasts. Barrows and others (1985, 1987) acknowledged that their total displacement for the San Andreas (102 km) is considerably less than the widely accepted displacement (240 km), and they pointed out that displacement on the Little Rock strand may be greater than the 21 km they were able to document. Their displacement estimate for the main San Andreas strand (21 km) also is considerably less than most workers would infer for the strand.

San Gabriel and Banning faults

San Gabriel fault. Geologic setting. The San Gabriel fault extends southeastward from Ridge Basin to the western San Gabriel Mountains, where most workers recognize north and south branches (Fig. 1, Plate IIIA). The north branch curves eastward through the San Gabriel Mountains to the southeastern part of the range, where its eastward continuation is obscured by a complex network of faults (Morton, 1975a). The south branch traverses the south flank of the San Gabriel Mountains, where the fault is obscured by the Quaternary frontal-fault zone that bounds the mountain front. Most workers suggest that the south and north branches rejoin in the vicinity of the eastern San Gabriel Mountains, but this structural reunion has not been documented and each worker has suggested a different arrangement of linking faults (contrast the views of Dibblee, 1968a, Fig. 1, 1982a, Fig. 1; Ehlig, 1973, Fig. 1, 1981, Fig. 10-2; Crowell, 1975a, Fig. 1, 1975c, p. 208–209, 1982a, Fig. 1). According to the generally accepted view, the combined south and north branches of the San Gabriel fault must work their way through the structural complexity of the eastern San Gabriel Mountains and continue eastward.

Continuity of the San Gabriel fault southeast and northwest of its mapped distribution is problematic. Matti and others (1985) proposed that the Banning fault forms the southeastward continuation of the San Gabriel fault; we elaborate this proposal below. Northwestward continuation of the San Gabriel fault beyond Ridge Basin is obscured by Quaternary thrust faults that carry crystalline rocks of the Frazier Mountain region southward over the San Gabriel fault and associated sedimentary rocks of Ridge Basin (Plate III). There, most workers follow Crowell (1950; 1975a; 1982a, Figs. 3, 4), who proposed that the San Gabriel fault at depth continues northwestward beneath the thrust sheets to join the San Andreas fault (Crowell, 1982a, p. 29). We interpret this junction as a meeting point where an older fault (the San Gabriel) intersects a younger fault (the San Andreas, sensu stricto), and we view the northwest end of the San Gabriel fault as a piercing point whose palinspastic position should be identifiable after right slip is restored on the San Andreas (sensu stricto). When we restore 160 km of right slip on the Mojave Desert segment of the San Andreas (Matti and others, 1986), the northwest end of the San Gabriel fault restores to the Cajon Pass region where we believe the Cajon Valley–Hitchbrook fault represents a former continuation of the San Gabriel fault (discussed below).

Displacement history. Estimates for right slip on the San Gabriel fault range from 0 to 70 km. Crowell (1952) first proposed large lateral displacements based on his recognition that distinctive Precambrian basement clasts in the upper Miocene Violin Breccia of Crowell (1954a) in Ridge Basin (Plate IIIA, sequence 4) and in upper Miocene marine beds of the Modelo Formation have been displaced 30 to 35 km from likely cross-fault sources in the Frazier Mountain region and western San Gabriel Mountains, respectively. However, Paschall and Off (1961) discounted the role of right slip on the San Gabriel fault and instead accounted for the distribution of basement clasts in the sedimentary units purely on the basis of vertical dip-slip movements. Crowell (1962, p. 39–40) reiterated his proposal that clasts in sedimentary rocks require about 32 km of displacement on the San Gabriel fault, and also suggested that basement rocks in the Frazier Mountain region may have been displaced by as much as 48 km from cross-fault counterparts in the western San Gabriel Mountains (Crowell, 1962, p. 41). Carman (1964) called for about 32 km of right slip on the fault based on his proposal that upper Miocene nonmarine deposits of the Caliente Formation in the Lockwood Valley area (Plate IIIA, sequence 13) were part of the same fluvial drainage that deposited the Mint Canyon Formation in Soledad basin (Plate IIIA, sequence 12); the two parts of the fluvial system since have been displaced by the San Gabriel fault. This proposal was expanded and refined by Ehlig and others (1975), who used stratigraphic patterns of distinctive volcanic clasts and basement-rock clasts to propose that the Caliente–Mint Canyon drainage was disrupted by as much as 56 to 65 km of right slip on the San Gabriel fault. Ehlert (1982b) subsequently increased this estimate to 70 km. Ehlig and Crowell (1982) subsequently refined the basement-rock correlations originally cited by Crowell (1962), and concluded that the San Gabriel fault has displaced these rocks by about 60 km. This figure has become widely accepted as the total displacement on the San Gabriel fault. In the San Gabriel Mountains, the 60-km displacement presumably is split between the north and south branches of the fault: about 22 km on the north branch, as shown by displaced crystalline rock units (Ehlig, 1968a), and the remaining 38 km presumably on the south branch, although this displacement has not been proven by cross-fault correlations. We propose that the San Gabriel fault has no more than about 44 km of displacement: 22 km on the north branch (Ehlig, 1973), and 22 km on the south branch based on our proposal that the fault has displaced the left-lateral Malibu Coast–Santa Monica–Raymond fault from the Evey Canyon–Icehouse Canyon fault in the southeastern San Gabriel Mountains (discussed below).

Crowell (1982a, Fig. 12) proposed that right slip on the San Gabriel fault was initiated in the late Miocene (about 10 m.y. ago) and largely was completed by the end of the Miocene (about 5 Ma; Plate IV). Late Miocene onset of faulting is interpreted from interfingering stratigraphic relations between the syntectonic Violin Breccia and marine sediments of late Miocene age. Termination of strike-slip displacements on the San Gabriel fault by earliest Pliocene time is interpreted from relations in Ridge Basin, where the Pliocene Hungry Valley Formation of Crowell (1950) has been mapped as a depositional cap that seals the main displacement history of the San Gabriel fault (Crowell, 1982c). The basal part of the Hungry Valley Formation (Plate IIIA, sequence 1) is about 5 m.y. old, and Crowell (1982a, b) concluded that right slip within the San Andreas transform system switched from the San Gabriel fault to the San Andreas fault at about this time. However, this conclusion has been questioned by some workers, most notably Weber (1982), who indicates that the Hungry Valley sequence is disrupted by the San Gabriel fault and that a few kilometers of right slip on the San Gabriel occurred after the Hungry Valley was deposited.

Banning fault. *Geologic setting.* The Banning fault is a strike-slip zone that can be identified or inferred over a distance of about 100 km between the Indio Hills and the San Jacinto fault (Plates III, V; Matti and others, 1985). During late Miocene and earliest Pliocene time, the ancestral Banning fault probably formed a single continuous trace throughout this 100-km extent. However, after early Pliocene time the fault ceased to be a major tectonic feature in southern California, and late Pliocene and Quaternary depositional and tectonic events have obscured the distribution and history of the ancestral structure.

The Banning fault zone consists of western, central, and eastern segments (Plates III, V), each having a unique geologic and geomorphic setting. The western segment—extending from the San Jacinto fault east to the San Gorgonio Pass region—has no surface expression because it is covered by Quaternary sediments, and the position of the fault must be inferred on the basis of gravity data (Willingham, 1971, 1981) and indirect geologic evidence. The central segment of the fault in the San Gorgonio Pass region is partly obscured by Neogene sedimentary deposits, and is enmeshed in reverse, thrust, and wrench faults of the Quaternary San Gorgonio Pass fault zone (Matti and Morton, 1982; Matti and others, 1985). The eastern segment of the fault traverses the Coachella Valley from the vicinity of Whitewater Canyon southeastward to the southern Indio Hills, where it merges with the San Andreas fault. Unlike the western and central segments, the Coachella Valley segment of the fault has produced right-lateral displacements in late Quaternary time (Matti and others, 1985).

Displacement history. Evidence for at least 16 to 25 km of right slip on the Banning fault is provided by geologic relations among Tertiary sedimentary rocks in the San Gorgonio Pass region (Plate IIIA, sequences 3 and 6). There, uppermost Miocene strata of the marine Imperial Formation south of the Banning fault have been displaced at least 11 km in a right-lateral sense from Imperial beds in the Whitewater area on the north side of the fault (Allen, 1957, p. 329). The 11-km separation is a minimum displacement, however, because the two sequences are not exact cross-fault counterparts: they differ in details of their physical stratigraphy and in their relations with underlying rocks. Moreover, their benthic foraminiferal assemblages suggest different paleogeographic settings for the two sequences (K. A. McDougall and J. C. Matti, unpublished data): Imperial beds south of the Banning fault represent a more offshore facies than beds north of the fault in the Whitewater area, which suggests that the southern sequence should be restored to a palinspastic position at least several kilometers farther offshore (southeast) from the more onshore Whitewater section. Thus, the 11-km separation between the two exposed Imperial sequences represents minimum displacement on the Banning fault.

In San Gorgonio Pass, the Painted Hill and Hathaway Formations (of Allen, 1957) provide additional evidence for right slip on the Banning fault. South of the fault, many conglomeratic beds in these units contain north-derived volcanic, plutonic, and gneissic clasts that could not have been derived from bedrock sources presently cropping out north of the fault (J. C. Matti and D. M. Morton, unpublished data). After about 140 km of right slip is restored on the Coachella Valley segment of the San Andreas fault (described below), restoration of 16 to 25 km of right slip on the Banning fault positions the Hathaway and Painted Hill Formations south of the southern Chocolate Mountains, where bedrock sources for some of their volcanic, plutonic, and metamorphic clasts can be found.

Available evidence restricts right slip on the Banning fault to the late Miocene and earliest Pliocene. Displacements occurred after about 8 Ma because sedimentary and volcanic rocks displaced by the fault in San Gorgonio Pass are that age and younger (D. M. Morton, J. C. Matti, and J. L. Morton, unpublished data). Right slip occurred after about 7 Ma, because the Banning fault has displaced the Imperial Formation 16 to 25 km and the unit appears to be about 7 m.y. old (Table 2, Plate IV). The late Miocene displacement episode may have extended into earliest Pliocene time, but recognition of Pliocene right slip on the Banning fault (if any) must await improved age control and facies analysis in the San Timoteo and Painted Hill Formations. In latest Quaternary time, 2 to 3 km of right slip has occurred on the Coachella Valley segment of the Banning fault (Matti and others, 1985). This right-slip history has not involved the central and western segments of the fault, however: neotectonic right slip on the Coachella Valley segment of the Banning is related geometrically and kinematically to right slip on the San Bernardino strand of the San Andreas fault, and has not carried over to the Banning fault west of San Gorgonio Pass (Matti and others, 1985). Thus, limited data suggest that the Banning fault generated at least 16 to 25 km of right slip between about 7.5 and 4 or 5 Ma, before the fault was abandoned as a right-lateral strand of the San Andreas transform system in earliest Pliocene time.

San Gabriel–Banning connection via Neogene faults in the southeastern San Gabriel Mountains. The San Gabriel and Banning faults are major right-lateral strike-slip structures

TABLE 2. DOCUMENTATION FOR K/Ar RADIOMETRIC AGE DATES FROM BASALT IN THE PAINTED HILL FORMATION, WHITEWATER AREA*

Sample	K_2O (%)	^{40}Ar (moles/gm)	$^{40}Ar/\Sigma^{40}Ar$ (%)	Age (Ma)
DG-1A	1.133, 1.121	9.813×10^{-12}	60	6.04 ± 0.18
DG-1B	1.095, 1.087	9.341×10^{-12}	66	5.94 ± 0.18

Constants used in age calculation:

$$\lambda\beta = 4.963 \times 10^{-10} \text{ yr}^{-1}$$
$$\lambda_e + \lambda_{e'} = 0.581 \times 10^{-10} \text{ yr}^{-1}$$
$$^{40}K/K_{total} = 1.167 \times 10^{-4} \text{ mole/mole}$$

*A basalt flow in the lower part of the Painted Hill Formation north of the Banning fault has yielded two concordant whole-rock K/Ar age determinations of 6.04 ± 0.18 and 5.94 ± 0.18 Ma. The analyses were performed by J. L. Morton at the U.S. Geological Survey. Both samples are fresh, fine-grained basalt with plagioclase phenocrysts as long as 0.5 cm. The whole-rock samples were crushed and sieved, then treated with hydrofluoric and nitric acid to reduce atmospheric argon contamination. Potassium was analyzed by flame photometry; argon was analyzed by standard isotope-dilution procedures, using a 60°, 15.2-cm radius, Neir-type mass spectrometer. The overall analytical uncertainty of the ages is approximately 3 percent and is a combined estimate of the precision of the argon and potassium measurements at 1 standard deviation.

that played significant roles in the late Miocene evolution of the San Andreas fault system in southern California. The two faults have several features in common that suggest a connection between them, including similarities in their amounts of right slip (a few tens of kilometers), their east-oriented regional strike, and their association with rocks of San Gabriel Mountains type. These similarities led Matti and others (1985) to propose that the San Gabriel and Banning faults once formed a single through-going right-lateral fault that since has been disrupted and obscured by younger tectonism. However, this conclusion does not follow obviously from what is known about the two faults because (1) the north and south branches of the San Gabriel fault cannot be traced easily through and beyond the eastern San Gabriel Mountains, (2) the Banning fault cannot be traced easily west of the San Gorgonio Pass region, and (3) Quaternary right slip on the San Jacinto fault has rearranged and obscured any throughgoing connection between the two older faults.

Rocks and structures in the southeastern San Gabriel Mountains provide insight into possible relations between the San Gabriel and Banning faults. In the southeasternmost part of the range, two distinct suites of crystalline rock occur (Fig. 3, Plate IIIA; Dibblee, 1968a, 1982a, Figs. 1, 2; Morton, 1975a, Fig. 1; especially see Ehlig, 1975, Fig. 1, and 1981, Fig. 10-2): (1) Pelona Schist and structurally overlying granitoid and gneissic rocks, which respectively, form lower and upper plates of the Vincent thrust; and (2) a suite of granitoid rocks and prebatholithic metasedimentary rocks of uncertain provincial affiliation. Lower and upper plate rocks of the Vincent thrust are typical of those elsewhere in the San Gabriel Mountains (Ehlig, 1975, Fig. 1; 1981, Fig. 10-2). However, granitoid and metasedimentary rocks in the southeasternmost part of the range have enigmatic provincial affinities and are structurally isolated from rocks of San Gabriel Mountains type. The two suites everywhere are separated by high-angle faults (Fig. 3): (1) on the east, the two suites rae separated by northwest-trending right-lateral faults traditionally assigned to the San Jacinto fault zone (Morton, 1975a); (2) on the north, the two suites are separated by an east-trending zone of strike-slip faults referred to as the Icehouse Canyon fault; (3) on the west, the two suites are separated by the poorly studied Evey Canyon and San Antonio faults that traverse San Antonio Canyon (Ehlig, 1975, Fig. 1; Morton, 1975a, Figs. 1, 2; Dibblee, 1982a, Fig. 1). We propose that these three fault zones and the granitic and metasedimentary terrane they enclose can be used to establish a connection between the San Gabriel and Banning faults.

Our correlation of the San Gabriel and Banning faults depends heavily on two inferences: (1) the enigmatic terrane of granitoid and metasedimentary rock in the southeastern San Gabriel Mountains is Peninsular Ranges–type rock that has been juxtaposed against San Gabriel Mountains–type rock (Matti and others, 1985), and (2) right-lateral displacement on a throughgoing San Gabriel–Banning fault contributed to this juxtaposition. In order to defend the San Gabriel–Banning connection, we must defend these two propositions.

Provincial affinities. Crystalline rocks enclosed by the San Jacinto, Icehouse Canyon, and Evey Canyon faults are broadly similar to rocks of Peninsular Ranges type. The granitoid rocks consist mainly of foliated biotite-hornblende quartz diorite and tonalite (Ehlig, 1975, 1981; Morton, 1975a; Evans, 1982; Morton and others, 1983; Morton and Matti, 1987, Plate 12.1, rock units Kqd and Kd; May and Walker, 1989) that locally are intruded by small bodies of monzogranite and garnetiferous muscovite-bearing monzogranite (Morton and Matti, 1987, Plate 12.1, unit Kqm). These granitoids probably all are Cretaceous, although emplacement ages of about 87 Ma (May and Walker, 1989) have been moderately to strongly reset by an early Tertiary thermal event (Miller and Morton, 1980). The quartz dioritic and tonalitic rocks are progressively more deformed southward toward the San Gabriel Mountain front, culminating in an east-trending zone of mylonite and mylonitic quartz diorite first studied by Alf (1948) and Hsu (1955) and later mapped by Morton (1975a, 1976; Morton and others, 1983; Morton and Matti, 1987, Plate 12.1). At the mountain front, crystalline rocks structurally beneath the mylonite belt consist of multiply deformed gneiss, quartz diorite, and garnetiferous hypersthene-bearing retrograded granulite rocks. Most workers assign a Precambrian age to these deformed rocks based on their unique structural and metamorphic history, but Ehlig (1975, Fig. 1; 1981, Fig. 10-2) believed them to be younger and grouped them with other granitoid rocks in this part of the southeastern San Gabriel Mountains that he inferred to be Mesozoic in age. Granulite-facies metamorphism appears to be early Cretaceous in

age on the basis of a 108-Ma U-Pb age obtained by May and Walker (1989) from a late-stage syntectonic pyroxene-plagioclase pegmatite associated with the retrograded granulite rocks.

Prebatholithic metasedimentary rocks intruded by the granitoids consist of amphibolite-grade marble, metaquartzite, and pelitic gneiss and schist that locally is graphitic. North of the main mylonite belt the metasedimentary rocks occur as large bodies and pendants; south of the mylonite belt the metasedimentary rocks occur only as thin septa and xenoliths. Most workers assign a late Proterozoic or early Paleozoic age to the sedimentary protoliths based on their general similarities to rocks known to be that age elsewhere in southern California.

Matti and others (1985, p. 3) speculated that the crystalline terrane enclosed by the San Jacinto, Icehouse Canyon, and Evey Canyon faults is part of the Peninsular Ranges block. We reiterate this proposal here, and support it with several lines of evidence. (1) Quartz diorite and tonalite in the southeasternmost San Gabriel Mountains are similar to quartz diorite and tonalite in the northeastern Peninsular Ranges block: the granitoids appear to have similar intrusive ages that have been reset by a younger thermal event (Miller and Morton, 1980, and unpublished K/Ar data), and they have similar major- and minor-element compositions (Baird and others, 1974, 1979). (2) The mylonitic belt of ductile deformation separating foliated granitoid rocks from highly deformed retrograded granulites in the San Gabriel Mountains is similar to the Eastern Peninsular Ranges mylonite zone of Sharp (1979), except that mylonitic lineations are oriented down-dip in most parts of the Eastern Peninsular Ranges mylonite zone (Erskine, 1985) but are oriented parallel to strike in the San Gabriel Mountains (Morton and Matti, 1987; May and Walker, 1989). This discrepancy could reflect either spatial differences in the orientation of the syntectonic strain field within the regionwide mylonite belt (May and Walker, 1989) or localized post-tectonic rotation of the mylonite fabrics around vertical and/or horizontal axes. Other workers have pointed out similarities between the two mylonite belts (Hsu, 1955; Ehlig, 1975, p. 183; Erskine, 1985). (3) Bodies of garnetiferous muscovite monzogranite structurally overlying the mylonite belt in the southeastern San Gabriel Mountains (Morton and Matti, 1987, Plate 12.1, units Kg and Kgc) are comparable to similar bodies structurally above and beneath the Eastern Peninsular Ranges mylonite belt in the Santa Rosa Mountains (Matti and others, 1983b, unit Mzlm). (4) Prebatholithic metasedimentary rocks in the southeasternmost San Gabriel Mountains are broadly similar to metasedimentary rocks structurally above and beneath the mylonite belt in the San Jacinto and Santa Rosa Mountains (Powell, 1982a; Matti and others, 1983b, units gsc, mq, and mc; Erskine, 1985).

Correlation of faults. If the enigmatic granitoid and metasedimentary rocks in the southeastern San Gabriel Mountains are rocks of Peninsular Ranges type, then their present tectonic juxtaposition against rocks of San Gabriel Mountains type must be explained by identifiable structures—ductile, brittle, or both. May and Walker (1989) proposed that the enigmatic suite (their

Cucamonga and San Antonio terranes) was juxtaposed against rocks of San Gabriel Mountains type in late Cretaceous or early Paleogene time by ductile oblique left-lateral convergence within a mylonite zone locally preserved along the boundary between the two terranes. Regional telescoping along Cretaceous or Paleogene ductile zones may well account for the primary structural geometry between the two crystalline terranes. However, we propose that their present juxtaposition in the San Gabriel Mountains resulted from brittle displacements on throughgoing Neogene faults whose dismembered segments now are represented by the San Jacinto, Icehouse Canyon, and Evey Canyon faults.

If these three faults are Neogene, and once were continuous with other Neogene strike-slip faults in southern California, then their regional counterparts should be nearby and should have a similar relation to rocks of San Gabriel Mountains type and Peninsular Ranges type. One candidate is the Banning fault, which in the San Gorgonio Pass region intervenes between rocks of San Gabriel Mountains type to the north and rocks of Peninsular Ranges type to the south. Likewise, in the southeastern San Gabriel Mountains, the San Jacinto and Icehouse Canyon faults intervene between rocks of San Gabriel Mountains type to the north and east and rocks we identify as Peninsular Ranges type to the south (Fig. 3). The structural position occupied by the San Jacinto and Icehouse Canyon faults thus is similar to that of the Banning fault (Fig. 3, Plate IIIA). We use this relation as a primary basis for our proposal that the Banning, San Jacinto and Icehouse Canyon faults once were continuous and shared common movement histories.

We complete the San Gabriel-Banning connection by projecting the Banning-San Jacinto-Icehouse Canyon trend westward beyond San Antonio Canyon. A likely candidate for this continuation is the east-oriented north branch of the San Gabriel fault—a structure that has about 22 km of late Miocene right slip (Ehlig, 1973, 1975, 1981) that compares well with 16 to 25 km of late Miocene right slip on the Banning fault in San Gorgonio Pass. Similarities in their movement histories suggest that the faults are related, and we propose that they once formed a single throughgoing right-lateral trend that since has been disrupted by about 25 km of Quaternary right slip on the San Jacinto fault and about 3 km of Quaternary left slip on the San Antonio fault. The former has displaced the east end of the Icehouse Canyon fault from the Banning fault; the latter has displaced the west end of the Icehouse Canyon fault from the north branch of the San Gabriel fault (Fig. 3, Plate IIIA).

This scenario provides a connection between the north branch of the San Gabriel fault and the Banning fault but leaves unresolved the connection (if any) between the south branch of the San Gabriel fault and the Banning. Most workers follow Crowell (1975a, b, 1981, 1982) and Ehlig (1975, 1981), who convey 38 km of right slip on the south branch eastward toward the San Andreas or Banning faults by way of connecting faults in the southeastern San Gabriel Mountains (Dibblee, 1968a, Fig. 1; Ehlig, 1973, Fig. 1, 1981, Fig. 10-2; Crowell, 1975a, Fig. 1, 1975c, p. 208–209, 1982a, Fig. 1). To these scenarios we add one

in which the San Gabriel fault (south branch) traverses the south front of the San Gabriel Mountains, is displaced about 3 km left laterally by the San Antonio fault, and continues east to the San Jacinto fault by way of the Stoddard Canyon fault and thence to the Banning fault (Fig. 3).

As developed to this point, our model for the San Gabriel-Banning connection accommodates three concerns: (1) it accounts for two of the three Neogene faults in the southeastern San Gabriel Mountains that separate Peninsular Ranges-type rock from San Gabriel Mountains-type rock; (2) it ties together similar right-slip displacement histories on the San Gabriel fault (north branch) and the Banning fault; and (3) it provides a testable hypothesis for a connection between the San Gabriel fault (south branch) and the Banning fault. This model also provides a testable solution for two other problems: the amount of right slip on the San Gabriel fault (south branch) and the regional distribution of the left-lateral Malibu Coast fault zone. We propose that a solution to both problems is provided by the Evey Canyon fault–the third Neogene fault in the southeastern San Gabriel Mountains that juxtaposes rocks of Peninsular Ranges type and San Gabriel Mountains type.

Pivotal to this analysis is the regional distribution and tectonic role of the Malibu Coast–Santa Monica–Raymond fault zone—a major left-lateral fault that trends easterly from the California coast to the south-frontal fault zone of the San Gabriel Mountains. Barbat (1958, Fig. 2, p. 64) originally suggested that the Santa Monica segment of the zone generated about 13 km of left slip. Subsequently, Yerkes and Campbell (1971), Jahns (1973), Campbell and Yerkes (1976), and Truex (1976) presented evidence for 60 to 90 km of left slip on the Malibu Coast–Santa Monica–Raymond trend during the middle Miocene (about 16 Ma to 12 Ma). Based on the premise that a left-lateral fault of this scale must be of regional extent, these workers extended the Malibu Coast–Raymond system eastward through the Cucamonga fault zone (Barbat, 1958, Fig. 1; Jahns, 1973, Fig. 5; Campbell and Yerkes, 1976, Fig. 1) and ultimately through San Gorgonio Pass by way of the Banning fault (Jahns, 1973, Fig. 6-9, p. 166).

We accept the premise that the Malibu Coast–Santa Monica–Raymond fault is a major left-lateral structure that should be recognizable east of the frontal fault zone of the San Gabriel Mountains. However, rather than merge the Malibu Coast system with the frontal fault zone and extend it east through the Cucamonga fault, we suggest that the Malibu Coast system may have been truncated by the south branch of the San Gabriel fault and displaced right laterally from a cross-fault counterpart located to the east. We propose that the Evey Canyon fault is the displaced continuation of the Malibu Coast–Santa Monica–Raymond trend, which requires that the Evey Canyon is a left-lateral fault that juxtaposed Peninsular Ranges-type rocks against San Gabriel Mountains-type rocks (Fig. 3; Matti and others, 1985). The once continuous middle Miocene Malibu Coast–Santa Monica–Raymond–Evey Canyon fault was truncated in late Miocene time by the San Gabriel fault (south branch), which has displaced the southwest end of the Evey Canyon fault about 22 km from the east end of the Raymond fault. This forms the basis for our displacement estimate of 22 km for the south branch, and underlies our conclusion that the San Gabriel fault (north and south branches) has no more than 44 km of right slip (22 km + 22 km).

Our conclusion that the Evey Canyon fault is a left-lateral structure that once was part of the Malibu Coast system has implications for the Icehouse Canyon, San Jacinto, and Banning faults. The northeast end of the Evey Canyon fault terminates against the north-trending San Antonio fault, a structure that has displaced the north and south branches of the San Gabriel fault about 3 km from their inferred cross-fault counterparts (the Icehouse Canyon and Stoddard Canyon faults, respectively; Fig. 3). Oblique convergence between the Evey Canyon and San Antonio faults precludes accurate determination of their geometric relations, and it is likely that left slip on the north-trending San Antonio fault has occurred by reactivation of north-trending segments of the Evey Canyon. Despite these complications, we propose that restoration of 3 km of left slip on the San Antonio fault would align the northeast end of the Evey Canyon fault with the Icehouse Canyon fault, which thereby is potentially continuous with the left-lateral Malibu Coast–Evey Canyon trend.

This implication appears to be at odds with our proposal that the Icehouse Canyon fault is a right-lateral component of the San Gabriel-Banning connection—a contradiction that becomes more profound if large left-lateral displacements proposed for the Malibu Coast-Raymond system are projected east along an Evey Canyon–Icehouse Canyon–San Jacinto trend that ultimately includes the Banning fault. This apparent contradiction can be resolved in two ways:

1. Fault zones established initially by middle Miocene left slip within a throughgoing Malibu Coast–Santa Monica–Raymond–Evey Canyon–Icehouse Canyon–San Jacinto–Banning trend could have been reactivated as zones of late Miocene right slip when the Banning–San Jacinto–Icehouse Canyon segment of the Malibu Coast system was incorporated into the San Gabriel-Banning system (a generic concept suggested to us by T. H. McCulloh).

2. May and Walker (1989, p. 1262–1263) suggested that kinematic models for tectonic rotations in southern California (Luyendyk and others, 1980; Hornafius and others, 1986) require that left slip on the Malibu Coast-Raymond system should decrease to the east, thereby eliminating the need to extend large left-lateral displacements throughout the entire length of the Malibu Coast system. May and Walker (1989) adopted our earlier suggestion (Matti and others, 1985) that the Evey Canyon fault is part of the Malibu Coast system, but suggest that large left-lateral displacements (60 to 90 km) proposed for western segments of the Malibu Coast–Evey Canyon system diminish to about 20 km in the southeastern San Gabriel Mountains. By this interpretation, left slip may not be significant on the Icehouse Canyon–San Jacinto–Banning segment of the Malibu Coast system, and may not even have extended all the way east on the Banning fault. This would obviate the need to extend large dis-

placements on the Malibu Coast system east through San Gorgonio Pass (Jahns, 1973), and would fit the lack of evidence for Neogene left slip on the Banning fault.

We cannot resolve uncertainties involving the regional distribution of the middle Miocene Malibu Coast system. However, rather than arbitrarily ending the Malibu Coast–Santa Monica–Raymond system at the northeast end of the Evey Canyon fault, our paleogeographic reconstructions (discussed below) incorporate modest left slip on the Evey Canyon–Icehouse Canyon–Banning segment of the Malibu Coast system. We estimate this left slip to be about 25 km based on our speculation that the Eastern Peninsular Ranges mylonite zone in the southeastern San Gabriel Mountains has been displaced about 25 km from a possible cross-fault counterpart between the Santa Monica Mountains and Verdugo Hills (Plate IIIA).

Summary. In the southeastern San Gabriel Mountains and San Gorgonio Pass region, several strike-slip faults traverse the boundary zone between rocks of Peninsular Ranges type and San Gabriel Mountains type. Our interpretation of movement histories for these faults requires that they are local segments of regionwide strike-slip fault trends that include (1) the middle Miocene Malibu Coast–Santa Monica–Raymond–Evey Canyon–Icehouse Canyon–San Jacinto–Banning left-lateral system, and (2) the late Miocene San Gabriel–Icehouse Canyon–Stoddard Canyon–San Jacinto–Banning right-lateral system. The Malibu Coast–Banning system produced about 25 km of left slip in middle Miocene time (14 to about 12 Ma). In late Miocene time (about 10 Ma), the Malibu Coast–Banning left-lateral system was dismembered and locally reactivated by displacements within the right-lateral San Gabriel–Banning system.

During late Miocene time the San Gabriel–Banning fault formed a single, throughgoing right-lateral structure that was part of the San Andreas transform system. The middle segment of this throughgoing structure, now located in the San Gabriel Mountains, developed two discrete strands that probably formed sequentially. To the northwest, these two strands coalesced to form a single strand (the San Gabriel fault) that traverses the west margin of Ridge Basin (Plate IIIA); to the southeast, the two strands coalesced to form a single strand (the Banning fault) that traverses the San Bernardino Valley, San Gorgonio Pass, and Salton Trough. Within the southeastern San Gabriel Mountains, remnants of the two San Gabriel–Banning fault strands are represented by the Stoddard Canyon, Icehouse Canyon, and San Jacinto faults.

The throughgoing San Gabriel–Banning system generated about 44 km of right-lateral displacement in late Miocene to earliest Pliocene time (10 to 4 or 5 Ma). The south branch of the San Gabriel fault probably is the older fault strand, and we propose that from about 10 Ma to about 7.5 Ma it generated about 22 km of displacement that displaced the Evey Canyon fault from the Raymond fault. This displacement must have extended along the Banning fault, but rocks of this age that would record right-lateral movements from 10 to 8 Ma are not exposed in San Gorgonio Pass. If the San Gabriel–Banning fault gradually was bowed into a convex-west arc during this early history, the curvature ultimately could have become too extreme to have accommodated efficient right-slip displacements. We propose that this happened about 7.5 m.y. ago, at which time the south branch was abandoned and right slip stepped inboard (east) to the north branch. From about 7.5 Ma to about 4 or 5 Ma the north branch generated 22 km of displacement (Ehlig, 1973, p. 174)—a displacement recorded by late Miocene sedimentary and volcanic rocks that are traversed by the Banning fault in San Gorgonio Pass. According to this model, total right slip on the combined north and south branches of the San Gabriel–Banning fault is about 44 km—16 km short of the 60 km proposed by Crowell (1982a) and by Ehlig and Crowell (1982). Major right-lateral activity on the throughgoing fault ceased about 4 or 5 m.y. ago (Crowell, 1982a, p. 35, Fig. 12) as the San Gabriel–Banning strand was bypassed by the San Andreas system and as strands of the San Andreas fault proper evolved to the east.

Our model for the distribution and history of Neogene faults in the southeastern San Gabriel Mountains has implications for the history of the San Jacinto fault. This model (Fig. 3, Plate V) requires that the fault trace traditionally identified as the San Jacinto fault in Lytle Creek Canyon not only has played a role in the Quaternary evolution of the San Andreas transform system (discussed above) but also played a role in the evolution of the evolution of older Neogene fault systems. In middle Miocene time the San Jacinto conveyed left slip from the Malibu Coast–Evey Canyon–Icehouse Canyon system eastward to the Banning fault; in late Miocene time it conveyed right slip from the Stoddard Canyon and Icehouse Canyon segments of the San Gabriel fault eastward to the Banning fault. The term "San Jacinto fault" in Lytle Creek Canyon thus has been applied to an ancient fault zone that has witnessed three separate episodes of strike-slip faulting, only the latest of which is related to Quaternary displacements on the San Jacinto fault that traverses the Peninsular Ranges Province to the southeast. This composite movement history is consistent with the broad crush zone that marks the trace of the San Jacinto fault between the Icehouse Canyon fault and the mountain front, and accounts for the pronounced contrast in basement-rock types on either side of the crush zone (Morton and Matti, 1987, Plate 12.1).

PALINSPASTIC RECONSTRUCTION OF THE SAN ANDREAS FAULT

Previous reconstructions

San Gabriel–San Andreas model. Most workers follow Crowell (1962, 1975a, 1979, 1981, 1982a) and Ehlig (1976, 1981, 1982), who have proposed that the late Miocene through Holocene history of the San Andreas fault system in southern California involved sequential right-lateral displacements on the San Gabriel and San Andreas faults. In this model the San Gabriel fault is an early strand of the San Andreas (sensu lato) that evolved during the late Miocene (about 10 Ma according to Crowell, 1982a, Fig. 12). At this time the San Andreas fault

(sensu stricto) had not yet evolved in southern California, although it was established in central California and experienced a prolonged late Miocene history (Huffman, 1972). The San Gabriel fault in southern California must have fed into the central California reach of the San Andreas (Crowell, 1982a, p. 35, Fig. 8; Ehlig, 1982, p. 375), and in this sense it is widely viewed as the first-generation strand of the San Andreas fault in southern California.

Crowell (1981) and Ehlig (1982) proposed that the San Andreas fault in southern California evolved about 5 m.y. ago as right slip on the San Gabriel fault dwindled or ceased. This event probably occurred because the broadly curved trace of the San Gabriel fault became misaligned within the strain field of the San Andreas transform system, and slip was transferred inboard (east) from the San Gabriel fault to the newly initiated San Andreas fault (sensu stricto). Inboard of the San Gabriel's curved trace the San Andreas formed separate and distinct structures; however, southeast of this curved segment the San Andreas and San Gabriel faults apparently coincided (Crowell, 1979, Fig. 1, 1981, Fig. 18-4; Ehlig, 1981, Fig. 10-4, 1982, Fig. 5), and the master trace formerly occupied by the San Gabriel fault was maintained by the younger San Andreas.

This two-stage tectonic model has provided a conceptual framework for reconstructing displacement on the San Andreas fault (sensu lato) in southern California. The cross-fault correlations originally proposed by Crowell (1962) for Precambrian and Oligocene and Miocene rocks have led to a palinspastic reconstruction for southern California (Ehlig, 1981, Fig. 10-4) that incorporates 60 km of displacement on the San Gabriel fault and 240 km of displacement on the San Andreas fault, for a total of about 300 km of displacement in the last 10 m.y. This model (Fig. 5) has become the accepted paradigm for interpreting the Neogene tectonic history of southern California.

Alternative views. Some observers have questioned premises that underlie the San Gabriel–San Andreas paradigm. Paschall and Off (1961) proposed that the clast assemblages cited by Crowell (1952, 1962) as evidence for right slip on the San Gabriel fault could best be explained by normal dip-slip movement on the fault. Woodford (1960) and Baird and others (1974) pointed to the apparent continuity of similar basement terranes across the Transverse Ranges and argued that these terranes have not been disrupted by large strike-slip displacements. Spittler and Arthur (1973, 1982) concluded that the Vasquez Formation of Soledad basin and the Diligencia Formation in the Orocopia Mountains differ in details of their physical stratigraphy and geochronology, and therefore cannot be exact cross-fault correlatives that require 210+ km of right slip on the San Andreas. Woodburne and Golz (1972) proposed that the late Miocene (Clarendonian and Hemphillian) Punchbowl Formation outboard (west) of the San Andreas fault near Valyermo was deposited in the same depositional regime as middle to late Miocene (Barstovian) strata inboard (east) of the San Andreas fault in Cajon Pass, leading Woodburne (1975, p. 20–21) to conclude that the San Andreas has displaced the two sequences by no more than 16 to 24 km of right slip since the Hemphillian.

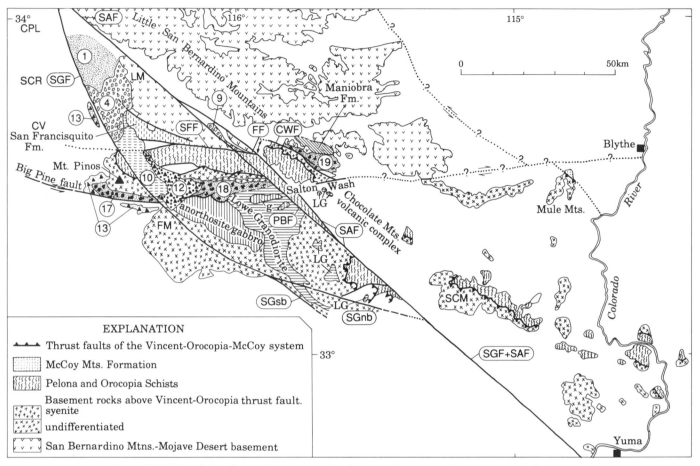

Figure 5. Widely cited palinspastic reconstruction for crystalline and sedimentary terranes in southern California based on restoration of 240 km of right slip on the San Andreas fault and 60 km of right slip on the San Gabriel fault; modified from Ehlig (1981, Fig. 10-4; also see Link, 1982c, Fig. 1, 1983, Fig. 5). In this reconstruction, upper Miocene and Pliocene sedimentary units accumulated farther southeast along the San Andreas fault than shown in our Plate IIIB: the upper Miocene Ridge Basin sequence (units 4, 10) and Punchbowl Formation (unit 9) accumulated adjacent to the southeastern Little San Bernardino Mountains, as did early Pliocene parts of the Hungry Valley Formation (unit 1)—an implication that derives from Figure 5 because older parts of the Hungry Valley Formation traditionally are thought to predate movement on the San Andreas fault (Ramirez, 1983, p. 43; Link, 1983). However, this palinspastic position for Hungry Valley beds is not in accord with Ramirez (1983) or Crowell (1982a, p. 39–40). Note misalignment between the San Francisquito–Fenner fault and the Clemens Well fault that Joseph and others (1982b, Fig. 6) inferred to have been continuous in late Oligocene and early Miocene time. In this figure we extend the reconstruction farther west than did Ehlig (1981, Fig. 10-4) in order to show the juxtaposition of middle and upper Miocene nonmarine sedimentary rocks of the Carrizo Plain (CPL), Cuyama Valley (CV), and southern Caliente Range (SCR) against the Liebre Mountain block (LM) after 60 km of right slip is removed on the San Gabriel fault. Paleogeographic implications of this juxtaposition are discussed in the text. CWF = Clemens Well fault; FF = Fenner fault; FM = Frazier Mountain region; LG = Lowe igneous pluton of Joseph and others (1982a); PBF = Punchbowl fault; SAF = San Andreas fault; SCM = southern Chocolate Mountains; SFF = San Francisquito fault; SGF = San Gabriel fault; SGnb = north branch San Gabriel fault; SGsb = south branch San Gabriel fault. 1, Hungry Valley Formation; 4, Ridge Route and Peace Valley Formations and Violin Breccia; 9, Punchbowl Formation; 10, Castaic Formation; 12, Mint Canyon Formation; 13, Caliente Formation; 17, Plush Ranch Formation; 18, Vasquez Formation; 19, Diligencia Formation.

An alternative reconstruction proposed by Powell (1981, Plate 6B, p. 307–339, 365–368; 1986) recognized the important role played by the San Andreas and San Gabriel faults in the San Andreas transform system, but required additional fault strands to account for the full history of right-lateral displacements in southern California. Powell's reconstruction incorporated only 240 km of displacement on the combined San Andreas and San Gabriel faults (our measurement from Powell, 1981, Plate 6B) and indicates that an older Miocene strand of the San Andreas fault is required in order to completely reassemble crystalline terranes and to provide the additional displacement (80 km) required if right slip on the San Andreas fault (sensu lato) in central and southern California are to match.

Powell assigned the missing 80 km to his Clemens Well–Fenner–San Francisquito fault—a highly dismembered fault zone that can be reassembled when offsets are restored on the San Andreas fault (Fig. 1, Plate IIIA). Earlier, D. P. Smith (1977) proposed a similar fault system, his San Juan–St. Francis–Clemens Well fault. However, Smith believed that the San Juan–Clemens Well fault was a late Oligocene and early Miocene structure that predated the late Miocene San Andreas system, and that 175 km of right-lateral displacement on the older fault were in addition to 300 km of post–middle Miocene displacement on the combined San Andreas and San Gabriel faults. Ehlig and Joseph (1977) and Joseph and others (1982b) presented isotopic evidence supporting the cross-fault correlations proposed by Smith (1977), and they shared his belief that the San Juan–Clemens Well fault preceded the San Andreas fault system. Powell (1981) observed that his Clemens Well–Fenner–San Francisquito fault probably was the same structure as Smith's. However, Powell viewed the fault as an integral part of the post–early Miocene San Andreas fault system that is necessary to completely reassemble displaced terranes and to match post–middle Miocene displacements in southern California with those in central California. We share Powell's view and incorporate his Clemens Well–Fenner–San Francisquito fault in this report.

A new reconstruction

Our palinspastic reconstruction for the San Andreas and San Gabriel faults in southern California (Plate IIIB) is based on three premises: (1) the 160-km displacement between the Mill Creek and Liebre Mountain Triassic megaporphyry bodies represents the total displacement on all strands of the San Andreas fault (sensu stricto) in southern California; (2) the San Gabriel–Banning fault has no more than 44 km of right slip; and (3) these displacements must be sequenced on various strands of the San Andreas system in a way that is compatible with the local history of these strands.

Restoration of specific fault displacements. The reconstruction in Plate IIIB arises from a specific sequence of restorations (Fig. 6), working backward from the most recent displacements along the neotectonic trace of the San Andreas fault to earliest displacements on the Banning–San Gabriel fault:

1. About 11 km of late Quaternary displacement is restored on the San Bernardino and Mill Creek strands of the San Andreas fault (Fig. 6A). This restoration utilizes the Mojave Desert segment of the San Andreas fault, and brings the Liebre Mountain megaporphyry body closer to the Mill Creek body.

2. About 20 km of displacement is restored on the San Jacinto fault utilizing the Mojave Desert segment of the San Andreas (Fig. 6B). This restoration occurs simultaneously with restorations on the San Bernardino and Mill Creek strands of the San Andreas, and brings the Liebre Mountain and Mill Creek megaporphyry bodies closer together. The San Andreas fault south of its junction with the San Jacinto fault is not involved in this 20-km restoration, however, because displacement on the San Jacinto occurs outboard (west) of the San Andreas. Thus, the San Andreas south of its junction with the San Jacinto must have less displacement than it does north of this junction. Plate IIIB incorporates this requirement.

3. About 89 km of Pliocene and early Pleistocene displacement is restored on the Mission Creek fault (Fig. 6C). This restoration brings the Liebre Mountain and Mill Creek megaporphyry bodies to within 40 km of each other, and approximately aligns the northwest termination of the Punchbowl fault (near Palmdale today) with the southeast termination of the Wilson Creek fault in the San Bernardino Mountains.

4. About 40 km of displacement restored on the Punchbowl–Wilson Creek fault juxtaposes the Liebre Mountain megaporphyry body against the San Bernardino Mountains near the Mill Creek body, and completes the sequence of restorations required to completely reconstruct the San Andreas fault (sensu stricto) in southern California (Fig. 6D).

5. About 44 km of right slip is restored on the Banning–San Gabriel fault (Fig. 6E). This figure reflects two components of slip: 22 km on the north branch of the San Gabriel fault (Ehlig, 1968a; 1975, p. 184), and 22 km on the south branch. By this interpretation, the north and south branches of the San Gabriel fault have a total displacement of about 44 km—not 60 km as proposed by Ehlig and Crowell (1982) and as cited widely by other workers. Whatever its total right slip turns out to be, displacement on the San Gabriel–Banning fault passes outboard (west) of the reconstructed San Andreas fault (sensu stricto), and does not affect its reconstruction.

6. About 58 km of left slip is restored on various left-lateral strike-slip faults of the eastern Transverse Ranges, including about 15 km on the Pinto Mountain fault (Dibblee, 1968b), about 8 km on the Mammoth Wash fault (Dillon, 1975, p. 329), and about 35 km on various left-lateral structures between the Pinto Mountain and Mammoth Wash faults (Powell, 1981, p. 315–333). Although Plate IIIB adopts these restorations, clockwise rotations interpreted to have accompanied left slip (Carter and others, 1987) have not been restored.

The palinspastic reconstruction that results from this sequence of restorations (Plate IIIB) yields a paleogeographic setting that would have existed about 10 m.y. ago—after displacements

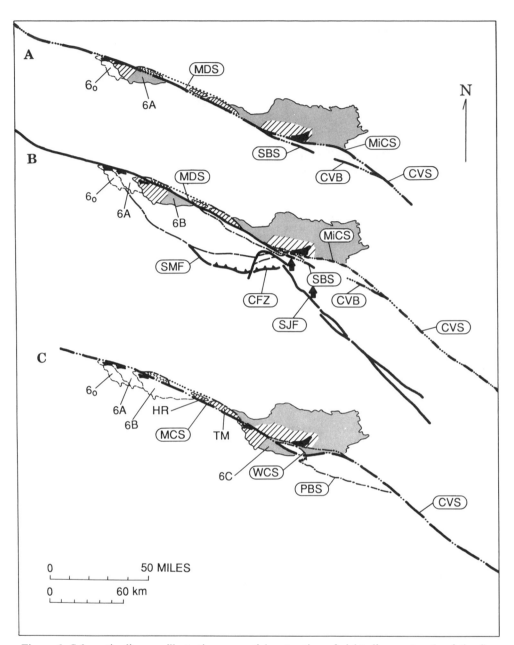

Figure 6. Schematic diagram illustrating sequential restoration of right slip on strands of the San Andreas fault system required to generate the palinspastic reconstruction of Plate IIIB. In each part of the figure (A-E), the Liebre Mountain Triassic megaporphyry body is restored to a palinspastic position closer to the Mill Creek megaporphyry body as a result of restored right slip on fault strands designated for each figure; preceding positions for the Liebre Mountain body are indicated by dashed outlies of the Liebre Mountain block. Rock and fault symbols same as Plates IIIA and B. A, The Liebre Mountain megaporphyry body is removed 11 km from its present position (6_o) by restoring displacements on the San Bernardino strand (SBS, 3 km) and Mill Creek strand (MiCS, 8 km), which merge with the Mojave Desert strand. B, The Liebre Mountain megaporphyry body is removed 20 km from its former position (6A) by restoring displacements on the San Jacinto fault (SJF); slip on the San Jacinto steps right onto the San Andreas fault and contributes to displacements on the Mojave Desert strand (MDS). CFZ = Cucamonga fault zone; CVB = Coachella Valley segment, Banning fault; CVS = Coachella Valley segment, San Andreas fault; MiCS = Mill Creek strand, San Andreas fault; SBS = San Bernardino strand, San Andreas fault; SMF = Sierra Madre fault. C, The Liebre Mountain megaporphyry body is removed 89 km from its former position (6B) by restoring displacements on the Mission Creek strand of the San Andreas fault (MCS), which merges with the Mojave Desert strand (MDS). This restoration approximately aligns the northwest end of the Punchbowl strand (PBS) with the southeast end of the

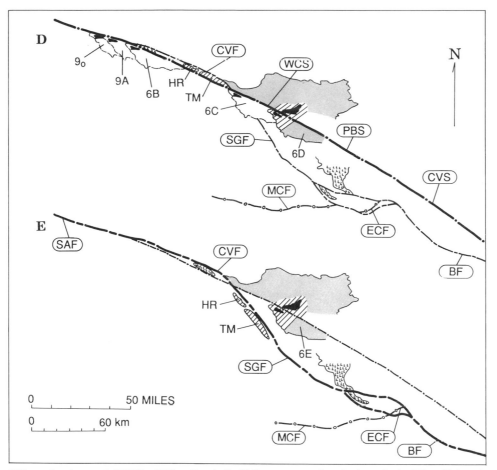

Wilson Creek strand (WCS). CVS = Coachella Valley segment, San Andreas fault; HR = Holcomb Ridge; TM = Table Mountain. D, The Liebre Mountain megaporphyry body is removed 40 km from its former position (6C) by restoring displacements on the continuous Punchbowl-Wilson Creek strand (PBS and WCS, respectively). This restoration reassembles the Liebre Mountain and Mill Creek occurrences of the Triassic megaporphyry body and completes the sequence of restorations required to reconstruct about 160 km of right slip on the San Andreas fault (sensu stricto) in southern California. Restoration of the Liebre Mountain and Mill Creek megaporphyry bodies approximately aligns the northwest end of the San Gabriel fault (SGF) with the southeast end of the Cajon Valley fault (CVF). BF = Banning fault; CVS = Coachella Valley segment, San Andreas fault; ECF = Evey Canyon fault; HR = Holcomb Ridge; MCF = Malibu Coast-Santa Monica-Raymond fault; TM = Table Mountain. E, About 22 km of right slip is removed on the north branch of the San Gabriel fault (SGF) by reuniting exposures of the Lowe igneous pluton as proposed by Ehlig (1973, 1975, 1981); approximately 22 km of right slip is removed on the south branch of the San Gabriel fault by matching the east end of the Malibu Coast–Santa Monica–Raymond fault (MCF) with the southwest end of the Evey Canyon fault (ECF). Reconstruction of 44 km of right slip on the San Gabriel fault completes the sequence of restorations required to generate the palinspastic reconstruction of Plate IIIB. BF = Banning fault; CVF = Cajon Valley fault; HR = Holcomb Ridge; SAF = San Andreas fault; TM = Table Mountain.

were completed on the Clemens Well–Fenner–San Francisquito fault of Powell (1981; this volume) but before displacements on the San Gabriel–Banning fault.

Implications

The reconstruction in Plate IIIB leads to a new way of viewing the restored positions of some well-known crystalline and sedimentary terranes in southern California.

Reunited crystalline rocks. *Liebre Mountain–San Bernardino Mountains.* The Liebre Mountain block is positioned adjacent to the San Bernardino and northern Little San Bernardino Mountains, rather than adjacent to the southern Little San Bernardino Mountains as in most reconstructions (Fig. 5). This juxtaposition yields an attractive comparison between batholithic and prebathoithic rocks: (1) It places the Liebre Mountain megaporphyry body within a Mojave Desert province of alkali-rich monzonitic rocks (Miller, 1978) that includes early Mesozoic alkalic rocks in the San Bernardino Mountains (Cameron, 1981; D. K. Smith, 1982) and that may include the quartz monzonite of Twentynine Palms (Trent, 1984; Brand and Anderson, 1982); (2) it juxtaposes extensive marble bodies in the Liebre Mountain block against comparable marble bodies in the San Bernardino Mountains in a way that is not accommodated if the Liebre Mountain block is placed against the southern Little San Bernardino Mountains (Fig. 5) where no extensive marble has been reported; and (3) it juxtaposes within a single region three bedrock occurrences of polka-dot granite—one in the Liebre Mountain block (Ehlert, 1982a) and one each in the southeastern San Bernardino Mountains and northern Little San Bernardino Mountains (Ehlig and Joseph, 1977).

San Gorgonio Pass–southern Chocolate Mountains. In Plate IIIB, crystalline rocks of the San Gorgonio Pass region between the Banning and Mission Creek faults are juxtaposed against crystalline rocks on the southwest flank of the southern Chocolate Mountains. This reconstruction compares favorably with Dillon's (1975, Fig. 70) reconstruction of these terranes following restoration of 180 ± 20 km on his north branch of the San Andreas fault. Dillon (1975, p. 59–60, 351–353) drew cross-fault comparisons between a variety of Precambrian orthogneisses and Mesozoic granitoid rock and orthogneiss, but two distinctive crystalline units provide strong cross-fault ties. (1) Small bodies of Triassic granitoid rocks of the Lowe igneous pluton (Joseph and others, 1982a) crop out in the San Gorgonio Pass terrane (Farley, 1979; Matti and others, 1983a) and along the south flank of the southern Chocolate Mountains (Dillon, 1975, p. 74–76, Fig. 14). When reconstructed as shown in Plate IIIB, the Lowe bodies in these two regions represent outliers of the regionally extensive Lowe igneous pluton that is preserved as large masses in the western San Gabriel Mountains but forms progressively smaller bodies to the east. (2) Bodies of distinctive gneissic porphyroblastic hornblende-biotite quartz monzonite with potassium-feldspar phenocrysts crop out in the San Gorgonio Pass terrane (saussuritized quartz monzonite unit of Allen, 1957, p. 319) and in the southern Chocolate Mountains and Cargo Muchacho Mountains (Dillon, 1975, p. 77–83, 343, Fig. 15). The Cargo Muchacho pluton yielded a Jurassic U-Pb age of 173 ± 1 Ma (Dillon, 1975), and Dillon believed that lithologically similar quartz monzonite bodies in the Chocolate Mountains and San Gorgonio Pass terranes are the same age.

Paleogeography of late Miocene depositional basins. The reconstruction in Plate IIIB provides a paleogeographic framework for various upper Miocene sedimentary sequences that have been dismembered and displaced by the San Andreas fault system. These sequences were deposited during all or part of the late Miocene, and today they are dispersed throughout southern California both inboard and outboard of the San Andreas fault (Plate IIIA, sequences 3, 4, 6, 7, 9–13). The reconstruction proposed by Ehlig (1981, Fig. 10-4; Fig. 5 of this report) provides a testable paleogeographic framework for three of the far-traveled sedimentary sequences (Ridge Basin Group, Castaic Formation of Link, 1982a, and the Punchbowl Formation; Plate IIIA, se-

quences 4, 10, and 9; see Link, 1983, Fig. 5), but it does not account for the palinspastic position of other far-traveled sequences that some workers claim to have originated in the region between the Orocopia and Cargo Muchacho Mountains (for example, the Mill Creek Formation of Gibson, 1971, upper Miocene units in the San Gorgonio Pass region, and the Coachella Fanglomerate; Plate IIIA, sequences 11, 6, and 7). In Plate IIIB, all of the upper Miocene sedimentary sequences known from within and adjacent to the San Andreas and San Gabriel fault zones are assembled within a paleogeographic setting that allows each sequence to be compared with the others and to be evaluated in the context of late Miocene tectonism and source areas.

Ridge Basin. Juxtaposition of the Liebre Mountain crystalline terrane against the southwestern margin of the San Bernardino Mountains positions upper Miocene sedimentary rocks of Ridge Basin against the San Bernardino Mountains (Plate IIIB, sequences 4 and 10) rather than farther south against the southern Little San Bernardino Mountains as in most reconstructions (Fig. 5; Link, 1982c, 1983, Fig. 5). Juxtaposition of Ridge Basin against the San Bernardino Mountains places the subsiding trough next to an upland that was undergoing orogenic uplift during the late Miocene (about 9 to 5 Ma according to Weldon, 1986; Alexander and Weldon, 1987; Meisling and Weldon, 1989). This uplift would have provided granitoid and gneissic debris to the Ridge Basin Group as it accumulated from 10 to 5 Ma. This paleogeographic reconstruction invites comparison between sedimentary units in Ridge Basin and sedimentary nits that occur in the San Bernardino Mountains. Candidates for comparison are the alluvial-fan facies of the Ridge Route Formation of Link (1982a, b) and Tertiary sedimentary rocks that fringe the western base of the San Bernardino Mountains (Plate IIIB, sequence 5; Morton and Miller, 1975, Fig. 1d-g, their unit of nonmarine conglomerate and sandstone that crops out between Cajon Pass and Mill Creek). The two sedimentary sequences have broadly similar lithologies and have several bedrock clast types in common. This speculative cross-fault correlation would allow the Ridge Route Formation and the unnamed sedimentary fringe to be coeval alluvial-fan deposits that lapped onto the reunited Liebre Mountain and San Bernardino Mountains blocks, which together were undergoing uplift and erosion.

Punchbowl and Mill Creek Formations. In Plate IIIB the upper Miocene Punchbowl Formation (Noble, 1954a) and the Mill Creek Formation of Gibson (1971) are positioned against the northern Little San Bernardino Mountains, together with Pelona Schist of Sierra Pelona and Oligocene and Miocene sedimentary and volcanic rocks of Soledad basin (Plate IIIB, sequences 9, 11, and 18). This palinspastic restoration places both the Punchbowl and Mill Creek Formations in positions where their paleocurrent data and clast compositions are compatible with possible source areas.

The Punchbowl Formation crops out today near Valyermo (Plate IIIA, sequence 9), where Noble (1954a) described a lower member containing granitoid and gneissic clasts together with polka-dot granite clasts (Pelka, 1971) and an upper member characterized by upwardly diminishing granitoid and gneissic clasts and progressively increasing numbers of volcanic clasts (Woodburne, 1975, Fig. 2; Barrows and others, 1987, recognized these two members plus a basal member rich in clasts of San Francisquito Formation). The volcanic clasts range from basaltic andesite to rhyolite in composition; metavolcanic clasts also are present. Reconstruction of the Punchbowl basin between Soledad basin and the northern Little San Bernardino Mountains provides appropriate source terranes for many of these clast types:

1. Volcanic clasts: The restoration in Plate IIIB places the Punchbowl Formation within 20 km of the east end of Soledad basin, and this juxtaposition raises the possibility that volcanic clasts in the Punchbowl fill were derived from Oligocene and Miocene volcanic rocks of the Vasquez Formation (Plate IIIB, sequence 18). This possibility also was raised by Barrows (1975) and by Barrows and others (1985, 1987), although they noted differences between the Punchbowl clasts and volcanic rocks exposed in the Soledad basin area. The source of metavolcanic clasts in the Punchbowl Formation is more certain. Robinson and Woodburne (1971) discussed their similarity to Triassic metavolcanic rocks of the Mojave Desert, and Woodburne (1975, p. 19–21, p. 63–65) suggested that if the clasts were derived from outcrops in the south-central Mojave Desert, the Punchbowl Formation may have been displaced no more than 16 to 24 km from its original depositional position in the Cajon Pass region. However, Barrows (1975, p. 201) indicated that the metavolcanic clasts probably have been recycled out of underlying Paleocene strata of the San Francisquito Formation. In that case, a paleogeographic connection between the Punchbowl Formation and metavolcanic rocks in the south-central Mojave Desert is not required.

2. Basement clasts: Gneissic and granitoid clasts in the Punchbowl Formation are compatible with crystalline rocks in the northern Little San Bernardino Mountains, including dikes and small bodies of polka-dot granite that occur there. The occurrence of polka-dot granite clasts in the Punchbowl Formation is the primary reason that Ehlert and Ehlig (1977) reconstructed the Punchbowl Formation to the vicinity of the northwestern Orocopia Mountains (Fig. 5), where polka-dot granite has been identified (Ehlig and Joseph, 1977; Ehlig, 1981, p. 279–280, Fig. 10-4). However, restoration of the Punchbowl Formation to the Orocopia Mountains region conflicts with the observation by Ehlig and Joseph (1977, p. 91) that clasts of polka-dot granite in the Punchbowl Formation are strikingly similar to polka-dot granite that crops out in the northern Little San Bernardino Mountains. Our reconstruction of the Punchbowl Formation near Morongo Valley is consistent with this observation.

The Mill Creek Formation crops out today in the southern San Bernardino Mountains (Plate IIIA, sequence 11), and has been the focus of interest ever since Gibson's (1964, 1971) proposal that the unit has been displaced 120 km from the Orocopia Mountains region—an area that Gibson proposed as a source for Pelona Schist clasts in the Mill Creek strata. The unit and underlying crystalline rocks are bounded on the west by the San Bernardino and Mission Creek strands of the San Andreas fault and

on the east by the Wilson Creek strand of the San Andreas (Matti and others, 1985). The Wilson Creek fault juxtaposes the Mill Creek strata against unnamed Tertiary sedimentary rocks that fringe the southern margin of the San Bernardino Mountains (Plate IIIA, sequence 5; Morton and Miller, 1975, Fig. 1d–g).

The Mill Creek Formation of Gibson (1971) consists of interfingering lithologies that have been mapped in many different ways (R. E. Smith, 1959; Gibson, 1964, 1971; Demirer, 1985; Sadler and Demirer, 1986; West, 1987; J. C. Matti, unpublished mapping, 1980–1988). Much of the formation consists of fluvial and lacustrine mudrock and fine-grained sandstone that forms a basin-axis facies. Conglomerate and coarse sandstone intertongue with the basin-axis facies from the north and northwest, west and southwest, and south and southeast. Conglomerate units that intertongue from the west through southeast are the most distinctive: these contain clasts derived from sources outboard (west) of the Mission Creek strand of the San Andreas fault, including Pelona Schist clasts (nongarnet-bearing) and volcanic clasts (quartz dacite, quartz latite, andesite, and basaltic andesitic).

Restoration of right slip on the Mission Creek and Wilson creek faults (Plate IIIA) should place the Mill Creek basin in a position where it could receive debris derived from outcrops of Tertiary volcanic rocks and Pelona Schist located to the west. The reconstruction in Plate IIIB achieves this requirement: the Mill Creek basin is positioned adjacent to the northeast end of Sierra Pelona (garnet-free outcrops of Pelona Schist) and within 10 to 15 km of intermediate to silicic Oligocene and Miocene volcanic rocks of Soledad basin that could have provided sources for Pelona Schist and volcanic clasts that occur in conglomerate tongues of the Mill Creek Formation. (Note: Gibson used the same Pelona Schist–bearing beds as evidence for sediment derivation from the Orocopia Mountains region on the east side of the San Andreas fault. We agree with Sadler and Demerer [1986, p. 132], who recognized that paleocurrent indicators within Pelona Schist-bearing Mill Creek beds are incompatible with an eastern source area.)

In Plate IIIB the Mill Creek and Punchbowl Formations are positioned together between the Little San Bernardino Mountains on the east and the Liebre Mountain–Sierra Pelona–Soledad basin region on the west. This juxtaposition implies a stratigraphic and paleogeographic relation between the two units that presently cannot be documented. Age relations between the two units have not been established, although they in part may be coeval (Plate IV): Woodburne (1975, Fig. 2) suggested that the Mill Creek Formation is latest Barstovian and Clarendonian in age (13 to about 10 Ma) and overlaps with the younger Punchbowl Formation that he considered to be of Clarendonian and Hemphillian age (12 to about 6 Ma). The physical stratigraphy of the two formations for the most part is very different; however, similarities between the upper member of the Punchbowl Formation and upper parts of the Mill Creek Formation suggest that these parts of the two units may have had a paleogeographic connection. Specifically, tongues of volcanic clast-bearing conglomerate and sandstone in the upper part of the Mill Creek Formation (Plate IV; unit 4 of Gibson, 1971) have bedding characteristics and clast compositions that are similar to volcanic clast-bearing conglomerate in the upper member of the Punchbowl Formation. We propose that these volcanic clast-bearing units were derived from a common source located southwest of both Formations—the Soledad basin in our reconstruction. Although the Mill Creek and Punchbowl Formations may not be entirely coeval and although their sedimentary fills may have evolved independently for much of their history, in late Clarendonian and early Hemphillian time, they may have developed enough paleogeographic continuity to have received volcanic-rich sediment derived from this common source.

Paleogeography of Santa Ana Sandstone. Juxtaposition of the Sierra Pelona–Soledad basin region near the southeastern San Bernardino Mountains during the period between 12 and 5 Ma may provide the correct time frame and paleoslope for Transverse Range–derived clasts reported in the Santa Ana Sandstone of Vaughan (1922) in the San Bernardino Mountains (Plate IIIA, sequence 16). Sadler and Demirer (1986, p. 133–136) reported garnet-bearing Pelona Schist, anorthosite, and intermediate volcanic clasts from poorly dated Santa Ana strata that could be as old as 15 Ma or as young as 3 or 4 Ma. Sadler and Demirer (1986) preferred the younger part of this time span, and derived the Transverse Ranges clasts from Sierra Pelona and the western San Gabriel Mountains as these terranes slid past the San Bernardino Mountains along the Pliocene San Andreas fault. If the part of the Santa Ana Sandstone containing the exotic clasts proves to be of late Miocene age rather than Pliocene, as we prefer, then juxtaposition of the Sierra Pelona–Soledad basin region near the southeastern San Bernardino Mountains in late Miocene time may provide a nearby source for the clasts.

Late Miocene paleogeography of San Gorgonio Pass. In Plate IIIB, two sedimentary units of upper Miocene strata that crop out today in the San Gorgonio Pass region are restored adjacent to the southern Chocolate Mountains: the Coachella Fanglomerate north of the Banning fault and the Hathaway Formation of Allen (1957) south of the Banning fault (Plates IIIA and B, sequences 7 and 8, respectively).

In Plate IIIB, the Coachella Fanglomerate is close to the position proposed by Dillon (1975, p. 342–347, Fig. 70). Dillon based his reconstruction on bedrock occurrences of porphyroblastic hornblende–biotite quartz monzonite and vein magnetite south of Mammoth Wash in the southern Chocolate Mountains that he believed to be sources for clasts of these rock types reported by Peterson (1975) in the Coachella Fanglomerate. In Plate IIIB, the Coachella Fanglomerate is positioned southwest of Mammoth Wash, and paleocurrent indicators recorded by Peterson (1975, Fig. 3) from the Coachella Fanglomerate are compatible with this position. However, Peterson (1975, p. 124–125) reconstructed the Coachella Fanglomerate farther south against candidate source rocks in the Cargo Muchacho Mountains—a restoration that requires 215 km of right slip on the San Andreas, rather than the 180 ± 20 km advocated by Dillon (1975). Our

placement of the Coachella Fanglomerate near Mammoth Wash positions the unit at the west end of a late Miocene depositional basin that included the coeval fanglomerate of Bear Canyon on the east (Dillon, 1975; Plate IIIB, sequence 8) and the Hathaway Formation on the south (Plate IIIB, sequence 6).

Restoration of 44 km of right slip on the Banning–San Gabriel fault in San Gorgonio Pass positions the Hathaway Formation of Allen (1957) between the San Jacinto and Santa Rosa Mountains on the south and the Chocolate Mountains on the north (Plate IIIB, sequence 6). This position is compatible with possible source areas for clasts in the Hathaway Formation (J. C. Matti and D. M. Morton, unpublished data). Some Hathaway conglomerate beds contain volcanic clasts that probably were derived from middle Tertiary volcanic units mapped by Dillon (1975) in the southern Chocolate Mountains as well as porphyroblastic biotite–hornblende quartz monzonite and other orthogneisses typical of basement terranes in San Gorgonio Pass and southern Chocolate Mountains. Other Hathaway conglomerate beds contain plutonic, cataclastic, mylonitic, and marble clasts that are similar to bedrock units in the upper and lower plates of the Eastern Peninsular Ranges mylonite zone of the San Jacinto and Santa Rosa Mountains mapped by Sharp (1979), Matti and others (1983b), and Erskine (1985).

Summary. The reconstruction in Plate IIIB restores to their original positions various upper Miocene sedimentary sequences that have been dismembered and displaced by the San Andreas fault system. Ridge Basin is juxtaposed adjacent to the San Bernardino Mountains at a time (9 to 5 Ma) when the range was undergoing uplift and erosion. Other late Miocene basins widely dispersed today are reassembled within a northwest-trending province that may represent initial subsidence of the Salton Trough in response to late Miocene displacements on the San Gabriel fault but precursory to the San Andreas fault (sensu stricto). Upper Miocene basin deposits accumulating in this trough included, from northwest to southeast, the Punchbowl and Mill Creek Formations, the Coachella Fanglomerate, the Hathaway Formation, and the fanglomerate of Bear Canyon.

Pliocene paleogeography. Hungry Valley Formation. In Plate IIIB, the Hungry Valley Formation, the Pliocene part of the sedimentary sequence that filled Ridge Basin (Link, 1982a), is juxtaposed against the southwestern San Bernardino Mountains (Plate IIIB, sequence 1). Both Crowell (1982a, b) and Ramirez (1983) suggested that the Hungry Valley Formation was deposited in the vicinity of the San Bernardino Mountains, but they restored the unit to different parts of the region: Crowell (1982a, p. 39–40) suggested that the Hungry Valley Formation was deposited in the vicinity of Cajon Pass and the western San Bernardino Mountains, whereas Ramirez (1983) suggested that the unit was deposited in the vicinity of the southeastern San Bernardino Mountains near Morongo Valley. Our reconstruction of the Hungry Valley basin falls between these two positions, and juxtaposes the unit against basement rocks that include extensive bodies of marble (Miller, 1979), which may have been the source for marble clasts emphasized by Ramirez (1983).

San Gorgonio Pass sequence. Uppermost Miocene and Pliocene deposits of the Imperial Formation and the Painted Hill Formation of Allen (1957) north of the Banning fault in San Gorgonio Pass are juxtaposed together with the underlying Coachella Fanglomerate against the southern Chocolate Mountains south of Mammoth Wash (Plate IIIB, sequence 3). This position is compatible with north-derived clasts in the Painted Hill Formation that include abundant volcanic clasts and various granitic and gneissic clasts, including porphyroblastic hornblende-biotite quartz monzonite and uncommon clasts of Lowe igneous pluton (J. C. Matti, unpublished data).

Paleogeology of the San Gabriel fault and Ridge Basin.
Reconstruction of the Liebre Mountain block against the southwestern margin of the San Bernardino Mountains may provide insight into the regional distribution of the San Gabriel fault and the paleogeology of Ridge Basin. If the San Gabriel fault is an early strand of the San Andreas fault system that was abandoned and subsequently offset to the northwest by the San Andreas (sensu stricto), then the northwest end of the San Gabriel must originally have joined the main trace of the San Andreas (sensu lato) well to the south of their present junction in Hill and Dibblee's (1953) Big Bend region. In our palinspastic reconstruction (Plate IIIB), the San Gabriel fault strikes toward Cajon Pass, which suggests that the San Gabriel–San Andreas junction originally was located in this region. There, on the east side of the San Andreas fault, the Cajon Valley fault (Woodburne and Golz, 1972; Dibblee, 1975b, Fig. 2) may represent a trace of the San Gabriel fault that was truncated by the San Andreas and left behind when the San Gabriel fault was offset to its present position in the Frazier Mountain region (Plates IIIA, B; also see Weldon, 1986, p. 360, Fig. 7.2). The Cajon Valley fault probably continues northwestward on the east side of Table Mountain and Holcomb Ridge, and may rejoin the San Andreas fault along the Hitchbrook fault of Wallace (1949). However, Ross (1984, p. 14) indicated that the Hitchbrook fault may not be a significant tectonic break, and although Plates IIIA and B adopt the speculative correlation between the Cajon Valley and Hitchbrook faults, this correlation is not essential to reconstructing rocks bounded by the Cajon Valley fault, so long as right slip on the Banning–San Gabriel–Cajon Valley fault came back into the San Andreas somewhere northwest of Cajon Pass.

If the Cajon Valley fault represents a former trace of the San Gabriel fault, then rocks bounded by this fault are not in place and must be restored to their original positions using offset estimates proposed for the San Gabriel fault. We adopt this proposal, and in Plate IIIB, crystalline rocks of the Table Mountain–Holcomb Ridge block are restored the full 44 km we attribute to the San Gabriel fault zone. This reconstruction has two implications:

1. Mesozoic batholithic rocks and prebatholithic marble and schist of the Table Mountain–Holcomb Ridge block are juxtaposed adjacent to the Liebre Mountain block, where generally similar batholithic and prebatholithic rocks occur. These two terranes in turn are positioned against the southwestern San Bernardino Mountains. The three terranes thus combined define a coherent province of generally similar batholithic and prebatholithic rocks—a notion anticipated by Ross (1972, p. 50), who noted that granitoid rocks of the Liebre Mountain region are indistinguishable petrologically and chemically from rocks in the Holcomb Ridge area. Dibblee (1968a, p. 263) also observed these similarities, but he discounted cross-fault correlation between the Liebre Mountains–Sawmill Mountain block and the Holcomb Ridge–Table Mountain block because the two blocks are only 80 to 95 km apart and thus have not been separated by the large displacements most workers attribute to the Mojave Desert segment of the San Andreas fault. Our reconstruction of the Holcomb Ridge–Table Mountain block against the Liebre Mountain block by restoring 44 km of right slip on the Cajon Valley–San Gabriel fault allows these compatible terranes to be juxtaposed but still permits large-scale displacements on the Mojave Desert segment of the San Andreas.

2. The restored position of the Table Mountain–Holcomb Ridge block has implications for the paleogeology of Ridge Basin relative to other Miocene basins. Ridge Basin was a trough whose subsiding depocenter migrated northwest from an initial position west of Soledad basin to a terminal position northwest of the Liebre Mountain massif. As the depocenter migrated northwestward, progressively younger deposits of the Ridge Basin Group overlapped the basin floor, with depocenter migration tracking the northwestward migration of the Frazier Mountain source area for the Violin Breccia (Crowell, 1982, Fig. 8, p. 32–35, 40). The late Miocene paleogeology of Ridge Basin traditionally is established by restoring about 60 km of right slip on the San Gabriel fault. This positions the Frazier Mountain source area initially southwest of the nascent Ridge Basin, and also juxtaposes middle and upper Miocene nonmarine deposits of the Caliente Formation of Cuyama Valley, the southern Caliente Range, and the Carrizo Plain against terrain inboard of the San Gabriel fault that later (in late Miocene time) would subside to receive the Ridge Basin fill (Fig. 5; implicit in Fig. 3 of Ehlig and others, 1975 and Fig. 10-4 of Ehlig, 1981; the region northwest of the "source area of Violin Breccia" in Crowell, 1982a, Fig. 8; Ehlert, 1983, Figs. 11A, B). This paleogeologic setting (Fig. 5) raises questions about the paleogeology and paleogeography of Ridge Basin prior to and during its evolution.

These questions focus on relations between Ridge Basin and the Caliente Formation: (1) Prior to late Miocene inception of Ridge Basin (about 10 Ma), did middle and upper Miocene sediments of the Caliente Formation deposited northwest of Mt. Pinos in Figure 5 lap east onto the basement terrane of the Liebre Mountain massif? If so, these sediments either must underlie beds of the Ridge Basin Group that subsequently accumulated in the northwest part of Ridge Basin or they were stripped away by erosion prior to deposition of these beds. (2) Alternatively, did middle and upper Miocene deposits of the Caliente Formation pinch out eastward against a paleogeographic barrier that prevented onlap of the Liebre Mountain basement? (3) Did Ridge Basin evolve in isolation from other late Miocene depositional basins in southern California? Paleogeographic and stratigraphic isolation seems to be required by the model proposed by Crowell (1982a). A paleogeographic barrier west of the future trace of the San Gabriel fault in Figure 5 not only would provide an eastward termination for the Caliente depositional province but would allow the late Miocene Ridge Basin to evolve in isolation from coeval upper Miocene beds of the Caliente Formation and from other late Miocene depositional basins in southern California.

The Frazier Mountain region generally is viewed as the paleogeographic high that flanked the west margin of the evolving Ridge Basin. However, Crowell (1982a, Fig. 8) has shown that the Frazier Mountain highland tracked northwest along the San Gabriel fault in parallel with the northwestward-migrating Ridge Basin depocenter. Therefore, during the early phase of basin development, any paleogeographic barrier that might have intervened between Ridge Basin and the Caliente depocenter could not have been formed by the Frazier Mountain terrane.

As a solution to this paleogeographic problem, we suggest that the Table Mountain–Holcomb Ridge block formed a paleogeographic highland of San Bernardino Mountain–type basement that flanked the entire northwest margin of the future Ridge Basin (Plate IIIB). This highland provided an eastern termination for the middle and late Miocene Caliente depositional province, and allowed the late Miocene Ridge Basin to evolve in isolation from upper Miocene deposits of the Caliente Formation. As Ridge Basin filled up with upper Miocene sediment, strike-slip displacements on the San Gabriel fault removed the Table Mountain–Holcomb Ridge highland and replaced it with the Frazier Mountain highland, which shed debris of the Violin Breccia eastward as both the highland and the migrating Ridge Basin depocenter tracked northwestward during the time frame 10 Ma to about 5 Ma (Crowell, 1982a).

Pliocene and Quaternary slip rates, and comparison of onshore strike-slip faulting with Gulf of California events. Plate IIIB incorporates 160 km of displacement on the San Andreas fault (sensu stricto) since its inception in the early Pliocene. This reconstruction yields a long-term slip rate of 32 to 40 mm/yr during the last 4 or 5 m.y.—a long-term rate that compares well with Holocene slip rates of about 35 mm/yr on the San Andreas fault north of Hill and Dibblee's Big Bend region (Sieh and Jahns, 1984) and comparable rates for the combined San Andreas and San Jacinto faults in southern California (Weldon and Humphries, 1986). By contrast, reconstructions for the San Andreas that require 240 km of right slip since the fault's inception 4 or 5 m.y. ago yield long-term slip rates of about 48 to 60 mm/yr. In order to accommodate the 35 mm/yr Holocene rates, the 240-km reconstruction must incorporate an early phase of rapid plate motion followed by a later phase of slower plate motion for at least the Holocene and probably the late Pleistocene. The 160-km reconstruction incorporates steady-state plate motion throughout the history of the San Andreas fault (sensu stricto).

The advent of the San Andreas fault (sensu stricto) in southern California has been linked temporally and genetically with the onset of sea-floor spreading in the Gulf of California. Most workers suggest that right slip on the San Andreas commenced with northwestward propagation of the East Pacific Rise and opening of the Gulf of California (Hamilton, 1961; Moore and Buffington, 1968; Larson and others, 1968; Atwater, 1970; Larson, 1972; Elders and others, 1972; Moore, 1973; Crowell, 1979, 1981; Curray and Moore, 1984). Curray and Moore (1984) proposed a two-phase model for the geologic history of the Gulf: an early phase of diffuse extension, crustal attenuation, and rifting that may have been accompanied by formation of oceanic crust without lineated magnetic anomalies, followed time-transgressively by a later phase of opening accompanied by formation of oceanic crust having lineated magnetic anomalies. According to this model, the early extensional phase commenced at about 5.5 Ma, but the actual opening phase commenced at about 4.9 Ma and culminated at about 3.2 Ma. Opening in the gulf has separated Peninsular Baja California from mainland Mexico by about 300 km.

Many workers link the offshore history of the Gulf of California with the onshore history of strike-slip faulting within the San Andreas transform system. Onshore, Crowell (1981, 1982a) and Ehlig (1981) proposed that the San Gabriel and San Andreas faults have generated about 300 km of right slip since about 10 Ma, with 240 km of this displacement occurring on the San Andreas fault in the last 5 m.y. Offshore, many workers propose that the gulf has opened about 300 km during the same time period. The onshore and offshore evidence are not completely compatible, however. For example, Curray and Moore (1984, p. 29) concluded that 300 km of combined slip on the San Gabriel and San Andreas faults corresponds with the 300-km separation of Baja California from mainland Mexico, and that the onshore and offshore events all commenced at about 5.5 Ma. However, this interpretation is in conflict with available onshore evidence for the late Miocene age of the San Gabriel fault (Crowell, 1982a, Fig. 12), and illustrates the lack of congruence between onshore and offshore histories.

This incongruence is even more pronounced in view of new information about the onshore history of the San Andreas fault presented by several chapters in this volume. If the San Andreas fault (sensu stricto) has only about 160 km of displacement since its inception 4 or 5 m.y. ago, then half the 300-km displacement required to open the Gulf of California since early Pliocene time is *not* recorded by the San Andreas fault (sensu stricto) onshore. Moreover, half of the 300-km displacement occurring on the San Andreas fault (sensu lato) onshore took place in the middle and late Miocene. These discrepancies between onshore and offshore tectonic records suggest that plate-tectonic models linking events in the Gulf of California to the San Andreas transform-fault system need to be reexamined.

DISASSEMBLY OF SOUTHERN CALIFORNIA: STEP-BY-STEP INSTRUCTIONS FOR REARRANGING A CONTINENTAL MARGIN

In this section we develop one possible interpretation of how the paleogeography of southern California evolved since the early Miocene (20 Ma to present). The reconstructions make no attempt to accommodate block rotations proposed for southern California on the basis of paleomagnetic data (Luyendyk and others, 1980, 1985; Luyendyk and Hornafius, 1987).

Paleogeography prior to disassembly (early Miocene [20 Ma], Fig. 7A)

Fault activity. The reconstruction in Figure 7A depicts the paleogeographic setting for southern California prior to major right slip within the San Andreas fault system. The reconstruction

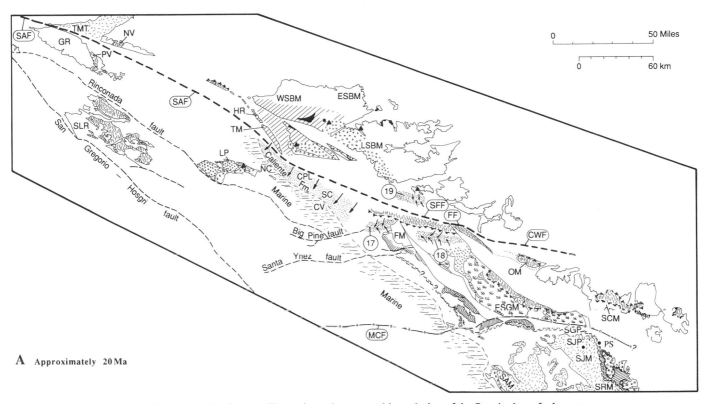

A Approximately 20 Ma

Figure 7A through K. Diagrams illustrating paleogeographic evolution of the San Andreas fault system in southern California from early Miocene to present. See text for discussion. Fault and rock symbols and sedimentary unit numbers same as in Plates IIIA and B. Heavy bold faults indicate structures that are active (dashed where soon to be initiated); thinner faults indicate structures whose displacement history is completed. Stippled pattern indicates active nonmarine sedimentation; short-dash pattern indicates active marine sedimentation; arrows indicate streamflow direction.

Figure 7A. Palinspastic reconstruction at about 20 Ma, prior to right-lateral displacements within the San Andreas fault system. About 50 km of right slip is restored on the Rinconda fault following Ross (1984), but we do not know how (or if) the Rinconda fault extended into southern California and affected the paleogeographic setting there. CPL = Carrizo Plain region; CWF = Clemens Well fault; CV = Cuyama Valley region; ESBM = eastern San Bernardino Mountains; ESGM = eastern San Gabriel Mountains; FF = Fenner fault; FM = Frazier Mountain region; GR = Gabilan Range; HR = Holcomb Ridge; LP = La Panza Range; LSBM = Little San Bernardino Mountains; MCF = Malibu Coast fault zone; NC = northern Caliente Range; NV = Neenach Volcanic field; OM = Orocopia Mountains; PS = Palm Springs; PV = Pinnacles Volcanic field; SAF = San Andreas fault; SAM = Santa Ana Mountains; SCM = southern Chocolate Mountains; SCR = southern Caliente Range; SFF = San Francisquito fault; SGM = San Gorgonio Mountain; SGP = San Gorgonio Pass; SJM = San Jacinto Mountains; SJP = San Jacinto Peak; SLM = Santa Lucia Mountains; SRM = Santa Rosa Mountains; TM = Table Mountain; TMT = Tehachapi Mountains; WSBM = western San Bernardino Mountains; WSGM = western San Gabriel Mountains.

obtains from Plate IIIB after 115 ± 10 km of right slip is restored on the Clemens Well–Fenner–San Francisquito–San Andreas fault, and about 25 km of left slip is restored on the Evey Canyon–Icehouse Canyon–San Jacinto–Banning segment of the Malibu Coast fault. We restore 50 km of right slip on the Rinconada fault following Ross (1984), but we do not extend this displacement into southern California. We also do not reconstruct left slip on the Big Pine and Santa Ynez faults, displacements that apparently accompany block rotations suggested by paleomagnetic data (Hornafius and others, 1986).

Basement patterns. Precambrian crystalline rocks of San Gabriel Mountains type in the Frazier Mountain region, northwestern San Gabriel Mountains, and Orocopia Mountains are reunited as originally proposed by Crowell and Walker (1962; Crowell, 1962), but in the arrangement proposed by Powell (1981; this volume). The Triassic Lowe igneous pluton forms a coherent pattern, with the main body in the western San Gabriel Mountains trailing east into remnants preserved in rocks of San Gabriel Mountains type in San Gorgonio Pass and in the northern and southern Chocolate Mountains. The 215-Ma Triassic megaporphyry body in the Liebre Mountain–San Bernardino Mountains block and associated rocks in the Twentynine Palms area occur north and mostly inboard of the Lowe pluton, and these two plutonic suites together form remnants of what must formerly have been an extensive late Permian and early Triassic plutonic province (Miller, 1978). Granitoid rocks of the La Panza Range are restored to the southern California region, although differently from the proposal by Smith (1977) and Joseph and others (1982b). These quartz monzonitic rocks form a western extension of similar quartz monzonite in the southern Little San Bernardino Mountains. Juxtaposition of the La Panza Range block west of the restored Liebre Mountain block positions polka-dot granite dikes in the eastern La Panza Range (Joseph and others, 1982b) at the northwest end of a province of polka-dot-bearing granitoid rocks that trends from the Orocopia Mountains northwestward through the Little San Bernardino and San Bernardino Mountains and into Liebre Mountain. We infer that the Eastern Peninsular Ranges mylonite zone occurs outboard (west) of all these crystalline terranes, and separates rocks of typical Peninsular Ranges type in the lower plate from an upper plate that includes rocks of atypical Peninsular Ranges type as well as rocks of La Panza Range type, San Bernardino Mountains type, and San Gabriel Mountains type. Onlapping Cretaceous marine rocks of the La Panza Range are coextensive with onlapping Cretaceous marine rocks in the Liebre Mountain region, and this regional stratigraphic marker coincides with onlapping Cretaceous marine rocks in the Santa Ana Mountains. Windows to Pelona Schist have not been opened, judging from the fact that clasts of Pelona Schist have not been identified in the Plush Ranch Formation of Carman (1964), Vasquez Formation, and Diligencia Formation.

Early Miocene depositional paleogeography. The early Miocene Neenach volcanics are reunited with the Pinnacles volcanics of central California as required by Matthews (1973, 1976). The

Diligencia, Vasquez, and Plush Ranch depositional basins are restored to a common region near the Orocopia Mountains as proposed by Crowell (1962, 1975a) and Bohannon (1975), but in the arrangement proposed by Powell (1981, this volume). Sediment in the Plush Ranch and Vasquez Formations was deposited from north- and west-flowing streams (Bohannon, 1975; Hendrix and Ingersoll, 1987); sediment in the Diligencia basin was deposited from north- and south-flowing streams (Bohannon, 1975; Spittler and Arthur, 1982).

Fault correlations. The Clemens Well, Fenner, and San Francisquito faults are connected as required by Powell (1981), although we position their junction farther northwest in the Salton Trough than did Powell (1981, Plate 6) because the Fenner and San Francisquito faults end up there when we restore the Liebre Mountain and Mill Creek Triassic megaporphyry bodies. Our palinspastic restoration of the San Francisquito and Fenner faults requires that we project a concealed extension of the Clemens Well fault for about 30 km along the southwestern flank of the Little San Bernardino Mountains. The Malibu Coast–Raymond system is connected to the Banning fault via the Evey Canyon, Icehouse Canyon, and San Jacinto faults. The Eastern Peninsular Ranges mylonite zone of the San Jacinto and Santa Rosa Mountains is continuous with similar mylonite in the southeastern San Gabriel Mountains; we infer that this ductile zone extends west to the Malibu Coast–Raymond Hill fault where it is displaced left-laterally and then continues to the northwest along a trend that projects the zone between the Santa Monica Mountains and Verdugo Hills.

Disassembly begins: Early displacements on the Clemens Well–Fenner–San Francisquito segment of the San Andreas fault (middle Miocene [16 Ma], Fig. 7B)

Fault activity. About 50 km of displacement has occurred on the Clemens Well–Fenner–San Francisquito fault zone. We interpret that this fault connected with the central California segment of the San Andreas fault to form a coextensive San Andreas–San Francisquito–Fenner–Clemens Well zone that lies outboard (west) of the Holcomb Ridge–Liebre Mountain–San Bernardino Mountains basement terrane. Left-lateral displacements are about to begin on the Malibu Coast–Raymond–Banning fault system.

Basement patterns. The Sierra Pelona–Blue Ridge–Orocopia Mountains window to Pelona Schist has opened as evidenced by Pelona Schist clasts in the Mint Canyon Formation, but there is no evidence that the San Bernardino and San Gorgonio windows of Pelona Schist are open. The Little San Bernardino Mountains–Liebre Mountain–La Panza Range belt of polka-dot-bearing granitoid rocks has been disrupted, along with the overlying Cretaceous marine onlap zone.

Middle Miocene depositional patterns. During early Barstovian time (16 Ma), two major domains of nonmarine deposition existed within intermontane southern California: (1) the Cajon, Crowder, and Santa Ana basins received sediments derived from Mojave Desert sources to the north (Meisling and Weldon, 1989; Sadler and Demirer, 1986); and (2) the Mint Canyon basin received sediments derived from the northern Chocolate Mountains to the east, the western San Gabriel Mountains to the south, and the Sierra Pelona to the north (Ehlig and others, 1975, and Ehlert, 1982b, although these authors place the Mint Canyon depocenter closer to and directly opposite from the Salton Wash area than we show in Fig. 7B). The Mint Canyon Formation passes westward into similar deposits of the Caliente Formation in the Cuyama Valley region.

Middle Miocene paleogeographic patterns following termination of Clemens Well–Fenner–San Francisquito fault (late Miocene 11 to 12 Ma], Fig. 7C)

Fault activity. Left-lateral displacements have ceased on the Malibu Coast–Banning fault system. Right-lateral displacements have ceased on the Clemens Well–Fenner–San Francisquito–San Andreas fault following 115 ± 10 km of displacement. A left step has developed in the fault at the latitude of the southeastern San Bernardino Mountains probably by oroflexural bending. The San Gabriel fault is about to become the active strand of the San Andreas fault system in southern California, and will evolve outboard (west) of the Clemens Well–Fenner–San Francisquito fault south of the Liebre Mountain region but inboard (east) of the San Andreas fault north of the major left step in the old San Andreas–Clemens Well system. The San Gabriel fault will generate right slip along a trace that includes the Hitchbrook–Cajon Valley–San Gabriel (south branch)–Stoddard Canyon–Banning segments. The Banning fault continues the San Gabriel trend into the Salton Trough, which by 11 Ma has begun to develop into a rift basin within the San Andreas system.

Basement patterns. The belt of granitoid basement rock

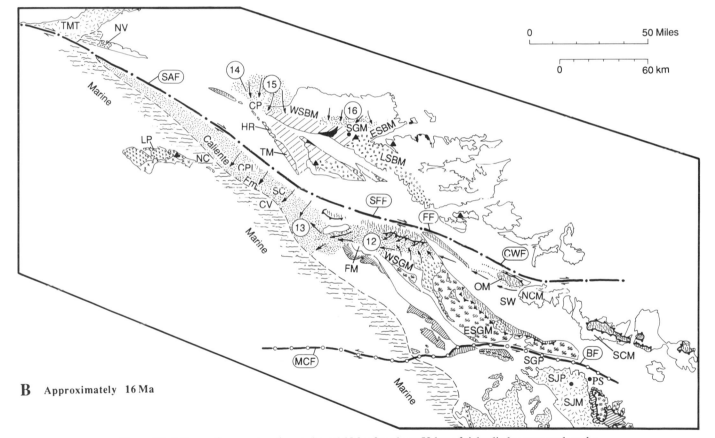

Figure 7B. Palinspastic reconstruction at about 16 Ma after about 50 km of right slip has occurred on the Clemens Well–Fenner–San Francisquito–San Andreas fault of Powell (1981, this volume). The left-lateral Malibu Coast-Banning fault system is about to be initiated. BF = Banning fault; CP = Cajon Pass region; CPL = Carrizo Plain region; CV = Cuyama Valley region; CWF = Clemens Well fault; ESBM = eastern San Bernardino Mountains; ESGM = eastern San Gabriel Mountains; FF = Fenner fault; FM = Frazier Mountain region; HR = Holcomb Rige; LP = La Panza Range; LSBM = Little San Bernardino Mountains; MCF = Malibu Coast fault zone; NC = northern Caliente Range; NV = Neenach Volcanic field; OM = Orocopia Mountains; PS = Palm Springs; SAF = San Andreas fault; SCM = southern Chocolate Mountains; SCR = southern Caliente Range; SFF = San Francisquito fault; SGM = San Gorgonio Mountain; SGP = San Gorgonio Pass; SJM = San Jacinto Mountains; SJP = San Jacinto Peak; SW = Salton Wash; TM = Table Mountain; TMT = Tehachapi Mountains; WSBM = western San Bernardino Mountains; WSGM = western San Gabriel Mountains.

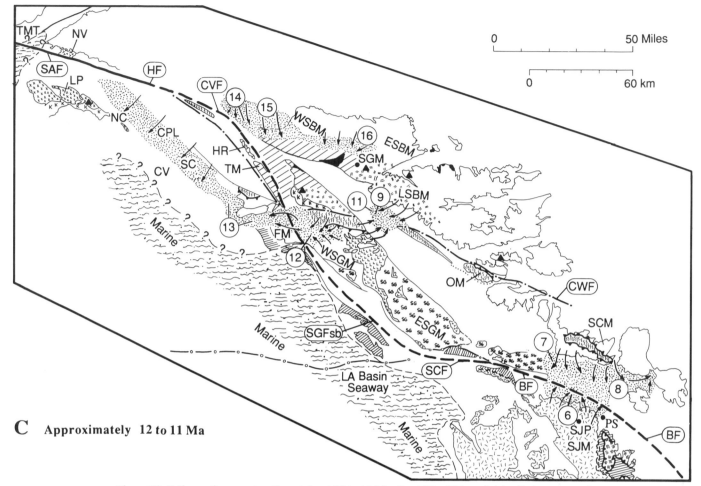

Figure 7C. Palinspastic reconstruction at about 12 to 11 Ma after 115 ± 10 km of right slip has occurred on the Clemens Well–Fenner–San Francisquito–San Andreas fault; the Clemens Well–Fenner–San Francisquito segment of this fault has been abandoned. Left slip on the Malibu Coast–Banning fault system is completed. The right-lateral San Gabriel–Banning fault is about to be initiated. Deposition of the Mint Canyon Formation continues, but we infer that the unit now is accumulating only in the western part of its depocenter. BF = Banning fault; CP = Cajon Pass region; CPL = Carrizo Plain region; CV = Cuyama Valley region; CVF = Cajon Valley fault; CWF, Clemens Well fault; ESBM, eastern San Bernardino Mountains; ESGM = eastern San Gabriel Mountains; FM = Frazier Mountain region; HF = Hitchbrook segment of San Gabriel–Cajon Valley fault; HR = Holcomb Ridge; LP = La Panza Range; LSBM = Little San Bernardino Mountains; NC = northern Caliente Range; MCF = Malibu Coast fault zone; NV = Neenach Volcanic field; OM = Orocopia Mountains; PS = Palm Springs; SAF = San Andreas fault; SCF = Stoddard Canyon fault; SCM = southern Chocolate Mountains; SCR = southern Caliente Range; SGFsb = San Gabriel fault (south branch); SGM = San Gorgonio Mountain; SJM = San Jacinto Mountains; SJP = San Jacinto Peak; TM = Table Mountain; WSBM = western San Bernardino Mountains; WSGM = western San Gabriel Mountains.

and overlapping Cretaceous has been completely disrupted by right-lateral displacement on the Clemens Well–San Andreas system. Crystalline rocks in the upper and lower plates of the Eastern Peninsular Ranges mylonite zone have been disrupted by about 25 km of left slip on the Malibu Coast–Banning fault.

Late Miocene depositional patterns. Along the Hitch-brook–Cajon Valley segment of the San Gabriel fault the nonma-rine Cajon and Crowder basins continue to receive sediment,

although deposits of this age (11 to 12 Ma) are not preserved beneath a regional unconformity in the Cajon Pass region (Plate IV; Woodburne and Golz, 1972; Woodburne, 1975; Meisling and Weldon, 1989). Farther to the east in the San Bernardino Mountains, the Santa Ana Sandstone continues to accumulate as an intermontane counterpart to the Crowder Formation (Meisling and Weldon, 1989). All of these nonmarine deposits continue to have a paleogeographic relation with upper Miocene

(Clarendonian) nonmarine sedimentary rocks of the Cuyama Valley and Caliente Range, which in turn pass outboard (west) into upper Miocene (Luisian to Mohnian) marine deposits. The Mint Canyon Formation continues to accumulate in this time frame, judging from the age of air-fall tuffs in the upper part of the unit that are 10.1 ± 0.8 and 11.6 ± 1.2 Ma (Obradovich and McCulloh, *in* Terres and Luyendyk, 1985, p. 12469). However, in Figure 7C we suggest that the locus of Mint Canyon deposition has shifted toward the Ridge Basin area from a larger area that formerly extended east throughout Soledad basin (Fig. 7B). We propose that eastern exposures of Mint Canyon Formation are being dissected at 11 Ma, with Mint Canyon sediment recycled into younger sedimentary basins.

Figure 7C depicts an intermontane nonmarine depocenter northwest of the Salton Trough that at 11 Ma is receiving Clarendonian sediments of the Mill Creek and Punchbowl Formations. Juxtaposition of the two sedimentary sequences derives from our restoration of right slip on the Mission Creek and Mill Creek faults followed by restoration of 40 km of right slip on the Punchbowl–Wilson Creek faults (discussed above), and leads us to propose that the two sequences formed opposing halves of a two-sided basin that evolved northwest of the early Salton Trough. The Mill Creek part of this basin received sediment from multiple sources: (1) granitoid rocks northwest of the basin that we propose was the southeast end of the Liebre Mountain block, (2) Pelona Schist southwest and south of the basin that we propose was Sierra Pelona, (3) sedimentary and volcanic debris from unconfirmed sources, and (4) granitoid sources to the north and northeast. The Punchbowl part of the two-sided basin received sediment from three sources: (1) metavolcanic and sedimentary clasts recycled from underlying Paleocene strata of the San Francisquito Formation (Barrows, 1975; Barrows and others, 1985, 1987), (2) gneissic and granitoic debris derived from eastern sources that included polka-dot granite (Pelka, 1971), and (3) volcanic debris from unconfirmed sources. In Figure 7C we show Soledad basin as the source for the volcanic debris in both the Mill Creek and Punchbowl Formations, with the volcanic clasts recycled out of the Vasquez Formation and/or the Mint Canyon Formation (although Barrows, 1975, p. 201, is uncomfortable with a Vasquez source for most of the Punchbowl volcanic clasts). Barrows (1987, Fig. 7A) proposed that volcanic clasts in the Punchbowl Formation were derived from a since-vanished volcanic field that formerly covered the east end of Sierra Pelona and Soledad basin.

In Figure 7C we show the Mill Creek Formation (of Gibson, 1971) accumulating partly on Pelona Schist, even though in most mappable exposures the unit rests noncomformably on crystalline rocks of the Wilson Creek block. Our reconstruction of the Mill Creek Formation adjacent to the restored Sierra Pelona and Blue Ridge windows would allow Mill Creek deposits to extend south onto Pelona Schist, and such an extension may be required to explain Pelona Schist outcrops that R. E. Smith (1959) mapped beneath or within the Mill Creek strata. We interpret Smith's Pelona Schist body as depositional basement beneath the Mill Creek Formation, and we interpret it as a remnant of either the Blue Ridge or Sierra Pelona windows. This interpretation raises the possibility that a segment of the San Francisquito Fenner–Clemens Well fault passes beneath the Mill Creek Formation, and thereby separates crystalline rocks of the Wilson Creek block on the north from Smith's (1959) Pelona Schist on the south (Plate IIIB).

Figure 7C depicts a major depocenter in the Salton Trough that received nonmarine sediments now represented by several rock units, including: (1) the fanglomerate of Bear Canyon (Dillon, 1975), derived from the southeastern Chocolate Mountains; (2) the Coachella Fanglomerate of Allen (1957), derived from the western part of the southern Chocolate Mountains (but contrast this interpretation with that of Peterson, 1975, who proposed that clasts in the Coachella Fanglomerate were derived primarily from the Cargo Muchacho Mountains); and (3) the Hathaway Formation of Allen (1957), derived both from the north (Chocolate Mountains) and the south (San Jacinto and Santa Rosa Mountains).

Paleogeography accompanying early displacements on San Gabriel fault (late Miocene [9 Ma], Fig. 7D)

Fault activity. About 10 km of right-lateral displacement has occurred on the Hitchbrook–Cajon Valley–San Gabriel (South Branch)–Stoddard Canyon–Banning fault.

Basement patterns. The Holcomb Ridge–Table Mountain basement sliver is being displaced northwest from the Liebre Mountain block by right slip on the Cajon Valley and Ridge Basin segments of the San Gabriel fault. Concurrently, the Frazier Mountain block of Precambrian crystalline rocks is being displaced northwest from its counterpart in the western San Gabriel Mountains, and is being slipped past the developing sag of Ridge Basin as required by Crowell (1952, 1962, 1975c, 1982b).

Depositional patterns. The Clarendonian paleogeographic setting at 9 Ma includes the following elements: (1) The Salton Trough depocenter continues to receive sediment transported southward from the Chocolate Mountains and northward from the San Jacinto and Santa Rosa Mountains. (2) The Punchbowl and Mill Creek depositional basins are receiving volcanic clast-bearing sediment from the south and southwest derived from volcanic rocks in Soledad basin. The locus of Clarendonian deposition at 9 Ma is in the Punchbowl basin; only minor sediment appears to have been deposited in the Mill Creek basin, judging from the minor component of volcanic material in the Mill Creek Formation (Plate IV). (3) Sediments of the Santa Ana Sandstone continue to accumulate in the San Bernardino Mountains, with sediment derived from Mojave Desert sources to the north (Sadler and Demirer, 1986). In addition, Figure 7D shows sediment derived from Soledad basin or from the western San Gabriel Mountains and transported northward into the Santa Ana basin. This paleogeographic scheme provides an origin for clasts of Pelona Schist (garnet-free variety), anorthosite, and volcanic rock described by Sadler and Demirer (1986) from the Santa Ana Sandstone, but requires that units bearing these clasts are Clarendonian and not Blancan as proposed by Sadler and Demirer (1986). An accurate age for these deposits presently cannot be determined.

West of the San Bernardino–Liebre Mountain block, Ridge Basin has begun to sag and receive marine and nonmarine sediment of the Ridge Basin Group (Link, 1982a, b). The northwest-aligned Ridge Basin is two-sided, and receives sediment from three sources: (1) crystalline basement rocks of the Frazier Mountain block that shed clasts east into the Violin Breccia (Crowell, 1952, 1982b), (2) crystalline basement rocks of Liebre Mountain that shed sediment west into the Ridge Route Formation, and (3) crystalline rocks of the Holcomb Ridge–Table Mountain slice that shed sediment down the southeast-dipping Ridge Basin paleoslope. Unnamed, poorly dated nonmarine sedimentary rocks

that fringe the southwest margin of the San Bernardino Mountains probably are related to the Ridge Basin fill.

Tectonic events. As the Clarendonian depositional regime evolved, the western San Bernardino Mountains region began to uplift (Meisling and Weldon, 1989), and the Barstovian and early Clarendonian Cajon and Crowder basins were disrupted. Uplift and dissection of the Cajon strata also may have contributed sediment to the Ridge Basin fill.

Paleogeographic transition between San Gabriel and San Andreas faults (latest Miocene [7 to 6 Ma], Fig. 7E)

Fault activity. The San Gabriel (south branch)–Stoddard Canyon–San Jacinto segment of the San Gabriel–Banning fault

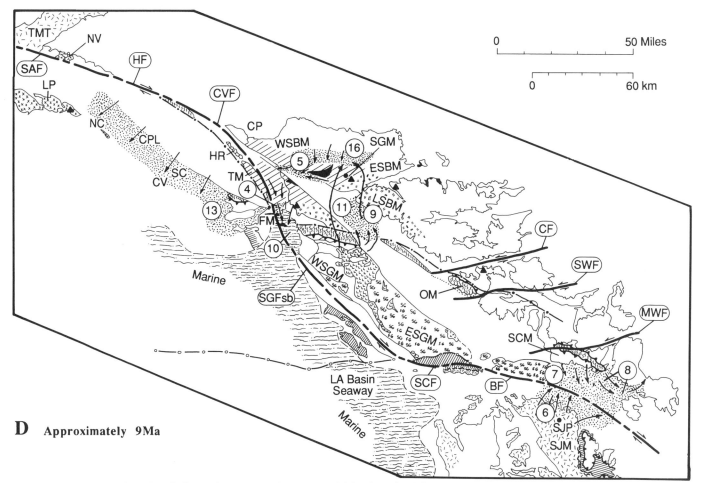

Figure 7D. Palinspastic reconstruction at about 9 Ma after a few kilometers of right slip has occurred on the San Gabriel–Banning fault. BF = Banning fault; CF = Chriaco fault; CP = Cajon Pass region; CPL = Carrizo Plain region; CV = Cuyama Valley region; CVF = Cajon Valley fault; ESBM = eastern San Bernardino Mountains; ESGM = eastern San Gabriel Mountains; FM = Frazier Mountain region; HF = Hitchbrook segment of San Gabriel–Cajon Valley fault; HR = Holcomb Ridge; LP = La Panza Range; LSBM = Little San Bernardino Mountains; MWF = Mammoth Wash fault; NC = northern Caliente Range; NV = Neenach Volcanic field; OM = Orocopia Mountains; PS = Palm Springs; SAF = San Andreas fault; SCF = Stoddard Canyon fault; SCM = southern Chocolate Mountains; SCR = southern Caliente Range; SGFsb = San Gabriel fault (south branch); SGM = San Gorgonio Mountain; SJM = San Jacinto Mountains; SJP = San Jacinto Peak; SWF = Salton Wash fault; TM = Table Mountain; WSBM = western San Bernardino Mountains; WSGM = western San Gabriel Mountains.

has generated 22 km of right slip; the segment was abandoned at about 7.5 Ma, and has been succeeded by the San Gabriel (north branch)–Icehouse Canyon–San Jacinto segment of the San Gabriel–Banning fault, which has generated about 10 to 15 km of right slip. These displacements carry northwest to the San Andreas fault by way of the Cajon Valley–Hitchbrook segment of the San Gabriel fault. We infer that left-lateral displacements are occurring east of the Salton Trough on the Mammoth Wash, Salton Wash, and Chiriaco faults at 7 Ma, and these inferred displacements may reflect region-wide deformation that led to abandonment of the south branch–Stoddard Canyon segment of the San Gabriel fault. By 6 Ma, left-lateral displacements have concluded on the Mammoth Wash and Salton Wash faults, but have continued on the Chiriaco fault. The Liebre–Squaw Peak thrust system is active across the Liebre Mountain and San Bernardino Mountain blocks. The Punchbowl–Wilson Creek strand

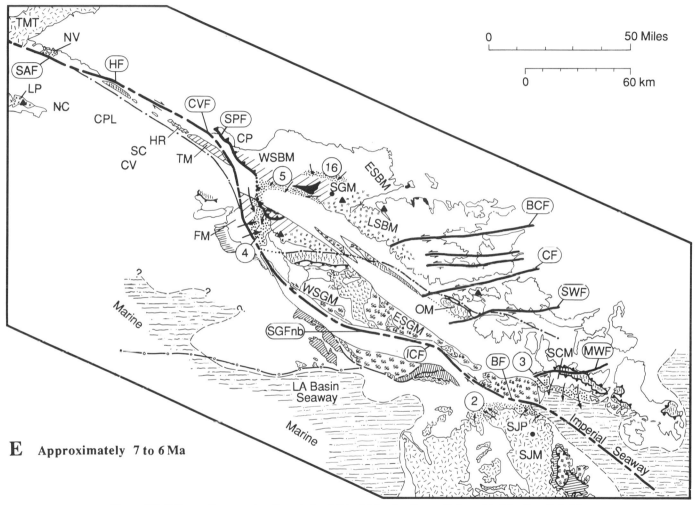

Figure 7E. Palinspastic reconstruction at about 7 to 6 Ma after the south branch of the San Gabriel–Banning fault has been abandoned following 22 km of right slip. The north branch of the San Gabriel–Icehouse Canyon–Banning fault has generated a few kilometers of right slip. Note that right slip on the Banning fault in this time frame occurs at the same time as marine deposition of the Imperial Formation in the Salton Trough. BCF = Blue Cut fault; BF = Banning fault; CF = Chiriaco fault; CP = Cajon Pass region; CPL = Carrizo Plain region; CV = Cuyama Valley region; CVF = Cajon Valley fault; ESBM = eastern San Bernardino Mountains; ESGM = eastern San Gabriel Mountains; FM = Frazier Mountain region; HF = Hitchbrook segment of San Gabriel–Cajon Valley fault; HR = Holcomb Ridge; ICF = Icehouse Canyon fault; LP = La Panza Range; LSBM = Little San Bernardino Mountains; MWF = Mammoth Wash fault; NC = northern Caliente Range; NV = Neenach Volcanic field; OM = Orocopia Mountains; PS = Palm Springs; SAF = San Andreas fault; SCR = southern Caliente Range; SGFnb = San Gabriel fault (north branch); SGM = San Gorgonio Mountain; SJM = San Jacinto Mountains; SJP = San Jacinto Peak; SWF = Salton Wash fault; TM = Table Mountain; WSBM = western San Bernardino Mountains; WSGM = western San Gabriel Mountains.

of the San Andreas fault has developed inboard (east) of the San Gabriel fault zone, and may have begun to generate right slip as early as 6 Ma.

Depositional patterns. In the Salton Trough, a marine incursion from the Gulf of California is depositing the Imperial Formation as far into the Salton Trough as the Whitewater area. Because these marine deposits occur as much as 100 m beneath a 6.0-Ma basalt flow (Plate IV), we suspect that marine sedimentation in this part of the Salton Trough occurred as early as 7 Ma—the older time frame of Figure 7E. Note that by 6 Ma the Imperial Formation has been displaced by movements on the San Gabriel–Banning fault. In the San Gorgonio Pass region, the marine deposits interfinger northward with nonmarine deposits of the Painted Hill Formation of Allen (1957) and probably southward with nonmarine deposits of the Hathaway Formation of Allen (1957; Plate IV). To the west, locally derived nonmarine sediments of the Potrero Creek deposits of Frick (1921) have begun to accumulate on the northern part of the Peninsular Ranges block in anticipation of a major depositional event that will follow in Pliocene time.

In the central San Bernardino Mountains and Ridge Basin, nonmarine sedimentation continues within a region that is increasingly affected by compressional tectonics. The Squaw Peak–Liebre Mountain thrust system has telescoped the Cajon and Crowder basins and emplaced a thin slice of crystalline rocks—including the Liebre Mountain Triassic megaporphyry body—atop uppermost Miocene beds of the Ridge Route Formation (Alexander and Weldon, 1987; Meisling and Weldon, 1989). Within this compressional regime locally derived sediments continue to accumulate within the Santa Ana Sandstone basin, and uppermost deposits of the Ridge Basin Group continue to fill up the Ridge Basin sag.

Initial paleogeography within San Andreas fault (sensu stricto) (latest Miocene and early Pliocene [6.0 Ma to 4.0 Ma], Fig. 7F

Fault activity. The San Gabriel fault has been abandoned after generating its full right-lateral displacement of 44 km. The San Andreas fault (sensu stricto) has evolved inboard (east) of the San Gabriel–Banning segment of the San Gabriel fault southeast of Ridge Basin, but outboard (west) of the Hitchbrook–Cajon Valley segment of the San Gabriel fault northwest of Ridge Basin. As a result, the Hitchbrook–Cajon Valley segment of the San Gabriel fault has been stranded against the Mojave Desert, and displacements on the San Andreas fault have displaced the truncated northwest end of the San Gabriel fault away from the southeast end of the Cajon Valley fault.

The Punchbowl–Wilson Creek fault, the earliest strand of the San Andreas fault (sensu stricto), evolved between 6 and 5 Ma and generated about 40 km of right slip. This slip episode disrupted the two-sided Punchbowl–Mill Creek basin and displaced the Mill Creek Formation and Wilson Creek block to the vicinity of the San Bernardino Mountains. In the early Pliocene,

compressional deformation of the Punchbowl–Wilson Creek fault terminated right slip on the strand, created its sinuous trace and west dips, and left the Wilson Creek block and Mill Creek Formation juxtaposed against the San Bernardino Mountains. Simultaneously, left slip on the Chiriaco and Blue Cut faults and associated structures projected the Little San Bernardino Mountains and the Punchbowl basin and associated Blue Ridge Window of Pelona Schist westward, allowing the succeeding Mission Creek strand of the San Andreas fault to evolve inboard (east) of the Punchbowl Formation and associated basement rocks. We propose that these events created the peculiar concave-east geometry of the Punchbowl fault observed today in the San Gabriel Mountains (Plate IIIA), and may explain the Nadeau north and south faults as early short-lived strands of the Mission Creek fault that evolved adjacent to the Little San Bernardino Mountains before the main Mission Creek trace prevailed. To the northwest, the Mission Creek strand has evolved outboard (west) of the Wilson Creek block and the Holcomb Ridge–Table Mountain block, leaving them and the Mill Creek Formation stranded against rocks of San Bernardino Mountains type. In the time frame of Figure 7F (6 to 4 Ma) the Mission Creek strand has just evolved, and will generate its full 89 km of right slip through about 1.2 Ma.

Depositional patterns. Between the San Gabriel Mountains and the San Bernardino and Little San Bernardino Mountains, nonmarine sediment of the Juniper Hills Formation of Barrows (1987) accumulates as displacements on the Punchbowl–Wilson Creek fault occur. This partly accommodates the requirements of Barrows (1987, Fig. 7), although he would have the Punchbowl fault continue to generate significant right slip during deposition of late Blancan beds of the Juniper Hills Formation (the time frame of our Fig. 7H; see Barrows, 1987, Fig. 8). In the Salton Trough, latest Miocene and earliest Pliocene progradation of the nonmarine Painted Hill Formation and correlative units has pushed the Imperial seaway farther southeast into the Salton Trough, where early Pliocene deposits of the Imperial Formation interfinger with nonmarine units in the Borrego area. The latter in part have been interpreted as distal deposits of the Colorado River displaced by the San Andreas and San Jacinto faults about 150 km from their former position on the southeast lobe of the Colorado River delta (Winker and Kidwell, 1986).

With the advent of the San Andreas fault (sensu stricto), two major nonmarine depocenters have evolved in the vicinity of the Transverse Ranges:

1. At the boundary between the Peninsular Ranges and Transverse Ranges, a large alluvial fan and braided stream complex consisting of sediment derived from the Transverse Ranges has begun to prograde southward onto the Peninsular Ranges block. This complex is preserved in deposits of the San Timoteo Badlands (of Frick, 1921, as used by Matti and Morton, 1975), but correlative sediments that must have been widespread to the west (the vanished fan of Morton and Matti, 1989) were stripped away in Quaternary time and can only be inferred from the distribution of relict clasts perched on the slopes of Peninsular

Ranges basement highs. As this alluvial fan complex accumulates, it gradually buries paleotopography on the northern part of the Peninsular Ranges.

2. In the vicinity of Cajon Pass, the Pliocene Hungry Valley Formation is receiving sediment derived from sources north of the San Andras fault. Meisling and Weldon (1989) proposed that sediment in the Hungry Valley Formation was recycled out of the Crowder Formation in Cajon Pass as that unit was uplifted in the upper plate of the Squaw Peak thrust. Paleogeographic linkage between Squaw Peak tectonism and Hungry Valley deposition has two implications: (1) emplacement of the Squaw Peak thrust, which uplifted the Crowder and displaced the unit southward toward headward-eroding streamflows that were feeding sediment into Ridge Basin, predated or accompanied deposition of the Hungry Valley Formation; and (2) initial displacements on

the Punchbowl–Wilson Creek strand of the San Andreas fault predated deposition of the Hungry Valley Formation. The second conclusion resolves an apparent conflict between Meisling and Weldon's (1989) requirement that the Hungry Valley Formation initially accumulated proximal to the Cajon Pass region and our requirement that the reconstructed Liebre Mountain and Mill Creek megaporphyry bodies place Ridge Basin and the site of the future Hungry Valley Formation farther southeast of the Cajon Pass region than Meisling and Weldon (1989) would prefer. To resolve this conflict, we propose that initial right slip on the San Andreas fault (Punchbowl–Wilson Creek strand) displaced Ridge Basin closer to Cajon Pass before initial deposits of the Hungry Valley Formation accumulated—that is, before 5 Ma, the widely accepted age for the basal Hungry Valley Formation (Plate IV; Link, 1982a, Fig. 1; note that Ensley and Verosub, 1982,

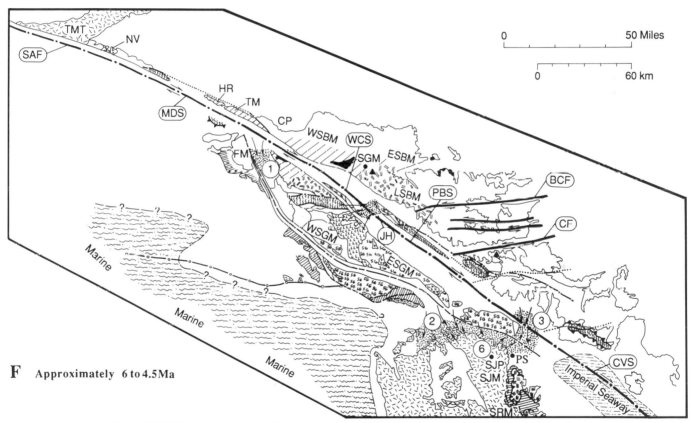

Figure 7F. Palinspastic reconstruction at about 6 to 4.5 Ma. The San Gabriel–Banning fault has been abandoned and the Punchbowl–Wilson Creek strand of the San Andreas fault has produced about 30 km of right slip. Separation of the Liebre Mountain and Mill Creek bodies of Triassic megaporphyry has begun. Sediment is shed east across the active Punchbowl fault into the lower Blancan Juniper Hills Formation (JH) as required by Barrows (1987). BCF = Blue Cut fault; CF = Chiriaco fault; CP = Cajon Pass region; ESBM = eastern San Bernardino Mountains; CVS = Coachella Valley segment, San Andreas fault; ESGM = eastern San Gabriel Mountains; FM = Frazier Mountain region; HR = Holcomb Ridge; LSBM = Little San Bernardino Mountains; MDS = Mojave Desert segment, San Andreas fault; NV = Neenach Volcanic field; PBS = Punchbowl strand, San Andreas fault; PS = Palm Springs; SAF = San Andreas fault; SGM = San Gorgonio Mountain; SJM = San Jacinto Mountains; SJP = San Jacinto Peak; SRM = Santa Rosa Mountains; TM = Table Mountain; WCS = Wilson Creek strand, San Andreas fault; WSBM = western San Bernardino Mountains; WSGM = western San Gabriel Mountains.

Fig. 7, inferred that the basal Hungry Valley may be as old as 5.2 or 5.3 Ma). This proposal suggests a start-up time for the San Andreas fault (sensu stricto) in southern California of about 5.5 to 6.0 Ma, and suggests that initial displacements on the Punchbowl–Wilson Creek strand overlapped with penultimate displacements on the San Gabriel–Banning fault (Figs. 7E, F).

Paleogeography dominated by the San Andreas fault (sensu stricto) (late Pliocene [4.0 to 2.5 Ma], Fig. 7G,H)

Fault activity. The Mission Creek strand of the San Andreas fault has generated about 50 km of displacement since its inception at 3.5 to 4.0 Ma. A left step is developing at the latitude of the left-lateral Pinto Mountain fault, which is active during this interval (Dibblee, 1968b). The north frontal-fault zone of the San Bernardino Mountains is active starting at about 2.5 Ma (Sadler,

1982a, b; May and Repenning, 1982; Meisling, 1984) leading to uplift of the range.

Basement patterns. The Lytle Creek–San Bernardino window of Pelona Schist is gradually beginning to open at about 2.5 to 2.0 Ma, judging from the age inferred for sediment containing clasts of Pelona Schist in the San Timoteo beds (Matti and Morton, 1975).

Depositional patterns. The alluvial fan complex formed by the San Timoteo beds and vanished fan have completely buried paleotopography on the northern part of the Peninsular Ranges block. The vertical distribution of basement clasts in the San Timoteo succession includes clasts of the upper plate of the Vincent thrust overlain by clasts of Pelona Schist (Matti and Morton, 1975)—a vertical pattern that records progressive unroofing of the San Bernardino and Lytle Creek windows of Pelona Schist during the time frame 4.0 to 2.0 Ma. After about 2.5 Ma, an

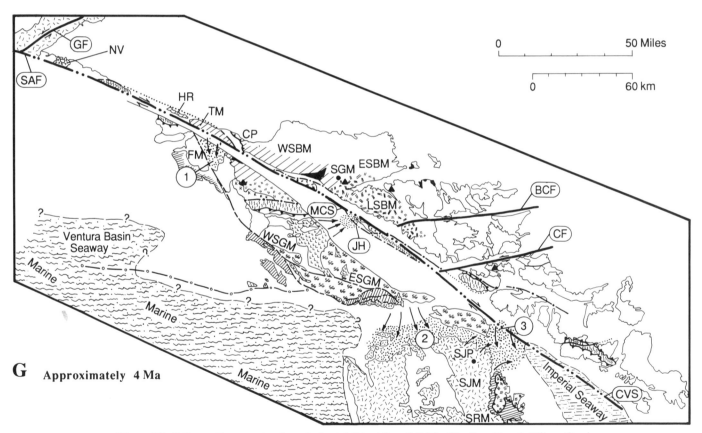

Figure 7G. Palinspastic reconstruction at about 4.5 to 4.0 or 3.5 Ma. The Punchbowl–Wilson Creek strand of the San Andreas fault has been abandoned after generating about 40 km of right slip and the Mission Creek strand of the San Andreas has been initiated. BCF = Blue Cut fault; CF = Chiriaco fault; CP = Cajon Pass region; CVS = Coachella Valley segment, San Andreas fault; ESBM = eastern San Bernardino Mountains; ESGM = eastern San Gabriel Mountains; FM = Frazier Mountain region; GF = Garlock fault; HR = Holcomb Ridge; JH = sediment of the Juniper Hills Formation of Barrows (1987); LSBM = Little San Bernardino Mountains; MCS = Mission Creek strand; San Andreas fault; MDS = Mojave Desert segment, San Andreas fault; NV = Neenach Volcanic field; PS = Palm Springs; SAF = San Andreas fault; SGM = San Gorgonio Mountain; SJM = San Jacinto Mountains; SJP = San Jacinto Peak; SRM = Santa Rosa Mountains; TM = Table Mountain; WSBM = western San Bernardino Mountains; WSGM = western San Gabriel Mountains.

orogenic sediment wedge (Old Woman Sandstone of Vaughan, 1922) is being shed northward from the rising San Bernardino Mountains (Sadler, 1982a, b; May and Repenning, 1982). Deposition of the Juniper Hills Formation continues east of the San Gabriel Mountains as required by Barrows (1987), although we believe that by the time frame of Figure 7H (2.5 Ma) these deposits postdate significant right slip on the Punchbowl fault (contrast with Barrows, 1987, Fig. 8).

Northwest of Cajon Pass, the Phelan Peak deposits of Meisling and Weldon (1989) is being shed north onto the Mojave Desert largely from sediment sources outboard (south) of the San Andreas fault. This event started at about 4.1 Ma (Meisling and Weldon, 1989; Weldon and others, this volume), and would have accelerated after the Hungry Valley depocenter was displaced to the northwest by the Mission Creek–Mojave Desert strand of the San Andreas.

Paleogeography reflecting complications within the San Andreas transform system (early Pleistocene [1.2 Ma], Fig. 7I)

Fault activity. The left step in the Mission Creek strand of the San Andreas fault in the San Gorgonio Pass region has become so pronounced that right slip along the strand has become difficult. This gives rise to the so-called San Gorgonio Pass knot—a slip barrier that has affected the entire San Andreas fault system in late Quaternary time. The San Andreas fault (sensu stricto) has developed a complex response to the San Gorgonio Pass knot, a response that led to the San Jacinto fault as well as to a variety of contractional and extensional fault complexes (Matti and others, 1985). We propose that the San Jacinto fault evolved about 1.2 m.y. ago as a way to transfer right slip around the San Gorgonio Pass knot (Morton and Matti, this volume).

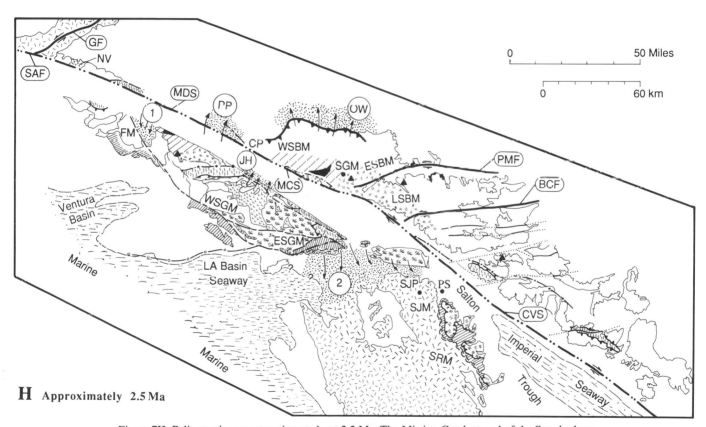

H Approximately 2.5 Ma

Figure 7H. Palinspastic reconstruction at about 2.5 Ma. The Mission Creek strand of the San Andreas fault has generated about 50 of its 89-km displacement. Initial left-lateral displacements on the Pinto Mountain fault (PMF) are beginning to develop a bend in the Mission Creek strand. BCF = Blue Cut fault; CP = Cajon Pass region; CVS = Coachella Valley segment, San Andreas fault; ESBM = eastern San Bernardino Mountains; ESGM = eastern San Gabriel Mountains; FM = Frazier Mountain region; GF = Garlock fault; LSBM = Little San Bernardino Mountains; MCS = Mission Creek strand, San Andreas fault; MDS = Mojave Desert segment, San Andreas fault; NV = Neenach Volcanic field; OW = Old Woman Sandstone of Vaughan (1922) as used by May and Repenning (1982) and Sadler (1982a, b); PP = sediments of Phelan Peak deposits of Meisling and Weldon (1989); PS = Palm Springs; SAF = San Andreas fault; SGM = San Gorgonio Mountain; SJM = San Jacinto Mountains; SJP = San Jacinto Peak; SRM = Santa Rosa Mountains; TM = Table Mountain; WSBM = western San Bernardino Mountains; WSGM = western San Gabriel Mountains.

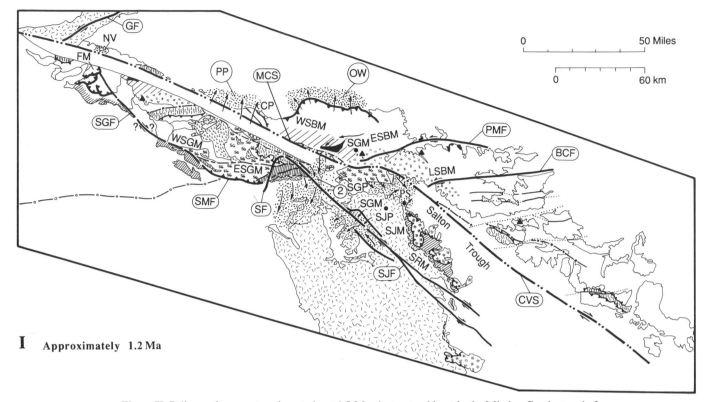

I   Approximately 1.2 Ma

Figure 7I. Palinspastic reconstruction at about 1.2 Ma. A structural knot in the Mission Creek strand of the San Andreas fault in San Gorgonio Pass has created a slip barrier that has led to initiation of the San Jacinto fault (SJF) to the west. BCF = Blue Cut fault; CP = Cajon Pass region; CVS = Coachella Valley segment, San Andreas fault; ESBM = eastern San Bernardino Mountains; ESGM = eastern San Gabriel Mountains; FM = Frazier Mountain region; GF = Garlock fault; LSBM = Little San Bernardino Mountains; MCS = Mission Creek strand, San Andreas fault; MDS = Mojave Desert segment, San Andreas fault; NV = Neenach Volcanic field; PMF = Pinto Mountain fault; PP = Phelan Peak deposits of Meisling and Weldon (1989); PS = Palm Springs; SAF = San Andreas fault; SF = San Antonio fault; SGF = San Gabriel fault in Ridge Basin region; SGM = San Gorgonio Mountain; SGP = San Gorgonio Pass; SJM = San Jacinto Mountains; SJP = San Jacinto Peak; SMF = Sierra Madre fault; SRM, Santa Rosa Mountains; TM = Table Mountain; WSBM = western San Bernardino Mountains; WSGM = western San Gabriel Mountains.

This conclusion follows from our interpretation of relations between the San Jacinto and San Andreas faults southeast and northwest of San Gorgonio Pass. Southeast of San Gorgonio Pass, the San Jacinto parallels the Coachella Valley segment of the San Andreas, and the two faults share whatever late Quaternary slip rate is assigned to them (commonly interpreted to be about 35 mm/yr). However, in the San Gorgonio Pass region, significant late Quaternary right slip on the San Andreas fault cannot be documented, so Matti and others (1985) proposed that right slip has stepped left from the Coachella Valley segment of the San Andreas fault to the San Jacinto fault, thereby accelerating slip on the San Jacinto northwest of San Gorgonio Pass. This left step involved regional compression that yielded two results: broad regional uplift of the San Gorgonio Pass region, including both the San Jacinto Mountain block and the San Bernardino Mountains block (Morton and Matti, this volume), and accelerated thrust faulting along the northern frontal-fault zone of the

San Bernardino Mountains (Meisling, 1984; Meisling and Weldon, 1989).

Initiation of the San Jacinto fault did not provide a simple and complete solution to the San Gorgonio Pass knot. If it had, the San Jacinto fault might be expected to merge cleanly with the San Andreas fault northwest of the San Gorgonio Pass knot, with right slip passing back to the San Andreas fault in a straightforward manner (Weldon and Humphreys, 1986). However, as discussed above, the San Jacinto at its northwest end cannot be shown to have a direct surface connection with the Mojave Desert segment of the San Andreas fault, and the intervening San Bernardino Valley and San Gabriel Mountains have experienced complex late Quaternary tectonic and alluvial histories. If right slip on the San Jacinto is transferred back to the San Andreas, this transfer must occur by mechanisms that are not easily documentable. We propose that right slip at the northwest end of the San Jacinto fault is transformed in one or more of three ways:

1. Slip steps right (east) onto the San Andreas fault between San Gorgonio Pass and Cajon Pass (Matti and others, 1985). This right step has created a normal fault-controlled late Quaternary depocenter that we term the San Bernardino basin (Morton and Matti, this volume). The San Bernardino basin is bounded on its margins by normal dip-slip faults like those of the Crafton Hills horst-and-graben complex (Matti and others, 1985); the basin has evolved in the last 750,000 yr after deposition of the San Timoteo beds terminated (Morton and Matti, this volume).

2. Right slip on the San Jacinto fault has been transformed into convergence and oblique left slip within the schuppen structure of the southeastern San Gabriel Mountains (Morton, 1975b; Morton and Matti, this volume). In the process, the southeastern corner of the San Gabriel Mountains between the San Antonio Canyon fault and the San Jacinto fault has been displaced northwestward relative to the rest of the range (Cramer and Harrington, 1987), resulting in about 3 km of left slip on the San Antonio fault and an equivalent amount of right slip within the San Jacinto fault zone. Right slip there has occurred as San Jacinto tectonism has reactivated the old Banning–San Gabriel fault and then broken away from that trend by extending northwest before transforming into east- and northeast-trending oblique left-lateral faults (Fig. 3; Morton, 1975a, b; Morton and Matti, this volume).

3. Geometric and kinematic complexities in the southeastern San Gabriel Mountains cannot represent a permanent sink for right slip within the San Andreas transform system. Ultimately, no matter how complex the local geometry is, right slip on the San Jacinto fault either must be transferred entirely to the Mojave Desert segment of the San Andreas fault (Weldon, 1984, 1986; Meisling and Weldon, 1989) or must be conveyed to faults elsewhere within the continental margin. We propose that geometric complexities within the southeastern San Gabriel Mountains may reflect a left step between the San Jacinto fault and structures to the west. The left step would require oblique right-lateral convergence on the south frontal fault zone of the San Gabriel Mountains (Sierra Madre fault), and would lead to reactivation of the San Gabriel fault, which Weber (1982) believed has had a Quaternary history, and/or to convergence within thrust faults of the western Transverse Ranges.

In this chapter we do not attempt to resolve these slip vectors at the northwest end of the San Jacinto fault; therefore, Figures 7I and J adopt all three types of right-slip transformation. Between 1.2 Ma and 750 ka, a left step between the San Jacinto fault and faults to the west may have been the predominant mechanism (Fig. 7I), whereas slip-transfer to the San Andreas fault may have prevailed during the last 750,000 yr (Fig. 7J). We cannot confirm these speculations, however.

Basement patterns. The Lytle Creek–San Bernardino window of Pelona Schist has been opened.

Depositional patterns. The alluvial fan depositional complex of the San Timoteo Formation and vanished-fan sequence is being uplifted and dissected on the southeast due to regional uplift in the San Gorgonio Pass region, but deposition continues on the northeast and northwest. This depocenter is receiving clasts derived from basement of San Bernardino Mountains type, including clasts of the 215-Ma Triassic megaporphyry. We interpret these San Bernardino Mountains–type clasts to be deposits of the ancestral Santa Ana River (Morton and others, 1986; Morton and Matti, this volume). Morton and others (1986) suggested that the San Bernardino Mountains source for the clasts is not much farther away from the San Timoteo depocenter today than when the sediments accumulated 1.2 m.y. ago, which implies that late Quaternary strands of the San Andreas fault (San Bernardino strand, Mill Creek strand, and youngest Mission Creek strand) that intervene between depocenter and source have not generated significant right slip in the last 1.2 m.y. (Morton and others, 1986; Morton and Matti, this volume; Matti and others, 1985, reached the same conclusion using different evidence). This leads us to conclude that the San Jacinto fault for much of its Quaternary history was the primary strand of the San Andreas system in this part of southern California.

Continued paleogeographic fragmentation in response to interaction between the San Andreas and San Jacinto faults (late Pleistocene [0.5 Ma], Fig. 7J)

Fault activity. The Mill Creek strand of the San Andreas fault has evolved inboard (east) of the abandoned Mission Creek strand, and will generate about 8 km of right slip between 0.5 Ma and 120 ka (Matti and others, 1985). The San Jacinto fault has generated about half its total 25-km displacement since 1.2 Ma, and will generate most of the balance between 0.5 Ma and 120 ka. Geometric and kinematic relations between the San Andreas and San Jacinto faults continue to be complex during this period. At the north end of the Salton Trough, the left step between the San Andreas and San Jacinto faults persists, but the amount of slip transferred from the San Andreas to the San Jacinto diminishes with time as right slip on the Mill Creek strand partly resolves the San Gorgonio Pass knot. At the northwest end of the San Bernardino Valley, right slip on the San Jacinto fault continues to be resolved into two components: a large component that steps right onto the San Andreas fault, thereby creating an extensional-strain field within the greater San Bernardino Valley region (Matti and others, 1985; Morton and Matti, this volume), and a small component of slip that is transformed within the southeasternmost San Gabriel Mountains by a combination of convergence along the Cucamonga fault zone (Morton and Matti, 1987) and left slip along east- and northeast-oriented faults of the southeastern San Gabriel Mountains (Morton, 1975a). Uplift along the north frontal fault zone of the San Bernardino Mountains largely has ceased during the time frame from 0.5 Ma to 120 ka (Meisling, 1984).

Basement patterns. The Eastern Peninsular Ranges mylonite zone has been displaced by the San Jacinto fault zone in the Borrego Valley area (Sharp, 1979).

Depositional patterns. Most of the vanished fan complex has been stripped off the northern part of the Peninsular Ranges

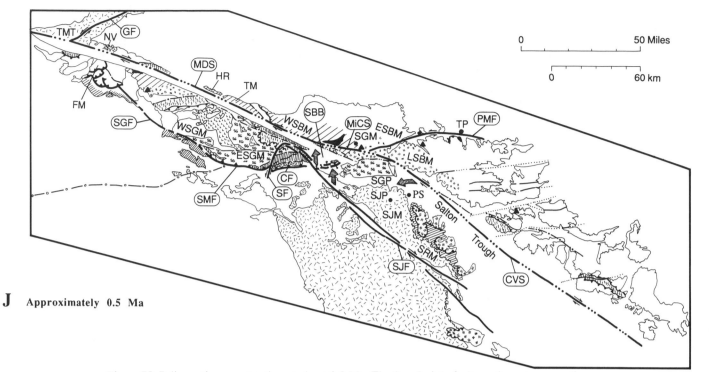

J   Approximately 0.5 Ma

Figure 7J. Palinspastic reconstruction at about 0.5 Ma. The San Jacinto fault continues to carry slip around the San Gorgonio Pass knot in the San Andreas zone, but the Mill Creek strand of the San Andreas (MiCS) has partly breached the San Gorgonio Pass slip barrier. Slip continues to step left between the San Andreas and San Jacinto faults in the San Gorgonio Pass region and right from the San Jacinto to the San Andreas in the San Bernardino Valley region (ruled arrows); the latter has led to extensional subsidence of the San Bernardino basin (SBB; Morton and Matti, this volume). CF = Cucamonga fault; CP = Cajon Pass region; CVS = Coachella Valley segment, San Andreas fault; ESBM = eastern San Bernardino Mountains; ESGM = eastern San Gabriel Mountains; FM = Frazier Mountain region; GF = Garlock fault; LSBM = Little San Bernardino Mountains; MDS = Mojave Desert segment, San Andreas fault; NV = Neenach Volcanic field; PMF = Pinto Mountain fault; PS = Palm Springs; SAF = San Andreas fault; SF = San Antonio fault; SGF = San Gabriel fault in Ridge Basin region; SGM = San Gorgonio Mountain; SGP = San Gorgonio Pass; SJF = San Jacinto fault; SJM = San Jacinto Mountains; SJP = San Jacinto Peak; SMF = Sierra Madre fault; SRM = Santa Rosa Mountains; TM = Table Mountain; WSBM = western San Bernardino Mountains; WSGM = western San Gabriel Mountains.

largely through downcutting by the Santa Ana River. Several climatically controlled cycles of late Quaternary alluvium fill the San Bernardino basin.

Modern paleogeographic setting (Holocene and latest Pleistocene [120 ka to present], Fig. 7K)

Fault activity. After generating about 8 km of right slip, the Mill Creek strand of the San Andreas fault has been abandoned by the San Andreas system, to be succeeded by the San Bernardino strand and the Coachella Valley segment of the Banning fault. These two strands are aligned with each other, but are separated by the late Quaternary San Gorgonio Pass fault zone, across which a throughgoing late Quaternary right-lateral fault is difficult to trace (Allen, 1957; Matti and others, 1985; but see an alternative viewpoint by Rasmussen and Reeder, 1986). The San Bernardino strand has evolved in the last 125,000 yr or so by reactivation of the old Mission Creek trace (Matti and others, 1985). As the San Bernardino strand has propagated southeastward toward the Coachella Valley segment of the Banning fault, its slip rate has increased through time, culminating in Holocene rates of 25 mm/yr (Harden and Matti, 1989) that are compatible with rates widely cited for the San Andreas fault southeast of Cajon Pass. Together, the San Bernardino strand and the Coachella Valley segment of the Banning fault represent modern neotectonic traces of the San Andreas fault zone in southern California. Both faults appear to have generated about 3 km of right slip in the last 125,000 yr (Matti and others, 1985), but this slip has accumulated in a complex way. Geometric and kinematic

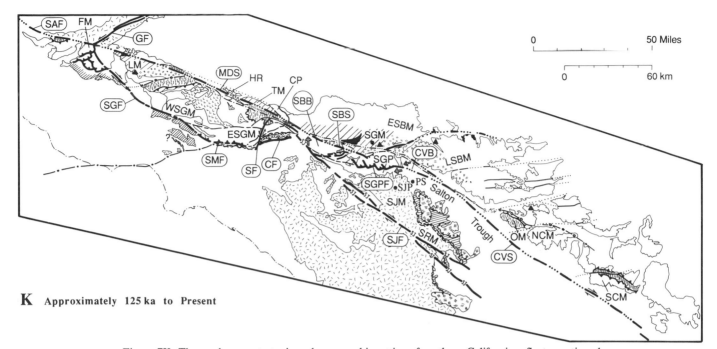

K Approximately 125 ka to Present

Figure 7K. The modern neotectonic and geomorphic setting of southern California reflects continued complexities within the San Andreas fault system. In a continuing attempt to breach the San Gorgonio Pass slip barrier, right slip on the San Andreas fault has reactivated the Banning fault in the Salton Trough to produce the neotectonic Coachella Valley segment of the Banning (CVB). Slip steps left from the Coachella Valley segment of the San Andreas (CVS) to the Banning fault (ruled arrows) and thence into San Gorgonio Pass, where convergence within the San Gorgonio Pass fault zone (SGPF) prevents a throughgoing connection between the Coachella Valley segment of the Banning fault and the newly evolved San Bernardino strand of the San Andreas (SBS). The San Bernardino basin (SBB) continues to subside as the right step (ruled arrow) between the San Jacinto fault and the San Bernardino strand creates extension within the San Bernardino Valley region. Complications created by integration of slip on the San Jacinto and San Andreas faults leads to convergence and uplift in the southeastern San Gabriel Mountains. CF = Cucamonga fault; CP = Cajon Pass region; CVS = Coachella Valley segment, San Andreas fault; ESBM = eastern San Bernardino Mountains; ESGM = eastern San Gabriel Mountains; FM = Frazier Mountain region; GF = Garlock fault; LSBM = Little San Bernardino Mountains; MDS = Mojave Desert segment, San Andreas fault; PMF = Pinto Mountain fault; PS = Palm Springs; SAF = San Andreas fault; SF = San Antonio fault; SGF = San Gabriel fault in Ridge Basin region; SGM = San Gorgonio Mountain; SGP = San Gorgonio Pass; SJF = San Jacinto fault; SJM = San Jacinto Mountains; SJP = San Jacinto Peak; SMF = Sierra Madre fault; SRM = Santa Rosa Mountains; TM = Table Mountain; WSBM = western San Bernardino Mountains; WSGM = western San Gabriel Mountains.

relations among the Coachella Valley segments of the San Andreas and Banning faults, the San Jacinto fault, and the San Bernardino strand of the San Andreas fault have produced a complex left- and right-stepping geometry by which right-lateral strain in the Salton Trough is passed around, not through, the structural knot in San Gorgonio Pass (Matti and others, 1985, Fig. 4). This geometry has created convergence in the San Gorgonio Pass fault zone and extension in the San Bernardino Valley region.

DISCUSSION

Our paleogeographic model for the late Miocene through recent evolution of southern California (Fig. 7A-K) is based on our palinspastic model for how crystalline and sedimentary ter-

ranes now dispersed throughout the San Andreas fault system (Plate IIIA) can be restored to their original positions (Plate IIIB). The reconstruction in Plate IIIB is based on two premises: (1) Bodies of Triassic megaporphyritic monzogranite that crop out on opposite sides of the San Andreas fault in the Liebre Mountain and Mill Creek regions are displaced counterparts that constrain right slip on the San Andreas fault (sensu stricto) to no more than 160 km; and (2) the two megaporphyry bodies must be reassembled using sequential fault restorations that conform to the timing and amount of displacement documented for individual strands of the San Andreas and San Gabriel faults. The first premise was evaluated by Frizzell and others (1986), who concluded that isotopic, petrologic, and geochronologic evidence directly support cross-fault correlation of the two megaporphyry bodies; in addition, Meisling and Weldon (1986, 1989) and Alexander and Weldon (1987) showed that similarities in structural setting between the Liebre Mountain megaporphyry outcrops and the western San Bernardino Mountains indirectly support cross-fault correlation between the two regions. The second premise is validated by the reconstruction itself, which incorporates or refines what is known about the local history of right-slip faults in the Ridge and Soledad basin regions (Crowell and Link, 1982), the San Gabriel Mountains (Barrows and others, 1987), the southeastern San Bernardino Mountains (Farley, 1979; Matti and others, 1985), and the Salton Trough (Dillon, 1975).

The reconstruction in Plate IIIB differs in important ways from the traditional reconstruction for southern California (Fig. 5), but it provides paleogeographic settings that are equally attractive or more so. For example, the palinspastic position of Ridge Basin adjacent to the San Bernardino Mountains in Plate IIIB allows the basin to evolve outboard (south and southwest) of a terrain that was undergoing orogenesis and uplift in late Miocene time; in Figure 5, similar linkage between the subsiding Ridge Basin and tectonic uplift to the north and northeast has not been documented. Similarly, palinspastic position of the Hungry Valley Formation adjacent to the western San Bernardino Mountains allows its early Pliocene paleogeographic evolution to occur in an orogenic setting not documented for its implied position in Figure 5. For basement rocks, Plate IIIB provides a coherent paleogeographic framework for far-flung, broadly similar crystalline slices like the Liebre Mountain block and the Holcomb Ridge and Table Mountain slices that are reunited against similar rocks in the San Bernardino Mountains, while simultaneously allowing crystalline rocks of the San Gorgonio Pass region to be restored against similar rocks in the southern Chocolate Mountains. The reconstruction in Plate IIIB also establishes a paleogeographic pattern for polka-dot-bearing granitoid rocks that differs from the pattern established by the traditional reconstruction (Fig. 5), but provides an equally attractive distribution of polka-dot granitoids and sedimentary basins containing clasts derived from such granitoids. In short, except for rocks of Soledad basin and the San Gabriel Mountains (the terranes originally cited by Crowell, 1962, as evidence for large right slip on the San Andreas fault), the palinspastic reconstruction in Plate IIIB does not violate anything that is known about the provincial affinities of crystalline rocks or the assumed source areas for clasts in Tertiary sedimentary basins—particularly if one accepts the premise that many crystalline terranes in southern California are so generalized and widespread that they do not constrain the exact position of reconstructed crystalline slices or the exact source for clasts in sedimentary basins.

The reconstruction in Plate IIIB has one major problem: it leaves sedimentary and crystalline rocks of the Soledad–San Gabriel region far short of their probable cross-fault counterparts in the Orocopia Mountains region, and it requires that the San Andreas and San Gabriel faults together account for only 205 km of displacement—about 110 km short of 315 ± 10-km post-early Miocene displacement proposed for the San Andreas fault

in central California. This problem must be resolved if displacements on the San Andreas fault (sensu lato) in central and southern California are about the same, but if the Liebre Mountain–Mill Creek Triassic megaporphyry is a valid piercing point for only 160 km of right slip on the San Andreas fault (sensu stricto) in southern California.

One solution to this problem was conceived by Weldon (1986; Meisling and Weldon, 1986; Alexander and Weldon, 1987; Meisling and Weldon, 1989) who accepted cross-fault correlation of the Liebre Mountain and Mill Creek megaporphyry bodies but proposed that their 160-km separation represents their displacement by the San Andreas fault (sensu stricto) only since the opening of the Gulf of California and the advent of the modern San Andreas. According to this model, the Triassic megaporphyry in the Liebre Mountain region is part of a thin crystalline sheet that was thrust across the trace of the San Andreas at the latitude of the San Bernardino Mountains and subsequently displaced 160 km to the northwest by Pliocene and Quaternary right slip on the San Andreas (sensu stricto; Fig. 7E, F). The Liebre Mountain thrust sheet was emplaced between about 9 and 5 m.y. ago, probably toward the end of right slip on the San Gabriel fault (Meisling and Weldon, 1986, 1989). Prior to these events, an early phase of right slip on the San Andreas (sensu lato) would have generated the 80-to 100-km displacements required to achieve the cross-fault matchups between the Orocopia Mountains and the Soledad– San Gabriel Mountains region and to accommodate 315 ± 10 km of displacement in both central and southern California. The thrust-fault model is appealing because it accommodates the Triassic megaporphyry as a piercing point for the young San Andreas fault while still allowing the San Andreas (sensu lato) to have large displacements comparable to those predicted by the traditional model (Fig. 5). However, the thrust-fault model requires that the movement history of the fault in southern California be extended much farther back into the Miocene than required by reconstructions that tie all 240 km of displacement on the San Andreas to opening of the Gulf of California 4 or 5 m.y. ago (Crowell, 1981, 1982a; Ehlig, 1981). Moreover, if the middle Miocene San Andreas occupied essentially the same trace as the Pliocene and Quaternary San Andreas, then displacements on the order of 240 km for the fault violate the cross-fault correlations proposed by Dillon (1975) and obviate the need for Powell's Clemens Well-Fenner-San Francisquito fault to have participated in the San Andreas system.

If the Liebre Mountain basement terrane ultimately is shown to be rock of San Bernardino Mountains type, despite local thrust faulting that may have redistributed the rocks, then the 160-km separation between the Liebre Mountain and Mill Creek megaporphyry bodies represents total displacement on the San Andreas fault in southern California, and some other right-lateral fault is required to make up the large difference between displacements on the fault in central and southern California. Powell's (1981) Clemens Well-Fenner-San Francisquito fault neatly resolves this problem by providing an additional 110 km of right slip during the Miocene. Restoration of this displacement on the Clemens Well system returns Precambrian and Oligocene and Miocene rocks of the Frazier Mountain–Soledad basin–San Gabriel Mountains region to the Orocopia Mountains region as originally recognized by Crowell (1962), but aligns these terranes in a different configuration than originally envisioned by Crowell (Powell, 1981, this volume). Powell's model for the Clemens Well–Fenner–San Francisquito fault has its own problems, however: restoration of middle Miocene right slip on this structure may well provide a better palinspastic alignment of basement terranes in southern California, but southeastward continuation of the Clemens Well fault and its geometric and kinematic relations (if any) to middle Miocene extension, rifting, and transform faulting in the Gulf of California have not been documented (see Powell, this volume; note that the reconstruction by Joseph and others, 1982b, Figs. 6-7, has a similar problem with the southeastward continuation of their early Miocene Clemens Well fault).

Although Plate IIIB does not reconstruct displacements on the Clemens Well–Fenner–San Francisquito fault, restoration of about 110 km on this fault as shown in Figure 7A would bring into southern California far-traveled basement terranes that are restored there by other workers. Granitoid rocks of the Gabilan Range are positioned against comparable rocks in the Tehachapi Mountains as proposed by Ross (1984, p. 8-16); this reconstruction also positions the Pinnacles Volcanic Formation against the Neenach Volcanic Formation as proposed by Matthews (1973, 1976). Granitoid rocks of the La Panza Range also are brought into southern California to a position near the restored Liebre Mountain block (Fig. 7A), although this position is not compatible with that proposed by Smith (1977) and Joseph and others (1982b), who restored the La Panza range by using 175 to 120 km of right slip on the Clemens Well–Fenner–San Francisquito–Chimeneas–San Juan–Red Hill fault proposed by Smith (1977). We favor cross-fault correlation between the La Panza Range granitoids and those in the southern Little San Bernardino Mountains; our restoration of the La Panza Range block adjacent to the Liebre Mountain block achieves this match, but requires that La Panza Range–type basement had an elongate geometry that would not require direct juxtaposition of the block against the southern Little San Bernardino Mountains as required by Smith (1977). Irrespective of the exact reconstruction, Smith (1977) and Joseph and others (1982b) favor an Oligocene or early Miocene time frame for offset of the La Panza Range, whereas Powell (1981) and we would have this displacement occur after the early Miocene as part of the San Andreas fault system. Thus, post–early Miocene displacements on the San Andreas fault (sensu lato) in southern California would total about 315 km, and would have been achieved by about 110 km of displacement on the Clemens Well–San Francisquito–San Andreas fault followed by sequential displacements of 44 km on the San Gabriel fault and 160 km on the San Andreas fault (sensu stricto). Thus, total right slip on the San Andreas fault in southern California would be fully comparable to post–early Miocene right slip documented for the fault in central California.

We presently cannot resolve whether the history of the San Andreas fault (sensu lato) in southern California involved a *long-lived San Andreas* that had a two-phase middle Miocene through Quaternary history punctuated by late Miocene right slip on the San Gabriel fault, or a *short-lived San Andreas* whose Pliocene and Quaternary history was preceded by late Miocene right slip on the San Gabriel fault and middle Miocene right slip on the Clemens Well–Fenner–San Francisquito fault. We adopt the latter model for our palinspastic reconstruction and paleogeographic interpretations (Plate IIIB, Fig. 7). However, the late Miocene through Quaternary paleogeographic evolution of southern California would be the same for either San Andreas scenario because, in our view, the cross-fault tie between Triassic megaporphyritic rocks at Liebre Mountain and in the San Bernardino Mountains leads to a late Miocene palinspastic reconstruction (Plate IIIB) that firmly constrains tectonic and depositional events between 10 Ma and today, irrespective of the middle Miocene (pre–San Gabriel fault) history of the San Andreas transform-fault system in southern California.

CONCLUSIONS

Distinctive bodies of alkalic porphyritic monzogranite and quartz monzonite of Triassic age that occur in the Mill Creek region of the San Bernardino Mountains and on the opposite side of the San Andreas fault at the northwest end of Liebre Mountain appear to be segments of a formerly continuous pluton that has been severed by the fault and displaced about 160 km. Reassembly of the megaporphyritic bodies by restoration of sequential right-lateral displacements on various strands of the San Andreas and San Gabriel fault zones leads to a palinspastic reconstruction for southern California that reassembles crystalline and sedimentary terranes differently from widely cited reconstructions.

Reassembled crystalline rocks establish three coherent patterns. (1) The reunited Liebre Mountain and Mill Creek Triassic megaporphyry bodies form western outliers of a province of Permian and Triassic alkalic granitoid rocks that occurs in the western Mojave Desert and San Bernardino Mountains. (2) Following sequential restorations of 160 km on the San Andreas fault and 44 km on the San Gabriel and Cajon Valley faults, the Table Mountain and Holcomb Ridge basement slices currently positioned along the west edge of the Mojave Desert near Wrightwood and Valyermo are reassembled with the Liebre Mountain and San Bernardino Mountains blocks. This restoration unites terranes of Mesozoic granitoid rock and Paleozoic(?) marble, metaquartzite, and pelitic gneiss that have strong lithologic and compositional similarities, and brings together within a single province quartz diorite and granodiorite that are hosts for three known bedrock occurrences of aluminous dike rocks (polka-dot granite) that previously have been used to assemble a different reconstruction for the San Andreas fault. (3) Crystalline rocks of the San Gorgonio Pass region, including Triassic monzogranite and granodiorite of the Lowe igneous pluton and Jurassic blastoporphyritic quartz monzonite, are juxtaposed adjacent to the southern Chocolate Mountains where similar rocks have been mapped.

The reconstructed crystalline rocks provide a paleogeographic framework for Pliocene and late Miocene sedimentary basins now scattered along the San Gabriel and San Andreas faults. Ridge Basin is juxtaposed adjacent to the southwestern San Bernardino Mountains in an orogenic setting compatible with stratigraphic relations, depositional fabrics, and clast compositions in the Ridge Basin fill. Synorogenic sediments of the upper Miocene Ridge Route and Peace Valley Formations accumulated in this setting at a time (9 to 5 Ma) when the western San Bernardino Mountains were undergoing uplift and erosion. The Pliocene Hungry Valley Formation was deposited toward the end of this orogenic pulse (5 to 4 Ma) when early movements on the San Andreas fault began to displace Ridge Basin away from the San Bernardino Mountains. Other late Miocene basins widely dispersed today are reassembled within an early Salton Trough that included, from northwest to southeast, the Punchbowl and Mill Creek basins, the Coachella Fanglomerate and Hathaway Formation, and the fanglomerate of Bear Canyon in the southern Chocolate Mountains.

The new reconstruction allows only about 205 km of displacement on the combined San Andreas and San Gabriel fault zones, which is about 110 km short of the 315 ± 10 km post–early Miocene offsets documented for the San Andreas in central California. This shortfall requires the existence of another Miocene strand of the San Andreas fault system in southern California. The Clemens Well–Fenner–San Francisquito fault system of Powell (this volume) provides this additional fault strand. Together, the Clemens Well, San Gabriel, and San Andreas faults generated about 315 ± 10 km of right slip comparable to displacements documented for the San Andreas fault in central California. The displacements in southern California have occurred in middle Miocene through Holocene time, and extend the history of late Cenozoic right-lateral faulting farther back into the Miocene than envisioned by traditional reconstructions for the San Andreas fault system in southern California.

ACKNOWLEDGMENTS

Early versions of this chapter were reviewed by V. A. Frizzell, P. L. Ehlig, T. H. McCulloh, and R. E. Powell. Thorough and constructive critiques were provided by Geological Society of America reviewers D. C. Ross and M. O. Woodburne. We have benefitted from stimulating discussions with A. G. Barrows, J. C. Crowell, R. E. Powell, P. R. Sadler, M. O. Woodburne, and R. E. Weldon regarding prospects and problems of reassembling bits and pieces of southern California. Pattie Lett entered the text into a word processor and cheerfully labored through endless revisions of the manuscript; thank you for your patience, Pattie. Geologic mapping and topical studies leading to the results of this work were sponsored by the U.S. Geological Survey's Mineral Resources of Public Lands Program, National Earthquake Hazards Reduction Program, and National Geologic Mapping Program.

REFERENCES CITED

Alexander, J., and Weldon, R., 1987, Comparison of structural style of the Liebre Mountain and Squaw Peak thrust systems, offset across the San Andreas fault, southern California: Geological Society of America Abstracts with Programs, v. 19, p. 353.

Alf, R. M., 1948, A mylonite belt in the southeastern San Gabriel Mountains, California: Geological Society of America Bulletin, v. 69, p. 1101–1120.

Allen, C. R., 1957, San Andreas fault zone in San Gorgonio Pass, southern California: Geological Society of America Bulletin, v. 68, p. 319–350.

Atwater, T., 1970, Implications of plate tectonics for the Cenozoic tectonic evolution of western North America: Geological Society of America Bulletin, v. 81, p. 3513–3535.

Baird, A. K., Morton, D. M., Baird, K. W., and Woodford, A. O.,1974, Transverse Ranges province; A unique structural-petrochemical belt across the San Andreas fault system: Geological Society of America Bulletin, v. 85, p. 163–174.

Baird, A. K., Baird, K. W., and Welday, E. E., 1979, Batholithic rocks of the northern Peninsular and Transverse Ranges, southern California, in Abbott, P. L., and Todd, V. R., eds., Mesozoic crystalline rocks; Peninsular Ranges batholith and pegmatites, Point Sal Ophiolite: San Diego, California, San Diego State University, p. 111–132.

Barbat, W. F., 1958, The Los Angeles Basin area, California, in Weeks, L. G., ed., Habitat of oil; A symposium: Tulsa, Oklahoma, American Association of Petroleum Geologists, p. 62–77.

Barrows, A. G., 1975, The San Andreas fault zone in the Juniper Hills Quadrangle, southern California, in Crowell, J. C., ed., San Andreas fault in southern California: California Division of Mines and Geology Special Report 118, p. 197–202.

—— , 1987, Geology of the San Andreas fault zone and adjoining terrane, Juniper Hills and vicinity, Los Angeles County, California, in Hester, R. L., and Hallinger, D. E., eds., San Andreas fault—Cajon Pass to Palmdale: Pacific Section, American Association of Petroleum Geologists Volume and Guidebook 59, p. 93–157.

Barrows, A. G., Kahle, J. E., and Beeby, D. J., 1985, Earthquake hazards and tectonic history of the San Andreas fault zone, Los Angeles County, California: California Division of Mines and Geology Open File Report 85-10LA, 139 p., scale 1:12,000.

Barrows, A. G., Kahle, J. E., and Beeby, D. J., 1987, Earthquake hazards and tectonic history of the San Andreas fault zone, Los Angeles County, California, in Hester, R. L., and Hallinger, D. E., eds., San Andreas fault—Cajon Pass to Palmdale: Pacific Section, American Association of Petroleum Geologists Volume and Guidebook 59, p. 1–92.

Berggren, W. A., Kent, D. V., Flynn, J. J., and Van Couvering, J. A., 1985, Cenozoic geochronology: Geological Society of America Bulletin, v. 96, p. 1407–1418.

Bilham, R., and Williams, P., 1985, Sawtooth segmentation and deformation processes on the southern San Andreas fault, California: Geophysical Research Letters, v. 12, no. 9, p. 557–560.

Bohannon, R. G., 1975, Mid-Tertiary conglomerates and their bearing on Transverse Range tectonics, southern California, in Crowell, J. C., ed., San Andreas fault in southern California: California Division of Mines and Geology Special Report 118, p. 75–82.

Brand, J. H., and Anderson, J. L., 1982, Mesozoic alkalic monzonites and peraluminous adamellites of the Joshua Tree National Monument, southern California: Geological Society of America Abstracts with Programs, v. 14, p. 151–152.

Buesch, D. C., and Ehlig, P. L., 1982, Structural and lower Miocene volcanic rock correlation between Soledad Pass and Salton Wash along the San Andreas fault: Geological Society of America Abstracts with Programs, v. 14, p. 153.

Cameron, C. S., 1981, Geology of the Sugarloaf and Delamar Mountain areas, San Bernardino Mountains, California [Ph.D. thesis]: Cambridge, Massachusetts Institute of Technology, 339 p.

Campbell, R. H., and Yerkes, R. F., 1976, Cenozoic evolution of the Los Angeles basin area; Relation to plate tectonics, *in* Howell, D. G., ed., Aspects of the geologic history of the California Continental borderland: Pacific Section, American Association of Petroleum Geologists Miscellaneous Publication 24, p. 541–558.

Carman, M. F., Jr., 1964, Geology of the Lockwood Valley area, Kern and Ventura Counties, California: California Division of Mines Special Report 81, 62 p.

Carter, B. A., 1982, Geology and structural setting of the San Gabriel anorthosite-syenite body and adjacent rocks of the western San Gabriel Mountains, Los Angeles County, California, Field trip 5, *in* Cooper, J. D., compiler, Geologic excursions in the Transverse Ranges, southern California; Geological Society of America Cordilleran Section 78th Annual Meeting, Anaheim, California, Guidebook: Geological Society of America, p. 1–53.

Carter, J. N., Luyendyk, B. P., and Terres, R. R., 1987, Neogene clockwise tectonic rotation of the eastern Transverse Ranges, California, suggested by paleomagnetic vectors: Geological Society of America Bulletin, v. 98, p. 199–206.

Clark, M. M., 1984, Map showing recently active breaks along the San Andreas fault and associated faults between Salton Sea and Whitewater River–Mission Creek, California: U.S. Geological Survey Miscellaneous Investigations Map I-1483, scale 1:24,000.

Cramer, C. H., and Harrington, J. M., 1987, Seismicity and tectonics of the Cucamonga fault and the eastern San Gabriel Mountains, San Bernardino County, California, *in* Morton, D. M., and Yerkes, R. F., eds., Recent reverse faulting in the Transverse Ranges, California: U.S. Geological Survey Professional Paper 1339, p. 7–26.

Crowe, B. M., Crowell, J. C., and Krummenacher, D., 1979, Regional stratigraphy, K-Ar ages, and tectonic implications of Cenozoic volcanic rocks, southeastern California: American Journal of Science, v. 279, p. 186–216.

Crowell, J. C., 1950, Geology of Hungry Valley area, southern California: American Association of Petroleum Geologists Bulletin, v. 34, p. 1623–1646.

——, 1952, Probable large lateral displacement on San Gabriel fault, southern California: American Association of Petroleum Geologists Bulletin, v. 36, no. 10, p. 2026–2035.

——, 1954a, Strike-slip displacement of the San Gabriel fault, southern California, *in* Jahns, R. H., ed., Geology of southern California: California Division of Mines Bulletin 170, p. 49–52.

——, 1954b, Geologic map of the Ridge Basin area, California: California Division of Mines Bulletin 170, Map Sheet 7.

——, 1960, The San Andreas fault in southern California: Report of the 21st International Geological Congress, Copenhagen, part 18, p. 45–52.

——, 1962, Displacement along the San Andreas fault, California: Geological Society of America Special Paper 71, 61 p.

——, 1975a, The San Andreas fault in southern California, *in* Crowell, J. C., ed., San Andreas fault in southern California: California Division of Mines and Geology Special Report 118, p. 7–27.

——, 1975b, Geologic sketch of the Orocopia Mountains, southeastern California *in* Crowell, J. C., ed., San Andreas fault in southern California: California Division of Mines and Geology Special Report 118, p. 99–110.

——, 1975c, The San Gabriel fault and Ridge Basin, southern California, *in* Crowell, J. C., ed., San Andreas fault in southern California: California Division of Mines and Geology Special Report 118, p. 208–219.

——, 1979, The San Andreas fault system through time: Geological Society of London Journal, v. 136, p. 293–302.

——, 1981, An outline of the tectonic history of southeastern California, *in* Ernst, W. G., ed., The geotectonic development of California; Rubey Volume 1: Englewood Cliffs, New Jersey, Prentice-Hall, p. 583–600.

——, 1982a, The tectonics of Ridge Basin, southern California, *in* Crowell, J. C., and Link, M. H., eds., Geologic history of Ridge Basin, southern California: Pacific Section, Society of Economic Paleontologists and Mineralogists Field Trip Guide and Volume, p. 25–42.

——, 1982b, Pliocene Hungry Valley Formation, Ridge Basin, southern California, *in* Crowell, J. C., and Link, M. H., eds., Geologic history of Ridge Basin, southern California: Pacific Section, Society of Economic Paleontologists and Mineralogists Field Trip Guide and Volume, p. 143–150.

——, 1982c, Geologic map of Ridge Basin, southern California, *in* Crowell, J. C., and Link, M. H., eds., Geologic history of Ridge Basin, southern California: Pacific Section, Society of Economic Paleontologists and Mineralogists Field Trip Guide and Volume.

Crowell, J. C., and Link, M. H., 1982, Geologic history of Ridge Basin, southern California: Pacific Section, Society of Economic Paleontologists and Mineralogists Field Trip Guide and volume, 304 p.

Crowell, J. C., and Ramirez, V. R., 1979, Late Cenozoic faults in southeastern California, *in* Crowell, J. C., and Sylvester, A. G., eds., Tectonics of the juncture between the San Andreas fault system and the Salton trough, southeastern California; A guidebook for fieldtrips, Geological Society of America Annual Meeting, San Diego, California: Santa Barbara, University of California, p. 27–39.

Crowell, J. C., and Suzuki, T., 1959, Eocene stratigraphy and paleontology, Orocopia Mountains, southeastern California: Geological Society of America Bulletin, v. 70, p. 581–592.

Crowell, J. C., and Walker, J.W.R., 1962, Anorthosite and related rocks along the San Andreas fault, southern California: Berkeley, University of California Publications in Geological Sciences, v. 40, p. 219–288.

Curray, J. R., and Moore, D. G., 1984, Geologic history of the mouth of the Gulf of California, *in* Crouch, J. K., and Bachman, S. B., eds., Tectonics and sedimentation along the California margin: Pacific Section, Society of Economic Paleontologists and Mineralogists, v. 38, p. 17–36.

Demirer, A., 1985, The Mill Creek Formation; A strike-slip basin filling in the San Andreas fault zone, San Bernardino County, California [M.S. thesis]: Riverside, University of California, 108 p.

Dibblee, T. W., Jr., 1954, Geology of the Imperial Valley region, California, *in* Jahns, R. H., ed., Geology of southern California: California Division of Mines Bulletin 170, p. 21–28, and plate 2, Chapter II.

——, 1964, Geologic map of the San Gorgonio Mountain Quadrangle, San Bernardino and Riverside Counties, California: U.S. Geological Survey Miscellenous Geologic Investigations Map I-431, scale 1:62,500.

——, 1967a, Areal geology of the western Mojave Desert, California: U.S. Geological Survey Professional Paper 522, 153 p.

——, 1967b, Geologic map of the Morongo Valley Quadrangle, San Bernardino County, California: U.S. Geological Survey Miscellaneous Geologic Investigations Map I-517, scale 1:62,500.

——, 1968a, Displacements on San Andreas fault system in San Gabriel, San Bernardino, and San Jacinto Mountains, southern California, *in* Dickinson, W. R., and Grantz, A., eds., Proceedings of Conference on Geologic Problems of San Andreas Fault System: Stanford, California, Stanford University Publications in Geological Sciences, v. 11, p. 260–278.

——, 1968b, Evidence of major lateral displacement on the Pinto Mountain fault, southern California, *in* Abstracts for 1967: Geological Society of America Special Paper 115, p. 322.

——, 1975a, Late Quaternary uplift of the San Bernardino Mountains on the San Andreas and related faults, *in* Crowell, J. C., ed., San Andreas fault in southern California: California Division of Mines and Geology Special Report 118, p. 127–135.

——, 1975b, Tectonics of the western Mojave Desert near the San Andreas rift, *in* Crowell, J. C., ed., San Andreas fault in southern California: California Division of Mines and Geology Special Report 118, p. 155–161.

——, 1982a, Geology of the San Gabriel Mountains, southern California, *in* Fife, D. L., and Minch, J. A., eds., Geology and mineral wealth of the California Transverse Ranges; Mason Hill Volume: Santa Ana, California, South Coast Geological Society Annual Meeting Symposium and Guidebook 10, p. 131–147.

——, 1982b, Geology of the San Bernardino Mountains, southern California, *in* Fife, D. L., and Minch, J. A., eds., Geology and mineral wealth of the California Transverse Ranges; Mason Hill Volume: Santa Ana, California, South Coast Geological Society Annual Meeting Symposium and Guidebook 10, p. 148–169.

Dickinson, W. R., and Snyder, W. S., 1979a, Geometry of triple junctions related to the San Andreas transform: Journal of Geophysical Research, v. 84, no. B2, p. 561–572.

——, 1979b, Geometry of subducted slabs related to San Andreas transform: Journal of Geology, v. 87, p. 609–627.

Dillon, J. T., 1975, Geology of the Chocolate and Cargo Muchacho Mountains, southeasternmost California [Ph.D. thesis]: Santa Barbara, University of California, 405 p.

Ehlert, K. W., 1982a, "Polka-dot" granite clasts in the Ridge Route Formation, Ridge Basin, southern California, *in* Crowell, J. C., and Link, M. H., eds., Geologic history of Ridge Basin, southern California: Pacific Section, Society of Economic Paleontologists and Mineralogists, Field Trip Guide and Volume, p. 203–204.

——, 1982b, Basin analysis of the Miocene Mint Canyon Formation, southern California, *in* Ingersoll, R. V., and Woodburne, M. O., eds., Cenozoic nonmarine deposits of California and Arizona: Pacific Section, Society of Economic Paleontologists and Mineralogists, p. 51–64.

Ehlert, K. W., and Ehlig, P. L., 1977, The "Polka-dot" granite and the rate of displacement of the San Andreas fault in southern California: Geological Society of America Abstracts with Programs, v. 9, p. 415–416.

Ehlig, P. L., 1968a, Displacement along the San Gabriel fault, San Gabriel Mountains, southern California, *in* Abstracts for 1967: Geological Society of America Special Paper 115, p. 55.

——, 1968b, Causes of distribution of Pelona, Rand, and Orocopia Schist along the San Andreas and Garlock faults, *in* Dickinson, W. R., and Grantz, A., eds., Proceedings of Conference on Geologic Problems of San Andreas Fault System: Stanford, California, Stanford University Publications in Geological Sciences, v. 11, p. 294–305.

——, 1973, History, seismicity, and engineering geology of the San Gabriel fault, *in* Moran, D. E., Slosson, J. E., Stone, R. O., and Yelverton, C. A., eds., Geology, seismicity, and environmental impact: Association of Engineering Geologists Special Publication, p. 247–251.

——, 1975, Basement rocks of the San Gabriel Mountains south of the San Andreas fault, southern California, *in* Crowell, J. C., ed., San Andreas fault in southern California: California Division of Mines and Geology Special Report 118, p. 177–186.

——, 1976, Magnitude and timing of displacement on San Andreas fault in southern California and its palinspastic implications: American Association of Petroleum Geologists Bulletin, v. 60, p. 668.

——, 1977, Structure of the San Andreas fault zone in San Gorgonio Pass, southern California: Geological Society of America Abstracts with Programs, v. 9, p. 416.

——, 1981, Origin and tectonic history of the basement terrane of the San Gabriel Mountains, central Transverse Ranges, *in* Ernst, W. G., ed., The geotectonic development of California; Rubey Volume 1: Englewood Cliffs, New Jersey, Prentice-Hall, p. 253–283.

——, 1982, The Vincent thrust; Its nature, paleogeographic reconstruction across the San Andreas fault, and bearing on the evolution of the Transverse Ranges, *in* Fife, D. L., and Minch, J. A., eds., Geology and mineral wealth of the California Transverse Ranges; Mason Hill Volume: Santa Ana, California, South Coast Geological Society Guidebook 10, p. 370–379.

Ehlig, P. L., and Crowell, J. C., 1982, Mendenhall Gneiss and anorthosite-related rocks bordering Ridge Basin, southern California, *in* Crowell, J. C., and Link, M. H., eds., Geologic history of Ridge Basin, southern California: Pacific Section, Society of Economic Paleontologists and Mineralogists Field Trip Guide and Volume, p. 199–202.

Ehlig, P. L., and Ehlert, K. W., 1972, Offset of Miocene Mint Canyon Formation from volcanic source along San Andreas fault, southern California: Geological Society of America Abstracts with Programs, v. 4, p. 154.

Ehlig, P. L., and Joseph, S. E., 1977, Polka dot granite and correlation of La Panza quartz monzonite with Cretaceous batholithic rocks north of Salton Trough, *in* Howell, D. G., Vedder, J. C., and McDougal, K. A., eds., Cretaceous Geology of the California Coast Ranges West of the San Andreas Fault: Pacific Section, Society of Economic Paleontologists and Mineralogists Field Guide 2, p. 91–96.

Ehlig, P. L., Ehlert, K. W., and Crowe, B. M., 1975, Offset of the upper Miocene Caliente and Mint Canyon Formations along the San Gabriel and San Andreas faults, *in* Crowell, J. C., ed., San Andreas fault in southern California: California Division of Mines and Geology Special Report 118, p. 83–92.

Elders, W. A., Rex, R. W., Meidav, T., Robinson, P. T., and Biehler, S., 1972, Crustal spreading in southern California: Science, v. 178, p. 15–24.

Ensley, R. A., and Verosub, K. L., 1982, Biostratigraphy and magnetostratigraphy of the southern Ridge Basin, central Transverse Ranges, California, *in* Crowell, J. C., and Link, M. H., eds., Geologic history of Ridge Basin, southern California: Pacific Section, Society of Economic Paleontologists and Mineralogists, p. 13–24.

Erskine, B. E., 1985, Mylonitic deformation and associated low-angle faulting in the Santa Rosa Mylonite zone, southern California [Ph.D. thesis]: Berkeley, University of California, 247 p.

Evans, J. G., 1982, The Vincent thrust system, eastern San Gabriel Mountains, California: U.S. Geological Survey Bulletin 1507, 15 p.

Farley, T., 1979, Geology of a part of northern San Gorgonio Pass, California [M.S. thesis]: Los Angeles, California State University, 159 p.

Farley, T., and Ehlig, P. L., 1977, Displacement on the Punchbowl fault based on occurrence of "Polka-dot" granite clasts: Geological Society of America Abstracts with Programs, v. 9, p. 419.

Frick, C., 1921, Extinct vertebrate faunas of the Badlands of Bautista Creek and San Timoteo Canon, southern California: University of California Publications in Geology, v. 12, no. 5, p. 277–424.

Frizzell, V. A., Mattinson, J. M., and Matti, J. C., 1986, Distinctive Triassic megaporphyritic monzogranite; Evidence for only 160 km offset along San Andreas fault, southern California: Journal of Geophysical Research, v. 91, no. B14, p. 14080–14088.

Gibson, R. C., 1964, Geology of a portion of the Mill Creek area, San Bernardino County, California [M.S. thesis]: Riverside, University of California, 50 p.

——, 1971, Nonmarine turbidites and the San Andreas fault, San Bernardino Mountains, California, *in* Elders, W. A., ed., Geological excursions in southern California: Riverside, University of California Campus Museum Contributions, no. 1, p. 167–181.

Hamilton, W. B., 1961, Origin of the Gulf of California: Geological Society of America Bulletin, v. 72, p. 1307–1318.

Harden, J. W., and Matti, J. C., 1989, Holocene and late Pleistocene slip rates on the San Andreas fault in Yucaipa, California, using displaced alluvial-fan deposits and soil chronology: Geological Society of America Bulletin, v. 101, p. 1107–1117.

Hendrix, E. D., and Ingersoll, R. V., 1987, Tectonics and alluvial sedimentation of the upper Oligocene/lower Miocene Vasquez Formation, Soledad Basin, southern California: Geological Society of America Bulletin, v. 98, p. 647–663.

Hill, M. L., 1981, San Andreas fault; History of concepts: Geological Society of America Bulletin, v. 92, p. 112–131.

Hill, M. L., and Dibblee, T. W., Jr., 1953, San Andreas, Garlock, and Big Pine faults, California: Geological Society of America Bulletin, v. 64, p. 443–458.

Hornafius, J. S., Luyendyk, B. P., Terres, R. R., and Kamerling, M. J., 1986, Timing and extent of Neogene tectonic rotation in the western Transverse Ranges, California: Geological Society of America Bulletin, v. 97, p. 1476–1487.

Howell, D. G., 1975, Early and Middle Eocene shoreline offset by the San Andreas fault, southern California, *in* Crowell, J. C., ed., San Andreas fault in southern California: California Division of Mines and Geology Special Report 118, p. 69–74.

Hsu, K. J., 1955, Granulites and mylonites of the region about Cucamonga and San Antonio Canyons, San Gabriel Mountains, California: Berkeley, University of California Publications in Geological Sciences, v. 30, p. 223–324.

Huffman, O. F., 1972, Lateral displacement of Upper Miocene rocks and the Neogene history of offset along the San Andreas fault in central California: Geological Society of America Bulletin, v. 83, p. 2913–2946.

Jahns, R. H., 1973, Tectonic evolution of the Transverse Ranges province as related to the San Andreas fault system, *in* Kovach, R. L., and Nur, A., eds., Proceedings of the Conference on Tectonic Problems of the San Andreas

Fault System: Stanford, California, Stanford University Publications in Geological Sciences, v. 13, p. 149–170.

Jennings, C. W., 1977, Geologic map of California: California Division of Mines and Geology Geologic Data Map 2, scale 1:750,000.

Johnson, C. E., and Hill, D. P., 1982, Seismicity of the Imperial Valley, *in* The Imperial Valley, California, earthquake of October 15, 1979: U.S. Geological Survey Professional Paper 1254, p. 15–24.

Joseph, S. E., Criscione, J. J., Davis, T. E., and Ehlig, P. L., 1982a, The Lowe igneous pluton, *in* Fife, D. L., and Minch, J. A., eds., Geology and mineral wealth of the California Transverse Ranges; Mason Hill Volume: Santa Ana, California, South Coast Geological Society Guidebook 10, p. 307–309.

Joseph, S. E., Davis, T. E., and Ehlig, P. L., 1982b, Strontium isotopic correlation of the La Panza Range granitic rocks with similar rocks in the central and eastern Transverse Ranges, *in* Fife, D. L., and Minch, J. A., eds., Geology and mineral wealth of the California Transverse Ranges; Mason Hill Volume: Santa Ana, California, South Coast Geological Society Guidebook 10, p. 310–320.

Keller, E. A., Bonkowski, M. S., Korsch, R. J., and Schlemon, R. J., 1982, Tectonic geomorphology of the San Andreas fault zone in the southern Indio Hills, Coachella Valley, California: Geological Society of America Bulletin, v. 93, p. 46–56.

Larson, R. L., 1972, Bathymetry, magnetic anomalies, and plate tectonic history of the mouth of the Gulf of California: Geological Society of America Bulletin, v. 83, p. 3345–3360.

Larson, R. L., Menard, H. W., and Smith, S. M., 1968, Gulf of California; A result of ocean-floor spreading and transform faulting: Science, v. 161, p. 781–784.

Link, M. H., 1982a, Stratigraphic nomenclature and age of Miocene strata, Ridge Basin, southern California, *in* Crowell, J. C., and Link, M. H., eds., Geologic history of Ridge Basin, southern California: Pacific Section, Society of Economic Paleontologists and Mineralogists, p. 5–12.

—— , 1982b, Introduction to the facies of the Ridge Route Formation, Ridge Basin, southern California, *in* Crowell, J. C., and Link, M. H., eds., Geologic history of Ridge Basin, southern California: Pacific Section, Society of Economic Paleontologists and Mineralogists, p. 99–104.

—— , 1982c, Provenance, paleocurrents, and paleogeography of Ridge Basin, southern California, *in* Crowell, J. C., and Link, M. H., eds., Geologic history of Ridge Basin, southern California: Pacific Section, Society of Economic Paleontologists and Mineralogists, p. 265–276.

—— , 1983, Sedimentation, tectonics, and offset of Miocene-Pliocene Ridge Basin, California, *in* Andersen, D. W., and Rymer, M. J., eds., Tectonics and sedimentation along faults of the San Andreas system: Pacific Section, Society of Economic Paleontologists and Mineralogists, p. 17–32.

Lonsdale, P., and Lawver, L. A., 1980, Immature plate boundary zones studied with a submersible in the Gulf of California: Geological Society of America Bulletin, Part 1, v. 91, p. 555–569.

Luyendyk, B. P., and Hornafius, J. S., 1987, Neogene crustal rotations, fault slip, and basin development in southern California, *in* Ingersoll, R. V., and Ernst, W. G., eds., Cenozoic basin development of coastal California; Rubey Volume 6: Englewood Cliffs, New Jersey, Prentice-Hall, p. 259–283.

Luyendyk, B. P., Kamerling, M. J., and Terres, R. R., 1980, Geometric model for Neogene crustal rotations in southern California: Geological Society of America Bulletin, v. 91, p. 211–217.

Luyendyk, B. P., Kamerling, M. J., Terres, R. R., and Hornafius, J. S., 1985, Simple shear of southern California during Neogene time suggested by paleomagnetic declinations: Journal of Geophysical Research, v. 90, no. B14, p. 12454–12466.

Matthews, V., 1973, Pinnacles-Neenach correlations; A restriction for models of the origin of the Transverse Ranges and the big bend in the San Andreas fault: Geological Society of America Bulletin, v. 84, p. 683–688.

—— , 1976, Correlation of Pinnacles and Neenach Volcanic Formations and their bearing on San Andreas fault problem: American Association of Petroleum Geologists Bulletin, v. 60, p. 2128–2141.

Matti, J. C., and Morton, D. M., 1975, Geologic history of the San Timoteo Badlands, southern California: Geological Society of America Abstracts with Programs, v. 7, p. 344.

—— , 1982, Geologic history of the Banning fault zone, southern California: Geological Society of America Abstracts with Programs, v. 14, p. 184.

Matti, J. C., and 6 others, 1982a, Mineral resource potential map of the Whitewater Wilderness Study Area, Riverside and San Bernardino Counties, California: U.S. Geological Survey Miscellaneous Field Studies Map MF-1478, scale 1:24,000.

Matti, J. C., Tinsley, J. C., McFadden, L. D., and Morton, D. M., 1982b, Holocene faulting history as recorded by alluvial history within the Cucamonga fault zone; A preliminary view, *in* Tinsley, J. C., McFadden, L. D., and Matti, J. C., eds., Late Quaternary pedogenesis and alluvial chronologies of the Los Angeles and San Gabriel Mountains areas, southern California: Geological Society of America Cordilleran Section 78th Annual Meeting, Anaheim, California, Guidebook, Field trip 12, p. 21–44.

Matti, J. C., Cox, B. F., and Iverson, S. R., 1983a, Mineral resource potential map of the Raywood Flat Roadless Area, San Bernardino and Riverside Counties, California: U.S. Geological Survey Miscellaneous Field Studies Map MF-1563-A, scale 1:62,500.

Matti, J. C., Cox, B. F., Powell, R. E., Oliver, H. W., and Kuizon, L., 1983b, Mineral resource potential map of the Cactus Spring Roadless Area, Riverside County, California: U.S. Geological Survey Miscellaneous Field Studies Map MF-1650-A, scale 1:24,000.

Matti, J. C., Morton, D. M., and Cox, B. F., 1985, Distribution and geologic relations of fault systems in the vicinity of the central Transverse Ranges, southern California: U.S. Geological Survey Open-File Report 85-365, 27 p., scale 1:250,000.

Matti, J. C., Frizzell, V. A., and Mattinson, J. M., 1986, Distinctive Triassic megaporphyritic monzogranite displaced 160 ± 10 km by the San Andreas fault, southern California; A new constraint for palinspastic reconstructions: Geological Society of America Abstracts with Programs, v. 18, p. 154.

May, D. J., and Walker, N. W., 1989, Late Cretaceous juxtaposition of metamorphic terranes in the southeastern San Gabriel Mountains, California: Geological Society of America Bulletin, v. 101, p. 1246–1267.

May, S. R., and Repenning, C. A., 1982, New evidence for the age of the Old Woman Sandstone, Mojave Desert, California, *in* Sadler, P. M., and Kooser, M. A., eds., Late Cenozoic stratigraphy and structure of the San Bernardino Mountains, *in* Cooper, J. D., compiler, Geologic excursions in the Transverse Ranges, southern California: Geological Society of America Cordilleran Section 78th Annual Meeting, Anaheim, California, Volume and Guidebook, Field trip 6, p. 93–96.

Meisling, K. E., 1984, Neotectonics of the north frontal fault system of the San Bernardino Mountains, southern California; Cajon Pass to Lucerne Valley [Ph.D. thesis]: Pasadena, California Institute of Technology, 394 p.

Meisling, K. E., and Weldon, R. J., 1982, The late Cenozoic structure and stratigraphy of the western San Bernardino Mountains, *in* Sadler, P. M., and Kooser, M. A., eds., Late Cenozoic stratigraphy and structure of the San Bernardino Mountains, *in* Cooper, J. D., compiler, Geologic excursions in the Transverse Ranges, southern California: Geological Society of America Cordilleran Section 78th Annual Meeting, Anaheim, California, Volume and Guidebook, Field trip 6, p. 75–82.

—— , 1986, Cenozoic uplift of the San Bernardino Mountains; Possible thrusting across the San Andreas fault: Geological Society of America Abstracts with Programs, v. 18, p. 157.

—— , 1989, Late Cenozoic tectonics of the northwestern San Bernardino Mountains, southern California: Geological Society of America Bulletin, v. 101, p. 106–128.

Miller, C. F., 1978, An early Mesozoic alkalic magmatic belt in western North America, *in* Howell, D. G., and McDougall, K. A., eds., Mesozoic paleogeography of the western United States: Pacific Section, Society of Economic Paleontologists and Mineralogists, p. 163–173.

Miller, F. K., 1979, Geologic map of the San Bernardino North 7.5′ Quadrangle: U.S. Geological Survey Open-File Report 79-770, scale 1:24,000.

Miller, F. K., and Morton, D. M., 1977, Comparison of granitic intrusions in the Pelona and Orocopia Schists, southern California: U.S. Geological Survey

Journal of Research, v. 5, no. 5, p. 643–649.

——— , 1980, Potassium-argon geochronology of the eastern Transverse Ranges and southern Mojave Desert, southern California: U.S. Geological Survey Professional Paper 1152, 30 p.

Moore, D. G., 1973, Plate-edge deformation and crustal growth, Gulf of California structural province: Geological Society of America Bulletin, v. 84, p. 1883–1906.

Moore, D. G., and Buffington, E. C., 1968, Transform faulting and growth of the Gulf of California since the late Pliocene: Science, v. 161, p. 1238–1241.

Morton, D. M., 1975a, Synopsis of the geology of the eastern San Gabriel Mountains, southern California, *in* Crowell, J. C., ed., San Andreas fault in southern California: California Division of Mines and Geology Special Report 118, p. 170–176.

——— , 1975b, Relations between major faults, eastern San Gabriel Mountains, southern California: Geological Society of America Abstracts with Programs, v. 7, p. 352–353.

——— , 1976, Geologic map of the Cucamonga fault zone between San Antonio Canyon and Cajon Creek, southern California: U.S. Geological Survey Open-File Report 76-726, scale 1:24,000.

Morton, D. M., and Matti, J. C., 1987, The Cucamonga fault zone; Geologic setting and Quaternary history, *in* Morton, D. M. and Yerkes, R. F., Recent reverse faulting in the Transverse Ranges, California: U.S. Geological Survey Professional Paper 1339, p. 179–203, scale 1:24,000.

——— , 1989, A vanished late Pliocene to early Pleistocene alluvial-fan complex in the northern Perris block, southern California, *in* Colburn, I. P., Abbott, P. L., and Minch, J., eds., Conglomerates in basin analysis; A symposium dedicated to A. O. Woodford: Pacific Section, Society of Economic Paleontologists and Mineralogists, v. 62, p. 73–80.

Morton, D. M., and Miller, F. K., 1975, Geology of the San Andreas fault zone north of San Bernardino between Cajon Canyon and Santa Ana Wash, *in* Crowell, J. C., ed., San Andreas fault in southern California: California Division of Mines and Geology Special Report 118, p. 136–146.

Morton, D. M., Cox, B. F., and Matti, J. C., 1980, Geologic map of the San Gorgonio Wilderness: U.S. Geological Survey Miscellaneous Field Studies Map MF-1164-A, scale 1:62,500.

Morton, D. M., Rodriguez, E. A., Obi, C. M., Simpson, R. W., Jr., and Peters, T. J., 1983, Mineral resource potential map of the Cucamonga Roadless Areas, San Bernardino County, California: U.S. Geological Survey Miscellaneous Field Studies Map MF-1646-A, scale 1:31,680.

Morton, D. M., Matti, J. C., Miller, F. K., and Repenning, C. A., 1986, Pleistocene conglomerate from the San Timoteo Badlands, southern California; Constraints on strike-slip displacements on the San Andreas and San Jacinto faults: Geological Society of America Abstracts with Programs, v. 18, p. 161.

Noble, L. F., 1926, The San Andreas rift and some other active faults in the desert region of southeastern California: Carnegie Institute of Washington Yearbook 25 (1925-1926), p. 415–435.

——— , 1932, The San Andreas rift in the desert region of southeastern California: Carnegie Institute of Washington Yearbook 31, p. 355–363.

——— , 1953, Geology of the Pearland Quadrangle, California: U.S. Geological Survey Geologic Quadrangle Map GQ-24, scale 1:24,000.

——— , 1954a, Geology of the Valyermo Quadrangle and vicinity, California: U.S. Geological Survey Geologic Quadrangle Map GQ-50, scale 1:24,000.

——— , 1954b, The San Andreas fault zone from Soledad Pass to Cajon Pass, California, *in* Jahns, R. H., ed., Geology of southern California: California Division of Mines Bulletin 170, p. 37–48, and Plate 5 of chapter IV, scale 1:125,000.

Paschall, R. H., and Off, T., 1961, Dip-slip versus strike-slip movement on the San Gabriel fault, southern California: American Association of Petroleum Geologists Bulletin, v. 45, p. 1941–1956.

Pelka, G. J., 1971, Paleocurrents of the Punchbowl Formation and their interpretation: Geological Society of America Abstracts with Programs, v. 3, p. 176.

Peterson, M. S., 1975, Geology of the Coachella fanglomerate, *in* Crowell, J. C., ed., San Andreas fault in southern California: California Division of Mines and Geology Special Report 118, p. 119–126.

Powell, R. E., 1981, Geology of the crystalline basement complex, eastern Transverse Ranges, southern California [Ph.D. thesis]: Pasadena, California Institute of Technology, 441 p.

——— , 1982a, Prebatholithic terranes in the crystalline basement complex of the Transverse Ranges, southern California: Geological Society of America Abstracts with Programs, v. 14, p. 225.

——— , 1982b, Crystalline basement terranes in the southern eastern Transverse Ranges, California, *in* Cooper, J. D., compiler, Geologic excursions in the Transverse Ranges, southern California: Geological Society of America Cordilleran Section 78th Annual Meeting, Anaheim, California, Guidebook, p. 109–136.

——— , 1986, Palinspastic reconstruction of crystalline-rock assemblages in southern California; San Andreas fault as part of an evolving system of late Cenozoic strike-slip faults: Geological Society of America Abstracts with Programs, v. 18, p. 172.

Ramirez, V. R., 1983, Hungry Valley Formation; Evidence for 220 kilometers of post Miocene offset on the San Andreas fault, *in* Anderson, D. W., and Rymer, M. J., eds., Tectonics and sedimentation along faults of the San Andreas system: Pacific Section, Society of Economic Paleontologists and Mineralogists, p. 33–44.

Rasmussen, G. S., and Reeder, W. A., 1986, What happens to the real San Andreas fault at Cottonwood Canyon, San Gorgonio Pass, California? *in* Kooser, M. A., and Reynolds, R. E., eds., Geology around the margins of the eastern San Bernardino Mountains: Publications of the Inland Geological Society, v. 1, p. 57–62.

Robinson, P. T., and Woodburne, M. O., 1971, Source of volcanic clasts in the Punchbowl Formation, Valyermo and Cajon Valley, California: Geological Society of America Abstracts with Programs, v. 3, p. 185–186.

Rogers, T. H., compiler, 1965, Santa Ana sheet of Geologic map of California: California Division of Mines and Geology, scale 1:250,000.

——— , 1967, San Bernardino sheet of Geologic map of California: California Division of Mines and Geology, scale 1:250,000.

Ross, D. C., 1972, Petrographic and chemical reconnaissance study of some granitic and gneissic rocks near the San Andreas fault from Bodega Head to Cajon Pass, California: U.S. Geological Survey Professional Paper 698, 92 p.

——— , 1984, Possible correlations of basement rocks across the San Andreas, San Gregorio–Hosgri, and Rinconada–Reliz–King City faults, California: U.S. Geological Survey Professional Paper 1317, 37 p.

Sadler, P. M., 1982a, An introduction to the San Bernardino Mountains as the product of young orogenesis, *in* Sadler, P. M., and Kooser, M. A., eds., Late Cenozoic stratigraphy and structure of the San Bernardino Mountains, *in* Cooper, J. D., compiler, Geologic excursions in the Transverse Ranges, southern California: Geological Society of America Cordilleran Section 78th Annual Meeting, Anaheim, California, 1982, Volume and Guidebook, Field trip 6, p. 57–65.

——— , 1982b, Provenance and structure of late Cenozoic sediments in the northeast San Bernardino Mountains, *in* Sadler, P. M., and Kooser, M. A., eds., Late Cenozoic stratigraphy and structure of the San Bernardino Mountains, *in* Cooper, J. D., compiler, Geologic excursions in the Transverse Ranges, southern California: Geological Society of America Cordilleran Section 78th Annual Meeting, Anaheim, California, 1982, Volume and Guidebook, Field trip 6, p. 83–92.

Sadler, P. M., and Demirer, A., 1986, Geology of upper Mill Creek and Santa Ana Canyon, southern San Bernardino Mountains, California, *in* Ehlig, P. L., compiler, Neotectonics and faulting in southern California: Geological Society of America Cordilleran Section, 82nd Annual Meeting, Los Angeles, California, 1986, Guidebook and Volume, Field trip 12, p. 129–140.

Sharp, R. V., 1967, San Jacinto fault zone in the Peninsular Ranges of southern California: Geological Society of America Bulletin, v. 78, p. 705–730.

——— , 1979, Some characteristics of the eastern Peninsular Ranges mylonite zone, *in* Analysis of actual fault zones in bedrock: U.S. Geological Survey Open-File Report 79-1239, p. 258–267.

——— , 1982, Tectonic setting of the Imperial Valley region, *in* The Imperial

Valley, California, earthquake of October 15, 1979: U.S. Geological Survey Professional Paper 1254, p. 1–14.

Sieh, K. E., and Jahns, R. H., 1984, Holocene activity of the San Andreas fault at Wallace Creek, California: Geological Society of America Bulletin, v. 95, p. 883–896.

Silver, L. T., 1971, Problems of crystalline rocks of the Transverse Ranges: Geological Society of America Abstracts with Programs, v. 3, p. 193–194.

Smith, D. K., 1982, Petrology of Mesozoic alkaline and calc-alkaline igneous rocks northwest of Holcomb Valley, San Bernardino County, California, *in* Fife, D. L., and Minch, J. A., eds., Geology and mineral wealth of the California Transverse Ranges; Mason Hill Volume: Santa Ana, California, South Coast Geological Society Guidebook 10, p. 321–329.

Smith, D. P., 1977, San Juan–St. Francis fault; Hypothesized major middle Tertiary right-lateral fault in central and southern California: California Division of Mines and Geology Special Report 129, p. 41–50.

Smith, R. E., 1959, Geology of the Mill Creek area, San Bernardino County, California [M.S. thesis]: Los Angeles, University of California, 95 p., scale 1:12,000.

Spittler, T. E., and Arthur, M. A., 1973, Post early Miocene displacement along the San Andreas fault in southern California, *in* Kovach, R. L., and Nur, A., eds., Proceedings of the Conference on Tectonic Problems of the San Andreas Fault System: Stanford, California, Stanford University Publications in Geological Sciences, v. 13, p. 374–382.

—— , 1982, The lower Miocene Diligencia Formation of the Orocopia Mountains, southern California; Stratigraphy, petrology, sedimentology, and structure, *in* Ingersoll, R. V., and Woodburne, M. O., eds., Cenozoic nonmarine deposits of California and Arizona: Pacific Section, Society of Economic Paleontologists and Mineralogists, p. 83–99.

Stewart, J. H., and Poole, F. G., 1975, Extension of the Cordilleran miogeosynclinal belt to the San Andreas fault, southern California: Geological Society of America Bulletin, v. 86, p. 205–212.

Terres, R. R., and Luyendyk, B. P., 1985, Neogene tectonic rotation of the San Gabriel region, California, suggested by paleomagnetic vectors: Journal of Geophysical Research, v. 90, no. B14, p. 12467–12484.

Trent, D. D., 1984, Geology of the Joshua Tree National Monument: California Geology, v. 37, p. 75–86.

Truex, J. N., 1976, Santa Monica and Santa Ana Mountains-relation to Oligocene Santa Barbara Basin: American Association of Petroleum Geologists Bulletin, v. 60, p. 65–86.

Vaughan, F. E., 1922, Geology of the San Bernardino Mountains north of San Gorgonio Pass: Berkeley, California University Publications in Geological Sciences, v. 13, p. 319–411.

Wallace, R. E., 1949, Structure of a portion of the San Andreas rift in southern California: Geological Society of America Bulletin, v. 60, p. 781–806.

Weber, F. H., Jr., 1982, New interpretation of structural elements in the vicinity of the northwestern terminus of the trace of the San Gabriel fault zone, Ventura County, California, *in* Fife, D. L., and Minch, J. A., eds., Geology and mineral wealth of the California Transverse Ranges; Mason Hill Volume: Santa Ana, California, South Coast Geological Society Guidebook 10, p. 321–329.

Weldon, R. J., 1984, Quaternary deformation due to the junction of the San Andreas and San Jacinto faults, southern California: Geological Society of America Abstracts with Programs, v. 16, p. 689.

—— , 1986, The late Cenozoic geology of Cajon Pass; Implications for tectonics and sedimentation along the San Andreas fault [Ph.D. thesis]: Pasadena, California Institute of Technology, 400 p.

Weldon, R. J., and Humphries, E., 1986, A kinematic model of southern California: Tectonics, v. 5, p. 33–48.

West, D. L., 1987, Geology of the Wilson Creek–Mill Creek fault zone; The north flank of the former Mill Creek basin, San Bernardino County, California [M.S. thesis]: Riverside, University of California, 94 p.

Willingham, R. C., 1971, Basement fault geometries in the San Bernardino Valley and western San Gorgonio Pass area, southern California: Geological Society of America Abstracts with Programs, v. 3, p. 217.

—— , 1981, Gravity anomaly patterns and fault interpretations in the San Bernardino Valley and western San Gorgonio Pass area, southern California, *in* Brown, A. R., and Ruff, R. W., eds., Geology of the San Jacinto Mountains: Santa Ana, California, South Coast Geological Society Annual Field Trip Guidebook 9, p. 164–174.

Winker, C. D., and Kidwell, S. M., 1986, Paleocurrent evidence for lateral displacement of the Pliocene Colorado River delta by the San Andreas fault system, southeastern California: Geology, v. 14, p. 788–791.

Woodburne, M. O., 1975, Cenozoic stratigraphy of the Transverse Ranges and adjacent areas, southern California: Geological Society of America Special Paper 162, 91 p.

Woodburne, M. O., and Golz, D. J., 1972, Stratigraphy of the Punchbowl Formation, Cajon Valley, southern California: Berkeley, University of California Publications in Geological Sciences v. 92, 73 p.

Woodford, A. O., 1960, Bedrock patterns and strike-slip faulting in southwestern California: American Journal of Science, v. 258A, p. 400–417.

Yerkes, R. F., and Campbell, R. H., 1971, Cenozoic evolution of the Santa Monica Mountains–Los Angeles Basin area; 1, Constraints on tectonic models: Geological Society of America Abstracts with Programs, v. 3, p. 222–223.

MANUSCRIPT ACCEPTED BY THE SOCIETY SEPTEMBER 6, 1990

Geological Society of America
Memoir 178
1993

Chapter 3

A speculative history of the San Andreas fault in the central Transverse Ranges, California

R. J. Weldon, II
Department of Geological Sciences, University of Oregon, Eugene, Oregon 97403, and U.S. Geological Survey, 345 Middlefield Road, Menlo Park, California 94025
K. E. Meisling
ARCO Oil and Gas Company, 2300 West Plano Parkway, Plano, Texas 75075
J. Alexander
U.S. Geological Survey, 345 Middlefield Road, Menlo Park, California 94025

ABSTRACT

It is generally accepted that the San Andreas fault formed between 4 and 5 Ma and that rocks west of it are now part of the Pacific plate, moving northwest relative to North America at 5 to 6 cm/yr. This model is inconsistent with the geologic record in the central Transverse Ranges.

Right-lateral shear began in the vicinity of the San Andreas fault system in early Miocene time. The San Andreas fault system in the central Transverse Ranges has since evolved through three major phases; this development has led to a generally simpler, more throughgoing main trace. Slip rates on the San Andreas system were about 1 cm/yr in the Miocene, increasing to their current level of 3.5 cm/yr between 4 and 5 Ma. The modern San Andreas fault still only accounts for just over half the current relative plate rate and retains kinematic complexities inherited from its earliest geometry.

The Early San Andreas transform system originated during early Miocene time in one of three transtensive zones that lay interior to the continent and east of the locus of transform motion between the Pacific and North American plates. The current three-fold division of motion in the plate boundary between the San Andreas fault, a coastal system, and an eastern California system dates to this time, as does the "anomalous" trend of the San Andreas fault through the Transverse Ranges. Basins and volcanic centers associated with this transtensive zone became dismembered as faults became integrated into a throughgoing system. Early motion led to juxtaposition of different rocks across faults now recognized as part of the Early San Andreas transform system, and to the development of sedimentary provincialism associated with uplift along the fault zone. Middle Miocene basins, including the Caliente, Cajon, Crowder, and Santa Ana basins that had previously received most of their sediments from sources far to the east, began to reflect local Transverse Ranges provenance. At least 100 km of slip is associated with the Early San Andreas transform system during early and middle Miocene time.

Slip across the geometrically complex late Miocene San Gabriel transform system—which includes the San Gabriel, Cajon Valley, and early Punchbowl faults—produced uplift in the proto–Transverse Ranges at a postulated restraining bend in the fault system. Compressional structures associated with this restraining bend include the

Weldon, R. J., II, Meisling, K. E., and Alexander, J., 1993, A speculative history of the San Andreas fault in the central Transverse Ranges, California, *in* Powell, R. E., Weldon, R. J., II, and Matti, J. C., eds., The San Andreas Fault System: Displacement, Palinspastic Reconstruction, and Geologic Evolution: Boulder, Colorado, Geological Society of America Memoir 178.

Squaw Peak and Liebre Mountain thrusts, related east-striking late Miocene reverse faults and folds, and, perhaps, northeast-striking left-lateral faults in the San Gabriel Mountains. Narrow fault-controlled basins formed during this period, including the Ridge basin, Devil's Punchbowl basin, Mill Creek basin, and part of the Santa Ana Sandstone basin. Offset of structures and relief associated with the proto–Transverse Ranges provides the best evidence for late Miocene restorations of the modern San Andreas fault. As much as 60 km of offset is associated with the late Miocene San Gabriel transform system.

Between 4 and 5 Ma, the modern San Andreas fault became the dominant member of the plate boundary system, cutting through the proto–Transverse Ranges and connecting more northerly striking traces to the north and south. The slip rate across the San Andreas fault system accelerated from 1 cm/yr to its current slip rate of 3.5 cm/yr prior to 4 Ma. The Pliocene rocks in the central Transverse Ranges do not contain evidence for relief as great as that of late Miocene or Quaternary time. The Pliocene trace of the modern San Andreas fault may have temporarily "solved" the geometric problem that led to late Miocene uplift. About 90 km of right-lateral displacement occurred on the modern San Andreas fault during Pliocene time.

During Quaternary time new regions of localized vertical deformation developed in the Transverse Ranges, apparently as the result of new geometric problems within the Pliocene solution to the restraining geometry of the fault system. Left-lateral motion on east-striking faults, probably due to a northward increase in Basin and Range extension, kinked the San Andreas fault at both ends of the Transverse Ranges, producing regions of extreme shortening and uplift. The development of young right-lateral faults through the Peninsular Ranges, including the San Jacinto and Elsinore faults, also contributed to renewed uplift in the Transverse Ranges. Sixty kilometers of right-lateral slip occurred across the San Andreas fault zone during Quaternary time.

INTRODUCTION

The slip history of the San Andreas fault system in southern California has long been inferred from correlation of rocks and structures in central California, the northwestern Transverse Ranges, the western San Gabriel Mountains, and desert regions of the eastern Transverse Ranges (Hill and Dibblee, 1953; Crowell, 1962). Late Cenozoic reconstructions of the San Andreas fault system in the central Transverse Ranges, defined here as the eastern San Gabriel Mountains, San Bernardino Mountains, and Cajon basin region (Fig. 1), have relied on externally imposed slip constraints rather than on specific offset criteria internal to the region itself. Though this region produced early ideas about offset on the San Andreas fault (Noble, 1926; Woodford, 1960) and is commonly incorporated in reconstruction models for southern California, the rocks of the central Transverse Ranges have only recently been documented well enough to provide critical constraints to models for evolution of the San Andreas fault system by providing specific offsets, rates of slip, and timing relationships. Recognition and documentation of significant offset constraints in the central Transverse Ranges was delayed by: (1) the scarcity of distinctive lithologic types within prebatholithic basement rocks and the lack of Oligocene to early Miocene sedimentary and volcanic rocks, (2) the general lack of detailed mapping and age control for distinctive units in the area, and (3) the difficulty in incorporating these rocks into externally derived models (e.g., Baird and others, 1974). In this chapter we present a new body of offset data that provide a consistent picture of the slip history of the San Andreas fault zone in the central Transverse Ranges. The data suggest that the history of the San Andreas fault in the Transverse Ranges is much longer and more complex than has been generally believed, and that rates of slip are lower than previously thought.

There are four key observations that motivate our discussion of the San Andreas fault in the central Transverse Ranges: (1) Quaternary slip rates are too low to account for the total cumulative slip required by existing plate tectonic models; (2) Miocene and Pliocene slip rates, while less certain, are not great enough to permit the deficit to be made up by increased rates in the past; (3) recently documented total cumulative offset between the San Bernardino and Liebre Mountains of ~150 km during the past 5 to 7 Ma are inconsistent with simple application of the widely accepted models for ~240 km of total slip on the modern San Andreas fault; and (4) middle to late Miocene pull-apart basins and right-lateral structures in the central Transverse Ranges suggest that motion on Miocene precursors to the modern San Andreas fault contributed slip to the widely cited total of ~240 km. These observations are briefly outlined here and discussed in detail in the body of the chapter. We combine these observations with evidence for the timing and magnitude of total plate motion across the Pacific–North American plate boundary to determine the role the San Andreas system has played in the evolving transform margin.

Well-documented Quaternary slip rates on the San Andreas

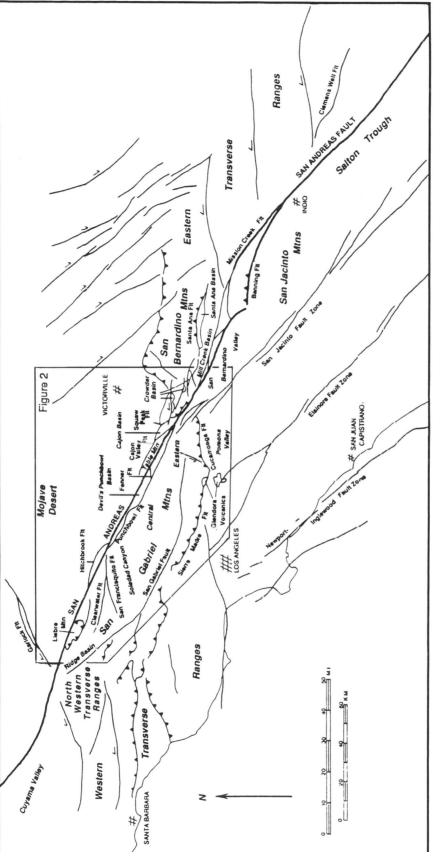

Figure 1. Index map of the San Andreas fault system in the central Transverse Ranges, southern California, showing the location of major physiographic and cultural features, principal regional faults, and key late Cenozoic sedimentary basins discussed in text. Box indicates location of regional geologic map, Figure 2.

fault are about half the rate required to accumulate the widely accepted value for total cumulative offset of 240 km in the past 4 to 5 Ma (Crowell, 1981; Ehlig, 1981). The widely accepted model, which implies that the San Andreas fault has accumulated all of the slip between the Pacific and North American plates, was a natural outgrowth of the recognition of the San Andreas fault as a plate margin transform boundary (Wilson, 1965). It has become increasingly apparent, from both plate-kinematic and slip-rate data, that all of the plate-boundary deformation is not occurring on the San Andreas fault alone, but is rather shared among many faults within a wide plate-boundary deformation zone, verifying the prediction of a broad "soft zone" of continental margin deformation (Atwater, 1970). An accounting of all the deformation that has occurred between the Pacific and North American plates during the past 4 to 5 Ma requires that the San Andreas fault has accrued just over half the plate total during this period, essentially the same proportion that is observed today.

Late Miocene, Pliocene, and Quaternary sedimentary rocks shed northeastward off the central and eastern San Gabriel Mountains into the San Bernardino Mountains and Mojave Desert, and southwestward of the San Bernardino Mountains into the northwestern Transverse Ranges, provide a relatively complete record of about 150 km of offset between these two evolving mountain masses across the modern San Andreas fault. While Tertiary sedimentary rocks do not provide the slip-rate resolution of Quaternary deposits, their inferred offsets and ages are consistent with the rates determined from the Quaternary units.

The proposed restoration of late Miocene structures in the western San Bernardino Mountains with those of the Ridge basin and Liebre Mountain region in the northwestern Transverse Ranges supports the correlation of a distinctive Triassic monzogranite found in both areas (Matti and others, 1985; Frizzell and others, 1986; Matti and Morton, this volume) for a total displacement of only ~150 km on the San Andreas fault during the past 5 to 7 Ma. Late Miocene juxtaposition of the western San Bernardino Mountains and Ridge basin provides a source for late Miocene and Pliocene sediments shed across the San Andreas fault into the Ridge basin.

Probable late Miocene rift basins along the San Andreas system are evidence for throughgoing right-lateral faults south and east of the San Gabriel fault prior to the 4 to 5 Ma initiation of the modern San Andreas fault. Whether they were formed by faults contemporaneous with or predating the San Gabriel fault is uncertain because of the inadequate age control of the sedimentary rocks in these basins. Evidence favors late activity on the integrated San Francisquito–Fenner–Clemens Well fault system (Powell, 1981, this volume) to produce these basins. Alternatively, the basins may have formed in an early phase of the southern San Andreas fault that was part of the San Gabriel transform system. Faults predating the modern San Andreas fault and San Gabriel transform system provide an additional ~100 km of displacement that, when combined with the ~150 km on the modern San Andreas fault and the ~60 km on the San

Gabriel transform system, yields the total offset of ~310 km across the San Andreas fault system in southern California (Crowell, 1962, 1981; Powell, 1981).

In this chapter we discuss evidence bearing on the history of the San Andreas fault system in order of increasing age and construct models to determine the role of the evolving system in the plate boundary. We then use this evidence to develop a speculative history of the San Andreas fault system in southern California from early Miocene to Quaternary time. The speculative history presented here is our interpretation of the available data for the central Transverse Ranges and our attempt to understand it within the context of the San Andreas system. Because the region we consider is too limited to include rocks that record the total offset of the system, examination of the earliest history necessarily focuses on timing and styles of deformation that can be interpreted from the rocks. Inferences about the amount of slip associated with these early structures are made from the literature, and the rates of deformation from kinematic models. We hope that the ideas contained in this chapter, which is intentionally broad in scope and speculative in nature, will stimulate and focus future studies on the slip history of the San Andreas fault in the central Transverse Ranges.

QUATERNARY SLIP HISTORY

Neotectonic and earthquake hazard studies have yielded a sizeable body of increasingly accurate Quaternary slip-rate data and kinematic models for active faulting across the western North American plate boundary. While short-term rates and geometric constraints are typically dismissed as being of limited use in palinspastic reconstructions, we present evidence that the Holocene slip rate for the modern San Andreas fault is about the same as the rate for the last 2 m.y. or longer. Quaternary rates of deformation are particularly important because they allow us to choose between various reconstruction models for Pliocene and earlier times that are less well constrained by dated offsets. Knowing the rates and total offset for even half the time that the modern San Andreas has been active allows us to place limits on what is available for its entire history. For the sake of simplicity we have rounded all slip rates in the following discussion to the nearest 0.5 cm/yr, since the accuracy of few slip-rate determinations warrants greater precision.

Holocene slip rates

In order to address the Holocene slip rate of the San Andreas fault in the central Transverse Ranges, it is first necessary to consider the manner in which slip is partitioned between the San Jacinto and San Andreas fault systems. The simplest model is that the sum of the slip on the San Andreas and San Jacinto faults south of their junction is equal to the slip or offset across the San Andreas fault north of the junction. While this simple model is attractive, and we adopt it here, we must recognize that other possibilities exist. Below we consider several model alternatives for partitioning slip between the San Jacinto and San Andreas faults.

Sharp (1981) has proposed that the San Andreas and San Jacinto faults have alternating periods of high activity, so that at any given time one or the other might accommodate a greater fraction of the combined slip rate. Sharp's model requires that combined slip rate will be accurate only if slip is calculated for both faults over the exact same time interval. Morton (1975) and Bird and Rosenstock (1984) suggested that much of the slip associated with the San Jacinto fault does not contribute to slip on the San Andreas fault to the north, but is taken up in thrusting and associated shortening across the southeastern range front of the San Gabriel Mountains. If so, slip on the San Andreas fault in the central Transverse Ranges only includes some fraction of the slip across the San Jacinto fault.

In this chapter we assume that most of the slip on the San Jacinto fault contributes to San Andreas fault motion north of the junction, which will maximize the possible slip rate on the San Andreas fault north of the San Gabriel Mountains. From the lack of crustal thickening beneath the San Gabriel Mountains and the style and rate of slip across the range-bounding thrust between the San Jacinto fault and northward projection of the Elsinore fault, we infer that the fraction of the slip on the San Jacinto fault consumed in uplift of the eastern San Gabriel Mountains is relatively small. Uplift along the southeastern range front of the San Gabriel Mountains can be satisfactorily explained by the restraining geometry of the junction alone (Weldon and Humphreys, 1986), eliminating the need for consumption of San Jacinto fault slip by thrusting. We therefore assume in the following discussion that slip on the San Andreas fault north of the San Jacinto fault is, to a reasonable approximation, the simple sum of the slip on the San Andreas and San Jacinto faults south of their junction. The details of how the motion is transferred across the complex junction spanning the San Bernardino Valley and the eastern San Gabriel Mountains are poorly resolved and are not addressed here.

Offsets of Holocene features across the San Andreas and San Jacinto faults support a combined slip rate of ~3.5 cm/yr. At Pallett Creek, north of the junction with the San Jacinto fault, Salyards (1988) measured distributed deformation across the San Andreas zone and added it to the brittle offset (~1 cm/yr) determined by Sieh (1984) for a total Holocene slip rate of 3.5 cm/yr. This rate is consistent with the combined slip rate of 3 to 4 cm/yr arrived at by summing average Holocene slip rate determinations for the San Andreas and San Jacinto faults south of their junction of 2 to 2.5 cm/yr (Weldon and Sieh, 1985) and 1 to 1.5 cm/yr (Rockwell and others, 1986; Merifield and others, 1987), respectively. This rate is also consistent with the well-documented Holocene slip rate of ~3.5 cm/yr for the San Andreas fault north of the Transverse Ranges (Sieh and Jahns, 1984). If Holocene slip rates for the San Andreas fault to the south of the Transverse Ranges of ~3 cm/yr (Sieh, 1986; Keller and others, 1982) are summed with slip rates for the San Jacinto fault of 1 to 1.5 cm/yr (Rockwell and others, 1986; Merifield and others, 1987), the

total is 4 cm/yr or greater. However, some of the motion south of the Transverse Ranges is transferred to the eastern California zone and does not enter the central Transverse Ranges. Geodetic data across the San Andreas fault in central California, and across the combined San Andreas and San Jacinto faults south of the Transverse Ranges, also yield a rate of about 3.5 cm/yr (Savage, 1983).

A higher slip rate of 4.5 to 6 cm/yr has been proposed by Rust (1982, 1986); however, analysis of his offsets and age control lead us to believe that his rates are maximums, not actual ranges (an assessment shared by Clark and others, 1984, and Barrows and others, 1985). Rust's study yields results that are so inconsistent with all of the other Holocene data from the area that we have not used them.

A lower slip rate for the San Andreas fault in the central Transverse Ranges of ~2 cm/yr has been proposed by Schwartz and Coppersmith (1984). This lower slip rate is within the range of preliminary slip-rate data near Littlerock (Schwartz and Weldon, 1986, 1987). This lower rate is consistent with a reinterpretation of the youngest offset of Weldon and Sieh (1985) by Sharp (*in* Schwartz and others, 1987), if it is combined with the assumption that the San Andreas fault has no change in slip rate across the northern termination of the San Jacinto fault. Geodetic data from north of the San Gabriel Mountains has been interpreted to support a rate of ~2 cm/yr (Savage, 1983; Savage and Prescott, 1986). However, more recent analysis (Savage, 1990; Savage and others, 1990; Eberhart-Phillips and others, 1990) leads to a preferred rate of 3.5 cm/yr. It is difficult to understand how the rate of slip on the San Andreas fault system could decrease to only 2 cm/yr in the Transverse Ranges where there are no diverging structures significant enough to accommodate the excess 1.5 cm/yr of slip that has been documented both to the north and south.

No compelling case can be made for a Holocene slip rate of 6 cm/yr, which would be the long-term rate required to accumulate the widely cited 240 km total offset in 4 m.y. (Crowell, 1981). We believe that the published data best support a Holocene slip rate of ~3.5 cm/yr; a Holocene slip rate of 2 cm/yr has some support. For the purpose of this chapter, which presents evidence for a longer and slower San Andreas fault slip history than generally thought, the lower slip rate of 2 cm/yr would only make our argument stronger.

Pleistocene slip rates

The Pleistocene slip rate for the San Andreas fault in the Transverse Ranges is better constrained than the Holocene slip rate. Sediments, shed northeastward off the central San Gabriel Mountains across the San Andreas fault and onto the floor of the western Mojave Desert, provide a continuous record of Pleistocene fault displacement at a rate of ~3.5 cm/yr.

Distinctive clast assemblages from the San Gabriel Mountains are found in the Pleistocene fanglomerates of the Harold Formation, Shoemaker Gravel, and older alluvium of the Victorville Fan (Noble, 1932, 1954; Weldon, 1984, 1986; Meisling and Weldon, 1989) now exposed in a narrow strip along the San Andreas fault northwest of the San Bernardino Mountains (Fig. 2). As the floor of the western Mojave Desert moved past the San Gabriel Mountains along the San Andreas fault, each major drainage system emerging from the range delivered a distinctive suite of clasts derived from source areas within the San Gabriel Mountains, laying down a vertical record of the lateral slip history (Fig. 3). This pattern of deposition was termed "hopper car" sedimentation by Crowell (1974) in his model for the deposition of the Violin Breccia in the Ridge basin.

An average Pleistocene slip rate for the San Andreas fault in the central Transverse Ranges can be determined by dating the clast-bearing strata and measuring their offsets from known source areas. Magnetostratigraphic data show that the Pleistocene units of the Victorville Fan are time-transgressive (Fig. 4). Dating, offsets, clast assemblages, and flow directions for these sediments are discussed in detail by Weldon (1984, 1986). Dated offsets are best fit by an average Pleistocene slip rate of ~3.5 cm/yr (Fig. 5).

Kinematic constraints on Quaternary slip rates

A Quaternary slip rate of 3.5 cm/yr is consistent with the kinematic constraints imposed by the relative motion for the Pacific–North American plate boundary in southern California and documented slip on other structures in the plate boundary deformation system. Although a detailed review of plate kinematic considerations is beyond the scope of this chapter, pertinent arguments are briefly summarized below because they place limits on the Quaternary slip rate for the San Andreas fault in the central Transverse Ranges. For a more detailed discussion of plate kinematics, as well as the slip rates, styles, and uncertainties for the major structures in the Transverse Ranges, see Humphreys and Weldon (1986, 1991) and Weldon and Humphreys (1986, 1987, 1989).

The San Andreas fault is only one of many Quaternary structures contributing slip to the plate boundary deformation system of southern California. Given a Quaternary motion vector between the Pacific and North American plates of 5.5 cm/yr (Minster and Jordan, 1984), the slip contribution by other Quaternary faults in southern California limits the slip rate on the San Andreas fault. Other structures in the plate boundary system contribute about 2 cm/yr of right-lateral slip, reducing the slip attributable to the San Andreas fault from the plate boundary total of ~5.5 to ~3.5 cm/yr.

Several vector diagrams illustrate the plate kinematic constraints (Fig. 6). Structures that contribute slip across the Pacific–North American plate boundary at the latitude of the Transverse Ranges can be divided into six categories: (1) the right-lateral San Andreas fault, which trends about 18° west of the Pacific–North American plate motion vector; (2) right-lateral faults east of the San Andreas fault in the Mojave Desert, which are parallel to the plate motion vector; (3) right-lateral faults west of the San Andreas fault in the Peninsular Ranges and continental borderland,

which also are parallel to the plate motion vector; (4) east-striking compressional structures, including thrusts, reverse faults, and folds in the western and central Transverse Ranges; (5) east-striking left-lateral faults throughout the Transverge Ranges; and (6) north-striking normal faults east of the San Andreas fault, and, possibly, in the offshore borderland. When added together, slip vectors on all these structures must equal the total motion vector across the plate boundary at the latitude of the Transverse Ranges (Fig. 6A).

Two principal factors affect the magnitude of slip available for the San Andreas fault at the latitude of the Transverse Ranges. First, northwest-striking right-lateral faults, north-striking normal faults, east-striking left-lateral faults, and block rotation all contribute at least a component of dextral shear in the direction and sense of the relative plate motion, thereby limiting the amount of right-lateral slip that can be accommodated by the San Andreas fault alone. The contribution of these structures, summed along a line across the plate boundary at the latitude of the central Transverse Ranges, is labeled "other" in Figure 6. Second, the more westerly motion vector for California west of the San Andreas fault, attributable to both the more westerly trend of the San Andreas fault through the Transverse Ranges and the westerly component of motion across the left-lateral and normal faults not parallel to the plate motion vector, must be restored to the trend of the Pacific plate motion vector by east-trending folds and thrust faults in the western Transverse Ranges. This can be readily understood by examining the vectors in Figure 6. If the slip rate of the San Andreas fault is increased beyond ~3.5 cm/yr, the north-south convergence associated with the east-trending structures will not close the loop. For example, if the San Andreas fault had a slip rate of 5 cm/yr, additional deformation (shown by the gap in Fig. 6B) must also be added for the total to match the plate vector. In this example, about 2 cm/yr of slip across north-northeast–striking reverse faults, west-northwest–striking left-lateral faults, or some oblique combination of these two end members would be required. Significant faults of these styles simply do not exist in the Transverse Ranges.

It should be noted that we use a plate motion vector that is about 8° more westerly and 0.5 cm/yr greater in magnitude than the currently accepted global value (NUVEL, N38°W, 4.9 cm/yr: DeMets and others, 1987). Figure 6C shows that the use of the accepted global value requires a slip rate of less than 3 cm/yr for the San Andreas fault through the Transverse Ranges and even higher rates of shortening across the east-striking reverse faults to close the vector circuit, given the geometry and approximate slip rates of the other faults in the plate boundary zone. It should also be noted that our model assumes little deformation away from the major faults and fold belts. If a significant fraction of the plate motion is absorbed by widely distributed minor structures, the amount of slip available for the major structures is correspondingly less. We submit that the distributed shear is quite minor, because the sum of all of the major faults approximately equals the relative plate motion. However, the distributed shear may help explain relatively low Holocene slip rates determined

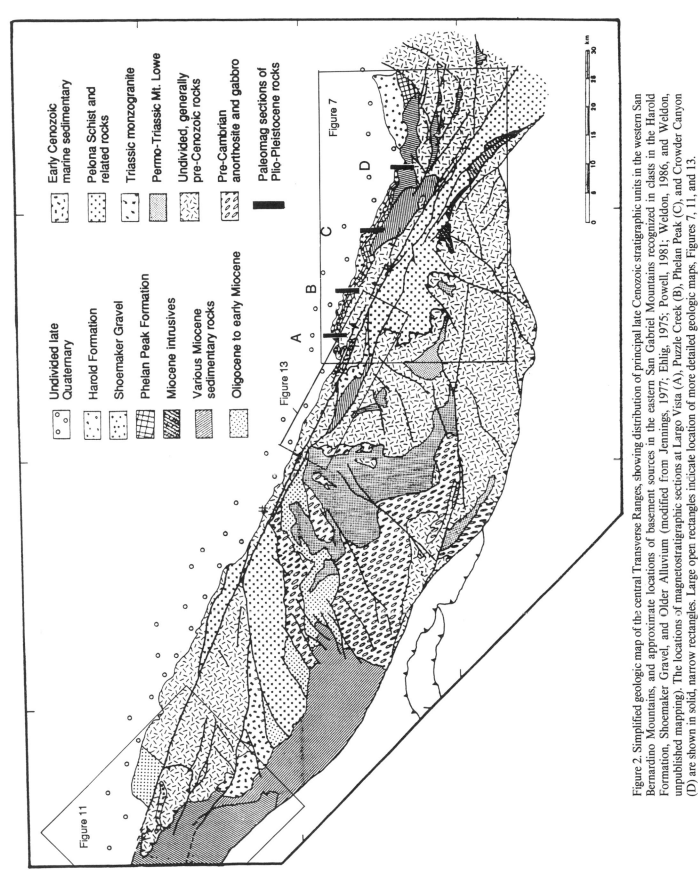

Figure 2. Simplified geologic map of the central Transverse Ranges, showing distribution of principal late Cenozoic stratigraphic units in the western San Bernardino Mountains, and approximate locations of basement sources in the eastern San Gabriel Mountains recognized in clasts in the Harold Formation, Shoemaker Gravel, and Older Alluvium (modified from Jennings, 1977; Ehlig, 1975; Powell, 1981; Weldon, 1986, and Weldon, unpublished mapping). The locations of magnetostratigraphic sections at Largo Vista (A), Puzzle Creek (B), Phelan Peak (C), and Crowder Canyon (D) are shown in solid, narrow rectangles. Large open rectangles indicate location of more detailed geologic maps, Figures 7, 11, and 13.

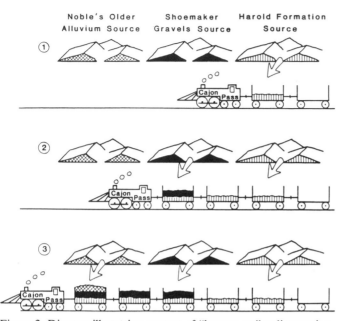

Figure 3. Diagram illustrating concept of "hopper car" sedimentation, originally proposed by Crowell (1974) for the Ridge basin, as applied to the deposition of the Harold Formation, Shoemaker Gravel, and Older Alluvium. Source terranes on the southwest side of the San Andreas fault, shown as mountains, shed distinctive sediments southward onto the floor of the Mojave Desert block, shown as a moving ore train, leading to a predictable vertical stacking of lithologically distinct sediment, reflected in the ordered contents of the "hopper cars." The diagram illustrates why the deposits shed across the San Andreas fault onto the moving floor of the Mojave Desert block are time-transgressive; variable amounts of time have elapsed since the filling of each successive hopper car, which represents a paleomagnetic section.

by measuring the offset of small features across a narrow zone, as evidenced by the results for the Pallett Creek site where both brittle and distributed shear were measured. Our Pleistocene and older offset features are wide enough to span the zone of faults near the San Andreas fault (e.g., Weldon and Springer, 1988) and therefore should incorporate the contribution of distributed shear.

To increase the share of the plate total that can be assigned to the San Andreas fault, it is tempting to decrease the slip rate on the other northwest-striking right-lateral faults or restore the San Andreas fault to a more favorable orientation during Pliocene time and argue that slip rates decreased during Quaternary time. Neither solution is supported by available data. Right-lateral shear east of the San Andreas fault at the latitude of the Transverse Ranges is currently 0.5 to 1 cm/yr (Sauber and others, 1986; Savage and others, 1990). Extrapolating these slip rates back to the beginning of the modern San Andreas fault at 4 to 5 Ma yields a total offset of 30 to 35 km, which is consistent with documented offsets across the northwest-striking right-lateral faults in the Mojave Desert (Dibblee, 1967; Powell, 1981; Dokka, 1983). Clockwise rotation in the western Transverse Ranges and the continental borderland (Luyendyk and others,

1980, 1985; Hornafius, 1985) and north-south shortening across the western Transverse Ranges of 53 km (Namson and Davis, 1988) indicate up to ~50 km of right-lateral shear and related shortening across the region west of the San Andreas fault during Pliocene time (Humphreys and Weldon, 1991). Only 250 km of right-lateral offset has occurred across the entire plate boundary since the onset of motion on the modern San Andreas fault at about 4.5 Ma. When the other deformation in the boundary zone is subtracted (250–35 km [northwest-trending Mojave faults] – 50 km [western Transverse Ranges rotation and shortening] – ? [undocumented distributed shear]), less than 160 km is left over for the San Andreas fault. The leftover amount is consistent with a maximum long-term slip rate of ~3.5 cm/yr.

If the San Andreas fault in the Transverse Ranges was initially oriented parallel to the Pacific plate motion vector, one could hypothesize that its slip rate has decreased as it has rotated into its currently unfavorable trend, implying higher Pliocene slip rates to accomplish a larger total offset. This hypothetical rotation would be due either to left-lateral shear associate with greater extension in the northern Basin and Range province (Garfunkel, 1974), or passive rotation in a right-lateral shear couple (Luyendyk and others, 1980). Even if such rotation of the San Andreas fault had occurred, the San Andreas fault could not necessarily assume a larger share of the total plate motion, because the rotation would be accompanied by broad right-lateral shear, which requires slip on other faults in the deforming system (Luyendyk and others, 1980), or rapid west-northwest–oriented extension within the Basin and Range province. Either process would contribute right slip to the overall motion of the Pacific plate relative to North America and therefore diminish the right-lateral shear available for the San Andreas fault alone. Rotation of the San Andreas fault requires the rotation of the blocks that bound it, especially the Mojave Desert and the San Gabriel blocks. There is evidence for counterclockwise rotation of the San Gabriel block (Terres and Luyendyk, 1985; Liu and others, 1988a). However, most authors agree that the rotation is due to the San Gabriel block rotating through the preexisting restraining geometry of the San Andreas fault in the central Transverse Ranges (Weldon and Humphreys, 1986; Terres and Luyendyk, 1985). Paleomagnetic data from the Mojave block suggest that little regional rotation has occurred during the time that the modern San Andreas fault was active (Weldon, 1986; Wells and Hillhouse, 1989). The rotation data suggest that the Mojave block has acted as a "backstop" for the rotating San Gabriel block, and that the San Andreas fault has always had its more westerly strike.

In summary, there is no evidence to support the hypothesis that the kinematics of the southern San Andreas fault were ever more favorable for right-lateral slip on the fault than they are today. In fact, faults and plate boundaries commonly simplify with increasing offset (e.g., Withjack and Jamison, 1986; Wesnouski, 1988), suggesting that the percentage of the plate motion accommodated by the San Andreas fault has increased, rather than decreased, through time. Such an increase is supported by our analysis of the fraction of the plate motion that the San

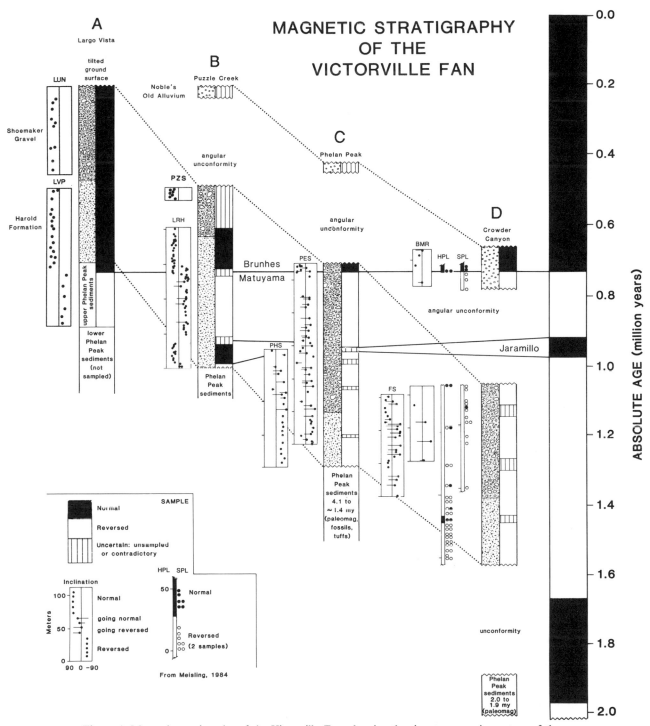

Figure 4. Magnetic stratigraphy of the Victorville Fan, showing the time-transgressive nature of the Harold Formation, Shoemaker Gravel, and Older Alluvium. See Figures 2 and 7 for locations of Largo Vista, Puzzle Creek, Phelan Peak, and Crowder Canyon magnetostratigraphic sites. Sample key shows symbols used for normal, reversed, and uncertain samples. Note that the Brunhes-Matuyama polarity transition occurs within the Older Alluvium at Crowder Canyon, within the Shoemaker Gravel at Phelan Peak, within the Harold Formation at Puzzle Creek, and within the Phelan Peak Formation at Largo Vista. Independent dating of the sections by vertebrate fossils, the presence of Pliocene ashes, and magnetostratigraphy in the underlying Phelan Peak Formation (Fig. 8) requires all sections to post-date the Olduvai magnetozone.

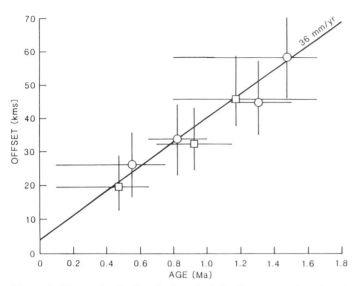

Figure 5. Slip rate for the San Andreas fault for Quaternary time, based on age and magnitude of offset from source drainages for distinctive Pleistocene stratigraphic units including the Harold Formation, Shoemaker Gravel, and Older Alluvium of Noble (see Weldon, 1986, for details). Least-squares fit to data indicates an average Quaternary slip rate of 36 mm/yr, which is consistent with documented Holocene rates from a variety of sources as well as plate-kinematic considerations.

Andreas fault has accommodated through time (see below). Given the presence and style of the other structures in the plate boundary, the San Andreas fault simply cannot account for more than 3.5 cm/yr of the total slip between the Pacific and North American plates.

PLIOCENE SLIP HISTORY

The Pliocene history of slip on the San Andreas fault and associated deformation in the central Transverse Ranges is recorded in the Pliocene deposits that lie north of the central Transverse Ranges. Pliocene evolution is less well constrained than either the Quaternary or Miocene history, due to the limited exposure and preservation of Pliocene strata around the flanks of the central Transverse Ranges. It is from this incomplete stratigraphic record that the Pliocene deformation history for the San Andreas fault system must be pieced together. Though widely separated, the Pliocene rocks that lie north and south of the central Transverse Ranges have strong lithologic similarities. There is no evidence in the Pliocene stratigraphic record for the tremendous relief near the San Andreas fault that characterizes the depositional setting of the Quaternary and late Miocene units. Because little has been formally published on the Pliocene rocks

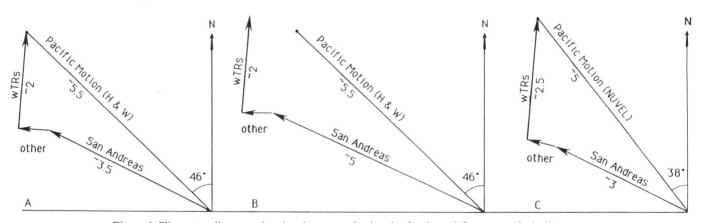

Figure 6. Slip vector diagrams showing the rate and azimuth of estimated Quaternary fault slip across major deformation systems in southern California relative to the Pacific–North American plate motion vector. Slip azimuth for major deformation systems, reflecting movement direction of western blocks relative to eastern blocks, is represented on diagram by vector azimuth relative to north and west axes of standard compass rose. Slip-rate magnitude, reflecting average rates of motion of southwestern blocks relative to northeastern blocks, is represented on diagram by vector lengths labeled in units of centimeters per year. Three different cases are shown: A, preferred solution in which San Andreas fault slip-rate vector is ~3.5 cm/yr, and slip vectors across western Transverse Ranges and other faults bring the plate-margin vector into agreement with the Pacific–North American plate motion vector as determined by Humphreys and Weldon (1987, 1991); B, solution in which the slip vector on the San Andreas fault is ~5 cm/yr, and the addition of slip for other major deformation provinces in California results in a significant mismatch of the total vector sum and the Pacific–North American plate vector; and C, solution illustrating that use of the more northerly NUVEL (DeMets and others, 1987) plate vector results in a lower slip rate on the San Andreas fault and a higher slip rate across the western Transverse Ranges. These three cases serve to illustrate that because the San Andreas fault in the Transverse Ranges has a more westerly orientation than the Pacific–North American plate-motion vector, any slip rate in excess of about 3.5 cm/yr on the San Andreas fault will have to be balanced by motion on other fault systems of a sense and magnitude that does not exist. See text for sources and discussion of data.

north of the central Transverse Ranges, we describe them in some detail here. Pliocene rocks that lie on the south flank of the central Transverse Ranges are discussed in detail elsewhere in this volume (Sadler and others, this volume; Morton and Matti, this volume).

Pliocene strata exposed along the San Andreas fault north of the San Gabriel Mountains (Figs. 2, 7) are transitional in character between overlying Quaternary fanglomerates and underlying Miocene fluvial sandstones. The Miocene fluvial sandstones were transported southwestward from sources in the Mojave Desert block into several distinctly different basins, whereas the Quaternary fanglomerates were transported northeastward in a broad alluvial apron from sources in the San Gabriel Mountains across the San Andreas fault. Intervening Pliocene strata locally contain coarse-grained facies derived from the south interbedded with fine-grained facies sourced from both north and south. Fine-grained facies are made up of lacustrine siltstone and claystone, indicating widespread conditions of restricted drainage (Barrows, 1979, 1980; Foster, 1980; Barrows and others, 1985; Meisling, 1984; Weldon, 1986; Meisling and Weldon, 1989). We therefore interpret the Pliocene strata as having been deposited in a series of restricted lacustrine basins, which we speculate were interconnected and elongated parallel to the San Andreas fault.

Phelan Peak Formation

The most complete and best-dated section of this transitional Pliocene stratigraphic interval is exposed at Phelan Peak at the northwestern end of Cajon Valley (Fig. 7). Fossil ages, fission track dates on volcanic ashes, and magnetostratigraphic correlations indicate that this section spans a time period from just before 4 Ma to sometime after 1.8 Ma (Fig. 8). The section is dominated by paleosol-rich distal fan and braided stream clastic facies, interbedded with coarse conglomerate exhibiting northward transport directions, and lacustrine siltstone and claystone. We chose this exposure as the type section for our Phelan Peak Formation, a distinctive stratigraphic unit that can be mapped for at least 30 km both east and west of Phelan Peak (Weldon, 1984; Meisling and Weldon, 1989).

Foster (1980) subdivided the Pliocene strata of the Phelan Peak Formation (his "western Crowder Formation") into three facies, consisting of a fine-grained member A, coarse-grained member B, and fine-grained member C (Fig. 8). He mapped these subdivisions for about 30 km west of Phelan Peak. Foster's facies subdivisions cannot be extended to the east, however, where we have mapped units lithologically equivalent to either member A or C that lack the conglomerate diagnostic of member B (Meis-

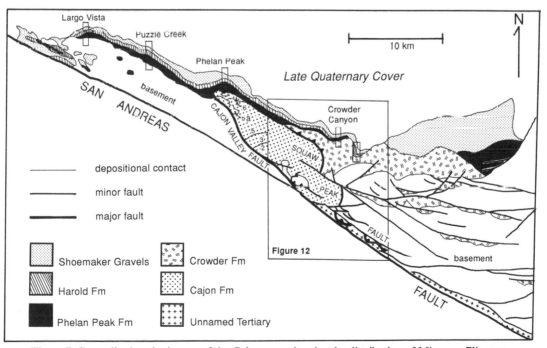

Figure 7. Generalized geologic map of the Cajon area, showing the distribution of Miocene, Pliocene, and Quaternary strata. Pliocene and Quaternary strata lie in profound angular unconformity on Miocene and older rocks, reflecting a major late Miocene deformation event associated with the Cajon Valley fault, the Squaw Peak fault, and related compressional structures (shown in more detail on Fig. 12). Pliocene and Quaternary strata in the Cajon area are now exposed along the north flank of a basement-cored antiform that developed during Quaternary uplift of the northwestern San Bernardino and northeastern San Gabriel Mountains. Subdivisions of the Cajon Formation are from Woodburne and Golz (1972), and are shown here only outside the area covered by Figure 12.

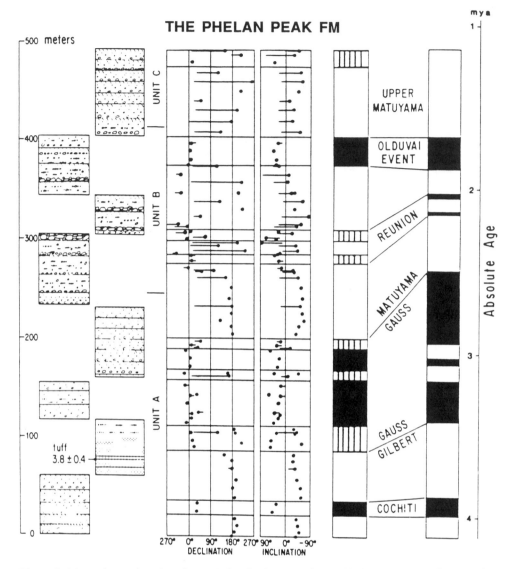

Figure 8. Magnetic stratigraphy of the Phelan Peak Formation at Phelan Peak (see Figs. 2, 7 for location). See Figure 4 for sample key showing symbols used for normal, reversed, and uncertain samples. Independent dating of the sections by vertebrate fossils and tephrachronology confirms their Pliocene to earliest Pleistocene age. Based on gradational contact with overlying Harold Formation, and depositional similarities to Quaternary units, we postulate that the Phelan Peak Formation is time-transgressive as well, so this section only represents the age at Phelan Peak. Preliminary results from paleomagnetic studies at Puzzle Creek, Crowder Canyon, and Largo Vista support a time-transgressive pattern of deposition for the Phelan Peak Formation, but it is not yet as well documented as the Quaternary units.

ling and Weldon, 1989). Poorer facies definition to the east may be explained by subdued relief and greater original distnace from southwestern sources of coarse clastic debris.

Juniper Hills Formation

West of Largo Vista, Barrows (1979, 1980; Barrows and others, 1985) described a very similar stratigrahic interval, called the Juniper Hills Formation. The Juniper Hills Formation differs

from the Phelan Peak Formation in its intimate involvement with deformation in the San Andreas fault zone, the presence of many distinct local facies, and the lack of organization into a simple continuous stratigraphic section like the type Phelan Peak Formation. Although the Phelan Peak Formation had a much simpler depositional and deformational history, the overall character of the two formations is similar enough that they may be lithologically correlative, and probably should be a single formation. We retain the distinct published usage in this chapter.

Preliminary paleomagnetic data (Fig. 4) from a section at Largo Vista (Fig. 2), which we and Foster (1980) have mapped as the western limit of the Phelan Peak Formation and Barrows and others (1985) have mapped as the easternmost limit of the Juniper Hills Formation, suggest that the Juniper Hills Formation is in part, if not entirely, younger than the type Phelan Peak section. The time line represented by the Bruhnes-Matuyama polarity reversal can be shown to cut across lithostratigraphic boundaries, occurring at the top of the Shoemaker Gravel at Phelan Peak and below the Harold Formation at Largo Vista (Fig. 4). Because of the gradational character of the contact at the base of the Harold Formation at Largo Vista, we infer that the reversal in the upper part of the underlying Phelan Peak Formation is the Bruhnes-Matuyama polarity reversal; this interpretation is supported by the long period of reversed polarity beneath the reversal. Given the demonstrated time-transgressive nature of the overlying Quaternary formations and the apparently conformable contact between the Phelan Peak and Quaternary formations in this area, it is likely that the Phelan Peak Formation and related units are also time transgressive.

The only unequivocal evidence of syntectonic deposition of Pliocene rocks is found within the Juniper Hills Formation, where the outcrop belt encounters the San Andreas fault zone (Barrows and others, 1985). Elsewhere north of the central Transverse Ranges, the Pliocene section overlies Miocene structures, is laterally continuous, and is no more deformed than overlying early Quaternary units. This suggests that Pliocene deformation on the San Andreas fault was concentrated in a relatively narrow zone, in contrast to the broad regions of associated faulting and warping that characterize Quaternary and late Miocene deformation.

Constraints on Pliocene slip history

Trails of Quaternary clastic debris allow us to track the displacement history of distinctive high-relief source areas past the western Mojave block along the San Andreas fault zone. Tracking Pliocene displacement is more difficult because there is little clastic debris that can be traced across the entire fault zone. Barrows and others (1985) documented 102 km of Plio-Pleistocene lateral slip on the San Andreas fault, and allowed for the possibility of a few "tens of kilometers of additional faulting on the Little Rock fault" (p. 215). They argued (p. 217) that significantly greater displacement was unlikely "if right-lateral displacement did not begin along the north side of the San Gabriel Mountains until the Pliocene."

The three-fold facies division of the Phelan Peak Formation may reflect a history of displacement of Pliocene paleotopographic features along the San Andreas fault zone. The abrupt transitions from fine- to coarse-grained clastic debris in the Phelan Peak Formation suggest the rapid appearance of drainage systems tapping areas of considerable topographic relief to the southwest. Clast imbrication data from coarse member B indicate

a flow direction of about N20E (Foster, 1980, p. 41), which is generally supported by our unpublished observations of the trends of paleochannels and clast imbrication in member B. These paleocurrent directions are essentially perpendicular to the northern San Gabriel rangefront. Since no angular or temporal unconformity exists between fine-grained member A and coarse-grained member B to indicate local uplift, we hypothesize that strike-slip motion on the San Andreas fault delivered and then removed a high-relief source area, resulting in the deposition of coarse-grained member B. Conversely, during deposition of fine-grained members A and C, we infer that there was no high-relief source terrane across the San Andreas fault from the site of deposition. As discussed below, the region between Liebre Mountain and the San Francisquito fault (Fig. 1) was uplifted during the late Miocene, and passed by the Phelan Peak area across the San Andreas fault during Pliocene or earliest Pleistocene time. By middle Pleistocene time the central San Gabriel Mountains faced the area, resulting in deposition of the Harold Formation and Shoemaker Gravel.

Foster (1980) has argued that the clasts in member B came exclusively from Table Mountain, a narrow sliver of basement rocks between the San Andreas fault and the Phelan Peak Formation outcrops (Figs. 1, 7). In support of his provenance interpretation, Foster (1980) noted the presence of abundant white marble clasts in member B of the "western Crowder Formation" near Phelan Peak, several kilometers down the paleocurrent direction from large marble outcrops at Table Mountain. There are several potential problems with Foster's interpretation: (1) the proposed source area at Table Mountain is fairly limited for such an extensive deposit; (2) unfolding of the broad antiform responsible for the unroofing of basement at Table Mountain removes the northward tilt of the Plio-Pleistocene strata and restores the basement outcrops at Table Mountain to a level topographically below the Phelan Peak Formation sediments and the relatively low-relief erosion surface upon which they were deposited; (3) the lack of marble clasts in member A, which lies below the marble-rich member B and directly on basement to the west (Fig. 7), argues against the availability of abundant marble clasts on the original unconformity surface; (4) no angular unconformity or disconformity has been found to support uplift following the deposition of member A and prior to the deposition of member B, which might have locally exposed marble outcrops; and (5) no Cajon Formation clasts are found above the base of member A (Foster, 1980, p. 45, confirmed by our observations), yet marble-rich basement was faulted against the Cajon Formation during late Miocene time and Pliocene uplift of the basement to create a local source of marble clasts should have provided Cajon Formation clasts as well.

We propose an alternate hypothesis in which the passage of the high-relief region southeast of Liebre Mountains was associated with the deposition of the coarse member B of the Phelan Peak Formation, whereas members A and C would have been deposited during passage of the low-relief Ridge basin and Soledad Canyon regions, respectively (Fig. 1). Basement rocks similar to those at Table Mountain exist in our proposed source region southeast of Liebre Mountain, which was part of the same basement assemblage prior to San Andreas faulting. Our model is highly speculative for several reasons. First, the marble clasts in member B of the Phelan Peak Formation can be derived from the strip of basement north of the San Andreas fault as proposed by Foster (1980). Second, the actual sources of clast types other than marble in the Phelan Peak Formation have not yet been established by direct comparison of diagnostic lithotypes. Finally, the fine-coarse-fine facies breakdown of the Phelan Peak Formation, which has only been documented in a 30-km strip in the vicinity of the type locality, differs from facies patterns in the lithostratigraphically correlative Juniper Hills Formation to the northwest and Phelan Peak Formation outcrops to the east. These regions must have migrated past the same topographic domains across the San Andreas fault, but unlike the Phelan Peak section, the Juniper Hills rocks are in the San Andreas fault zone and reflect local sources and syndepositional internal deformation.

If our model is correct, the Phelan Peak Formation provides a continuous record of Pliocene slip on the San Andreas fault zone. Simply matching the topographic domains with their associated deposits yields a slip rate of 3 to 4 cm/yr, essentially the same as the Quaternary rate. More detailed study of sources and clasts should help prove or disprove the model and refine the slip estimate. Our model is consistent with a total Plio-Pleistocene displacement across the San Andreas fault zone of about 150 km, which is in reasonable agreement with the work of Barrows and others (1985) and the better constrained Miocene offsets discussed below.

LATE MIOCENE SAN GABRIEL TRANSFORM SYSTEM AND PROTO-TRANSVERSE RANGES

The late Miocene history of the San Andreas fault system is recorded in a diverse group of structures, unconformities, and sedimentary rocks that can be understood within the context of the rapidly evolving San Gabriel transform system and development of a proto–Transverse Ranges. If we extrapolate the Pliocene through Quaternary slip rate of ~3.5 cm/yr back to earliest Pliocene time, the rugged ancestral San Bernardino Mountains are restored 150 km on the San Andreas fault to a location opposite the Ridge basin. This restoration is in reasonable agreement with the proposal by Matti and others (1985; Frizzell and others, 1986; Matti and Morton, this volume) for correlation of a distinctive Triassic megaporphyritic monzogranite body in the Liebre Mountain area to a location opposite similar rocks in the central San Benardino Mountains, yielding a total cumulative offset of 160 km on the modern San Andreas fault. We favor a 150-km restoration, which aligns the southern end of the Cajon Valley fault with the northern projection of the San Gabriel fault to form the late Miocene San Gabriel transform system (Matti and others, 1985; Weldon, 1986), and juxtaposes the coeval Liebre Mountain and Squaw Peak thrust systems to form the proto–Transverse Ranges.

The proposed 150-km correlation between the Ridge basin and Western San Bernardino Mountains areas is supported by the similarity in timing of structural events in the two areas (Fig. 9). Development of the modern San Andreas fault was preceded by late Miocene dextral displacement of ~60 km on the San Gabriel–Cajon Valley fault system and shortening on the Squaw Peak–Liebre Mountain thrust system. The San Gabriel fault was active between about 11 and 5 Ma (Crowell, 1982b). The Cajon Valley fault was active sometime between the end of deposition of the Cajon Formation (≤13.5 Ma) and the beginning of deposition of the Phelan Peak Formation (4 Ma; Meisling and Weldon, 1989). The Squaw Peak thrust was active between the end of Crowder Formation deposition (≤9.5 Ma) and the onset of deposition of the Phelan Peak Formation (4 Ma; Meisling and Weldon, 1989); activity on the Liebre Mountain thrust has been bracketed between 7.3 and 5.0 Ma (Crowell, 1982a, b; Ensley and Verosub, 1982). Widespread erosion associated with uplift of the proto–Transverse Ranges along the Squaw Peak–Liebre Mountain thrust system is interpreted to have supplied sediment to fill the Ridge basin.

Parts of central Transverse Ranges that currently lie within the San Gabriel and San Bernardino Mountains were uplifted during late Miocene time, whereas other parts subsided rapidly producing narrow fault-controlled basins, filled with coarse, locally derived debris. Although paleotopography during late Miocene time probably did not rival that of the Quaternary central Transverse Ranges, it does appear that local relief was greater than in Pliocene time. We propose that uplift of the proto–Transverse Ranges was a consequence of the restraining geometry of the late Miocene San Gabriel transform system—which was dominated by the San Gabriel fault—and that this late Miocene style of uplift and deposition terminated with development of the modern San Andreas fault.

Uplift on the Squaw Peak–Liebre Mountain thrust system

We interpret late Miocene uplift in the western San Bernardino Mountains to be the result of thrusting and reverse faulting associated with the southwest-directed Squaw Peak thrust system (Meisling and Weldon, 1989). We relate all of the late Miocene east-striking reverse faults of the western and central San Bernardino Mountains to this system, which we believe extended to the east to include the uplift and reverse faulting associated with the potentially contemporaneous Santa Ana thrust system. We also extend this system to the west across the San Andreas fault to the Liebre Mountain thrust system, and speculate that contemporaneous uplift in the central San Gabriel Mountains was also part of a late Miocene proto–Transverse Ranges (Fig. 10).

The similarities in structural style and timing between the Ridge basin region and the western San Bernardino Mountains can be seen by comparing the structural styles on Figures 11 and 12, and timing on Figure 9. The Liebre Mountain and Squaw Peak thrust systems have many features in common (Alexander, 1985; Alexander and Weldon, 1987; Miller and Weldon, 1989a,

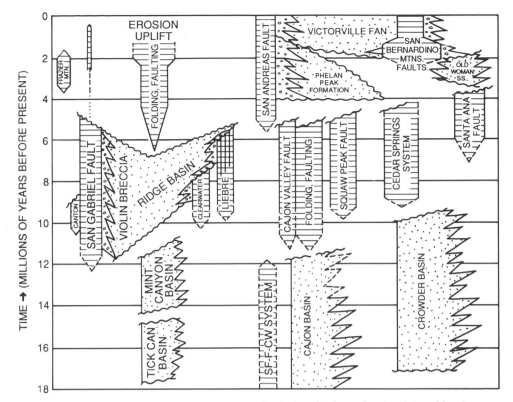

Figure 9. Tectonostratigraphic correlation chart showing timing of deformational and depositional events in the western San Bernardino Mountains (modified from Meisling and Weldon, 1989) and their counterparts in the Ridge basin (from Crowell, 1982b) plotted on a linear time scale. Dot patterns indicate depositional events; horizontal lines indicate deformational events. Note that motion on the Squaw Peak fault system, the Cedar Springs reverse fault system, and structures in the Cajon basin are contemporaneous with the movement on the Liebre thrust system and related structures in the Ridge basin. Note also that motion on the Cajon Valley fault is contemporaneous with motion on the San Gabriel fault in the Ridge basin. Miocene stages in the tectonostratigraphic evolution of the two areas are remarkably similar, and coincident with key periods in the development of the San Andreas–San Gabriel plate-margin fault system.

1989b; 1992). In addition to the obvious similarity in age, both systems display a zig-zag surface expression, with northwest-striking low-angle segments alternating with north-striking high-angle segments. Both fault zones are characterized by a distinctive style of lower plate deformation in which Miocene sedimentary rocks are tightly folded into northwest-striking, south-vergent, anticline-syncline pairs interpreted to be due to forced folding over northeast-dipping high-angle reverse faults at depth. In contrast, the upper plate basement rocks and overlying sedimentary cover are only weakly deformed, and cut by northwest-striking arcuate reverse faults interpreted as imbricate splays off a master detachment at depth. Both fault systems truncate and produce intense deformation within the Tertiary basins they encounter, yet neither shows particularly distinctive basement contrast (Ehlig, 1988, for the Squaw Peak fault; Crowell, 1982b; Alexander, 1985, for the Liebre Mountain fault).

We also have proposed that the Santa Ana thrust system is part of this family of east-striking late Miocene structures (Meis-

ling and Weldon, 1989). This suggestion is at best speculative because of the poor age control in the Santa Ana basin. Sadler (1985, this volume) has documented middle Miocene vertebrate fossils (~15 Ma) at the base of the western facies of the Santa Ana Sandstone and late Miocene basalt (~6.2 Ma) within the eastern facies. The presence of basalt clasts in the upper part of the central facies of the Santa Ana Sandstone has led Sadler (1982a, 1985, this volume) to conclude that the central facies is younger than the eastern, perhaps 4 to 6 Ma. He has acknowledged, however, that deposition of the central facies could extend from the middle Miocene age of the vertebrate fossils at the base to the age of the Quaternary uplift of the San Bernardino Mountains.

We are impressed by the similarity of depositional and structural settings between the western facies of the Santa Ana Sandstone and the eastern outcrops of the Crowder Formation. We do not argue for direct correlation of these units, but favor a similar age and structural setting. If the central facies of the Santa Ana

Sandstone is more closely related in age to the fossils in the western facies, and the basalt clasts in the upper part of the central facies are derived from outcrops other than those in the eastern facies (see Neville and Chambers, 1982; Neville, 1986), then the western and central facies could be essentially contemporaneous with the Crowder Formation. In this case the eastern facies would be the youngest unit, and all deposition would stop with the onset of motion on the Santa Ana thrust between 6 and 4 Ma. This

model would allow timing of motion on the Santa Ana thrust to be contemporaneous with motion on the Squaw Peak–Liebre Mountain thrust system in the western San Bernardino Mountains.

Inclusion of the western end of the Santa Ana basin in our model for the late Miocene evolution of the western San Bernardino Mountains is not crucial to our argument; it simply extends the late Miocene regional uplift eastward. In fact, such inclusion

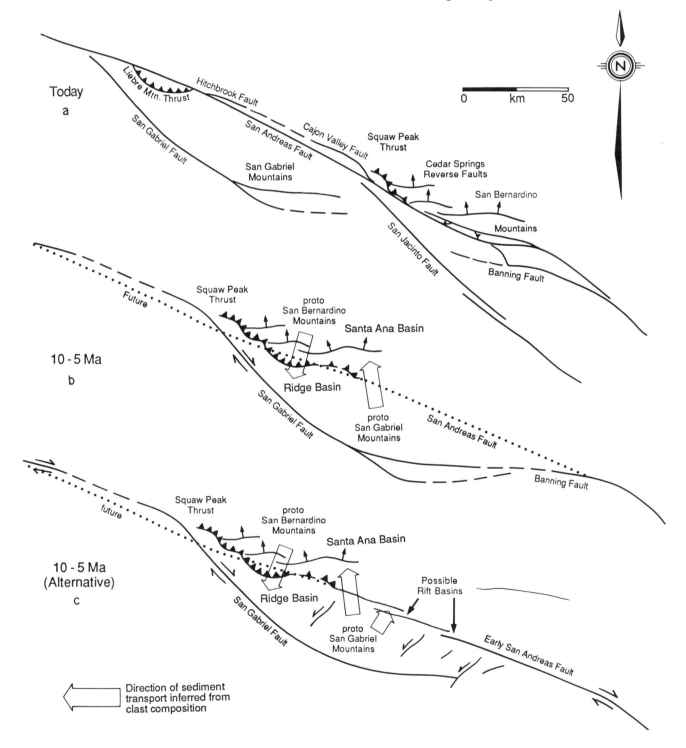

requires a different depositional model and offset history for the San Andreas fault than that of Sadler (1982a, 1985, this volume). Until better age control is available for the crucial central facies of the Santa Ana Sandstone, we favor the interpretation of the Santa Ana Sandstone as a middle to late Miocene unit.

Eastern Sediment Sources for the Ridge Basin

Crowell (1982a, b) proposed that most of the late Miocene to Pliocene Ridge Route and Hungry Valley Formations of the Ridge basin were derived from sources across the San Andreas fault to the northeast, hypothesizing (Crowell, 1982b, p. 40) that rocks in the westernmost San Bernardino Mountains are the most likely source terrane for the Ridge basin fill. We concur with this interpretation and suggest that (1) the Hungry Valley Formation was sourced from recycled sediment eroded from the Crowder Formation and underlying basement rocks during uplift of the ancestral western San Bernardino Mountains, (2) the Ridge Route Formation is composed of sediment derived from the basement terrane that lies generally to the east of the Crowder basin, and (3) the "megaconglomerate" of Ramirez (1982) was derived from recycled sediment and underlying basement rocks of the westernmost edge of the Santa Ana basin. A comparison of clast types in these three groups of rocks supports this correlation.

A profound angular unconformity separates Pliocene strata from underlying Miocene strata in the Cajon Pass and western San Bernardino Mountains region (Meisling and Weldon, 1989). Deposition of both the Cajon and Crowder Formations ended in

Figure 10. Proposed restorations of the San Andreas fault to its configuration between about 10 and 5 Ma, based on correlation of late Miocene structural and stratigraphic elements between the Ridge basin and the western San Bernardino Mountains areas. a, distribution of key faults as they appear today; b and c, two alternate interpretations of a reconstruction in which the Liebre Mountains thrust system restores to the leading edge of the Squaw Peak thrust system during late Miocene uplift of the proto–San Bernardino Mountains. In both interpretations, the Ridge basin is filled by sediment shed southwestward off the rising proto–San Bernardino Mountains, and the Santa Ana basin is filled by sediment shed northeastward off the rising proto–San Gabriel Mountains. This reconstruction is supported by the occurrence in each of these basins of distinctive clast types that can be tied to source terranes in correlative areas across the San Andreas fault. In b, the San Gabriel fault connects with the Cajon fault on the north and the Banning fault on the south, creating a restraining bend in the late Miocene San Gabriel fault system that may explain the observed similarities between late Miocene and present-day tectonic patterns in the central Transverse Ranges. Alternatively, c shows the central Transverse Ranges as a step-over zone from the San Gabriel fault to an early San Andreas fault system with Pliocene rift basins and left-lateral faults perhaps forming a portion of the kinematic connection between the two zones.

the beginning of late Miocene time, at about 12.5 and 9.5 Ma, respectively. The angular unconformity, which omits any record of deposition in the Cajon Pass area between 9.5 and about 4 Ma, is overlain by the basal beds of the Phelan Peak Formation. Development of the angular unconformity coincides with the proposed uplift of the ancestral San Bernardino Mountains in the Cajon Pass and western San Bernardino Mountains areas (Meisling and Weldon, 1989). Erosion at the unconformity stripped the Crowder Formation and significant amounts of underlying bedrock from much of the western San Bernardino Mountains and shed the resulting clastic debris to the southwest, hypothetically into Ridge basin.

The Hungry Valley Formation (Crowell, 1982a; Ramirez, 1982) is, in part, a second-generation deposit reworked from a white arkose containing clasts of rock types from the Victorville area of the Mojave Desert. The Crowder Formation is the only available source with the right composition, age, and uplift history to supply recycled sediment to the Hungry Valley Formation. Ramirez (1982) proposed that the Hungry Valley Formation came across the San Andreas fault from the far eastern end of the San Bernardino Mountains, now offset about 220 km from the site of deposition. His model requires that the bulk of the sediment came from the central Mojave Desert region and mixed with local rock types underlying the erosion surface in the eastern San Bernardino Mountains.

The Ramirez proposal is untenable for two reasons. (1) Ramirez appealed to a source region in the easternmost San Bernardino Mountains that did not develop positive relief until Pleistocene time (Sadler, 1981, 1982a, b, 1985, this volume; Meisling and Weldon, 1989), long after the late Miocene or Pliocene deposition of the Hungry Valley Formation. An extensive middle Miocene erosion surface that underlies much of the easternmost San Bernardino Mountains is locally capped by basalt flows of late Miocene age (Neville, 1986), the extensive preservation of which precludes erosion of significant volumes of material from the region since middle Miocene. (2) The rock types described by Ramirez (1982) from the Hungry Valley Formation do not exist in the easternmost San Bernardino Mountains and must be derived from the central Mojave block or some unit earlier derived from the Mojave block.

The Ridge Route Formation (Fig. 11), widely exposed in the Ridge basin to the southeast of the Hungry Valley Formation, is principally composed of first-generation sediment derived from a crystalline terrane that included a variety of gneissic and plutonic rock types. We propose that the sediment for the Ridge Route Formation was derived primarily from the region of uplifted basement that lies between and south of the Crowder and the Santa Ana basins.

We infer that the "megaconglomerate" Ramirez (1982), which rests above the Liebre Mountain thrust, came from the western edge of the Santa Ana basin or the eastern edge of the Crowder basin. While Ramirez (1982) emphasized the occurrence of large marble clasts and rare small basalt clasts, we also note the presence of a variety of volcanic, plutonic, and meta-

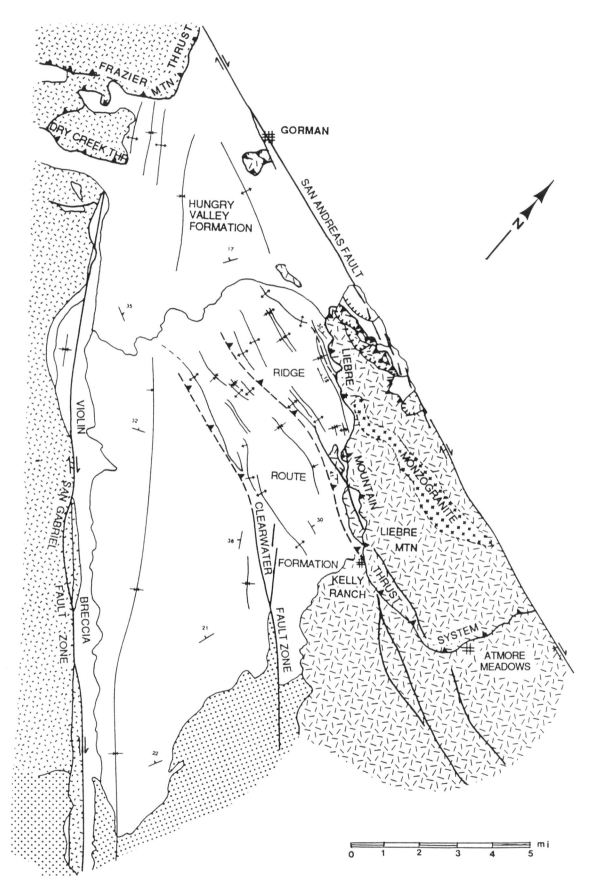

morphic clasts that appear to have been reworked from an older, in part Mojave-derived, sedimentary unit. We found several large blocks of lithic arkose containing similar clasts, confirming our hypothesis that the source includes an older sedimentary unit. We suggest that the best source for this deposit is the western facies of the Santa Ana Sandstone and its underlying basement. This source region is on the eastern edge of the uplifted ancestral San Bernardino Mountains and the lithic arkose is much like the clasts found in the megaconglomerate. Furthermore, the lower Santa Ana Sandstone is in part middle Miocene in age and thus old enough to be the source. Large marble clasts described by Ramirez (1982) may have come from roof pendants of miogeoclinal rocks within the basement rocks that underlie the Santa Ana Sandstone, and the basalts may have been reworked from the upper part of the Santa Ana Sandstone itself or derived directly from flows some distance to the north or east (Sadler, 1982b, 1985, this volume; Sadler and Reeder, 1983).

We interpret the "megaconglomerate" deposit as having been shed off an escarpment at the site of an active fault, perhaps a reverse fault in the Liebre Mountain thrust system. This origin would explain why the megaconglomerate deposits are uncommon in the Ridge basin. If the conglomerate was first shed off a scarp onto the Liebre Mountain block, which originated some distance from the basin, it later could have been carried southwest across the future trace of the San Andreas fault and over the basin sediments on the upper plate of the Liebre Mountain thrust sheet. The scarp deposit would have been isolated when the San Andreas fault offset the Liebre Mountain block from the ancestral San Bernardino Mountains.

Restraining geometry of the San Gabriel transform system

The San Gabriel transform system is the middle to late Miocene phase of the San Andreas fault system, which included the San Gabriel–Cajon Valley fault system, as well as its northern and southern continuations. Motion on the San Gabriel fault was contemporaneous with (1) uplift on the Squaw Peak and Liebre Mountain thrust systems; (2) uplift along the thrust systems within the Ventura basin (Crowell, 1982b), (3) uplift in the source area for the Violin Breccia (Crowell, 1974, 1982b), and (4) uplift in the central San Gabriel Mountains, which provided the source of sediments to both the eastern Ridge basin (e.g., Ehlert, 1982) and central Santa Ana Sandstone basin (Sadler and

Demirer, 1986; Sadler and others, this volume). When offset on the modern San Andreas fault is restored, the San Gabriel fault has a more easterly trend in the central Transverse Ranges than contemporary faults in central California to the north or the Salton Trough to the south (Fig. 10). We hypothesize that widespread late Miocene uplift was a response to convergence in a restraining bend in the San Gabriel transform system.

One possible geometry for the proposed late Miocene restraining bend would be a smaller version of that in the San Andreas fault of the Transverse Ranges today (Fig. 10B). Alternatively, the restraining bend in the San Gabriel transform system could have had a step-over geometry in the central Transverse Ranges (Fig. 10C). We favor the latter case, in which the San Gabriel fault decreased in offset to the southeast as slip was transferred across the zone of convergence and uplift of the proto–Transverse Ranges to a southern continuation of the San Gabriel transform system. The hypothetical southern fault segment would presumably have displayed a similar northwestward decrease in offset and either terminated at the late Miocene ancestral San Bernardino Mountains or connected in some unknown way with the northern continuation of the late Miocene transform system in central California. The step-over model may explain the presence of northeast-striking left-lateral faults in the San Gabriel Mountains as part of a kinematic connection between two parallel right-lateral systems (Fig. 10C).

Slip on the San Gabriel fault decreases from a maximum of 60 to ~22 km southeast of the Ridge basin in the central San Gabriel Mountains (Ehlig, 1981). While Ehlig (1981) suggested that the extra slip was taken up on a southern branch of the San Gabriel fault (see Fig. 10B), his hypothesis is unlikely because of the lack of major offset of the Glendora Volcanics (Shelton, 1955; Fig. 1) that lie across the proposed southern branch. If the San Gabriel fault does indeed decrease in offset to the southeast, another parallel fault would be required to take up the missing slip in the San Gabriel transform system.

There are several late Miocene pull-apart basins along the San Andreas fault system that are too old to be associated with the modern San Andreas fault. The existence of these late Miocene basins may also require the existence of a southern continuation of the San Gabriel transform system. The location of this inferred southern fault does not allow a direct connection with the southern end of the San Gabriel fault, giving further credence to the presence of a major parallel southern fault zone. Alternatively, the pull-apart basins may represent the waning phase of deformation on Powell's San Francisquito–Fenner–Clemens Well fault system (Powell, 1981, this volume), which would predate the development of the San Gabriel transform system.

Correlations across the San Andreas fault south of the Ridge basin indicate about 180 km of total cumulative offset (Dillon, 1975; Dillon and Ehlig, this volume; Powell, 1981, this volume), whereas correlation of the Ridge basin and San Bernardino Mountains regions only allows about 150 km of cumulative offset. The extra 30 km could be due to slip on the postulated late Miocene southern continuation of the San Gabriel transform sys-

---

Figure 11. Geology of the Liebre Mountain–Ridge basin area, simplified and modified from Crowell (1982b) and Alexander (1985). Note the structural style near the Liebre Mountain thrust, indicating syndepositional contraction, and the zig-zag pattern of outcrop, like the Squaw peak fault (Figs. 7, 12). We adopt the shape of the monzogranite and the outcrop trace of the Liebre Mountain thrust east of Kelly Ranch from Alexander (1985), not Crowell (1982b). The monzogranite in Liebre Mountain is completely surrounded by this thrust fault of unknown displacement.

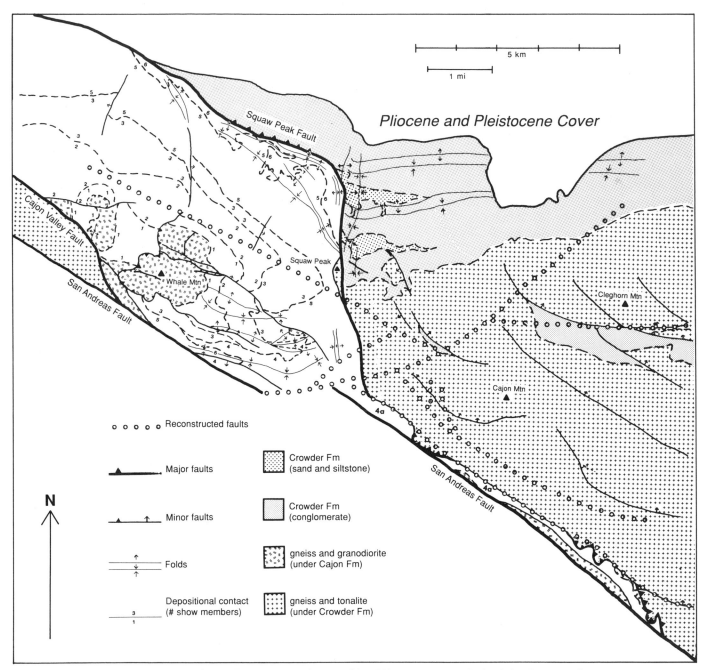

Figure 12. Simplified palinspastic geologic map of late Miocene structures in the Cajon Pass area. See Weldon (1986) and Meisling and Weldon (1989) for current distribution of reconstructed structures. Several kilometers of left-lateral slip were restored across the Cleghorn fault (decreasing from east to west to match offset structures), about 1 km of right-oblique offset is restored across a northeast-striking fault zone that offsets the Squaw Peak fault but offset by the Cleghorn, and lesser amounts of right-oblique slip is restored across two faults parallel to the San Andreas. Once restored, the Squaw Peak fault forms a large Z shape and is low angle where it strikes northwest and high angle where it strikes north. Note that it separates the Cajon and Crowder basins and truncates the large antiform that wraps around Whale Mountain in the Cajon Formation. Note that the east-trending folds occur as pairs, suggestive of forced folding over north-dipping reverse faults, like those underlying the Crowder and associated basement to the east.

tem south of the point at which slip is shared with the San Gabriel fault (essentially the model shown in Fig. 10C).

There are many other possible explanations for this apparent 30-km mismatch. Slip associated with faults within the Mojave Desert region contributes to slip on the San Andreas fault system and could account for much of the 30-km difference (Powell, 1981). A small amount of right-lateral slip across the Liebre Mountain fault zone (Crowell, 1982b) and subparallel faults like the Clearwater fault could also contribute to the deficit. Another possibility, suggested by Meisling and Weldon (1986, 1989), is that thrusting associated with the Liebre Mountain fault crossed an early southern branch of the San Andreas fault that had already accumulated some offset. Subsequent motion on the modern San Andreas fault could then have cut this thrust flap, which contains the Triassic monzogranite, and translated it an additional 150 km. This scenario yields a total offset of ~180 km for rocks not carried across the postulated early fault strand in the upper plate of the thrust. Current reconstructions are not sufficiently precise to account for the discrepancy, and additional work will be necessary to resolve between these several possibilities.

Tectonic setting of the Ridge basin

In light of evidence for widespread late Miocene compression within the proto–Transverse Ranges in response to a restraining bend in the San Gabriel transform system, it is worth reconsidering the proposed origin of the Ridge basin as an extensional or pull-apart feature. Crowell (1974) explained both subsidence of the Ridge basin and uplift of source areas to the northwest as related phenomena caused by an S-shaped restraining bend in the San Gabriel fault system. Crowell (1974, 1982b) postulated that compression was localized within the restraining portion of the bend, with extension developing at the releasing point where the fault returned to its original trend. Crowell's model offers no explanation for the Liebre Mountain thrust system, a compressional feature active during deposition, on the eastern margin of the Ridge basin during alleged extensional subsidence in the releasing portion of the postulated bend (Fig. 10).

Link (1982) extended Crowell's model to include the Liebre Mountain thrust, but it is clear that Crowell (1974) did not view the Liebre Mountain thrust as the result of compression in the bend prior to opening of the basin. Liebre Mountain is located where the basin pulls away from the northeastern side in Crowell's model (Crowell, 1974, Fig. 5, also 1982b, Fig. 8). Although Crowell (1982a, b) mapped the Liebre Mountain fault as a thrust at the surface, in cross section he shows it steepening with depth to become a vertical structure. Our mapping indicates that the low angle structures extend completely around Liebre Mountain, and that the Liebre Mountain thrust does not steepen with depth. Our observations suggest that the Ridge basin extends beneath Liebre Mountain at least as far as the San Andreas fault.

The proposed correlation of the Liebre Mountain and Squaw Peak thrusts suggests that the Ridge basin formed at the leading edge of a regionally extensive thrust system. We propose

(Fig. 10) that the Ridge basin may have originated in a compressional setting, perhaps in response to shortening and flexural downwarping of the crust between the toe of the Squaw Peak–Liebre Mountain thrust system and the San Gabriel fault. This interpretation is supported by the geometry of secondary structures that lie between the San Gabriel and Liebre Mountain faults (Fig. 11). In Crowell's restraining bend model (Crowell, 1974), compression should precede extensional subsidence. However, evidence of intense syn- and postdepositional compressional deformation in the Ridge Route and Hungry Valley Formations demonstrates that the sediments were shortened by the Liebre Mountain thrust system both during and after deposition.

A compressional origin for the Ridge basin is consistent with the setting in which we find basins in the Transverse Ranges today. In fact, Crowell (1974) used the San Bernardino basin as a modern example of this basin-forming process. Just as the modern San Andreas fault traverses the modern Transverse Ranges, the San Gabriel fault could have traversed the late Miocene Transverse Ranges. By analogy with the San Andreas fault of today, the San Gabriel fault would have been characterized by strike-slip motion, with compression on secondary thrusts producing downwarped basins (Fig. 10B).

Pull-apart basins in the late Miocene San Gabriel transform system

During late Miocene time deposition took place in several small depressions within the southern continuation of the San Gabriel transform system, along the future trace of the modern San Andreas fault. The two basins that are best preserved from this period are the Devil's Punchbowl and the Mill Creek basins (Fig. 1). Both lie completely within the San Andreas fault zone, are structurally controlled and elongate parallel to the bounding faults, are deep relative to their size, and have been folded into synclines that closely follow the original axes of deposition. Spatially associated with each of these well-defined basins are long slivers of Tertiary rock that may be genetically related as well.

Mill Creek basin. The Mill Creek basin of Gibson (1971; Sadler and Demirer, 1986; Sadler and others, this volume) is a small, lens-shaped body of late Miocene sedimentary rock within the San Andreas fault system. The Mill Creek basin has many of the characteristics of a pull-apart basin (Sadler and Demirer, 1986; Sadler and others, this volume). It lies within a strike-slip system, is elongate in the direction of the strike-slip faults, and has been deformed into a synform parallel to the original basin margins. The stratigraphic section is very thick relative to the small size of the basin, displaying abrupt lateral facies changes and breccia at the fault-bounded margins. The basin center is dominated by finer grained, more mature axial facies that interfinger with coarser marginal facies displaying paleocurrent indicators that point to the basin center. All of these characteristics invite comparison with the other Miocene pull-apart basins in the region. Sadler and others (this volume) have proposed that the Mill Creek basin formed within the San Francisquito–Fenner–Clemens Well system of Powell (1981, this volume).

The Mill Creek basin is spatially associated with a 50-km-long discontinuous chain of slivers of similar rocks to the northwest between the Wilson Creek and Mill Creek strands of the San Andreas fault (Matti and others, 1985). Early workers generally included all of the Tertiary rocks north of San Bernardino Valley in one formation (e.g., Potato Sandstone of Vaughan, 1922). Sadler and Demirer (1986; Sadler and others, this volume) note that there are stratigraphic reasons for including all of the Tertiary rocks within the Mill Creek Formation, but also add (Sadler and others, this volume) that there are some structural and stratigraphic problems with integrating the northwestern slivers into the tectonic setting of the main Mill Creek basin.

Matti and others (1985) called the slivers the "Tertiary; fringe" of the San Bernardino Mountains, and suggested that the slivers are completely unrelated to the Mill Creek basin. They separated the slivers from the Mill Creek basin with the Wilson Creek fault, which they inferred has large displacement. They hypothesized that the "Tertiary fringe" was shed off the ancestral San Bernardino Mountains southwestward into the San Gabriel Mountains, and therefore could be equivalent to the Ridge Route Formation in the Ridge basin. Meisling and Weldon (1989) noted that the northern boundary of the Tertiary slivers is commonly a thrust and that their structural setting is similar to that of the Cajon Formation, suggesting the possibility that these rocks could be middle Miocene in age.

Although several plausible interpretations have been proposed for the origin and correlation of the Tertiary rocks associated with the Mill Creek basin within the San Andreas fault zone, none of the existing models simply explains their structural and stratigraphic evolution. Better age control and more structural and stratigraphic analysis will be required before their affinity can be resolved.

Devil's Punchbowl basin. Devil's Punchbowl basin is a late Miocene pull-apart basin that lies between the San Andreas and Punchbowl faults. The age, stratigraphy, structure, and provenance of the Devil's Punchbowl basin have been extensively documented by previous workers, including Noble (1926, 1932, 1953, 1954), Pelka (1971), Dibblee (1967, 1975, 1987), Woodburne (1975), Woodburne and Golz (1972), Farley and Ehlig (1977), Barrows and Barrows (1975), and Barrows and others (1985).

Like the Mill Creek basin, the Devil's Punchbowl basin is lens-shaped, elongate in a northwesterly direction, and forms a sharp syncline that follows the original axis of deposition. The stratigraphic section in the Devil's Punchbowl basin is very thick relative to its areal extent, with a finer, more mature central facies, deposited by axial flow, that interfingers with a marginal breccia that extends southeast along the Punchbowl fault. The Punchbowl Formation is in depositional contact with the Paleocene San Francisquito Formation to the northeast, is truncated to the southwest by the Punchbowl fault, and is bounded on the northwest by the southern Nadeau and Holmes faults (Fig. 13). The only original fault-controlled margin, preserved along the Punchbowl fault, is complicated by later reactivation. Here, we

focus on the depositional setting and deformation of the southwest margin of the basin, particularly the fault-related breccia (Figs. 13 and 14).

The lithic arkose of the Punchbowl Formation was deposited into a northwest-striking trough that follows the strike of the structures in the underlying Paleocene rocks. Individual beds (see Fig. 14) onlap the margins of this northwest-striking depression, and have been folded along the axis of the preexisting trough. Paleocurrent indicators within the sandstone suggest flow directions from east to west (Pelka, 1971). Given the inferred shape of the basin from the buttressing relationships shown in Figure 14, the flow was essentially axial to the basin.

The lithic arkose of the Punchbowl Formation in the Devil's Punchbowl area interfingers with, and overlies, a coarse fault-related breccia that extends for 11 km to the southeast along the Punchbowl fault zone, never exceeding 0.5 km in width (Fig. 14). The base of the Punchbowl Formation, where it rests unconformably on Paleocene sedimentary rocks (Fig. 13), is marked by a breccia that is both cut by, and overlies, strands of the Punchbowl fault, suggesting contemporaneous deposition and fault movement.

The breccia is everywhere truncated on its southwestern margin by the reactivated main strand of the Punchbowl fault, but there is no evidence that this truncating fault was active during deposition of the breccia. Marginal facies are not present near the southern Nadeau fault or the western end of the Punchbowl fault. These faults offset formations younger than the Punchbowl Formation, and we infer that the northwestern boundary of the Devils's Punchbowl basin lay beyond these faults to the northwest.

Little, if any of the breccia can be lithologically linked to rocks that currently lie southwest of the Punchbowl fault. However, rocks that lie along the fault to the north, including the Pelona Schist of Blue Ridge and the Paleocene San Francisquito Formation, are well represented in the breccia. Locally, the breccia contains clasts of several far-traveled rock types that have been juxtaposed as slivers within the Punchbowl fault zone, including Pelona Schist, Oligocene(?) sandstones and volcanic rocks, and gneissic and granitic rocks of unknown affinity. These observations further support the interpretation that the Punchbowl fault was active immediately before, during, and after deposition of the breccia.

Farley and Ehlig (1977) proposed that the Devil's Punchbowl basin was part of the Ridge basin, due to the presence of clasts of "polka dot granite" and west-directed paleocurrents common to both formations. The distribution of the "polka dot granite" is much more widespread than originally thought, and includes outcrops in the basement ridge that is interpreted to have separated the Ridge and Devil's Punchbowl basins prior to their offset according to Ehlert (1982). We can find no compelling evidence to support the correlation of Farley and Ehlig (1977), and conclude, based on our mapping, that the Devil's Punchbowl was a small isolated pull-apart basin that onlapped relief to the southeast.

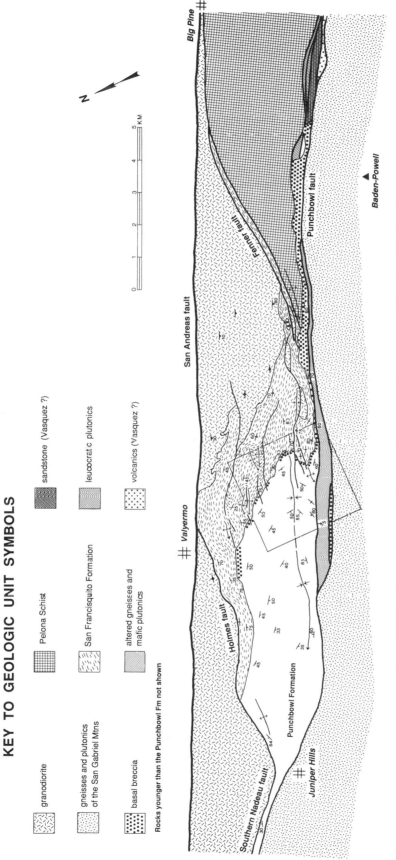

Figure 13. Simplified geologic map of the rocks between the San Andreas and Punchbowl faults in the north-central San Gabriel Mountains, showing key stratigraphic and structural relationships. Note the lens shape of the Punchbowl basin and its relationship to the bounding faults. Also, note the slivers of exotic rock in the Punchbowl zone that extend more than 10 km to the southeast of the basin. The location of Figure 14 is shown in the large rectangles.

KEY TO GEOLOGIC UNIT SYMBOLS

granodiorite

gneisses and plutonics
of the San Gabriel Mtns

basal breccia

Pelona Schist

San Francisquito Formation

altered gneisses and
mafic plutonics

sandstone (Vasquez ?)

leucocratic plutonics

volcanics (Vasquez ?)

Rocks younger than the Punchbowl Fm not shown

Origin and evolution of the Punchbowl fault

The Punchbowl fault is interpreted to have originated as part of the fault system that created the Devil's Punchbowl depression. The basin may have formed at an S-shaped bend in the Punchbowl fault (Fig. 13), or displacement could have stepped from the Punchbowl fault to a trace to the northwest. Significant right-lateral offset occurred along the Punchbowl fault some time after deposition of the Punchbowl Formation (Dibblee, 1967, 1987; Barrows and others, 1985). The magnitude of this later offset along the Punchbowl fault is best constrained by the separation of about 48 km between the Fenner fault and San Francis-

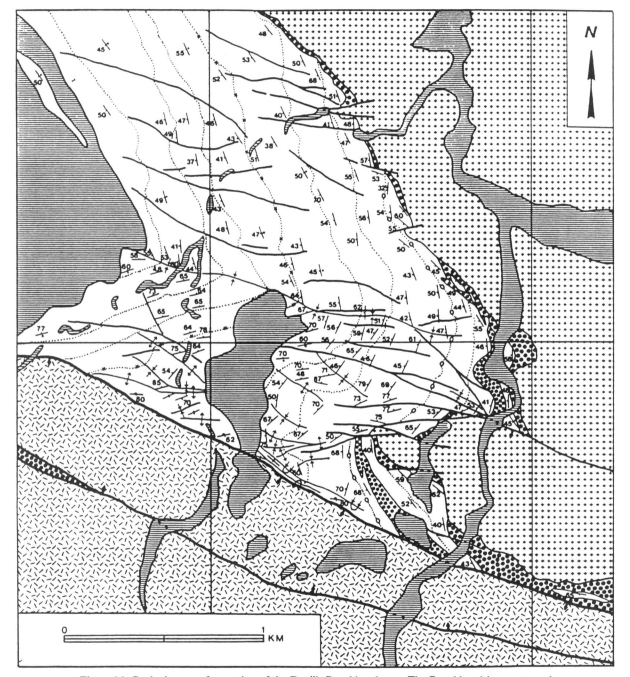

Figure 14. Geologic map of a portion of the Devil's Punchbowl area. The Punchbowl is unpatterned, basement is stipple patterned, the Paleocene has small pluses, and Quaternary deposits are striped. Note the spotted basal breccia that occurs at the base of the Punchbowl, the margins, and along the Punchbowl fault zone. The dotted contacts show marker beds that can be traced through the map; note that the sediments buttress on the edges of the basin and are thickest at the axis of the syncline, which follows the trough into which the sediments were deposited.

quito fault (Dibblee, 1967). This offset interpretation is consistent with the inferred location of the Devil's Punchbowl basin at the time of its deposition as deduced from clast provenance (Farley and Ehlig, 1977; Dibblee, 1975, 1987; Barrows and others, 1985), and, to a lesser degree, the distribution of debris shed across the Punchbowl fault into the Juniper Hills Formation (Barrows and others, 1985). The Punchbowl fault may be interpreted as part of the postulated southern fault zone of the San Gabriel transform system (Fig. 10C).

An important distinction should be made between the offset correlation of the Fenner–San Francisquito fault system, which is widely accepted, and the offset correlation of the source area for the Devil's Punchbowl basin, which is less well constrained. The offset of the Fenner fault records displacement on the Punchbowl fault both before and after deposition of the Punchbowl Formation. The basal breccia of the Punchbowl Formation overlies the Fenner fault and is spatially associated with the early phase of activity on the Punchbowl fault zone, but this does not require that the offset of the Fenner fault and the Punchbowl basin be the same, as is generally inferred. It is possible that some displacement accumulated on the early Punchbowl fault prior to deposition of the breccia. If the Punchbowl basin were offset the same distance as the Fenner fault, there would be no offset left over for the early Punchbowl fault to juxtapose diverse slivers of rocks, and create the elongate trough into which the basal breccia was shed.

It is possible that the early Punchbowl fault was connected to the San Francisquito fault, as an anastomosing strand of the San Francisquito–Fenner–Clemens Well system (Powell, 1981) that locally supplanted and postdated the Fenner fault. Rocks between the Fenner and Punchbowl faults—the Pelona Schist of Blue Ridge—constitute a sliver that floats in the reconstruction of Powell (1981), depending on how the motion on the two faults are split. If some of the deformation on the San Francisquito–Fenner zone caused by early slivering and trough development, the later Punchbowl fault may have accumulated the entire 48 km of offset. A possible test of this hypothesis may come from the age control on the other members of the San Francisquito–Fenner–Clemens Well system. The basal breccia is intimately involved with the main basin-filling arkose, implying that the early Punchbowl fault must have immediately predated or have been contemporaneous with the lower basin sediments. The age range of these sediments can be compared with other parts of Powell's (1981) system.

It is also possible that the early Punchbowl fault did not connect to any other fault, but was an isolated or preexisting fault that was later integrated into the throughgoing system. Restoring the Fenner fault to a position east of the San Francisquito fault (Fig. 1) approximately aligns the western Punchbowl fault with the Vincent thrust bounding the Pelona Schist at Sierra Pelona (southern margin of the Pelona Schist in Fig. 2). The slices in the Punchbowl fault zone are reminiscent of slices of rock along that thrust, so perhaps the fault started as a deformed part of the regional Vincent thrust.

Alternatively, the total offset on the Punchbowl fault may be the offset between the Fenner and San Francisquito faults. In this case, the offset of the Punchbowl basin must be less than the offset between these faults, because some motion on the early system is required to displace the exotic slivers and create the trough and breccia. This alternative also requires an early fault of the San Andreas system that lay northwest of the San Francisquito fault. Such an early San Andreas fault could be consistent with the model presented in Figure 10C, where displacement on the San Gabriel fault decreases to the southeast and is matched by increasing displacement on a southern continuation of the San Gabriel transform system. This southern fault would likely have relatively little displacement in the vicinity of the Devil's Punchbowl basin, which would restore closer to the northwestern end of the San Gabriel fault.

Barrows and others (1985) suggested that debris shed from the Devil's Punchbowl into the Juniper Hills Formation could be used to track all but the first 13 km of offset of the Punchbowl fault. If this is the case, slip on the pre–Punchbowl basin phase of the Punchbowl fault must have been less than this amount. More work is required to unravel the complex history of the Punchbowl fault zone.

EARLY TO MIDDLE MIOCENE SAN ANDREAS TRANSFORM SYSTEM AND EARLY TRANSVERSE RANGES PROVINCIALISM

The exact timing of onset of motion on the earliest faults of the San Andreas system is difficult to document. Powell (1981, this volume) discussed the timing of the various elements of the San Francisquito–Fenner–Clemens Well fault system, which he interpreted to be the earliest phase of the San Andreas transform system. In this section we use the sedimentary, volcanic, and structural record along the San Andreas fault zone in the central Transverse Ranges to infer the age of onset and subsequent switching of strands within the San Francisquito–Fenner–Clemens Well transform system.

Middle Miocene basins and Transverse Ranges provenance

In the central Transverse Ranges, several middle Miocene basins that were contemporaneous with the transition from the San Francisquito–Fenner–Clemens Well transform system to the San Gabriel transform system are well preserved. In the San Bernardino Mountains, the Cajon, Crowder, and, perhaps, the western Santa Ana basins began accumulating sediments within a few million years of each other in latest early Miocene time (~18 to 20 Ma). The Caliente–Tick Canyon–Mint Canyon basins were deposited within the same time period, prior to the onset of motion on the San Gabriel fault (Ehlig and others, 1975). All of these middle Miocene basins are characterized by: (1) southwest-directed paleocurrents; (2) a preponderance of clasts from the eastern California desert region, including unique lithologic types inferred to have been transported great distances; and (3) relatively broad, homogeneous distribution of internal facies. Although most of the debris in the middle Miocene basins was shed southwestward from the Mojave Desert region, several basins record evidence of the development of local relief by the end of middle Miocene time, reflecting the onset of the stratigraphic provincialism that characterizes the late Miocene proto–Transverse Ranges.

Cajon Formation. The majority of the Cajon Formation (Members 1 through 5) in the Cajon basin (Fig. 1) is predominantly derived from sources to the northeast within the Mojave block (Woodburne and Golz, 1972; Meisling and Weldon, 1989). The uppermost unit of the Cajon Formation (Member 6), however, contains abundant volcanic, red sandstone, and green sandstone clasts that are unlike clasts in the older part of the formation. The matrix of Member 6 is buff to white in color and lacks the abundant pink arkosic component that characterizes the rest of the formation; paleocurrent directions have not been reported from Member 6. This member cannot be derived from the northeast because a suitable source terrane simply does not exist within the Mojave Desert region. The character of the red sandstone clasts, particularly associated with mafic volcanics, is suggestive of Oligocene redbeds such as the Vasquez Formation. We speculate that the clasts in the upper Cajon Formation came from south or west across the San Andreas fault, suggesting relief near the San Andreas fault and disruption of regional early Miocene southwestward-flowing drainage systems.

Members 4 and 5a of the Cajon Formtion may also suggest some relief near the San Andreas fault. Member 5a was shed east or northeast across the Cajon Valley fault (Woodburne and Golz, 1972), and Member 4 was apparently shed west or southwest across the Squaw Peak fault, from the vicinity of its intersection with the San Andreas fault (Meisling and Weldon, 1989). Members 4 and 5a are thus inferred to be locally developed marginal facies of the more typical and widely distributed Member 5, and may indicate the onset of disruption of southwesterly flowing drainages that ultimately resulted in the complete reversal of drainage direction by the time Member 6 was deposited. The uppermost Cajon Formation is tentatively dated at 13.5 Ma by microvertebrate fossils (Reynolds and Weldon, 1988) and magnetostratigraphic correlation (Liu and others, 1988b). We conclude that provincial drainage systems, and by inference activity on the early San Andreas fault, was well established by this time.

Caliente Formation. Deposition of the Caliente Formation persisted well into the late Miocene, ending at 7 Ma. The uppermost late Miocene unit differs significantly from middle Miocene units. Carman (1964) noted that the late Miocene Member 3 of the Caliente Formation contains predominantly locally derived clasts, speculating that this change in source character was due to the onset of movement of the San Gabriel fault. Ehlert (1982) developed this hypothesis with a paleogeographic argument for movement on the San Gabriel fault that separated the Mint Canyon and lower Caliente Formations, causing a shift to local provenance in both basins.

Crowder Formation. The Crowder Formation is separated into lower (Members 1, 2, 3) and upper (Members 4, 5) parts on

the basis of a similar change in the depositional basin (Foster, 1980; Winston, 1985; Weldon, 1984, 1986; Meisling and Weldon, 1989). The lower Crowder Formation was deposited in a broad, braided stream environment, with conglomerate clasts derived almost exclusively from the Mojave Desert area to the northeast and with paleocurrent directions consistently to the southwest (Foster, 1980; Winston, 1985). In contrast, the upper Crowder Formation records a greater diversity of paleocurrent directions (Foster, 1980) and a different clast suite reflecting local source contributions. This is consistent with an evolution toward a more provincial and less throughgoing depositional system. Magnetostratigraphic data (Weldon, 1984, 1986; Weldon and others, 1984; Winston, 1985) and microvertebrate paleontologic studies (Reynolds, 1984, 1985; Reynolds and Weldon, 1988) demonstrate that the change in depositional patterns occurred about 11.5 Ma. They also show that the sedimentation rate in the upper part of the Crowder Formation was at least four times greater than that of the lower part.

Santa Ana Sandstone. Additional evidence for late middle Miocene relief and the onset of provincialism in the central Transverse Ranges can be found in the Santa Ana Sandstone. Sadler (1985, Sadler and Reeder, 1983, this volume) described clast suites in the lower part of the central facies of the Santa Ana Sandstone that were clearly derived from the central San Gabriel Mountains. Clast populations include distinctive rock types such as Pelona Schist, marine sandstone from the San Francisquito Formation, and "polka dot granite." Sadler interpreted the sediments containing these distinctive clast types to be Pliocene in age, but he acknowledged that they are only constrained to be younger than middle Miocene vertebrate fossils at the base of the western facies and older than the Quaternary onset of uplift of the San Bernardino Mountains. We prefer a late middle to late Miocene age for the Santa Ana Sandstone, and present our arguments in the next section.

Sadler (this volume) proposes a model in which the strata containing the distinctive clasts were shed from the Sierra Pelona near the point where it intersects the San Andreas fault (Fig. 2). Sadler's model requires offset on the Punchbowl fault system to occur prior to deposition of the Santa Ana Sandstone, in order to block the flow of other unusual clast types from the central San Gabriel Mountains. His model also requires that ~120 km of displacement has occurred on the San Andreas fault since the deposition of these sediments. Sadler argues that these requirements support a Pliocene age for the strata bearing distinctive clasts, because the modern San Andreas system would have already accumulated considerable offset by the time of their deposition.

An alternative model for the age and provenance of the Santa Ana Sandstone

We propose an alternative model for the deposition of the Santa Ana Sandstone, which allows these sediments to completely predate movement on the modern San Andreas fault and be either late middle or late Miocene in age. Our hypothesis is motivated by the fact that the entire Santa Ana Sandstone section is involved in deformation along the Santa Ana thrust fault, which we consider to be part of the late Miocene Squaw Peak–Liebre Mountain thrust system that was active in the western San Bernardino Mountains before 4 Ma (Meisling and Weldon, 1989). If our hypothesis is correct, the entire Santa Ana Sandstone section would be older than 4 Ma, and Sadler's western and central facies would be closer in age to the middle Miocene fossils at the base of the western facies. In our model, deposition of the Santa Ana Sandstone would predate the entire ~150 km of displacement associated with the modern San Andreas fault, and the diagnostic clast assemblage could not have been derived from the eastern end of the Sierra Pelona as Sadler (1985, this volume) has proposed.

We hypothesize that the exotic clasts were derived from the western end of the Sierra Pelona outcrop belt (Figs. 2, 10) during late Miocene time and were carried by streams flowing north-northwest across the San Andreas fault just southeast of Liebre Mountain (proposed path shown with a sediment transport arrow in Fig. 10). This path, which was specifically chosen to allow Santa Ana Sandstone deposition to be consistent with the full ~150 km of offset of the modern San Andreas fault rather than the 120 km proposed by Sadler (this volume), has several potential advantages. First, the proposed drainage path, which originates in the Sierra Pelona, crosses outcrop belts of both the San Francisquito Formation and the basement terrane containing documented occurrences of "polka dot granite." These outcrop belts were probably source areas in the late Miocene (Ehlert, 1982), and could have supplied suitable rock types for deposition in the Santa Ana Sandstone. Second, by deriving the Pelona Schist clasts from the western end of the Sierra Pelona outcrop belt, there is no need to invoke activity on the Punchbowl fault to divert drainages bearing distinctive rock types from the central San Gabriel Mountains away from the Santa Ana basin as proposed by Sadler (this volume).

Our model is in conflict with the interpretation of Sadler (1985, this volume) that the San Gabriel clasts in the Santa Ana Sandstone came from only ~5 km away, based on the northward decrease in clast size and obvious fan shape of facies containing the San Gabriel clast assemblage. In our view, these relationships are also consistent with a model in which a fairly long stream debouched onto a depositional plain where the mouth of the canyon, rather than the source, was ~5 km away. The decrease in size of the Pelona Schist clasts in the Santa Ana Sandstone would be similar to that which is observed on the modern fans exiting the central San Gabriel Mountains, where the rapid decrease in clast size is primarily a function of distance from the mouth of the canyon rather than distance from the source of clasts. Our proposed stream would not have been any longer, nor would it have carried clasts that were any larger, than the streams that drain the Transverse Ranges today.

Our proposed streams carried clasts north to Santa Ana basin rather than northeast, perpendicular to the San Andreas fault, as proposed by Sadler. Without specific paleogeographic evidence, however, one cannot assume that all clasts shed across the San Andreas fault were eroded directly off the fault scarp and transported perpendicular to the fault trace. Neither is it reasonable to assume that paleocurrent directions in depositional basins along the San Andreas fault can be linearly extrapolated all the way back to the source area. These simplifying assumptions are common to many models for sedimentation across the San Andreas fault, including our own proposals for transport of Miocene and younger sediments across the San Andreas fault, and those of Ehlig and others (1975) for transport of clasts from an inferred source in the Chocolate Mountains in the eastern Transverse Ranges into the Mint Canyon basin. Such simplifying assumptions can be misleading, especially when the drainage systems originate in mountain ranges of unknown paleotopography and cross active strike-slip faults.

Need for middle Miocene slip on an earlier transform system

Our model for the early to middle Miocene evolution of the Early San Andreas transform system in the central Transverse Ranges is similar to early models for the offset history of the San Andreas fault in central California. Tertiary rocks of late early Miocene and older age consistently record about 315 km of total cumulative offset across the San Andreas fault zone in central California (e.g., Matthews, 1973). Younger units are offset less across the San Andreas fault zone, implying middle Miocene onset of activity on the San Andreas system in central California (Hill and Dibblee, 1953; Hill, 1971; Dickinson and others, 1972; Huffman, 1972; Huffman and others, 1973). The timing of motion that led to these middle Miocene offsets is critical in determining whether they were contemporaneous with motion on the San Gabriel transform system or earlier faults of the Early San Andreas transform system in southern California.

Hill and Dibblee (1953; also Hill, 1971; Huffman, 1972; Dickinson and others, 1972; Stanley, 1987b) document an offset of middle Miocene rocks in central California that is up to 50 km less than the 315 km of total cumulative offset across the San Andreas fault. The offset rocks are dated as middle Miocene based on marine fauna that includes early Mohnian benthic forams and Santa Margaritan molluscs, and vertebrate fauna assigned to the latest Barstovian or earliest Clarendonian stages of the vertebrate time scale. These rocks correlate with the Mint Canyon Formation in the Ridge basin, which was deposited prior to activity on the San Gabriel fault (Crowell, 1982b). Even if the faunal time scales are stretched to allow the central California offsets to be contemporary with the onset of activity on the San Gabriel fault (during deposition of the Castaic Formation), the rocks in central California were already offset by an amount nearly equal the total cumulative displacement of ≤60 km across the San Gabriel fault by the time the San Gabriel fault began moving. However, the San Gabriel fault did not accumulate its total offset until Pliocene time (Crowell, 1982b). Even within the errors of the various relative time scales, there is little doubt that the San Andreas fault in central California started well before the San Gabriel fault, and was therefore associated with an earlier system to the south. Crowell (1973) acknowledged this discrepancy and suggested that earlier faults, such as the Clemens Well fault, could account for the difference.

San Francisquito–Fenner–Clemens Well fault system

Powell (1981) was the first to completely elucidate the possibility that the San Francisquito–Fenner–Clemens Well fault system was the Early San Andreas transform system. We concur that the San Francisquito–Fenner–Clemens Well system was the first major integrated fault system of the transform boundary in the Transverse Ranges, forming in an area that was already undergoing transtension and was only part of the overall right-lateral shear between the plates.

We believe that the San Francisquito–Fenner–Clemens Well fault system (Powell, 1981, this volume) was the dominant fault within the continental portion of the transform boundary during the middle Miocene. Powell (1981, this volume) has proposed about 100 km of middle Miocene offset on this fault system to account for the bedrock distribution before the onset of activity on the San Gabriel and modern San Andreas faults. Sadler and others (this volume) have suggested that the Mill Creek pull-apart basin was also formed in this system; we have suggested above that the early phase of activity on the Punchbowl fault may be part of this system, but other models are possible. Other than these two speculative suggestions, and the existence of the Fenner fault for which no independent offset data exist, there is little evidence preserved in the central Transverse Ranges to characterize the activity of the Early San Andreas transform system.

We know that there must be an early phase of transform activity to make our observations consistent with the total offset of rocks across the modern San Andreas system. We prefer Powell's model over other possibilities—such as greater offset across the modern San Andreas fault—because it requires that the early transform system pass south of the Liebre and San Bernardino Mountains prior to their offset by the modern San Andreas fault. It is a key requirement of our model that these rock masses remain together until Pliocene time. We find Powell's (1981) basement correlations compelling, in particular his requirement that the rocks in the central and northwestern San Gabriel Mountains initially moved in a more westerly direction than the later offset on the modern San Andreas fault.

Silver (1982, 1986, personal communication, 1987) also began his reconstruction of the rocks along the San Andreas system by moving several thrust sheets in the region of the Transverse Ranges westward relative to the rocks to the north or south. In fact, he required that the basement of the central and eastern San Gabriel Mountains move west relative to the rocks in the southeastern and northwesternmost Transverse Ranges, effectively solving the problem that motivates Powell's (1981) reconstruction. Silver's model requires low-angle structures that as yet

have not been recognized. Silver called on core-complex style extension in southeastern California as the mechanism for the westward motion of the region that was to become the Transverse Ranges province, whereas Powell invoked early transform-related shear.

We do not view these models as completely contradictory. We speculate that the earliest right-lateral shear in the region probably followed the pattern of deformation seen in most oblique rifts (e.g., Withjack and Jamison, 1986). Lateral and extensional structures coexisted in a zone of distributed deformation that became progressively integrated into what we recognize as the Early San Andreas transform system with time. The spatial and temporal relationships of the middle Miocene basins with the San Francisquito–Fenner–Clemens Well fault system suggest that it was transtensive in nature. An analog might be found in the zone of transtensive right-lateral shear in eastern California, that includes the Owens Valley to Death Valley region and faults crossing the Mojave Desert today. Like the San Francisquito–Fenner–Clemens Well fault system, which lay inboard of the locus of middle Miocene transform motion offshore, the transtensive zone in eastern California is inboard of the modern San Andreas fault and accommodates both right-lateral shear and extension. The transtensive zone in eastern California accounts for about 10 to 20 percent of the plate boundary deformation. As discussed below, this is about the same percentage that the San Francisquito–Fenner–Clemens Well system accommodated in the middle Miocene plate boundary.

Origin and significance of Oligocene to early Miocene volcanic centers

The San Andreas fault system offsets a number of areally restricted Oligocene to early Miocene volcanic centers, including the Pinnacles-Neenache, Tecuya–San Juan Batista, and Vasquez–Delihencia–Plush Ranch eruptive complexes (summarized by Stanley, 1987a). These and other relatively restricted volcanic centers and basins have proven very useful in reconstructing the total offset across the San Andreas and other faults. In fact, it seems unlikely that the San Andreas fault system would have truncated these centers simply by chance. Stanley (1987a) proposed that these and other late Oligocene to early Miocene volcanic centers record a period of transtensional activity on the California plate margin. He inferred based on spatial relationships with the San Andreas and Hosgri faults that volcanism was contemporaneous with the onset of fault motion. The often noted spatial relationship of the location of the San Andreas fault and early exposures of deep seated Pelona Schist (e.g., Powell, 1981; Ehlig, 1981) is also consistent with early transtension along the zone that would later become the San Andreas fault system.

Stanley's (1987a) arguments suggest that the crust may have been extended, thinned, and weakened along a transtensional zone of initially discontinuous fault segments that subsequently became integrated to form a throughgoing system. The volcanics would have been emplaced in response to extension within this early zone of thinned crust. If we accept that the throughgoing San Andreas fault system that offset the volcanic centers ~315 km grew within the same tectonic context that created these volcanic centers, we can use ~23 Ma as the time of onset of motion along the San Andreas system. Because of the presence of similar aged volcanic centers along the coastal Hosgri fault system (Stanley, 1987a), and within transtensive basins in the eastern California and the Mojave Desert (Dokka and Travis, 1990), we infer that all of these systems began at about the same time and may have contributed slip at rates equal to that of the Early San Andreas transform system until the acceleration in slip rate that accompanied development of the modern San Andreas fault in latest Miocene or Pliocene time.

KINEMATIC CONSTRAINTS ON MIOCENE SLIP RATES

It is difficult to determine from direct observations what role the San Andreas system played in the overall plate boundary deformation during the early stages of its development. However, a range of models can be constructed to take advantage of the constraints afforded by available offset data, the relationship between different fault zones in the system, and externally imposed plate boundary conditions. By placing portions of the slip across the various major structures into time intervals for which the data are good, and requiring that the sum of deformation always add up to the plate total, we can make inferences about the poorly defined intervals.

For example, if a large fraction of the offset across a boundary is taken up in a particular time interval, then little of it is available for other time intervals, and other structures in the plate boundary must have lesser displacement during the interval that a single structure dominates. Essentially, we are applying a simplified kinematic analysis like that applied above for the Quaternary, but also requiring that both rates and total displacements add up at any given time. Here we consider only the components of motion in the direction of slip across the plate boundary, and round values to the nearest 0.5 cm/yr, to illustrate the tradeoffs.

We take the San Andreas fault to have ~315 km of offset accumulated over ~23 Ma, divided into 150 km on the modern San Andreas fault during the past 4 to 5 Ma, 60 km on the San Gabriel transform system during the period from 12–11 to 4–5 Ma, and 105 km on the Early San Andreas transform system prior to 12 Ma. Note that Powell (this volume) uses a lower value for offset on the San Gabriel fault, but for the purposes of this simple analysis it does not matter. We take faults west of the San Andreas fault, including the Hosgri, San Gregorio, and Nacimiento, to accommodate ~115 km of offset (Grahm and Dickinson, 1978) that does not contribute to the San Andreas fault system in southern California. This system also started around 23 Ma (Stanley, 1987a).

For simplicity we represent all deformation to the west of the continent as one boundary zone at the continental margin, and all deformation to the east of the San Andreas system as another; eastern deformation consists mainly of extension within the Basin and Range, strike slip across the Mojave block, and, perhaps, strike slip across southeastern California. Although the magnitudes of right-lateral offset across these two outer boundary zones are somewhat problematic, two approaches allow us to place limits on displacement across these zones.

Limits can be placed on the deformation east and west of coastal California by considering the northward migration of the Mendocino triple junction (Fig. 15). In principle, the Mendocino triple junction moves relative to stable North America at the relative Pacific plate rate. By comparing the rate of migration of the triple junction relative to the California coast, one can infer what portion of the plate rate was occurring west of the coast and what fraction was internal to the continent.

Fox and others (1985) were the first to document the rate of northern migration of the triple junction by tracing the volcanism they inferred was related to the passage of the triple junction and the creation of a slab-free window (Dickinson and Snyder, 1979). Fox and others (1985) concluded that the San Andreas and faults to the west account for ~3.5 cm/yr averaged over the past 23 Ma. Weigand and Thomas (1987) have supported this inference. Stanley (1987a) noted that the volcanics discussed by Fox and others (1985) consisted of two groups, those clustered around 22 to 24 Ma and those that young to the northwest, apparently following the triple junction. If the position of the triple junction controlled the early volcanism, then Stanley (1987a) concurred with Fox and others (1985) that the triple junction has migrated at ~3.5 cm/yr relative to coastal California. Because this rate is not the entire plate motion, Fox and others (1985) inferred that the excess was east of the San Andreas system, particularly in the Basin and Range Province. Given the amount of slip that has occurred across the plate boundary during the time period of interest here, 5 to 6 cm/yr for ~23 Ma, one can divide the motion across the four major boundaries: 370 km west of the coast, 115 km across coastal faults, 315 km across the San Andreas, and 400 km east of the San Andreas fault.

While Fox and others (1985) offered evidence in support of this large value east of the San Andreas fault, the actual offset is probably lower, so we decrease the large amount of right-lateral shear in the Basin and Range to 200 km and increase the offshore to 570 km (to keep the plate total the same). Such a model is supported by reconstructions of the right-lateral shear across the western Basin and Range (e.g., Stewart and others, 1968) and paleomagnetic data that can be interpreted to suggest as much as 200 km of northward transport of the Sierra block relative to North America (Kanter and McWilliams, 1982; Frei and others, 1984; Frei, 1986; Kanter, 1988). This value is approximately two times the amount that can be inferred from the work of Dokka and Travis (1990) in the Mojave and the work of Wernicke and others (1989) at the latitude of Death Valley, but the work includes only late Miocene and younger activity and may be minimum values. Reducing the value east of the San Andreas may violate the inference of Fox and others (1985) for the origin of the Oligocene and early Miocene volcanics in central California be-

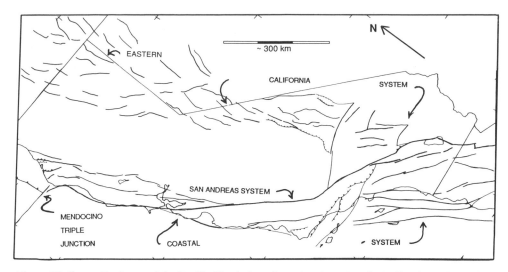

Figure 15. Generalized map of the Pacific–North American plate boundary in California. Deformation within the continent can be divided into three broad zones, a coastal system, the San Andreas system, and an eastern California system. The relative role of these three main systems is explored in Figure 16.

cause it places some of these volcanics north of the triple junction. However, it does allow the later volcanics to track the migration of the triple junction, as preferred by Stanley (1987a). We consider 200 km a reasonable compromise for the purpose of this simple model.

An alternative to adding the motion taken from the Basin and Range to the offshore boundary is to simply eliminate it from the total. DeMets and others (1987) suggested that their relatively low rate for the past few million years (~5 cm/yr) extrapolates beyond the past few million years, and over tens of millions of years could reduce the total by several 100 km. This possible refinement does not change the conclusions discussed here so we do not include it.

Figure 16 summarizes some possible models and illustrates the tradeoffs involved. Each line represents an offset path that a hypothetical rock has taken relative to the triple junction, which is assumed to move northwest relative to cratonic north America at the plate rate of ~5.5 cm/yr (Minster and Jordan, 1978). Each block is assumed to move faster than the one to the west, relative to the Pacific plate (i.e., the boundaries are all assumed to have a component of right-lateral motion). Effectively, this means that the rock's paths shown in Figure 16 cannot cross. For example, offset across faults near the California coast moves all rocks to the east in a right-lateral sense relative to the Pacific plate, the San Andreas fault moves rocks east of it even more, and any motion on boundaries to the east will add yet a third component to North America's motion relative to the Pacific plate. At all times the sum of motions on all structures must be the plate rate and by today the rocks must be in their current positions relative to the triple junction.

To illustrate the usefulness of such models we consider two unrealistic end members. In model A, we require that all of the boundaries have had constant rates for the past 23 Ma. To have

the displacement observed, model A requires the slip rate be partitioned as 2 cm/yr at the continental margin, 0.5 cm/yr across western California, 1.5 cm/yr across the San Andreas, and 1 cm/yr to the east. Obviously, the San Andreas has waxed during this time, and the offshore waned. Model A simply serves to illustrate an end-member model in which structures always have their average rate.

In model B, we have incrementally moved the total plate-boundary slip from outboard to more inboard structures. Initially, the offshore has the full 5.5 cm/yr of the displacement, then the coastal faults, then the San Andreas, and now (by this model) the region to the east. Obviously, this is not the case and it only serves to illustrate another end member. However, this sequential development is often called on for the jump from offshore to the San Andreas about 4.5 Ma (e.g., Curray and Moore, 1984). The model illustrates the point that if the San Andreas ever has the total plate motion, all other boundaries must be inactive during that time interval. Also, the other boundaries must play much larger roles during other intervals to gain the correct cumulative totals. Clearly, this has not been the case, though it is implied by widely cited models.

Model C is the simplest model that is consistent with the current kinematics of the plate boundary, the history of the San Andreas, and the division of the total offset east and west of California, as discussed above. Because the faults within the continent currently account for all of the plate motion (Weldon and Humphreys, 1987) the transform at the continental boundary is essentially dead, and probably has been for the past 4 to 5 m.y. Since this boundary must accumulate all of its ~570 km of offset in the previous 18 to 19 m.y., it must average ~3 cm/yr from 23 to 5 Ma, leaving only 2 to 2.5 cm/yr for all boundaries within the continent. Since the San Gabriel phase of the San Andreas system accumulated ~60 km between 11 and 5 Ma, it must have ~1

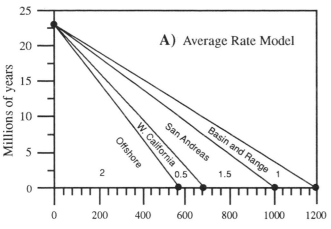

Distance (in kilometers) from the triple junction

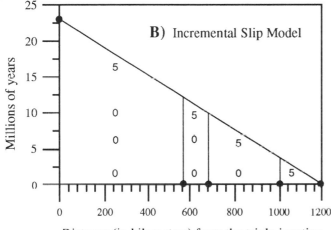

Distance (in kilometers) from the triple junction

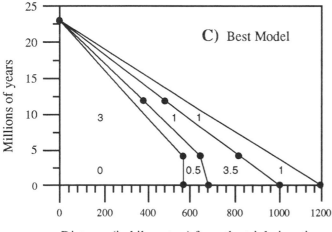

Distance (in kilometers) from the triple junction

cm/yr, leaving only 1 to 1.5 cm/yr for the coastal faults and eastern California during that interval. The division between the others is difficult to determine, but neither can surpass the San Andreas without taking essentially all motion from the other.

Between 23 and 12 Ma the San Andreas accumulated ~105 km of offset for an average rate of ~1 cm/yr. Again, given the amount of motion that must be accommodated off the continent, this leaves only 1 to 1.5 cm/yr for the other fault zones, suggesting that the San Andreas has been the dominant fault zone since it formed. It is possible that for short periods of time other faults could dominate but only if the other zones slow down. For example, if the coastal zone accumulated all of its 115 km of offset between 23 and 12 Ma, to equal the San Andreas's 105 km of offset during this time, the eastern zone would have had to have gone to essentially zero or the offshore boundary would have had to have temporarily decreased its rate to only 2 cm/yr. Either case would require that the zone that slowed down speed up by at least 1 cm/yr for the next 10 m.y. to generate enough cumulative offset today. The only model that precludes such varying rates between zones is one in which the continental zones share ~2 cm/yr of slip from ~23 to 5 Ma, with the San Andreas having almost half of the total, and the eastern zone taking most of the rest. When the locus of sea-floor spreading in the East Pacific Rise shifted into the Gulf of California (Atwater, 1970; Moore and Curray, 1982), the San Andreas accelerated to its currently high rates, without much change in the rate in the other two zones. The San Andreas system appears to have dominated the other two zones, despite its unfavorable geometry through the Transverse Ranges, because it provided the most direct connection between the Gulf of California to the south and the Mendocino Triple Junction to the north. While this scenario is not necessarily the only solution, it does explain the major features of the development of the plate boundary. It also illustrates that both the total and cumulative rate on faults within the boundary provide important constraints on the possible level of activity on all other faults within the boundary.

Figure 16. Models illustrating the tradeoffs involved in partitioning slip between principal structures of the southern California plate margin during the migration of the triple junction. All models start ~23 m.y. ago, when all three continental boundaries first became active (see text). The sum of rates across all zones must always equal the plate rate, and the offset across the individual zones must have their proper values by today. In the first model, A, all boundaries have the average rate required to offset the rocks. This clearly underestimates the San Andreas today and the offshore boundaries earlier. In model B, the motion incrementally moves into the continent, and only one boundary is active at a time; this also cannot be correct. Model C is the simplest model consistent with the data (see text). In it the onshore faults start out slowly, with the San Andreas slightly faster than the others; between 4 and 5 Ma the San Andreas accelerated, mainly at the expense of the offshore, which must now be essentially inactive. Given the amount of slip partitioned across all of the boundaries, all zones must be relatively active for most of the time since the middle Miocene, or faults must turn on and off, sequentially having very high rates.

MODEL FOR SLIP HISTORY OF THE SAN ANDREAS FAULT SYSTEM

The evidence and inferences discussed above lead us to propose a very different model for the evolution of the San Andreas fault than that which is widely cited: (1) the modern San Andreas fault in the central Transverse Ranges has accumulated about 150 km of right-lateral offset during the past 4 to 5 Ma; (2) as much as 60 km accumulated on the San Gabriel transform system during late Miocene time between 11 and 5 Ma, perhaps in part shared with early activity on the Punchbowl fault; and (3) at least 100 km of offset occurred during early to middle Miocene time, a slip history summarized in Figure 17.

A constant slip rate is assumed for the last 4 to 5 Ma of motion on the modern San Andreas fault, based on well-documented Quaternary offsets and arguments for total cumulative offset of distinctive rocks in the Liebre Mountain and San Bernardino Mountains areas. Average slip rates for the San Gabriel transform system and the San Francisquito–Fenner–Clemens Well fault system are more difficult to determine accurately because of uncertainty in the age of these structures. If we infer that deformation began ~23 Ma, and distribute the total slip for each structure within its appropriate time interval, the three points of control for the Miocene could fall on a line. The data could also be interpreted to represent a smoothly increasing slip rate, as has been suggested for similar plots of slip history for the San Andreas fault in central California (e.g., Hill, 1971; Dickinson and others, 1972; Huffman, 1972). However, Quaternary data do not require any systematic increase in southern California, and can be simply satisfied by a marked change between 4 and 5 Ma. Also, geologic evidence for long periods characterized by constant depositional and structural styles, separated by relatively brief transitions from activity on one strand to another, suggest major periodic reorganizations of the tectonic regime that are likely to produce discrete changes in the slip rates.

The earliest evidence of deformation along the San Andreas

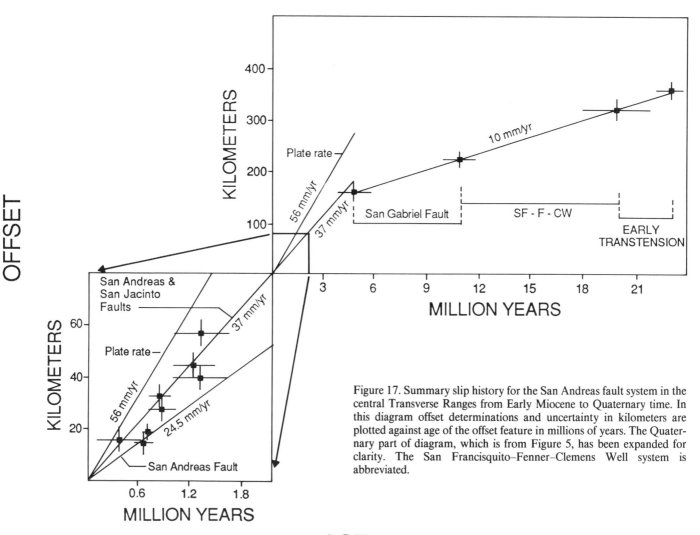

Figure 17. Summary slip history for the San Andreas fault system in the central Transverse Ranges from Early Miocene to Quaternary time. In this diagram offset determinations and uncertainty in kilometers are plotted against age of the offset feature in millions of years. The Quaternary part of diagram, which is from Figure 5, has been expanded for clarity. The San Francisquito–Fenner–Clemens Well system is abbreviated.

fault system is the creation of the late Oligocene to early Miocene basins and associated volcanic centers along the future trace of the fault zone (Stanley, 1987a). In the central Transverse Ranges this period of time is represented in the rock record by the Vasquez Formation and small slivers of similar rock along the Punchbowl zone; regionally, this time period includes development of the Pinnacles-Neenach, Tecuya–San Juan Batista, and Vasquez–Delihencia–Plush Ranch centers. Progressive shear across this transtensive zone probably produced many faults that were initially poorly integrated and had a diversity of structural styles. Faults active at this time include northwest-striking right-lateral faults like the Russell fault (Yeats and others, 1989) which was active by 23 Ma in the Cuyama Valley region (Fig. 1), and northwest- through northeast-striking oblique-normal faults that bound the early Miocene basins (e.g., Ehlig, 1981; Yeats and others, 1989). All of the deformation documented for this time period, including the clockwise rotations of Luyendyk and others (1985), contributed to northwestward motion of the rocks west of the zone of deformation. We estimate the cumulative deformation rate across this zone to have been about 1 cm/yr.

We view the San Francisquito–Fenner–Clemens Well system as the product of integration within the transtensive zone, which may solve some of the problems that have plagued acceptance of this system as an early version of the San Andreas fault. Powell (1986) pointed out that this system coexisted with west-striking left-lateral faults, producing a "zig-zag pattern" of deformation through California. Contemporaneous development in the Transverse Ranges of middle Miocene basins, northwest-striking right-lateral faults, and west- to southwest-striking left-lateral faults suggests that the early deformation was transtensive in style. For example, the San Francisquito–Fenner–Clemens Well system heads east at its southern end and has no obvious connection back to the plate boundary in the Gulf of California. If the San Francisquito–Fenner–Clemens Well system was simply an integrated portion of a broad zone of transtension that included the region that became the San Andreas and the Gulf of California, there is no reason that the fault zone must directly connect to the plate boundary.

In the modern Transverse Ranges there are many faults, basins, and zones of distributed shear not directly connected to the plate boundary that play the same kinematic role as the San Francisquito–Fenner–Clemens Well system would have played in early to middle Miocene time. Another analogy can be made with the right-lateral shear crossing eastern California today. Individual faults, like the Death Valley–Furnace Creek fault, account for a fraction of the right-lateral shear in the Pacific–North American plate boundary even though they do not directly connect to the triple junctions to the north and south. Any fault that contributes right-lateral shear in a northwesterly direction is a kinematic player in the plate margin.

With the onset of activity on the San Gabriel fault at about 11 Ma, deposition in the middle Miocene basins changed dramatically or came to an end, and the region developed the stratigraphic provincialism that characterized late Miocene time. The onset of activity on the San Gabriel fault also coincided with development of the proto–Transverse Ranges, which included uplift and erosion of the ancestral San Bernardino Mountains (Meisling and Weldon, 1989), uplift of the central San Gabriel Mountains, and related tectonism elsewhere in the Transverse Ranges. It is not known whether this change was simply a consequence of increased offset across the geometrically complex San Gabriel transform system or was due to some external cause such as a change in plate motion direction. Based on available data and kinematic modeling, we interpret the slip rate across the San Andreas fault system to have been the same in late Miocene time as it was in middle Miocene time, about 1 cm/yr.

In the western San Gabriel Mountains, the Mint Canyon and lower Caliente Formations were superseded by strata reflecting separation and provincial source areas, culminating in deposition of the late Miocene Ridge basin. The Mill Creek and Punchbowl Formations are both late Miocene basins within the San Andreas system that share many characteristics of strike-slip pull-apart basins. These basins were either associated with the waning phase of motion on the San Francisquito–Fenner–Clemens Well system or incipient deformation along an early southern San Andreas fault, perhaps linked to early motion on the Punchbowl fault. The Mill Creek and Punchbowl basins also reflect the development of local relief that characterizes the late Miocene basins all along the fault zone. This relief was the consequence of the restraining geometry on the San Gabriel transform system where it traversed the proto–Transverse Ranges.

Sometime between 4 and 5 Ma, the modern San Andreas fault became the dominant member of the San Andreas system, and the slip rate accelerated to 3.5 cm/yr. This increase in slip coincided with the shift in the locus of spreading along the East Pacific Rise to the Gulf of California (Atwater, 1970). Whether the rate change was also associated with increased convergence due to a change in the direction of Pacific–North American plate motion is controversial (e.g., Engebretson and others, 1985). Despite the increase in rate and possibly greater component of compression across the plate boundary, there is little evidence for major uplift in the central Transverse Ranges along the San Andreas fault during Pliocene time. Relatively minor deposits of coarse debris in the Hungry Valley Formation in the Ridge basin (Crowell, 1982a; Ramirez, 1982), the Phelan Peak Formation north of the San Gabriel Mountains, and the San Timoteo Formation south of the Transverse Ranges (Matti and Morton, 1975; Morton and Matti, 1979) can all be attributed to relief that originated with the late Miocene proto–Transverse Ranges. This is surprising, in light of the observation that the San Andreas must have had its "anomalous" trend by Pliocene time, and was accumulating slip faster then than it did during Miocene time.

It appears that strand-switching from the San Gabriel to the San Andreas fault temporarily relieved much of the local convergence associated with the restraining geometry of the San Gabriel transform system. However, the modern San Andreas fault is only slightly more favorably oriented than the San Gabriel fault and a marked increase in slip rate accompanied the switch, but there appears to have been an initial decrease in uplift rates. It is possible that the counterclockwise rotation of the San Gabriel block, which currently transfers most of the convergence across the central Transverse Ranges to the western Transverse Ranges (Weldon and Humphreys, 1986), began at this time, and the local geometric complexities that are currently associated with uplift in different parts of the central Transverse Ranges had not yet developed.

During Quaternary time, new regions of localized vertical deformation in the Transverse Ranges evolved that appear to be associated with developing problems in the Pliocene solution to the restraining geometry of the system. Left-lateral motion on east-striking faults, caused by the northward increase in Basin and Range extension, has kinked the fault at both ends of the Transverse Ranges segment and produced local regions of extreme shortening and uplift. The development of new right-lateral faults crossing the Peninsular Ranges has also led to renewed uplift in the central and western Transverse Ranges. Sixty kilometers of right-lateral slip has occurred on the San Andreas fault system during the Quaternary.

CONCLUSIONS

Although our "speculative history" has focused on activity along the San Andreas fault zone, the roles of other zones within the plate boundary are equally important. The concept of the San Andreas fault as a simple transform revolutionized our thinking, resulting in the integration of many diverse observations into a unifying tectonic model. The concept of the San Andreas fault as a simple transform in which all of the plate motion is expressed on a single fault has its limitations, however, and it has become an impediment to our understanding.

The San Andreas fault is the most important fault in a broad transform system. It is the natural product of 23 m.y. of evolution of the transform boundary. The San Andreas fault did not form full-blown at 4.5 Ma, or even, for that matter, at 11 Ma. Although it is developing toward a simple transform, it is still not there and may never be. The complexity that is present in the central Transverse Ranges plate-boundary zone may frustrate our efforts to make simple models, yet the central Transverse Ranges contain a wealth of information that is too often ignored because the "San Andreas problem" was so convincingly "solved" 15 yr ago. The fact that the San Andreas system in the Transverse Ranges has had so many different strands and phases of deformation offers us far more insight into the evolution of the transform boundary than would be possible if it were one simple fault. We hope that by presenting our speculative evolutionary model, with all the alternatives and uncertainties that still exist for the San Andreas fault in the central Transverse Ranges, we will motivate people to reconsider what we really know about the history of the San Andreas system. Much work will be necessary if we are to take the next step in understanding the evolution of the San Andreas fault system in southern California.

ACKNOWLEDGMENTS

We thank Bob Powell, Jon Matti, and Gene Humphreys for important intellectual contributions to this work. Mike Rymer, Bob Powell, and Doug Morton reviewed the latest manuscript, and their recommendations led to substantial improvements. Kerry Sieh, Clarence Allen, Lee Silver, and Joe Kirschvink supervised much of the work that went into this study, while two of us (R.J.W. and K.E.M.) were students at the California Institute of Technology. Important conceptual improvements came from the 1986 meeting from which this volume grew; we appreciate the contributions of all participants. Most of the work was financially supported by grants from the U.S. Geological Survey.

REFERENCES CITED

Alexander, J., 1985, A test of the correlation of the Liebre Mountain thrust with the Squaw Peak fault, 150 km to the southeast, on the north side of the San Andreas fault [B.A. thesis]: Los Angeles, California, Occidental College, 18 p., 1 plate.

Alexander, J., and Weldon, R., 1987, Comparison of structural style of the Liebre Mountain and Squaw Peak thrust systems, offset across the San Andreas fault, southern California: Geological Society of America Abstracts with Programs, v. 19, p. 353.

Atwater, T., 1970, Implications of plate tectonics for the Cenozoic tectonic evolution of western North America: Geological Society of America Bulletin, v. 81, p. 3513–3536.

Baird, A. K., Morton, D. M., Woodford, A. O., and Baird, K. W., 1974, Transverse Ranges province; A unique structural-petrochemical belt across the San Andreas fault system: Geological Society of America Bulletin, v. 85, p. 163–174.

Barrows, A. G., 1979, Geology and fault activity of the Valyermo segment of the San Andreas fault zone, Los Angeles County, California: California Division of Mines and Geology Open-File Report 79-1 LA, 49 p.

——, 1980, Geologic map of the San Andreas fault zone and adjoining terrane, Juniper Hills and vicinity, Los Angeles County, California: California Division of Mines and Geology Open-File Report 80-2 LA.

Barrows, A. G., Kahle, J. E., and Beeby, D. J., 1985, Earthquake hazards and tectonic history of the San Andreas fault zone, Los Angeles County, California: California Division of Mines and Geology Open-File Report 85-10 LA, 236 p., 21 plates.

Barrows, K. J., and Barrows, A. G., 1975, Comparison of lithology and provenance of cobbles in the western and type facies of the Punchbowl Formation: Geological Society of America Abstracts with Programs, v. 7, p. 295–296.

Bird, P., and Rosenstock, R. W., 1984, Kinematics of present crust and mantle flow in southern California: Geological Society of America Bulletin, v. 95, p. 946–957.

Carmen, M. F., Jr., 1964, Geology of the Lockwood Valley area, Kern and Ventura Counties: California Division of Mines and Geology Special Report 81, 62 p.

Clark, M. M., and 12 others, 1984, Preliminary slip-rate table and map of late-Quaternary faults of California: U.S. Geological Survey Open-File Report 84-106.

Crowell, J. C., 1962, Displacement along the San Andreas fault, California: Geological Society of America Special Paper 71, 62 p.

——, 1973, Problems concerning the San Andreas fault system in southern California, *in* Proceedings of the Conference on Tectonic Problems of the San Andreas Fault System: Stanford, California, Stanford University Publications in the Geological Sciences, v. 13, p. 368–373.

——, 1974, Sedimentation along the San Andreas fault, California, *in* Dott,

R. H., ed., Modern and ancient geosynclinal sedimentation: Society of Economic Paleontologists and Mineralogists Special Publication 19, p. 292–303.

——, 1981, An outline of the tectonic history of southeastern California, *in* Ernst, W. G., ed., The geotectonic development of California; Rubey Volume 1: Englewood Cliffs, New Jersey, Prentice-Hall, p. 583–600.

——, 1982a, Pliocene Hungry Valley Formation, Ridge basin, southern California, *in* Crowell, J. C., and Link, M. H., eds., Geologic history of Ridge basin, southern California: Pacific Section, Society of Economic Paleontologists and Mineralogists, p. 143–149.

——, 1982b, The tectonics of the Ridge basin, southern California, *in* Crowell, J. C., and Link, M. H., eds., Geologic history of Ridge basin, southern California: Pacific Section, Society of Economic Paleontologists and Mineralogists, p. 25–42.

Curray, J. R., and Moore, D. G., 1984, Geologic history of the mouth of California, *in* Crouch, J. K., ed., Tectonics and sedimentation along the California margin: Pacific Section, Society of Economic Paleontologists and Mineralogists, v. 38, p. 17–36.

DeMets, C., Gordon, R. G., Stein, S., and Argus, D. F., 1987, A revised estimate of Pacific–North America motion and implications for western North America plate boundary zone tectonics: Geophysical Research Letters, v. 14, p. 911–914.

Dibblee, T. W., Jr., 1967, Areal geology of the western Mojave Desert, California: U.S. Geological Survey Professional Paper 522, 153 p.

——, 1975, Late Quaternary uplift of the San Bernardino Mountains on the San Andreas and related faults, *in* Crowell, J. C., ed., San Andreas fault in southern California; A guide to San Andreas fault from Mexico to Carrizo Plain: California Division of Mines and Geology Special Report 118, p. 127–135.

——, 1987, Geology of the Devil's Punchbowl, Los Angeles County, California, *in* Hill, M. L., ed., Cordilleran Section of the Geological Society of America: Boulder, Colorado, Geological Society of America Centennial Field Guide 1, p. 207–210.

Dickinson, W. R., and Snyder, W. S., 1979, Geometry of subducted slabs related to San Andreas transform: Journal of Geology, v. 87, p. 609–627.

Dickinson, W. R., Cowan, D. S., and Schweichert, R. A., 1972, Discussion of "Test of the new global tectonics": American Association of Petroleum Geologists Bulletin, v. 56, p. 375–384.

Dillon, J. T., 1975, Geology of the Chocolate and Cargo Muchacho Mountains, southeasternmost California [Ph.D. thesis]: Santa Barbara, University of California, 405 p.

Dokka, R. K., 1983, Displacements on late Cenozoic strike-slip faults of the central Mojave Desert, California: Geology, v. 11, p. 305–308.

Dokka, R. K., and Travis, C. J., 1990, Late Cenozoic strike-slip faulting in the Mojave Desert, California: Tectonics, v. 9, p. 311–340.

Eberhart-Phillips, D., Lisowski, M., and Zoback, M. D., 1990, Crustal strain near the Big Bend of the San Andreas fault; Analysis of the Los Padres–Tehachapi triangulation networks, California: Journal of Geophysical Research, v. 95, p. 1139–1153.

Ehlert, K. W., 1982, "Polka-dot" granite clasts in the Ridge Route Formation, Ridge basin, southern California, *in* Crowell, J. C., and Link, M. H., eds., Geologic history of Ridge basin, southern California: Pacific Section, Society of Economic Paleontologists and Mineralogists, Field Trip Guide and Volume, p. 203–204.

Ehlig, P. L., 1975, Basement rocks of the San Gabriel Mountains, south of the San Andreas fault, southern California, *in* Crowell, J. C., ed., San Andreas fault in southern California; A guide to San Andreas fault from Mexico to Carrizo Plain, California Division of Mines and Geology, Special Report 118, p. 177–186.

——, 1981, Origin and tectonic history of the basement terrane of the San Gabriel Mountains, central Transverse Ranges, *in* Ernst, W. G., ed., The geotectonic development of California; Rubey Volume 1: Englewood Cliffs, New Jersey, Prentice-Hall, p. 253–283.

——, 1988, Characteristics of basement rocks exposed near the Cajon Pass scientific drill hole: Geophysical Research Letters, v. 15, p. 949–952.

Ehlig, P. L., Ehlert, K. W., and Crowe, B. M., 1975, Offset of the upper Miocene Caliente and Mint Canyon Formations along the San Gabriel and San Andreas faults, *in* Crowell, J. C., ed., San Andreas fault in southern California; A guide to San Andreas fault from Mexico to Carrizo Plain: California Division of Mines and Geology Special Report 118, p. 83–92.

Engebretson, D. C., Cox, A., and Gordon, R. G., 1985, Relative motions between oceanic and continental plates in the Pacific Basin: Geological Society of America Special Paper 206, 59 p.

Ensley, R. A., and Verosub, K. L., 1982, Biostratigraphy and magnetostratigraphy of the southern Ridge basin, central Transverse Ranges, California, *in* Crowell, J. C., and Link, M. H., eds., Geologic history of Ridge basin, southern California: Pacific Section, Society of Economic Paleontologists and Mineralogists, p. 13–24.

Farley, T., and Ehlig, P. L., 1977, Displacement on the Punchbowl fault based on occurrence of "polka-dot" granite clasts: Geological Society of America Abstracts with Programs, v. 9, p. 419.

Foster, J. H., 1980, Late Cenozoic tectonic evolution of Cajon Valley [Ph.D. thesis]: Riverside, University of California, 242 p.

Fox, K. F., Jr., Fleck, R. J., Curtis, G. H., and Meyer, C. E., 1985, Implications of the northwestwardly younger age of the volcanic rocks of west-central California: Geological Society of America Bulletin, v. 96, p. 647–654.

Frei, L. S., 1986, Additional paleomagnetic results from the Sierra Nevada; Further constraints on Basin and Range extension and northward displacement in the western United States: Geological Society of America Bulletin, v. 97, p. 840–849.

Frei, L. S., Magill, J. R., and Cox, A., 1984, Paleomagnetic results from the central Sierra Nevada; Constraints on reconstructions of the western United States: Tectonics, v. 3, p. 157–177.

Frizzell, V. A., Jr., Mattison, J. M., and Matti, J. C., 1986, Distinctive Triassic megaporphyritic monzogranite; Evidence for only 160 kilometers of offset along the San Andreas fault, southern California: Journal of Geophysical Research, v. 91, p. 14080–14088.

Garfunkel, Z., 1974, Model for the late Cenozoic tectonic history of the Mojave Desert, California, and its relation to adjacent regions: Geological Society of America Bulletin, v. 85, p. 1931–1944.

Gibson, R. C., 1971, Non-marine turbidites and the San Andreas fault, San Bernardino Mountains, California, *in* Elders, W. A., ed., Geological excursions in southern California: Riverside, University of California Campus Museum Contributions 1, p. 167–181.

Graham, S. A., and Dickinson, W. R., 1978, Evidence for 115 kilometers of right slip on the San Gregorio-Hosgri fault trend: Science, v. 199, p. 179–181.

Hill, M. L., 1971, A test of new global tectonics; Comparisons of northeast Pacific and California structures: American Association of Petroleum Geologists Bulletin, v. 55, p. 3–9.

Hill, M. L., and Dibblee, T. W., Jr., 1953, San Andreas, Garlock, and Big Pine faults, California: A study of the character, history, and tectonic significance of their displacement: Geological Society of America Bulletin, v. 64, p. 443–458.

Hornafius, J. S., 1985, Neogene tectonic rotation of the Santa Ynez Range, western Transverse Ranges, California, suggested by paleomagnetic investigation of the Monterey Formation: Journal of Geophysical Research, v. 90, no. B14, p. 12503–12522.

Huffman, O. F., 1972, Lateral displacement of upper Miocene rocks and the Neogene history of offset along the San Andreas fault in central California: Geological Society of America Bulletin, v. 83, p. 2913–2946.

Huffman, O. F., Turner, D. L., and Jack, R. N., 1973, Offset of late Oligocene-Early Miocene volcanic rocks along the San Andreas fault in central California, *in* Proceedings of the Conference on Tectonic Problems of the San Andreas Fault System: Stanford, California, Stanford University Publications in the Geological Sciences, v. 13, p. 368–373.

Humphreys, E., and Weldon, R., 1986, Pacific–North America relative motions in southern California [abs.]: EOS American Geophysical Union Transactions, v. 67, p. 905.

Humphreys, E. D., and Weldon, R. J., 1991, Kinematic constraints on the rifting of Baja California, chapter 12, in Dauphin, J. P., and Simoneit, B.R.T., eds., The Gulf and Peninsular Provinces of the Californias: American Association of Petroleum Geologists Memoir 47, p. 217–229.

Jennings, C. W., 1977, Geologic map of California, California Division of Mines and Geology, scale 1:750,000.

Kanter, L. R., 1988, Paleolatitude of the Butano Sandstone, California, and its implications for the kinematic histories of the Salinian terrane and the San Andreas fault: Journal of Geophysical Research, v. 93, p. 11699–11710.

Kanter, L. R., and McWilliams, M. O., 1982, Rotation of the southernmost Sierra Nevada, California: Journal of Geophysical Research, v. 87, p. 3819–3830.

Keller, E. A., Bonkowski, M. S., Korsch, R. J., and Shlemon, R. J., 1982, Tectonic geomorphology of the San Andreas fault zone in the southern Indio Hills, Coachella Valley, California: Geological Society of America Bulletin, v. 93, p. 46–56.

Link, M. H., 1982, Sedimentation, tectonics, and offset of Miocene-Pliocene Ridge Basin, California, in Anderson, D. W., and Rymer, M. J., eds., Tectonics and sedimentation along the San Andreas system: Pacific Section, Society of Economic Paleontologists and Mineralogists, p. 17–31.

Liu, W., Kirschvink, J. L., and Weldon, R., 1988a, Paleomagnetic study of the Punchbowl Formation, California [abs.]: EOS Transactions of the American Geophysical Union, v. 69, p. 1164.

Liu, W., Weldon, R. J., and Kirschvink, J. L., 1988b, Paleomagntism of sedimentary rocks from and near the DOSECC Cajon Pass well, southern California: Geophysical Research Letters, v. 15, p. 1065–1068.

Luyendyk, B. P., Kamerling, M. J., and Terres, R. R., 1980, Geometric model for Neogene crustal rotations in southern California: Geological Society of America Bulletin, v. 91, p. 211–217.

Luyendyk, B. P., Kamerling, M. J., Terres, R. R., and Hornafius, J. S., 1985, Simple shear of southern California during Neogene time suggested by paleomagnetic declinations: Journal of Geophysical Research, v. 90, no. B14, p. 12454–12466.

Matthews, V., 1973, Pinnacles-Neenach correlation; A restriction for models of the origin of the Transverse Ranges and the Big Bend in the San Andreas fault: Geological Society of America Bulletin, v. 84, p. 683–688.

Matti, J. C., and Morton, D. M., 1975, Geologic history of the San Timoteo Badlands, southern California: Geological Society of America Abstracts with Programs, v. 7, p. 344.

Matti, J. C., Morton, D. M., and Cox, B. F., 1985, Distribution and geologic relations of fault systems in the vicinity of the central Transverse Ranges, southern California: U.S. Geological Survey Open-File Report 85-865, 27 p.

Meisling, K. E., 1984, Neotectonics of the north frontal fault system of the San Bernardino Mountains; Cajon Pass to Lucerne Valley, California [Ph.D. thesis]: Pasadena, California Institute of Technology, 394 p.

Meisling, K. E., and Weldon, R. J., 1986, Cenozoic uplift of the San Bernardino Mountains: possible thrusting across the San Andreas fault: Geological Society of America Abstracts with Programs, v. 18, p. 157.

—— , 1989, Late Cenozoic tectonics of the northwestern San Bernardino Mountains, southern California: Geological Society of America Bulletin, v. 101, p. 106–128.

Merifield, P. M., Rockwell, T. K., and Loughman, C. C., 1987, Slip rate on the San Jacinto fault zone in the Anza seismic gap, southern California: Geological Society of America Abstracts with Programs, v. 19, p. 431–432.

Miller, M. G., and Weldon, R. J., 1989a, Evolution of a lateral ramp in the Squaw Peak thrust, Cajon Pass, CA: Geological Society of America Abstracts with Programs, v. 21, p. 118.

—— , 1989b, Structural evolution of part of the north-trending segment of the Squaw Peak fault, Cajon Pass, California: U.S. Geological Survey Open-File Report, 89-189, 29 p., 2 plates.

—— , 1992, A lateral ramp origin for the north-trending segment of the Squaw Peak fault, Cajon Pass, California: Journal of Geophysical Research, v. 97, p. 5153–5165.

Minster, J. B., and Jordan, T. H., 1978, Present-day plate motions: Journal of Geophysical Research, v. 83, p. 5331–5354.

—— , 1984, Vector constraints on Quaternary deformation of the western United States east and west of the San Andreas fault, in Crouch, J. K., and Bachman, S. B., eds., Tectonics and sedimentation along the California margin: Pacific Section, Society of Economic Paleontologists and Mineralogists, v. 38, p. 1–16.

Moore, D. G., and Curray, J. R., 1982, Geologic and tectonic history of the Gulf of California, in Curray, J. R., and others, eds., Initial Reports of the Deep Sea Drilling Project: Washington, D.C., U.S. Government Printing Office, v. 64, p. 1279–1296.

Morton, D. M., 1975, Synopsis of the geology of the eastern San Gabriel Mountains, southern California, in Crowell, J. C., ed., San Andreas fault in southern California; A guide to San Andreas fault from Mexico to Carrizo Plain: California Division of Mines and Geology Special Report 118, p. 170–176.

Morton, D. M., and Matti, J. C., 1979, Evidence for a vanished post–middle Miocene pre–late Pleistocene alluvial fan complex in the northern Perris Block, Southern California: Geological Society of America Abstracts with Programs, v. 11, p. 118.

Namson, J., and Davis, T. L., 1988, Structural transect of the western Transverse Ranges, California; Implications for lithospheric kinematics and seismic risk evaluation: Geology, v. 16, p. 675–679.

Neville, S. L., 1986, Late Miocene alkaline volcanism, Ruby Mountain, San Bernardino County, California, in Kooser, M. A., and Reynolds, R. E., eds., Geology around the margins of the eastern San Bernardino Mountains: Inland Geological Society, v. 1, p. 95–99.

Neville, S. L., and Chambers, J. M., 1982, Late Miocene alkaline volcanism, northeastern San Bernardino Mountains and adjacent Mojave Desert, in Cooper, J. D., ed., Geologic excursions in the Transverse Ranges; Geological Society of America Annual Meeting Guidebook: Fullerton, California State University, p. 103–106.

Noble, L. F., 1926, The San Andreas rift and some other active faults in the desert region of southeastern California: Carnegie Institution of Washington Yearbook 25, p. 415–422.

—— , 1932, The San Andreas rift in the desert region of southeastern California: Carnegie Institution of Washington Yearbook 31, p. 355–363.

—— , 1953, Geology of the Pearland Quadrangle, California: U.S. Geological Survey Geologic Quadrangle Map GQ-24, scale 1:24,000.

—— , 1954, Geology of the Valyermo Quadrangle and vicinity, California: U.S. Geological Survey Quadrangle Map GQ-50, scale 1:24,000.

Pelka, G. J., 1971, Paleocurrents of the Punchbowl Formation and their interpretation: Geological Society of America Abstracts with Programs, v. 3, p. 176.

Powell, R. E., 1981, Geology of the crystalline basement complex, eastern Transverse Ranges, southern California; Constraints on regional tectonic interpretations [Ph.D. thesis]: Pasadena, California Institute of Technology, 441 p.

—— , 1986, Palinspastic reconstruction of crystalline-rock assemblages in southern California; San Andreas fault as part of an evolving system of late Cenozoic conjugate strike-slip faults: Geological Society of America Abstracts with Programs, v. 18, no. 172.

Ramirez, V. R., 1982, Hungry Valley Formation; Evidence for 220 km of post Miocene offset on the San Andreas fault, in Anderson, D. W., and Rymer, M. J., eds., Tectonics and sedimentation along the San Andreas system: Pacific Section, Society of Economic Paleontologists and Mineralogists, p. 33–44.

Reynolds, R. E., 1984, Miocene faunas in the lower Crowder Formation, Cajon Pass, California; A preliminary discussion, in Hester, R. L., and Hallinger, D. E., eds., San Andreas fault, Cajon Pass to Wrightwood: Pacific Section, American Association of Petroleum Geologists Volume and Guidebook 55, p. 17–21.

—— , 1985, Tertiary small mammals in the Cajon Valley, San Bernardino County, California, in Reynolds, R. E., ed., Geologic investigations along Interstate 15; Cajon Pass to Manix Lake, California: Western Association of Vertebrate Paleontologists 60th Meeting Field Trip Guidebook: Redlands, California, San Bernardino County Museum, p. 49–58.

Reynolds, R. E., and Weldon, R. J., 1988, Vertebrate paleontologic investigation, DOSECC Deep Hole project, Cajon Pass, San Bernardino County, California: Geophysical Research Letters, v. 15, p. 1073–1076.

Rockwell, T. K., Merifield, P. M., and Shepard, F. P., 1986, Holocene activity of the San Jacinto fault in the Anza seismic gap, southern California: Geological Society of America Abstracts with Programs, v. 18, p. 177.

Rust, D. J., 1982, Radiocarbon dates for the most recent large prehistoric earthquake and for late Holocene slip rates; San Andreas fault in part of the Transverse Ranges north of Los Angeles: Geological Society of America Abstracts with Programs, v. 14, p. 229.

——, 1986, Neotectonic behavior of the San Andreas zone in the Big Bend: Geological Society of America Abstracts with Programs, v. 18, p. 179.

Sadler, P. M., 1981, Structure of the northeast San Bernardino Mountains: California Division of Mines and Geology Open-File Report and Maps.

——, 1982a, Late Cenozoic stratigraphy and structure of the San Bernardino Mountains, *in* Cooper, J. D., ed., Geologic excursions in the Transverse Ranges; Geological Society of America Annual Meeting Guidebook: Fullerton, California State University, p. 55–106.

——, 1982b, Provenance and structure of late Cenozoic sediments in the northeast San Bernardino Mountains, *in* Cooper, J. D., ed., Geologic excursions in the Transverse Ranges; Geological Society of America Annual Meeting Guidebook: Fullerton, California State University, p. 83–91.

——, 1985, Santa Ana Sandstone; Its provenance and significance for the late Cenozoic history of the Transverse Ranges, *in* Reynolds, R. E., ed., Geologic investigations along Interstate 15, Cajon Pass to Manix Lake, California: Western Association of Vertebrate Paleontologists Field Trip Guidebook, 60th Meeting: Redlands, California, San Bernardino County Museum, p. 69–78.

Sadler, P. M., and Demirer, A., 1986, Pelona schist clasts in the Cenozoic of the San Bernardino Mountains, southern California, *in* Ehlig, P., ed., Neotectonics and faulting in southern California; Geological Society of America Cordilleran Section Meeting Guidebook: Los Angeles, California State University, p. 129–146.

Sadler, P. M., and Reeder, W. A., 1983, Upper Cenozoic, quartzite-bearing gravels of the San Bernardino Mountains, southern California, Recycling and mixing as a result of transpressional uplift, *in* Anderson, D. W., and Rymer, M. J., eds., Tectonics and sedimentation along faults of the San Andreas system: Pacific Section, Society of Economic Paleontologists and Mineralogists, p. 45–57.

Salyards, S. L., 1988, Dating and characterizing late Holocene earthquakes using paleomagnetics [Ph.D. thesis]: Pasadena, California Institute of Technology, 217 p.

Sauber, J., Thatcher, W., and Solomon, S., 1986, Geodetic measurements of deformation in the central Mojave Desert, California: Journal of Geophysical Research, v. 91, p. 12683–12693.

Savage, J. C., 1983, Strain accumulation in western United States: Annual Reviews of Planetary Science, v. 11, p. 11–43.

——, 1990, Equivalent strike-slip earthquake cycles in half-space and lithosphere-asthenosphere earth models: Journal of Geophysical Research, v. 95, p. 4873–4879.

Savage, J. C., and Prescott, W. H., 1986, Strain accumulation in southern California: Journal of Geophysical Research, v. 91, p. 7455–7494.

Savage, J. C., Lisowski, M., and Prescott, W. H., 1990, An apparent shear zone trending north-northwest across the Mojave Desert, into Owens Valley, eastern California: Geophysical Research Letters, v. 17, p. 2113–2116.

Schwartz, D. P., and Coppersmith, K. J., 1984, Fault behavior and characteristic earthquakes; Examples from the Wasatch and San Andreas faults: Journal of Geophysical Research, v. 89, p. 5681–5698.

Schwartz, D. P., and Weldon, R., 1986, Late Holocene slip rate on the Mojave segment of the San Andreas fault zone, Littlerock, California; Preliminary results [abs.]: EOS Transactions of the American Geophysical Union, v. 67, p. 906.

——, 1987, San Andreas slip rates; Preliminary results from the 96th St. site near Littlerock, California: Geological Society of America Abstracts with Pro-

grams, v. 19, p. 448.

Schwartz, D. P., Sharp, R. V., and Weldon, R. J., 1987, San Andreas segmentation, Cajon Pass to Wallace Creek; National Earthquake Hazard Reduction Program Summaries of Technical Reports, Volume 23: U.S. Geological Survey Open-File Report 87-63, p. 576–578.

Sieh, K., 1984, Lateral offsets and revised dates for earthquakes at Pallett Creek, California: Journal of Geophysical Research, v. 89, p. 7641–7670.

——, 1986, Slip rate across the San Andreas fault and prehistoric earthquakes at Indio, California [abs.]: EOS Transactions of the American Geophysical Union, v. 67, p. 1200.

Sieh, K., and Jahns, R. H., 1984, Holocene activity of the San Andreas fault at Wallace Creek: Geological Society of America Bulletin, v. 95, p. 883–896.

Sharp, R. V., 1981, Variable rates of late Quaternary strike slip on the San Jacinto fault zone, southern California: Journal of Geophysical Research, v. 86, p. 1754–1762.

Shelton, J. S., 1955, Glendora volcanic rocks, Los Angeles basin: Geological Society of America Bulletin, v. 66, p. 45–89.

Silver, L. T., 1982, Evidence and a model for west-directed early- to mid-Cenozoic basement overthrusting in southern California: Geological Society of America Abstracts with Programs, v. 14, p. 617.

——, 1986, Southern California geology; A product of evolution in marginal environments: Geological Society of America Abstracts with Programs, v. 18, p. 184–185.

Stanley, R. G., 1987a, Implications of the northwestwardly younger age of the volcanic rocks of west-central California; Alternative interpretation: Geological Society of America Bulletin, v. 98, p. 612–614.

——, 1987b, New estimates of displacement along the San Andreas fault in central California based on paleobathymetry and paleogeography: Geology, v. 15, p. 171–174.

Stewart, J. H., Albers, J. P., and Poole, F. G., 1968, Summary of regional evidence for right-lateral displacement in the western Great Basin: Geological Society of America Bulletin, v. 79, p. 1407–1414.

Terres, R. R., and Luyendyk, B. P., 1985, Neogene tectonic rotation of the San Gabriel region, California, suggested by paleomagnetic vectors: Journal of Geophysical Research, v. 90, p. 2467–2484.

Vaughan, F. E., 1922, Geology of the San Bernardino Mountains north of San Gorgonio Pass: Berkeley, University of California Department of Geological Sciences Bulletin, v. 13, p. 319–411.

Weigand, P. W., and Thomas, J. M., 1987, Geochemical and age data from Miocene volcanic rocks form the San Emigdio/San Juan Bautista and Neenach/Pinnacles areas, California; Implications for early movement on the San Andreas fault: Geological Society of America Abstracts with Programs, v. 19, p. 462.

Weldon, R. J., 1984, Implications of the age and distribution of late Cenozoic stratigraphy in Cajon Pass, southern California, *in* Hester, R. L., and Hallinger, D. E., eds., San Andreas fault, Cajon Pass to Wrightwood: Pacific Section, American Association of Petroleum Geologists, Volume and Guidebook 55, p. 9–16.

——, 1986, The late Cenozoic geology of Cajon Pass; Implications for tectonics and sedimentation along the San Andreas fault [Ph.D. thesis]: Pasadena, California Institute of Technology, 400 p.

Weldon, R., and Humphreys, E., 1987, Plate model constraints on the deformation of coastal southern California north of the Transverse Ranges: Geological Society of America Abstracts with Programs, v. 19, p. 462.

Weldon, R. J., and Humphreys, E., 1986, A kinematic model of southern California: Tectonics, v. 5, p. 33–48.

——, 1989, Comment on "Structural transect of the western Transverse Ranges, California: Implications for lithospheric kinematics and seismic risk evaluation" by J. Namson and T. Davis: Geology, v. 17, p. 769–770.

Weldon, R. J., and Sieh, K. E., 1985, Holocene rate of slip and tentative recurrence interval for large earthquakes on the San Andreas fault in Cajon Pass, southern California: Geological Society of America Bulletin, v. 96, p. 793–812.

Weldon, R. J., and Springer, J. E., 1988, Active faulting near the Cajon Pass well, southern California; Implications for the stress orientation near the San Andreas fault: Geophysical Research Letters, v. 15, p. 993–996.

Weldon, R. J., Winston, D. S., Kirschvink, J. L., and Burbank, D. W., 1984, Magnetic stratigraphy of the Crowder Formation, Cajon Pass, southern California: Geological Society of America Abstracts with Programs, v. 16, p. 689.

Wells, R. E., and Hillhouse, J. W., 1989, Paleomagnetism and tectonic rotation of the lower Miocene Peach Springs Tuff, Colorado Plateau, Arizona, to Barstow, California, Geological Society of America Bulletin, v. 101, p. 846–863.

Wernicke, B., Axen, G. J., and Snow, J. K., 1989, Basin and Range extensional tectonics at the latitude of Las Vegas, Nevada, Geological Society of America Bulletin, v. 100, p. 1738–1757.

Wesnouski, S. G., 1988, Seismological and structural evolution of strike-slip faults: Nature, v. 335, p. 340–343.

Wilson, J. T., 1965, A new class of faults and their bearing on continental drift: Nature, v. 207, p. 343–347.

Winston, D. S., 1985, Magnetic stratigraphy of the Crowder Formation, southern California [M.S. thesis]: Los Angeles, University of Southern California, 100 p.

Withjack, M. O., and Jamison, W. R., 1986, Deformation produced by oblique rifting: Tectonophysics, v. 126, p. 99–124.

Woodburne, M. O., 1975, Cenozoic stratigraphy of the Transverse Ranges and adjacent areas, southern California: Geological Society of America Special Paper 162, 91 p.

Woodburne, M. O., and Golz, D. J., 1972, Stratigraphy of the Punchbowl Formation, Cajon Valley, southern California: Berkeley, University of California Publications in Geological Sciences, v. 92, 57 p.

Woodford, A. O., 1960, Bedrock patterns and strike slip faulting in south-western California: American Journal of Science, v. 258A, p. 400–417.

Yeats, R. S., Calhoun, J. A., Nevins, B. B., Schwing, H. F., and Spitz, H. M., 1989, Russell Fault of California Coast Ranges: American Association of Petroleum Geologists Bulletin, v. 73, p. 1089–1102.

MANUSCRIPT ACCEPTED BY THE SOCIETY SEPTEMBER 6, 1990

Geological Society of America
Memoir 178
1993

Chapter 4

Displacement on the southern San Andreas fault

John T. Dillon*
Division of Geological and Geophysical Surveys, Alaska Department of Natural Resources, Fairbanks, Alaska 99701
Perry L. Ehlig
Department of Geology, California State University, Los Angeles, California 90032

ABSTRACT

The pre-Quaternary geology of the southern Chocolate and Cargo Muchacho mountains correlates with that exposed in San Gorgonio Pass between the Mission Creek and Banning branches of the San Andreas fault. Matching features include: (1) the southernmost exposures of the Triassic Mount Lowe Granodiorite on opposite sides of the San Andreas; (2) the presence in both areas of distinctive melanocratic quartz monzonite, dated as mid-Jurassic in the southern area; (3) the occurrence in both areas of kyanite schist formed by hydrothermal leaching of granitic and gneissic rocks; (4) the correlation of the Bear Canyon fanglomerate and associated late Miocene basalt flows in the southern area with the Coachella Fanglomerate and associated late Miocene andesite and basalt flows in the northern area; and (5) the correlation of the northern limit of the late Miocene to Pliocene marine Bouse Formation in the Yuma area east of the Salton Trough, with the northern limit of the late Miocene to Pliocene marine Imperial Formation adjacent to the Mission Creek fault in San Gorgonio Pass. These correlations require 185 ± 20 km of right slip on the Mission Creek–Coachella Valley segment of the San Andreas fault since late Miocene. An additional right slip of about 30 km has occurred in San Gorgonio Pass along the Banning–Coachella Valley segment of the San Andreas fault, as indicated by the displacement of the northern limit of the Imperial Formation. A total right slip of 90 ± 20 km is needed on the Banning fault to be consistent with other data presented by us, but the displacement cannot be demonstrated because basement terrane north of the Banning fault is unrelated to that south of the Banning fault.

The Salton Creek fault, located east of the San Andreas fault between the Orocopia Mountains and the Chocolate Mountains, correlates with an unnamed fault in Soledad Pass, located west of the San Andreas at the west end of the San Gabriel Mountains. In both areas the correlative faults have volcanic necks along them, and separate Precambrian syenite and related rocks on the northwest from Triassic Mount Lowe Granodiorite overlain by late Oligocene to early Miocene volcanic rocks on the southeast. The displacement of these correlative faults is younger than about 10 Ma, since it postdates deposition of conglomerate within the middle to late Miocene Mint Canyon Formation in Soledad basin. The conglomerate contains clasts of unusual volcanic rocks whose only known source is in the northern Chocolate Mountains. This correlation requires 240 km of right slip on the San Andreas fault, including displacement contributed by the San Jacinto fault.

---

*Deceased.

Dillon, J. T., and Ehlig, P. L., 1993, Displacement on the southern San Andreas fault, *in* Powell, R. E., Weldon, R. J., II, and Matti, J. C., eds., The San Andreas Fault System: Displacement, Palinspastic Reconstruction, and Geologic Evolution: Boulder, Colorado, Geological Society of America Memoir 178.

The San Gabriel fault is an abandoned branch of the San Andreas fault located southwest of Soledad Pass. It probably connected to the Banning fault prior to development of the Coachella Valley segment of the San Andreas fault. The San Gabriel fault displaces middle Miocene and older rocks 60 km to the right. The addition of this displacement to the Soledad Pass–Salton Creek offset yields a total right slip of 300 km on the San Andreas fault in southern California since middle Miocene time.

Our determination of the total offset on the San Andreas fault is consistent with that obtained by Matthews (1976) from the correlation of the Neenach Volcanic Formation, located east of the San Andreas fault in the central Transverse Ranges, with the Pinnacles Volcanic Formation, located west of the fault in the central Coast Ranges. However, there is a conflict between the evidence presented here for 240 km of right slip on the San Andreas fault between Salton Creek and Soledad Pass and the evidence presented by Frizzell and others (1986) and Matti (this volume) for 160 km of right slip on the San Andreas between the central San Bernardino Mountains and Liebre Mountain, based on their correlation of distinctive Triassic monzogranite in the two areas. This conflict needs to be resolved.

Our data are consistent with the hypothesized origin of the San Andreas as a transform fault caused by crustal extension in the Gulf of California. It requires displacement to have started by about 10 Ma. Only about 60 km of right slip needs to have occurred prior to about 5 Ma; 240 Ma of right slip has probably occurred since then.

INTRODUCTION

The role of the modern San Andreas fault as a right-slip transform fault along the boundary between the Pacific and North American crustal plates is a generally accepted fact. It links the northern end of the Gulf of California spreading system with the Mendocino triple junction and is among the simpler transform faults within continents (Allen, 1981). However, the continental crust is complexly deformed and fragmented on both sides of the San Andreas fault as a result of tectonism, both antecedent to and contemporaneous with development of the San Andreas fault. Because of this complexity and the vagarious nature of surface exposures that provide most of our geologic information, work on the history of the San Andreas fault has progressed slowly.

This chapter deals with the pre-Quaternary history of the San Andreas fault in southeasternmost California and the Transverse Ranges, and addresses three questions. Did the San Andreas fault originate in association with the opening of the Gulf of California or does it have an earlier history? How much displacement has occurred along the San Andreas fault in southern California, and what is the timing of that displacement? Does palinspastic unslipping of the San Andreas fault produce an orderly arrangement of terranes along this segment of the Pacific margin of North America? Our answers to some parts of these questions are firm, and others are tentative. More precise and accurate answers should evolve as more is learned about the terranes bordering the San Andreas fault.

Correlations have been established between terranes southwest of the San Andreas fault in the central Transverse Ranges and those east of the Salton Trough in southeasternmost California. The initial correlation by Crowell (1962) is based on similarities in the kinds and histories of rocks exposed in the western San Gabriel Mountains and adjoining Soledad basin and those exposed in the Orocopia Mountains (Fig. 1). The keystone to his correlation is the occurrence of distinctive Precambrian anorthosite, syenite, and related rocks in both areas, as documented by Crowell and Walker (1962). Crowell (1962) concluded that the two areas adjoined one another prior to middle Miocene time but were subsequently separated by about 210 km of right slip on the San Andreas fault. Crowell (1975a) increased his estimate of the displacement to 240 km based on new information. The proposed displacement is the horizontal separation between the correlative terranes and is not constrained by an offset piercing point.

It is difficult to establish unequivocal piercing points on the San Andreas fault because of four factors: (1) the extensive occurrence of Quaternary deposits concealing bedrock along and adjacent to the fault zone; (2) the presence of numerous fault slices within the fault zone, which create gaps between correlative piercing points on either side of the fault zone; (3) the complex nature of deformations affecting terranes in southern California, including the juxtaposition of rocks derived from different crustal levels; and (4) the repetitious nature of rock types within some granitic and gneissic terranes bordering the fault. These difficulties are partially overcome by establishing a series of correlations, each independent of the others, that collectively constrain piercing points.

Evidence is presented in support of two piercing points. One is based on a series of correlations that match the geology of the Cargo Muchacho and southern Chocolate Mountains with that south of the Mission Creek fault in San Gorgonio Pass. It indicates that the two areas were connected as recently as latest Miocene. It is based mainly on studies by Dillon (1975), Peterson (1975), and Farley (1979). This piercing point is poorly con-

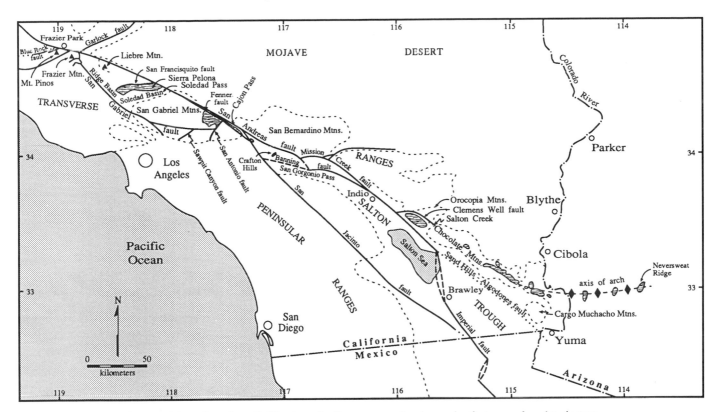

Figure 1. Map of southern California and adjacent areas showing major features referred to in text. Squiggle pattern designates present exposures of Pelona-Orocopia schist.

strained but yields a post-Miocene right slip of 185 ± 20 km on the Mission Creek–Coachella Valley branch of the San Andreas fault. The displacement on the Banning fault must be added to that of the Mission Creek fault to get the total displacement on the San Andreas fault in San Gorgonio Pass.

The second piercing point is a precise correlation between the Salton Creek fault and an unnamed fault in Soledad Pass. Data indicate the two were connected as recently as late Miocene. This piercing point is based on correlations by Ehlig and Ehlert (1972), Ehlig and others (1975), Joseph and Davis (1977), Ehlig (1981), and Buesch and Ehlig (1982). It is compatible with Crowell's data and yields the same right slip as Crowell's (1975a) revised estimate.

The San Gabriel fault, located southwest of the San Andreas fault in the central Transverse Ranges (Fig. 1), has 60 km of right slip along it and is interpreted to have been a segment of the main branch of the San Andreas fault during late Miocene time (Crowell, 1962, 1975b, 1982). Evidence for the amount and timing of its displacement is reviewed here because of its relevancy to the history of the San Andreas fault.

We also discuss two other published piercing points. One proposed by Matti and others (1986), Frizzell and others (1986), and Matti (this volume), correlates porphyritic monzogranite exposed northeast of the San Andreas fault in the San Bernardino

Mountains with similar monzogranite exposed southwest of the fault on Liebre Mountain (Fig. 1). This correlation yields a right slip of only 160 km on the San Andreas fault and is in apparent conflict with the correlations presented by us. The other correlation, between the Neenach Volcanic Formation, exposed northeast of the fault near Leibre Mountain, and the Pinnacles Volcanic Formation, exposed northwest of the fault in the central Coast Ranges, yields a right slip of about 300 km on the San Andreas fault (Matthews, 1976). This agrees with the offsets obtained from our correlations, including the 60 km of right slip on the San Gabriel fault.

SALTON TROUGH AND THE ACTIVE SAN ANDREAS FAULT

The Salton Trough is a deep topographic and structural depression formed by late Cenozoic extension associated with the opening of the Gulf of California (Elders and others, 1972). It is a northern continuation of the Gulf of California that was separated from the gulf by growth of the Colorado River delta (Sharp, 1982). During the late Miocene and early Pliocene, marine waters of the gulf extended to the northern limit of the present Salton Trough, as shown by the presence of the marine Imperial Formation in San Gorgonio Pass (Allen, 1957). The greatest thickness of sedimentary fill is in the south-central part of the trough, where unmetamorphosed sedimentary rocks extend to a depth as great as 6 km, and metasedimentary rocks may extend to depths of 10 to 16 km (Fuis and others, 1982).

We once assumed that the active San Andreas fault extended along the entire eastern margin of the Salton Trough and entered Mexico in the desert west of Yuma, Arizona. We now believe this is incorrect. The Salton Trough segment of the active San Andreas fault, referred to as the Coachella Valley segment by Matti and others (1985), extends southeastward from San Gorgonio Pass to the southeast corner of the Salton Sea. Here, the San Andreas fault appears to join the northeast corner of the Brawley seismic zone, the most northerly spreading axis in the system of short spreading axes and interconnected transform faults that form the divergent plate boundary in the Gulf of California (Elders and others, 1972; Johnson and Hadley, 1976; Johnson and Hill, 1982; Clark, 1984). As defined by earthquake epicenters during the period from 1973 through 1978 (Johnson and Hill, 1982, Fig. 7), the Brawley seismic zone trends about N20°W and extends for a distance of 75 km between the northern part of the Imperial fault and the southern terminus of the present San Andreas fault. Although the precise nature of the Brawley seismic zone is not fully understood, for our purpose it can be treated as a spreading axis 30 km long when measured perpendicular to the San Andreas fault.

South of the Brawley seismic zone, the Imperial fault occupies a medial position within the Salton Trough. Thick sedimentary fill extends east of the Imperial fault to the vicinity of the Sand Hills and Algodones faults, but neither of these faults appears to be active (Fuis and others, 1982; Olmsted and others, 1973). Therefore, the spreading axis at the Brawley seismic zone is probably migrating northwestward at the half spreading rate, as is occurring along the East Pacific Rise at the mouth of the Gulf of California (Larson and others, 1968). If this is the case, the Sand Hills and Algodones faults are fossil transform faults where minor compactional and isostatic adjustments may be occurring.

The part of the Salton Trough north of the Brawley seismic zone spreading axis, including the area occupied by the Salton Sea and Coachella Valley, is being translated northward along the west side of the San Andreas fault, presumably at the full spreading rate, while the spreading axis is translated northward at half the spreading rate. South of Indio, the San Andreas fault trends about N45°W, roughly parallel to relative plate motion. North of Indio, the San Andreas fault bends westward and splits into the Mission Creek and Banning branches. Because these faults trend more westerly than relative plate motion, the northern part of the Salton Trough is being compressed, particularly in San Gorgonio Pass where basement terrane is being thrust southward over sedimentary deposits (Allen, 1957; Matti and others, 1985; Blanck, 1987). The thrust faulting has disrupted the surface continuity of the San Andreas fault within San Gorgonio Pass, and it is unclear how Holocene right slip has been accommodated within the pass.

From San Gorgonio Pass to Cajon Pass the active trace of the San Andreas fault follows the southern margin of the San Bernardino Mountains and has an average trend of about N65°W. In the vicinity of Cajon Pass the fault trace bends to the right and then to the left, displacing the San Andreas fault 2.5 km to the right in a distance of about 20 km. This deflection coincides with the convergence of the San Jacinto fault with the San Andreas fault. The San Jacinto fault trends about N45°W and presumably joins the San Andreas fault at depth, although the two lack a visible surface connection. Westward from Cajon Pass to the Garlock fault the active San Andreas fault has a relatively straight trace trending N65W.

According to the rigid plate model of Minster and Jordan (1978, 1987), 56 ± 3 mm/yr of right slip should be occurring along this part of the Pacific and North American plate boundary. Only 20 to 35 mm/yr is occurring on the Coachella Valley segment of the San Andreas fault (Keller and others, 1982). The remainder is occurring farther west along the San Jacinto fault (10 to 15 mm/yr; Merifield and others, 1987), the Elsinore fault (about 4 mm/yr; Pinault and Rockwell, 1984) and more westerly faults, and by crustal deformation between major faults.

EASTERN LIMIT OF THE SAN ANDREAS FAULT SYSTEM

Although the active San Andreas fault emanates from the north end of the system of spreading centers in the Gulf of California and the Salton Trough, uncertainty surrounds its history. Is it an exclusive product of relative plate motion associated with the opening of the gulf or has it functioned in other capacities? One aspect of this question is whether the San Andreas fault has inactive branches east of its presently active trace within southern California. The answer appears to be no, except for abandoned fault segments within a few kilometers of the active trace.

No branches of the San Andreas fault have been recognized in the Mojave Desert north of the central Transverse Ranges or in the eastern Transverse Ranges to the east of the San Andreas fault in Cajon Pass. In the area east of the Salton Trough, Miocene and older bedrock crops out in a semicontinuous belt extending through the Orocopia, Chocolate, and Cargo Muchacho mountains to the California-Mexico border. Quaternary alluvium conceals bedrock directly east of the active trace of the San Andreas fault in the Salton Trough, but in most places, bedrock exposures occur within 15 km of the east edge of the thick sedimentary fill in the Salton Trough. Although numerous minor faults occur east of the Salton trough, including some northwest-trending ones with right separations along them, only one, the Clemens Well fault, is a major fault that may be genetically related to the San Andreas fault.

As mapped by Crowell (1962), the Clemens Well fault diverges eastward from the San Andreas fault in the Mecca Hills and transects the Orocopia Mountains where it separates unrelated basement terranes. Crowell (1975a) considered it a major right-slip fault of unknown displacement. The correlation of Pelona-Orocopia Schist bodies across the San Andreas fault (Dibblee, 1968; Ehlig, 1968, 1981) requires the Clemens Well fault to be an eastward continuation of the Fenner and San Francisquito faults in the central Transverse Ranges. Smith (1977) correlated these faults with faults in the southern Coast Ranges and presented evidence for 175 km of Oligocene and early Miocene right slip along them. Powell (1981; this volume) offered an alternative reconstruction that requires about 110 km of right slip on these faults. Smith (1977) and Powell (this volume) suggest that the Clemens Well and correlative faults are part of the San Andreas fault system. This may be correct, depending on one's definition of the San Andreas fault system; however, the east end of the Clemens Well fault is older than, and does not connect to, the system of faults and spreading centers in the Gulf of California. We would prefer to restrict the usage of San Andreas fault system to faults that are directly associated with the boundary between the Pacific and North American plates. We consider it most likely that the Clemens Well and correlative faults are genetically related to the formation of Oligocene basins by crustal extension (Bohannon, 1975) and are tear faults bounding an extensional terrane.

Neither the Clemens Well fault nor major branches of the San Andreas fault transect the terrane southeast of the Orocopia Mountains. This is shown by the continuity of a mid-Tertiary arch along which the Vincent–Chocolate Mountain thrust and underlying Pelona-Orocopia Schist are exposed in a series of windows extending from the central Chocolate Mountains on the east side of the Salton Trough to Neversweat Ridge in Arizona (Haxel and Dillon, 1973; Dillon, 1975; Haxel, 1977).

OVERVIEW OF CORRELATIVE TERRANES

Basement terranes east of the Salton Trough and correlatives west of the San Andreas fault in the central Transverse Ranges contain rock types and structural features that distinguish them from all other terranes along the western margin of North America. An important structural feature is the arch that brings the Vincent–Chocolate Mountain thrust and underlying Pelona-Orocopia Schist to the surface in a series of windows extending westward from Neversweat Ridge in Arizona to the southern Chocolate Mountains in California and hence northwestward to the west end of the Orocopia Mountains where the arch is truncated by the San Andreas fault (Fig. 1). The arch occurs west of the San Andreas in the central Transverse Ranges but its continuity is disrupted by faulting. The arch involves great uplift and denudation. The Vincent–Chocolate Mountain thrust is inferred to be a fossil subduction zone of Late Cretaceous to early Paleocene age (Dillon, 1986; Ehlig, 1981). The underlying Pelona-Orocopia Schist is a supracrustal oceanic terrane of late Mesozoic age, whereas basement at the base of the thrust's upper plate is of midcrustal cratonic origin.

The Pelona-Orocopia Schist consists mainly of gray mica–albite–quartz schist derived from graywacke and argillite of continental derivation. Greenschist and amphibolite derived from basalt are locally abundant. Thin beds of metamorphosed chert and limestone are scattered through much of the section, particu-

larly in association with metabasalt. Exotic masses of serpentinite have isolated occurrences. Metamorphism to greenschist and lower amphibolite facies occurred during subduction beneath the overlying continental crust. For greater detail, refer to Ehlig (1958, 1968, 1981), Dillon (1975), Haxell (1977), Haxel and Dillon (1978), Haxel and others (1987), Jacobson (1980, 1983), and Jacobson and others (1988).

Studies by Graham and Powell (1984) indicated that metamorphism of the schist exposed on Sierra Pelona occurred at pressures of 8 to 9 kbar, a value equivalent to a depth of about 30 km. The stripping away of the upper continental crust to expose the schist was probably accomplished tectonically by mid-Tertiary detachment faulting, with surface erosion playing a subordinant role. The importance of detachment faulting in this terrane has only recently been recognized (Frost and others, 1986), and details have not been worked out. Some detachment faults have their soles at the top of the Pelona-Orocopia Schist and place rocks from different structural levels against the schist, greatly complicating the geology.

In areas where detachment faults do not extend to the top of the Pelona-Orocopia Schist, the schist is overlain by mylonite and retrograde metamorphic rocks formed by ductile flow and recrystallization within the overlying cratonic rocks. The mylonite truncates the structure of the overlying cratonic basement in a manner that indicates the cratonic rocks originally extended to greater depth but have had their roots removed during thrusting.

The cratonic basement overlying the thrust contains several distinctive rock units whose limited distribution make them useful indicators of strike-slip displacement on the San Andreas fault. These include the Precambrian granulite-facies Mendenhall Gneiss, a Precambrian anorthosite-syenite-gabbro complex, the Triassic Mount Lowe Granodiorite, an unnamed Jurassic quartz monzonite, and a kyanite-bearing gneiss and schist. In addition to the above units, the upper plate basement includes rock types whose wide distribution make them less easily used for determining displacement on the San Andreas fault. These include Precambrian quartzofeldspathic banded gneiss and migmatite of uncertain protolith, amphibolite and dioritic gneiss of metaigneous origin, augen gneiss derived from porphyritic granite, and middle to late Mesozoic batholithic rocks ranging from diorite to granite in composition.

The oldest dated units within the cratonic basement include layered gneisses and amphibolite with an inferred protolith age of 1,710 ± 40 Ma and 1,670 ± 15 Ma augen gneiss (Silver, 1966, 1971). These are widely distributed. Part of this terrane was subjected to granulite-facies metamorphism at 1,440 ± 15 Ma (Silver and others, 1963; Silver, 1971), creating a distinctive assemblage of rocks. Subsequent amphibolite-facies metamorphism has largely destroyed the granulite-facies minerals, but the rocks are easily recognized by replacement textures and some relict minerals, particularly perthite and blue to violet quartz. This assemblage was named the Mendenhall Gneiss where it crops out north of the San Gabriel fault in the western San Gabriel Mountains (Oakeshott, 1958). The Mendenhall Gneiss crops out southwest of the San Gabriel fault in the area west of Ridge basin (Ehlig and Crowell, 1982). Similar rock types also occur in association with anorthosite and syenite in the northwestern San Gabriel Mountains and Orocopia Mountains. Powell (1981, 1982) has identified similar rocks in several ranges north and east of the Orocopia Mountains, but none have been recognized in the Chocolate Mountains or ranges to the south and east.

The 1,220 ± 10–Ma (Silver and others, 1963) anorthosite-gabbro-syenite complex of the western San Gabriel Mountains was most recently described by Carter (1980, 1982) and Carter and Silver (1972). The correlation between the complex in the San Gabriel Mountains and that in the Orocopia Mountains is documented by Crowell and Walker (1962). Minor occurrences are also present in association with correlatives of the Mendenhall Gneiss southwest of the San Gabriel fault (Ehlig and Crowell, 1982) and in ranges north and east of the Orocopia Mountains (Powell, 1981, 1982). Correlatives appear to be absent in the Chocolate Mountains and ranges to the south and east.

The main occurrence of the Triassic Mount Lowe Granodiorite is in the San Gabriel Mountains to the southwest of the San Andreas fault. Here, it occurs as a zoned pluton that is divided into four facies by Ehlig (1975, 1981). Each facies has a distinctive composition and texture, such that the four facies of the Mount Lowe Granodiorite are easily recognized in large hand specimens and readily distinguished from other granitoid rocks in southern California. One facies contains stout prismatic to ovoid hornblende phenocrysts 5 to 20 mm long, set in a white matrix of andesine, minor quartz, and K-feldspar. Another has a spotted "Dalmatian" texture formed by pink K-feldspar phenocrysts,

usually 20 to 70 mm long, and black hornblende phenocrysts set in a white matrix of andesine and minor quartz. A third facies contains scattered garnet phenocrysts as much as 20 mm in diameter, along with abundant large K-feldspar phenocrysts and sparse hornblende phenocrysts. The fourth facies consists mainly of oligoclase or albite, K-feldspar, and minor quartz and biotite. The chemistry of the Mount Lowe Granodiorite is also distinctive (Joseph and others, 1982; Barth, 1985; Barth and Ehlig, 1988). The Mount Lowe Granodiorite has been dated at 220 ± 10 Ma by zircon U/Pb method (Silver, 1971) and 208 ± 14 Ma by Rb/Sr whole-rock method (Joseph and others, 1982).

Outside the San Gabriel Mountains, the only known exposures of Lowe Granodiorite on the west side of the San Andreas fault zone are between the Banning fault and active trace of the San Andreas fault to the west of San Gorgonio Pass (P. L. Ehlig, unpublished data), and between the Banning and Mission Creek branches of the San Andreas fault in San Gorgonio Pass (Farley, 1979). East of the San Andreas, Lowe Granodiorite occurs in the Chocolate Mountains (Crowell, 1973; Dillon, 1975), the Little Chuckwalla Mountains (Powell, 1982), and the Mule Mountains (Tosdal, 1988).

Melanocratic quartz monzonite crops out in the southern Chocolate Mountains just south of Mammoth Wash (Fig. 2), and related quartz monzonite crops out in the Cargo Muchacho Mountains where it has been dated at 173 Ma (Dillon, 1975). Part of that in the Chocolate Mountains contains abundant pink K-feldspar phenocrysts, typically 10 to 20 mm wide, embedded in a dark greenish gray matrix. Metamorphism has recrystallized mafic minerals to fine-grained felted aggregates of biotite and

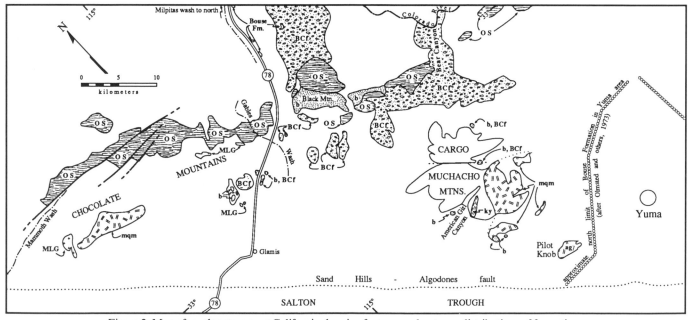

Figure 2. Map of southeasternmost California showing features and outcrop distributions of formations referred to in text. Key: ag = augen gneiss; MLG = Mount Lowe Granodiorite; mqm = melanocratic quartz monzonite; ky = kyanite-bearing rocks; OS = Orocopia Schist; BCf = Bear Canyon fanglomerate; b = basalt. Geology mainly from Dillon (1975).

minor hornblende. These black minerals project into the K-feldspar phenocrysts, creating fuzzy boundaries. Plagioclase is partially altered to epidote. Lenticular diorite inclusions, typically several centimeters long, are abundant within much of this quartz monzonite. These features create a rock that is easily distinguished from other rocks in the region. Clasts of this quartz monzonite are abundant in part of the Miocene Bear Canyon fanglomerate in the vicinity of the southern Chocolate and Cargo Muchacho mountains. In San Gorgonio Pass, similar quartz monzonite is exposed south of the Mission Creek fault (Fig. 3) (mapped as monzogranite by Matti and others, 1982), and occurs as clasts in the overlying Miocene Coachella Fanglomerate (Peterson, 1973, 1975).

Kyanite is abundant in veins and schist along the south side of the entrance to American Girl Canyon and occurs locally in the area northwest of American Girl Canyon in the southwest Cargo Muchaco Mountains (Fig. 2) (Henshaw, 1942; Dillon, 1975). It is typically blue and forms thin bladed crystals, some more than 10 cm long. Quartz is the principal associated mineral in veins, and muscovite and quartz are the dominant minerals in schist. Biotite, staurolite, and tourmaline also occur in some schist. The kyanite appears to have formed by the high-temperature hydrothermal leaching of Na, Ca, Mg, and some K from preexisting quartzofeldspathic gneiss and granitic rocks. All gradations are

present between the unaltered host rocks and kyanite-bearing schist. No other major bedrock occurrences of kyanite have been observed in the surrounding area, although thin lenses of kyanite-garnet-muscovite-quartz schist occur in retrograde schist along the Chocolate Mountain thrust in the southern Chocolate Mountains. At San Gorgonio Pass, kyanite has been found at two locations in the basement between the Mission Creek and Banning faults. In both places the kyanite-bearing rocks appear to have formed by high-temperature hydrothermal leaching of compositionally banded quartzofeldspathic gneiss. A steeply inclined northwest-trending band of kyanite-bearing schist has been traced for about 650 m in the area between two branches of Mias Canyon about 80 m south of the Gandy Ranch fault (Blanck, 1987). The kyanite occurs with muscovite, biotite, and quartz. A few pods and small veins of kyanite and quartz are also present. The kyanite is blue and occurs in irregularly oriented bladed crystals as much as 10 cm long. The second location is on the east side of Banning Canyon, about 1.5 km northwest of the other location (Farley, 1979). Here, the kyanite is difficult to see in hand specimen. An adjacent schist contains abundant crystals of red-brown staurolite that are readily visible.

Cenozoic volcanic and sedimentary rocks are also present within the correlative terranes but are only described as they relate to fault displacement.

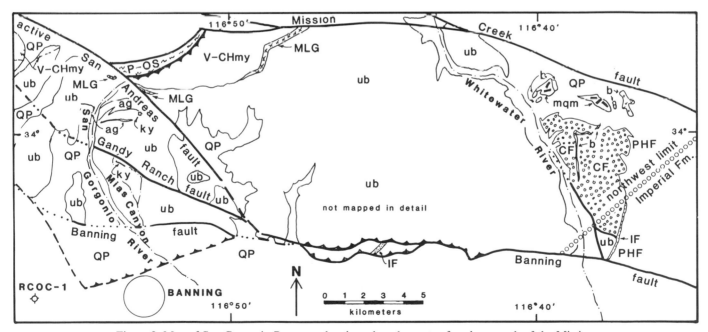

Figure 3. Map of San Gorgonio Pass area showing selected aspects of geology south of the Mission Creek branch of the San Andreas fault. Key: ag = augen gneiss; MLG = Mount Lowe Granodiorite; mqm = melanocratic quartz monzonite; ky = kyanite-bearing rocks; ub = undifferentiated basement; P-OS = Pelona-Orocopia Schist; V-CHmy = = mylonitic rocks of the Vincent–Chocolate Mountain thrust; CF = Coachella Fanglomerate; b = basalt; IF = Imperial Formation; PHF = Painted Hill Formation; QP = mainly late Pliocene and Quaternary alluvial deposits; RCOC-1 = wildcat well of Riverside County Oil Company. Data from Allen (1957), Peterson (1973), Farley (1979), Matti and others (1982), Blanck (1987), and unpublished mapping by P. Ehlig.

CARGO MUCHACHO AND SOUTHERN CHOCOLATE MOUNTAINS–SAN GORGONIO PASS CORRELATIONS

Similarities in the rock types and formations exposed east of the San Andreas fault in the Cargo Muchacho and southern Chocolate mountains and those exposed between the Mission Creek and Banning branches of the San Andreas fault in San Gorgonio pass suggest that the two terranes were formed close to each other and subsequently displaced by the San Andreas fault. In order to measure the magnitude of the displacement, it is necessary to find displaced geologic lines that form piercing points on opposite sides of the San Andreas fault. Thus far, no precise piercing points have been located, but five imprecise piercing points have been established: (1) the most southern occurrence of Triassic Mount Lowe Granodiorite on each side of the fault; (2) the limited occurrence in both areas of distinctive melanocratic porphyritic quartz monzonite; (3) the limited occurrence in both areas of kyanite schist formed by hydrothermal leaching; (4) the occurrence in both areas of upper Miocene fanglomerate containing basalt flows and similar clast assemblages; and (5) the northwest limit on each side of the fault of late Miocene and early Pliocene marine strata of the Bouse and Imperial Formations. Collectively, the five constrain the right slip on the Mission Creek branch of the San Andreas fault to 185 ± 20 km since late Miocene.

Southern limit of Mount Lowe Granodiorite

The main body of the Mount Lowe Granodiorite is essentially intact in the northwest and central San Gabriel Mountains. The eastern and southeastern parts were fragmented and distended during emplacement of younger Mesozoic granitoid plutons into it. The Mount Lowe Granodiorite may be distended by detachment faulting in southeasternmost California.

East of the San Andreas fault, the most southerly exposures of Mount Lowe Granodiorite are in the southern Chocolate Mountains north of State Highway 78 (Fig. 2). Strongly foliated Lowe Granodiorite crops out on two small knolls surrounded by alluvium about 2 km west of the open pit of the Mesquite Gold Mine. On one knoll the granodiorite contains hornblende phenocrysts; on the other it contains hornblende and K-feldspar phenocrysts. About 8.5 km farther north, biotite-bearing Mount Lowe Granodiorite crops out in a narrow lens about 1 km long and consists of the biotite-bearing facies. It is underlain by mylonitic rocks of the Chocolate Mountain thrust and overlain by compositionally banded gneiss. A larger body, 1.5 km long and about 100 m thick, occurs 5 km farther northwest. It contains hornblende and K-feldspar phenocryst.

The most southerly exposure of Lowe Granodiorite west of the Coachella branch of the San Andreas fault is between the Mission Creek and Banning faults in San Gorgonio Pass (Fig. 3). Here, it is interlayered with gneiss immediately above mylonitic rocks of the Vincent–Chocolate Mountain thrust. The largest exposed body is 2 km long and about 150 m thick. It contains ovoid phenocrysts of hornblende and K-feldspar aligned parallel to foliation in a white matrix of oligoclase to andesine and minor quartz.

There is a strong similarity between the Mount Lowe Granodiorite in San Gorgonio Pass and the one that contains hornblende and K-feldspar phenocrysts in the southern Chocolate Mountains. The associated banded gneiss is similar in both areas, and in both areas the Mount Lowe Granodiorite is immediately above the Vincent–Chocolate Mountains thrust. In San Gorgonio Pass, the most southeasterly exposure of the Mount Lowe Granodiorite is truncated by the Mission Creek fault (Fig. 3). In the southern Chocolate Mountains the most southerly exposure of the Mount Lowe Granodiorite is separated from the Sand Hills–Algodones fault by several kilometers of sand dunes and alluvium (Fig. 2). However, when projected into the fault at a right angle, the most southerly exposure of Mount Lowe Granodiorite in the Chocolate Mountains is about 200 km from that in San Gorgonio Pass.

Melanocratic porphyritic quartz monzonite

This correlation is restricted to the previously described distinctive melanocratic porphyritic quartz monzonite that contains abundant diorite inclusions. In San Gorgonio Pass, this rock type is exposed in an area about 4 km long, which is located 1 to 2 km south of the Mission Creek fault (Fig. 3). In the Chocolate Mountains this rock type forms part of a quartz monzonite pluton that underlies an area of at least 22 km^2 to the south of Mammoth Wash (Fig. 2). Several kilometers of sand and alluvium separate the exposures in the Chocolate Mountains from the San Hills–Algodones fault. When projected into the fault at a right angle, the occurrence of this rock type in the Chocolate Mountains is about 165 to 170 km from its occurrence in San Gorgonio Pass. This correlation yields a smaller separation than that obtained from the most southerly occurrence of Mount Lowe Granodiorite. There are three plausible explanations for the discrepancy. The western edge of the quartz monzonite in the southern Chocolate Mountains may trend southward at a low angle to the Sand Hills–Algodones fault and intersect the fault many kilometers south of its exposure in the Chocolate Mountains. Alternatively, the pluton in the southern Chocolate Mountains may not extend ot the fault, and the offset equivalent to the pluton that is exposed in San Gorgonio Pass may be buried beneath alluvium to the south of the pluton exposed in the Chocolate Mountains. A pluton of similar quartz monzonite crops out over an area of about 40 km^2 in the Cargo Muchacho Mountains (Fig. 2), but it lacks exposures of the distinctive rock type found in the Chocolate Mountains. The third possibility is that the southern limit of the Mount Lowe Granodiorite may trend southeastward at an oblique angle to the Sand Hills–Algodones fault such that its separation is less than the 200 km obtained by projecting the southernmost exposure of the Mount Lowe Granodiorite into the fault at a right angle.

Occurrences of kyanite

The presence of kyanite-bearing rocks of similar appearance and origin in the Cargo Muchacho Mountains and south of the Mission Creek fault in San Gorgonio Pass provides supporting evidence for the proposed correlation of the two terranes. Kyanite is rare in southern California; we are aware of no other occurrences of similar appearance and origin. When projected at a right angle into the Sand Hills–Algodones fault, the main occurrence of kyanite in the Cargo Muchacho Mountains is 215 to 220 km from the two known occurrences of kyanite in San Gorgonia Pass. We do not regard this, however, as a valid piercing point for measuring fault displacement. The local presence of kyanite-bearing mylonitic rocks along the Vincent–Chocolate Mountains thrust in the southern Chocolate Mountains suggests that the kyanite may have a wide, though spotty, distribution in this terrane.

Correlation of Bear Canyon fanglomerate with Coachella Fanglomerate

The Bear Canyon fanglomerate, informally named by Dillon (1975) for exposures in Bear Canyon between the Cargo Muchacho Mountains and the Colorado River, crops out over a broad area east of the Salton Trough (Fig. 2). It unconformably overlies late Oligocene to early Miocene volcanic and older basement rocks and is locally overlain by the Pliocene Bouse Formation, Colorado River gravel, and locally derived Pliocene and younger alluvium. The Bear Canyon fanglomerate has not been studied in detail, but the distribution of distinctive clast types and observations of clast imbrication indicate that it was transported northeastward from a source adjacent to the Salton Trough near the southern Chocolate and Cargo Muchacho mountains. Distinctive clast types include the melanocratic porphyritic quartz monzonite, kyanite-bearing rocks and titanium-poor magnetite rock. These do not occur in exposed bedrock outside the proposed source area. The magnetite clasts, as much as 30 cm in diameter, are probably derived from replacement deposits that occur in the melanocratic quartz monzonite and chloritized diorite in the southern Chocolate Mountains (Sampson and Tucker, 1942; Dillon, 1975).

Blocky flows of basalt, and in one case andesite, are interbedded with Bear Canyon fanglomerate within a limited area in the southern Chocolate and Cargo Muchaco mountains. The most extensive exposures are in the vicinity of Black Mountain where olivine basalt flows lap onto basement rocks. One of these flows has been dated at 13.1 ± 2.5 Ma by the K-Ar method (Crowe and others, 1979). Flows also form erosionally resistant caps on several small mesas in the southern Chocolate and Cargo Muchacho mountains. Olivine basalt, which may be part of the same series of flows, is present beneath the Bouse Formation, as revealed by a hole drilled south of the Cargo Muchacho Mountains (Olmsted and others, 1973). The flows do not occur in the Bear Canyon fanglomerate as far north as Mammoth Wash in the southern Chocolate Mountains or as far east as the Colorado River. The typical basalt contains phenocrysts of olivine in a groundmass of labradorite to calcic andesine, augite, and minor olivine and opaque minerals.

The Miocene Coachella Fanglomerate is exposed south of the Mission Creek fault on the east side of San Gorgonio Pass. It has been most extensively studied by Allen (1957) and Peterson (1973, 1975). The fanglomerate rests unconformably on basement rocks and is unconformably overlain by the late Miocene to early Pliocene Imperial and Painted Hill Formations. Andesite from a sequence of basalt flows and andesite breccias located 250 m above the base of the Coachella Fanglomerate has been dated at 10.0 ± 1.2 Ma by the K-Ar method (Peterson, 1975).

On the basis of clast imbrication, Peterson (1973, 1975) concluded that most of the fanglomerate was derived from a source area on the opposite side of the Mission Creek fault. The fanglomerate contains clasts of varied composition, including gneissic, granitic, and volcanic rocks. Some beds contain abundant clasts of distinctive quartz monzonite similar to that previously described. Although similar quartz monzonite is exposed between the fanglomerate and the Mission Creek fault, this source should have been covered during deposition of most of the fanglomerate. No similar rocks are present in terrane north of the Mission Creek fault in the vicinity of San Gorgonio Pass. The most likely source is to the east of the Salton Trough. Peterson (1973, 1975) compared clasts of the quartz monzonite from the Coachella Fanglomerate with similar clasts from the Bear Canyon fanglomerate and found them to have essentially identical petrologic and chemical characteristics.

An uncommon but distinctive clast type in the Coachella Fanglomerate is titanium-poor magnetite rock. Its clasts are as much as a meter in diameter. Similar magnetite rock occurs with the melanocratic porphyritic quartz monzonite south of the Mission Creek fault, but it is not a likely source for clasts except those near the base of the fanglomerate. There are no other known sources for large clasts of magnetite in the vicinity of San Gorgonio Pass. Sources do exist in the anorthosite-syenite terrane in the Orocopia Mountains, but the magnetite is titaniferous. The most likely source area for the magnetite clasts is in the southern Chocolate and Cargo Muchacho Mountains.

The olivine basalt flows and augite andesite breccia lens out in a southward direction within the Coachella Fanglomerate. Dikes near the Mission Creek fault are a potential source, but the flows and breccias most likely came from across the Mission Creek fault. They are petrologically and chemically similar to their counterparts in the Bear Canyon fanglomerate (Dillon, 1975). Thus, it is likely that the basalt and andesite came from the same magma source.

Although the evidence is not conclusive, it strongly suggests that the Bear Canyon fanglomerate and the Coachella Fanglomerate were deposited at about the same time from a common source terrane east of the San Andreas fault in the vicinity of the southern Chocolate and Cargo Muchacho mountains. Detailed geochemical and isotopic analyses should be able to confirm or refute this correlation.

Correlation of the Bouse and Imperial Formations

The Pliocene Bouse Formation occurs in basins along the Colorado River and adjacent desert areas of southeastern California and western Arizona. Its most southerly occurrence is beneath Colorado River deposits in the vicinity of Yuma, Arizona. The type area of the Bouse Formation is within the large basin that encompasses the Parker-Blythe-Cibola area of Arizona and California. Here, its basinal facies consists of a basal coquinoid marl overlain by subhorizontally bedded clay, silt, and fine sand (Metzger, 1968; Winterer, 1975). Erosion has removed the topographically highest parts of the Bouse Formation in most areas, but erosionally resistant remnants of stomatolithic travertine are draped across rocky parts of the basin margin. The highest occurrence of travertine marks the Bouse shoreline in many areas.

Based on the presence of a limited marine fauna, Metzger (1968) concluded that the Bouse Formation of the Parker-Blythe-Cibola area is a marine- to brackish-water formation deposited in an embayment of the Gulf of California. Because the Bouse Formation crops out northeast of the drainage divide between Milpitas Wash and the Salton Trough, Metzger (1968, p. 133) concluded that a seaway extended through the divide between the two areas. However, one of us (Ehlig) is skeptical of this conclusion. Although remnants of the Bouse Formation crop out immediately northeast of the divide within the Colorado River drainage area, none has been found at a similar elevation on the Salton Trough side of the divide. Dillon (1975) was of the opinion that calcareous sand and silt, which overlie or are interbedded with the Bear Canyon fanglomerate in exposures along Ogilby Road 10 km south of the drainage divide, might represent the basal part of the Bouse Formation. Dillon (1975), however, found no fossils in these beds. During subsequent reconnaissance in this area, Ehlig has found no beds that he considers lithologically typical of the Bouse Formation. Since Dillon completed his work in 1975, an unpublished regional reconnaissance study by Ehlig indicates that travertine deposits of the Bouse Formation are widespread at elevations between 400 and 1,000 ft above sea-level on both sides of the Colorado River throughout the Parker-Blythe-Cibola basin, but none have been found on the Salton Trough side of the drainage divide. This has led Ehlig to conclude that the Bouse Formation of the Parker-Blythe-Cibola area was probably deposited in a saline lake and was not an embayment of the Gulf of California. Such a lake could have formed when water from the Colorado River inundated desert basins and could have been similar to the present Salton Sea, except that its surface may have been several hundred feet above sea level. The limited marine fauna may have been introduced by birds.

In the Yuma area of Arizona, California, and Mexico, the Bouse Formation is widespread in the subsurface, but surface exposures are restricted to small area southeast of Imperial Dam (Olmsted and others, 1973). Data from drillholes and geophysical studies indicate that the Bouse Formation extends westward across the Algodones fault into the Salton Trough (Olmsted and others, 1973; Mattick and others, 1973). The fauna within the Bouse Formation and an unnamed underlying marine unit of this area is more varied than that of the Parker-Blythe-Cibola area and indicates marine open-shelf conditions (Winterer, 1975). Thus, this area was connected to the Gulf of California. In this area the northern margin of the Bouse Formation intersects the eastern margin of the Salton Trough (defined by the Algodones fault) near the California-Mexico border west of Pilot Knob (Mattick and others, 1973).

An ash bed near the base of the Bouse Formation in the area west of Cibola has yielded a K/Ar age of 5.4 ± 0.2 Ma (Lucchitta, 1972). Based on its marine fauna, Winterer (1975, Fig. 24) has rejected this age as too old. She considers the Bouse Formation of the Parker-Blythe-Cibola basin to be entirely late Pliocene in age (2 to 3 Ma). The Yuma basin contains older marine strata, the oldest of which she believes are early Pliocene in age (4 to 5 Ma).

The upper Miocene to lower Pliocene marine Imperial Formation crops out in the southwestern and northern parts of the Salton Trough and is present in the subsurface throughout most of the Salton Trough. There are no reports of it east of the Coachella Valley and Mission Creek branches of the San Andreas fault. Its northernmost occurrence is in San Gorgonio Pass. On the east side of the pass it crops out for a distance of 2 km, extending northward from the Banning fault. Here it consists of a fossiliferous neritic facies that extends unconformably across a fault contact between basement and the Coachella Fanglomerate and an olivine basalt dike (Allen, 1957). On the north it interfingers with the basal part of the nonmarine Painted Hill Formation. The projected northwest limit of the Imperial Formation is shown in Figure 3. It also crops out 15 km farther west in a fault sliver along the Banning fault (Allen, 1957). The most westerly reported occurrence is in Riverside County Oil Company Well Number 1, drilled south of the Banning fault on the west side of the pass (well RCOC-1, Fig. 3) (Blanck, 1987). The Imperial Formation is interpreted to have a thickness of about 600 m in this well (Blanck, 1987). The distribution of the Imperial Formation to the south of the Banning fault appears to require at least 15 km of right slip on the Banning fault since deposition of the Imperial Formation. The actual displacement may be as much as 35 km if Blanck (1987) has correctly interpreted the subsurface distribution of the Imperial Formation.

Although fossils point to a Pliocene age for most of the Imperial Formation (Ingle, 1973, 1974; Winterer, 1975), Matti and others (1985) reported K-Ar whole-rock age determinations of 5.94 ± 0.18 and 6.04 ± 0.18 Ma on an intercalated basalt flow in the Painted Hill Formation that overlies the Imperial Formation in San Gorgonio Pass. This indicates the Imperial Formation is no younger than late Miocene along its northwestern limit in San Gorgonio Pass. However, it does not necessarily negate the conclusions of Ingle (1973, 1974) and Winterer (1975) that most of the Imperial Formation was deposited during the Pliocene.

Ingle (1973, 1974) and Winterer (1975) have correlated the Bouse Formation of the Yuma area with the Imperial Formation of the Salton Trough based on similarities in the fossil contents and depositional environments of the two formations. For this reason, and because the bedrock south of the Mission Creek fault in San Gorgonio Pass correlates with that east of the Salton Trough in the southern Chocolate and Cargo Muchacho mountains, Dillon (1975) proposed that the northern limit of the Imperial Formation in San Gorgonio Pass is the offset westward continuation of the northern limit of the Bouse Formation to the east of Salton Trough. In his reconstruction, Dillon (1975, Fig. 70) assumed that the shoreline of the Bouse Formation extended along the west side of the Cargo Muchacho Mountains, subparallel to the San Andreas fault, and crossed the San Andreas fault adjacent to the south end of the Chocolate Mountains. In his reconstruction, the offset along the San Andreas fault is about 170 km. His reason for extending the shoreline so far north was to permit a seaway to extend eastward across the drainage divide between Salton Trough and Milipitas Wash.

An alternative interpretation preferred by Ehlig is that the northern limit of the Bouse Formation crosses the Algodones fault west of Pilot Knob near the California-Mexico border, essentially as shown in Figure 15 of Olmsted and others (1973). Assuming that the Algodones fault is the fossil trace of the San Andreas fault, this yields a right slip of about 200 km along the Mission Creek–Coachella Valley branch of the San Andreas fault. However, this may be too great a distance. The Algodones fault may be several kilometers east of the original San Andreas transform fault that extended from the northernmost spreading center in the gulf during latest Miocene time. If the shoreline was oblique to the Salton Trough so as to embay northward into the Salton Trough, the offset piercing points would be closer together (Dillon, 1975, Fig. 65). We consider a right slip of 185 ± 20 km to encompass a reasonable range of projections of the northern limits of temporally equivalent parts of the two formations. The age of the displacement is poorly constrained because of uncertainty in the age of the offset parts of the Imperial and Bouse formations, but displacement probably occurred during the last 6 m.y.

CORRELATION OF SALTON CREEK FAULT WITH UNNAMED FAULT IN SOLEDAD PASS

Crowell (1962) and Crowell and Walker (1962) made the initial correlation between the geology of the Orocopia Mountains to the north of the Salton Creek fault and that of the easternmost Soledad basin and western San Gabriel Mountains. The correlation was based on similarities in the types and features of the entire assemblage of rocks exposed in the two areas but focused on the distinctive suite of anorthosite, syenite, and related rocks present in the two areas (Figs. 4, 5). Crowell (1962) concluded that the two areas were offset by 210 km of post–earliest Miocene right slip on the San Andreas fault. His estimate of the offset was based on gross similarities between the two areas and lacked a unique piercing point. Silver (1971) strengthened the correlation by showing that the rocks in the two areas have identical Precambrian isotopic ages.

The first approximation of a piercing point was obtained from studies of clast assemblages in conglomerate within the lower part of the upper Miocene Mint Canyon Formation in Soledad basin and the delineation of sources of the clasts (Ehlig and Ehlert, 1972; Ehlig and others, 1975; Ehlert, 1982). The Mint Canyon Formation was previously known to contain volcanic clasts exotic to the local area, but the clasts were assumed to have come from the western Mojave Desert (Oakeshott, 1958). The studies showed that clast assemblages along the central axis of the westward-trending Mint Canyon trough are predominantly volcanic (more than 90 percent volcanic in many places). The assemblages contain an abundance of volcanic types not present in volcanic terranes of the western Mojave Desert, but these types are present in the northwestern Chocolate Mountains, south of the Salton Creek fault. The study focused on a distinctive group of quartz latite porphyry clasts characterized by abundant glomeroporphyritic phenocrysts of feldspar in which potassium feldspar mantles plagioclase and vice versa (rapakivi texture). A variety of such clasts from the Mint Canyon Formation were matched with identical rocks from outcrops in the Chocolate Mountains. A subsequent isotopic study (Joseph and Davis, 1977; Joseph, 1981) dated one of the dikes in the Chocolate Mountains (Fig. 4) at 27.0 ± 0.9 Ma and showed that clasts and outcrop samples have identical initial $^{87/86}$Sr ratios of 0.7059 ± 0.0002; thereby confirming the correlation. The offset yielded by this correlation depends on the location of the channel that transported volcanic clasts from the Chocolate Mountains to the Mint Canyon Formation. Constraints are provided by the nonvolcanic clast assemblages in Mint Canyon conglomerate. On the southeast side of the trough axis, conglomerate is rich in clasts of the biotite facies of the Mount Lowe Granodiorite along with lesser amounts of other facies of the Mount Lowe Granodiorite and anorthosite. The main potential source of the biotite facies clasts is the area surrounding Mount Emma and Mount Pacifico in the San Gabriel Mountains (Fig. 5). On the northwest side of the trough, conglomerate lacks clasts of Mount Lowe Granodiorite but contains abundant clasts of syenite, Pelona-Orocopia Schist, and, in the more westerly exposures, brown sandstone from the Paleocene San Francisquito Formation. In order to account for the change in clast types on either side of the trough axis, the drainage system that transported the clasts must have crossed the San Andreas fault near the present-day location of Soledad Pass. This suggests that Soledad Pass and Salton Creek wash were adjacent to each other prior to displacement on the San Andreas fault.

A comparison of the geology of Soledad Pass with that of Salton Creek wash indicates that an unnamed fault on the northwest side of Soledad Pass is the offset westward extension of the Salton Creek fault (Buesch and Ehlig, 1982). In both areas the subject fault places syenite on the northwest against Mount Lowe Granodiorite overlain by late Oligocene to early Miocene volcanic rocks on the southeast. Both faults have volcanic necks along them, indicating they were present during volcanism. Basaltic andesite predominates at Soledad Pass, but silicic rocks con-

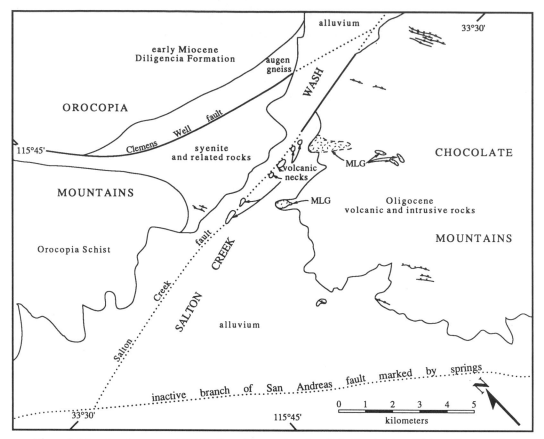

Figure 4. Map showing general distribution of exposed rock units in the vicinity of Salton Creek wash. Key: MLG = Mount Lowe Granodiorite; --x-- = dikes of quartz latite porphyry with phenocrysts of mantled feldspar. Data from Crowell and Walker (1962) and unpublished mapping by P. Ehlig.

taining sanidine and quartz occur in tuff-breccia and partially welded tuff. The volcanic suite adjacent to the Salton Creek fault includes the rock types present in Soledad Pass but contains a greater abundance and variety of silicic rocks. The differences between the two volcanic suites are most likely the result of unexposed lateral changes within a segment of the fault that fits between the two exposed segments. At Soledad Pass, bedrock exposures extend to the San Andreas fault zone, but at Salton Creek wash, alluvium conceals bedrock for a distance of about 10 km northeast of the active San Andreas fault. Part of the intervening bedrock may also exist as concealed fault slivers along the San Andreas fault.

The correlation of the two faults yields a precise piercing point, but only an approximate displacement, because of the concealed segment of Salton Creek fault that must fit between the two exposed segments. Assuming the Salton Creek fault continues along its exposed trend to the San Andreas fault, and that any segments lost as fault slivers are narrow, the correlation requires about 240 km of right slip on the intervening segment of the San Andreas fault within about the last 10 m.y. since deposition of the volcanic conglomerate member of the Mint Canyon Formation.

SAN GABRIEL FAULT AND ITS RELATION TO THE SAN ANDREAS FAULT

The displacement of Soledad Pass from its original position opposite Salton Creek wash does not account for all displacement on the southern San Andreas fault. Soledad Pass is within a structural block between the San Gabriel and the San Andreas faults. The San Gabriel fault has about 60 km of right slip along it and is believed to have been part of the main trace of the San Andreas fault during the period from about 12 to 5 Ma (Crowell, 1962, 1975, 1982; Ehlig and others, 1975). The exposed trace of the San Gabriel fault is 130 km long. Its northwest end is truncated by the San Andreas fault, but the truncation is concealed beneath the Frazier Mountain and Dry Creek thrusts (Crowell, 1954). The San Gabriel fault splits into north and south branches within the San Gabriel Mountains. The south branch extends to the south-central front of the range, where it is truncated by the Sierra Madre fault. The north branch extends eastward to the eastern part of the range, where it is truncated by the San Antonio fault.

In the San Gabriel Mountains, the nearly vertical contact between the Mendenhall Gneiss and Mount Lowe Granodiorite

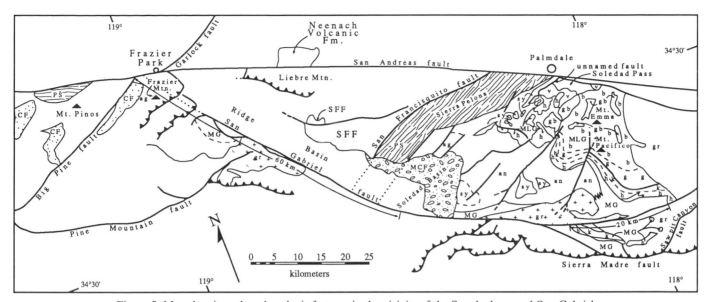

Figure 5. Map showing selected geologic features in the vicinity of the San Andreas and San Gabriel faults in the central Transverse Ranges. Key: ag = augen gneiss; MG = Mendenhall Gneiss; an = mostly anorthosite; sy = mostly syenite; gb = mostly gabbro; MLG = Mount Lowe Granodiorite; gr = late Mesozoic granitic rocks; PS = Pelona Schist; SFF = San Francisquito Formation; MCF = Mint Canyon Formation; CF = Caliente Formation. Patterns are "h" for hornblende phenocrysts, "k" for K-feldspar and hornblende phenocrysts, "g" for garnet present, and "b" for biotite present. Modified after the Los Angeles Sheet of the Geologic Map of California by Jennings and Strand (1969), Ehlig (1981), and Ehlig and Crowell (1982).

is displaced about 20 km to the right along the north branch of the San Gabriel fault (Ehlig, 1966). No correlations have been established across the south branch. From the western San Gabriel Mountains to the west end of the fault, all rock units older than about 12 Ma are displaced about 60 km to the right (Crowell, 1975b, 1982; Ehlig and others, 1975; Ehlig and Crowell, 1982; Ehlert, 1982). The net slip represents the amount of displacement that must be removed in order to have rock units southwest of the fault match their counterparts northeast of the fault. Reconstructions are complicated by deformation that post-dates displacement on the San Gabriel fault. Leucocratic granitic rocks, Mendenhall Gneiss, and a small body of anorthosite to the west of Ridge basin correlate with similar rocks in the western San Gabriel Mountains. Augen gneiss on Frazier Mountain correlates with that exposed in Soledad basin. The Caliente Formation to the west of Frazier Mountain is a westward continuation of the Mint Canyon Formation in Soledad basin. The Pelona Schist of Mount Pinos is the offset equivalent of the Pelona Schist of Sierra Pelona. The Blue Rock fault to the north of Mount Pinos is a major fault that correlates with the San Francisquito fault on the north side of Sierra Pelona (Bohannon, 1975; Smith, 1977).

Strike-slip displacement on the San Gabriel fault began near the end of middle Miocene time during deposition of the nonmarine Mint Canyon Formation in Soledad Basin and its correlative, the Caliente Formation, in the area west of Frazier Mountain. Basin analyses of the two formations indicate the lower part of the Mint Canyon Formation was deposited in a westward-descending alluvial wash that opened onto an alluvial fan of the Caliente Formation (Ehlig and others, 1975; Ehlert, 1982). About 10 to 12 Ma, displacement on the San Gabriel fault blocked the drainage system, causing a lake to form in the Soledad basin (Ehlert, 1982). The San Gabriel fault was continuously active during the late Miocene, as shown by analyses of formations in Ridge basin, a deep sedimentary basin formed by a releasing bend in the San Gabriel fault (Crowell, 1950, 1975b, 1982).

OTHER PUBLISHED DISPLACEMENTS

There are two published values of net slip on the San Andreas fault that bear directly on the values presented here. They are 315 km of right slip, based on correlation of the Pinnacles and Neenach volcanic formations (Matthews, 1976), and 160 km of right slip, based on the correlation of two bodies of distinctive Triassic monzogranite (Matti and others, 1985, 1986; Frizzell and others, 1986; Matti, this volume).

The Neenach Volcanic Formation overlies granitic basement northeast of the San Andreas fault in the southwest corner of the Mojave Desert (Fig. 5). The Pinnacles Volcanic Formation unconformably overlies granitic basement west of the San Andreas fault in the Gabilan Range of the southern Coast Ranges. A detailed comparison of the stratigraphy, petrography, and chemistry of the two formations indicates they are essentially identical and originated adjacent to each other (Matthews, 1976). This correlation requires about 315 km of right slip on the San Andreas fault since emplacement of the volcanics at 23.5 Ma. The displacement is in good agreement with a right slip of about 300 km for the combined displacements of the San Andreas and San Gabriel faults.

A monzogranite containing large phenocrysts of K-feldspar crops out north of the Mill Creek branch of the San Andreas fault in the central San Bernardino Mountains. A body of similar monzogranite crops out south of the San Andreas fault on Liebre Mountain, directly across the fault from the Neenach Volcanic Formation. Frizzell and others (1986) presented petrographic, chemical, and isotopic data indicating the two rocks are the same and have similar ages of 215 Ma. If the two originated adjacent to each other, only 160 km of right slip has occurred along the segment of the San Andreas fault that lies between them (Frizzell and others, 1986; Matti and others, 1985, 1986; Matti, this volume). This is about half of the displacement proposed by Matthews (1976) based on the correlation of the Neenach and Pinnacles volcanic formations. However, because the San Gabriel fault is southwest of Liebre Mountain, about 60 km can be subtracted from the difference. This leaves 95 km of right slip unaccounted for by correlation of the two bodies of monzogranite. The discrepancy is 80 km with respect to the Salton Creek–Soledad Pass correlation. The cause for these differences in apparent displacement needs to be resolved.

Each of the previously described correlations of rock units appears to be well founded. Therefore, there must be an explanation for the smaller apparent displacement of the monzogranite bodies. A simple explanation would be that the monzogranite of Liebre Mountain has not experienced total displacement on the San Andreas fault. Meisling and Weldon (1986) have postulated that the monzogranite at Liebre Mountain is in the upper plate of a thrust that originated in the San Bernardino Mountains and was displaced across the San Andreas fault during the late Miocene after the San Andreas fault had accumulated the missing displacement. This hypothesis is supported by a comparison of the structural styles and timing of thrusts in the two areas (Alexander and Weldon, 1987). Such a thrust might develop under conditions similar to those currently present in San Gorgonio Pass.

The thrust hypothesis is attractive, but there are other plausible hypotheses. One would be that Liebre Mountain is a fault sliver that was broken from the San Bernardino Mountains after 80 km of right slip occurred on an unrecognized branch of the San Andreas fault bounding Liebre Mountain on the southwest. The identity of such a fault might be obscured by subsequent transpressional tectonics. In this case, and in the case of the hypothesized thrust fault, field work needs to demonstrate that the monzogranite of Liebre Mountain is in fault contact with adjoining granitic and gneissic rocks to the southeast. Basement terrane 5 km southeast of Liebre Mountain contains distinctive "polka dot" granite similar to that found in the eastern Transverse Ranges north of the Salton trough (Joseph, 1981). The distribu-

tion of polka dot granite does not constrain a piercing for displacement on the San Andreas fault but does favor a displacement significantly larger than 160 km.

Other plausible explanations for the smaller apparent displacement of the two monzogranite bodies are that the two bodies were derived from the same magma source but were emplaced as separate plutons tens of kilometers apart, or that they are parts of a single pluton that have been separated by thrust faulting, detachment faulting, or some other type of faulting during the time interval between their crystallization in the Triassic and the start of displacement along the San Andreas fault.

PALINSPASTIC RECONSTRUCTION AND DISCUSSION

Figure 6 is a schematic palinspastic reconstruction showing the relative positions of selected rock units after removal of the following fault slips: 185 km on the Mission Creek and Coachella Valley branches of the San Andreas fault to approximately align previously correlated features; 240 km on the San Andreas fault to restore Soledad Pass to a position opposite Salton Creek; 25 km on the San Jacinto fault (based on Sharp, 1967) to restore terrane of the eastern San Gabriel Mountains against that between the Banning and San Andreas faults, and 20 km on the north branch and 60 km on the main branch of the San Gabriel fault. The north branch of the San Gabriel fault is assumed to have extended east of its present truncation by the San Antonio fault (Fig. 1) and connected to the Banning fault. The south branch of the San Gabriel fault is assumed to have extended eastward along the southern edge of the San Gabriel Mountains and connected to the Banning fault prior to displacement on the San Jacinto fault, or to have bent back into the central San Gabriel Mountains along the Sawpit Canyon fault, as shown by Dibblee (1968, Fig. 1), and rejoined the north branch of the San Gabriel fault.

In order to account for all of the right slip that must pass through San Gorgonio Pass, it is necessary to assume that the Banning fault has a total right slip of 90 ± 20 km, of which about 60 km must have occurred while it was connected to the San Gabriel fault and about 30 km must have occurred after the San Andreas fault established its present position northeast of the San Gabriel fault. Both assumptions are reasonable. As previously described, basement terrane north of the Banning fault is related to that of the San Gabriel Mountains and to the Chocolate and Cargo Muchacho Mountains but is significantly different from the basement terrane of the Peninsular Ranges that is exposed to the south of the Banning fault. If the late Miocene–early Pliocene Imperial Formation is present in the abandoned oil well drilled west of Banning (well RCOC-1 on Fig. 3), as interpreted by Blanck (1987), the northwest limit of the Imperial Formation may have been displaced by as much as 30 km along the Banning fault. During this displacement the Banning fault had to have connected to the present San Andreas fault in the area east of the San Gabriel Mountains in order to be part of the 240 km of offset shown by the Soledad Pass–Salton Creek correlation.

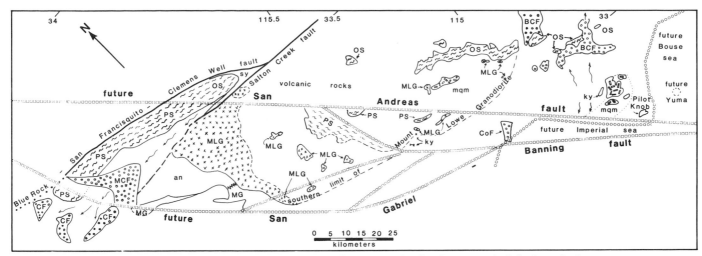

Figure 6. Simplified map showing distribution of selected rock units after removal of slip from the San Andreas, Mission Creek, and San Gabriel faults. Area northeast of the San Andreas fault is assumed to have remained fixed as shown by present-day latitude, longitude, and north arrow, except for bending of the Clemens Well fault to match with the San Francisquito fault. Key: MG = Mendenhal Gneiss; an = dominantly anorthosite; sy = dominantly syenite; MLG = Mount Lowe Granodiorite; mqm = melanocratic quartz monzonite; ky = kyanite-bearing rocks; PS = Pelona Schist; OS = Orocopia Schist; MCF = Mint Canyon Formation; CF = Caliente Formation; BCf = Bear Canyon fanglomerate; CoF = Coachella Fanglomerate. Wiggly arrows indicate direction of alluvial flow during deposition of Mint Canyon and Caliente Formations and Bear Canyon and Coachella fanglomerates.

The reconstruction shown in Figure 6 requires rotation and straightening of terrane from the central Transverse Ranges so that it fits against the terrane east of the Salton Trough. It also requires the west end of the Clemens Well fault to be rotated counterclockwise to join with the San Francisquito fault. The misalignment of the two faults is attributed to drag of Pelona-Orocopia Schist along the San Andreas fault and the insertion of fault slices between the active trace of the San Andreas fault and the Clemens Well fault (Ehlig, 1981). The Pelona-Orocopia Schist is much weaker and more prone to be affected by drag than other basement rocks along the San Andreas fault.

Based on our reconstruction, slip on the Mission Creek and Banning branches of the San Andreas fault may account for all displacement on the San Andreas fault and its branches except the 25 km of right slip on the San Jacinto fault. The approximately 215 km of right slip within San Gorgonio Pass that postdates deposition of the most northerly part of the Imperial-Bouse Formation has probably occurred within the last 4 to 6 m.y. If this is correct, the average slip rate of the San Andreas fault in San Gorgonio Pass and adjoining Coachella Valley has been between 3.5 and 5.5 cm/yr during the past 4 to 6 m.y. This is significantly faster than the current rate of about 2.5 cm/yr along the Coachella Valley segment of the San ANdreas fault reported by Keller and others (1982). The apparent shifting of motion away from the Coachella Valley branch of the San Andreas fault is probably the result of resistance caused by structural complications in the San Gorgonio Pass area and progressive westward

deflection of the San Andreas fault within the Transverse Ranges. The San Jacinto fault is probably taking up plate motion that used to be accommodated by slip on the Coachella Valley segment of the San Adnreas fault.

Based on our reconstruction, the San Gabriel fault was part of the San Andreas fault during the period prior to about 5 Ma. Its south end probably connected with the east side of a spreading center a short distance south of the California-Mexico border. It extended along the present axis of the Salton Trough before the Salton Trough existed and merged with the modern trace of the San Andreas fault in the vicinity of San Gorgonio Pass. The Coachella Valley segment of the modern San Andreas fault developed to the east of the San Gabriel segment about 5 Ma and has subsequently translated the San Gabriel fault northwestward as a part of the Pacific plate. Based on this reconstruction, the terrane that presently borders the west side of the San Andreas fault for a distance of 60 km beyond the northern terminus of the San Gabriel fault was originally along the southwest side of the San Gabriel fault. Our reconstruction also requires all of the southern Coast Ranges south of the Gabilan Range to have been adjacent to the Mojave Desert during middle Miocene time.

Our palinspastic reconstruction indicates crustal rocks of the Transverse Ranges were organized differently during the middle Miocene than they are now. However, it does not alter the anomalous absence of Mesozoic accretionary and arc terrane similar to that present farther south in the Peninsular Ranges or farther north in the Sierra Nevada. Our reconstruction indicates that a scallop-shaped marine embayment existed in the vicinity of the present Salton Trough during middle Miocene time if the region east of the San Andreas fault had its present configuration. However, the scallop and the bend in the San Andreas will probably disappear if late Miocene and younger crustal extension are removed from the Basin and Range Province.

ACKNOWLEDGMENTS

This work was in preparation at the time of John Dillon's unfortunate death. John had a more encompassing knowledge of terrane east of the Salton Trough than anyone else. A part of that knowledge was not in the literature or in the prepared part of the manuscript. It is hoped that this chapter does justice to his knowledge. Both authors owe much to John Crowell, who chaired both our Ph.D. dissertations, and who, through field trips and discussions, introduced us to problems of the San Andreas fault and basement terranes in southern California. Many other individuals have contributed to our work, as indicated by citations in this chapter. We especially thank Gordon Haxel, who often worked with John and freely shared ideas and information, and Robert Powell and Ray Weldon, who reviewed this manuscript and provided helpful suggestions.

REFERENCES CITED

Alexander, J., and Weldon, R., 1987, Comparison of structural style of the Liebre Mountain and Squaw Peak thrust systems, offset across the San Andreas fault, southern California: Geological Society of America Abstracts with Programs, v. 19, p. 353.

Allen, C. R., 1957, San Andreas fault zone in San Gorgonio Pass, southern California: Geological Society of America Bulletin, v. 68, p. 315–349.

——, 1981, The modern San Andreas fault, *in* Ernst, W. G., ed., The geotectonic development of California; Rubey Volume 1: Englewood Cliffs, New Jersey, Prentice-Hall, p. 511–534.

Barth, A. P., 1985, Petrology of the Mount Lowe intrusion, San Gabriel Mountains, California [M.S. thesis]: Los Angeles, California State University, 82 p.

Barth, A. P., and Ehlig, P. L., 1988, Geochemistry and petrogenesis of the marginal zone of the Mount Lowe intrusion, central San Gabriel Mountains, California: Contributions to Mineralogy and Petrology, v. 100, p. 192–204.

Blanck, E. L., 1987, Geologic structure of the Beaumont-Banning area from gravity profiles [M.S. thesis]: Los Angeles, California State University, 78 p.

Bohannon, R. G., 1975, Mid-Tertiary conglomerates and their bearing on Transverse Range tectonics, southern California, *in* Crowell, J. C., ed., San Andreas fault in southern California: California Division of Mines and Geology Special Report 118, p. 75–82.

Buesch, D. C., and Ehlig, P. L., 1982, Structural and lower Miocene volcanic rock correlation between Soledad Pass and Salton Wash along the San Andreas fault: Geological Society of America Abstracts with Programs, v. 14, p. 153.

Carter, B. A., 1980, Structure and petrology of the San Gabriel anorthosite-syenite body, Los Angeles County, California [Ph.D. thesis]: Pasadena, California Institute of Technology, 393 p.

——, 1982, Field petrology and structural development of the San Gabriel anorthosite-syenite body, Los Angeles County, California, *in* Cooper, J. D., compiler, Geologic excursions in the Transverse Ranges, southern California; Geological Society of America Annual Meeting Guidebook: Fullerton, California State University, p. 1–53.

Carter, B. A., and Silver, L. T., 1972, Structure and petrology of the anorthosite-syenite body, California: 24th International Geological Congress, Section 2, p. 303–311.

Crowe, B. M., Crowell, J. C., and Krummenacher, D., 1979, Regional stratigraphy, K-Ar ages, and tectonic implications of Cenozoic volcanic rocks, southeastern California: American Journal of Science, v. 279, p. 186–216.

Crowell, J. C., 1950, Geology of the Hungry Valley area, southern California: American Association of Petroleum Geologists, v. 34, p. 1623–1646.

——, 1962, Displacement along the San Andreas fault, California: Geological Society of America Special Paper 71, 61 p.

——, 1973, Problems concerning the San Andreas system in southern California, *in* Kovach, R. L., and Nur, A., eds., Proceedings of the Conference on Tectonic Problems of the San Andreas Fault System: Stanford, California, Stanford University Publications in the Geological Sciences, v. 13, p. 125–135.

——, 1975a, Geologic sketch of the Orocopia Mountains, southeastern California, *in* Crowell, J. C., ed., San Andreas fault in southern California: California Division of Mines and Geology Special Report 118, p. 99–110.

——, 1975b, The San Gabriel fault and Ridge Basin, southern California, *in* Crowell, J. C., ed., San Andreas fault in southern California: California Division of Mines and Geology Special Report 118, p. 208–221.

——, 1982, The tectonics of Ridge Basin, southern California, *in* Crowell, J. C., and Link, M. H., eds., Geologic history of Ridge Basin, southern California: Pacific Section, Society of Economic Paleontologists and Mineralogists, p. 25–42.

Crowell, J. C., and Walker, J.W.R., 1962, Anorthosite and related rocks along the San Andreas fault, southern California: Berkeley, University of California Publications in the Geological Sciences, v. 40, p. 219–288.

Dibblee, T. W., Jr., 1968, Displacements on the San Andreas fault system in the San Gabriel, San Bernardino, and San Jacinto Mountains, southern California, *in* Dickinson, W. R., and Grantz, A., eds., Proceedings of the Conference on Geologic Problems of the San Andreas Fault System: Stanford, California, Stanford University Publications in the Geological Sciences, v. 11, p. 294–305.

Dillon, J. T., 1975, Geology of the Chocolate and Cargo Muchacho Mountains, southeasternmost California [Ph.D. thesis]: Santa Barbara, University of California, 405 p.

——, 1986, Timing of thrusting and metamorphism along the Vincent–Chocolate Mountain thrust system, southern California: Geological Society of America Abstracts with Programs, v. 18, p. 101.

Ehlert, K. W., 1982, Basin analysis of the Miocene Mint Canyon Formation, southern California, *in* Ingersoll, R. V., and Woodburne, M. O., eds., Cenozoic nonmarine deposits of California and Arizona: Pacific Section, Society of Economic Paleontologists and Mineralogists, p. 51–64.

Ehlig, P. L., 1958, Geology of the Mount Baldy region of the San Gabriel Mountains, California [Ph.D. thesis]: Los Angeles, University of California, 192 p.

——, 1966, Displacement along the San Gabriel fault, San Gabriel Mountains, southern California [abs.]: Geological Society of America 1966 Annual Meetings Program, p. 60.

——, 1968, Causes of distribution of Pelona, Rand, and Orocopia Schist along the San Andreas and Garlock faults, *in* Dickinson, W. R., and Grantz, A., eds., Proceedings of the Conference on Geologic Problems of San Andreas Fault System: Stanford, California, Stanford University Publications in the Geological Sciences, v. 11, p. 294–305.

——, 1975, Basement rocks of the San Gabriel Mountains, south of the San Andreas fault, southern California, *in* Crowell, J. C., ed., San Andreas fault in southern California: California Division of Mines and Geology Special Report 118, p. 177–186.

——, 1981, Origin and tectonic history of the basement terrane of the San Gabriel Mountains, central Transverse Ranges, *in* Ernst, W. G., ed., The geotectonic development of California; Rubey Volume 1: Englewood Cliffs, New Jersey, Prentice-Hall, p. 253–283.

Ehlig, P. L., and Crowell, J. C., 1982, Mendenhall Gneiss and anorthosite-related rocks bordering Ridge Basin, southern California, *in* Crowell, J. C., and Link, M. H., eds., Geologic history of Ridge basin, southern California:

Pacific Section, Society of Economic Paleontologists and Mineralogists, p. 199–202.

Ehlig, P. L., and Ehlert, K. W., 1972, Offset of Miocene Mint Canyon formation from volcanic source along the San Andreas fault, southern California: Geological Society of America Abstracts with Programs, v. 4, p. 154.

Ehlig, P. L., Ehlert, K. W., and Crowe, B. M., 1975, Offset of Upper Miocene Caliente and Mint Canyon Formations along the San Gabriel and San Andreas faults, *in* Crowell, J. C., ed., San Andreas fault in southern California: California Division of Mines and Geology Special Report 118, p. 83–92.

Elders, W. A., Rex, R. W., Meidav, T., Robinson, P. T., and Biehler, S., 1972, Crustal spreading in southern California: Science, v. 178, p. 15–24.

Farley, T., 1979, Geology of a part of northern San Gorgonio Pass, California [M.S. thesis]: Los Angeles, California State University, 159 p.

Frizzell, V. A., Jr., Mattinson, J. M., and Matti, J. C., 1986, Distinctive Triassic megaporphyritic monzogranite; Evidence for only 160 km offset along the San Andreas fault, southern California: Journal of Geophysical Research, v. 91, p. 14080–14088.

Frost, E., Drobeck, P., and Hillemyer, B., 1986, Geologic setting of gold and silver mineralization in southeastern California and southwestern Arizona, *in* Cenozoic stratigraphy, structure, and mineralization in the Mojave Desert; Geological Society of America Cordilleran Section 82nd Annual Meeting Guidebook and Volume: Los Angeles, Geology Department, California State University, unpublished report, p. 71–119.

Fuis, G. S., Mooney, W. D., Healey, J. H., McMechan, G. A., and Lutter, W. J., 1982, Crustal structure of the Imperial Valley region, *in* The Imperial Valley, California, earthquake of October 15, 1979: U.S. Geological Survey Professional Paper 1254, p. 15–24.

Graham, C. M., and Powell, R., 1984, A garnet-hornblende geothermometer; Calibration, testing, and application to the Pelona Schist, southern California: Journal of Metamorphic Geology, v. 2, p. 13–31.

Haxel, G. B., 1977, The Orocopia Schist and Chocolate Mountain thrust, Picacho–Peter Kane Mountain area, southeasternmost California [Ph.D. thesis]: Santa Barbara, University of California, 277 p.

Haxel, G., and Dillon, J., 1973, The San Andreas fault system in southeasternmost California, *in* Kovach, R. L., and Nur, A., eds., Proceedings of the Conference on Tectonic Problems of the San Andreas Fault System: Stanford, California, Stanford University Publications in the Geological Sciences, v. 13, p. 322–333.

—— , 1978, The Pelona-Orocopia Schist and Vincent–Chocolate Mountain thrust system, southern California, *in* Howell, D. G., and McDougall, K. A., eds., Mesozoic paleogeography of the western United States: Pacific Section, Society of Economic Paleontologists and Mineralogists, p. 453–469.

Haxel, G. B., and 5 others, 1987, Geochemistry of the Orocopia Schist, and southeastern California; Summary, *in* Dickinson, W. R., and Klute, M. A., eds., Mesozoic rocks of southern Arizona and adjacent areas: Arizona Geological Society Digest, v. 18, p. 49–64.

Henshaw, P. C., 1942, Geology and mineral resources of the Cargo Muchacho Mountains, Imperial County, California: California Division of Mines Report 38, p. 147–196.

Ingle, J. C., Jr., 1973, Neogene marine history of the Gulf of California; Foraminiferal evidence: Geological Society of America Abstracts with Programs, v. 5, p. 62.

—— , 1974, Paleobathymetric history of Neogene marine sediments, northern Gulf of California, *in* Gastil, G., and Lillegraven, J., eds., Geology of the Peninsular California: Pacific Section, American Association of Petroleum Geologists Guidebook, p. 121–138.

Jacobson, C. E., 1980, Deformation and metamorphism of the Pelona Schist beneath the Vincent thrust, San Gabriel Mountains, California [Ph.D. thesis]: Los Angeles, University of California, 231 p.

—— , 1983, Structural geology of the Pelona Schist and Vincent thrust, San Gabriel Mountains, California: Geological Society of America Bulletin, v. 94, p. 753–767.

Jacobson, C. E., Dawson, M. R., and Postlethwaite, C. E., 1988, Structure,

metamorphism, and tectonic significance of Pelona, Orocopia, and Rand schists, southern California, *in* Ernst, W. G., ed., Metamorphism and crustal evolution of the western United States; Rubey Volume 7: Englewood Cliffs, New Jersey, Prentice-Hall, p. 976–997.

Jennings, C. W., and Strand, R. G., 1969, Geologic map of California; Los Angeles sheet: California Division of Mines and Geology, scale 1:250,000.

Johnson, C. E., and Hadley, D. M., 1976, Tectonic implications of the Brawley earthquake swarm, Imperial Valley, California, January 1975: Seismological Society of America Bulletin, v. 66, p. 1133–1144.

Johnson, C. E., and Hill, D. P., 1962, Seismicity of the Imperial Valley, *in* The Imperial Valley, California, earthquake of October 15, 1979: U.S. Geological Survey Professional Paper 1254, p. 15–24.

Joseph, S. E., 1981, Isotopic correlation of the La Panza Range granitic rocks with similar rocks in the central and eastern Transverse Ranges [M.S. thesis]: Los Angeles, California State University, 77 p.

Joseph, S. E., Criscione, J. J., Davis, T. E., and Ehlig, P. L., 1982, The Lowe igneous pluton, *in* Fife, D. L., and Minch, J. A., eds., Geology and mineral wealth of the California Transverse Ranges: Santa Ana, California, South Coast Geological Society, p. 307–309.

Joseph, S. E., and Davis, T. E., 1977, $^{87}Sr/^{86}Sr$ correlation of Rapakivi-textured porphyry to measure offset on the San Andreas fault: Geological Society of America Abstracts with Programs, v. 9, p. 443.

Keller, E. A., Bonkowski, M. S., Lorsch, R. J., and Shlemon, R. J., 1982, Tectonic geomorphology of the San Andreas fault zone in the southern Indio Hills, Coachella Valley, California: Geological Society of America Bulletin, v. 93, p. 46–56.

Larson, R. L., Menard, H. W., and Smith, S. M., 1968, Gulf of California; A result of ocean-floor spreading and transform faulting: Science, v. 161, p. 781–784.

Lucchitta, I., 1972, Early history of the Colorado River in the Basin and Range Province: Geological Society of America Bulletin, v. 83, p. 1933–1948.

Matthews, V., III, 1976, Correlation of Pinnacles and Neenach Volcanic Formations and their bearing on San Andreas fault problem: American Association of Petroleum Geologists Bulletin, v. 60, p. 2128–2141.

Matti, J. C., and 6 others, 1982, Mineral resource potential map of the Whitewater Wilderness Study Area, Riverside and San Bernardino Counties, California: U.S. Geological Survey Miscellaneous Field Studies Map MF-1478-A, scale 1:24,000.

Matti, J. C., Morton, D. M., and Cox, B. F., 1985, Distribution and geologic relations of fault systems in the vicinity of the central Transverse Ranges, southern California: U.S. Geological Survey Open-File Report 85-365, 23 p.

Matti, J. C., Frizzell, V. A., and Mattinson, J. M., 1986, Distinctive Triassic megaporphyritic monzogranite displaced 160 ± 10 km by the San Andreas fault, southern California: Geological Society of America Abstracts with Programs, v. 18, p. 154.

Mattick, R. E., Olmsted, F. H., and Zody, A. R., 1973, Geophysical Studies in the Yuma area, Arizona and California: U.S. Geological Survey Professional Paper 726-D, 36 p.

Merifield, P. M., Rockwell, T. K., and Lughman, C. C., 1987, Slip rate on the San Jacinto fault zone in the Anza seismic gap, southern California: Geological Society of America Abstracts with Programs, v. 19, p. 431–432.

Metzger, D. G., 1968, The bouse Formation (Pliocene) of the Parker-Blythe-Cibola area, Arizona and California: U.S. Geological Survey Professional Paper 600-D, p. D126–D136.

Minster, J. B., and Jordan, T. H., 1978, Present-day plate motions: Journal of Geophysical Research, v. 83, p. 5331–5354.

—— , 1987, Vector constraints on western U.S. deformation from space geodesy, neotectonics, and plate motions: Journal of Geophysical Research, v. 92, p. 4798–4804.

Oakeshott, G. B., 1958, Geology and mineral deposits of the San Fernando Quadrangle, Los Angeles County, California: California Division of Mines Bulletin 172, 147 p.

Olmsted, F. H., Loeltz, O. J., and Irelan, B., 1973, Geohydrology of the Yuma area, Arizona and California: U.S. Geological Survey Professional Paper 486-H, 227 p.

Peterson, M. S., 1973, Geology of the Coachella Fanglomerate, San Gorgonio Pass, California [M.A. thesis]: Santa Barbara, University of California, 114 p.

—— , 1975, Geology of the Coachella Fanglomerate, *in* Crowell, J. C., ed., San Andreas fault in southern California: California Division of Mines and Geology Special Report 118, p. 119–126.

Pinault, C. T., and Rockwell, T. K., 1984, Rates and sense of Holocene faulting on the southern Elsinore fault; Further constraints on the distribution of dextral shear between the Pacific and North American plates: Geological Society of America Abstracts with Programs, v. 16, p. 624.

Powell, R. E., 1981, Geology of the crystalline basement complex, eastern Transverse Ranges, southern California [Ph.D. thesis]: Pasadena, California Institute of Technology, 441 p.

—— , 1982, Crystalline basement terranes in the southern eastern Transverse Ranges, California, *in* Cooper, J. D., compiler, Geologic excursions in the Transverse Ranges; Geological Society of America Annual Meeting Guidebook: Fullerton, California State University, p. 109–136.

Sampson, R. J., and Tucker, W. B., 1942, Mineral resources of Imperial County, California: California Division of Mines Report 38, p. 105–146.

Sharp, R. V., 1967, San Jacinto fault zone in the Peninsular Ranges of southern California: Geological Society of America Bulletin, v. 78, p. 705–730.

—— , 1982, Tectonic setting of the Imperial Valley region, *in* The Imperial Valley, California, earthquake of October 15, 1979: U.S. Geological Survey Professional Paper 1254, p. 15–24.

Silver, L. T., 1966, Preliminary history for the crystalline complex of the central Transverse Ranges, Los Angeles, California: Geological Society of America Program 1966 Annual Meeting, p. 201–202.

—— , 1971, Problems of crystalline rocks of the Transverse Ranges: Geological Society of America Abstracts with Programs, v. 3, p. 193–194.

Silver, L. T., McKinney, C. R., Deutsch, S., and Bolinger, J., 1963, Precambrian age determinations in the western San Gabriel Mountains, California: Journal of Geology, v. 71, p. 196–214.

Smith, D. P., 1977, San Juan–St. Francis fault; Hypothesized major middle Tertiary right-lateral fault in central and southern California, *in* Short contributions to California geology: California Division of Mines and Geology Special Report 129, p. 41–50.

Tosdal, R. M., 1988, Mesozoic rock units along the Late Cretaceous Mule Mountains thrust system, southeastern California and southwestern Arizona [Ph.D. thesis]: Santa Barbara, University of California, 365 p.

Winterer, J. I., 1975, Biostratigraphy of Bouse Formation; A Pliocene Gulf of California deposit in California, Arizona, and Nevada [M.S. thesis]: Long Beach, California State University, 132 p.

MANUSCRIPT ACCEPTED BY THE SOCIETY SEPTEMBER 6, 1990

Geological Society of America
Memoir 178
1993

Chapter 5

Extension and contraction within an evolving divergent strike-slip fault complex: The San Andreas and San Jacinto fault zones at their convergence in southern California

Douglas M. Morton
U.S. Geological Survey, Riverside, California 92521
Jonathan C. Matti
U.S. Geological Survey, Department of Geological Sciences, University of Arizona, Tucson, Arizona 85721

ABSTRACT

A variety of extensional and contractional structures is produced by strike slip faulting. The variety and extent of the structures are directly related to the kind and extent of geometric complexities of the fault zone or system. The area of convergence of the San Andreas fault zone and the much younger San Jacinto fault zone in the eastern Transverse Ranges is exquisitely complex. We propose that the San Jacinto fault zone formed in response to a structural knot in San Gorgonio Pass probably within the past 1.5 Ma. In the area of their convergence we propose that slip is transferred both east and west from the San Jacinto fault zone northward to the San Andreas fault zone over a 60- to 70-km band that extends northwestward from the south end of the San Bernardino basin to the east end of the San Gabriel Mountains. We further propose several structural adjustments as a consequence of onset or acceleration of lateral movement on the San Jacinto fault zone: accelerated uplift of the eastern San Gabriel Mountains, development or accentuation of an arcuate schuppen-like structure in the eastern San Gabriel Mountains, inception of the San Bernardino basin, cessation of deposition in the present-day San Timoteo badlands area, inception of the San Jacinto basin, and an increase in compression and uplift in the San Gorgonio Pass area. We interpret the uplift and compression in San Gorgonio Pass to result from two formerly disparate structural blocks—the eastern San Bernardino and San Jacinto blocks—becoming a relatively coherent block, and the San Gorgonio Pass area constituting a left step between the San Andreas fault zone in the Coachella Valley area and the San Jacinto fault zone in the San Jacinto Valley area. The compression and uplift led to the formation of the San Gorgonio Pass thrust faults and disruption of any through-going San Andreas strands, at least at the surface.

In partitioning slip between the San Andreas and San Jacinto fault zones, consideration should be given to the bandwidth over which horizontal strain has accumulated. The average slip rate of the northern part of the San Jacinto fault zone during the past 1.5 m.y. may have been about 20 mm/yr and about 15 mm/yr on the San Andreas. South of the San Bernardino basin, current strain accumulation based on repeated geodetic surveys is nearly equally divided between the San Jacinto and San Andreas fault zones.

Morton, D. M., and Matti, J. C., 1993, Extension and contraction within an evolving divergent strike-slip fault complex: The San Andreas and San Jacinto fault zones at their convergence in southern California, *in* Powell, R. E., Weldon, R. J., II, and Matti, J. C., eds., The San Andreas Fault System: Displacement, Palinspastic Reconstruction, and Geologic Evolution: Boulder, Colorado, Geological Society of America Memoir 178.

INTRODUCTION

As it evolves, a complex segment of a strike-slip fault system concurrently and sequentially generates a great variety of contrasting structures. A single fault zone can have a complex history of strand development, abandonment, and reactivation (e.g., Matti and others, 1985). Irregularities in strike of a fault can result in restraining (compressional) and/or releasing (tensional) bends (Crowell, 1974) and their attendant structural features such as uplift and thrusting or development of pull-apart basins (e.g., Sharp, 1967). A new fault developed outside an existing fault zone will consist of a series of en echelon step-bounded strands with a zone of compression or extension within the area of a step in addition to possible restraining and/or releasing bends along the length of a strand. The frequency of steps per unit length of a fault is apparently a function of cumulative offset: the less the offset, the greater the frequency (Wesnousky, 1988).

If a new fault develops in a direction that is divergent from the existing fault zone, interaction between the divergent faults yields broad domains of tension and compression, both between and outside the area of the divergent faults. Complexes of normal and thrust faults can be produced within the domains resulting in pull-apart basins and in complex arches or domes, in addition to those produced at steps and bends in the newly formed fault.

Additional complexities can arise from the presence of preexisting faults, active and/or inactive, in the area of the divergent faults. These preexisting faults, which may have developed in a stress field quite different from the present stress field, can be accelerated, reactivated, and/or deformed to further complicate the structure in the vicinity of the divergent faults.

Transference of slip from one fault zone to another can likewise produce a domain of compression or tension. Depending on the distance over which the slip is transferred, the resultant subsidence or uplift can be local or of regional scale.

Based on our mapping of the San Andreas and San Jacinto fault zones in the area of the central Transverse Ranges of southern California, we have deduced an ongoing history of faulting that includes strand development, abandonment, and reactivation; development of structures associated with bends along strike; evolution of divergent fault zones and their attendant features, including acceleration and reactivation of slip of preexisting faults, and regional uplift resulting from transference of slip from one fault zone to another. We describe here mainly features produced by the continuing interaction of the relatively young San Jacinto fault zone in the area of its divergence from the older San Andreas fault zone, expanding and modifying our earlier interpretations (Morton, 1975a, b; Matti and others, 1985).

We first describe the geologic setting of the area of the divergent faults, the San Bernardino basin and eastern San Gabriel Mountains, then describe the late Cenozoic faults, present our interpretations, and conclude with some speculations. We propose a model that includes slip transference between the San Jacinto and San Andreas fault zones, termination of the San Jacinto fault in the eastern San Gabriel Mountains, a young age

for the San Jacinto fault zone, and finally, consider some slip rates on both the San Jacinto and San Andreas fault zones. As with other models we know of for this part of southern California, we consider our model to be permissive, but not required by the data.

The interrelationship between the northern end of the San Jacinto fault zone and the San Andreas fault involves structures of both the eastern San Gabriel Mountains and the adjacent San Bernardino basin (Fig. 1). Both of these structural subprovinces are triangular-shaped with slightly overlapping apices. The eastern San Gabriel Mountains is an area of compression and uplift, sharply contrasting with the San Bernardino basin, an area of tension and subsidence.

SAN BERNARDINO BASIN

We informally define the "San Bernardino basin" as the triangular-shaped part of the upper Santa Ana Valley between the San Andreas and San Jacinto fault zones northwest of the Crafton Hills (Fig. 2). The basin's narrow northwest apex is overlapped by the adjacent eastern San Gabriel Mountains. This northwest apex is coincident with the mouth of Cajon Canyon, near the Glen Helen, Peters, and Tokay Hill faults.

Most of the basin surface is covered by alluvial fans derivative from the San Bernardino Mountains, and our knowledge of the age and thickness of the basin fill and of structures within the basin is both skeletal and indirect. Surface exposures and water well data indicate that the basin contains sedimentary fill which is locally more than 300 m thick (e.g., Dutcher and Garrett, 1963). Models based on gravity data suggest that the sedimentary fill in the western part of the basin is 1 to 2.5 km thick (Willingham, 1968; Lambert, 1987). The only age determinations for valley fill are limited to a depth of 17 m. Nine ^{14}C ages obtained from organic material in alluvium at depths of 3 to 17 m below the surface of the west-central part of the basin (Carson and others, 1986) indicate depositional rates of 1 to 11.5 mm/yr for the alluvium.

Faulting is recognized at both ends of the San Bernardino basin (Morton and Miller, 1975; Matti and others, 1985; Weldon and Matti, 1986; Weldon and Springer, 1988). The Peters and Tokay Hill faults are located at the northwest apex and the Crafton Hills fault complex forms the southeastern margin (Fig. 3). The Tokay Hill fault appears to be a reverse fault; the Peters fault scarp faces north toward the San Bernardino Mountains, but its dip is unknown. There is no evidence of any surface faulting within the central part of the basin. Based on groundwater levels, the presence of some faults have been inferred marginal to the basin, and parallel to the basin margins (Dutcher and Garrett, 1963). Hills underlain by Pelona Schist in the northern part of the basin are elongate subparallel to the San Jacinto and San Andreas fault zones. Their direction of elongation suggests they might be fault-bounded. Alluvium exposed in a 6-m-deep trench across the ends of some of these hills appeared to be unfaulted (D. M. Morton, unpublished observations).

Interpretations of seismicity within the basin area are con-

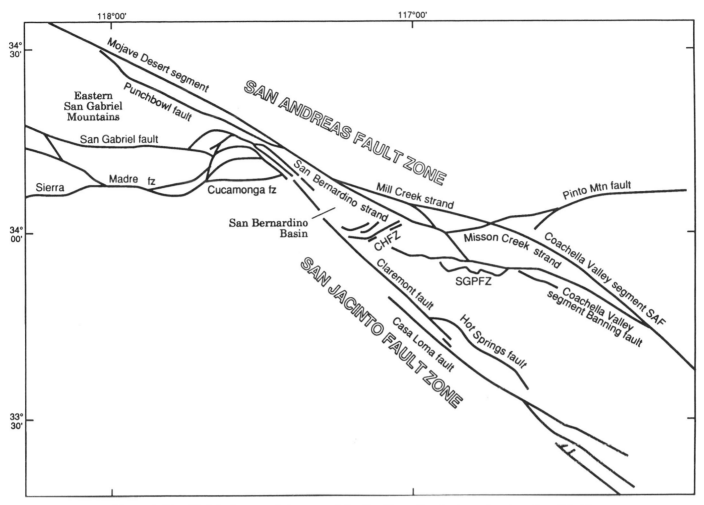

Figure 1. Simplified fault map of the area of the San Bernardino basin and eastern San Gabriel Mountains. CHFZ = Crafton Hills fault zone; SGPFZ = San Gorgonio Pass fault zone; SAF = San Andreas fault; fz = fault zone.

flicting. Based on single-event fault plane solutions, Jones (1988) has interpreted the seismicity as mostly resulting from normal faulting. Nicholson and others (1986a, b), on the other hand, has made use of composite event focal mechanisms and interpreted the seismicity chiefly as a result of left slip and rotation on northeast oriented faults. We prefer Jones's interpretation indicating normal faulting based on single-event fault plane solutions because it best fits our knowledge of the geology. Northeast-striking left-lateral fault solutions have been determined for seismicity occurring along northeast-oriented seismic zones in the valley area south of the eastern San Gabriel Mountains and west of the San Bernardino basin (Hadley and Combs, 1974; Cramer and Harrington, 1987; Yerkes and others, 1987), where documented northeast-striking faults exist.

Within the San Bernardino basin the exposed basement rock indicates that the basin was formerly a topographic high. Hills underlain by the Pelona Schist rise above the alluvial valley surface in the northern part of the basin (Fig. 2). The Pelona Schist

here is white mica–albite–quartz schist. In the Crafton Hills (J. C. Matti and others, unpublished mapping, 1977) and eastern San Gabriel Mountains (Ehlig, 1958; D. M. Morton, unpublished mapping, 1974), the Pelona Schist is structurally overlain by 60 to 600 m of mylonitic rock. The mylonitic rock is separated from the Pelona Schist by the Vincent thrust (Fig. 2). In both the eastern San Gabriel Mountains and the Crafton Hills, the Pelona Schist immediately below the Vincent thrust consists largely of greenstone. The Pelona Schist, with common to abundant greenstone, is 250 to several thousand meters thick in the eastern San Gabriel Mountains (Ehlig, 1958; D. M. Morton, unpublished mapping, 1974) and more than 100 m thick in the Crafton Hills where the base of the greenstone is not exposed (J. C. Matti and others, unpublished mapping, 1977). Below the greenstone in the San Gabriel Mountains is white mica–albite–quartz schist. The presence of white mica-albite-quartz schist and the lack of greenstone outcrops within the San Bernardino basin (D. M. Morton, unpublished mapping, 1976) suggests the exposed Pelona Schist

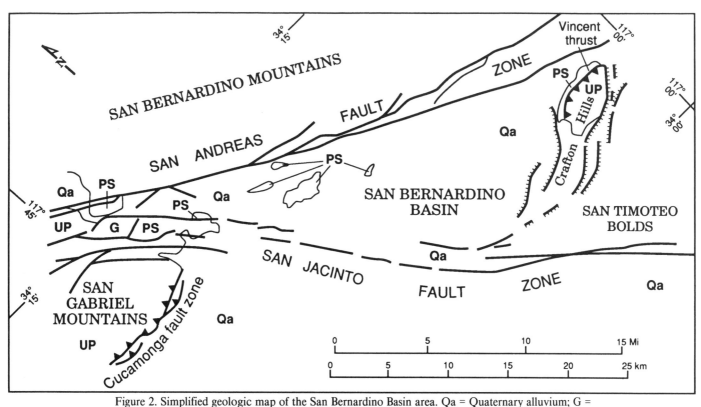

Figure 2. Simplified geologic map of the San Bernardino Basin area. Qa = Quaternary alluvium; G = Miocene granitic rock; PS = Pelona Schist; UP = upper plate rocks.

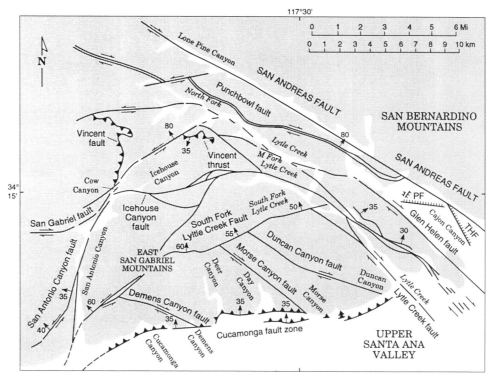

Figure 3. Map showing major physiographic features and principal fault zones in the eastern San Gabriel Mountains. PF = Peters fault; THF = Tokay Hill fault.

in the basin was located several hundred to a thousand or more meters beneath the Vincent thrust. The implication is that, in prebasin time, the San Bernardino basin area was a positive physiographic feature.

Restoration of 24 km of right-lateral displacement, the amount determined for the San Jacinto fault zone in the northern Peninsular Ranges (Sharp, 1967), places the formerly positive physiographic San Bernardino basin area as an eastward continuation of the present high-standing eastern San Gabriel Mountains.

San Timoteo badlands

Immediately south of the San Bernardino basin is the San Timoteo badlands (Fig. 2). The badlands, underlain by a thick section of continental sedimentary rocks, extend about 35 km southeastward from the southwestern margin of the San Bernardino basin. The lower part of this section, the Mount Eden Formation (Fraser, 1931), consists of sandstone and mudstone that are chiefly fluvial and partly lacustrine in origin (Matti and Morton, 1975). A fossil vertebrate assemblage called the Mount Eden local fauna, from the lower part of the Mount Eden Formation, is about 5 to 5.4 Ma (May and Repenning, 1982; Repenning, 1987).

Overlying and interfingering with the upper part of the Mount Eden Formation is the fluvial San Timoteo Formation (Fraser, 1931), mainly coarse-grained sandstone and conglomeratc (J. C. Matti, unpublished mapping, 1973). The youngest dated part of the San Timoteo Formation contains the 1.3 ± 0.3-Ma Shutt Ranch local fauna (Repenning, 1987). The Shutt Ranch fauna is in turn underlain by strata that contain the El Casco local fauna whose age is estimated to be 1.5 ± 0.3 Ma (Repenning, 1987). Also in the upper part of the San Timoteo Formation is a conglomeratic section containing clasts derived from the San Bernardino Mountains, as well as clasts from rock units lying above and below the Vincent thrust. Clasts from above the Vincent thrust include those derived from a variety of mylonites and the Mount Lowe Granodiorite of Miller (1934); clasts derived from units lying below the thrust consist primarily of greenstone of the Pelona Schist. Clasts derived from the San Bernardino Mountains include those derived from Precambrian quartzite, the monzogranite of Keller Park, and a very distinctive megaporphyritic hornblende monzogranite (Frizzell and others, 1986, Matti and Morton, this volume); presently, all these lithologic units crop out together only in the Santa Ana River drainage in the San Bernardino Mountains northeast of the exposures of the sedimentary rocks (Morton and others, 1986). The Olive Dell local fauna, collected from near the top of this conglomeratic section, is presently considered to be about 1.5 Ma (Repenning, 1987). Based on a rodent tooth identified as *Pitymys meadensis,* this part of the section was previously considered to be about 700 ka (Morton and others, 1986); this tooth has been reidentified as *Microtus Californicus* (C. A. Repenning, personal communication, 1986) indicating an age of about 1.5 Ma for this part of the

section. Soil developed on an alluvial surface at the top of the San Timoteo Formation is tentatively estimated to have an age of about 500 ka (L. D. McFadden, personal communication, 1988).

In summary, clasts in the lower part of the San Timoteo Formation were derived from lithologies known only from above the Vincent thrust; none are known from Pelona Schist rock types from below the thrust. Approximately in the middle of the formation the clast assemblage includes clasts of mylonite and Mount Lowe Granodiorite derived from structurally above the Vincent thrust and clasts of the Pelona Schist derived from structurally below the Vincent thrust. Greenstone clasts derived from structurally high Pelona Schist are a common lithic constituent of the upper part of the San Timoteo Formation.

The basement inferred to have been eroded from the San Bernardino basin prior to basin development, the presently exposed basement, and the distribution of rock types as clasts within the sedimentary rocks of the San Timoteo badlands suggest that during the interval 3.5 to 4(?) Ma to about 0.8 to 1.3 Ma the area of the present-day San Bernardino basin was relatively high standing. The eroding basement shed coarse clastic debris westward and southward into the San Timoteo badlands depositional area. Within the San Timoteo Formation, the transition from rock types derived from above the Vincent thrust to rock types derived from both above and below the Vincent thrust marks the unroofing of the Vincent thrust. The occurrence of San Bernardino Mountains rocks in sedimentary rocks dated at 1.5 Ma, is the first known record of a San Bernardino Mountains drainage extending across the San Bernardino basin area.

Structurally high rocks, above the Vincent thrust, occur southeast and east of the San Bernardino basin, from Crafton Hills eastward into San Gorgonio Pass. The presence of these structurally high rocks suggests the area of their occurrence was probably relatively topographically low when the San Bernardino basin area was topographically high.

EASTERN SAN GABRIEL MOUNTAINS

Basement rocks of the eastern San Gabriel Mountains can be subdivided into two general groups separated by the Icehouse Canyon fault. The northern basement rock group consists of the Pelona Schist and structurally overlying mylonite, both intruded by middle Tertiary granitic bodies. The southern basement rock group consists of an assemblage that progresses southward through marble, schist, quartzite, Cretaceous tonalitic rocks, and granulitic rocks, mylonitized and retrogressively metamorphosed to greenschist and amphibolite facies, intruded by charnockitic rocks. The eastern San Gabriel Mountains are bounded on the north by the San Andreas fault and on the south by the Cucamonga fault (Fig. 3).

Within the northern group of basement rocks, the Pelona Schist between the San Andreas fault and the Punchbowl fault is upper greenschist to lower amphibolite grade. South of the Punchbowl fault the schist is greenschist grade. Southward, the Pelona Schist descends beneath the Vincent thrust. The Vincent thrust separates the Pelona Schist from the structurally overlying mylonitic and gneissic complex. The Vincent thrust, of Late Cretaceous or early Tertiary age (Conrad and Davis, 1977; Ehlig, 1981), is a regionally developed zone of decoupling that probably underlies much of southern California (Haxel and Dillon, 1978; Ehlig, 1981). Mylonite overlying the Vincent thrust ranges from about 60 m thick east of San Antonio Canyon to as much as 600 m or more west of San Antonio Canyon. This extreme variation in thickness suggests that deformation that postdates thrusting on the Vincent thrust may have removed variable thicknesses of mylonite. South of the Punchbowl fault, both the Pelona Schist and mylonite have been intruded by a pluton and numerous dikes and sills of Miocene granodiorite (Hsu and others, 1963; Miller and Morton, 1977).

Included in the southern group of rocks is the Placerita metasedimentary suite (Powell, this volume). It consists of amphibolite-grade quartzite, marble and biotite, sillimanite, and graphitic schist, all intruded by Cretaceous-age tonalitic rock. Much of the tonalitic rock is mylonitized. The degree of mylonitic deformation increases southward toward the mountain front where there is a 300-m-thick, relatively uniform mylonite.

South of the Placerita metasedimentary suite rocks and the tonalitic rock is a heterogeneous lithologic metamorphic assemblage characterized by retrogressively metamorphosed and mylonitized granulite (Hsu, 1955; May, 1986; D. M. Morton unpublished mapping, 1974). The rocks include biotite schist, amphibolite, marble, and calc-silicate rock, intruded by small bodies of charnockitic rocks. The age of the metamorphic protolith is unknown, but it has been commonly thought to be either Precambrian(?) (largely based on the granulite grade of the metamorphism) or Paleozoic age (based on the occurrence of marble). The charnockitic rocks have been dated as 70 Ma (Walker and May, 1986). For a recent structural analysis of the basement rocks of the southeastern San Gabriel Mountains, see May (1986).

LATE CENOZOIC FAULTS

San Andreas fault zone

The San Andreas fault zone contrasts sharply with the San Jacinto throughout most of its length in that it is a relatively continuous break appropriately identified as the San Andreas fault. A major exception is that part of the San Andreas fault zone that extends eastward from the upper Santa Ana Valley, through the San Gorgonio Pass area into the Coachella Valley. Descriptions of San Andreas fault complexities in the San Gorgonio Pass area were published as early as 1908 by Fairbanks and in 1922 by Vaughan. Here, the San Andreas fault zone is characterized by a much more complex surface geometry (Allen, 1957; Clark, 1984; Matti and others, 1985) than occurs elsewhere along its trace, or indeed, along the northern part of the San Jacinto fault zone. The San Andreas fault north of the Salton Sea to the Indio Hills area of the Coachella Valley (latitude 33°45′) parallels the

San Jacinto fault zone (Fig. 1), whereas northward from the Indio Hills it branches into the Coachella Valley segment of the San Andreas fault and the Coachella Valley segment of the Banning fault (Matti and others, 1985), both of which strike progressively more westward toward San Gorgonio Pass (Fig. 1). In the area of San Gorgonio Pass and the southeastern San Bernardino Mountains (Allen, 1957; Matti and others, 1985), the complex San Andreas zone is oriented about N80°W. Thus, westward from San Gorgonio Pass the San Jacinto fault and that part of the San Andreas fault zone that defines the northern boundary of the San Jacinto Mountains structural block converge at an angle of about 35° (Fig. 1).

Northwestward from San Gorgonio Pass, along the south side of the San Bernardino Mountains, the San Bernardino strand is the currently active strand of the San Andreas fault zone (Matti and others, 1985). This nearly linear trace strikes N60°W and converges with the San Jacinto fault zone at an angle of about 15° (Fig. 1). The San Bernardino strand is subtly concave northward, seeming to sag southward against the north side of the upper Santa Ana Valley (Fig. 2). North of the San Gabriel Mountains, the nearly linear Mojave Desert segment of the San Andreas fault is slightly concave southward, as if it had been pushed northward into the Mojave Desert (Fig. 3) (Weldon, 1984, 1986; Weldon and Humphrys, 1986; Weldon and Matti, 1986; Weldon and Springer, 1988).

The Punchbowl fault zone, located just south of the San Andreas fault (Figs. 1, 3), is a deformed and faulted early strand of the San Andreas fault. It consists of two closely spaced faults separated at most places by a sliver of intensely deformed rocks. In the southern extent of the fault zone the sliver consists of gneiss and tonalitic rock. In the northern extent it includes sedimentary and volcanic rocks in addition to basement rock (Barrows and others, 1985; R. J. Weldon, personal communication, 1989). In the eastern San Gabriel Mountains, the northern of the two faults juxtaposes the relatively high-grade part of the Pelona Schist with highly deformed gneiss and locally tonalite. The southern fault juxtaposes the gneiss with the lower grade part of the Pelona Schist. In many places recognizable gneiss is missing and the two grade types of the Pelona Schist are separated by thoroughly sheared basement rock of uncertain parentage.

Secondary faults in the area of the San Gabriel Mountains commonly have reverse slip (Weldon and Matti, 1986), and analysis of current seismicity gives a mix of reverse and strike-slip fault plane solutions (Jones, 1988).

Cucamonga fault zone

The Cucamonga fault defines the southern margin of the eastern San Gabriel Mountains and marks the eastern end of the frontal fault system of the San Gabriel Mountains (Figs. 1, 3). The Cucamonga fault zone consists of numerous anastomosing east-striking north-dipping thrusts that separate the crystalline basement of the eastern San Gabriel Mountains from the alluvium of upper Santa Ana Valley to the south as well as thrust faults entirely within alluvium (Morton and Matti, 1987). Slickensides in the basement are consistently oriented downdip, indicating the most recent displacements along the Cucamonga zone have been pure thrust. We project the Cucamonga fault zone downdip 13 km to merge with the San Andreas fault.

Faulting has occurred throughout the Quaternary with individual faulting events estimated at about 6.7 M with a recurrence of about 625 yr for the past 13,000 yr (Morton and Matti, 1987). In contrast, the recurrence interval for major earthquakes along the frontal fault zone of the San Gabriel Mountains to the west in the Pasadena area is estimated at more than 5,000 yr (Crook and others, 1987). The average north-south convergence across the Cucamonga fault zone is estimated to have been in the range of 3 (Weldon, 1986) to 5 mm/yr (Matti and others, 1985; Morton and Matti, 1987).

San Jacinto fault zone

Peninsular Ranges. The San Jacinto fault zone is the most seismically active member of the San Andreas fault system in southern California (Thatcher and others, 1975). Northwestward from Borrego Valley, the San Jacinto fault zone consists of a series of relatively widely spaced, en echelon faults characterized by a relatively consistent N45°W strike (Sharp, 1967, 1975) (Fig. 1). No single strand within this zone extends over a great enough distance to merit identification as the San Jacinto fault (Sharp, 1975). Indeed, there is less lateral continuity and relatively greater width across the en echelon breaks within this zone than is generally associated with transcurrent fault zones (Sharp, 1975).

San Gabriel Mountains. The San Jacinto fault zone loses its identity as an easily mappable throughgoing strike-slip fault at its northern end, where it enters the eastern Gabriel Mountains. In the San Gabriel Mountains area, the fault that is labeled on many maps as the San Jacinto fault has been simply depicted as either branching directly off the San Andreas fault on the north side of the San Gabriel Mountains or merging with the inactive Punchbowl fault. Mapping by Dibblee (1968, 1982) and Morton (1975a, b) indicates that interpretations of this sort are not only simplistic but clearly incorrect. According to Dibblee (1982), the San Jacinto retains its northwest strike and simply dies out within the San Gabriel Mountains. Mapping by Morton (1975b), on the other hand, indicates that the San Jacinto fault zone interacts in some fashion with east- to northeast–striking faults in the interior of the eastern San Gabriel Mountains.

The feature generally identified as the "San Jacinto fault zone," where it penetrates the southeastern corner of the San Gabriel Mountains near the mouth of Lytle Creek (Fig. 3), is a 300-m-wide zone of three near-vertical faults. From west to east these three faults bound separately mappable blocks of biotite gneiss, cataclastic leucogranite, Pelona Schist, and Miocene granodiorite. The fault with the greatest width of crushed rock is overlain by apparently unfaulted alluvium thought to be 200 to 500 ka old (Morton and Matti, 1987). Four kilometers into the

range, the San Jacinto fault zone consists of a relatively homogeneous zone of gouge and crushed rock, 200 to 300 m thick, bordered on the east by a thrust fault. Apparently unfaulted alluvium capped by soils whose degree of weathering appears to be the same as the S2 soil stage of McFadden (1982), considered to be 200 to 500 ka (Morton and Matti, 1987), overlies the broad fault zone, but is offset along the eastern edge by the thrust fault.

Two faults, the Lytle Creek and the Glen Helen (Fig. 3), commonly thought to be branches of the San Jacinto fault zone, parallel the fault zone near the mountain front. The Glen Helen fault, exposed along the west side of Cajon Canyon, is the only fault within the eastern San Gabriel Mountains characterized by a variety of more-or-less continuous youthful fault features, such as sag ponds and scarps. These fault features are developed in alluvium capped by soils whose degree of development appears the same as the S4 or S5 soil stage of McFadden (1982) considered to be between 4 and 70 ka (Morton and Matti, 1987). The Lytle Creek fault offsets alluvial deposits forming scarps capped by soils, the younger of which are considered by Morton and Matti to also belong to S4 or S5 soil stage of McFadden (1982). These younger soils offset by the Lytle Creek fault are apparently the same as the ones considered to be 50 to 60 ka by Mezger and Weldon (1983).

Six kilometers northwest of the mountain front, the 200- to 300-m-wide San Jacinto zone coalesces with three north-dipping faults, each along a separate fork of Lytle Creek (Fig. 3). West of this junction these three faults display a progressively counterclockwise change in strike until they are all oriented in a northeast direction. Branching of most of these faults occurs in some measure within the mountains, and many converge to the west near the mountain front at the mouth of San Antonio Canyon. Just west of San Antonio Canyon, two northern branches seem to coincide with the San Gabriel fault zone in Cow Canyon (Fig. 3).

Based on the distribution of basement rocks, suggested separation on the northwest-striking faults is oblique-right-reverse, that on the east-striking faults appears to be thrust, and that on the northeast-striking faults is oblique-left-reverse. This is similar to the generalized sense of displacement determined from a recent microearthquake study (Cramer and Harrington, 1987). The overall geometry and sense of displacement of the faults is an antiformal schuppen-like structure.

Located north of the Cucamonga fault between Lytle Creek and San Antonio Canyon are three northwest-striking right-lateral faults, Duncan Canyon, Morse Canyon, and Demens Canyon faults (Fig. 3). They appear to be terminated on the south by the Cucamonga fault zone and on the north by the South Fork Lytle Creek fault. All three show right-lateral separation, the 1.5-km separation along Duncan Canyon being the greatest.

Nature of fault junctions in the San Gabriel Mountains

The locations of abrupt changes in strike of all those faults within the eastern San Gabriel Mountains included within the schuppen-like structure are under alluvium or otherwise obscured. Most of the changes in strike of these faults can be explained as abrupt curving, branching, one fault offsetting another, or some form of conjugate faulting.

Low-angle fault junctions, such as in the upper reaches of the Middle Fork of Lytle Creek (Fig. 3), appear to be simple branches. Larger angle junctions, such as that at the east end of the South Fork Lytle Creek fault (Fig. 3), suggest termination of an older fault against a younger throughgoing fault. A clear case of a northeast-striking left-slip fault offsetting a northwest-striking right-slip fault is in the upper reaches of the North fork of Lytle Creek (Fig. 3). Here, the older Punchbowl fault is offset by a northeast-striking left-slip fault.

Age. Although rocks offset by the San Jacinto fault zone poorly constrain the age of its inception, Sharp (1967) argued that this fault zone might be as young as Pliocene.

The model developed in this chapter proposes accelerated elevation of the eastern San Gabriel Mountains, inception of the San Bernardino basin, cessation of major deposition in the area of the San Timoteo badlands, inception of the San Jacinto basin, and uplift of the San Gorgonio Pass area. All are attributable to or coincident with the onset of right-lateral slip on the San Jacinto fault zone.

The start of uplift of the eastern San Gabriel Mountains is estimated at 2 Ma in a model proposed by Meisling and Weldon (1989). The age of inception of the San Bernardino basin and cessation of deposition of the San Timoteo Formation are interdependent. The Shutt Ranch local fauna (1.3 ± 0.3 Ma) is from the youngest dated assemblage contained within the San Timoteo Formation. The time involved in the deposition of the part of the San Timoteo Formation overlying the beds of the Shutt Ranch fauna is estimated at 0.1 to 0.3 Ma. This suggests that major deposition of the San Timoteo Formation ceased about 1 ± 0.3 Ma. This is also a minimum age for the structural inception of the San Bernardino basin; the area of the basin would remain a topographic positive area for some time after basin inception.

An independent crude estimate of the age of the inception of faulting can be obtained from the age of the San Jacinto basin, a pull-apart basin located at a major right step in the San Jacinto fault zone between the Claremont and Casa Loma faults (Fig. 1). Because both faults bounding the basin record 24-km offsets, the age of the basin should equate with the time required for the accumulation of 24 km of lateral slip. Radiocarbon ages of wood recovered from depths of 150 m indicate tectonic subsidence rates for the basin of 3 to 11 mm/yr, and average about 5 mm/yr (Morton, 1977, and unpublished data). Geophysical surveys suggest basin fill of at least 3 km (Fett, 1968; Shawn Biehler, personal communication, 1988). Unconstrained linear extrapolation of the average rate of subsidence from the upper 150 m to depths of 3.8 km suggests an age as young as 0.6 Ma for the basin's inception.

Slip rates. Constant slip-rate models for the San Andreas and San Jacinto faults are generally based on long-term values of 10 mm/yr on the San Jacinto and 25 mm/yr on the San Andreas (e.g., Weldon, 1986; Weldon and Sieh, 1985, Prentice and others, 1986, Rockwell and others, 1986). Based on a constant 10 mm/yr slip-rate model for the San Jacinto, the derived 2.4-Ma age of the fault challenges our independently developed determinations and we think exaggerates the actual age by about 1 Ma.

A slip rate of 10 to 15 mm/yr on the San Jacinto fault in the Anza area, 60 km southeast of the San Bernardino basin, has been determined for the past 15 to 20 ka (Rockwell and others, 1986). A *minimum* slip rate of 8 to 12 mm/yr was earlier determined by Sharp (1981) for the fault in the same general area. We suggested in an earlier report that slip steps from the San Andreas fault zone in the northern Coachella Valley area westward to the San Jacinto fault zone in the area of the San Jacinto Valley (Matti and others, 1985). Thus, the slip rate for the San Jacinto at Anza may be different than the slip rate for the area of the San Bernardino basin.

In the Reche Canyon area of the northern part of the San Timoteo badlands, the eastern part of some dissected alluvial surfaces abut the San Jacinto fault zone (D. M. Morton, unpublished mapping, 1976). Based on soil development, surfaces of at least two different ages are present. The age of the younger surface is tentatively estimated at about 50 ka; the age of the older at about 500 ka (L. D. McFadden, personal communication, 1988). Streams incised into the younger(?) surface appear to be offset 0.6 to 0.7 km (Prentice and others, 1986), which yield an average slip rate of 12 to 14 mm/yr. Soils of both ages contain clasts derived from the conglomerate containing the Olive Dell local fauna, from the east side of the San Jacinto fault. Clasts in the older soil are offset a minimum of 4 km from their source on the east side of the fault; clasts in the younger soil are offset about 1 km from their source (D. M. Morton, unpublished mapping, 1976). Assuming the age estimates for the soils are approximately correct, the offset older clast-bearing soil yields a *minimum* slip rate of 8 mm/yr and for the younger soil a slip rate of about 20 mm/yr.

Existing data are consistent with an interpretation that right slip on the San Jacinto fault zone began about 1.5 Ma. Throughout much of this 1.5-m.y. period, the San Jacinto fault zone accommodated much of the slip that had previously been taken up on the San Andreas fault zone. Current strain accumulations on the San Andreas and San Jacinto faults, based on geodetic resurveys, are about equal. Our model of relatively long-term average slip rates is compatible with a short-period variable slip-rate model for the San Jacinto fault zone such as proposed by Sharp (1981).

INTERPRETATIONS

The San Andreas zone through that part of California between the Cuyama and Coachella Valleys (Fig. 4A) can be viewed as a left-stepping right-slip fault (e.g., Ehlig, 1981). This large-scale left step of about 90 km (Fig. 4B) results in compression coincident with the central and eastern parts of the present-day tectonic Transverse Ranges province. This compression produces the seismicity with reverse fault-plane solutions

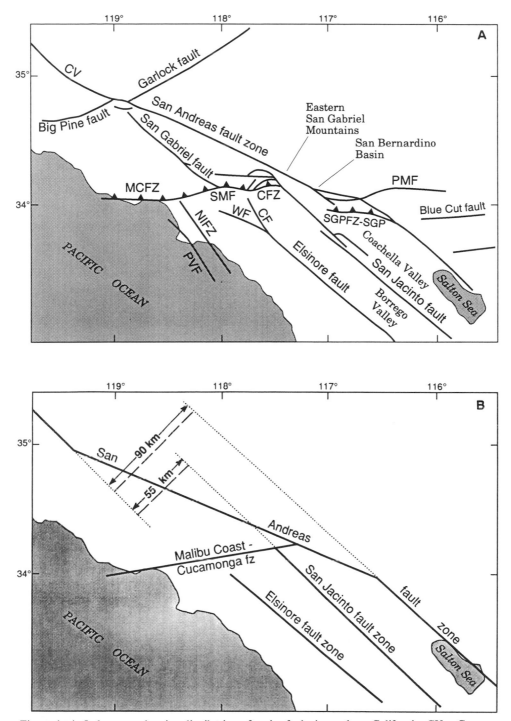

Figure 4. A, Index map showing distribution of major faults in southern California. CV = Cuyama Valley; SGP = San Gorgonio Pass; MCFZ = Malibu Coast fault zone; SMF = Sierra Madre fault zone; CFZ = Cucamonga fault zone; PVF = Palos Verdes fault; NIFZ = Newport-Inglewood fault zone; WF = Whittier fault; CF = Chino fault; SGPFZ = San Gorgonio Pass fault zone; PMF = Pinto Mountain fault. B, Schematic map showing 90-km left step in the San Andreas fault zone and 55-km left step between the San Jacinto fault zone and the San Andreas fault zone. fz = fault zone.

throughout this length of the San Andreas fault zone (Jones, 1988). Contained within this regionally developed contractional province, or "rhombohump," are a large number of smaller sized compressional and tensional structural domains, several of which have persisted through much of late Cenozoic time.

A large structural domain, dominated by contractional strain attributable to a 55-km left step (Fig. 4B), extends from the San Bernardino basin northwest to the "big bend" of the San Andreas in the area of Cuyama Valley (Fig. 4A). Two smaller structural subdomains, produced by the discordance in azimuth between the San Jacinto and San Andreas faults, are the eastern San Gabriel Mountains and the San Bernardino basin. The 15° convergence of the San Andreas fault zone with the San Jacinto fault zone has produced a compression vector in the area west of the projection of the San Jacinto fault zone into the eastern San Gabriel Mountains (Weldon and Humphrys, 1986). Similarly, the southwestward divergence and right-stepping of the San Jacinto fault zone from the San Andreas fault zone results in tension in the area between the two faults producing the San Bernardino basin. Our model proposes progressive transfer of slip from the San Jacinto to the San Andreas zone from well within the southeastern San Gabriel Mountains through the San Bernardino basin, a distance of 60 to 70 km (Fig. 1).

Thus, the topographically high eastern San Gabriel Mountains are a product of the impingement of the area west of the San Jacinto fault, moving N45°W, against the area to the east that is moving N60°W. We suggest that deformation accommodated within the schuppen-like structure in the narrow eastern San Gabriel Mountains is distributed over a much broader area to the west. We interpret the 5-km north step along the south side of the San Gabriel Mountains east of San Antonio Canyon to result from a minimum of 5 km of convergence along the south side of the schuppen-like structure. The Cucamonga fault zone, bounding the south side of the San Gabriel Mountains, has been displaced 5 km farther north than the mountain front to the west of San Antonio Canyon.

The San Gabriel Mountains area east of the longitude of the San Antonio Canyon has a high level of seismicity; the level of seismicity abruptly decreases west of San Antonio Canyon area both within the mountains and along the mountain front (e.g., Cramer and Harrington, 1987; Yerkes and others, 1987). Fault-plane solutions from microseismic events in the eastern San Gabriel Mountains are basically right slip for northwest-striking faults, thrust for east-striking faults, and left slip for northeast-striking faults (Cramer and Harrington, 1987). The localization of seismicity in the area of the schuppen-like structure suggests the deformation responsible for the activity is likewise concentrated in that area.

Reconnaissance of the San Gabriel fault zone west of San Antonio Canyon suggests that the San Gabriel fault is devoid of youthful fault features. Similarly, west of San Antonio Canyon, for at least 40 km to the Pasadena area, the frontal fault zone of the San Gabriel Mountains, the Sierra Madre, generally lacks evidence of Holocene activity (Morton, 1973; Crook and others,

1987) and also lacks, or has a low level of, current seismic activity (Cramer and Harrington, 1987; Yerkes and others, 1987). For the Sierra Madre fault zone in the Pasadena area, the recurrence interval is estimated as more than 5,000 yr, and may be as long as 11,000 yr, (Crook and others, 1987), considerably longer than the estimated recurrence interval of about 600 yr for the central part of the Cucamonga fault zone (Morton and Matti, 1987). The youngest recognized fault in the Pasadena area, the Raymond fault, marked by scarps and closed depressions, diverges southwest from the mountain front and the Sierra Madre fault zone (Crook and others, 1987). The recurrence interval for the Raymond fault may be of the order of 3,000 yr (Crook and others, 1987).

The impingement of the San Gabriel Mountains against the San Andreas is the same as or similar to the model of Weldon and Humphrys (1986) and Weldon and Springer (1988). Weldon (1984, 1986) suggested that the geometric configuration of convergence and the stepping of slip from the San Jacinto to the San Andreas requires a zone of compression, an interpretation with which we concur. We interpret that the compression formed some new faults and reactivated and deformed some old faults. Much of the neotectonic displacement is probably taken up on faults produced in an earlier stress regime, such as the Cucamonga fault zone and the Icehouse Canyon fault. We interpret the northward concavity of the trace of the San Andreas on the north side of the San Gabriel Mountains to result from the compression in the mountains; it suggests, if seen at depth, that the San Andreas would dip northward to a depth of about 13 km where we interpret the Cucamonga fault zone to merge with the San Andreas. The concave north surface trace is the same as Weldon's (1984, p. 689) "marked right-lateral deflection where the projection of the San Jacinto would join it" (the San Andreas fault), an explanation that has since been expanded and illustrated (Weldon, 1986, p. 383).

Southeast of the San Gabriel Mountains, additional slip is transferred to the San Andreas fault from the San Jacinto fault zone across the San Bernardino basin. The Glen Helen fault steps right from the San Andreas fault zone at the northern end of the San Bernardino basin. The right step produces extension marked by normal focal mechanisms (Jones, 1988). Extensional faults also occur north of the basin within the western San Bernardino Mountains (Matti and others, 1985; Weldon, 1986; Weldon and Springer, 1988) accommodating additional transfer of slip between the San Jacinto and San Andreas faults.

Searching for a "San Jacinto fault" within the eastern San Gabriel Mountains is akin to searching for the continuation of an en echelon break in the area of structural overlap between two breaks defining an idealized rhombochasm. The northernmost surface evidence of a neotectonic break along the clearly defined San Jacinto fault zone occurs 9 km southeast of the San Gabriel Mountains; linear extrapolation of this break puts it on line with the Lytle Creek fault. In the area of convergence between the San Andreas and the San Jacinto faults, the Glen Helen fault is the northernmost fault characteristic of the faults within San Jacinto fault zone in the Peninsular Ranges. Indeed, the distribution of surface breaks along the western edge of the San Bernardino basin suggests segments of the San Jacinto fault zone step progressively right toward the San Andreas fault zone. It is in this sense, accordingly, that the fault features characteristic of San Jacinto fault zone end with the Glen Helen fault.

Faults within the San Gabriel Mountains, which predate the inception of movement of the San Jacinto fault have been reactivated or accelerated in their slip rate by accommodating strain resulting from stress associated with the convergence between the San Jacinto and San Andreas faults. Preexisting faults oriented at high angles to the San Jacinto fault zone, such as the Cucamonga fault, have not been considered as potential candidates as continuations of the San Jacinto fault zone, even though they may have served as slip sumps. On the other hand, the northwest-striking broad shear zone at the mountain front we interpret to be the Icehouse Canyon fault reactivated with the onset of displacement on the San Jacinto fault, constitutes in a sense a composite Icehouse Canyon–San Jacinto fault zone. Moreover, this zone is likely the composite of the south (Sierra Madre fault) and north branches of the San Gabriel fault and the San Antonio Canyon fault, as well as accommodating strain associated with the San Jacinto fault zone (Figs. 1, 3).

SOME SPECULATIONS

The confluence between the San Jacinto–San Andreas fault zones is marked by an intimate intertwining of faults, although the relation of one to the other is less clear—certainly more complex than implied by some.

Crowell (1981) concluded from the linearity and high seismicity of the San Jacinto fault zone that it represents the most active "strand" of the San Andreas. Implicit in his suggestion is that much of the plate motion is accommodated by movement on the San Jacinto, a conclusion that should be testable based on slip rates for those faults comprising the San Andreas system in southern California. However, the uncertainties associated with determined slip rates along both the San Jacinto and San Andreas faults are challenging (e.g., Harden and others, 1986; Harden and Matti, 1989; Keller and others, 1982; Prentice and others, 1986; Rasmussen, 1982; Rockwell and others, 1986; Rust, 1982, 1986; Schwartz and Weldon, 1986; Sharp, 1981, 1986; Sieh, 1984; Sieh and Jahns, 1984; Weldon, 1986; Weldon and Sieh, 1985), and presently seem to preclude the development of an unequivocal slip-rate model.

Partitioning current strain accumulation across the San Andreas fault system appears less ambiguous than the derived Quaternary slip rates. Across the San Jacinto fault zone, in San Jacinto–Moreno Valley, strain accumulation (right-lateral shear), determined from geodimeter measurements for the period 1969 to 1975, is 0.4 ± 0.1 μrad/yr (engineering strain). This is consistent with a ratio of slip rate to depth of the locked zone of 1.2×10^{-6}/yr (Savage and Prescott, 1976). Hypocenters of the deeper earthquakes along the San Jacinto fault zone in the area of

the geodimeter measurements average about 17 km (Sanders, 1990). If the San Jacinto is locked to 17 km, the slip rate is 20 + 5 mm/yr in the San Jacinto–Moreno Valley area.

Parts of the San Andreas and the San Jacinto fault zones, south and west from the area of Palm Springs, are included in an extensive trilateration network. Resurveys of this network disclosed shear-strain maxima γmax = 0.35 ± 0.02 μrad/a and γ/max = 0.40 ± 0.02 μrad/a centered on the San Jacinto and San Andreas fault zones (King and Savage, 1983). These surveys suggest a near-equal division of shear strain accumulation between the San Jacinto and the San Andreas.

Models partitioning slip between the San Andreas and San Jacinto fault zones should consider the distance over which slip is being transferred between the two fault zones (Matti and others, 1985). Slip-rate figures used to partition slip between the two fault zones should be derived from the San Andreas fault north of the eastern San Gabriel Mountains and from the San Jacinto fault southeast of the Crafton Hills fault complex.

The recent evolution of the San Andreas system between Cajon Pass and Coachella Valley was characterized by a complex history of fault strand development, strand switching, and strand abandonment related to the creation of the structural knot in the San Gorgonio Pass area (Matti and others, 1985). Successively developed strands were apparently created in response to those mechanical considerations that tended to facilitate continuing transcurrent movement between plates. The effect has been development of a progressively more complex geometric structural knot. We consider two major structural features resulted from progressive development of the structural knot. For the San Andreas fault zone it resulted in the apparent lack of a throughgoing fault, at least at the surface. In terms of strand development, the San Jacinto fault zone represents the most recent phase in this mechanistic evolution and simply bypasses the San Gorgonio knot. Thus, it is likely that an active strand of the San Andreas fault does not extend through San Gorgonio Pass (Allen, 1957; Matti and others, 1985), and an active strand has not, perhaps, extended through San Gorgonio Pass since soon after the onset of right slip on the San Jacinto fault zone. Thus with the onset of displacement on the San Jacinto, two formerly disparate structural blocks, the San Jacinto and eastern San Bernardino, became a relatively coherent block forming a left step in the San Andreas system that underwent compression producing a regional uplift. The uplift elevated both the pass and adjacent San Jacinto and San Bernardino Mountains on the order of 700 m. The compression results in shortening across the San Gorgonio Pass thrust complex, estimated at 10 mm/yr (Matti and others, 1985) and deep seismicity, with hypocenters to more than 20 km (Jones, 1988; Nicholson and others, 1986a,b).

CONCLUSIONS

Although the two may merge at depth, the San Jacinto fault does not presently coalesce with the San Andreas fault at the surface. Contraction and uplift occurring in the eastern San Ga-

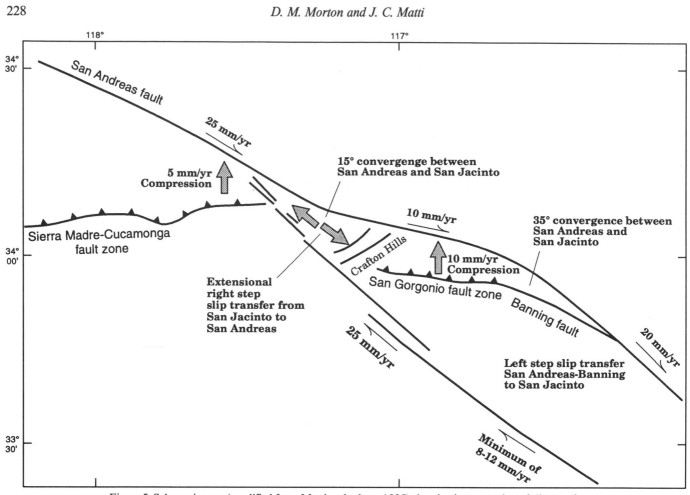

Figure 5. Schematic map (modified from Matti and others, 1985) showing interpretation of slip transfer between some of the faults in the area of the San Bernardino basin and southeastern San Gabriel Mountains.

briel Mountains, between the Cucamonga fault zone and the San Andreas fault, is interpreted as strain that has accumulated in response to the transfer of slip between the San Jacinto and San Andreas fault zones (Fig. 5). Additional slip is transferred between the two fault zones southeast of the San Gabriel Mountains through the tensional San Bernardino basin (Fig. 5). We consider the amount of slip that can be accommodated in the San Bernardino basin without development of a throughgoing surface fault is at least 25 km. This amount of slip is by analogy with 25 km of slip transferred from the Claremont fault to the Casa Loma fault across the San Jacinto graben without the development of a throughgoing surface fault. Slip transferring across the San Bernardino basin is a combination of both right slip across the San Jacinto fault zone and that resulting from the 15° convergence between the San Jacinto fault zone and the San Andreas (Fig. 5). Slip transfer between the Banning fault and the San Jacinto fault zone results in compression and uplift in San Gorgonio Pass and adjacent mountains (Fig. 5).

The age of the San Jacinto fault zone is estimated at about

1.5 Ma. Accelerated uplift in the eastern San Gabriel Mountains, development of the San Bernardino and San Jacinto basins, cessation of deposition of the San Timoteo Formation, and accelerated uplift in the San Gorgonio Pass area are considered to be the result of the inception of slip on the San Jacinto fault zone.

We propose a model of slip distribution for the past 1.5 m.y. that has much of the combined San Andreas–San Jacinto slip occurring on the San Jacinto. Geodetic measurements indicate near-equal division of present-day shear strain accumulation between the San Jacinto and San Andreas fault zones.

ACKNOWLEDGMENTS

We appreciate thoughtful and critical reviews by R. O. Castle, R. E. Powell, and R. F. Yerkes of the U.S. Geological Survey; by L. H. Cohen and M. O. Woodburne of the University of California, Riverside; and by R. J. Weldon of the University of Oregon.

REFERENCES CITED

Allen, C. R., 1957, San Andreas fault zone in San Gorgonio Pass, southern California: Geological Society of America Bulletin, v. 68, p. 319–350.

Allen, C. R., and Sieh, K. E., 1983, Creep and strain studies in southern California, *in* Summaries of technical reports, Earthquake Hazards Reduction Program, v. 17: U.S. Geological Survey Open-File Report 83-918, p. 199–202.

Barrows, A. G., Kahle, J. E., and Beeby, D. J., 1985, Earthquake hazards and tectonic history of the San Andreas fault zone, Los Angeles County, California: California Division of Mines and Geology Open-File Report 85-10 LA, 236 p.

Carson, S. E., Matti, J. C., Throckmorton, C. K., and Kelley, M. M., 1986, Stratigraphic and geotechnical data from a regional drilling investigation in the San Bernardino valley, California: U.S. Geological Survey Open-File Report 86-225, 78 p.

Clark, M. M., 1984, Map showing recently active breaks along the San Andreas fault and associated faults between Salton Sea and Whitewater River–Mission Creek, California: U.S. Geological Survey Miscellaneous Investigations Map I-1483, scale 1:24,000.

Conrad, R. L., and Davis, T. E., 1977, Rb-Sr geochronology of cataclastic rocks of the Vincent thrust, San Gabriel Mountains, southern California: Geological Society of America Abstracts with Programs, v. 9, p. 403–404.

Cramer, C. H., and Harrington, J. M., 1987, Seismicity and tectonics of the Cucamonga fault and the eastern San Gabriel Mountains, San Bernardino County, California, *in* Recent reverse faulting in the Transverse Ranges, California: U.S. Geological Survey Professional Paper 1339, p. 7–26.

Crook, R., Jr., Allen, C. R., Kamb, B., Payne, C. M., and Proctor, R. J., 1987, Quaternary geology and seismic hazard of the Sierra Madre and associated faults, western San Gabriel Mountains: U.S. Geological Survey Professional Paper 1339, p. 179–203.

Crowell, J. C., 1974, Origin of Late Cenozoic basins in southern California, *in* Dickinson, W. R., ed., Tectonics and sedimentation: Society of Economic Paleontologists and Mineralogists Special Publication 22, p. 190–204.

——, 1981, An outline of the tectonic history of southeastern California, *in* Ernst, W. G., ed., The geotectonic development of California; Rubey Volume 1: Englewood Cliffs, New Jersey, Prentice-Hall, p. 583–600.

Dibblee, T. W., Jr., 1968, Displacements on San Andreas fault system in San Gabriel, San Bernardino, and San Jacinto Mountains, southern California, *in* Dickinson, W. R., and Grantz, A., eds., Proceedings of Conference on Geologic Problems of San Andreas Fault System: Stanford, California, Stanford University Publications in Geological Sciences, v. 11, p. 269–278.

——, 1982, Geology of the San Gabriel Mountains, southern California, *in* Fife, D. L., and Minch, J. A., eds., Geology and mineral wealth of the California Transverse Ranges; Mason Hill Volume: Santa Ana, California, South Coast Geological Society Annual Symposium and Guidebook 10, p. 131–147.

Dutcher, L. C., and Garrett, A. A., 1963, Geologic and hydrologic features of the San Bernardino area California: U.S. Geological Survey Water-Supply Paper 1419, 114 p.

Ehlig, P. L., 1958, Geology of the Mount Baldy region of the San Gabriel Mountains, California [Ph.D. thesis]: Los Angeles, University of California, 153 p.

——, 1981, Origin and tectonic history of the basement terrane of the San Gabriel Mountains, central Transverse Ranges, *in* Ernst, W. G., ed., The geotectonic development of California; Rubey Volume 1: Englewood Cliffs, New Jersey, Prentice-Hall, p. 253–283.

Fairbanks, H. W., 1908, The San Andreas rift as a geomorphic feature; Pajaro River to the north end of the Colorado Desert, *in* Lawson, A. C., chairman, The California earthquake of April 18, 1906; Report of the state earthquake investigation commission: Carnegie Institution of Washington, v. 1, p. 38–52.

Fett, J. D., 1968, Geophysical investigation of the San Jacinto Valley, Riverside County, California [M.A. thesis]: Riverside, University of California, 87 p.

Fraser, D. M., 1931, Mining in California: California Division of Mines, v. 27,

no. 4, p. 511–514.

Frizzell, V. A., Jr., Mattinson, J. M., and Matti, J. C., 1986, Distinctive Triassic megaporphyritic monzogranite; Evidence for only 160 km offset along the San Andreas fault, southern California: Journal of Geophysical Research, v. 91, no. B14, p. 14080–14088.

Hadley, D., and Combs, J., 1974, Microearthquake distribution and mechanisms of faulting in the Fontana–San Bernardino area of southern California: Bulletin of the Seismological Society of America, v. 64, no. 5, p. 1477–1499.

Harden, J. W., and Matti, J. C., 1989, Holocene and late Pleistocene slip rates on the San Andreas fault in Yucaipa, California, using displaced alluvial-fan deposits and soil chronology: Geological Society of America Bulletin, v. 101, p. 1107–1117.

Harden, J. W., Matti, J. C., and Terhune, C., 1986, Late Quaternary slip rates along the San Andreas fault near Yucaipa, California, derived from soil development on fluvial terraces: Geological Society of America Abstracts with Programs, v. 18, p. 113.

Haxel, G. B., and Dillon, J., 1978, The Pelona-Orocopia Schist and Vincent–Chocolate Mountain thrust system, southern California, *in* Howell, D. G., and McDougall, K. A., eds., Mesozoic paleogeography of the western United States: Pacific Section, Society of Economic Paleontologists and Mineralogists Pacific Coast Paleogeography Symposium 2, p. 453–469.

Hsu, K. J., 1955, Granulites and mylonites of the region about Cucamonga and San Antonio Canyons, San Gabriel Mountains, California: Berkeley, University of California Publications in Geological Sciences, v. 30, no. 4, p. 223–352.

Hsu, K. J., Edwards, G., and McLaughlin, W. A., 1963, Age of intrusive rocks of the southeastern San Gabriel Mountains, California: Geological Society of America Bulletin, v. 74, p. 507–512.

Jones, L. M., 1988, Focal mechanisms and the state of stress on the San Andreas fault in southern California: Journal of Geophysical Research, v. 93, no. B8, p. 8869–8891.

Keller, E. A., Bonkowski, M. S., Korsch, R. J., and Shlemon, R. J., 1982, Tectonic geomorphology of the San Andreas fault zone in the southern Indio Hills, Coachella Valley, California: Geological Society of America Bulletin, v. 93, p. 46–56.

King, N. E., and Savage, J. C., 1983, Strain-rate profile across the Elsinore, San Jacinto, and San Andreas faults near Palm Springs, California, 1973–81: Geophysical Research Letters, v. 10, no. 1, p. 55–57.

Lambert, D. W., 1987, A geophysical survey of a contaminated aquifer in Redlands, California [M.S. thesis]: Riverside, University of California, 126 p.

Matti, J. C., and Morton, D. M., 1975, Geologic history of the San Timoteo badlands, southern California: Geological Society of America Abstracts with Programs, v. 7, p. 344.

Matti, J. C., Morton, D. M., and Cox, B. F., 1985, Distribution and geologic relations of fault systems in the vicinity of the Central Transverse Ranges, southern California: U.S. Geological Survey Open-File Report 85-365, 23 p.

May, D. J., 1986, Amalgamation of metamorphic terranes in the southeastern San Gabriel Mountains, California [Ph.D. thesis]: Santa Barbara, University of California, 325 p.

May, S. R., and Repenning, C. A., 1982, New evidence for the age of the Mount Eden fauna, southern California: Journal of Vertebrate Paleontology, v. 2, no. 1, p. 109–113.

McFadden, L. D., 1982, The impacts of temporal and spatial climatic changes on alluvial soils genesis in southern California [Ph.D. thesis]: Tucson, University of Arizona, 430 p.

Meisling, K. E., and Weldon, R. J., 1989, Late Cenozoic tectonics of the northwestern San Bernardino Mountains, southern California: Geological Society of America Bulletin, v. 101, p. 106–128.

Mezger, L., and Weldon, R. J., 1983, Tectonic implications of the Quaternary history of lower Lytle Creek, southeast San Gabriel Mountains: Geological Society of America Abstracts with Programs, v. 15, p. 418.

Miller, F. K., and Morton, D. M., 1977, Comparison of granitic intrusions in the Pelona and Orocopia Schists, southern California: U.S. Geological Survey Journal of Research, v. 5, no. 5, p. 643–649.

Miller, W. J., 1934, Geology of the western San Gabriel Mountains of California: Berkeley, University of California Publications in Mathematics and Physical Science, v. 1, p. 1–114.

Morton, D. M., 1973, Geology of parts of the Azusa and Mt. Wilson Quadrangles, Los Angeles County, California: California Division of Mines and Geology Special Report 105, 21 p.

——, 1975a, Relations between major faults, eastern San Gabriel Mountains, southern California: Geological Society of America Abstracts with Programs, v. 7, p. 352–353.

——, 1975b, Synopsis of the geology of the eastern San Gabriel Mountains, southern California, *in* Crowell, J. C., ed., San Andreas fault in southern California: California Division of Mines and Geology Special Report 118, p. 170–176.

——, 1977, Surface deformation in part of the San Jacinto Valley, southern California: U.S. Geological Survey Journal of Research, v. 5, no. 1, p. 117–124.

Morton, D. M., and Matti, J. C., 1987, The Cucamonga fault zone; Geologic setting and Quaternary history: U.S. Geological Survey Professional Paper 1339, p. 179–203.

Morton, D. M., and Miller, F. K., 1975, Geology of the San Andreas fault zone north of San Bernardino between Cajon Canyon and Santa Ana Wash, *in* Crowell, J. C., ed., San Andreas fault in southern California: California Division of Mines and Geology Special Report 118, p. 136–146.

Morton, D. M., Matti, J. C., Miller, F. K., and Repenning, C. A., 1986, Pleistocene conglomerate from the San Timoteo badlands, southern California; Constraints on strike-slip displacements on the San Andreas and San Jacinto faults: Geological Society of America Abstracts with Programs, v. 18, p. 161.

Nicholson, C., Seeber, L., Williams, P., and Sykes, L. R., 1986a, Seismicity and fault kinematics through the eastern Transverse Ranges, California; Block rotation, strike-slip faulting, and low-angle thrusts: Journal of Geophysical Research, v. 91, no. B5, p. 4891–4908.

——, 1986b, Seismic evidence for conjugate slip and block rotation within the San Andreas fault system, southern California: Tectonics, v. 5, no. 4, p. 629–648.

Prentice, C. S., Weldon, R. J., and Sieh, K. E., 1986, Distribution of slip between the San Andreas and San Jacinto faults near San Bernardino, southern California: Geological Society of America Abstracts with Programs, v. 18, p. 172.

Rasmussen, G. S., 1982, Geologic features and rate of movement along the south branch of the San Andreas fault, San Bernardino, California, *in* Cooper, J. D., compiler, Neotectonics in southern California: Geological Society of America Cordilleran Section, 78th Annual Meeting, Anaheim, California, Volume and Guidebook, Field Trip 4, p. 109–114.

Repenning, C. A., 1987, Biochronology of the microtine rodents of the United States, *in* Woodburne, M. O., ed., Cenozoic mammals of North America: Berkeley, University of California Press, p. 236–268.

Rockwell, T. K., Merrifield, P. M., and Loughman, C. C., 1986, Holocene activity on the San Jacinto fault in the Anza seismic gap, southern California: Geological Society of America Abstracts with Programs, v. 18, p. 177.

Rust, D. J., 1982, Radiocarbon dates for the most recent large prehistoric earthquake and for late Holocene slip rates; San Andreas fault in part of the Transverse Ranges north of Los Angeles: Geological Society of America Abstracts with Programs, v. 14, p. 229.

——, 1986, Neotectonic behavior of the San Andreas fault zone in the big bend: Geological Society of America Abstracts with Programs, v. 18, p. 179.

Sanders, C. O., 1990, Earthquake depths and the relation to crustal strength and stress in southern California: Journal of Geophysical Research, (in press).

Savage, J. C., and Prescott, W. H., 1976, Strain accumulation on the San Jacinto fault near Riverside, California: Seismological Society of America Bulletin,

v. 66, p. 1749–1754.

Schwartz, D. P., and Weldon, R. J., 1986, *in* Schwartz, D. P., Sharp, R. V., and Weldon, R. J., San Andreas segmentation; Cajon Pass to Wallace Creek, *in* National Earthquake Hazards Reduction Program; Summaries of technical reports, volume 23: U.S. Geological Survey Open-File Report 87-63, p. 576–578.

Sharp, R. V., 1967, San Jacinto fault zone in the Peninsular Ranges of southern California: Geological Society of America Bulletin, v. 78, p. 705–730.

——, 1975, En echelon fault patterns of the San Jacinto fault zone, *in* Crowell, J. C., ed., San Andreas fault in southern California: California Division of Mines and Geology Special Report 118, p. 147–152.

——, 1981, Variable rates of late Quaternary strike slip on the San Jacinto fault zone, southern California: Journal of Geophysical Research, v. 86, no. B3, p. 1754–1762.

——, 1986, *in* Schwartz, D. P., Sharp, R. V., and Weldon, R. J., San Andreas segmentation; Cajon Pass to Wallace Creek, *in* National Earthquake Hazards Reduction Program; Summaries of technical reports, volume 23: U.S. Geological Survey Open-File Report 87-63, p. 576–578.

Sieh, K. E., 1984, Lateral offsets and revised dates of large prehistoric earthquakes at Pallet Creek, southern California: Journal of Geophysical Research, v. 89, no. B9, p. 7641–7670.

Sieh, K. E., and Jahns, R. H., 1984, Holocene activity of the San Andreas fault at Wallace Creek, California: Geological Society of America Bulletin, v. 95, p. 883–896.

Thatcher, W., Hileman, J. A., and Hanks, T. C., 1975, Seismic slip distribution along the San Jacinto fault zone, southern California, and its implications: Geological Society of America Bulletin, v. 86, p. 1140–1146.

Vaughan, F. E., 1922, Geology of San Bernardino Mountains north of San Gorgonio Pass: Berkeley, University of California, Bulletin of the Department of Geological Sciences, v. 13, no. 9, p. 319–411.

Walker, N. W., and May, D. J., 1986, U-Pb zircon ages from the SE San Gabriel Mountains, California; Evidence for Cretaceous metamorphism, plutonism, and mylonitic deformation predating the Vincent thrust: Geological Society of America Abstracts with Programs, v. 18, p. 195.

Weldon, R. J., 1984, Quaternary deformation due to the junction of the San Andreas and San Jacinto faults, southern California: Geological Society of America Abstracts with Programs, v. 16, p. 689.

Weldon, R. J., II, 1986, The late Cenozoic geology of Cajon Pass; Implications for tectonics and sedimentation along the San Andreas fault [Ph.D. thesis]: Pasadena, California Institute of Technology, 382 p.

Weldon, R. J., and Humphrys, E., 1986, A kinematic model of southern California: Tectonics, v. 5, p. 33–48.

Weldon, R. J., and Matti, J. C., 1986, Geologic evidence for segmentation of the southern San Andreas fault: EOS Transactions of the American Geophysical Union, v. 67, p. 905–906.

Weldon, R. J., II, and Sieh, K. E., 1985, Holocene rate of slip and tentative recurrence interval for large earthquakes on the San Andreas fault, Cajon Pass, southern California: Geological Society of America Bulletin, v. 96, p. 793–812.

Weldon, R. J., II, and Springer, J. E., 1987, Active faulting near the Cajon Pass well, southern California: Implications for the stress orientation near the San Andreas fault: Geophysical Research Letters, v. 15, no. 9, p. 993–996.

Wesnousky, S. G., 1988, Seismological and structural evolution of strike-slip faults: Nature, v. 335, no. 6188, p. 340–344.

Willingham, C. R., 1968, A gravity survey of the San Bernardino valley, southern California [M.A. thesis]: Riverside, University of California, 96 p.

Yerkes, R. F., Levine, P., and Lee, W.H.K., 1987, Contemporary tectonics along south boundary of central and western Transverse Ranges, southern California: EOS Transactions of the American Geophysical Union, v. 68, no. 44, p. 1510.

MANUSCRIPT ACCEPTED BY THE SOCIETY SEPTEMBER 6, 1990

Geological Society of America
Memoir 178
1993

Chapter 6

Chronology of displacement on the San Andreas fault in central California: Evidence from reversed positions of exotic rock bodies near Parkfield, California

John D. Sims
U.S. Geological Survey, MS 977, Middlefield Road, Menlo Park, California 94025

ABSTRACT

This chapter presents a synthesis of data pertaining to post–early Miocene slip on the San Andreas fault in central California and suggests a three-phase evolition of the San Andreas system. The cricial evidence that supports the three phases of evolution comes from the reversed positions of two exotic rock fragments in the vicinity of Parkfield, California. The three-phase evolution of the San Andreas is also supported by the correlation of other exotic fragments, the basement rocks on which they lie, overlying Tertiary stratigraphic sequences, and distinctive Miocene strata derived from these fragments during their transport along the fault.

The 40-km-long section of the San Andreas fault near Parkfield is characterized by exotic blocks composed of Cretaceous hornblende quartz gabbro at Gold Hill and lower Miocene volcanic rocks in Lang Canyon. The gabbro is correlated petrographically with similar rocks near Eagle Rest Peak, 145 km to the southeast, and near Logan, 165 km to the northwest. The lower Miocene volcanic rocks, informally termed the volcanic rocks of Lang Canyon, are correlated with the Neenach Volcanics 220 km to the southeast and the Pinnacles Volcanics 95 km to the northwest. All three fragments of volcanic rocks are unconformably overlain by similar successions of Tertiary sedimentary rocks.

The original positions of the bodies of gabbro and volcanic bodies and their overlying sedimentary cover may be reconstructed from these exotic fragments that now lie along the San Andreas fault between San Juan Bautista and the northwestern Mojave Desert. The original undeformed gabbroic body was composed of the hornblende quartz gabbro of Eagle Rest Peak, Gold Hill, and Logan. In its initial prefaulted position, the original gabbroic body lay about 55 km northwest of the early Miocene volcanic assemblage. The undeformed volcanic assemblage was composed of the Neenach Volcanics, Pinnacles Volcanics, and volcanic rocks of Lang Canyon. The original spatial relationship between the undeformed gabbro and volcanic assemblage and their sedimentary cover is preserved in the present position of the gabbro of Logan and the Pinnacles Volcanics. However, in the Parkfield segment of the San Andreas, the gabbro of Gold Hill lies east of the main trace of the San Andreas fault, and the volcanic rocks of Lang Canyon lie 2 km west of the fault. The reversed relative positions of the gabbro of Gold Hill and the volcanic rocks of Lang Canyon suggest a complex history of movement on the San Andreas fault.

Consequently, plainspastic reconstruction of these bodies and their overlying sedimentary cover is constrained by the unusual distribution of exotic blocks near Parkfield.

Sims, J. D., 1993, Chronology of displacement on the San Andreas fault in central California: Evidence from reversed positions of exotic rock bodies near Parkfield, California, *in* Powell, R. E., Weldon, R. J., II, and Matti, J. C., eds., The San Andreas Fault System: Displacement, Palinspastic Reconstruction, and Geologic Evolution: Boulder, Colorado, Geological Society of America Memoir 178.

The resulting proposed history of movement is divided into three stages that begins with the eruption of the early Miocene volcanic rocks about 24 Ma. The Neenach-Pinnacles Volcanics, erupted after passage of the Mendocino triple junction, were soon cut by the growing San Andreas transform system.

During the first phase of movement the Salinian block, which contains the Pinnacles and Logan godies, was detached from the Mojave and Sierran blocks. The Pinnacles and Logan bodies were transported about 95 km northwest from the Neenach Volcanics and the gabbro of Eagle Rest Peak. At the end of the first phase, the Logan and Pinnacles fragments lay adjacent to the west side of what is now the San Joaquin Valley. Concurrently, fan-deltas deposited debris that was derived from the Gabilan Range. the fan-deltas spread across the San Andreas fault into the middle Miocene sea in the San Joaquin trough.

During the second phase of movement, the San Andreas—at least locally—stepped eastward and detached a second fragment from the Neenach Volcanics. This fragment consists of the volcanic rocks of Lang Canyon. Slip was transferred to the new trace of the San Andreas fault, and the older trace became completely or largely inactive. After transferral of slip to the new trace of the San Andreas fault, the volcanic rocks of Lang Canyon and the Pinnacles Volcanics remained about 95 km apart on the Salinian Block west of the San Andreas fault.

During the third phase, the Gold Hill fragement was slivered off the Logan fragment and was tectonically emplaced on the east side of the San Andreas fault when the Logan fragment lay at the latitude of Gold Hill. The process of slivering off of the Gold Hill fragment was accomplished by deformation of the San Andreas in an eastward bend along what is now the Jack Ranch fault. Bending of the fault was stimulated by the presence of highly sheared Franciscan rocks that crop out near the San Andreas and extend to great depth. Eventually the San Andreas bent to such a degree that slip could not be conducted around the bend, and a new, stable, straight segment was formed. The straightening of the fault resulted in slivering of the Gold Hill fragment from the Logan fragment.

After detachment of the Gold Hill fragment, the Salinian block containing the gabbro of Logan, the Pinnacles Volcanics, and the volcanic rocks of Lang Canyon was transported an additional 160 km northwest to its present position. This reconstruction honors the current positions of all the related exotic fragments of gabbro, volcanics, and sedimentary rocks. The timing of the sequence of movements required to reconstruct the original bodies suggests that the three phases of evolution of the San Andreas fault in central California are characterized by increasing slip rates. The rate for the first phase probably averaged about 10 mm/yr over a period of about 8 m.y. The rate for the second phase averaged about 8 mm/yr over a period of about 7 m.y. The rate rate for the third phase averaged about 33 mm/yr over a period of about 5 m.y.

INTRODUCTION

Large-scale right-lateral movement of the San Andreas fault on the order of hundreds of kilometers was first demonstrated in the seminal paper by Hill and Dibblee (1953). They used the lithologic and faunal similarities of widely separated rock bodies on opposite sides of the San Andreas fault to show that younger strata are displaced shorter distances than older strata. They also suggested that the total offset since late Oligocene to early Miocene time was on the order of 280 km. Later studies built on this theme (for example, Bazeley, 1961; Addicott, 1968; Turner, 1968, 1970). However, all these cross-fault correlations were based on broadly defined features such as ancient shorelines,

latitudinally restricted faunal distributions, and paleoisobaths offset by movement along the San Andreas. The point of intersection of these broadly defined features with the San Andreas fault required the straightline projection of their trends over distances between about 2 to 20 km. Projection of these kinds of features over such distances considerably reduced the precision of the offset determination. These imprecise estimates of offset prompted a search for more areally restricted and distinctive rock units that are offset by the San Andreas. Several distinctive volcanic units and crystalline basement rocks were focused on as likely candidates for the more precise cross-fault correlations that were needed.

One early estimate of the magnitude of large-scale strike-slip movement along the San Andreas fault was based on the presence of distinctive flow-banded rhyolite clasts in the Santa Margarita Formation of the southwestern Temblor Range (Berry and others, 1968; Fletcher, 1962, 1967). Flow-banded rhyolite from the Pinnacles Volcanics[1] was identified as the source of these distinctive clasts because both the clasts in the Santa Margarita Formation and the rocks of the Pinnacles are petrographically similar and have similar potassium-argon ages (Turner, 1968, p. 67). This work was further expanded and the Pinnacles Volcanics were demonstrated to be correlative with the Neenach Volcanics[2] (Turner and others, 1970; Matthews, 1973a), which are now separated by a distance of 315 km along the strike of the San Andreas fault.

At about the same time that studies of the volcanic rocks were being conducted, the gabbroic basement in the San Emigdio Mountains was correlated with similar rocks near San Juan Bautista. Later, the sedimentary and volcanic rocks that unconformably overlay these volcanic rocks were also correlated (Fig. 1, Table 1). The gabbroic basement consists of distinctive hornblende quartz gabbro of Cretaceous and Jurassic age that is unconformably overlain by Eocene to Miocene sedimentary and volcanic rocks. These bodies, which are petrographically and geochemically similar (Ross, 1970; James and others, 1986), lie about 55 km northwest of the Pinnacles and Neenach Volcanics, respectively. The two bodies, the gabbro of Logan near San Juan Bautista and the gabbro of Eagle Rest Peak in the San Emigdio Mountains, and their overlying sedimentary rocks, are separated by about 315 km along the San Andreas fault (Turner and others, 1970; Ross, 1970, 1984, Matthews, 1973a; Nilsen, 1984).

Slivers of rocks similar to the Neenach and Pinnacles Volcanics and the gabbros of Logan and Eagle Rest Peak also occur near Parkfield, an area intermediate between the offset pairs (Dickinson, 1966; Turner, 1968; Ross, 1970; Sims, 1988, 1990). The displaced tectonic fragments of the Parkfield segment of the San Andreas fault lie just northwest of the 1-km right stepover of the fault in Cholame Valley and about 1.5 km northwest of the hamlet of Parkfield (Fig. 1B). Here, the volcanic rocks of Lang Canyon and the gabbro of Gold Hill are correlated with similar distinctive rocks 95 and 150 km to the northwest on the west side of the fault and 145 and 220 km to the southeast on the east side of the fault. The gabbro of Gold Hill is overlain by strata of early Eocene age and the volcanic rocks of Lang Canyon are unconformably overlain by sedimentary rocks of middle to late Miocene age. The relative positions of the volcanic and gabbro fragments are reversed compared to their correlatives to the northwest and southeast. The unusual reversed relative positions of the volcanic rocks of Lang Canyon and gabbro of Gold Hill suggests a more complex history of faulting along the San Andreas fault than has been previously proposed.

The correlations of the Pinnacles Volcanics with the Neenach Volcanics and the gabbro of Logan with the gabbro of Eagle Rest Peak are generally accepted. However, several correlations of displaced rock units that support the correlation of the volcanics and gabbro are less well known, although no less well established. As these additional correlations are crucial to the interpretation of the detailed movement history of the San Andreas fault and the palinspastic reconstruction I present, I first briefly describe pertinent correlations and their interrelationships.

SUMMARY OF CORRELATIVE ROCK UNITS

Gabbroic rocks overlain by Eocene through lower Miocene strata

Three bodies of distinctive and areally restricted gabbroic rocks lie along the San Andreas fault at Logan, near Eagle Rest Peak, and at Gold Hill (Fig. 1). Correlation of the gabbro of Gold Hill with other similar distinctive hornblende quartz gabbro bodies along the San Andreas fault was first suggested by Ross (1970). He correlated the gabbro of Gold Hill with the gabbro of Eagle Rest Peak 145 km to the southeast, and with the gabbro of Logan 170 km to the northwest (Ross, 1970). The gabbros of Logan and Eagle Rest Peak also have identical U/Pb zircon ages of 161 Ma and similar Pb and initial Sr ratios (James and others, 1986). Ross (1970, p. 3058) concluded, based on the petrographic and chemical characteristics of the rocks, "it seems virtually inescapable" that the Logan and Gold Hill masses were once a single body.

All three gabbro bodies are unconformably overlain by conglomerate of Eocene age, and distinctive and correlative Eocene to Miocene stratigraphic successions overlie the gabbroic bodies in the Logan and Eagle Rest Peak areas (Fig. 2). Strata younger than Eocene are absent on the Gold Hill block. The Eocene to

---

[1]Andrews (1936) first introduced the name Pinnacles Formation for the sequence of volcanic rocks exposed in Pinnacles National Monument; he did not designate a type for the unit. In a doctoral dissertation, Matthews (1973b) more fully described and mapped this unit, which he called the Pinnacles Volcanic Formation. He described it as consisting of a sequence of calc-alkaline andesite, dacite, and rhyolite flows interbedded with pyroclastic and volcaniclastic rocks. He divided the formation into seven informal members with an overall thickness of at least 2,185 m. He noted that the unit occurs in several patches north of the Pinnacles National Monument (Matthews, 1973b, Fig. 18) and that it rests unconformably on Cretaceous granitic rocks and is overlain unconformably by unnamed middle Miocene shale. He cited K/Ar dates ranging from 23.9 ± 1.2 Ma to 21.5 ± 3.2 Ma as evidence for an early Miocene age. I accept Matthews's usage of the name Pinnacles and his suggested type section along High Peaks Trail in the Pinnacles National Monument (SW¼SE¼Sec35 to SW¼SE¼Sec.34, T.16S., R.4E., North Chalone Peak 7½-minute Quadrangle). However, his term, Pinnacles Volcanic Formation, does not conform to the North American Stratigraphic Code (North American Commission on Stratigraphic Nomenclature, 1983); thus the unit is here renamed the Pinnacles Volcanics, a term also used by Huffman (1972). Although the Pinnacles Volcanics is not a widespread unit, its extremely important status in California regional geology justifies its retention as a formational rank unit. The retention of formational rank requires the designation of a type section.

[2]The Neenach Volcanic Formation (Dibblee, 1967) is here renamed Neenach Volcanics to conform with the recommendation of the North American Stratigraphic Code (North American Commission on Stratigraphic Nomenclature, 1983). The Neenach Volcanics is here considered to be early Miocene in age, based on correlation with the Pinnacles Volcanics (Turner, 1968; Turner and others, 1970; Matthews, 1973a, 1976).

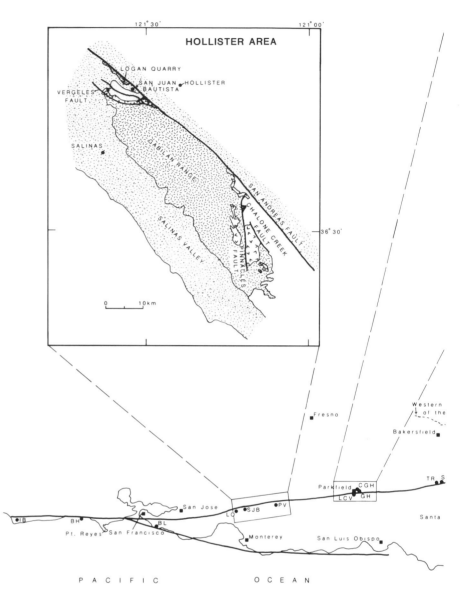

Miocene sedimentary and volcanic rocks of the Logan and Eagle Rest Peak areas were first suggested to be correlative by Turner (1968). Later work in the San Emigdio Mountains (Nilsen and others, 1973; Nilsen, 1987) and in the Juan Bautista area (Dibblee and others, 1979; Nilsen, 1984) supports this correlation. The Miocene strata at Logan and Eagle Rest Peak contain andesite, dacite, and basalt flows interbedded with nearshore marine and continental sedimentary rocks. The Miocene volcanic rocks in both areas are petrographically and geochemically similar, and have similar stratigraphic positions and potassium/argon ages (Thomas, 1986). The three associations of petrographically similar gabbroic basement that are overlain by Eocene to Miocene strata and Miocene volcanic units are the basis of correlation of the three units.

Thus, units ranging in age from Jurassic to early Miocene are offset a constant amount—315 km. Younger rocks are offset lesser amounts (Hill and Dibblee, 1952; Addicott, 1968). The correlation of the Neenach Volcanics with the Pinnacles Volcanics, in addition to the correlation of gabbroic basement rocks at Logan near with the gabbro of Eagle Rest Peak precisely and dramatically illustrate the magnitude of the total offset along the San Andreas fault in central California.

San Juan Bautista area. Near the town of San Juan Bautista lies a 10-km-long fault-bounded fragment of basement composed of hornblende quartz gabbro and anorthositic gabbro. This unit, the gabbro of Logan, is bounded on the east by the San Andreas fault (Ross, 1970). Aeromagnetic data suggests that the fragment, 1 to 2 km thick, has a vertical western boundary considered to be a fault (R. H. Jachens, oral communication, 1987). The gabbro is well exposed in Logan quarry, and is unconform-

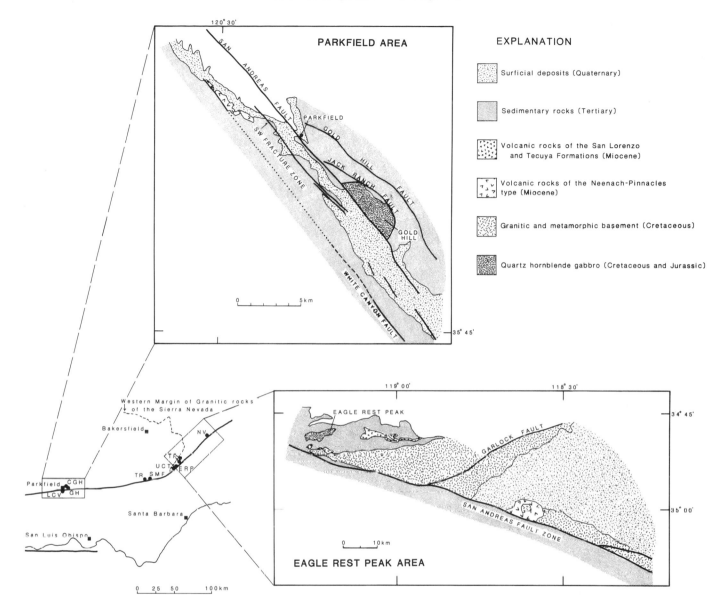

Figure 1. California and the location of the San Andreas fault and sites used to restore on the San Andreas and San Gregorio–Hosgri faults (see Table 1). Sites are keyed to Table 1 and from northwest to southeast are: IB = Iversen Basalt; BH = Bodega Head; BL = tonalite of Ben Lomond; LO = gabbro of Logan; SJB = rhyolite, andesite, and dacite of the San Juan Bautista area; PV = Pinnacles Volcanics; LCV = volcanic rocks of Lang Canyon; GH = hornblende quartz gabbro of Gold Hill; CGH = Eocene Uvas Conglomerate Member of Tejon Formation of the Gold Hill area (Sims, 1988); TR = Santa Margarita Formation–bearing Sr-isotope–dated Crasostrea in the southern Temblor Range; SMF = volcanic clasts of Neenach-Pinnacles–type of the Santa Margarita Formation in southern Temblor Range; STR = volcanic-rich facies in the Temblor Formation of the southern Temblor Range; UCT = Uvas Conglomerate Member of the Tejon Formation; ERP = hornblende-quartz gabbro of Eagle Rest Peak; TF = volcanic rocks in the Tecuya Formation; NV = Neenach Volcanics. A, Geologic map of the Gabilan Range and vicinity. Illustrates the spatial relationship of the gabbro of Logan (Ross, 1970), the Pinnacles Volcanics, and volcanic rocks of the San Lorenzo Formation. After Allen (1946), Dibblee (1971), and Dibblee and others (1979). B, Geologic map of the Parkfield area illustrating the spatial relationship between the gabbro of Gold Hill, Eocene sedimentary rocks deposited on the Gold Hill body, and the volcanic rocks of Lang Canyon (Sims, 1988). C, Geologic map of the Eagle Rest Peak–Neenach Volcanics are illustrating the spatial relationship between the gabbro of Eagle Rest Peak Eocene sedimentary rocks in the San Emigdio Mountains, and the Neenach Volcanics.

**TABLE 1. ROCK BODIES AND STRATIGRAPHIC UNITS THAT DEMONSTRATE OFFSET
ALONG THE SAN ANDREAS AND THE SAN GREGORIO–HOSGRI FAULT
IN POST–LATE MIOCENE TIME***

Name of Unit	Age (m.y.)[†]	Distance from Parent Body (km)	References[‡]
1. Neenach Volcanics (NV)	19.0 - 24.1[§]	Parent	1
2. Volcanic rocks of Lang Canyon (LCV)	22.3 - 23.6	220	2, 4, 9
3. Pinnacles Volcanics (PV)	24.2 ± 0.5	315	2
4. Volcanic clasts in Santa Margarita Formation, southern Temblor Range (SMF)	23.3 ± 0.7 to 24.2 ± 0.7	....	2
5. Sr age of Ostera from Santa Margarita Formation, southern Temblor Range (STR)	13.6**	....	5
6. Dacite flow in Tecuya Formation (TF)	22.2 ± 0.3	Parent	3
7. Volcanic rocks near San Juan Bautista (SJB)	22.2 ± 0.7	315	2, 12
8. Gabbro of Eagle Rest Peak (ERP)	137 ± 4 to 212 ± 10	Parent	6
9. Gabbro of Gold Hill (GH)	147 ± 7	145	6, 8
10. Tejon Formation near Gold Hill (CGH)	Eocene "Tejon" stage		
11. Gabbro of Logan (LO)	159 ± 8	315	6
12. Uvas Conglomerate Member of Tejon Formation, San Emigdio Mountains (UC)	Eocene "Capay" to "Tejon" stage	Parent	10, 11

*Ages and distances used in the palinspastic reconstruction of rock bodies across the San Andreas (see Fig. 1 for locations). Letters in parentheses refer to the designations on Figure 1.

[†]Ages converted to conform to dates determined by new IUGS constants (Dalrymple, 1979).

[§]Minimum age, probable age is 23.7 Ma.

**No error stated.

[‡]1 = D. L. Turner, unpublished data cited by Matthews (1973a); 2 = Turner (1968, p. 67); 3 = Turner (1970); 4 = Matthews (1973a); 5 = Hornafius (written communication, 1987); 6 = Ross (1970); 7 = Sims (1986); 8 = Sims (1988); 9 = Sims (1990); 10 = Nilsen (1984); 11 = Nilsen (1987); 12 = Thomas (1986).

ably overlain by Eocene to Miocene sedimentary and volcanic rocks (Fig. 1A).

The Eocene strata that unconformably overlie the gabbro of Logan have been studied by a number of authors (Kerr and Schenck, 1925; Hill and Dibblee, 1953; Clark and Reitman, 1973; Nilsen, 1984). These strata, considered to be equivalent to the lower part of the San Juan Bautista Formation of Kerr and Schenck (1925), are assigned to an unnamed formation by Dibblee and others (1979). The basal member of the unnamed formation consists of sandstone, conglomerate, and breccia (Fig. 2). The cobbles, boulders, and blocks in the conglomerate have the same lithology as the underlying gabbroic basement rocks. The basal conglomerate member is laterally discontinuous and grades upward into a shaley section that contains foraminifers diagnostic of Ulatisian to Narizian age (Nilsen, 1984). Conformably overlying the unnamed formation are the deep-water fine-grained deposits of the upper part of the San Juan Bautista and Pinecate Formations of Kerr and Schenck (1925), assigned by Dibblee and others (1979) to the San Lorenzo and Vaqueros Formations. These strata are correlated by Nilsen (1984) to the San Emigdio

and lowermost Pleito Formations in the San Emigdio Mountains (Fig. 2).

The uppermost unit of the Cenozoic section in the San Juan Bautista area, assigned to the lower Miocene Zayante Sandstone by Dibblee and others (1979), is considered to be the equivalent of the "red beds" and the "volcanic rocks with interbedded sandstone" of Kerr and Schenck (1925). Interbedded with these strata is a volcanic member composed of andesite, dacite, and rhyolite (Thomas, 1986). Radiometric dates from the volcanic member range from 21.3 ± 1.3 to 23.5 ± 1.4 Ma and average 22.2 ± 0.6 Ma (Table 2, Appendix I). This unit is correlated with the upper part of the Tecuya Formation of the San Emigdio Mountains (Nilsen, 1984).

Eagle Rest Peak area. The gabbro of Eagle Rest Peak and related rocks form the westernmost basement outcrops in the San Emigdio Mountains (Fig. 1C). These rocks are markedly different from the dominantly granitic and gneissic basement of the San Emigdio Mountains and are correlated with the gabbro of Eagle Rest Peak (Ross, 1970).

The Eocene Tejon Formation unconformably overlies the

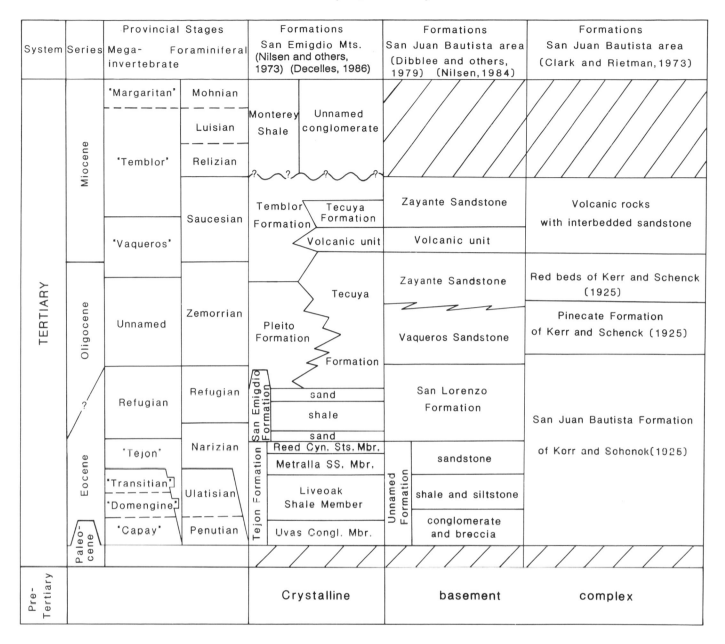

Figure 2. Stratigraphic correlation of Eocene through Miocene strata in the San Emigdio Mountains with those near San Juan Bautista. Modified from Nilsen (1984) and Bishop and Davis (1984).

gabbro of Eagle Rest Peak. The Tejon Formation consists of four members in this area, the basal Uvas Conglomerate Member, the Liveoak Shale Member, the Metralla Sandstone Member, and the uppermost Reed Canyon Siltstone Member. The Uvas Conglomerate Member contains boulder conglomerate composed of clasts derived from the underlying basement, pebble and cobble conglomerate derived from more distant sources, and angular fragments formed in place by weathering of the gabbro (Nilsen, 1987). This unit is oldest in the western part of the San Emigdio Mountains and is progressively younger eastward.

The Eocene marine rocks in the San Emigdio Mountains are conformably overlain by continental sedimentary and volcanic rocks of the Eocene to Miocene Tecuya Formation (Nilsen and others, 1973; T. H. Nilsen, oral communication, 1987). The Tecuya interfingers westward with the marine Eocene and Oligocene Pleito Formation and the Oligocene and Miocene Temblor Formation (Fig. 2). The upper part of the Tecuya contains distinctive andesite and dacite flows overlain by basalt flows. Both volcanic subunits are laterally extensive and extend westward into the Temblor Formation (Nilsen and others, 1973). Radio-

**TABLE 2. POTASSIUM-ARGON AGE DATES OF VOLCANIC ROCKS OFFSET BY MOVEMENT
ON THE SAN ANDREAS FAULT IN CENTRAL CALIFORNIA***

Analysis Number	Unit Dated	Area	Age (Ma)	Reference[†]
KA2215	Tecuya Formation	Eagle Rest Peak	18.1 ± 7.6	1, p. 16
KA2175	Tecuya Formation	Eagle Rest Peak	22.9 ± 0.7	1, p. 16
KA2166	Tecuya Formation	Eable Rest Peak	22.4 ± 0.7	1, p. 15
KA2164	Tecuya Formation	Eagle Rest Peak	24.6 ± 7.2	1, p. 15
KA2162	Tecuya Formation	Eagle Rest Peak	25.2 ± 2.9	1, p. 15
KA2114	Tecuya Formation	Eagle Rest Peak	22.1 ± 0.6	1, p. 13
KA2115	Tecuya Formation	Eagle Rest Peak	22.5 ± 0.7	1, p. 13
	Pooled Mean Age		22.5 ± 0.1	See Appendix
KA2157	Volcanic rocks near San Juan Bautista	Logan	22.2 ± 0.7	1, p. 60[§]
SBJ-1	Volcanic rocks near San Juan Bautista	Logan	23.5 ± 1.4	3
SJB-2	Volcanic rocks near San Juan Bautista	Logan	21.3 ± 1.3	3
	Pooled Mean Age		22.2 ± 0.3	See Appendix
KA2079	Pinnacles Volcanics	Pinnacles	22.1 ± 3.2	1, p. 74
KA2087	Pinnacles Volcanics	Pinnacles	24.1 ± 0.7	1, p. 74
KA2091R	Pinnacles Volcanics	Pinnacles	24.3 ± 0.7	1, p. 74
KA2092	Pinnacles Volcanics	Pinnacles	24.5 ± 1.2	1, p. 74
	Pooled Mean Age		24.2 ± 0.2	See Appendix
KA2102	Volcanic rocks of Lang Canyon	Parkfield	23.6**	1, p. 74
KA2103	Volcanic rocks of Lang Canyon	Parkfield	22.3**	1, p. 74
....	Volcanic rocks of Lang Canyon	Parkfield	23.8 ± 0.7	2
....	Neenach Volcanics	Neenach	24.1**	4
....	Neenach Volcanics	Neenach	23.0**	4
....	Neenach Volcanics	Neenach	22.5**	4

*All dates are corrected for the new IUGS constants (Dalrymple, 1979).

[†]1 = Turner (1968); 2 = D. L. Turner (oral communication, 1987); 3 = Thomas (1986); 4 = D. L. Turner, unpublished data cited *In* Matthews (1973a).

[§]Sample number is incorrectly stated as KA2153 *In* Turner (1968, p. 60); however, KA2153 is described as the Modelo Tuff in the Appendix (p. 80). The correct sample number for the San Juan Bautista dacite is KA2157 (Turner, 1968, p. 80).

**Minimum age.

metric dates from the dacites in the lower part of the volcanic member of the Tecuya Formation have an average potassium-argon age of 22.3 ± 0.6 Ma (Turner, 1970; Thomas, 1986). This section is similar in all respects to the one near San Juan Bautista (Fig. 2).

Gold Hill area. The gabbro of Gold Hill is composed of hornblende quartz gabbro that is surrounded by late Cenozoic strata near Parkfield (Ross, 1970). The gabbro of Gold Hill is bounded on the west by the generally straight trace of the San Andreas fault and on the east by the curving trace of the Jack Ranch fault (Fig. 1B). Aeromagnetic data suggest that the gabbro is about 1 to 2 km thick, 5 to 6 km long, and has vertical

boundaries to at least 1 km depth (R. H. Jachens, oral communication, 1987). The gabbro of Gold Hill is unconformably overlain by strata of the Eocene Tejon Formation and locally derived Quaternary gravel (Sims, 1988). The gabbro and the overlying strata of the Tejon Formation together comprise a fault-bounded, lozenge-shaped tectonic fragment that lies east of the San Andreas fault (Fig. 1B). A notable aspect of the geologic structure of the area east of the Gold Hill fragment is that the fold axes that lie between the Jack Ranch and Gold Hill faults all have a curvature similar to that of the Jack Ranch fault (Fig. 3).

The Eocene sedimentary rocks, correlated with the Tejon Formation of the San Emigdio Mountains (Sims, 1988), outcrop

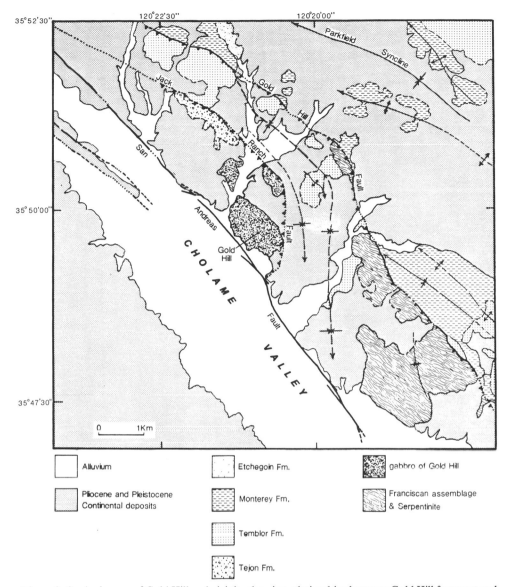

Figure 3. Geologic map of Gold Hill and vicinity showing relationships between Gold Hill fragment and concentric bending of fold axes in the Temblor Formation and the Gold Hill fault. Relationship of Tejon Formation to gabbro of Gold Hill also shown.

in a narrow band adjacent to the Jack Ranch fault on the northeast side of the fragment. This correlative of the Tejon Formation consists of sandstone and conglomerate that contain pebbles, cobbles, and boulders of the gabbro of Gold Hill. Well-preserved molluscan fossils are locally abundant in sandstone of the Tejon Formation north-northwest of Gold Hill (USGS locality M8901). These strata yield the gastropods *Turritella uvasana* sp. (cf. *T. u. sargenti*) and *Neverita globosa* and the bivalves *Pitaria* sp. (cf. *P. uvasana*) and *Yoldia* sp. (cf. *Y. tenuissima*), which are diagnostic of the middle Eocene "Tejon" Stage (Sims, 1988). These Eocene rocks are correlated with the Uvas Conglomerate Member of the Tejon Formation, which contains a larger fauna but includes *T. uvasana, P. uvasana,* and *Pitaria* sp.

Lower Miocene volcanic rocks and derivative strata of the upper Miocene Santa Margarita Formation

Three bodies of distinctive lower Miocene volcanic rocks lie along and near the San Andreas fault at the Pinnacles, the Neenach Volcanics in the northwestern Mojave desert, and in Lang Canyon near Parkfield (Fig. 1). Each unit is unconformably overlain by middle to upper Miocene strata assigned to the Santa Margarita Formation. The sedimentary strata consist of conglomerate and sandstone derived from similar volcanic and granitic sources. Similar sedimentary strata, not now associated with volcanic and granitic basement rocks, crop out in the western Temblor Range northeast of the San Andreas fault (Fig. 1).

Both the Pinnacles and Neenach Volcanics are complex assemblages of flow-banded rhyolite, andesite, dacite, and pyroclastic deposits that overlie granitic basement rocks (Matthews, 1973a; Ross, 1984). These volcanic rocks have similar stratigraphic successions and major and trace element chemistry. The Neenach Volcanics have no reliable potassium/argon dates; however, those given for the Neenach are compatible with dates from the Pinnacles and its correlatives (Matthews, 1973a, b, 1976; Turner, 1968). In some sections both the volcanic and basement rocks are unconformably overlain by marine and nonmarine conglomeratic rocks of the Santa Margarita Formation, which were derived in part from the underlying volcanic and basement rocks (Fletcher, 1967; Wiese, 1950; Sims, 1986). A sliver of similar volcanic rocks and overlying early Miocene conglomeratic strata is also present in the Lang Canyon area 95 km southeast of the Pinnacles (Fig. 1). A fourth occurrence of conglomeratic Santa Margarita Formation, not now associated with lower Miocene volcanic rocks or granitic basement, lies east of the San Andreas fault in the southern Temblor Range (Ryder and Thomson, 1989).

Strata of the Santa Margarita Formation in the southwestern Temblor Range, as well as those that overlie the Pinnacles and Neenach Volcanics, have similar sources of granitic and Miocene volcanic rocks (Huffman, 1972). In addition Ross (1984) demonstrated the similarity of the granitic and tonalitic basement rocks of the Gabilan Range and in the Western Mojave, on which the volcanic rocks rest. This evidence supports the conclusion that the middle Miocene conglomeratic rocks, commonly assigned to the Santa Margarita Formation in central California, were derived from debris eroded from the Neenach and Pinnacles Volcanics and their underlying granitic basement. The volcanic rocks west of the San Andreas fault, the Pinnacles Volcanics and the volcanic rocks of Lang Canyon, are tectonically transported fragments of the same initial volcanic field. The relationships between these volcanics and the sedimentary strata derived from them suggest that the initial volcanic field, which includes the Neenach Volcanics, was repeatedly cut by the evolving San Andreas fault system to produce these fragments.

Pinnacles area. The Pinnacles Volcanics are exposed in and near the Pinnacles National Monument in the Gabilan Range (Fig. 1A). This formation consists of rhyolite, dacite, andesite, rhyolite agglomerate, rhyolite breccia, and pumice lapilli-tuff. The formation rests unconformably on the granitic basement of the Gabilan Range and is unconformably overlain by upper Miocene strata assigned to the Santa Margarita Formation (Matthews, 1973a). Potassium/argon ages from rocks of the Pinnacles Volcanics range from 25.2 ± 2.9 to 22.1 ± 0.6 m.y. (Turner, 1969) and average 24.2 ± 0.5 m.y. (Table 2). The Miocene strata that unconformably overlie the granitic basement and Pinnacles Volcanics consist of boulder conglomerate and interbedded lithic-feldspathic arenites, diatomaceous shale, and pyroclastic breccia assigned to the Santa Margarita Formation (Fletcher, 1967). Basal conglomerate that lacks rhyolite clasts unconformably rests on diatomaceous shale of the Monterey Formation and granitic

basement. The basal conglomerate is overlain by conglomerate that contains clasts of granite and flow-banded rhyolite that is interbedded with agglomerate of the Pinnacles Volcanics (Fletcher, 1967, p. 79).

Neenach area. The Neenach Volcanics crop out in a 40-km^2 area adjacent to and east of the San Andreas fault along the western edge of the Mojave Desert about 55 km southeast at the Eagle Resti Peak area (Fig. 1C). The formation consists of rhyolite, dacite, andesite, rhyolite agglomerate, and pumice lapilli-tuff. The formation rests unconformably on the Cretaceous granitic basement and is unconformably overlain by upper Miocene strata assigned to the Santa Margarita Formation (Wiese, 1950). Potassium/argon analyses of rocks from the Neenach Volcanics yield minimum ages that range from 22.5 to 24.1 Ma. (Turner, 1969 cited *in* Matthews, 1973b). The upper age limit is preferred, owing to the similarity of these rocks with those of the Pinnacles Volcanics (Matthews, 1973b).

The Miocene strata, here assigned to the Santa Margarita Formation[3], are of two facies; maroon sandstone, conglomerate, and mudstone and gray to buff sandstone and conglomerate. The maroon facies overlies and is in part conformable with the Neenach Volcanics (Wiese, 1950). The maroon strata interfinger with light gray arkosic facies of the Santa Margarita Formation (Wiese, 1950, p. 34). The light gray arkosic facies of the more typical Santa Margarita Formation is largely composed of granitic debris that is similar in composition to the granitic rocks present in the San Emigdio Mountains and on Portal Ritter Ridge. Volcanic material is rare in the lower part of the light gray facies but increases in the upper one-half of the unit. In the uppermost part of the light gray facies of the Santa Margarita volcanic debris is about equal to the granitic debris. Volcanic clasts in the Santa Margarita are identical to rocks of the Neenach Volcanics and attest to local derivation (Wiese, 1950, p. 32–33).

Lang Canyon area. The volcanic rocks that crop out in Lang Canyon near Parkfield were first correlated with the Pinnacles Volcanics on the basis of similar K/Ar dates and field petrographic relationships (Turner, 1968). These volcanic rocks were later correlated with the Neenach Volcanics (Turner and others, 1970; Huffman, 1972; Matthews, 1973b). The volcanic rocks of

---

[3]Gray to buff conglomerate and sandstone that bear fossils and outcrop in the Neenach and Lebec Quadrangles, Los Angeles and Kern Counties, California were referred to the Santa Margarita Formation by Wiese (1950). An unnamed unit of Miocene(?) continental deposits of similar composition but maroon in color was also described by Wiese (1950) and Crowell (1952). These strata were reassigned to the Quail Lake Formation and the Oso Canyon Formation, respectively, by Dibblee (1967). A comparison of clast types between the two units and with the Santa Margarita Formation at its type locality and in the southern Temblor Range (J. G. Vedder, oral communication, 1987) and near Parkfield and in the Gabilan Range shows that all these strata are strikingly similar in clast lithology. Strata exposed in the excavation for the West Branch of the California Aqueduct near Quail Lake shows that lentils of gray conglomerate and sandstone of Santa Margarita–type occur in maroon conglomerate and sandstone, thus suggesting that the two colors of sandstone and conglomerate are facies of the same formation. Therefore, Quail Lake Formation and the Oso Canyon Formation are both hereby abandoned and their strata reassigned to the Santa Margarita Formation.

Lang Canyon (Sims, 1988), a body of flow-banded rhyolite, obsidian, and volcanic breccia, lie 16 km northwest of Gold Hill (Fig. 1B). The volcanic rocks crop out in a narrow northwest-southeast elongated belt about 6 km long and 1 km wide about 2 km west of the main trace of the San Andreas fault, but are best exposed in Lang Canyon 4 km northwest of Parkfield. The volcanic rocks dip steeply to the northeast in contrast with the gently southwest-dipping Santa Margarita Formation that unconformably overlies the fragment on the southwest. The fragment is considered to be fault-bounded although the bounding faults are poorly exposed (Sims, 1990). The southwest-bounding fault is in part covered by strata of the late Miocene Santa Margarita Formation. The northeast boundary of the fragment is overlapped by alluvial deposits of Cholame Creek. Sedimentary rocks of the upper Miocene Santa Margarita Formation are of two distinct types in the Parkfield area. The upper Santa Margarita consists of light-colored, well-sorted, fossiliferous sandstone. The lower Santa Margarita Formation consists of conglomerate and conglomeratic sandstone that is dominantly composed of coarse granitic debris. Sandstone of the upper Santa Margarita unconformably overlaps the west edge of the volcanic rocks of Lang Canyon. These white to pale gray, coarse-grained, fossiliferous sandstone strata represent a shallow, nearshore, marine environment. Well-sorted calcareous and fossiliferous arenites lap onto the volcanic rocks of Lang Canyon in sections 20 and 29, T.23S., R.14E. in the Parkfield 7½-minute quadrangle. Megafossils collected from the white calcareous sandstone suggest a late Miocene age for the rocks (J. W. Durham, written communication, 1985). Strata of this type are not present in the Santa Margarita of the Neenach and Pinnacles area, but are similar to the Santa Margarita in its type area (Sims, 1990).

The lower Santa Margarita outcrops principally about 1 km west of the volcanic rocks of Lang Canyon and is unconformably(?) overlain by sandstone of the upper Santa Margarita. The lower Santa Margarita consists of interbedded granitic debris, boulder conglomerate, and interbedded quartzose arenite, mudstone, and claystone. The conglomerate contains volcanic clasts composed of andesite, flow-banded rhyolite, and rare purple amygdaloidal andesite (Sims, 1990). The andesite and purple andesite clasts, common in the Neenach Volcanics, are not present in the volcanic rocks of Lang Canyon (Wiese, 1950, p. 31; Crowell, 1952, p. 11–12; Matthews, 1973b, p. 43). The presence of purple amygdaloidal andesite clasts in the lower part of the Santa Margarita Formation suggests that these strata were, in part, derived from the Neenach or Pinnacles Volcanics.

Granitic and metamorphic basement rocks

Basement rocks that underlie the Neenach and Pinnacles Volcanics and the sedimentary rocks derived from them also show striking similarities to each other (Ross, 1984). Correlations of these granitic and metamorphic rocks together with correlations of late Cenozoic volcanic and sedimentary rocks suggests that the Gabilan Range lay opposite the tail of the Sierra and the

western Mojave Desert prior to initiation of movement on the San Andreas fault.

The granodiorite of Natividad in the Gabilan Range closely resembles the granodiorite of Lebec in the San Emigdio Mountains (Fig. 4). Both are medium grained, and contain distinctive coarse biotite flakes and scattered hornblende crystals with red cores of skeletal clinopyroxene crystals. Coarse-grained biotite granite, the granites of Fremont Peak and Brush Mountain, typical and common "low melting trough"-type granites, are also associated with both the granodiorite of Natividad and Lebec (Ross, 1984). The tonalite and granodiorite of Johnson Canyon, which crops out near the Pinnacles Volcanics in the Gabilan Range, are similar lithologically, modally, and chemically to the granodiorite of Fairmont Reservoir which crops out near the Neenach Volcanics (Fig. 4). Both the Johnson Canyon and Fairmont Reservoir bodies contain distinctive, euhedral sphene crystals (Ross, 1984, p. 13, 33). The Johnson Canyon, Bickmore Canyon, Burnt Peak, and Fairmont Reservoir bodies are also noteworthy for the presence of small pink K-feldspar phenocrysts, as much as 2 cm long, in them. The granite of Bickmore Canyon is also strikingly similar to the felsic variant of the Fairmont Reservoir body. Another similarity between the basement rocks that underlie the Pinnacles and Neenach Volcanics is the abundance of alaskite, aplite, and pegmatite. Ross (1984) further suggested the correlation of the schist of Sierra de Salinas, in the Gabilan Range, to the schist of Portal-Ritter Ridge (Fig. 4). The strong physical and chemical similarity of these two schists is well established (Ross, 1976).

REGIONAL PATTERN OF CORRELATIVE ROCKS

North to south arrangement

Fragments of gabbroic basement, lower Miocene volcanic rocks, and Eocene to upper Miocene sedimentary rocks are arrayed on both sides of the San Andreas fault over a distance of about 370 km from near San Juan Bautista in the north to the western Mojave Desert in the south (Fig. 1). The gabbro of Logan, the northernmost fragment, lies 55 km northwest of the Pinnacles Volcanics on the west side of the San Andreas fault. The gabbro of Logan is adjacent to the San Andreas, but the Pinnacles lie about 2 km west of the San Andreas. The volcanic rocks of Lang Canyon lie 95 km southeast of the Pinnacles and 16 km northwest of the Gold Hill fragment. The Lang Canyon fragment lies about 2 km west of the San Andreas, and the Gold Hill fragment is adjacent to the San Andreas but on the east side of it. The gabbro of Eagle Rest Peak lies 145 km southeast of Gold Hill and 55 km northwest of the Neenach Volcanics, the southernmost fragment. The Eagle Rest Peak fragment lies about 3 km east of the San Andreas, and the Neenach Volcanics lies adjacent to and east of the San Andreas.

The gabbro and volcanic rocks of both the Logan and Pinnacles and Eagle Rest Peak and Neenach fragments are 55 km apart. The volcanic rocks lie to the southeast in both cases. How-

EXPLANATION

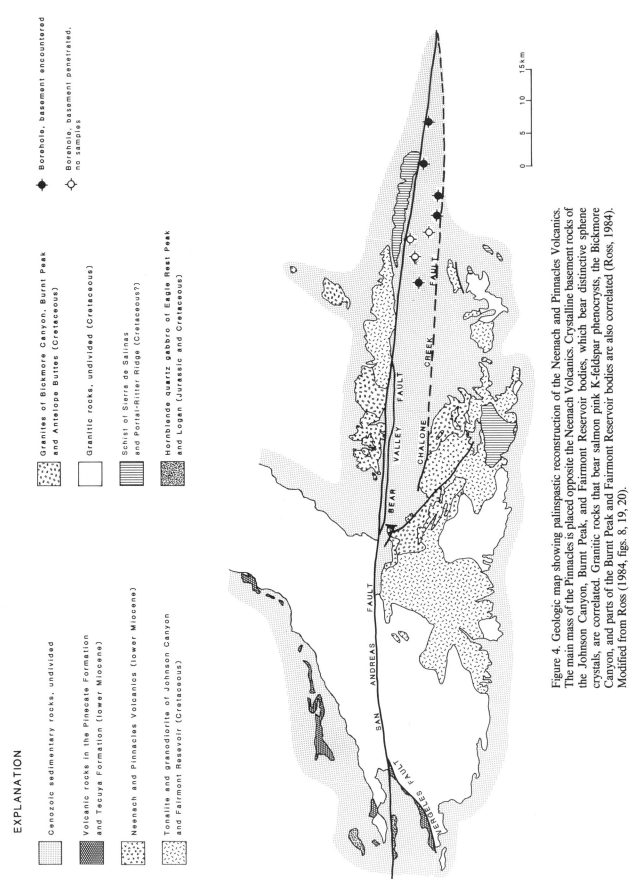

Figure 4. Geologic map showing palinspastic reconstruction of the Neenach and Pinnacles Volcanics. The main mass of the Pinnacles is placed opposite the Neenach Volcanics. Crystalline basement rocks of the Johnson Canyon, Burnt Peak, and Fairmont Reservoir bodies, which bear distinctive sphene crystals, are correlated. Granitic rocks that bear salmon pink K-feldspar phenocrysts, the Bickmore Canyon, and parts of the Burnt Peak and Fairmont Reservoir bodies are also correlated (Ross, 1984). Modified from Ross (1984, figs. 8, 19, 20).

ever, the Lang Canyon and Gold Hill fragments are 16 km apart and the volcanic rocks lie to the northwest (Sims, 1988, 1990). This anomaly provides the clue to how these bodies attained their present positions.

Reversed positions of fragments near Parkfield

The relative positions of the volcanics and gabbro in the Parkfield area are reversed with respect to their correlative rock bodies and they lie only 16 km apart, not 55 km, as the correlations with their parent bodies predict. Both the Lang Canyon and Gold Hill fragments are fault-bounded (Sims, 1988, 1990). The fact of the reversed positions of these two fragments requires a complex series of movements on their respective bounding faults to achieve their present locations.

A two-stage sequence of slivering the original Neenach/Pinnacles body explains the present position of the Lang Canyon fragment as well as the presence of east- and west-bounding faults. In the initial episode of strike-slip faulting, the Pinnacles Volcanics body was slivered off the parent block. The separation between the two Pinnacles and the Neenach bodies increased with time to about 95 km. At that point a second episode of strike-slip faulting occurred on a new trace of the San Andreas transform to produce the Lang Canyon fragment. The present position of the Lang Canyon fragment, intermediate between its correlative bodies, is best explained by cessation of movement on the first strike-slip fault and transfer of motion to the newly formed fault. With the first fault inactive, the relative position and distance between the Pinnacles and Lang Canyon fragments is maintained, and the lateral separation between the Lang Canyon and Neenach fragment is also explained.

The position of the Gold Hill fragment east of the San Andreas fault and just north of the 1-km right step in the San Andreas is anomalous because its southeastern parent fragment is also on the east side of the San Andreas. The Gold Hill fragment is bounded on the southwest by the San Andreas and on the northeast by the arcuate Jack Ranch fault. Arcuate fold axes in the Temblor Formation are nearly parallel to the Jack Ranch fault. Farther eastward the arcuate trace of the Gold Hill fault also parallels the Jack Ranch, but to a lesser degree. This unusual family of arcuate structures coincides with a similarly curved topographic configuration of Cholame Valley (Sims, 1988). The similar curvature of the faults and folds and the broadening of Cholame Valley around Gold Hill have prompted speculation that the Jack Ranch fault is a warped and abandoned older trace of the San Andreas fault (R. W. Simpson, written communication, 1986). Highly sheared Franciscan melange and serpentinite basement outcrops to the east of the Jack Ranch fault in a wedge-shaped mass bounded on the east by the Gold Hill fault. South of Gold Hill, Franciscan melange and serpetinite are less abundant and are unknown more than about 10 to 15 km to the south (Dibblee, 1971). This evidence suggests that, as the parent mass of the Gold Hill fragment, the Logan block, approached the latitude of Gold Hill, the easily deformable Franciscan melange

gave way and a bend in the San Andreas fault developed. As the Logan block moved closer to the present position of Gold Hill the curvature of the San Andreas increased to a point that configuration of the fault was no longer stable (Fig. 5). The unstable configuration was resolved by formation of a new straight segment that cut through the Logan fragment. The reason the new straight segment of the San Andreas passed through the gabbro body rather than around it is unclear. The straightened segment may have occupied previously formed shear planes or other zones of weakness in the gabbro. Formation of the new fault segment resulted in the slivering off of the Gold Hill block and the freeing of the remainder of the Logan fragment to be transported to its present position. Farther eastward, bending of the San Andreas fault after slivering off of the Gold Hill fragment probably did not occur because of the buttressing effect of the steeply dipping Parkfield syncline, west-directed thrusting on the Table Mountain thrust fault (Sims, 1988, 1990), and convergence across the San Andreas fault.

PALINSPASTIC RECONSTRUCTION

Palinspastic reconstruction of rock bodies cut and displaced by the San Andreas fault in central California is largely constrained by the location and relative position of the various fragments of tectonically transported exotic rock bodies that now lie between the San Juan Bautista and Neenach areas. The reversed fragments of exotic rocks in the Parkfield area are critical to this reconstruction. Correlation of the fault-bounded fragments near Parkfield with rocks that lie adjacent to or near the San Andreas to the northwest and southeast is well established. It is their relative position in the Parkfield area that implies a time sequence for tectonic transport and emplacement.

Simple removal of the 315 km of accumulated slip on the San Andreas fault places the gabbro of Logan, the Pinnacles Volcanics, and the overlying Eocene to Miocene strata opposite the gabbro of Eagle Rest Peak, Neenach Volcanics, and overlying Eocene to Miocene strata. Reconstruction of these bodies also juxtaposes the basement rocks on which they lie, the tonalite of Johnson Canyon, with the granodiorites of Fairmont Reservoir and Burnt Peak. The reconstruction also places the Schist of Sierra de Salinas opposite the schist of Portal Ritter Ridge (Ross, 1984) (Fig. 2). No time history for the San Andreas, other than that determined from the ages of the rocks that are cut, can be implied.

Restoration of the exotic blocks of the Parkfield area to their respective parent blocks requires a more complicated sequence of movements. These movements must both remove the reversed relative positions and restore the Gold Hill block to the west side of the San Andreas fault. Reconstruction of these fragments and with their basement rocks, and associated derivative sedimentary rocks is accomplished in three phases (Fig. 6). Ultimately these phases of movement are related to the movement of the Pacific and North American plates and the migration of the two triple junctions that developed following the subduction of the Farallon plate (Atwater, 1970).

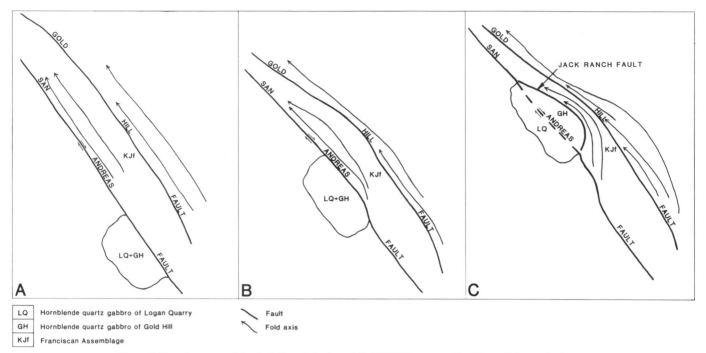

LQ Hornblende quartz gabbro of Logan Quarry

GH Hornblende quartz gabbro of Gold Hill

KJf Franciscan Assemblage

⟨fault symbol⟩ Fault

⟨fold axis symbol⟩ Fold axis

Figure 5. Development of the Jack Ranch fault and Gold Hill Fragment. See Figure 3 for geologic map of Gold Hill and vicinity. See Figure 3 for geologic map of Gold Hill and vicinity. A, Initial stage of tectonic transport of the Logan Quarry–Gold Hill gabbro body into the Parkfield area along the San Andreas fault. B, Second stage of tectonic transport of the Logan Quarry–Gold Hill gabbro body in the Parkfield area. Sheared Franciscan rocks yield to northeast-directed convergence across the San Andreas fault. Compression causes the San Andreas fault to bend. Fold axes and existing faults are also warped by northeast movement of the rigid gabbro body. C, Final stage of tectonic transport of the Logan Quarry–Gold Hill gabbro body in the Parkfield area. The San Andreas fault is now warped greatly enough that slip around the bend is difficult and a new path is sought for release of strain. The new path of strain release splits off the Logan Quarry fragment to be transported northwestward. The Gold Hill body remains in the Parkfield area 16 km southeast of the volcanic Rocks of Lang Canyon (Sims, 1990).

The present positions of the northernmost fragments suggest that they were separated from the parent masses first and those that now lie in the Parkfield area were separated later (Fig. 7A). The sequence of motions required to restore all of the exotic blocks to their initial positions is constrained to begin with the reversal of slip that has accumulated on the most recent trace of the San Andreas fault. I assume that in the first step of this sequence of reverse motions the Logan, Pinnacles, and Lang Canyon fragments remain fixed with respect to each other on the Salinian block.

1. Reverse movement of 165 km of the Logan fragment from its present position places it opposite the Gold Hill fragment. The Pinnacles and Lang Canyon fragments, having moved the same amount, then lie 55 and 140 km, respectively, to the southeast of Gold Hill (Fig. 7B). The Logan and Gold Hill bodies are joined and in further movements will move together.

2. Strike-slip movement is transferred to the Jack Ranch fault, and strike-slip movement and the east-directed compressional deformation of the fold axes to the east of Gold Hill are removed. The position of the amalgamated Logan and Gold Hill fragments fixes their position with respect to the Pinnacles and Lang Canyon fragments.

3. Reverse movement of an additional 55 km carries the Lang Canyon fragment to a position opposite the Neenach (Fig. 7C). The volcanic rocks of Lang Canyon are joined to the Neenach Volcanics fragment.

4. Following the amalgamation of the Lang Canyon fragment with the Neenach fragment strike-slip motion is transferred to the White Canyon–Red Hills–San Juan–Chimineas fault zone. The remaining 95 km of deformation is removed along this western boundary fault of the Lang Canyon fragment (Fig. 7D). Movement along this fault zone joins the Pinnacles with the

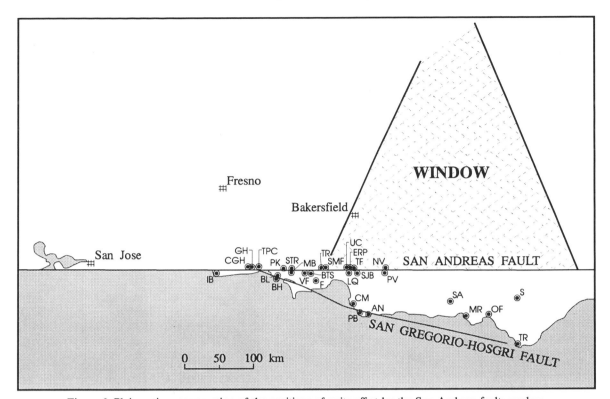

Figure 6. Plainspatic reconstruction of the positions of units offset by the San Andreas fault; modern coastline shown for reference. Map constructed by moving block bounded by San Andreas fault and San Gregorio–Hosgri fault 150 km southeast; rotation of the Tehachapi and San Emigdio Mountains were proposed by McWilliams and Li (1986) and Plescia and Calderone (1986) and the restoration of left-lateral movement on the Garlock fault serves to straighten out the trace of the San Andreas. Restoration of right-lateral slip on the San Gregorio–Hosgri fault based on James and Mattinson (1985). Sites are keyed to Table 1, and from northwest to southeast are: IB = Iversen Basalt; BH = Bodega Head; BL = tonalite of Ben Lomond; LO = gabbro of Logan; SJB = rhyolite, andesite, and dacite of the San Juan Bautista area; PV = Pinnacles Volcanics; LCV – volcanic rocks of Lang Canyon; GH = hornblende quartz gabbro of Gold Hill; CGH = Eocene Uvas Conglomerate Member of Tejon Formation of the Gold Hill area (Sims, 1988); TR = Santa Volcanics clasts of Neenach-Pinnacles–type of the Santa Margarita Formation in southern Temblor Range; STR = volcanic-rich facies in the Temblor Formation of the southern Temblor Range; UCT = Uvas Conglomerate Member of the Tejon Formation; ERP = hornblende-quartz gabbro of Eagle Rest Peak; TF = volcanic rocks in the Tecuya Formation; NV = Neenach Volcanics. Geometry of no-slab window in subducted slab after Dickinson and Snyder (1978a,b).

previously amalgamated Neenach and Lang Canyon body. This final movement also joins the Logan and Gold Hill fragments with the Eagle Rest Peak fragment.

DISPLACEMENT HISTORY ON THE SAN ANDREAS FAULT SYSTEM

Sequence, magnitude, and timing

Following subduction of the Farallon plate, the collision of the Pacific plate with the North American plate resulted in the formation of two triple junctions (Atwater, 1970). The northern transform-transform-trench Mendocino triple junction moved northwest with the Pacific plate, and the southern transform-ridge-trench Rivera triple junction moved southeast. As the separation of the two triple junctions increased, the San Andreas transform system formed. The San Andreas lengthened in response to migration of the triple junctions. As the San Andreas fault system developed, volcanic rocks erupted along it to form a broad, nearly linear, northwest-trending belt of upper Tertiary to Quaternary volcanic rocks (Fox and others, 1985). Two volcanic centers, represented by the Neenach and Pinnacles Volcanics and the volcanic member of the Tecuya and San Juan Bautista Formations, are among the oldest extrusive rocks thus formed. Both

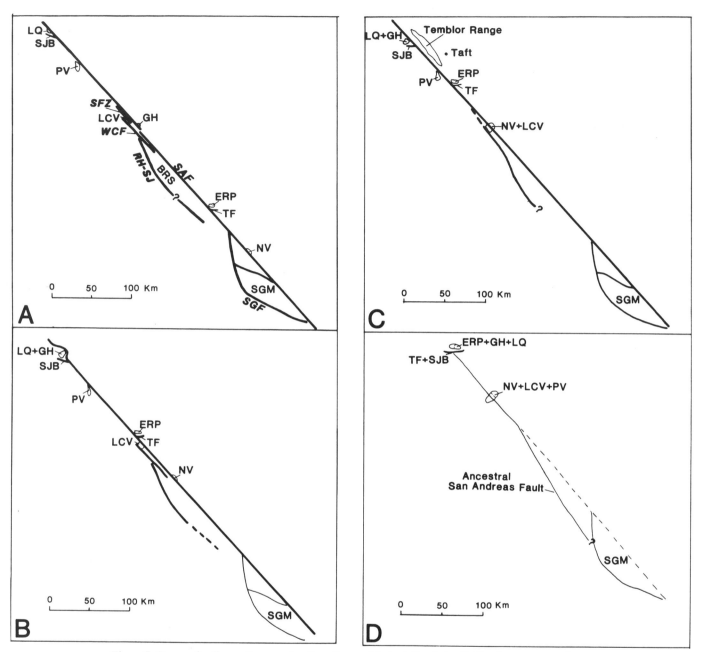

Figure 7. Stages of palinspastic reconstruction discussed in text. Symbols of offset fragments are, from northwest to southeast; LO = gabbro of Logan; SJB = volcanic rocks near San Juan Bautista; PV = Pinnacles Volcanics LCV = volcanic rocks of Lang Canyon; GH = gabbro of Gold Hill; BRS = Barrett Ridge Slice; ERP = gabbro of Eagle Rest Peak; TF = volcanic members of Tecuya Formation; NV = Neenach Volcanis; SGM = San Gabriel Mountains. Faults (in italics): *SAF* = San Andreas Fault; *SFZ* = Southwest fracture zone; *WCF* = White Canyon fault; *RH-SJ* = Red Hills–San Juan–Chimineas fault; *SGF* = San Gabriel fault.   A, Present configuration of exotic fragments along the San Andreas fault. B, First stage of back slip. The slip removed is 165 km. The gabbro of Gold Hill is moved along the present trace of the San Andreas fault to lie opposite the gabbro of Gold Hill. C, Second stage of back slip. The slip removed is an additional 55 km. The volcanic rocks of Lang Canyon are moved to lie opposite the Neenach Volcanics. D, Third stage of back slip. The amount of slip removed is an additional 95 km and completes the slip removal. The Pinnacles Volcanics are moved to lie opposite the Neenach Volcanics, and the combined gabbros of Logan and Gold Hill lie opposite the Gabbro of Eagle Rest Peak.

volcanic masses were cut and offset by the San Andreas fault in central California (Fig. 1).

The age of the Neenach and Pinnacles Volcanics is revealed by K/Ar dating to be 24.2 ± 0.5 Ma (Table 2). Lower Miocene volcanic rocks of similar composition, the Tecuya Formation of the San Emigdio Mountains and their correlatives in the San Juan Bautista formation, are K/Ar dated at 22.6 ± 0.3 Ma and 22.2 ± 0.6 (Table 2). The Tecuya and San Juan Bautista units are offset the same amount as the Neenach and Pinnacles Volcanics—315 km. Thus, development of the San Andreas fault northwest of the San Emigdio Mountains postdates the eruption of both of these volcanic centers (Thomas, 1986; Peter Weigand, oral communication, 1987). The amount of time by which the establishment of the San Andreas fault postdates these volcanic units is imprecisely known. A minimum age of 18 Ma for initiation of movement on the San Andreas fault is based on evidence from basaltic volcanic debris deposited in the Temblor Formation of the western San Joaquin Valley (O'Day and Sims, 1986). An extensional tectonic regime is suggested between 24 and 20 Ma in western California from the LaHonda Basin in the northwest and the Diligencia Basin in the southeast (Tennyson, 1989). These suggestions are at variance with other suggestions that initial movement on the San Andreas fault occurred in late Oligocene time (Stanley, 1987; Addicott, 1968). On the basis of present information, strike-slip movement on the San Andreas transform appears to have begun between 22 and 18 Ma.

The evolution of the San Andreas fault in central California is defined on the basis of the location of one or more exotic fragments along the trace of the fault system. I divide the evolution into three phases. The events that mark the beginning and end of phases are: the eruption of the Neenach and Pinnacles Volcanics and the volcanic rocks in the Tecuya and San Juan Bautista Formations, deposition of flow-banded rhyolite clasts in the Santa Margarita Formation of the southern Temblor Range, and slivering off of the gabbro of Gold Hill from the gabbro of Logan.

Phase 1. The event that marks the initiation of recognizable movement on the San Andreas fault in central California is the eruption of the Neenach and Pinnacles Volcanics, and the volcanic members of the Tecuya and the San Juan Bautista Formations between 24.2 ± 0.5 and 22.2 ± 0.6 Ma (Fig. 7A). Eruption of these volcanics postdates the migration of the Menodicino triple junction through the area and is associated with expansion of the no-slab window as the subducted slab of Pacific Plate continued to pass beneath the North American plate (Fig. 8). Following eruption of the volcanics, the San Andreas continued to extend to cut the volcanic rocks, the gabbros of Logan and Eagle Rest Peak, the overlying Eocene to lower Miocene sedimentary rocks, and their associated basement. The precise time that the San Andreas cut the volcanic rocks is unknown, but for present purposes tentatively taken at less than 22 Ma. Right-lateral movement continued during this phase and the San Andreas transform system extended as the Mendocino triple junction migrated farther northwest. The Salinian block, on which lay the

Pinnacles and Logan fragments, moved northwestward away from the Neenach Volcanics and gabbro of Eagle Rest Peak until the Pinnacles fragment lay at the latitude of the southern Temblor range (Fig. 9). Little or no deformation occurred on the Salinian block since the initiation of strike-slip movement on the San Andreas system. This is attested to by the fact that the distance between the Pinnacles and Logan fragments is the same as the Neenach and Eagle Rest Peak fragments.

Arrival of the Pinnacles Volcanics in the vicinity of the southern Temblor Range, about 95 km northwest of the Neenach Volcanics (Fig. 9), is marked by the presence of abundant clasts of flow-banded rhyolite in member C of the Santa Margarita Formation (Ryder and Thomson, 1989). The rhyolite clasts have the same lithology and K/Ar age as the Pinnacles Volcanics (Fletcher, 1967; Turner, 1968). Members A, B, and D of the Santa Margarita Formation in the Temblor Range are characterized by coarse detritus derived from granitic and metamorphic basement rocks that are free of volcanic rock types. The members of the Santa Margarita record the passage of the exposed basement of the Gabilan Range through the latitude of the southern Temblor Range. The time of the passage is placed at about 14 Ma on the basis of Sr-isotope dating of fossil molluscan shells from member B of the Santa Margarita (S. Hornafius, written communication, 1987). Member C, which carries the flow-banded rhyolite clasts, conformably overlies member B. Member D of the Santa Margarita Formation is overlain by latest Miocene and Pliocene strata. The time of first deposition of rhyolite clasts in member C cannot be directly determined. However, their deposition began after 14 Ma but before late Miocene time, about 10 Ma. Therefore, I suggest a tentative date of 12 Ma for the initiation of the deposition of flow-banded rhyolite clasts in the Santa Margarita Formation of the western Temblor Range.

The end of the first phase of movement of the Andreas fault is placed at the cessation of the deposition of rhyolite clasts in the upper part of the Santa Margarita Formation of the southern Temblor Range. Strike-slip movement up to this time occurred on a fault that is now represented by segments of the San Juan, Chimineas, Red Hills, and White Canyon faults. The northwestward extension of the White Canyon fault connects with the west boundary of the Lang Canyon fragment. Abandonment of this fault occurred at the time the Lang Canyon fragment was slivered off of the Neenach fragment. Abandonment of the fault is chosen at the end of phase 1 because the Lang Canyon fragment lies 95 km southwest of the Pinnacles, and the Pinnacles and Logan fragments have remained a constant distance apart since they were slivered off their respective parent bodies at the beginning of the phase. The reason for abandonment of the old fault and realignment of the San Andreas to sliver off another fragment of the Neenach Volcanics is unclear. The realignment may have resulted from the development of a bend in the fault that was an unstable configuration. The bend may have resulted from rotation in the tail of the Sierra Nevada (McWilliams and Li, 1986; Plescia and Calderone, 1986) and accompanying bending of the fault.

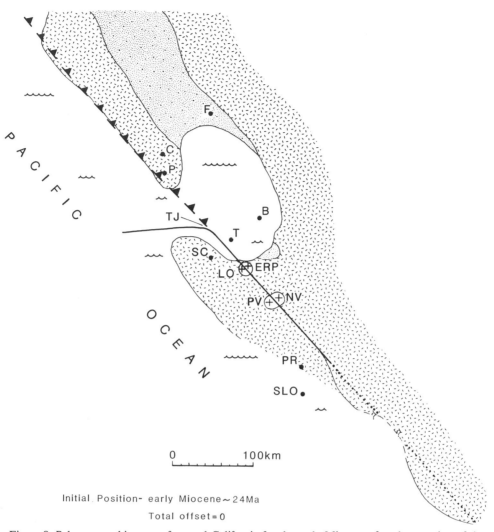

Figure 8. Paleogeographic map of central California for the early Miocene after the eruption of the Neenach/Pinnacles and Tecuya/San Juan Bautista Volcanics. Shorelines and depositional units taken from Bartow (1987), Perkins (1987) and Graham and others (1989). C = Coalinga; B = Bakersfield; ERP = gabbro of Eagle Rest Peak; F = Fresno; LO = gabbro of Logan; NV = Neenach Volcanics; P = Parkfield; PR = Paso Robles; PV = Pinnacles Volcanics; QSV = Quien Sabe Volcanics; SC = Santa Cruz; SF = San Francisco; SLO = San Luis Obispo; SV = Sonoma Volcanics; TJ = Mendocino triple junction.

Phase 2. The beginning of phase 2 is marked by the end of deposition of flow-banded rhyolite clasts in the Santa Margarita Formation of the southern Temblor Range and an eastward jump in the location of the San Andreas fault (Fig. 9). The eastward realignment of the San Andreas fault resulted in a second fragment, the volcanic rocks of Lang Canyon, being sliced from the Neenach Volcanics. The old trace of the San Andreas, represented by the San Juan–Chimineas–Red Hills–White Canyon fault zone, became wholly or largely inactive. The exact length of the abandoned trace of the San Andreas is in doubt. However, the fault segments that bound the Barrett Ridge slice including the fault segment that forms the west boundary of the Lang Canyon fragment were most likely abandoned (Fig. 10).

The precise length of the newly activated segment of the San Andreas fault in phase 2 is uncertain. The segment certainly extended at least from just northwest of the Lang Canyon body to some distance southeast of the Neenach Volcanics. This fault is now represented, in part, by the Southwest fracture zone of the San Andreas fault near Parkfield (Brown and others, 1967; Sims, 1990). The Southwest fracture zone is thought to connect with the west trace of the San Andreas fault, which lies on the west side of Cholame Valley to the southwest (Fig. 1B). The northwest extension of the Southwest fracture zone is not clearly defined, but the trend of its en echelon segments suggests that it connects with the San Andreas on Middle Mountain (Sims, 1990). The Chalone Creek fault, which lies to the northwest of Middle

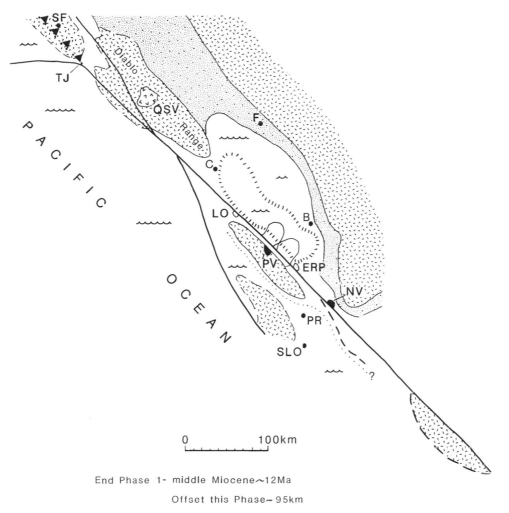

Figure 9. Paleogeographic map of central California for the early Miocene after the eruption of the Neenach/Pinnacles and Tecuya/San Juan Bautista Volcanics. Shorelines and depositional units taken from Bartow (1987), Perkins (1987) and Graham and others (1989). C = Coalinga; B = Bakersfield; ERP = gabbro of Eagle Rest Peak; F = Fresno; LO = gabbro of Logan; NV = Neenach Volcanics; P = Parkfield; PR = Paso Robles; PV = Pinnacles Volcanics; QSV = Quien Sabe Volcanics; SC = Santa Cruz; SF = San Francisco; SLO = San Luis Obispo = SV = Sonoma Volcanics; TJ = Mendocino triple junction. Phase 1 of the evolution of the San Andreas fault began with strike-slip movement following eruption of the (composite) Neenach-Pinnacles Volcanics and the volcanic rocks in the Tecuya Formation. The end of Phase 1 is placed at the time flow-banded rhyolite clasts first occur in the Santa Margarita Formation in the southern Temblor Range. The age of the Santa Margarita is here considered to be ≈12 Ma. The time span of Phase 1 is about 10 m.y. The amount of slip is 95 km. Thus the minimum slip rate for Phase 1 is 9.5 mm/yr.

Mountain is considered by Matthews (1973a) to be an old segment of the Andreas fault. Matthews hypothesized that the Chalone Creek fault is connected to a fault that lies on the west side of Peachtree Valley, here referred to as the Peachtree Valley fault (Fig. 2). Matthews considered the combined Chalone and Peachtree Valley faults to predate the San Andreas fault and to have a comparable amount of offset. Matthews's suggestion was later questioned by Ross (1984), who noted that the basement rocks that lie between the Chalone Creek fault and the San Andreas fault are similar to the granitic rocks and schist of the Gabilan Range. However, Ross (1984) also showed that the rocks of the Gabilan are strikingly similar to those of Portal-Ritter Ridge. Thus, Ross's argument that large-scale offset on the Chalone Creek fault comparable to the San Andreas cannot be demonstrated is not a convincing one. There is also abundant evidence that the active trace of the San Andreas has not remained fixed

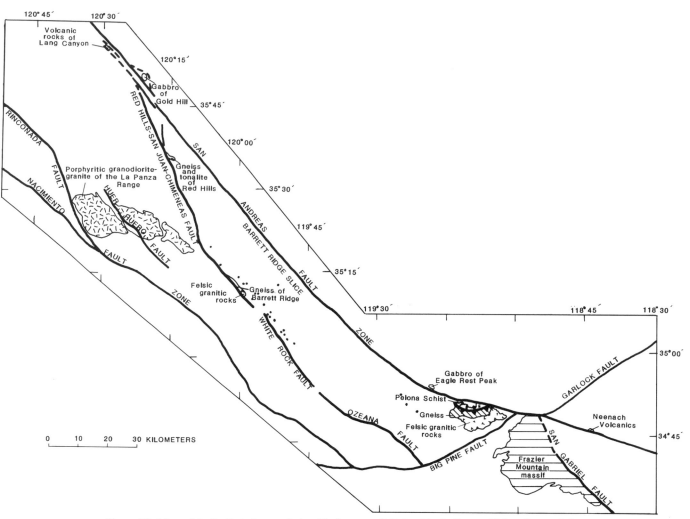

Figure 10. Map of faults that shows relationship between Salinian block, Barrett Ridge slice, and the sliver between the Chalone Creek fault and San Andreas fault. Wells that reach basement of the Barrett Ridge slice also shown. Modified from Ross (1984).

through time, as shown here and by Matti and others (1986) and Powell (1986).

The end of phase 2 is marked by the separation of the Gold Hill fragment from the Logan fragment (Fig. 11). The location of the gabbro of Gold Hill constrains the amount of lateral offset that accumulated in phase 2 (Fig. 5). The timing of the split off the Gold Hill fragment is poorly constrained but can be estimated by the age of associated late Cenozoic units involved in deformation that accompanied the positioning of Gold Hill. Units older than Pleistocene(?) and younger than Eocene are absent on the Gold Hill fragment (Fig. 1B). The absence of the upper Miocene and Pliocene Etchegoin Formation on the Gold Hill fragment is significant; because these formations crop out within 2 km of exposures of the gabbro of Gold Hill (Sims, 1988). Fossil molluscs from these strata are all of late Miocene Relizian and younger age (E. M. Moore, written communication, 1985, 1986). Clasts in conglomerate of the Etchegoin Formation are composed

only of Franciscan debris and clasts of basement rock derived from the Logan or Gold Hill fragments are absent. The Etchegoin Formation in the San Joaquin Valley ranges in age from about 4 to 10 Ma (Perkins, 1987). Additionally, remnant patches of upper Miocene sedimentary rocks of the Polonio Sandstone Tongue of the Monterey Formation of Marsh (1960) are also present throughout the area southeast of the Jack Ranch fault, but none of these rocks is present on the Gold Hill fragment (Sims, 1988). The absence of these rocks on the Gold Hill fragment suggests that Gold Hill arrived after their deposition between about 5 and 10 Ma. Last, the average Holocene and late Pleistocene slip rate of the San Andreas fault is between 27 and 33 mm/yr (Sieh and Jahns, 1984; Sims, 1987). Such a rate may also be representative of the Pleistocene and Pliocene. If this rate can be extended to the Pliocene, then the 165 km of separation of the Gold Hill and Logan fragments could be accounted for in about 5 Ma (see discussion of phase 3 below). Thus, I tentatively place the end of phase 2, marked by the slivering off of the Gold Hill fragment, at about 5 Ma.

Phase 3. The final phase of movement of the San Andreas follows the separation of the Gold Hill fragment from the Logan body, the abandonment of the Jack Ranch fault, and establishment of the present trace of the San Andreas in the Parkfield area (Fig. 11). At the beginning of this phase, the Pinnacles Volcanics lay about 55 km southeast of Gold Hill. Thus, the granitic and metamorphic basement of the Gabilan Range lay exposed from a point a few kilometers to more than 50 km southeast of Gold Hill. During this phase, the distance between the Pinnacles Volcanics and the volcanic rocks of Lang Canyon continued to re-

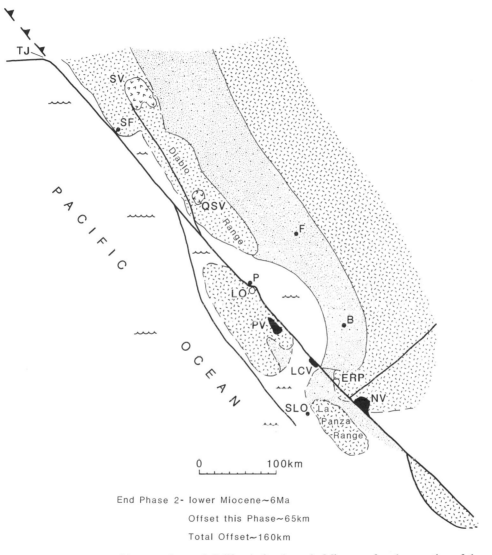

End Phase 2- lower Miocene~6Ma

Offset this Phase~65km

Total Offset~160km

Figure 11. Paleogeographic map of central California for the early Miocene after the eruption of the Neenach/Pinnacles and Tecuya/San Juan Bautista Volcanics. Shorelines and depositional units taken from Bartow (1987) and Perkins (1987). C = Coalinga; B = Bakersfield; ERP = gabbro of Eagle Rest Peak; F = Fresno; LO = gabbro of Logan; NV = Neenach Volcanics; P = Parkfield; PR = Paso Robles; PV = Pinnacles Volcanics; QSV = Quien Sabe Volcanics; SC = Santa Cruz; SF = San Francisco; SLO = San Luis Obispo; SV = Sonoma Volcanics; TJ = Mendocino triple junction. Phase 2—strike-slip movement is transferred to a new segments of the San Andreas and the older segment occupied during Phase 1 becomes inactive. The new segment severs the sliver composed of the now designated volcanic rocks of Lang Canyon from the Neenach Volcanics. The distance between the Pinnacles Volcanics and the volcanics of Lang Canyon remains constant. The end of Phase 2 is chosen as the time of slivering of the Gold Hill sliver from Logan body about 7 m.y. ago. The distance between the Logan unit and Gold Hill unit at the end of stage 1 is 65 km. This stage is estimated at about 7 m.y. long. Thus the minimum slip rate for Phase 2 is 8 mm/yr.

main constant, and the position of the San Andreas fault was probably similar to its present configuration. During phase 3 the Pinnacles Volcanics fragment was transported 160 km from about 55 km southeast of the Gold Hill fragment to its present position (Fig. 11). Passage of the Gabilan Range through the Parkfield area is recorded by the deposition of the Varian Ranch beds of Dickinson (1966). Debris was eroded off the Gabilan Range and was deposited in a small basin now occupied by the Parkfield syncline. Formation of this basin and the subsequent

Parkfield syncline may have been in response to the change in absolute motion of the Pacific plate and the associated noncollisional orogeny at about 4 Ma (Harbet and Cox, 1989). The Varian Ranch beds unit is restricted to the structural trough formed by the Parkfield Syncline. The unit is poorly exposed but does not contain flow-banded rhyolite clasts (Sims, 1990). Thus, they were probably derived from the northern Gabilan Range prior to passage of the Pinnacles Volcanics through the Parkfield area (Fig. 12). A second indicator that the northern Gabilan

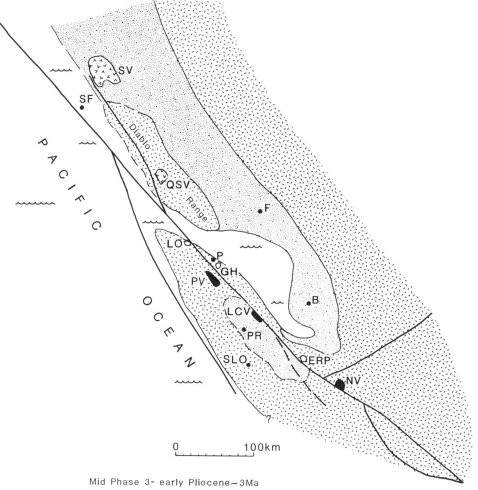

Mid Phase 3- early Pliocene~3Ma

Offset this Phase~25km

Total Offset~185km

Figure 12. Paleogeographic map of central California for the early Miocene after the eruption of the Neenach/Pinnacles and Tecuya/San Juan Bautista Volcanics. Shorelines and depositional units taken from Bartow (1987), and Perkins (1987). C = Coalinga, B = Bakersfield, ERP = gabbro of Eagle Rest Peak, F = Fresno, LO = gabbro of Logan, NV = Neenach Volcanics, P = Parkfield, PR = Paso Robles, PV = Pinnacles Volcanics, QSV = Quien Sabe Volcanics, SC = Santa Cruz, SF = San Francisco, SLO = San Luis Obispo, SV = Sonoma Volcanics, TJ = Mendocino triple junction. Phase 3—follows the slivering off of the Gold Hill block. Right-lateral movement continues to the present time. The distance between the Pinnacles Volcanics and the volcanic rocks of Lang Canyon remains the same as does the distance between the gabbro of Logan and the Pinnacles Volcanics. The distance from the Gold Hill unit to the present location the gabbro of Logan is 165 km, which is also the distance required to bring the volcanic rocks of Lang Canyon to its present position. This phase is estimated as being about 5 m.y. Thus the slip rate for Phase 3 is 33 mm/yr. See text for discussion.

Range passed through this area are allochthonous granite and marble slabs and the large blocks along and near the northeast side of the San Andreas fault in the Parkfield area. These slabs and blocks overlie Late Miocene strata of the Monterey and Etchegoin Formations, indicating late Miocene to early Pliocene deposition. The exact relationship of these allochthonous blocks with the Varian Ranch beds is largely obscured by younger surficial deposits. However, the lithology of the allochthons, as well as that of the coarse detritus in the Varian Ranch beds, strongly suggests derivation from basement rocks of the Gabilan Range. These basement rocks could not have arrived at the latitude of Parkfield until after the Gold Hill fragment was slivered off the Logan fragment. The Varian Ranch beds overlie upper Miocene age strata of the Etchegoin Formation (E. M. Moore, written communication, 1985, 1986). The Etchegoin Formation, as noted above, contains coarse detritus derived from the Franciscan assemblage and lacks debris derived from gabbroic or granitic basement.

Most of the slip during phase 3 occurred on the present-day San Andreas fault, although a lesser amount may have occurred on the Southwest fracture zone in the Parkfield area. However, there is little evidence for the amount of movement on the Southwest fracture zone because of the lack of distinctive offset markers. The southwest fracture zone is considered to be an extension of the San Andreas fault to the northwest of the right stepover in Cholame Valley (Sims, 1988) (Fig. 10). The segment of the San Andreas that on lies the west side of Cholame Valley may be extended through several young fault scarps on the northwest side of Cholame Valley to connect with the southernmost mapped exposure of the Southwest fracture zone (Sims, 1988).

Average slip rates

Incremental offset along the San Andreas fault since early Miocene time may be estimated from the positions of exotic fragments and sedimentary facies that contain clasts derived from these exotic rocks. Estimates of post–early Miocene average slip across the San Andreas fault zone have long been based on the 315-km offset the Pinnacles and Neenach Volcanics and on the assumption that slip began about the same time as the eruption of the Neenach and Pinnacles (Huffman, 1972; Matthews, 1973a). These data and assumptions yield an average slip rate of about 12 to 14 mm/yr on the San Andreas for the entire period from early Miocene time to present, a maximum of about 22 Ma. This rate is at variance with slip rates determined from offset Holocene and Pleistocene strata as well as geodetic measurements of rates of slip. Holocene and late Pleistocene slip rates range from 22 ± 4 mm/yr near Hollister (Perkins and others, 1989), to 27 ± 4 mm/yr near Parkfield (Sims, 1987), to 33.9 ± 1.9 mm/yr on the Carrizo Plain (Sieh and Jahns, 1984). Geodetic rates are reported at 28 to 33 mm/yr for seismogenic depths in central California (Segal and Harris, 1986; Harris, 1986).

The reconstruction and slip history proposed here suggests varying slip rates for the three phases of post–early Miocene

evolution of the San Andreas fault in central California. Each phase is characterized by the movement of specific exotic fragments along the fault, and by a different rate of average slip (Fig. 13). Evidence from transported tectonic fragments of crystalline basement, lower Miocene volcanic rocks, and displaced formations and facies of sedimentary rocks suggest that a more detailed sequence of events may be determined. This sequence of events suggests a step-like rate of movement on the San Andreas system since early Miocene time. Based on the data presented here, the minimum slip rate was about 10 mm/yr for phase 1, about 8 mm/yr for phase 2, and about equal to the present rate of 33 mm/yr for phase 3 (Table 3).

The slip rate for phase 1 is highly dependent on fixing the time of initiation of movement along the San Andreas fault. This is commonly assumed to be shortly after the eruption of the volcanic rocks. However, the relationship between the eruption of volcanic rocks, the passage and location of the triple junction, and the lengthening of the San Andreas transform fault is not precisely understood. Thus, placing the initiation of movement to before, during, or after eruption of the two volcanic units will change the slip rate slightly. The other constraint on the slip rate for phase 1 is the end of the phase. I place the end of phase 1 at the cessation of deposition of the flow-banded rhyolite clasts in the Santa Margarita Formation in the southern Temblor Range. Again, the age of this event is not well defined and may have occurred as much as 2 Ma on either side of the time here chosen.

The average slip rate for phase 2 is similarly dependent on the timing of the events that are used to define the phase. However, the beginning and end of phase 2 are perhaps better constrained by specific events—the cessation of deposition of flow-banded rhyolite clasts in the southern Temblor Range and the slivering off of the Gold Hill fragment. The time of initiation of phase 2, in sharing a common event with the end of phase 1,

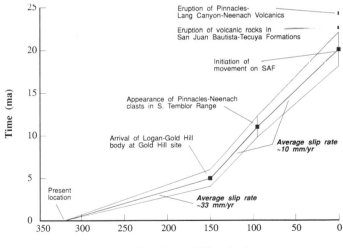

Figure 13. Slip history of the San Andreas transform system in central California since inception of strike-slip movement less than 20 m.y. ago. The dates of eruption of the Neenach/Lang Canyon/Pinnacles Volcanics and the volcanic member of the San Juan Bautista/Tecuya Formation are shown for reference.

TABLE 3. NET INCREMENTAL SLIP AND AVERAGE SLIP RATE FOR EACH PHASE OF THE EVOLUTION OF THE SAN ANDREAS FAULT IN CENTRAL CALIFORNIA*

Phase	Net Slip (km)	Period of Movement (Ma)	Period Length (Ma)	Evidence[†]	Slip Rate (mm/yr)
1	95	24 - 10	14	A	7
1	95	22 - 12	10	A	10
1	95	18 - 10	8	A	12
2	55	12 - 6	6	B	9
2	55	12 - 5	7	B	8
2	55	10 - 6	4	B	14
3	165	6 - present	6	C	27
3	165	5 - present	5	C	33
3	165	4 - present	4	C	41

*Three estimates of period of movement and the subsequent time length and average slip rate is given for each phase. The three estimates for each phase of movement are the maximum (upper value), preferred (middle value), and minimum (lower value).

[†]A = Clasts of flow-banded rhyolite in Santa Margarita Formation of southern Temblor Range. Eruption of Neenach and Pinnacles Volcanics and the volcanic rocks near San Juan Bautista and in the Tecuya Formation.

B = End of deposition of flow-banded clasts in the Santa Margarita Formation in the southern Temblor Range. Slivering off of the volcanic rocks of Lang Canyon and the ancestral San Andreas fault is abandoned.

C = Slivering off of gabbro of Gold Hill from gabbro of Logan and the Jack Ranch segment of the San Andreas fault system is abandoned. Etchegoin Formation is widely distributed in the Gold Hill area but are not present on Gold Hill.

Temblor Formation and Polonio Sandstone member of Monterey Formation are widely distributed in Gold Hill area but not present on Gold Hill.

Varian Ranch "beds" of Dickinson (1966) contain eroded granitic debris similar to the basement rocks of the Gabilan Range but do not contain flow-banded rhyolite clasts derived from the Pinnacles Volcanics.

shares the uncertainty associated with the time that event. The end of phase 2 is marked by the slivering off of Gold Hill from the Logan fragment. The time of this event is further constrained by the distribution of upper Miocene and Pliocene strata in the Parkfield area and the differing style and amount of folding in them. Upper Miocene and Pliocene(?) strata of the Etchegoin Formation and the Varian Ranch beds of Dickinson (1966) lie in the core of the Parkfield syncline. The Parkfield syncline is slightly asymmetrical with steeply dipping limbs. It lies northeast of Gold Hill and is subparallel to the San Andreas fault. The structural style of the fold is similar to the faults and minor folds northeast of Gold Hill in that their axes are curved in a similar manner to the Jack Ranch and Gold Hill faults (Sims, 1990). The curvature of the fold axes and faults decreases to the northeast away from Gold Hill (Fig. 3). The similar but decreasing curvature of the faults and fold axes suggests that they were deformed by northeast-directed translation of the Gold Hill–Logan fragment prior to slivering off of the Logan body (Fig. 5). Thus, the uncertainty in timing of end of phase 2 largely controlled by the age of the Miocene Pliocene boundary and the age of the Varian Ranch beds of Dickinson. This uncertainty is estimated at about 2 Ma (Sims, 1990). However, the uncertainty in the endpoints of phase 2 does not produce a remarkable change in the overall slip rate (Fig. 10).

The uncertainty in the timing of phase 3, which ends at the present, is primarily in its starting time. The slip rate proposed for phase 3 is similar to the present rate and similar to the rate for the last 5 Ma proposed by Weldon and Meisling (1986) for southern California. Most important, however, is that the rates proposed for phases 1 and 2 are similar and about one-third to one-fourth of those proposed for phase 3—the past 5 Ma. The start of phase 3 roughly coincides with the change in direction and speed of the Pacific plate at about 5 Ma (Minster and Jordan, 1973; Stock and Molnar, 1989). The realignment of the Pacific plate with respect to the North American plate resulted in transpression that deformed the rocks lying along the boundary between the two plates (Harbert and Cox, 1989). Deformation associated with this episode of transpression in the Parkfield area resulted in folding and uplift of the Middle Mountain area, closure of the Parkfield Syncline, and eastward translation of the Gold Hill fragment (Sims, 1988, 1990). This transpressional episode was accompanied by an increase in velocity of the Pacific plate and thus the San Andreas fault.

ACKNOWLEDGMENTS

Many people helped and encouraged me in the course of this work. Foremost, I thank my father, John R. Sims, for his encouragement and support and for instilling a sense of wonder and curiosity about the natural universe in his children. I also thank D. C. Ross, R. C. Jachens, J. C. Matti, R. E. Powell, and J. C. Crowell for their numerous conversations and forbearance in answering my questions. The manuscript was greatly improved by the comments and criticism of D. C. Ross, J. A. Bartow, R. J. Weldon, R. E. Powell, C. S. Prentice, and Peter Weigand. Drafting and editing by J. C. Hamilton, D. E. Meier, and A. J. Foss are gratefully acknowledged. Special thanks are due to the Hearst Corporation, owners, and to Larry Ford, foreman, of the Jack Ranch for permission to roam freely over the ranch in the course of this work. Thanks also to Mr. and Mrs. Arthur Claasen for access to their ranch. My work would have been impossible without the goodwill and assistance of these and other ranch owners.

APPENDIX I

Method of Combining K/Ar Dates

The procedures of Wilson and Ward (1978) were used to examine the K/Ar age dates for the Pinnacles Volcanics, volcanic rocks in the Tecuya Formation of the San Emigdio Mountains, and the volcanic rocks in the San Juan Bautista area. The procedure requires the calculation of a pooled mean date, A_p, from the following:

$$A_p = \frac{\Sigma \, (A_i/E_i^2)}{\Sigma(1/E_i^2)} \, .$$

The pooled mean date A_p is then used in

$$T = \Sigma[(A_i-A_p)^2/E_i^2],$$

which has a χ^2 distribution on n-1 degrees of freedom. If $T < \chi^2$, the null hypothesis is accepted and the dates may be combined. If $T > \chi^2$, the null hypothesis is rejected and combination of the dates is not valid. When the null hypothesis is accepted the pooled standard deviation, σ_p, of A_p is calculated from

$$\sigma_p = 1/\Sigma(1/E_i^2)^{\frac{1}{2}}.$$

The dates for the Pinnacles Volcanics are:

22.1 ± 3.2 (KA 2079)
24.1 ± 0.7 (KA 2087)
24.3 ± 0.7 (KA 2091R)
24.5 ± 1.2 (KA 2192).

The calculated statistical values are $A_p = 24.2$, $T = 0.53$, df = 3, $\chi^2 = 7.81$ at p = 0.05, and $\sigma_p = 0.21$. Thus, the combined age date is 24.2 ± 0.2 Ma.

Dates for the volcanic member of the Tecuya Formation in the San Emigdio Mountains are:

22.9 ± 0.7 (KA 2175)
22.4 ± 0.7 (KA 2166)
24.6 ± 7.2 (KA 2169)
25.2 ± 2.9 (KA 2162)
22.1 ± 0.6 (KA 2114)
22.5 ± 0.7 (KA 2115)
18.1 ± 7.6 (KA 2215).

The calculated statistical values are $A_p = 22.5$, $T = 2.08$, df = 5, $\chi^2 = 11.07$ at p = 0.05, and $\sigma_p = 0.11$. Thus the combined date is 22.5 ± 0.1 Ma.

Dates for the volcanic member of the Zayante Sandstone of Dibblee and others (1979) near San Juan Bautista are:

22.2 ± 0.7 (KA 2157)
23.5 ± 1.4 (SJB-1)
21.3 ± 1.3 (SJB-2).

The calculated statistical values are $A_p = 22.2$, $T = 1.35$, df = 2, $\chi^2 = 5.99$ at p = 0.05, and $\sigma_p = 0.32$. Thus the combined date is 22.2 ± 0.3 Ma.

REFERENCES CITED

Addicott, W. O., 1968, Mid-Tertiary zoogeographic and paleogeographic discontinuities across the San Andreas fault, California, *in* Dickinson, W. R., and Grantz, A., eds., Proceedings of a Conference on Geologic Problems of the San Andreas Fault System: Stanford, California, Stanford University Publications in the Geological Sciences, v. 11, p. 144–165.

Allen, J. E., 1946, Geology of the San Juan Bautista Quadrangle, California: Division of Mines and Geology Bulletin 133, p. 9–76.

Andrews, P., 1936, Geology of the Pinnacles National Monument: California University Department of Geological Sciences Bulletin, v. 24, no. 1, p. 1–38.

Atwater, T., 1970, Implications of plate tectonics for the Cenozoic evolution of western North America: Geological Society of America Bulletin, v. 81, p. 3513–3536.

Bartow, J. A., 1987, Cenozoic nonmarine sedimentation in the San Joaquin basin, central California, *in* Ingersoll, R. V., and Ernst, W. G., Cenozoic basin development of coastal California; Rubey Volume 6: Englewood Cliffs, New Jersey, Prentice-Hall, p. 146–171.

Bazeley, W.J.M., 1961, 175 miles of lateral movement along the San Andreas fault since Lower Miocene: Pacific Petroleum Geologist, v. 15, p. 2–3.

Berry, F.A.F., Huffman, O. F., and Turner, D. L., 1968, Post-Miocene movement along the San Andreas fault, California, *in* Abstracts for 1966: Geological Society of America Special Paper 101, p. 15.

Bishop, C. C., and Davis, J. F., compilers, 1984, Central California province correlation chart, *in the collection,* Correlation chart series; Correlation of stratigraphic units in North America: Tulsa, Oklahoma, American Association of Petroleum Geologists, sheet 2.

Brown, R. D., Jr., and 8 others, 1967, The Parkfield-Cholame California, earthquakes of June-August 1966; Surface geologic effects, water resources aspects, and preliminary seismic data: U.S. Geological Survey Professional Paper 579, 66 p.

Clark, J. C., and Reitman, J. D., 1973, Oligocene stratigraphy, tectonics, and paleogeography southwest of the San Andreas fault, Santa Cruz and Gabilan Range, California Coast Ranges: U.S. Geological Survey Professional Paper 783, 18 p.

Crowell, J. C., 1952, Geology of the Lebec Quadrangle: California Division of Mines and Geology Special Report 24, 23 p.

Dalrymple, G. B., 1979, Critical tables for conversion of K-Ar ages from old to new constants: Geology, v. 7, p. 558–560.

Decelles, P. E., 1986, Middle Tertiary depositional systems of the San Emigdio Range, southern California: Society of Economic Paleontologists and Mineralogists, Pacific Section Guidebook 47, 32 p.

Dibblee, T. W., Jr., 1967, Areal geology of the western Mojave Desert: U.S. Geological Survey Professional Paper 522, 153 p.

—— , 1971, Geologic maps of 17 15-minute quadrangles along the San Andreas fault in the vicinity of King City, Coalinga, Panoche Valley and Paso Robles, with index map: U.S. Geological Survey Open-File Report 71-87, scale 1:62,500.

Dibblee, T. W., Jr., Nilsen, T. N., and Brabb, E. E., 1979, Geologic map of the San Juan Bautista Quadrangle, San Benito and Monterey Counties, California: U.S. Geological Survey Open-File Report 79-375, scale 1:24,000.

Dickinson, W. R., 1966, Structural relationship of San Andreas fault system, Cholame Valley and Castle Mountain Range, California: Geological Society of America Bulletin, v. 77, p. 707–736.

Dickinson, W. R., and Snyder, W. S., 1979a, Geometry of subducted slabs related to San Andreas Transform: Journal of Geology, v. 87, p. 609–627.

—— , 1979b, Geometry of triple junctions related to San Andreas Transform: Journal of Geophysical Research, v. 84, no. B2, p. 561–572.

Fletcher, G. L., 1962, The Recruit Pass area of the Temblor Range, San Luis Obispo and Kern Counties, California, *in* Geology of the Carrizo Plains and San Andreas fault: San Joaquin Geological Society, Pacific Sections, American Association of Petroleum Geologists, and Society of Economic Paleontologists and Mineralogists Field Guidebook, p. 16–20.

—— , 1967, Post–late Miocene displacement along the San Andreas fault zone, central California: Pacific Sections, American Association of Petroleum Geologists and Society of Economic Paleontologists and Mineralogists Guidebook, p. 74–80.

Fox, K. F., Jr., Fleck, R. J., Curtis, G. H., and Meyer, C. E., 1985, Implications of the northwestwardly younger age of the volcanic rocks of west-central California: Geological Society of America Bulletin, v. 96, p. 647–654.

Graham, S. A., Stanley, R. G., Bent, J. V., and Carter, J. B., 1989, Oligocene and Miocene paleogeography of central California and displacement along the San Andreas fault: Geological Society of America Bulletin, v. 100, p. 711–730.

Harbert, W., and Cox, A., 1989, Late Neogene motion of the Pacific plate: Journal of Geophysical Research, v. 94, no. B3, p. 3052–3064.

Harris, R. A., 1986, Slip at depth along the San Andreas fault south of Parkfield: EOS Transactions of the American Geophysical Union, v. 67, p. 1215–1216.

Hill, M. L., and Dibblee, T. W., Jr., 1953, San Andreas, Garlock, and Big Pine faults, California: Geological Society of America Bulletin, v. 64, p. 443–458.

Huffman, O. F., 1972, Lateral displacement of upper Miocene rocks and the Neogene history of offset along the San Andreas fault in central California: Geological Society of America Bulletin, v. 83, p. 2913–1946.

James, E. W., and Mattinson, J. M., 1985, Evidence for 160-km post–mid Cretaceous slip on the San Gregorio–Hosgri fault, coastal California: EOS Transactions of the American Geophysical Union, v. 66, p. 1093.

James, E. W., Kimborough, D. L., and Mattinson, J. M., 1986, Evaluation of Tertiary piercing points along the northern San Andreas fault using zircon dating, initial Sr, and common Pb isotopic ratios: Geological Society of America Abstracts with Programs, v. 18, p. 121.

Kerr, P. F., and Schenck, H. G., 1925, Active thrust faults in San Benito County, California: Geological Society of America Bulletin, v. 36, p. 465–494.

Mash, O. T., 1960, Geology of the Orchard Peak area, California: California Division of Mines and Geology Special Report 62, 42 p.

Matthews, V., III, 1973a, Pinnacles-Neenach correlation; A restriction for models of the origin of the Transverse Ranges and the big bend of the San Andreas fault: Geological Society of America Bulletin, v. 84, p. 683–688.

—— , 1973b, Geology of the Pinnacles Volcanic Formation and the Neenach Volcanics Formation and their bearing on the San Andreas fault problem [Ph.D. thesis]: Santa Cruz, University of California, 214 p.

—— , 1976, Correlation of the Pinnacles and Neenach Volcanic Formations and their bearing on the San Andreas fault problem: American Association of Petroleum Geologists Bulletin, v. 60, p. 2128–2141.

Matti, J. C., Frizzell, V. A., and Mattinson, J. M., 1986, Distinctive Triassic megaporphyritic monzogranite displaced 106 ± 10 km by the San Andreas fault, southern California; A new constraint for palinspastic reconstructions: Geological Society of America Abstracts with Programs, v. 18, p. 154.

McWilliams, M., and Li, Y., 1986, Oroclinal bending of the southern Sierra Nevada batholith: Science, v. 230, p. 172–175.

Minster, J. B., and Jordan, T. H., 1978, Present-day plate motions: Journal of Geophysical Research, v. 83, no. B11, p. 5331–5354.

Nilsen, T. H., 1984, Offset along the San Andreas fault of Eocene strata from the San Juan Bautista area and western San Emigdio Mountains, California: Geological Society of America Bulletin, v. 95, p. 599–609.

—— , 1987, Stratigraphy and sedimentology of the Eocene Tejon Formation, San Emigdio and western Tehachapi Mountains, California: U.S. Geological Survey Professional Paper 1268, 110 p.

Nilsen, T. H., Dibblee, T. W., Jr., and Addicott, W. O., 1973, Lower and middle Tertiary stratigraphic units of the San Emigdio and western Tehachapi Mountains, California: U.S. Geological Survey Bulletin 1372-H, 23 p.

North American Stratigraphic Commission, 1983, North American stratigraphic code: American Association of Petroleum Geologists Bulletin, v. 67, p. 841–875.

O'Day, P. A., and Sims, J. D., 1986, Sandstone compositions and paleogeography of the Temblor Formation in central California; Evidence for early middle Miocene right-lateral displacement on the San Andreas fault System: Geological Society of America Abstracts with Programs, v. 18, p. 165.

Perkins, J. A., 1987, Provenance of the upper Miocene and Pliocene Etchegoin Formation; Implications for paleogeography of the late Miocene of central California: U.S. Geological Survey Open-File Report 87-167, 86 p.

Perkins, J. A., Sims, J. D., and Sturgess, S. S., 1989, Late Holocene movement along the San Andreas fault at Melendy Ranch; Implications for the distribution of fault slip in central California: Journal of Geophysical Research, v. 94, no. B8, p. 10217–10230.

Plescia, J. B., and Calderone, G. J., 1986, Paleomagnetic constraints on the timing and extent of rotation of the Tehachapi Mountains, California: Geological Society of America Abstracts with Programs, v. 18, p. 171.

Powell, R. E., 1986, Palinspastic reconstruction of crystalline-rock assemblages in southern California; San Andreas fault as part of an evolving system of late

Cenozoic conjugate strike-slip faults: Geological Society of America Abstracts with Programs, v. 18, p. 172.

Ross, D. C., 1970, Quartz gabbro and anorthositic gabbro; Markers of offset along the San Andreas fault in the California Coast Ranges: Geological Society of America Bulletin, v. 81, p. 3647–3662.

—— , 1976, Metagraywacke of the Salinian block, central Coast Ranges, California; A possible correlative across the San Andreas fault: U.S. Geological Survey Journal of Research, v. 4, p. 683–696.

—— , 1984, Possible correlations of basement rocks across the San Andreas, San Gregorio–Hosgri, and Rinconada–Reliz–King City faults, California: U.S. Geological Survey Professional Paper 1317, 37 p.

Ryder, R. T., and Thomson, A., 1989, Tectonic controlled fan delta and submarine fan sedimentation of late Miocene age, southern Temblor California: U.S. Geological Survey Professional Paper 1442, 59 p.

Segal, P., and Harris, R., 1986, Slip deficit on the San Andreas fault, Parkfield, California, as revealed by an inversion of geodetic data: Science, v. 233, p. 1409–1413.

Sieh, K. E., and Jahns, R. H., 1984, Holocene activity of the San Andreas fault at Wallace Creek, California: Geological Society of America Bulletin, v. 94, p. 883–896.

Sims, J. D., 1986, The Parkfield shuffle; Displaced rock bodies as a clue to the post-Eocene history of movement on the San Andreas fault in central California: Geological Society of America Abstracts with Programs, v. 18, p. 185.

—— , 1987, Late Holocene slip along the San Andreas fault near Cholame, California: Geological Society of America Abstracts with Programs, v. 19, p. 415.

—— , 1988, Geologic map of the San Andreas fault zone in the Cholame Valley and Cholame Hills Quadrangles, Monterey and San Luis Obispo Counties, California: U.S. Geological Survey Miscellaneous Field Studies Map MF-1995, scale 1:24,000.

—— , 1990, Geologic map of the San Andreas fault zone in the Parkfield 7½-minute Quadrangle, Monterey and Fresno Counties, California: U.S. Geological Survey Miscellaneous Field Studies Map, scale 1:24,000.

Stanley, R. G., 1987, Implications of the northwestwardly decreasing age of the volcanic rocks of west-central California; Alternative interpretation: Geological Society of America Bulletin, v. 98, p. 612–614.

Stock, J., and Molnar, P., 1989, Uncertainties and implications of the late Cretaceous and Tertiary position of North America relative to the Farallon, Kula, and Pacific plates: Tectonics, v. 7, no. 6, p. 1339–1384.

Tennyson, M. E., 1989, Pre-transform early Miocene extension in western California: Geology, v. 17, no. 9, p. 792–796.

Thomas, J. M., 1986, Correlation and petrogenesis of the Miocene volcanic rocks in the San Emigdion and San Juan Bautista areas, California [M.S. thesis]: Northridge, California State University, 94 p.

Turner, D. L., 1968, Potassium argon dates concerning the Tertiary foraminiferal time scale and San Andreas fault displacement [Ph.D. thesis]: Berkeley, University of California, 99 p.

—— , 1970, Potassium argon dating of Pacific coast Miocene foraminiferal stages, *in* Bandy, O. L., ed., Radiometric dating and paleontologic zonation: Geological Society of America Special Paper 124, p. 91–129.

Turner, D. L., Curtis, G. H., Berry, F.A.F., and Jack, R. N., 1970, Age relationship between the Pinnacles and Parkfield felsites and felsite clasts in the southern Temblor Range, California; Implications for San Andreas fault displacement: Geological Society of America Abstracts with Programs, v. 2, p. 154.

Weldon, R. J., and Meisling, K. E., 1986, Rates of slip on the San Andreas fault in the Transverse Ranges during the last 4 million years: Geological Society of America Abstracts with Programs, v. 18, p. 197.

Wiese, J. H., 1950, Geology and mineral resources of the Neenach Quadrangle, California: California Division of Mines and Geology Bulletin 153, p. 53.

Wilson, G. K., and Ward, S. R., 1978, Procedures for comparing and combining radiocarbon age determinations; A critique: Archaeometry, v. 20, p. 19–31.

MANUSCRIPT ACCEPTED BY THE SOCIETY SEPTEMBER 6, 1990

Geological Society of America
Memoir 178
1993

Chapter 7

Evaluation of displacements of pre-Tertiary rocks on the northern San Andreas fault using U-Pb zircon dating, initial Sr, and common Pb isotopic ratios

Eric W. James*, David L. Kimbrough*, and James M. Mattinson
Department of Geological Sciences, University of California, Santa Barbara, California 93106

ABSTRACT

The Eagle Rest Peak igneous complex in the San Emigdio Mountains, and Logan gabbro, near San Juan Bautista, California, crop out on opposite sides of the San Andreas fault. They have identical U-Pb zircon ages of 161 Ma and similar Sr initial isotopic ratios. These data support previous correlations of these rocks (Ross, 1970) and require 305 km of post-Jurassic slip on the northern San Andreas fault, a figure equal to the total slip on the southern segment.

The Eagle Rest Peak complex has previously been proposed as the source for gabbro clasts in the Upper Cretaceous Gualala Formation near Point Arena (Ross, 1970; Ross and others, 1973). This implies >440 km of post–Late Cretaceous slip on the San Andreas fault and the existence of a proto–San Andreas fault. Also, granitic to quartz dioritic clasts in the Gualala Formation have been interpreted as detritus derived from the Salinian block in the Cretaceous.

However, new U-Pb zircon data from Gualala Formation gabbroic clasts indicate minimum ages of 163 and 165 Ma, slightly older than the 161-Ma age of the Eagle Rest Peak complex and Logan gabbro. Published K-Ar ages also suggest the Gualala cobbles are older than the Eagle Rest Peak complex. These data and the presence of alternate sources for the Gualala cobbles indicate that the Gualala–Eagle Rest Peak tie is not suitable for determining slip on the San Andreas fault.

Despite the differences between specific areas, the Eagle Rest Peak complex, Gold Hill and Logan gabbros, and the Gualala gabbro clasts are similar in age and lithology to mafic-ultramafic complexes that form a widespread part of the Jurassic Sierran-Klamath arc. Although Gualala gabbro clasts cannot be uniquely matched to the Eagle Rest Peak complex and Logan rocks, they probably are derived from similar rocks cropping out in the Sierra foothills or buried in the Great Valley.

U-Pb–age, Sr, and Pb isotopic data from a single Gualala Formation granodiorite clast do not support a Salinian provenance. The clast is older than 154 Ma, older than the 80- to 120-Ma Salinian granites. The clast also has less radiogenic Pb and Sr isotopic ratios than plutonic rocks from the Salinian block. A reevaluation of paleocurrent data and clast types from the Gualala Formation also suggests a non-Salinian source. The Gualala area is apparently not part of the Salinian block.

*Present addresses: James, Bureau of Economic Geology, University of Texas at Austin, Austin, Texas 78713; Kimbrough, Department of Geological Sciences, San Diego State University, San Diego, California 92182.

James, E. W., Kimbrough, D. L., and Mattinson, J. M., 1993, Evaluation of displacements of pre-Tertiary rocks on the northern San Andreas fault using U-Pb zircon dating, initial Sr, and common Pb isotopic ratios, *in* Powell, R. E., Weldon, R. J., II, and Matti, J. C., eds., The San Andreas Fault System: Displacement, Palinspastic Reconstruction, and Geologic Evolution: Boulder, Colorado, Geological Society of America Memoir 178.

INTRODUCTION

Cretaceous conglomeratic strata on the northern California coast near Gualala and three outcrops of Mesozoic gabbroic basement along the San Andreas fault provide key evidence to help decipher the movement history of the northern San Andreas fault (Fig. 1). These rocks comprise the only generally accepted pre-Eocene correlation across the San Andreas fault north of the California Transverse Ranges (Ross, 1970; Nilsen, 1978). It has

been established that the Eocene through Miocene rocks north of the Transverse Ranges and Precambrian through Miocene rocks south of Transverse Ranges are offset about 300 km (Crowell and Walker, 1962; Clarke and Nilsen, 1973; Powell, this volume). In contrast, the correlation of gabbro outcrops near Eagle Rest Peak at the southern end of the San Joaquin Valley with gabbroic cobbles in Cretaceous sediments near Gualala suggests as much as 560 km of pre-Eocene slip on the northern part of the San Andreas fault (Wentworth, 1966; Ross, 1970; Ross and others,

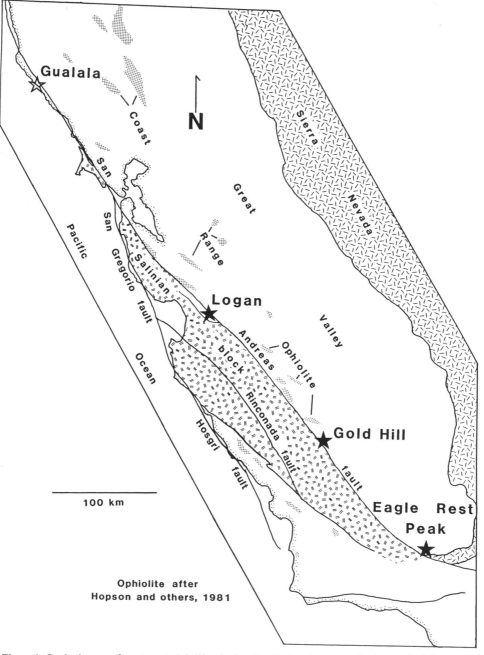

Figure 1. Geologic map of west-central California showing the locations of major faults, Gualala, Logan, Gold Hill, Eagle Rest Peak complex, Coast Range ophiolite, and Sierra Nevada batholith.

1973). Recognition of faults that feed slip into the northern San Andreas fault—the San Gregorio–Hosgri fault and the Rinconada fault (Graham and Dickinson, 1978; Dibblee, 1976; Ross, 1984)—reduces the mismatch of offset, but still leaves a 100-km discrepancy.

Coupled with this dilemma is the problematic paleogeographic relationship of the Salinian block to the rest of California. Interbedded with the gabbroic conglomerates at Gualala are conglomerates made up of granitic to tonalitic clasts that were considered by Wentworth (1966, 1968) to have a source within the Salinian block. If the provenance determinations for both the gabbroic and granitic clasts are correct (Ross and others, 1973), then the Salinian block was in contact with rocks east of the San Andreas fault in the late Cretaceous, i.e., Campanian-Maastrichtian time.

Further complications in the paleogeography are suggested by paleomagnetic evidence that indicate possible large displacements (2,500 to 4,100 km) of parts of the California Coast Ranges (McWilliams and Howell, 1982; Vedder and others, 1983; Champion and others, 1984; Kanter and Debiche, 1985). These interpretations of the paleomagnetic data bear directly on the paleogeography of the Gualala and San Emigdio Mountains areas and are examined below.

The possible correlation of pre-Eocene rocks across the northern San Andreas fault is thus an important link in several views of the paleogeography of California. This chapter presents new isotopic and age data on rocks from the Eagle Rest Peak complex, Gold Hill, Logan, and Gualala, and discusses their implications.

PREVIOUS WORK

Correlation of Cretaceous sedimentary rocks near Gualala with sources in the Salinian block and areas south of the Diablo Range east of the San Andreas fault (Wentworth, 1966, 1968; Ross, 1970; Ross and others, 1973) is the most important evidence for large (440 to 560 km) displacements on the San Andreas fault system. These models have been extrapolated by others to indicate the presence of a pre-Tertiary "proto–San Andreas fault" and to include the Gualala area in the Salinian block. Ross (1970) and Ross and others (1973) also interpreted outcrops of gabbroic rocks along the San Andreas fault at Logan, near San Juan Bautista, and Gold Hill, near Cholame, as correlative with basement rocks in the San Emigdio Mountains (Fig. 1). A review of the data and interpretations that led to these correlations, as well as several more recently proposed correlations, follows.

Gualala Formation

The Gualala Formation is situated on the north coast of California, west of the San Andreas fault (Fig. 1). It is composed of Upper Campanian to Lower Maastrichtian mudstone, sandstone, and conglomerate (Wentworth, 1966, 1968; Ross and others, 1973) (Fig. 2). These rocks structurally overlie the spilite of

Black Point. This fault-bound exposure of spilitized pillow basalt is the only exposure of "basement" in the Gualala block. The Gualala Formation is overlain conformably by the Paleocene to Eocene German Rancho Formation, which is in turn overlain by younger sedimentary rocks and the Miocene Iverson Basalt.

Wentworth (1966) divided the Gualala Formation into two members based on composition. The Stewarts Point Member contains potassium feldspar arkose (~20 percent K-feldspar) with conglomerates containing clasts of plutonic rocks ranging from granite to tonalite in composition. These rocks interfinger with, and are overlain by, the Anchor Bay Member, which contains plagioclase arkose (<5 percent K-feldspar) and clasts of quartz gabbro. The overlying Paleocene and Eocene German Rancho Formation is primarily K-feldspar arkose containing cobbles of granite and granodiorite.

Paleocurrents and Paleogeography. Paleocurrent indicators are very scarce in the K-feldspar–bearing Stewarts Point Member. This greatly complicates determining its provenance. The overlying Anchor Bay Member contains ripples and sole marks that indicate flow primarily to the west and northwest. Paleocurrents in the German Rancho Formation flowed to the northwest with subsidiary trends to the northeast and southeast. These paleocurrents were interpreted (Wentworth, 1966, 1968; Ross and others, 1973) to indicate deposition in a northwest-trending basin with a gabbroic, plagioclase-rich source on the northeast and a granitic K-feldspar-rich source on the southwest (Figs. 2, 3). Because the Cretaceous Stewarts Point Member con-

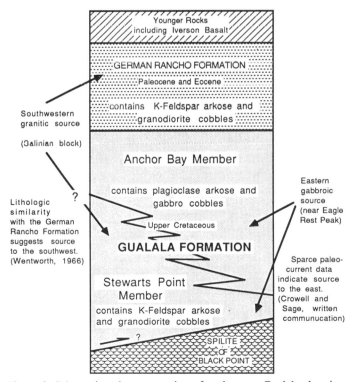

Figure 2. Schematic columnar section of rocks near Gualala showing hypothesized sediment sources.

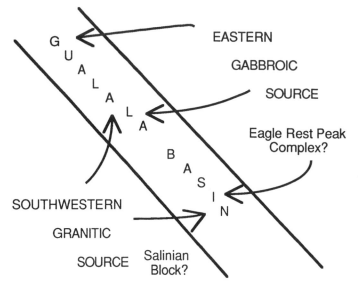

After Wentworth (1966) and Ross and others (1973)

Figure 3. Schematic paleogeographic basin model of Wentworth (1966) from Ross and others (1973).

tains granitic clasts similar to those in the Paleocene to Eocene German Rancho Formation, their source was interpreted to be the same, from the southwest.

Paleogeographic reconstruction based on these data and interpretations (Fig. 3) ties the Gualala Formation to the Salinian block and to areas east of the San Andreas fault (Wentworth 1966, 1968; Ross and others, 1973). The reconstruction places the Salinian block to the southwest of the Gualala basin to supply granitic detritus to the Stewarts Point Member and the German Rancho Formation (Wentworth, 1966, 1968; Ross and others, 1973). The validity of this reconstruction rests on the interpretation that the occurrence of granitic cobbles and K-feldspar arkose in the Stewarts Point Member of the Gualala Formation and in the German Rancho Formation indicates that they have the same source terrane.

Paleocurrent data for the gabbro-clast-bearing Anchor Bay sediments indicate they were derived from an eastern source. Wentworth (1966, 1968) noted the lack of an exposed monolithologic source of mafic, K-feldspar-poor detritus in the Franciscan and Great Valley rocks east of the San Andreas fault or in the Salinian block west of the San Andreas fault. Yet the volume and coarseness of the Anchor Bay sediments required a nearby source "several hundred square miles" in area (Wentworth, 1968, p. 136). Given these constraints, Wentworth (1968) speculated that the nearest source is now no closer than 270 miles (440 km) and no farther than 350 miles (570 km) southeast on the other side of the San Andreas fault.

Gabbro bodies along the San Andreas fault

Ross (1970) and Ross and others (1973) suggested the possibility of a significantly limited source area for the Gualala gabbro cobbles east of the San Andreas fault in the San Emigdio Mountains. This location of the source for cobbles in the Anchor Bay Member of the Gualala Formation, the Eagle Rest Peak igneous complex, together with the interpreted Salinian source for the Stewarts Point Member, suggests the existence of a proto–San Andreas fault. A proto–San Andreas fault explains the juxtaposition of western granitic and eastern gabbroic terranes prior to Late Campanian–Early Maastrichtian time.

They also suggested that slices of Eagle Rest Peak complex quartz gabbro have been left dispersed along the San Andreas fault at Logan, near San Juan Bautista and at Gold Hill, north of Cholame. Ross (1970) and Ross and others (1973) have suggested that these rocks are ophiolitic in character and possibly part of the Coast Range ophiolite.

Eagle Rest Peak complex. The Eagle Rest Peak complex has been mapped and described by Hammond (1958), Ross (1970), Ross and others (1973), Dibblee and Nilsen (1973), and Reitz (1983, 1986). It is composed of a complex, but cogenetic, suite of pyroxenites, gabbros, and quartz diorites. Syn-plutonic deformation has formed complicated intrusive relations between different phases (Reitz, 1983, 1986). The rocks are unconformably overlain and surrounded by Eocene and younger sediments (Hammond, 1958; Nilsen, 1984). Similar gabbros are penetrated by a well to the east, adjacent to the San Andreas fault (C. Hopson and A. Reitz, personal communication, 1983).

Logan and Gold Hill gabbros. The rocks at Logan and Gold Hill are quartz gabbros containing roughly equal amounts of green hornblende and calcic plagioclase, with as much as 20 percent quartz. The rocks also contain clinopyroxene and rare traces of orthopyroxene (Ross, 1970, 1972). The outcrops are adjacent to the San Andreas fault and are separated from other basement outcrops by Tertiary cover (Hay, 1963; Dickinson, 1966; Nilsen, 1984).

Ross (1970) stated that petrography and location along the San Andreas fault leave little doubt about the correlation of the Gold Hill sliver and the Logan gabbro. The correlation of these bodies with the Eagle Rest Peak complex is also reasonable and persuasive.

Correlation of overlying sedimentary rocks. Nilsen (1984) has added considerable support to the Eagle Rest Peak–Logan correlation. He determined that the stratigraphy of the Eocene through Miocene rocks that unconformably overlie the Eagle Rest Peak complex matches that of the sedimentary rocks unconformably overlying the Logan gabbro. This ~300-km offset is in accord with other slip estimates from Tertiary rocks along the northern San Andreas fault (Huffman, 1972; Matthews, 1973) and matches the offsets of Precambrian through Miocene features south of the Transverse Ranges (Crowell and Walker,

1962). The sedimentary rocks that overlie the Gold Hill sliver are also very similar to the Logan and San Emigdio sections (Hay, 1963; Dickinson, 1966).

K-Ar and Sr initial ratios

The petrographic correlations of gabbro from the Eagle Rest Peak complex, Logan, and Gold Hill, and clasts from the Gualala Formation were further tested by Ross (1970) and Ross and others (1973), who presented hornblende K-Ar dates. These dates range from 187 to 134 Ma and are summarized in Table 1. There is considerable overlap in the dates from the different areas, generally substantiating the correlations. The general similarity between the K-Ar dates and the U-Pb dates determined in this study is treated below.

A further similarity between the Eagle Rest Peak complex, Logan, and Gold Hill bodies and Gualala gabbroic cobbles is indicated by their initial Sr isotopic ratios (Kistler and others, 1973; Johnson and O'Neil, 1988; and this study). New Sr data, and those of Kistler and others (1973), are listed in Table 2. They range between 0.7031 and 0.7048. Ratios measured by Johnson and O'Neil (1988) are as low as 0.7020. These contrast with the relatively radiogenic ratios found in the Salinian block (e.g., 0.706 to 0.708, Kistler and others, 1973; Mattinson, 1978; James, 1986) and the central and eastern Sierra Nevada (e.g., >0.706, Kistler and Peterman, 1973).

Paleomagnetic studies

Recent studies have emphasized that the western coast of North America is made up of a collage of geologically unrelated fragments, some of which appear far-traveled (Coney and others, 1980; Jones and others, 1982). These rocks record shallow paleomagnetic inclinations that may be indicative of magnetization at low paleolatitudes. Recent paleomagnetic studies may be important in understanding the history of the Gualala and Eagle Rest Peak areas.

McWilliams and Howell (1982) and Vedder and others (1983) have presented geologic and paleomagnetic data that suggest that the Salinian block and a composite terrane, consisting of the Coast Range ophiolite, Great Valley sequence, and Francis-

TABLE 1. PUBLISHED K/Ar HORNBLENDE AGES OF ROCKS FROM NEAR GUALALA, EAGLE REST PEAK, LOGAN, AND GOLD HILL

Rocks	Sample No.	Ages (Ma)
Cobbles from Gualala Formation, strata of Anchor Bay		
1. Hornblende quartz gabbro	(ABc-1)	175 ± 8
2. Hornblende quartz gabbro	(ABc-4)	141 ± 4
3. Hornblende quartz diabase	(ABi-1)	186 ± 7
Cobble from German Rancho Formtion		
4. Biotite granodiorite	(OC-3)	82 ± 3
Engles Rest Peak area		
5. Hornblende quartz gabbro	(DR-1169A)	134 ± 4
6. Hornblende quartz gabbro	(DR-1182C)	165 ± 6
7. Hornblende–plagioclase–quartz pegmatoid	(DR-1189C)	207 ± 10*
Logan area		
8. Hornblende quartz gabbro	(DR-998)	156 ± 8
9. Hornblende quartz gabbro		171 ± ?
Gold Hill area		
10. Altered hornblende gabbro		142 ± ?
11. Hornblende quartz gabbro	(DR-1062)	144 ± 7

*Results of analysis are suspect, see text. Samples 1 through 7 and 9 from Ross and others, 1973; 8 and 11 from Ross, 1970; 10 from Hay, 1963.

TABLE 2. FELDSPAR COMMON Pb AND Sr INITIAL RATIOS

Sample	$\frac{^{206}Pb}{^{204}Pb}$	$\frac{^{207}Pb}{^{204}Pb}$	$\frac{^{208}Pb}{^{204}Pb}$	$\frac{^{87}Sr}{^{86}Sr_i}$
Eagles Rest Peak				
ERP2	18.57	15.45	38.06	
ERP3	18.63	15.52	37.99	
DR-1182				0.7039*
DR-761B				0.7031*
DR-761 plag.				0.7034*
Gold Hill				
DR-121				0.7038*
Logan				
LC1	18.30	15.61	38.15	
DR-589B				0.7039*
Gualala Cobbles				
AB-1	18.25	15.51	38.07	0.70384
SP-1	18.25	15.52	37.87	0.70327
ABC-1				0.7042*
SR-3				0.7048*
Analytical uncertainty at 95 percent level				
	±0.06%	±0.12%	±0.13%	±0.004% or ±0.016%*

*Recalculated from Kistler and others (1973), assuming an age of 161 Ma rather than 200 Ma.

can assemblage, were sutured together at a low latitude during Late Cretaceous time. Interestingly, this hypothesis rests on the interpretation that clasts in equivalents of Great Valley sequence southwest of the Salinian block were derived from the Salinian block. If this model is applicable to the Gualala area, the granitic detritus in the Gualala Formation may record the amalgamation of the Salinian block with the Gualala block.

Kanter (1983) and Kanter and Debiche (1985) measured anomalously shallow paleomagnetic inclinations in the spilite of Black Point and the German Rancho Formation and interpreted the results to indicate translations of 4,100 ± 300 and 1,810 ± 200 km for these units. These interpretations fit with the model of McWilliams and Howell (1982) and Vedder and others (1983) outlined above if the Gualala area and the Salinian block are part of a composite terrane amalgamated prior to the Late Cretaceous.

However, nontectonic causes for shallow paleomagnetic inclinations must be considered as possible alternatives to large-scale northward transport. These include depositional and compactional flattening in sediments (Coe and others, 1985; Blow and Hamilton, 1978; Creer, 1974; Liddicoat and Coe, 1979), failure of the dipolar field hypothesis (Coupland and Van der Voo, 1980; Livermore and others, 1984; Sager, 1984), remagnetization after tilting, and failure to remove all traces of the present-day magnetic field.

There are also specific uncertainties in the geologic and paleomagnetic data from the Gualala area that weaken the case for large-scale movements. First, the determination of paleohorizontal in pillow lavas, such as those at Black Point, is difficult. The Black Point lavas contain no intercalated sediments that would aid in orienting samples and are further complicated by shattering and deformation. These rocks may not have been horizontal when they were last magnetized. Spilites do not retain their original basaltic mineralogy and may not retain their original magnetization. Normal faulting, tilting, and hydrothermal alteration prior to their last magnetization are distinct possibilities. Furthermore, the lack of age control for the spilite means there is no way to select an accurate reference paleomagnetic pole. Finally, the paleomagnetic results from the Black Point spilite have very limited applicability. The spilite is separated from the Gualala Formation by faults (Wentworth, 1966, 1968; Ross and others, 1973) and, except for its limited extent, could be considered a terrane by itself. There is no evidence that its "transport history" is applicable to neighboring rocks.

The interpretation of the paleomagnetic data is obviously a difficult and controversial matter. The geologic data and interpretations in this chapter indicate a much less mobile history for rocks at Gualala. Indeed, the geologic evidence suggests close association of the Gualala strata with the Klamath-Sierran province since at least the late Cretaceous.

ANALYTICAL METHODS

Samples of gabbroic and granitic clasts from the Gualala Formation, samples from all the major rock units of the Eagle Rest Peak complex, and samples of the Logan and Gold Hill gabbros were collected and cleaned to remove surface contaminants and then crushed. Zircon, feldspar, and apatite were separated using the Wilfley table, heavy liquids, and magnetic methods. Zircon was dissolved and U and Pb separated using techniques slightly modified from Krogh (1973). Common Pb compositions were determined on separates of feldspar (±minor quartz). Initial Sr-isotopic compositions were determined on apatite separates. Pb and Sr were run on the Finnigan-MAT multi-collector mass spectrometer at the University of California, Santa Barbara (UCSB). Uranium was determined on the UCSB 35-cm AVCO single-collector mass spectrometer. Details of analytical procedures are contained in James (1986).

RESULTS

The sampling locations, field settings, and petrography of the samples are summarized in Table 3. Zircon U-Pb analytical data and calculated ages are listed in Table 4. Common Pb and Sr initial ratios are listed in Table 2.

Interpretation of discordant U-Pb results

Because the U-Pb and Pb-Pb ages of some individual zircon fractions are not concordant and because there are differences in calculated ages between fractions of zircons from the same sample, determining the age of crystallization of the samples requires some interpretation. Several different explanations for discordance of zircon are commonly invoked and allow derivation of a crystallization age. Episodic or continuous loss of Pb (or gain of U) (Wetherill, 1956; Tilton, 1960) and the incorporation of older zircon into the melt during magmagenesis or emplacement (Pasteels, 1964; Grauert, 1973) have been demonstrated in many cases. The distribution of discordant analyses on the concordia diagram usually implicates one (or several) of these causes of discordance. Commonly, where discordance is due to Pb loss, U-Pb ages of zircon fractions are roughly inversely proportional to their U content (Silver, 1963; Silver and Deutsch, 1963).

The samples analyzed in this study form linear arrays on the concordia diagram that indicate Pb-loss (Fig. 4). Unfortunately, the near linearity of the concordia curve in the age range in question, the small spread in ages of zircon fractions from each sample, and the closeness of all analyses to the concordia curve mean that a concordia intercept interpretation is not possible for most of the samples. However, given a Pb-loss explanation for the discordance, lower limits may be placed on the age of crystallization of each sample. In these samples, the $^{206}Pb^*/^{238}U$ age of the oldest fraction is a conservative minimum age for the sample. It is important to note that these samples are in no way unusual in their style of discordance. They are comparable to analyses from similar rocks from other areas (cf. Saleeby, 1982).

An alternate approach to relying on minimum ages is to utilize the relationship between U content of zircons and their susceptibility to Pb loss (Silver and Deutsch, 1963; Silver, 1964). There should be a correlation between apparent U-Pb age and U

TABLE 3. SAMPLE SITE LOCATIONS AND PETROGRAPHY
FOR GUALALA, LOGAN, AND EAGLE REST PEAK

AB-1 (zircon fraction 9) Gualala 7.5' quadrangle, latitude 38°48'4.5"N, longitude 123°34'57"W. 40-cm well-rounded cobble from conglomerate bed in beach cliff. Anchor Bay Member of Gualala Formation. Augite hornblende quartz gabbro, An_{55-85} plagioclase, late magmatic green hornblende replacing clinopyroxene, interstitial quartz.

AB-2 (zircon fractions 10 and 11) Gualala 7.5' quadrangle, latitude 38°48'7.6"N, longtitude 123°34'57"W. 30-cm well-rounded cobble from conglomerate bed in beach cliff. Anchor Bay Member of Gualala Formation. Augite hornblende quartz-bearing gabbro, An_{50-80} plagioclase, interstitial quartz. Very similar to AB-1.

AB-3 (zircon fractions 12 and 13) Gualala 7.5' quadrangle, latitude 38°47'7.7"N, longtitude 123°34'57"W. 30-cm well-rounded cobble from conglomerate bed in beach cliff. Anchor Bay Member of Gualala Formation. Very altered quartz gabbro composed of quartz, chlorite, and epidote.

SP-1 (zircon fractions 14 and 15) Stewarts Point 7.5' quadrangle, latitude 38°42'46.5"N, longitude 123°27'18"W. 40-cm well-rounded cobble from conglomerate bed in beach cliff. Stewarts Point Member of Gualala Formation. Biotite-bearing granodiorite. Slightly sericitized An_{40-55} plagioclase, quartz, orthoclase, and biotite.

L 1 (zircon fractions 7 and 8) San Juan Bautista 15' quadrangle, latitude 36°52'29.9"N, longitude 121°34'27"W. Roadcut on Highway 101, near DR 998 of Ross (1970). Hornblende quartz gabbro, An_{75} plagioclase and green hornblende.

GH 1 (zircon fraction 16) Cholame Valley 7.5' quadrangle, latitude 35°50'N, longitude 120°21'W. Gabbro collected by G. Van Kooten.

ERP 1 (zircon fraction 1) Eagle Rest Peak 7.5' quadrangle, latitude 34°56'30.4"N, longitude 119°13'16.8"W. Dike cutting gabbro in canyon wall exposure. Aplitic hornblende-bearing leuco tonalite, An_{60} plagioclase, quartz, and hornblende.

ERP 2 (zircon fractions 2, 3, 4, and 5) Eagle Rest Peak 7.5' quadrangle, latitude 34°56'41.3"N, longitude 119°13'16.8"W. 3 x 5 m felsic segregation in gabbro. Dike cutting gabbro in canyon wall exposure. Quartz, plagioclase, hornblende pegmatoid. Coarse hornblende to 3 cm, An_{55-75} plagioclase, and quartz.

ERP 3 (zircon fraction 6) Eagle Rest Peak 7.5' quadrangle, latitude 34°55'52.3"N, longitude 119°13'0"W. 50-cm segregation in pyroxenelte collected by A. Reitz and C. Hopson. Hornblende-bearing leuco tonalite, turbid plagioclase, quartz, and chloritized hornblende.

content that allows a better age estimate. Figure 5 shows the U contents versus $^{206}Pb*/^{238}U$ ages of the samples in this study. The U–age relationship allows an extrapolation to zero lead-loss conditions and a crystallization age for the Eagle Rest Peak samples. Several uncertainties in the method should be noted. The data do not form perfect linear arrays, indicating the U content is not the sole variable causing lowered ages. Also, the relationship between U and lowered ages may not be linear. Perhaps the lines on Figure 5 should be convex-up curves, indicating less lead loss at low U contents rather than a strictly linear model (Mattinson, 1978). Interconnection of damaged domains in a crystal and susceptibility to daughter product loss might be expected to increase nonlinearly with radiation damage. Unfortunately, the exact shape of the curve is unknown. The straight-line model is a reasonable approximation given the scatter in the data. In any case, the very low U contents allow good age determinations.

Ages of the Eagle Rest Peak complex, Logan, and Gold Hill

The samples from Eagle Rest Peak complex are concordant and near concordant. They show the effect of lead loss clearly, partly because they have a wide range of U contents. Concordance of fractions 2 and 7 (Fig. 4) and the age–U content relationship indicate that the crystallization age of the complex is 161 Ma (Fig. 5).

The Logan samples have the same U-Pb systematics as the Eagle Rest Peak samples, concordant to very slightly discordant, and show the same 161-Ma age.

A single, small (<1 mg) fraction of zircon from the Gold Hill gabbro yields an apparently concordant date of 143 Ma. This could conceivably be its crystallization age, but the large uncertainty in the $^{207}Pb*/^{206}Pb*$ age makes it equally probable that the age represents lead loss. The possibility of lead loss and the recov-

TABLE 4. U-Pb AGES OF ZIRCON FROM EAGLE REST PEAK COMPLEX, LOGAN, GUALALA FORMATION COBBLES, AND GOLD HILL

Sample*	Composition§					Ages (Ma)**		
	^{238}U (ppm)	$^{206}Pb^*$ (ppm)	$\frac{^{208}Pb}{^{206}Pb}$	$\frac{^{207}Pb}{^{206}Pb}$	$\frac{^{206}Pb}{^{204}Pb}$	$\frac{^{206}Pb^*}{^{238}U}$	$\frac{^{207}Pb^*}{^{235}U}$	$\frac{^{207}Pb^*}{^{206}Pb^*}$
Eagle Rest Peak complex								
1. ERP1 b	1111	23.3	0.2325	0.05201	6109	154.5	155.9	177 ± 2
2. ERP2 c	202	4.40	0.1433	0.05070	9921	160.1	160.0	159 ± 3
3. ERP2 c	192	4.14	0.1330	0.05058	5394	158.6	158.8	162 ± 6
4. ERP2 c ab	270	5.90	0.1771	0.05575	2289	160.2	160.5	165 ± 2
5. ERP2 f	181	3.90	0.1324	0.05148	7215	157.9	158.7	169 ± 2
6. ERP3 f	725	15.27	0.2152	0.05364	3455	155.1	155.9	168 ± 2
Logan Gabbro								
7. L1 c	55.3	1.21	0.08431	0.05368	3367	161.0	161.2	164 ± 4
8. L1 c	58.4	1.27	0.09387	0.05108	8547	160.1	160.4	165 ± 3
Gualala Formation cobbles								
Anchor Bay Member gabbro cobbles								
9. AB1 b	157	3.47	0.3918	0.06543	913	162.7	163.2	170 ± 7
10. AB2 c	85.2	1.88	0.2061	0.05242	4873	162.0	162.3	168 ± 6
11. AB2 f	99.3	2.22	0.2480	0.05785	1793	164.6	165.6	180 ± 13
12. AB3 c	254	5.22	0.1205	0.05388	3271	150.7	151.7	168 ± 6
13. AB3 f	287	5.71	0.1148	0.05448	3164	146.5	149.0	189 ± 4
Stewarts Point Member granodiorite cobble								
14. SP1 c	184	3.84	0.1096	0.05147	6868	153.7	154.3	164 ± 2
15. SP1 f	142	2.83	0.09981	0.05226	5073	146.8	147.9	165 ± 18
Gold Hill gabbro								
16. GH1 b	3504‡	67.69‡	0.10855	0.51541	5957	143.3	143.4	144 ± 30

Pb* = radiogenic Pb.

†All fractions are nonmagnetic on a Franz isodynamic separator at 1.6 amps, 1° side tilt except for those labeled b for bulk. C = coarse, F = fine, ab = abraded.

§Corrected for mass fraction of 0.12 percent ±0.03 per mass unit.

**λ ^{238}U = 1.5515 x 10^{-10}/y, λ^{235}U = 9.8485 x 10^{-10}/y (Jaffey and others, 1971), $^{238}U/^{235}U$ = 137.88. Stated $^{207}Pb^*/^{206}Pb^*$ age uncertainty is a worst-case sum of analytical uncertainties in $^{206}Pb/^{204}Pb$ at the 95 percent level and the conservative assumption of a 0.1 uncertainty in $^{207}Pb/^{204}Pb$ common lead corrections. $^{206}Pb^*/^{238}U$ age ±0.4 percent based on replicate analyses. $^{207}Pb^*/^{235}U$ age <±0.5 percent for Pb/Pb age uncertainties less than 10 Ma.

‡Uncertainties in U and Pb concentrations are large due to uncertainties inherent in weighing very small samples (~0.2 mg). Use of a mixed ^{205}Pb-^{235}U spike ensures ages are not affected.

ery of only enough zircon for a single analysis makes 143 Ma a minimum age for the Gold Hill gabbro.

When the age and uranium content data for the Gold Hill analysis are plotted, it appears that the Gold Hill sample falls along the same trend with the Logan and Eagle Rest Peak complex analyses. However, the U content of the Gold Hill zircons is not well known because of the uncertainty in weighing very small samples. The apparent U and Pb concentrations are an order of magnitude greater than most other samples also suggesting a weighing error. Fortunately, the use of a mixed U-Pb spike means that these errors cancel in age calculations.

Ages of the Gualala cobbles

Determining crystallization ages for the Gualala cobbles is hampered by the limited number of zircon fractions available for analysis and their limited range of U content. Sample AB-1 and the coarse fraction of AB-2 are near concordant at about 162 to 163 Ma. However, the fine fraction of AB-2 has slightly older U-Pb ages of 165 to 166 Ma, which represents a minimum age for this sample. Zircon fractions from the altered sample AB-3 are discordant and have younger U-Pb ages ranging from about 147 to 152 Ma. These, too, must be considered minimums. Lead loss

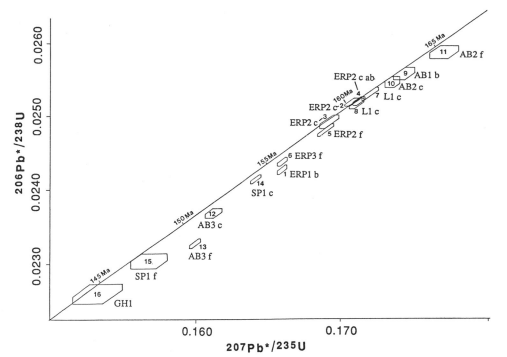

Figure 4. Concordia diagram of Eagle Rest Peak complex, Logan, Gualala, and Gold Hill zircon analyses. Fraction numbers refer to Tables 2 and 4. Boxes indicate 95 percent confidence level of measurement. See Table 2 caption for explanation.

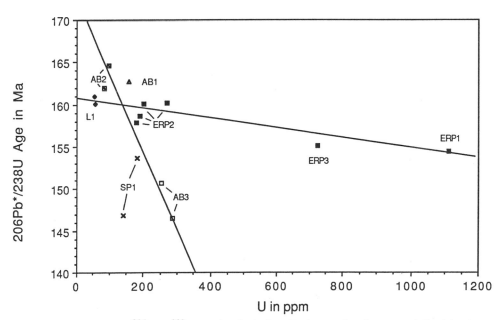

Figure 5. U content vs. $^{206}Pb*/^{238}U$ age for Eagle Rest Peak complex, Logan and Gualala zircon samples. Filled symbols indicate Logan and Eagle Rest Peak; open symbols and crosses indicate Gualala cobble samples. Gold Hill sample is not plotted because of the large uncertainty in the U concentration.

and consequent lower ages are very likely related to the severe alteration of this sample (Table 3).

The granodiorite cobble, SP-1, from the K-feldspar arkose has also experienced lead loss as shown by the spread in ages between the coarse and fine zircon fractions. The coarse fraction is near concordant at about 154 Ma (a minimum crystallization age), but the fine fraction is slightly less concordant at about 148 Ma.

Comparison of U-Pb and K-Ar ages

Published K-Ar dates (Table 1) generally agree with the age interpretations of the U-Pb data within the stated uncertainties or yield younger dates reflecting the lower closing temperature of the K-Ar hornblende system, or slight argon loss. The K-Ar dates of the Gold Hill gabbro agree particularly well with the U-Pb date. The two dates with the greatest discrepancies are the 171-Ma date from Logan (Oakeshott, quoted in Ross and others, 1973) and the 207-Ma date from Eagle Rest Peak complex (Ross and others, 1973). The 171-Ma date is difficult to assess because no analytical uncertainty is given and the analytical data are unpublished. The 171-Ma age may in fact agree with the U-Pb dates within analytical uncertainty. The 207-Ma date from Eagle Rest Peak complex is probably incorrect. Ross (personal communication) believes the sample may have been impure. There is also a possibility that this sample contains excess argon. J. Saleeby (personal communication, 1987) has noted a number of cases where hornblende pegmatites yield erroneous older K-Ar ages in comparison to well-determined U-Pb ages. The two older dates on gabbroic cobbles from the Anchor Bay Member of the Gualala Formation (175 and 186 Ma) are older than the U-Pb dates from Eagle Rest Peak complex, Logan, or Gold Hill, and support the contention that the U-Pb dates on the Gualala cobbles (162, 165, and 155 Ma) are indeed minimums.

Common Pb and Sr initial systematics

The common Pb isotopic ratios for the basement outcrops and all the cobbles are relatively nonradiogenic. When $^{207}Pb/^{204}Pb$ and $^{208}Pb/^{204}Pb$ are plotted against $^{206}Pb/^{204}Pb$, all the points, with the exception of Logan, lie below the curve of Stacey and Kramers (1975) for average crust (Fig. 6). The higher $^{208}Pb/^{204}Pb$ and $^{207}Pb/^{204}Pb$ (or low $^{206}Pb/^{204}Pb$) of the Logan sample are the only noticeable difference between the Logan and Eagle Rest Peak complex rocks.

Two determinations of initial Sr isotopic composition (Sr_i) were made on apatite from samples AB-2 and SP-1 from the Gualala Formation (Table 2). These ratios are nonradiogenic, 0.70308 and 0.70327, respectively. The 0.70308 value from AB-2 is lower than the 0.7041 and 0.7048 values determined by Kistler and others (1973) for other gabbroic cobbles from the Anchor Bay Member of the Gualala Formation. Sample SP-1 is from the K-feldspar–bearing Stewarts Point Member and is also notably nonradiogenic.

DISCUSSION

This study concurs with Ross (1970), Ross and others (1973), and Kistler and others (1973) in the petrographic, isotopic, and geochronologic similarity of rocks from Gualala, Logan, and the Eagle Rest Peak complex. Ross (1970, p. 3659) concluded that, "The similarity of the anorthositic gabbro of Gold Hill, Logan, and Eagle Rest Peak to some clasts in the conglomerate beds at Gualala suggests they originated in a similar environment, . . . but not necessarily an origin from the same mass." The new data and arguments allow us to narrow this statement further and preclude an origin for the cobbles from the same mass, yet reinforce the idea that they are products from a similar magmatic and tectonic setting.

Age differences

There are small, but significant, age differences between the gabbroic cobbles from the Gualala Formation and the Eagle Rest Peak complex. Cobble sample AB-2 is at least 165 Ma, but the Eagle Rest Peak complex is no older than 161 Ma. The differences, in fact, could be larger because the U-Pb ages of the Gualala

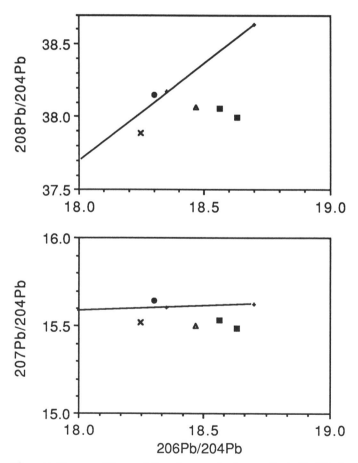

Figure 6. Common Pb correlation diagrams for Logan, Eagle Rest Peak complex, and Gualala cobbles. Lines indicate average crust from Stacey and Kramers (1975). Symbols as in Figure 5.

cobbles are minimums. The published K-Ar dates (Table 1) reinforce this conclusion with ages of 175 and 186 Ma for gabbro and diabase clasts from Gualala.

The U concentration vs. ^{206}Pb*/^{238}U-age relationships in the two suites are also different (Fig. 5). The Logan and Eagle Rest Peak complex zircon fractions form a coherent linear trend. The cobble samples, individually or grouped, have steeper, less coherent trends. The trends for the cobbles suggest older ages and greater Pb loss. Different postcrystallization histories might account for the steep trends indicative of lead loss, but there is no evidence for when this hypothetical disturbance took place.

Alternate Correlations

The Eagle Rest Peak complex is probably part of a suite of rocks that stretches the length of California. There are many Jurassic gabbroic plutons in California that represent possible sources for the mafic cobbles at Gualala. Two major groups of Late Jurassic plutons are logical candidates: the Coast Range ophiolite, as suggested by Ross (1970), and a unique suite of ultramafic to intermediate plutons that make up part of the Jurassic arc (Snoke and others, 1981; Wright and Sharp, 1982; Harper and Wright, 1984).

The Coast Range ophiolite (Bailey and others, 1970; Hopson and others, 1981) stretches through most of the California Coast Ranges on both sides of the San Andreas fault and has U-Pb ages ranging from 153 to 169 Ma (Hopson and others, 1981). Part of the ophiolite, gabbros that crystallized in the upper levels of the magma chamber, are petrographically identical to many of the Gualala samples (Table 3, cf. Hopson and others, 1981, p. 485–486). They are hypidiomorphic gabbros and diorites containing late magmatic green hornblende surrounding earlier crystallized clinopyroxene. Interstitial quartz may make up several percent of the rock.

There are several major objections to deriving the Gualala cobbles from the Coast Range ophiolite. The lack of cobbles in the Gualala Formation representing the rest of the ophiolite pseudostratigraphy and its cover presents an obstacle. Wentworth (1968) first noted the scarcity of ultramafic clasts in the section. The Gualala conglomerates contain few or no clasts of cumulate gabbro, ultramafic cumulates or tectonites, basalt, or chert. Many diabase cobbles from Gualala contain quartz, whereas quartz-bearing diabase is relatively scarce in the Coast Range ophiolite.

Besides the lack of the expected rock types, there is the difficulty of eroding detritus from the Coast Range ophiolite in the Cretaceous to feed into the Gualala basin. The Coast Range ophiolite is capped by upper Jurassic sediments throughout most of southern Coast Ranges, and it is unlikely that it was exposed during the Cretaceous. In the northern Coast Ranges, parts of the ophiolite were apparently exposed to plutonic levels during the Late Jurassic. There is a possibility that a Cretaceous unconformity lies to the east of the southern exposures of ophiolite beneath other Great Valley sediments, but this is special pleading. Although the age and petrography of the Coast Range ophiolite

make it an attractive source for the Gualala gabbro cobbles, the stratigraphic problems and lack of a wide range of ophiolitic rock types in the Gualala Formation weigh against the correlation.

The second possible source of plutonic detritus is the Jurassic arc of California. If one mentally subtracts the vast volume of Cretaceous granite from California, the Jurassic plutonic arc becomes a major entity. Its proximity to the western edge of the continent makes it a logical source for Cretaceous sediments.

The rocks from the Jurassic arc that most resemble the Gualala cobbles are zoned mafic to ultramafic plutonic complexes. These peridotitic to dioritic complexes are found throughout California (Snoke and others, 1982; Wright and Sharp, 1982) from the Klamath Mountains (Wright, 1981; Harper and Wright, 1984) south through the Sierra Nevada (Saleeby and Sharp, 1980; Saleeby, 1982). They have calc-alkaline differentiation trends with rock types ranging through dunite, wehrlite, olivine-hornblende-clinopyroxenite, olivine gabbro, gabbronorite, hornblende gabbro, hornblende quartz gabbro, and quartz diorite. Their ages are appropriate, ranging from 150 to 200 Ma, with many dates clustering around 165 Ma.

The Eagle Rest Peak complex is very probably part of this group of plutons (A. Reitz and C. Hopson, personal communication, 1984). The location along strike, rock types, differentiation trend, complicated intrusive relations, and age of the Eagle Rest Peak complex all support its inclusion with the group of mafic to ultramafic complexes.

It is possible that Great Valley sequence and younger sedimentary rocks conceal the southern continuation of mafic-ultramafic Jurassic intrusives in the western Sierra Nevada foothills until they surface again in the San Emigdio Mountains. The gravity and magnetic anomalies that follow the axis of the Great Valley may be an expression of these buried mafic to ultramafic plutons (Reitz, 1983), although the anomalies have also been attributed to Lower Cretaceous gabbros (Saleeby and Williams, 1978). Well data show that there is a substantial amount of gabbro in the subsurface (Saleeby, 1975, 1981).

The Coast Range ophiolite and Jurassic peridotitic to dioritic complexes have been suggested as two possible sources for the gabbroic cobbles in the Gualala Formation. They may in fact be two parts of a single magmatic response to the Late Jurassic tectonic regime (Saleeby, 1981). Extension and rifting of the Jurassic arc can account for the peridotitic to dioritic complexes in the arc and the Coast Range ophiolite out-board of the arc. The position of the two end members on either side of the Great Valley invites speculation that there are intermediate members hidden in between.

A final complication must be dealt with. Offset on the Hosgri–San Gregorio fault feeds into the San Andreas fault well south of Gualala. If the Gold Hill rocks are in place, rather than a sliver in the San Andreas fault zone, then the combination of slip could have moved the Gualala rocks from a position opposite Gold Hill. However, the Gold Hill area has been mapped as a fault sliver by many investigators (Hay, 1963; Dickinson, 1966; Ross, 1970). Furthermore, the sedimentary section at Gold Hill is comparable to Logan and Eagle Rest Peak areas but is very different from the adjacent area east of the San Andreas fault and different from Gualala (cf. Dickinson, 1966, with Nilsen, 1984). Although the age data do not resolve this question, the sedimentary rocks and structural settings do not support derivation of Gualala clasts from the Gold Hill area.

Assessment of correlations

In summary, several lines of evidence preclude the use of the Eagle Rest Peak complex and the Gualala sediments as an accurate indication of displacement on the San Andreas fault: (1) the difference in crystallization ages between the Eagle Rest Peak complex gabbro and the Gualala cobbles; (2) the wide distribution of Late Jurassic ophiolite and pyroxenitic to dioritic complexes parallel to the San Andreas fault; and (3) the voluminous younger sedimentary cover in the Great Valley that may conceal more exact correlatives.

Evaluating the Gualala-Salinia link

Although the data are limited, they suggest a second negative conclusion. They do not support the hypothesis that the K-feldspar arkose of the Gualala Formation was derived from the Salinian block. The U-Pb age of 153 Ma (a minimum) is much older than any rock thus far dated in the block (Mattinson and James, 1985). The common Pb and Sr_i ratios are also considerably less radiogenic than any pluton in the Salinian block. It might be argued that the Jurassic cobbles came from a presently unexposed or missing part of the Salinian block. There is, however, no known evidence to substantiate this argument.

Further evidence against a Salinian source includes a reinvestigation of paleocurrents in the Gualala Formation (J. Crowell and O. Sage, written communication, 1984). These new, but sparse, measurements indicate westward sediment transport in the Stewarts Point Member (K-feldspar arkose), rather than eastward transport as inferred from the younger German Rancho Formation. It seems better to base the paleotransport direction on these sparse data taken directly from the Stewarts Point K-feldspar arkose rather than relying on the lithologic link with the Tertiary German Rancho Formation.

The lack of metamorphic clasts in the Cretaceous granitic conglomerates and their occurrence in the overlying Tertiary sediments (Ross and others, 1973) also argues against a Salinian source for the Stewarts Point member. It seems probable that the two members of the Gualala Formation were derived from the east and that the overlying Paleocene through Eocene German Rancho Formation, with its more northerly current directions and metamorphic clasts, were derived from a different, probably Salinian source. A single K-Ar age from a German Rancho Formation cobble of 82 ± 3 Ma (Ross and others, 1973) reinforces this possibility.

Paleogeographic models

The data presented here support conventional models of displacement on the northern San Andreas system. In particular, our results are in accord with a model developed by Ross (1984, Fig. 14A), which reconstructs several plutonic and metamorphic bodies cut by the San Andreas fault. Our model involves about 300 km of displacement between Logan and the Eagle Rest Peak complex, with the Gold Hill fault sliver being picked up on the way.

The Gualala strata are more difficult to place and require restoring slip on other faults. Correlations of Salinian Tertiary strata and basement rocks at Point Reyes with rocks near Monterey (Clark and others, 1984) and correlations of basement rocks near Tomales Bay with rocks on Ben Lomond (James and Mattinson, 1985) suggest approximately 150 km of slip on the northern San Gregorio fault, possibly representing combined slip on the Rinconada and southern San Gregorio–Hosgri systems. If San Gregorio fault slip is removed, the Gualala rocks move back to just north of Montara Mountain but are not adjacent to similar rocks or sediment sources. Backtracking slip on the San Andreas fault 300 km still does not place the Gualala block opposite equivalent rocks. In short, we can match Jurassic and younger rocks across the San Andreas fault as far north as Logan, but Cretaceous strata at Gualala do not fit any simple model of San Andreas system movement. This dilemma may be approached in several ways.

First, the Gualala block may not have traveled along the San Gregorio–Hosgri fault a full 150 km or along the San Andreas fault all of 300 km. This ignores the arguments of Wentworth (1966) that there are no source terranes for these rocks north of the Transverse Ranges. Despite the fact that there is no known *source* for the sediments, there are very similar Upper Cretaceous sedimentary rocks (whose source is also a mystery) east of the San Andreas fault. Strata similar to the Gualala Formation resting structurally or depositionally on Franciscan rocks lie east of the San Andreas fault and west of the Sargent fault in the Santa Cruz Mountains. This block contains Upper Cretaceous sandstones, some with 5 to 20 percent K-feldspar, and conglomerates with granitic clasts (Bailey and Everhart, 1964). These strata are in fault contact with spilites in several places (Bailey and Everhart, 1964; Dibblee and Brabb, 1980). If these strata are correlative to the Gualala Formation, then the Gualala block has not traveled 300 km along the San Andreas fault and a once-active strand of the San Andreas fault must lie offshore west of Gualala. A systematic comparison of the Santa Cruz Mountains assemblage with the Gualala rocks is needed to access this speculative correlation.

Second, westward emplacement of the Salinian block along low-angle faults during the latest Cretaceous or Tertiary (Silver, 1982, 1983) may have greatly complicated the paleogeography. These faults may be responsible for moving the Gualala Formation away from its original position or may have covered the source terrane. The change from an eastern sediment source derived from the Jurassic arc in the Late Cretaceous to a Salinian source to the south in the Paleocene may reflect the same cause. This possibility replaces the "proto–San Andreas fault" with one or more low-angle faults that moved west or northwest.

SUMMARY

The link between the Eagle Rest Peak complex and gabbroic cobbles from the Gualala Formation is not unique. Despite similar petrography and Pb and Sr isotopic signatures, differences in U-Pb dates and the presence of several widespread gabbro sources make these outcrops unsuitable for measuring slip on the San Andreas fault. Two suites of rocks that might have supplied the gabbroic clasts to the Gualala Formation are the Coast Range ophiolite or peridotitic to dioritic complexes in the Jurassic arc. East of the San Andreas fault, Cretaceous sedimentary rocks broadly similar to the Gualala Formation crop out in the Santa Cruz Mountains.

On the other hand, correlation of the Eagle Rest Peak complex with the Logan gabbro (Ross, 1970) is reinforced by the U-Pb zircon age data. These age data, coupled with the correlation of the overlying sediments at Logan and the Eagle Rest Peak complex (Nilsen, 1984) and the correspondence of the amount of offset with determinations south of the Transverse Ranges and post-Eocene slip north of the Transverse Ranges, suggest that post-Jurassic slip on the San Andreas fault is about 300 km at least as far north as Logan. The U-Pb date from a single zircon fraction from the Gold Hill gabbro is insufficient to confirm or reject a correlation with the Eagle Rest Peak complex and Logan gabbros, but other workers (Ross, 1970; Ross and others, 1973) believe this is the best petrographic match with the Eagle Rest Peak complex rocks.

The limited data on the K-feldspar arkose in the Gualala Formation suggest that these sediments do not have a Salinian source. One cobble typical of the conglomerates in the K-feldspar arkose has too old a U-Pb age and has Sr and Pb isotopic characteristics incompatible with a Salinian source. A source in the Jurassic arc is more likely. A reappraisal of the paleocurrents and clast types in these strata also suggests a non-Salinian source. Therefore, tectonic models are not constrained to amalgamate Salinia and the Gualala block before Campanian-Maastrichtian time. There is no reason to make the Gualala area part of the Salinian block. However, the data do suggest the close proximity of the Gualala block to Jurassic arc rocks at this time. More significantly, because the pyroxenitic to dioritic complexes of the California Jurassic arc intrude most of the lithotectonic belts of the Klamath Mountains and Sierra Nevada, association of the Logan, Gold Hill, Eagle Rest Peak complex, and Gualala rocks with these intrusive complexes suggests the close geographic relationship of these terranes since the Late Jurassic.

ACKNOWLEDGMENTS

We thank T. Davis and A. Reitz for help in the field. Reviews by D. Ross, J. Saleeby, R. Powell, and B. Globerman are much appreciated. The research was supported by National Science Foundation grant EAR-80-08215 (to J.M.M.).

REFERENCES CITED

Bailey, E. H., Blake, M. C., Jr., and Jones, D. L., 1970, On-land Mesozoic oceanic crust in California Coast Ranges, *in* Geological Survey Research, 1970: U.S. Geological Survey Professional Paper 700-C, p. C70–C81.

Bailey, E. H., and Everhart, D. L., 1964, Geology and quicksilver deposits of the New Almeden district, Santa Clara County, California: U.S. Geological Survey Professional Paper 360, 206 p.

Blow, R. A., and Hamilton, N., 1978, Effect of compaction on the acquisition of a detrital remnant magnetization in fine grain sediments: Geophysical Journal of the Royal Geophysical Society, v. 52, p. 13–23.

Champion, D., Howell, D. G., and Gromme, S., 1984, Paleomagnetic and geologic data indicating 2,500 km of northward displacement for the Salinian and related terranes, California: Journal of Geophysical Research, v. 89, p. 7736–7752.

Clark, J. C., Brabb, E. E., Greene, H. G., and Ross, D. C., 1984, Geology of the Point Reyes peninsula and implications for San Gregorio fault history, *in* Crouch, J. C., and Bachman, S. B., eds., Tectonics and sedimentation along the California margin: Pacific Section, Society of Economic Paleontologists and Mineralogists, p. 67–86.

Clarke, S. H., Jr., and Nilsen, T. H., 1973, Displacement of Eocene strata and implications for the history of offset along the San Andreas fault, central, and northern California, *in* Kovach, R. L., and Nur, A., eds., Proceedings of the Conference on Tectonic Problems of the San Andreas Fault System: Stanford, California, Stanford University Publications in the Geological Sciences, v. 13, p. 358–367.

Coe, R. S., Globerman, B. R., Plumley, P. W., and Thrupp, G. A., 1985, Paleomagnetic results from Alaska and their tectonic implications, *in* Howell, D. G., ed., Tectonostratigraphic terranes of the Circum-Pacific region: Houston, Texas, Circum-Pacific Council for Energy and Mineral Resources, p. 85–108.

Coney, P. J., Jones, D. L., and Monger, J.W.H., 1980, Cordilleran suspect terranes: Nature, v. 288, p. 329–333.

Coupland, D. H., and Van der Voo, R., 1980, Long-term nondipole components in the geomagnetic field during the last 130 m.y.: Journal of Geophysical Research, v. 85, p. 3529–3548.

Creer, K. M., 1974, Geomagnetic variations for the interval 7,000-25,000 yr BP as recorded in a core of sediment from station 1474 of the Black Sea cruise of "Atlantis II": Earth and Planetary Science Letters, v. 23, p. 34–42.

Crowell, J. C., and Walker, W. R., 1962, Anorthosite and related rocks along the San Andreas fault, southern California: Berkeley, California University Publications in the Geological Sciences, v. 40, no. 4, p. 219–287.

Dibblee, T. W., Jr., 1976, The Rinconada and related faults in the southern Coast Ranges, California, and their tectonic significance: U.S. Geological Survey Professional Paper 981, 55 p.

Dibblee, T. W., Jr., and Brabb, E. E., 1980, Preliminary geologic map of the Loma-Prieta Quadrangle, Santa Cruz and Santa Clara Counties, California: U.S. Geological Survey Open-File Report 80-944, scale 1:24,000.

Dibblee, T. W., Jr., and Nilsen, T. H., 1973, Geologic map of the San Emigdio and western Tehachapi Mountains, California, *in* Vedder, J. G., chairman, Sedimentary facies changes in Tertiary rocks; California Transverse and southern Coast Ranges: Society of Economic Paleontologists and Mineralogists Annual Meeting Field Trip Guidebook 2, scale 1:62,500.

Dickinson, W. R., 1966, Structural relationships of the San Andreas fault, Cholame Valley and Castle Mountain Range, California: Geological Society of America Bulletin, v. 77, p. 707–726.

Graham, S. A., and Dickinson, W. R., 1978, Evidence for 115 kilometers of right slip on the San Gregorio–Hosgri fault trend: Science, v. 199, p. 179–181.

Grauert, B., 1973, U-Pb isotopic studies of zircons from the Gunpowder Granite, Baltimore County, Maryland: Department of Terrestrial Magnetism Year Book, v. 72, p. 288–290.

Hammond, P. E., 1958, Geology of the lower Santiago Creek area, San Emigdio Mountains, Kern County, California [M.S. thesis]: Los Angeles, University of California, 108 p.

Harper, G. D., and Wright, J. E., 1984, Middle to Late Jurassic tectonic evolution of the Klamath Mountains, California-Oregon: Tectonics, v. 3, p. 759–772.

Hay, E. A., 1963, Age and relationships of the Gold Hill pluton, Cholame Valley, California, *in* Geology of Salinas Valley and the San Andreas fault: Pacific Section, American Association of Petroleum Geologists Guidebook, p. 113–115.

Hopson, C. A., Mattinson, J. M., and Pessangno, E. A., Jr., 1981, Coast Range ophiolite, western California, *in* Ernst, W. G., ed., The geotectonic development of California; Rubey Volume 1: Englewood Cliffs, New Jersey, Prentice-Hall, p. 419–510.

Huffman, O. F., 1972, Lateral displacement of upper Miocene rocks and the Neogene history of offset along the San Andreas fault in central California: Geological Society of America Bulletin, v. 83, p. 2913–2946.

Jaffey, A. H., Flynn, K. F., Glendenin, L. E., Bentley, W. C., and Essling, A. M., 1971, Precision measurement of half-lives and specific activities of ^{235}U and ^{238}U: Physical Review, v. C4, p. 1889–1906.

James, E. W., 1986, Age, isotopic characteristics, and paleogeography of parts of the Salinian block of California [Ph.D. thesis]: Santa Barbara, University of California, 176 p.

James, E. W., and Mattinson, J. M., 1985, Evidence for 160 km post mid-Cretaceous slip on the San Gregorio fault, coastal California: EOS Transactions of the American Geophysical Union, v. 66, p. 1093.

Johnson, C. M., and O'Neil, J. R., 1988, Constraints of pre-Tertiary movement on the San Andreas fault system (SAF); Stable and radiogenic isotope and trace element data from Jurassic gabbros: Geological Society of America Abstracts with Programs, v. 20, p. A381.

Jones, D. L., Silberling, N. J., Gilbert, W., and Coney, P., 1982, Character, distribution, and tectonic significance of accretionary terranes in the central Alaska Range: Journal of Geophysical Research, v. 87, p. 3709–3717.

Kanter, L. R., 1983, Paleomagnetic constraints on the motion of Salinia [Ph.D. thesis]: Stanford, California, Stanford University, 156 p.

Kanter, L. R., and Debiche, M., 1985, Modeling the motion histories of the Point Arena and central Salinia terranes, *in* Howell, D. G., ed., Tectonostratigraphic terranes of the Circum-Pacific region: Houston, Texas, Circum-Pacific Council for Energy and Mineral Resources, p. 226–238.

Kistler, R. W., and Peterman, Z. E., 1973, Variations in Sr, Rb, K, Na, and initial ^{87}Sr/^{86}Sr in Mesozoic granitic rocks and intruded wall rocks in central California: Geological Society of America Bulletin, v. 84, p. 3489–3512.

Kistler, R. W., Peterman, Z. E., Ross, D. C., and Gottfried, D., 1973, Strontium isotopes and San Andreas fault, *in* Kovach, R. L., and Nur, A., eds., Proceedings of the Conference on Tectonic Problems of the San Andreas Fault System: Stanford, California, Stanford University Publications in the Geological Sciences, v. 13, p. 339–347.

Krogh, T. E., 1973, A low-contamination method for hydrothermal decomposition of zircon and extraction of U and Pb for isotopic age determinations: Geochimica et Cosmochimica Acta, v. 37, p. 485–494.

Liddicoat, J. C., and Coe, R. S., 1979, Mono Lake geomagnetic excursion: Journal of Geophysical Research, v. 84, p. 261–271.

Livermore, R. A., Vine, F. J., and Smith, A. G., 1984, Plate motions and the geomagnetic field; 2, Jurassic to Tertiary: Geophysical Journal of the Royal Astronomical Society, v. 79, p. 939–961.

Matthews, V. M., III, 1973, Pinnacles-Neenach correlation; A restriction for models of the origin of the Transverse Ranges and the big bend in the San Andreas fault: Geological Society of America Bulletin, v. 84, p. 683–688.

Mattinson, J. M., 1978, Age, origin, and thermal histories of some plutonic rocks from the Salinian block of California: Contributions to Mineralogy and Petrology, v. 67, p. 233–245.

Mattinson, J. M., and James, E. W., 1985, Salinian block U-Pb age and isotopic variations; Implications for origin and emplacement of the Salinian terrane, *in* Howell, D. G., ed., Tectonostratigraphic terranes of the Circum-Pacific region: Houston, Texas, Circum-Pacific Council for Energy and Mineral Resources, p. 215–226.

McWilliams, M. O., and Howell, D. G., 1982, An exotic origin for terranes of western California: Nature, v. 297, p. 215–217.

Nilsen, T. H., 1978, Late Cretaceous geology and the problem of the proto–San Andreas fault, *in* Howell, D. G., and McDougall, K. A., eds., Mesozoic paleogeography of the western United States: Pacific Section, Society of Economic Paleontologists and Mineralogists Paleogeography Symposium 2, p. 559–573.

—— , 1984, Offset along the San Andreas fault of Eocene strata from the San Juan Bautista area and western San Emigdio Mountains, California: Geological Society of America Bulletin, v. 95, p. 599–609.

Pasteels, P., 1964, Measures d'ages sur les zircons de quelques roches des Alpes: Schweiwerische Mineralogische und Petrologische Mitteilungen, v. 44, p. 519–541.

Reitz, A., 1983, San Emigdio Mountains plutonic sequence; Remnants of a pre–late Jurassic arc complex: Geological Society of America Abstracts with Programs, v. 15, p. 411.

—— , 1986, The geology and petrology of the northern San Emigdio plutonic complex, San Emigdio Mountains, southern California [M.S. thesis]: Santa Barbara, University of California, 80 p.

Ross, D. C., 1970, Quartz gabbro and anorthositic gabbro markers of offset along the San Andreas fault in the California Coast Ranges: Geological Society of America Bulletin, v. 81, p. 3647–3662.

—— , 1972, Petrographic and chemical reconnaissance study of some granitic and gneissic rocks near the San Andreas fault from Bodega Head to Cajon Pass, California: U.S. Geological Survey Professional Paper 698, 92 p.

—— , 1984, Possible correlations of basement rocks across the San Andreas, San Gregorio–Hosgri, and Rinconada–Reliz–King City faults, California: U.S. Geological Survey Professional Paper 1317, 37 p.

Ross, D. C., Wentworth, C. M., and McKee, E. H., 1973, Cretaceous mafic conglomerate near Gualala offset 350 miles by San Andreas fault from oceanic crustal source near Eagle Rest Peak, California: U.S. Geological Survey Journal of Research, v. 1, p. 45–52.

Sager, W. W., 1984, Paleomagnetism of the Abbott Seamount and implications for the latitudinal drift of the Hawaiian hot spot: Journal of Geophysical Research, v. 89, p. 6271–6284.

Saleeby, J. B., 1975, Structure, petrology, and geochronology of the Kings-Kaweah mafic-ultramafic belt, southwestern Sierra Nevada foothills, California [Ph.D. thesis]: Santa Barbara, University of California, 286 p.

—— , 1981, Ocean floor accretion and volcano plutonic arc evolution of the Mesozoic Sierra Nevada, *in* Ernst, W. G., ed., The geotectonic development of California; Rubey Volume 1: Englewood Cliffs, New Jersey, Prentice-Hall, p. 957–972.

—— , 1982, Polygenetic ophiolite belt of the California Sierra Nevada; Geochronological and tectonostratigraphic development: Journal of Geophysical Research, v. 87, p. 1823–1824.

Saleeby, J. B., and Sharp, W. D., 1980, Chronology of the structural and petrologic development of the southwest Sierra Nevada foothills, California: Geological Society of America Bulletin, v. 91, part 2, p. 1416–1535.

Saleeby, J. B., and Williams, H., 1978, Possible origin for California Great Valley gravity-magnetic anomalies [abs.]: EOS Transactions of the American Geophysical Union, v. 59, p. 1189.

Silver, L. T., 1963, The relation between radioactivity and discordance in zircons, *in* Nuclear geophysics: Washington, D.C., National Academy of Sciences–National Research Council Nuclear Science Series Repository 38, Publication 1075, p. 34–39.

—— , 1982, Evidence and a model for west directed Early to Mid-Cenozoic basement overthrusting in southern California: Geological Society of America Abstracts with Programs, v. 14, p. 617.

—— , 1983, Paleogene overthrusting in the tectonic evolution of the Transverse Ranges, Mojave and Salinian regions, California: Geological Society of America Abstracts with Programs, v. 15, p. 438.

Silver, L. T., and Deutsch, S., 1963, Uranium-lead isotopic variations in zircons; A case study: Journal of Geology, v. 71, p. 721–758.

Snoke, A. W., Sharp, W. D., Wright, J. E., and Saleeby, J. B., 1982, Significance of mid-Mesozoic peridotitic to dioritic intrusive complexes, Klamath Mountains–western Sierra Nevada, California: Geology, v. 10, p. 160–166.

Stacey, J. S., and Kramers, G. D., 1975, Approximation of terrestrial lead isotopic evolution by a two-stage model: Earth and Planetary Science Letters, v. 26, p. 207–221.

Tilton, G. R., 1960, Volume diffusion as a mechanism for discordant lead ages: Journal of Geophysical Research, v. 65, p. 2933–2945.

Vedder, J. G., Howell, D. G., and McLean, H., 1983, Stratigraphy, sedimentation, and tectonic accretion of exotic terranes, southern Coast Ranges, California: American Association of Petroleum Geologists Memoir 34, p. 471–496.

Wentworth, C. M., Jr., 1966, The Upper Cretaceous and lower Tertiary rocks of the Gualala area, northern Coast Ranges, California [Ph.D. thesis]: Stanford, California, Stanford University, 197 p.

—— , 1968, Upper Cretaceous and lower Tertiary strata near Gualala, California, and inferred large right-lateral slip on the San Andreas fault, *in* Dickinson, W. R., and Grantz, A., eds., Proceedings of a Conference on Geologic Problems of the San Andreas Fault System: Stanford, California, Stanford University Publications in the Geological Sciences, v. 11, p. 130–143.

Wetherill, G. W., 1956, Discordant uranium-lead ages: Transactions of the American Geophysical Union, v. 37, p. 302–326.

Wright, J. E., 1981, Geology and uranium-lead geochronology of the western Paleozoic and Triassic subprovince, southwestern Klamath Mountains [Ph.D. thesis]: Santa Barbara, University of California, 300 p.

Wright, J. E., and Sharp, W. D., 1982, Mafic-ultramafic intrusive complexes of the Klamath-Sierran region, California; Remnants of a Middle Jurassic arc complex: Geological Society of America Abstracts with Programs, v. 14, p. 245–246.

MANUSCRIPT ACCEPTED BY THE SOCIETY SEPTEMBER 6, 1990

Geological Society of America
Memoir 178
1993

Chapter 8

Whole-rock K-Ar ages and geochemical data from middle Cenozoic volcanic rocks, southern California: A test of correlations across the San Andreas fault

Virgil A. Frizzell, Jr.
Office of Earthquakes, Volcanoes and Engineering, U.S. Geological Survey, MS 905, Reston, Virginia 22092
Peter W. Weigand
Department of Geological Sciences, California State University, Northridge, California 91330

ABSTRACT

Potassium-argon determinations (n = 19) and whole-rock trace-element analyses (n = 9) on volcanic rocks from the Plush Ranch, Vasquez, and Diligencia Formations located along the San Andreas fault system in southern California support earlier correlations that suggest palinspastic proximity of the transtensional basins in which the lavas erupted. Volcanic rocks of the Plush Ranch and Vasquez Formations, now located southwest of the San Andreas fault, were extruded contemporaneously at about 23.1 to 26.5 Ma and 23.6 to 25.6 Ma, respectively, whereas those of the Diligencia Formation, now located northeast of the fault, erupted about 20.6 to 23.6 Ma. The rocks yield trace-element and isotopic ratios that define a petrologic suite unique in southern California. These subalkaline and calc-alkaline rocks range from medium-potassium basalt to high-potassium dacite. Vasquez samples are generally higher in SiO_2 than are those from either the Plush Ranch or Diligencia Formations. Similar rare-earth-element patterns exhibit moderate light-REE and flat heavy-REE enrichment. Spider diagrams for selected incompatible major and trace elements have relatively tight and featureless patterns. Initial whole-rock $^{87}Sr/^{86}Sr$ ratios range from 0.7048 to 0.7062 and plagioclase $\delta^{18}O$ ranges from 6.2 to 7.7; both values correlate positively with SiO_2. These data indicate that the volcanic rocks were derived from similar magmas, probably incorporating different amounts of crustal component. Our data indicate that the volcanic rocks formed at only slightly different times, had similar petrogenetic histories, and have been separated along the San Andreas fault system. We cannot distinguish, however, either the magnitude or timing of movement on the various strands of the system in southern California that resulted in their disruption.

INTRODUCTION

A cumulative offset of approximately 300 km has been substantiated for the San Andreas fault in central California (Hill and Dibblee, 1953; Clarke and Nilsen, 1973; Stanley, 1987). In southern California, an equivalent offset has been distributed among several faults of the San Andreas fault system (Crowell, 1962)—the San Andreas (sensu stricto), San Gabriel, San Jacinto, Punchbowl, and Clemens Well–Fenner–San Francisquito faults, among others. Although various palinspastic reconstructions rely on differing amounts of offset for various components of the San Andreas fault system, most ultimately arrive at a proximal pre–San Andreas cross-fault juxtapositioning of middle Tertiary nonmarine volcanic and sedimentary sequences that now crop out in Lockwood Valley, the Soledad basin, and the Orocopia Mountains (Crowell, 1962; Bohannon, 1975; Ehlig,

Frizzell, V. A., Jr., and Weigand, P. W., 1993, Whole-rock K-Ar ages and geochemical data from middle Cenozoic volcanic rocks, southern California: A test of correlations across the San Andreas fault, *in* Powell, R. E., Weldon, R. J., II, and Matti, J. C., eds., The San Andreas Fault System: Displacement, Palinspastic Reconstruction, and Geologic Evolution: Boulder, Colorado, Geological Society of America Memoir 178.

1981; Powell, this volume). These nonmarine units, the Plush Ranch Formation of Carman (1964) and the Vasquez Formation located southwest of the San Andreas (Fig. 1), and the Diligencia Formation of Crowell (1975) located northeast of the fault, share many similarities. However, their mutual correlation has proved difficult to document, because fossils and K-Ar determinations are scarce and discrepant, and because of apparent differences in the chemical composition of the interbedded volcanic rocks.

Middle Cenozoic volcanic rocks occur at many localities in southern California (Crowe and others, 1979; Weigand, 1982). Weigand (1982) and Johnson and O'Neil (1984) presented data that roughly divide Cenozoic volcanic rocks west of the San Andreas into two main provinces. They are distinguishable on the basis of major elements, trace elements, isotopic values, and age and are thought to have originated from fundamentally different tectonic environments—a coastal province (i.e., Conejo and Santa Cruz Islands volcanic rocks) and a northern province (i.e., Neenach, Pinnacles, and Sonoma volcanics). Geochemical data suggest that the volcanic rocks of the Plush Ranch, the Vasquez, and the Diligencia Formations form a distinct subgroup within the northern province (Weigand, 1982).

This chapter reviews and reevaluates the physical stratigraphy and the fossil, K-Ar, and paleomagnetic data from these three units, which are of great interest for making reconstructions along the San Andreas fault. It also reviews briefly some of the previous correlations and reconstructions, described elsewhere in this volume, and presents new whole-rock K-Ar determinations and trace-element and isotopic analyses that support the notion that the lavas formed nearly synchronously in a lower Miocene volcanic province disrupted by the San Andreas fault system.

STRATIGRAPHIC AND STRUCTURAL SETTING OF THE VOLCANIC ROCKS

Plush Ranch Formation

The Plush Ranch Formation of Carman (1964) is exposed chiefly on the north side of the Big Pine fault in the western Transverse Ranges (Fig. 1). Eocene marine rocks underlie the Plush Ranch, and Miocene nonmarine conglomeratic rocks overlie it (Carman, 1964, Plate 1), probably with unconformable relations (Carman, 1964, p. 26). Crystalline rocks form the depositional basement for the Eocene marine rocks that underlie the Plush Ranch. One- and two-mica monzogranite, biotite granodiorite, and tonalite (Ross, 1972) intrude layered gneiss on Mt. Pinos and Sawmill Mountain north of the Plush Ranch outcrops. This plutonic terrane tectonically overlies mica schist, greenschist, and quartzite correlative with the "Pelona-Orocopia schist" of Haxel and Dillon (1978; see also Ziony, 1958) along a relatively shallow southward-dipping fault exposed on the north slope of Mt. Pinos.

Conglomerate, sandstone, shale, and fresh-water limestone predominate in the 1,800-m-thick Plush Ranch Formation (Fig. 2). Carman (1964, Table 1) divided the formation into six interfingering members; the fourth member contains a volcanic se-

quence at its top. This member is composed predominantly of borate-bearing lacustrine deposits, including gypsiferous shale, sandstone, and tuffaceous sandstone, and has a maximum thickness of about 380 m. The volcanic rocks are 200 m thick in the east and thin to 90 m in the west (Carman, 1964, p. 29).

Two published K-Ar analyses on plagioclase indicate ages of 17.9 ± 3.7 and 20.1 ± 1.1 Ma (Fig. 3; Crowell, 1973, Table 1; these and other K-Ar determinations reported in the literature have been converted by using revised constants after Dalrymple, 1979); the Plush Ranch has not yielded datable fossils. These two K-Ar determinations apparently overlap the age of the unconformably overlying Caliente Formation—also a nonmarine unit—that is locally Hemingfordian in age (about from 16.5 to 20.0 Ma, Tedford and others, 1987, Fig. 6.2) at its base (James, 1963, Fig. 2).

Breccias in different members of the Plush Ranch indicate that marginal faults paralleled the present elongate outcrop pattern (Carman, 1964, p. 37; Bohannon, 1976). If an ambiguous counterclockwise rotation of $6°$ ($\pm 12°$; Terres, 1984; Luyendyk and others, 1985) is corrected, the basin could have formed during a north-south extensional regimen, normal to that which appears to have existed during the evolution of the Vasquez and Diligencia basins, as discussed below.

Vasquez Formation

The Vasquez Formation is exposed in three fault-bounded sub-basins in the northwestern part of the San Gabriel Mountains. It overlies crystalline basement (Fig. 1; Ehlig, 1981, Fig. 10-2) and is unconformably overlain by nonmarine clastic rocks of the Tick Canyon Formation. From south to north, these are the Vasquez Rocks (= Soledad), the Texas Canyon, and the Charlie Canyon sub-basins. The Vasquez Rocks sub-basin is bounded by the Soledad and Mint Canyon faults and is broken by several others. Sedimentary rocks were deposited upon Precambrian banded and augen gneiss and the Mendenhall Gneiss, and upon anorthosite, a hornblende-rich phase of the Triassic Mt. Lowe intrusion, and other Mesozoic granitoids. The Vasquez in the Texas Canyon sub-basin, between the Vasquez Canyon and Pelona faults, was also deposited on Precambrian gneiss and Mesozoic granitoids. A fault separates the Pelona Schist from the Vasquez along the eastern edge of the sub-basin (Bohannon, 1976). The San Francisquito fault separates the Charlie Canyon sub-basin from the Sierra Pelona, the type area of the Pelona Schist where the schist forms a west-plunging antiform. The Bee Canyon fault separates the Vasquez there from the stratigraphically lower marine San Francisquito Formation, which was deposited upon basement consisting of Precambrian gneiss intruded by Mesozoic granitoids. Because the clasts in, and sedimentary history of, the Charlie Canyon sub-basin differ from those in the two southern basins (Bohannon, 1976; Hendrix and Ingersoll, 1987), its relation to them is enigmatic.

The Vasquez is thickest in its easternmost sub-basin (Muehlberger, 1958), where it attains a thickness of about 5,500 m (Hendrix and Ingersoll, 1987). An approximately 1,300-m-

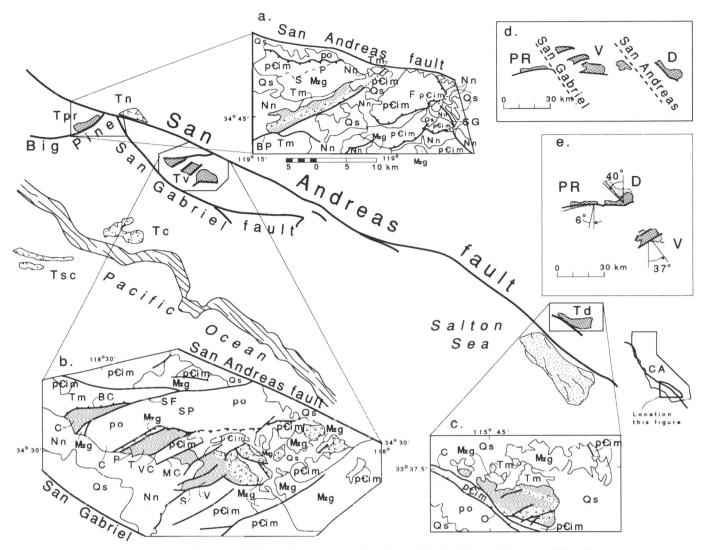

Figure 1. Map of southern California showing present locations of the Plush Ranch Formation (Tpr) of Carman (1964), Vasquez Formation (Tv), and Diligencia Formation (Td) of Crowell (1975), as well as other selected volcanic suites in the region south of and adjacent to the San Andreas fault; Tc = Conejo Volcanics; Tn = Neenach Volcanics; Tsc = Santa Cruz Island Volcanics. a, Sketch map of Sawmill and Frazier Mountains region (after Frizzell and Vedder, 1986; see Carman, 1964, for details) showing the undivided part of the Plush Ranch (dots) and its interbedded volcanic rocks (v's); BP = Big Pine fault; F = Frazier Mountain; P = Mt Pinos; S = Sawmill Mountain; SG = San Gabriel fault; Qs = Quaternary surficial deposits; Nn = Neogene nonmarine sedimentary rocks; Tm = lower Tertiary marine sedimentary rocks; Mzg = Mesozoic granitoids; po = "Pelona-Orocopia schist" of Haxel and Dillon (1978); pЄim = Precambrian igneous and metamorphic suite. b, Sketch map of part of San Gabriel Mountains (after Jennings and Strand, 1969; Ehlig, 1981; Hendrix and Ingersoll, 1987) showing the undivided part of the Vasquez Formation (dots) and its interbedded volcanic rocks (v's); symbols same as in a, except BC = Bee Canyon fault; C = Charlie sub-basin; MC = Mint Canyon fault; P = Pelona fault; S = Soledad fault; SF = San Francisquito fault; SP = Sierra Pelona; T = Texas sub-basin; V = Vasquez Rocks sub-basin; VC = Vasquez Canyon fault. c, Sketch map of Orocopia Mountains area (after Jennings, 1967; Spittler and Arthur, 1982) showing the undivided part of the Diligencia (dots) and its interbedded volcanic rocks (v's); symbols same as in a, except C = Clemens Well fault; O = Orocopia thrust. d, Simplified reconstruction of Bohannon (1975, Fig. 3). e, Schematic palinspastic reconstruction after Powell (1981) with basins restored (solid boundaries) using paleomagnetic data discussed in text.

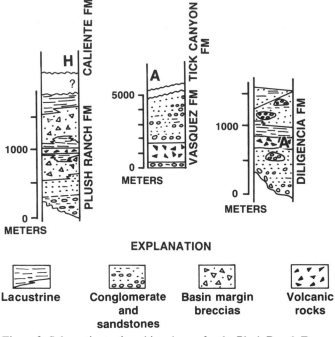

Figure 2. Schematic stratigraphic columns for the Plush Ranch Formation of Carman (1964), Vasquez Formation, and Diligencia Formation of Crowell (1975), after Carman (1964) and Bohannon (1976), Muehlberger (1958), Spittler (1974), and Hendrix and Ingersoll (1987), and Spittler and Arthur (1982). Relative vertical position adjusted according to probably ages of volcanic units; note different vertical scales. A = Arikareen fauna; H = Hemingfordian fauna.

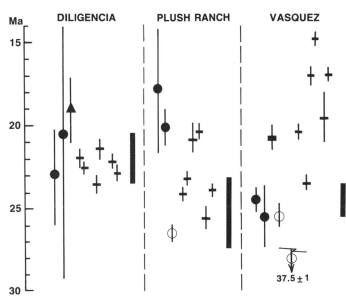

Figure 3. Potassium-argon determinations for the Plush Ranch Formation of Carman (1964), Vasquez Formation, and Diligencia Formation of Crowell (1975). Filled circles indicate data of Crowell (1973); triangle, Spittler (1974); rectangle, Woodburne (1975); open circles, P. E. Damon (written communication, 1981; Table 1); bars, this study (Table 1). Bold vertical bars represent pools of most reliable determinations; volcanic rocks probably erupted during these intervals.

thick sequence of andesite and basalt flows, shallow intrusions, and minor interbedded conglomerate overlies about 300 m of basal conglomerate and is in turn overlain by 3,900 m of conglomerate and sandstone. Four upward-fining megacycles in the upper conglomerate and sandstone sequence probably formed in response to tectonic activity on the Soledad fault (Hendrix and Ingersoll, 1987). Near the Mint Canyon fault, borate-rich, gypsum-bearing, and ripple-marked beds indicate deposition in ephemeral lake/fan delta environments (Muehlberger, 1958; Hendrix and Ingersoll, 1987).

Three K-Ar determinations on plagioclase (Fig. 3) from the volcanic section of 20.7 ± 0.8 (Woodburne, 1975, p. 59) and 24.5 ± 0.8 and 25.6 ± 2.1 Ma (Crowell, 1973, Table 1) indicate a latest Oligocene to early Miocene age; no fossil ages have been reported from the Vasquez. The two older determinations seem compatible with deformation that occurred before deposition of the overlying Tick Canyon Formation, which contains a sparse late Arikareean (20 to 23 Ma, Tedford and others, 1987) fauna (Whistler, 1967; Woodburne and Whistler, 1973).

Volcanic flows in the Vasquez Formation show an apparent clockwise rotation of 37° (±12°; Terres and Luyendyk, 1985), although orientations differ significantly near faults. Data from the stacked sequence consisting of the Vasquez, Tick Canyon, and Mint Canyon Formations provide evidence that this apparent rotation may have resulted from an initial clockwise rotation of 53° and a subsequent counterclockwise rotation of 16° (Terres and Luyendyk, 1985). The clockwise rotation apparently occurred in concert with left-lateral faulting after deposition of the unconformably overlying blanket of Tick Canyon Formation, but before deposition of the Mint Canyon Formation, which shows counterckockwise rotation after its deposition. Thus, the Vasquez may have been deposited in restricted northeast-oriented fault-bounded sub-basins that formed as a result of northwest/southeast extension.

Diligencia Formation

The Diligencia Formation of Crowell (1975) is exposed northeast of the San Andreas fault (Fig. 1) and consists of a nonmarine clastic sequence and interbedded volcanic flows (Spittler and Arthur, 1982, Fig. 3). The Diligencia basin is underlain by Mesozoic monzogranite and Eocene marine rocks on the north and east, and a complex of Precambrian crystalline rocks on the south (Crowell and Walker, 1962). On the southwest, a steeply dipping fault separates the Diligencia from this crystalline terrane, which in turn tectonically overlies, along a relatively shallow northeast-dipping fault, an antiformal body of the Orocopia Schist on its southwest.

The Diligencia is at least 1,500 m thick, may locally be 2,000 m thick (Fig. 2; Spittler and Arthur, 1982, p. 88), and it comprises four members. The 330-m-thick basal member was deposited in a closed basin and consists of coarse-grained conglomerate (Arthur, 1974, who indicated the clasts are predominantly the underlying Mesozoic monzogranite) that grades upward into thin-bedded, laterally extensive, and varicolored siltstone, mudstone, and limestone. This basal unit is overlain by a 160-m-thick sequence of lava flows and rare intrusive rocks (Spittler and Arthur, 1982, p. 90). The upper flows interfinger with overlying fine-grained evaporite beds and lacustrine sedimentary rocks that are as much as 230 m thick. The uppermost two units consist of 450 m of poorly sorted and bedded conglomerate, well-bedded sandstone, siltstone and mottled sandstone, and fine-grained sandstone, siltstone, and mudstone. Squires and Advocate (1982) recognized alluvial fan/braided-fluvial, shoreline, fluvial-deltaic, and lacustrine depositional environments in the Diligencia.

Woodburne and Whistler (1973, p. 908) have suggested that an oreodont fragment from the Diligencia is "best attributed" to the late Arikareean (20 to 23 Ma, Tedford and others, 1987). Three K-Ar ages on plagioclase have been determined (Fig. 3). Crowell (1973, Table 1) reported ages of 23.0 ± 2.9 and 20.6 ± 8.9 Ma, and Spittler (1974, Table 1) reported one of 19.1 ± 1.9 Ma.

Volcanic rocks of the Diligencia Formation exhibit up to 200° of apparent clockwise rotation (Terres, 1984; Luyendyk and others, 1985). Much of this rotation is no doubt local, because nearby Cretaceous dikes and lava flows of late Miocene and Pliocene age all indicate regional clockwise rotation of about 40° since late Miocene time (Terres, 1984; Carter and others, 1987; Luyendyk and others, 1985).

The Diligencia Formation appears to have been deposited in a structurally controlled basin. Two syndepositional normal faults marginal to the current outcrop area (and probably to the paleobasin as well) indicate extensional tectonics early in Diligencia history (Spittler and Arthur, 1982, p. 95). The approximate east-west orientation of the basin and its marginal faults led Spittler and Arthur to propose a north-south extensional direction; based on the paleomagnetic data, this orientation apparently resolves to east-west extension (Terres, 1984).

CORRELATION OF THE PLUSH RANCH, VASQUEZ, AND DILIGENCIA FORMATIONS

As outlined above, the Plush Ranch, Vasquez, and Diligencia Formations share many similarities. They have similar basement rocks and structure, similar physical stratigraphy, similar age, and similar volcanic interbeds. The similarity of the nonmarine sequences and associated volcanic rocks in two of the units, now called the Diligencia and Vasquez Formations, prompted Crowell (1962, p. 38) to suggest that they had been separated by the San Andreas fault. He (1962, p. 41) also used similarities between the later-named Vasquez and Plush Ranch Formations to strengthen a correlation of the basement terrane on which those two units were deposited.

Carman (1964, p. 35–37) also noted similarities, such as similar clasts in a sedimentary breccia, that suggested correlation of the Plush Ranch with the Vasquez. Spittler and Arthur (1973), however, argued that, on a fine scale, the sedimentary and volcanic rocks in the Vasquez and Diligencia were not similar, that

the basins were a different age, and thus the rocks were not correlative. Nevertheless, based on composition of the conglomerates, sedimentologic data, and previous suggestions concerning offset along faults in the San Andreas fault zone, Bohannon (1975, Fig. 3) drew a palinspastic diagram (simplified in Fig. 1d) that brought the three units into close proximity and produced an east-west belt of Pelona-Orocopia Schist. This restoration of the Pelona with the Orocopia was first suggested by Hill and Dibblee (1953), who worked mostly farther north along the San Andreas.

Reconstructions by Hill and Dibblee (1953), Crowell (1962), and Bohannon (1975; Fig. 1d) juxtaposed, from west to east, the volcanic and sedimentary sequence of the Lockwood Valley area (Plush Ranch Formation) against similar sequences from the Soledad basin (Vasquez Formation). These sequences in turn were juxtaposed against the sequence in the Orocopia Mountains (Diligencia Formation) and against the northern Chocolate Mountains. More recently, identification of a hitherto unrecognized strand of the San Andreas fault—the Clemens Wells–Fenner–San Francisquito fault—led Powell (1981; this volume) to suggest that these previous reconstructions may be intermediate to an alternative pre–San Andreas starting point of basement rocks that places the overlying basins in different relative positions (cf. Fig. 1e; Powell, 1981). This reconstruction resulted in the volcanic and sedimentary sequences of the Plush Ranch and those of the Diligencia being clumped together a little northwest of the Vasquez sequences, which were juxtaposed against volcanic rocks in the southern Chuckwalla and the Little Chuckwalla Mountains (Crowe and others, 1979). We find this alternative attractive because it requires only some 160 to 180 km of offset for part of the San Andreas fault in southern California; offset of this magnitude has recently been suggested by the cross-fault correlation of megaporphyritic monzogranite (Frizzell and others, 1986). Matti and others (1986) explored alternative reconstructions based on this cross-fault correlation.

PETROGRAPHY OF THE VOLCANIC ROCKS

The volcanic rocks of the Plush Ranch, Vasquez, and Diligencia Formations are calc-alkaline, ranging from medium-potassium basalt to high-potassium dacite. Textures are variable, and phenocrysts of plagioclase and clino- and orthopyroxene are present in various proportions. Olivine occurs in the more mafic rocks. A few samples from each formation contain plagioclase phenocrysts that exhibit narrow rims filled with masses of inclusions (Carman, 1964, p. 32; Spittler, 1974; Weigand, 1982).

Vitrophyric to holocrystalline basalt in the Plush Ranch Formation is characterized by strongly and oscillatorily zoned labradorite laths with subordinate amounts of olivine, and large and small crystals of sub- and euhedral augite and hypersthene. Groundmass pyroxene makes up about a third of the rocks (Carman, 1964). Andesite of the Vasquez Formation has intersertal to hyalopilotaxitic textures and contains oscillatory-zoned plagioclase, subhedral orthopyroxene and clinopyroxene, and rare embayed olivine, in a finely crystalline matrix (Spittler and Arthur, 1973, p. 379; Weigand, 1982). The volcanic rocks in the Diligencia Formation form two distinct petrographic groups (Spittler and Arthur, 1982, p. 95): basalt with phenocrysts of olivine, augite, and plagioclase in an intergranular texture; and porphyritic, 2-pyroxene andesite with orthopyroxene, clinopyroxene, and andesine in an intersertal to hyalophitic matrix. Evidence for alteration in the volcanic rocks includes "calcite, zeolite, and serpentine filling the cavities" of amygdaloidal zones of Plush Ranch volcanic rocks (Carman, 1964, p. 31), and clays derived from glass in the matrix of selected samples from the Vasquez and Diligencia (Spittler, 1974, Appendix I).

POTASSIUM-ARGON DATING

Published K-Ar determinations for the three volcanic suites range from 17.9 to 25.6 Ma, with ages from each suite having somewhat smaller discrepancies (Fig. 3). Because the age of the units remains a critical problem to their correlation, we conducted a series of K-Ar experiments using consistent methods and equipment.

Three samples of volcanic rock from each formation were qualitatively selected from more than a dozen on the basis of relative lack of alteration as determined by microscopic examination of thin sections, though neither the species of alteration products nor their amount was determined. All the samples were altered, and thus were not particularly good candidates for K-Ar analysis. Nevertheless we elected to proceed with the extraction experiments. Relative stratigraphic position of the samples is known only for those from the Diligencia. Hand samples were crushed and sieved to accumulate a 32- to 60-mesh fraction, which was then etched with a 14-percent HNO_3 solution for approximately 30 min. The etched samples were split with a microsplitter, and one fraction was treated further with 5 percent solution of HF for approximately 2 min. Both the HNO_3 and HNO_3-HF acid-treated fractions were further split to yield four fractions for K_2O analysis by flame photometry. A minimum of two samples were used for conventional argon extractions for analysis by isotope dilution on a mass spectrometer. Our 19 determinations and three other unpublished determinations (P. E. Damon, written communication, 1981) are presented in Table 1 and Figure 3.

Acid treatment was used to reduce the amounts of altered minerals and glass from whole-rock basalt samples. In one study (Tabor and others, 1984, p. 40), this treatment tended to increase relative K_2O content, as well as age. Six of our splits etched only in HNO_3 yielded ages equal to or older than the split etched both in HNO_3 and HF. Interestingly, the samples analyzed by P. E. Damon (written communication, 1981), lightly etched in HF only, yielded relatively older determinations. Although analytical error overlapped on the splits of only three samples (PR 11, DF 4, and DF 30), two extractions from the same sample (VF 23B) yielded similar ages.

A common method of assigning the "age" to a volcanic unit

TABLE 1. NEW POTASSIUM-ARGON DETERMINATIONS ON WHOLE-ROCK SAMPLES FROM MIDDLE TERTIARY VOLCANIC ROCKS, SOUTHERN CALIFORNIA*

Sample Number	K_2O (%)	$^{40}Ar_{rad}$ (m/gm x 10⁻)	$^{40}Ar_{rad}$ (%)	Age (Ma)
Plush Ranch Formation				
PR 6A	1.820, 1.917, 1.846, 1.835	6.516	79.1	24.2 ± 0.4
PR 6B	1.742, 1.742, 1.784, 1.770	5.897	81.5	23.1 ± 0.3
PR 11A	1.363, 1.354, 1.346, 1.344	4.096	44.9	20.9 ± 0.9
PR 11B	1.425, 1.409, 1.440, 1.429	4.220	34.4	20.4 ± 0.9
PR 13A	1.413, 1.420, 1.426, 1.426	5.210	58.1	25.3 ± 0.6
PR 13B	1.430, 1,419, 1.412, 1.435	4.940	87.5	23.9 ± 0.3
PR F6†	1.724, 1.718	6.62	86.6	26.5 ± 0.5
Vasquez Formation				
VF 7A	2.624, 2.601, 2.647, 2.619	7.806	67.1	20.6 ± 0.4
VF 7B	2.580, 2.589, 2.605, 2.585	8.841	78.7	23.6 ± 0.4
VF 23A	1.571, 1.623, 1.548, 1.615	3.909	49.2	17.0 ± 0.5
VF 23B†	1.619, 1.616, 1.644, 1.535	3.406	49.1	14.7 ± 0.4
VF 23B₁	do	3.531	48.3	15.2 ± 0.4
VF 25A₂	1.722, 1.730, 1.732, 1.729	4.906	18.6	19.6 ± 1.5
VF 25B	1.697, 1.731, 1.751, 1.754	4.318	41.7	17.2 ± 0.4
VF L3	3.374, 3.392, 3.367	12.43	94.9	25.4 ± 0.6
VF L6**	6.806, 6,729, 6.797	36.98	49.7	37.5 ± 1.0
Diligencia Formation				
DF 4A	1.250, 1.284, 1.276, 1.270	4.051	60.3	22.0 ± 0.5
DF 4B	1.362, 1.333, 1.326, 1.320	4.364	80.3	22.6 ± 0.3
DF 21A	1.358, 1.376, 1.392, 1.357	4.681	62.0	23.6 ± 0.5
DF 21B	1.427, 1.427, 1.424, 1.421	4.401	50.5	21.3 ± 0.6
DF 30A	1.297, 1.306, 1.263, 1.279	4.134	59.3	22.2 ± 0.5
DF 30B	1.327, 1.301, 1.312, 1.332	4.383	60.1	22.9 ± 0.5

*A samples etched in HNO_3; B samples etched in HNO_3 and HF. Ages calculated using IUGS decay and abundance constants; errors based on variation in replicate K_2O and argon analyses or expected variation derived from empirical formula of Tabor and others, 1985.

†P. E. Damon (written communication, 1981) mean of three extractions from same sample of andesite; groundmass etched lightly in HF.

§P. E. Damon (written communication, 1981) mean of five extractions from same sample of dacite; groundmass etched lightly with HF.

**P. E. Damon (written communication, 1981) mean of five extractions from same sample of biotite from tuff with detrital components.

yielding discrepant determinations is to assume that the oldest ages approximate a "minimum" that is assumed to represent the approximate time of volcanic activity. Since the HNO_3-etched splits tended to produce older determinations, these may best approximate the "minimum," but we considered all numbers when we assigned probable ranges. When assigning these ranges, we also considered gaps between determinations and gaps in analytical error. In addition, we applied a test (the Critical Value test of Dalrymple and Lanphere, 1969, p. 120) to determine if, at 95-percent confidence, any real difference in age had been detected. These ranges, then, consist of pools of what we consider to be the most reliable determinations, and we believe that eruption of the volcanic rocks probably took place during the time represented by the ranges. Finally, we examined the range in light of the sparse fossil record. The number of determinations discarded during the assignment process emphasizes the importance of performing multiple extractions on different samples when using altered rock to determine the age of a given volcanic suite.

Plush Range Formation

We report seven new determinations for the Plush Ranch Formation (Table 1, Fig. 3). The five oldest (PR 6A, B; PR 13A, B; PR F6) range from 23.1 to 26.5 Ma.

One sample yielded two determinations (PR 11A, B) significantly younger than five older determinations and similar in value to the oldest of the two determinations already in the literature (20.1 Ma, Crowell, 1973). A 1-m.y. gap separates the analytical error bars of these younger determinations from the error bar of the next oldest determination (23.1 Ma, PR 6B). The difference between the two younger determinations reported here and the older ones exceeds the critical value, indicating that a real difference has been detected. No real difference exists between the average of the splits for PR 6 and PR 13. No difference exists between PR 13A and PR F6, although comparisons of analyses between labs may be questionable. Based on these considerations, we believe that the eruption of these volcanic rocks probably occurred approximately 23.1 to 26.5 Ma, making the Plush Ranch about 3 m.y. older than previously reported.

Although we prefer this range for the Plush Ranch Formation, the Hemingfordian fauna (James, 1963) in the unconformably overlying Caliente Formation may support some of the younger determinations that we disregarded, because the Hemingfordian-Arikareean boundary is about 20 m.y. old (Tedford and others, 1987). Further work with more carefully selected samples may help resolve this issue.

Vasquez Formation

New determinations (n = 9) from the Vasquez Formation give the most discrepant results (Table 1, Fig. 3). Our oldest determination of 23.6 Ma (VF 7B) is younger than the oldest published determinations (24.5 and 25.6 Ma, Crowell, 1973), as well as younger than an unpublished 25.4-Ma determination (P. E. Damon, written communication, 1981). However, error bars overlap slightly with the youngest of the older published determinations.

We therefore accept Crowell's 24.5- and 25.6-Ma determinations and the 25.4-Ma determination by Damon, because they pass the critical-value test. We also reluctantly accept our oldest determination of 23.6 Ma (VF 7B), which is about 1 m.y. younger, because it passes the critical-value test when compared to Crowell's two determinations. However, we repeat that comparisons between gas samples extracted in different labs may be questionable.

The balance of our samples fail to meet the requirements of a closed system, fail the critical-value test when compared to the older determinations, and are probably not useful for determining the age of the volcanism. Three of our determinations (VF 23A, 23B, and 25B) are younger than allowed by the late Arikareean fauna from the unconformably overlying Tick Canyon Formation. Oreodonts from this unit are represented in the Mojave Desert in units assigned to the late Arikareean, which there is considered to be 21 to 22 Ma (Miller, 1980, p. 59; Woodburne and others, 1982; late Arikareen ranges from 20 to 23 Ma throughout North America, Tedford and others, 1987). Two of our determinations (20.6 and 19.6, VF 7A and 25A, respectively) and that of Woodburne (1975) cannot be excluded because of the fossil evidence, since the Arikareean ranges up to 20 Ma, even though Woodburne (1975, p. 59) indicated that his 20.7-Ma age appeared too young. Our figures fail the critical-value test with the next older samples, however, and we believe that they also are too young to represent Vasquez volcanism.

Although low $^{40}K_{rad}$ does not necessarily indicate alteration, the relatively low ages from many of the extractions from Vasquez samples, and the relatively low average $^{40}Ar_{rad}$ compared to the other two units discussed here, may indicate significant postdeposition alteration and argon loss, despite the fact that the samples were selected for qualitatively less alteration. In some cases, an assemblage of altered minerals may include more atmospheric argon than the original assemblage, although we determined neither the quantity nor the species of altered minerals. Alteration of these samples may be indicated by comparing the percentage $^{40}Ar_{rad}$ with the difference between oxygen isotopes from whole rocks and from plagioclase: higher percentage $^{40}Ar_{rad}$ (possibly less altered sample, cleaner extraction, or both) roughly corresponds with smaller differences between the two isotopic values (indicating, perhaps, relatively less alteration). Further study with stratigraphically controlled samples may help us understand how the possible alteration of volcanic rocks from the Vasquez Formation relates to the young determinations derived from them.

The 37.5-Ma determination (P. E. Damon, written communication, 1981) probably is beyond a reasonable figure for the Vasquez. The determination was made on biotite from a tuff with detrital components. Because this determination fails the critical-value test with the next youngest sample, and because a small amount of detrital biotite from the nearby crystalline basement could significantly alter a Miocene determination, we reject it as representing the age of eruption for the volcanic rocks of the Vasquez. If further work should corroborate the determination, ideas about southern California paleogeography would require drastic revision.

Therefore, we consider the range 23.6 to 25.6 Ma to represent the eruptive period of the Vasquez Formation.

Diligencia Formation

Our six determinations for the Diligencia Formation cluster from 21.3 to 23.6 Ma (Fig. 3, Table 1). With the exception of DF 21B, the determinations stack in proper stratigraphic order (T. E. Spittler, written communication, 1984), but no real difference exists among the averages of DF 4, DF 21, and DF 30. The difference between the DF 21 splits does slightly exceed the critical value. No difference exists, however, between the younger splits of DF 4 and DF 30 and of DF 21A—the older of the DF 21 splits. It is interesting to note that the range defined by these determinations approximates the 21- to 22-Ma range assigned the Arikareean oreodonts in the Mojave (Miller, 1980, p. 59; Woodburne and others, 1982).

The youngest determination in the literature (19.1 Ma, Spittler, 1974) is probably too young to represent the age of eruption; it is 2 m.y. younger than our youngest determination and younger than the 20-Ma Hemingfordian/Arikareean boundary (Tedford and others, 1987). It does pass the critical-value test, though, because of the rather large error (10 percent). The younger of Crowell's (1973) two determinations (20.6 Ma) appears also to be a bit young compared to ours. It passes the critical value test because of the rather large error, but that error and the relatively low amount of K_2O (0.280 percent) compared to Crowell's other sample (0.530 percent) and ours (Table 1) make it difficult to evaluate. Nevertheless, because the Arikareean ranges up to 20 Ma, we must include that determination in the range representing Diligencia volcanism, but we omit the 19.1-Ma determination.

Therefore, the 20.6 to 23.6-Ma determinations probably represent the period during which the volcanic rocks erupted.

GEOCHEMISTRY

A combination of major- and trace-element data, and relations and isotopic compositions of the Plush Ranch, Vasquez, and Diligencia Formations indicate that these formations constitute a distinct petrologic subgroup within the northern province of volcanic rocks in western California (Weigand, 1982, p. 178, Fig. 25). The volcanic rocks under study here are subalkaline on the basis of their silica-alkali contents and calc-alkaline on the basis of their AFM ratios (Weigand, 1982, Figs. 14, 15; Spittler and Arthur, 1982, Table 3). These rocks are also classed as calc-alkaline on the basis of the SiO_2-FeO/MgO criterion of Miya-

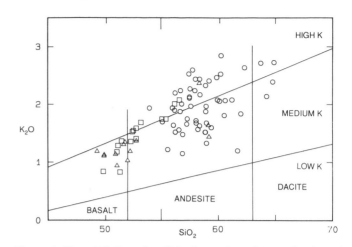

Figure 4. Plot of K_2O against SiO_2 for analyses from volcanic rocks interbedded in the Plush Ranch Formation of Carman (1964) (squares) (Weigand, 1982), Vasquez Formation (circles) (Spittler, 1974; Weigand, 1982 and unpublished data), and Diligencia Formation of Crowell (1975) (triangles) (Spittler and Arthur, 1982), superposed on nomenclatural scheme of Gill (1981).

shiro (1974), the Al_2O_3-normative plagioclase criterion of Irvine and Baragar (1971), the alkali/lime index of Peacock (1931), and the trace-element criterion of Pearce and Cann (1973). A plot of SiO_2 against K_2O (Fig. 4) shows that the rocks range from medium-potassium basalt to high-potassium dacite. The Plush Ranch and Diligencia rocks are medium-potassium basalt with subordinate amounts of medium- and high-potassium andesite; the Vasquez rocks are medium- and high-potassium andesite with subordinate amounts of dacite. Additional analyses seem to indicate that the bimodality in K_2O for the Vasquez volcanic rocks described by Weigand (1982) is a sampling artifact.

The biggest difference in major oxides (Fig. 5) is the lower SiO_2 content of the Plush Ranch and Diligencia rocks compared to those from the Vasquez. This difference, in fact, was the major geochemical criterion that Spittler and Arthur (1973) used to show that the Diligencia and Vasquez volcanic rocks are not correlative. The other major oxides exhibit systematic variation with SiO_2 on Harker diagrams with moderate amounts of scatter (Fig. 5).

Trace elements (Table 2) also exhibit systematic variations with SiO_2. In the case of some elements (La, Zr, Hf, Cs, Th Hf), there is considerable overlap in concentration between the three suites. For other elements (Ni, Cr, U, Sb), concentrations of elements in Plush Ranch and Diligencia samples overlap and differ from those in the Vasquez samples. Differences in SiO_2 usually emphasize this relation. For still other elements (Ba, Sc, La, Co), however, there are nonoverlapping concentration ranges among the three suites. Compared to other Neogene volcanic centers in western California (Weigand, 1982; Johnson and O'Neil, 1984), these three suites are generally lower in Ni and Co and higher in Sr, Rb, Ba, Ta, and U at a given content of SiO_2. In contrast to K_2O relations (Weigand, 1982), the three suites are

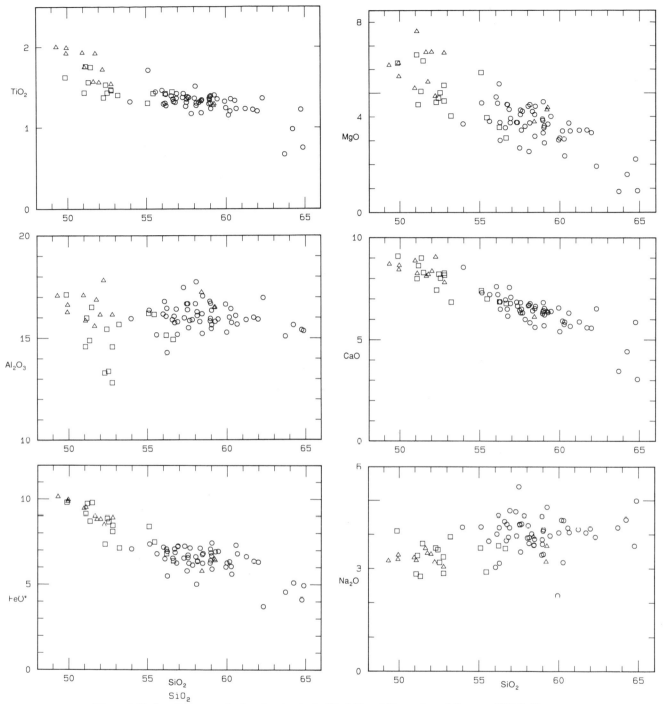

Figure 5. Harker diagrams of volcanic suites in the Plush Ranch Formation of Carman (1964), Vasquez Formation, and Diligencia Formation of Crowell (1975). Sources and symbols same as in Figure 4.

not intermediate in trace-element content between the more mafic rocks in the coastal area (the Conejo and Santa Cruz Island Volcanics) and the more felsic volcanic rocks farther to the north (i.e., Neenach, Pinnacles, and Sonoma Volcanics). On many plots of trace elements and ratios of trace elements against SiO₂, points for volcanic rocks from the Plush Ranch, the Vasquez, and the

Diligencia form relatively discrete linear arrays within large irregular fields representing the two provinces of Cenozoic volcanic rocks from western California. In plots of Th, U, and Hf against Ta, however, our samples form clusters distinct from the large fields representing the coastal and northern provinces (Fig. 6).

Rare-earth element (REE) patterns (Fig. 7) are very similar

TABLE 2. MAJOR- AND TRACE-ELEMENT AND ISOTOPIC ANALYSES FOR SAMPLES OF VOLCANIC ROCKS FROM THE PLUSH RANCH, VASQUEZ, AND DILIGENCIA FORMATIONS, SOUTHERN CALIFORNIA*

	Diligencia Formation			Plush Ranch Formation			Vasquez Formation		
	DF 4	DF 21	DF 30	PR 6	PR 11	PR 13	VF 7	VF 23	VF 25
SiO_2[†]	52.24	52.79	51.66	55.09	52.30	51.48	57.48	56.10	59.04
Ni[§]	70	90	90	85	85	90	35	35	<20
Cr	139	216	184	229	241	209	63.5	90.9	40.6
Co	40.5	37.0	38.5	30.2	29.7	36.7	21.5	25.6	19.4
Zn	70.5	69.7	70.6	63.4	61.2	73	52	59.8	59.1
Zr	160	152	175	178	173	206	180	176	196
Hf	3.96	4.05	3.86	4.35	4.48	4.19	4.83	4.42	5.37
Sc	24.8	22.5	22.9	19.9	19.6	22.7	16.4	18.0	14.5
Sb	0.11	0.14	0.17	0.16	0.32	0.14	0.62	0.44	0.41
Ta	1.46	1.38	1.48	1.58	1.61	1.81	1.43	1.39	1.67
Rb	28.3	32.5	36.2	63.8	71.6	39.1	92.4	89.2	54.5
Sr	525	456	459	402	424	475	414	454	444
Ba	349	401	364	501	525	441	651	630	711
Cs	0.6	1.47	0.963	1.67	3.28	0.498	1.80	5.77	8.07
U	1.05	1.38	1.32	2.13	2.31	1.24	2.60	2.47	2.57
Th	4.17	5.25	5.03	7.27	7.76	4.98	8.10	7.68	10.0
La	23.1	26.0	26.3	26.3	27.3	27.5	28.3	28.1	33.4
Ce	47.9	50.6	52.5	51.9	56.9	53.6	58.5	57.3	65.1
Nd	24.4	23.8	24.4	22.3	25.5	25.6	26.0	27.1	29.2
Sm	5.13	4.93	4.92	5.11	5.47	5.63	5.65	5.44	5.64
Eu	1.77	1.63	1.70	1.49	1.47	1.81	1.57	1.49	1.54
Gd	5.39	5.14	5.56	5.02	5.43	5.81	5.23	4.85	5.13
Tb	0.812	0.714	0.817	0.757	0.769	0.927	0.794	0.704	0.862
Tm	0.399	0.382	0.351	0.371	0.364	0.404	0.388	0.381	0.359
Yb	2.42	2.30	2.21	2.12	2.09	2.50	2.34	2.15	2.28
Lu	0.340	0.315	0.336	0.283	0.293	0.387	0.334	0.304	0.331
$\delta^{18}Op$**	nd	nd	nd	7.2	7.2	6.2	7.6	7.2	7.7
$\delta^{18}Owr$	7.9	8.2	7.4	8.6	9.9	8.8	10.4	10.2	10.8
$^{87}Sr/^{86}Sri$[‡]	0.70564	0.70527	0.70475	0.70509	0.70514	0.70478	0.70558	0.70538	0.70619

*Values in parts per million, except SiO_2 reported in weight percent and oxygen isotopes reported in per mil.
[†]SiO_2 data from Weigand (1982) for Vasquez and Plush Ranch; Spittler and Arthur (1982) for Diligencia. Reported in weight percent.
[§]Trace- and rare earth elements analyzed by J. R. Budahn and R. J. Knight, U.S.G.S., Lakewood, Colorado, using instrumental neutron activation analysis. Reported in parts per million.
**Oxygen isotope analyses by Ivan Barnes and L. D. White, U.S.G.S., Menlo Park, California. Reported in per mil; p = data from plagioclase separates; wr = data from whole-rock samples.
[‡]Strontium isotope analyses by R. W. Kistler and A. C. Robinson, U.S.G.S., Menlo Park, California.

for the nine analyzed samples. They have a moderate light REE enrichment (La ~85 × chondrite) and a flat heavy REE enrichment (Tb_N/Yb_N = 1.1). Diligencia samples have no Eu anomaly, whereas samples from the other two suites exhibit small negative Eu anomalies; Eu*/Eu ranges to 1.16 for the Vasquez and to 1.22 for the Plush Ranch. The La_N/Lu_N ratio is about 8.5, not 85 as reported in Weigand and Frizzell (1986).

We used spider diagrams (Fig. 8) to depict incompatible major and trace elements normalized to chondritic abundances.

The order of the elements was chosen to yield a systematic upward, left-to-right pattern for mid-ocean ridge basalt samples (Thompson and others, 1984, Fig. 2). We eliminated eight rare earth elements, normally found on such plots, from our Figure 8 to reduce redundancy with Figure 7. Basalt and andesite samples from this study (Fig. 8a) form a tight group characterized by a relative depletion of Ba relative to Rb and a fairly monotonic decrease from Rb to Ti. For comparison, Conejo Volcanics (Fig. 8b) are characterized by a large range in concentration for most

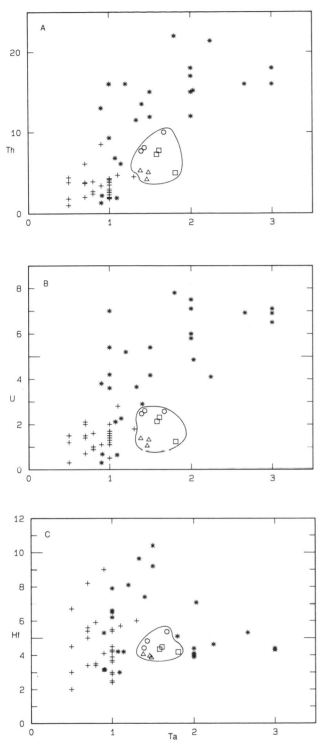

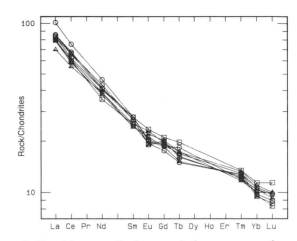

Figure 7. Chondrite-normalized rare-earth element patterns for volcanic rocks from the Plush Ranch Formation of Carman (1964), Vasquez Formation, and Diligencia Formation of Crowell (1975). Data are from Table 2; symbols same as Figure 4.

Figure 6. Trace-element plots comparing data from volcanic rocks in the Plush Ranch, Vasquez, and Diligencia Formations with data from coastal and northern volcanic provinces in western California. Symbols same as Figure 4. A, Th against Ta. B, U against Ta. C, Hf against Ta. Data for the coastal (crosses) and northern (stars) volcanic provinces are largely from Johnson and O'Neil (1984) and Weigand (1982, and unpublished).

of the elements and by a depletion in P relative to Sr and Zr. Overall, absolute abundances are lower than in samples from this study. Andesite samples from Santa Cruz and Anacapa Islands (Fig. 8c) show patterns similar to those of samples from this study, including the relative Ba depletion. In detail, they have a larger concentration range and are depleted in P relative to Zr. Early Miocene volcanic rocks (Fig. 8d) from San Juan Bautista, San Emigdio, Pinnacles, and Neenach are closely related in age and composition (Weigand and Thomas, 1989). San Emigdio basalt samples have lower concentrations in all elements. Andesite samples from these four areas have patterns with steep slopes and a trough for Sr and P.

Initial whole-rock $^{87}Sr/^{86}Sr$ ratios range from 0.7048 to 0.7062. Plagioclase $\delta^{18}O$ values for Plush Ranch and Vasquez samples, which approximate primary magmatic values, range from 6.2 to 7.7 (Fig. 9). Wider $\delta^{18}O$ ranges (Fig. 9) presented by Weigand and Frizzell (1986) were obtained from analyses of whole-rock samples; the data confirm secondary alteration, evident in all samples studied, and increases of $\delta^{18}O$ values by as much as several per mil due to that alteration (Johnson and O'Neil, 1984). These isotopic ratios show a systematic increase with SiO_2, both within suites and between suites. As a group, the samples have higher $^{87}Sr/^{86}Sr$ ratios than Los Angeles–area Miocene volcanic rocks, and lower $\delta^{18}O$ values than Neogene volcanic areas farther to the north (Fig. 10).

DISCUSSION

We believe that the similarities in age ranges, in trace-element relations, and in isotopic compositions of the volcanic rocks we studied indicate that they formed nearly contemporaneously in very similar volcanic provinces. We believe that these similarities allow the Plush Ranch, the Vasquez, and the Diligencia Formations—three cousins estranged from each other by the

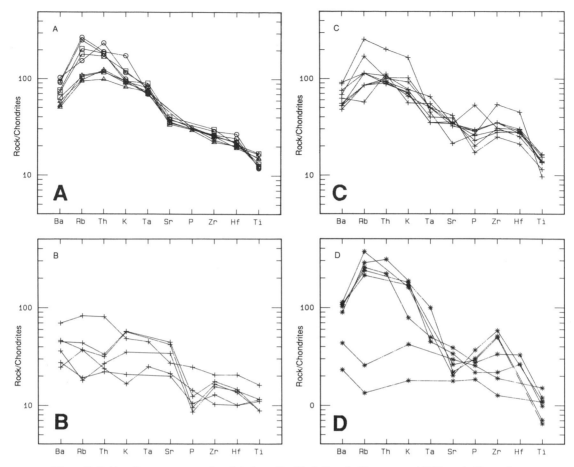

Figure 8. Spider diagrams comparing data from the Plush Ranch, Vasquez, and Diligencia Formations with data from other middle Cenozoic volcanic areas in western California. A, Samples from this study; symbols same as Figure 4. B, Basalt and andesite samples from the Conejo volcanics; data from Weigand (1982 and unpublished). C, Andesite samples from Santa Cruz and Anacapa Islands; data from Weigand (1990). D, Basalt and andesite samples from San Juan Bautista, San Emigdio, Pinnacles, and Neenach; data from Weigand and Thomas (1989).

San Andreas fault system—to be reunited by palinspastic reconstruction of that fault system in southern California.

Volcanic rocks are sparse east of, and adjacent to, the San Andreas fault in southern California (Jennings, 1977). Matthews (1973) correlated the northernmost body east of the fault in southern California, the Neenach volcanics (Fig. 1), with rocks west of the fault at the Pinnacles National Monument, 314 km north. Progressing south along the east side of the fault, the next volcanic rocks encountered are those found in the Diligencia Formation.

These Diligencia rocks occur at the northwest apex of a wedge-shaped region of southeastern California in which early and middle Cenozoic volcanic rocks with various temporal and chemical affinities crop out. However, the Diligencia volcanics occupy a position north of the Salton Creek fault—the northern boundary of the volcanic and plutonic province described by Crowe and others (1979). The volcanic rocks in the Diligencia could be correlative with, and partly fill a time gap in, the "cap-

ping" flows—the youngest in a regional tripartite sequence ranging in age from 36 to 13 Ma (Crowe and others, 1979). The use of sparse chemical data from these volcanic rocks and their intrusive equivalents (Crowe and others, 1979, Table 2; Miller and Morton, 1977, Table 1, nos. 4 through 6) would extend trends in Harker diagrams (Fig. 5) toward more silicic values.

No mafic representatives of the capping unit described by Crowe and others (1979) crop out near the Diligencia basin; instead, the unit is represented in the northwest Chocolate Mountains by dacite domes. Most of the volcanic centers that existed along the length of the range have been removed by erosion, exposing probable cogenetic plutonic affiliates (Crowe and others, 1979, p. 193, Fig. 3-6). These intrusive rocks yield 20- to 24-Ma K-Ar determinations (Miller and Morton, 1977). Thus, if we use the reconstructions of Hill and Dibblee (1953), Crowell (1962), and Bohannon (1975) to reunite the volcanic rocks of the Plush Ranch and Vasquez Formations with an estranged cousin, the volcanic rocks of the Diligencia represent the only petrologic, chemical, and temporal possibility. We cannot rule out, however, the possibility that other similar extrusive rocks have been eliminated by erosion or that additional correlates will be discovered as more data become available.

Alternatively, Powell's (1981, this volume) reconstruction reorders the reunited basins differently. He placed the volcanic rocks of the Vasquez basin adjacent to the southwestern Chuckwalla Mountains and Little Chuckwalla Mountains. There, according to Crowe and others (1979), olivine- and pyroxene-bearing lava flows, representing the three members of the regional tripartitie sequence, yield 26-Ma K-Ar determinations. The volcanic rocks found in the Vasquez Formation are thus similar in composition and age to those found in the Chuckwallas, and our data allow the reconstruction proposed by Powell (1981, this volume).

Our preferred ranges for the extrusion of the lava for the three units indicate that no large disparity in age exists between the three middle Cenozoic basins. The Diligencia, though, is somewhat younger than the Plush Ranch and Vasquez Formations, according to both K-Ar and fossil data.

It might be argued that the K-Ar data also indicate an additional episode of volcanic activity from 20 to 21 Ma for the Plush Ranch and Vasquez Formations. However, we do not believe that the data support such an interpretation. We think that geologic or analytical problems account for these "young" determinations.

The following lines of geochemical evidence suggest a close relation between Vasquez, Plush Ranch, and Diligencia lavas: (1) several trace elements have unique concentrations relative to SiO_2 and define unique fields when plotted against SiO_2 or each other and compared to data from other volcanic rocks of similar age (Fig. 6); (2) the REE patterns are virtually identical (Fig. 7); (3) patterns for other incompatible elements are also virtually identical and differ when compared to other volcanic rocks of similar age from California (Fig. 8); and (4) isotopic ratios of O and Sr are similar, show systematic variation with SiO_2, and are unique compared to other volcanic areas of similar age in southwestern California (Figs. 9, 10).

The moderately high values for Sr and O isotopic ratios for these three suites (and a low Nd-isotopic ratio measured on a single Vasquez sample; Weigand, 1982) indicate the presence of a continental crustal component in these rocks, and the positive correlations between Sr and O isotopic ratios and SiO_2 suggest a systematic increase of crustal interaction. Geochemical data show that Vasquez lavas are more highly evolved than the Diligencia or Plush Ranch rocks. Perhaps the continental crust that interacted with Vasquez magmas was higher in Sr and O isotopes or was thicker, thus providing a longer time for crust-magma interaction. This possibility favors Powell's (1981) reconstruction, which places the Vasquez basin somewhat inboard of the Plush Ranch and Diligencia basins. On the other hand, perhaps Vasquez

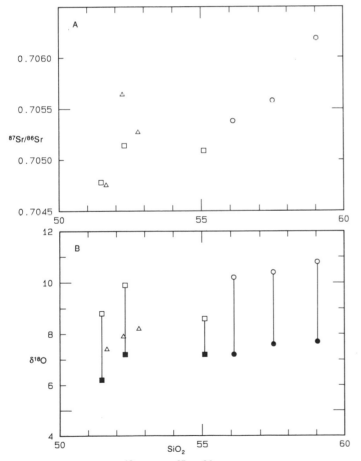

Figure 9. Plot of $\delta^{18}O$ and $^{87}Sr/^{86}Sri$ vs. SiO_2. Analyses on whole-rock samples selected for dating, as well as samples from the Plush Ranch Formation of Carman (1964) and the Vasquez Formation. Symbols same as Figure 4, except filled symbols indicate plagioclase values.

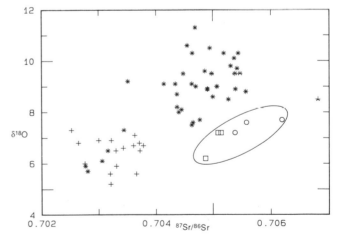

Figure 10. Plot of $\delta^{18}O$ vs. $^{87}Sr/^{86}Sri$, comparing data from the Plush Ranch Formation of Carman (1964) (squares) and the Vasquaz Formation (circles) with those from other Tertiary volcanic suites in California. The numbers of samples from each suite is shown in parentheses after the abbreviation of the suite name. Eastern Coast Range volcanic centers (stars) include Pinnacles (3), Quien Sabe (11), Tolay (3), Sonoma (15), and Clear Lake (8). Southern California centers (plus) include Santa Cruz Island (4), Conejo (9), El Modeno (1), and San Joaquin Hills (1). Data largely from Johnson and O'Neil (1984); some are from Weigand (1982) and Hurst (1983).

magmas simply had longer residence time in the crust and thus a longer time for interaction, relative to the other magmas.

We present no new palinspastic reconstruction, but instead refer to Powell's reconstruction (1981) as a possible "starting point." This basement configuration clumps the overlying estranged nonmarine units and their interbedded volcanic rocks together (Fig. 1e), but does not represent a piercing point. Although Powell's reconstruction seems most attractive, our data only indicate that such clumping of the Oligocene and Miocene volcanic rocks is possible.

Indeed, geochemical and geochronologic data cannot prove the correlation. Neither do our data allow us to distinguish the magnitude and timing of the displacements on various strands of the San Andreas fault system. Instead, we describe the rocks more fully and constrain their time of eruption more closely in order to advance our understanding of early Miocene paleogeography in southern California.

ACKNOWLEDGMENTS

Discussions with T. E. Spittler and R. R. Terres improved our understanding of these rocks. Volcanic rocks analyzed from the Diligencia were collected by Spittler. K. M. Ort made some of the K-Ar analyses. R. G. Bohannon, B. P. Hausback, R. E. Wells, and M. G. Sawlin provided helpful reviews, and suggestions by the editors of this volume were elucidating.

REFERENCES CITED

Arthur, M. A., 1974, Stratigraphy and sedimentation of Lower Miocene nonmarine strata of the Orocopia Mountains; Constraints for late Tertiary slip on the San Andreas fault system in southern California [M.S. thesis]: Riverside, University of California, 200 p.

Bohannon, R. G., 1975, Mid-Tertiary conglomerates and their bearing on Transverse Range tectonics, southern California, *in* Crowell, J. C., ed., The San Andreas fault in southern California: California Division of Mines and Geology Special Report 118, p. 75–82.

—— , 1976, Mid-Tertiary nonmarine rocks along the San Andreas fault in southern California [Ph.D. thesis]: Santa Barbara, University of California, 309 p.

Carman, M. F., Jr., 1964, Geology of the Lockwood Valley area, Kern and Ventura Counties, California: California Division of Mines and Geology Special Report 81, 62 p.

Carter, J. N., Luyendyk, B. P., and Terres, R. R., 1987, Neogene clockwise tectonic rotation of the eastern Transverse Ranges, California, suggested by paleomagnetic vectors: Geological Society of America Bulletin, v. 98, p. 199–206.

Clarke, S. H., Jr., and Nilsen, T. H., 1973, Displacement of Eocene strata and implications for the history of offset along the San Andreas fault, central and northern California, *in* Kovach, R. L., and Nur, A., eds., Proceedings of the Conference on Tectonic Problems of the San Andreas Fault System: Stanford, California, Stanford University Publications in the Geological Sciences, v. 13, p. 358–367.

Crowe, B. M., Crowell, J. C., and Krummenacher, D., 1979, Regional stratigraphy, K-Ar ages, and tectonic implications of Cenozoic volcanic rocks, southeastern California: American Journal of Science, v. 279, p. 186–216.

Crowell, J. C., 1962, Displacement along the San Andreas fault, California: Geological Society of America Special Paper 71, 61 p.

—— , 1973, Problems concerning the San Andreas fault system in southern California: Stanford, California, Stanford University Publications in the Geological Sciences, v. 13, p. 125–135.

—— , 1975, Geologic sketch of the Orocopia Mountains, southeastern California, *in* Crowell, J. C., ed., San Andreas Fault in southern California: California Division of Mines and Geology Special Report 118, p. 99–110.

Crowell, J. C., and Walker, J.W.R., 1962, Anorthosite and related rocks along the San Andreas fault, southern California: Berkeley, University of California Publications in Geological Sciences, v. 40, p. 219–288.

Dalrymple, G. G., 1979, Critical tables for conversion of K-Ar ages from old to new constants: Geology, v. 7, p. 558–560.

Dalrymple, G. G., and Lanphere, M. A., 1969, Potassium-argon dating: San Francisco, California, W. H. Freeman, 258 p.

Ehlig, P. L., 1981, Origin and tectonic history of the basement terrane of the San Gabriel Mountains, central Transverse Ranges, *in* Ernst, W. G., ed., The geotectonic development of California; Rubey Volume 1: Englewood Cliffs, New Jersey, Prentice-Hall, p. 253–283.

Frizzell, V. A., Jr., and Vedder, J. G., 1986, Geologic map of roadless areas and the Santa Lucia wilderness in the Los Padres National Forest, southwestern California: U.S. Geological Survey Miscellaneous Field Studies Map MF-1655-A, scale 1:250,000.

Frizzell, V. A., Jr., Mattinson, J. M., and Matti, J. C., 1986, Distinctive Triassic megaporphyritic monzogranite; Evidence for only 160 km offset along the San Andreas fault, southern California: Journal of Geophysical Research, v. 91, no. B14, p. 14080–14088.

Gill, J. B., 1981, Orogenic andesites and plate tectonics: Berlin, Springer-Verlag, 390 p.

Haxel, G., and Dillon, J., 1978, The Pelona-Orocopia Schist and the Vincent–Chocolate Mountain thrust system, southern California, *in* Howell, D. G., and McDougall, K. A., eds., Mesozoic paleography of the western United States: Pacific Section, Society of Economic Paleontologists and Mineralogists Pacific Coast Paleogeography Symposium 2, p. 453–469.

Hendrix, E. D., and Ingersoll, R. V., 1987, Tectonics and alluvial sedimentation of the Upper Oligocene/lower Miocene Vasquez Formation, Soledad basin, southern California: Geological Society of America Bulletin, v. 98, p. 647–663.

Hill, M. L., and Dibblee, T. W., 1953, San Andreas, Garlock, and Big Pine faults; A study of the character, history, and significance of their displacements: Geological Society of America Bulletin, v. 64, p. 443–458.

Hurst, B. H., 1983, Volcanogenesis contemporaneous with mid-ocean ridge subduction and translation, *in* Augustithis, S. S., ed., The significance of trace elements in solving petrogenetic problems and controversies: Athens, Theophrastus Publications, p. 197–213.

Irvine, T. N., and Baragar, W. R., 1971, A guide to the chemical classification of the common igneous rocks: Canadian Journal of Earth Sciences, v. 8, p. 523–548.

James, G. T., 1963, Paleontology and nonmarine stratigraphy of the Cuyama Valley Badlands, California: Berkeley and Los Angeles, University of California, Department of Geological Sciences Bulletin, v. 45, 170 p.

Jennings, C. W., 1967, Geologic map of California, Salton Sea sheet: California Division of Mines and Geology, scale 1:250,000.

—— , 1977, Geologic map of California: California Division of Mines and Geology, scale 1:750,000.

Jennings, C. W., and Strand, R. G., 1969, Geologic map of California, Los Angeles sheet: California Division of Mines and Geology, scale 1:250,000.

Johnson, C. M., and O'Neil, J. R., 1984, Triple junction magmatism; A geochemical study of Neogene volcanic rocks in western California: Earth and Planetary Science Letters, v. 71, p. 241–262.

Luyendyk, B. P., Kamerling, M. J., Terres, R. R., and Hornafius, J. S., 1985, Simple shear of southern California during Neogene time suggested by paleomagnetic declinations: Journal of Geophysical Research, v. 90, p. 12454–12466.

Matthews, V., III, 1973, Pinnacles-Neenach correlation; A restriction for models of the origin of the Transverse Ranges and the Big Bend in the San Andreas fault: Geological Society of America Bulletin, v. 84, p. 683–688.

Matti, J. C., Frizzell, V. A., and Mattinson, J. M., 1986, Distinctive Triassic megaporphyritic monzogranite displaced 160 ± 10 km by the San Andreas fault, southern California; A new constraint for palinspastic reconstructions: Geological Society of America Abstracts with Programs, v. 18, p. 154.

Miller, F. K., and Morton, D. M., 1977, Comparison of granitic intrusions in the Pelona and Orocopia Schists, southern California: U.S. Geological Survey Journal of Research, v. 5, no. 5, p. 643–649.

Miller, S. T., 1980, Geology and mammalian biostratigraphy of a part of the northern Cady Mountains, California: U.S. Geological Survey Open-File Report 80-978, 121 p.

Miyashiro, A., 1974, Volcanic rock series in island arcs and active continental margins: American Journal of Science, v. 274, p. 321–355.

Muehlberger, W. R., 1958, Geology of northern Soledad basin: American Association of Petroleum Geologists Bulletin, v. 42, p. 1812–1844.

Peacock, M. A., 1931, Classification of igneous rock series: Journal of Geology, v. 39, p. 54–67.

Pearce, J. A., and Cann, J. R., 1973, Tectonic setting of basic volcanic rocks determined using trace element analyses: Earth and Planetary Science Letters, v. 19, p. 290–300.

Powell, R. E., 1981, Geology of the crystalline basement complex, eastern Transverse Ranges, southern California; Constraints on regional tectonic interpretation [Ph.D. thesis]: Pasadena, California Institute of Technology, 441 p.

Ross, D. C., 1972, Petrographic and chemical reconnaissance of some granitic and gneissic rocks near the San Andreas fault from Bodega Head to Cajon Pass, California: U.S. Geological Survey Professional Paper 698, 92 p.

Spittler, T. E., 1974, Volcanic petrology and stratigraphy of nonmarine strata, Orocopia Mountains; Their bearing on Neogene slip on the San Andreas fault, southern California [M.S. thesis]: Riverside, University of California, 115 p.

Spittler, T. E., and Arthur, M. A., 1973, Post early Miocene displacement along the San Andreas fault in southern California, *in* Kovach, R. L., and Nur, A., eds., Proceedings of the Conference on Tectonic Problems of the San Andreas Fault System: Stanford, California, Stanford University Publications in the Geological Sciences, v. 13, p. 374–382.

—— , 1982, The lower Miocene Diligencia Formation of the Orocopia Mountains, southern California; Stratigraphy, petrology, sedimentology, and structure, *in* Ingersoll, R. V. and Woodburne, M. O., eds., Cenozoic nonmarine deposits of California and Arizona: Pacific Section, Society of Economic Paleontologists and Mineralogists, p. 83–89.

Squires, R. L., and Advocate, D. M., 1982, Sedimentary facies of the nonmarine lower Miocene Diligencia Formation, Canyon Spring area, Orocopia Mountains, southern California, *in* Ingersoll, R. V. and Woodburne, M. O., eds., Cenozoic nonmarine deposits of California and Arizona: Pacific Section, Society of Economic Paleontologists and Mineralogists, p. 83–89.

Stanley, R. G., 1987, New estimates of displacement along the San Andreas fault in central California based on paleobathymetry and paleogeography: Geology, v. 15, p. 171–174.

Tabor, R. W., Frizzell, V. A., Jr., Vance, J. A., and Naeser, C. W., 1984, Ages and stratigraphy of lower and middle Tertiary sedimentary and volcanic rocks of the central Cascades, Washington; Application to the tectonic history of the Straight Creek fault: Geological Society of America Bulletin, v. 95, p. 26–44.

Tabor, R. W., Mark, R. K., and Wilson, R. H., 1985, Reproducibility of the K-Ar ages of rocks and minerals; An empirical approach: U.S. Geological Survey Bulletin 1654, 5 p.

Tedford, R. H., and 8 others, 1987, Faunal succession and biochronology of the Arikareean through Hemphillian intervals (late Oligocene through earliest Pliocene epochs) in North America, *in* Woodburne, M. O., ed., Cenozoic mammals of North America: Los Angeles, University of California Press, p. 153–210.

Terres, R. R., 1984, Paleomagnetism and tectonics of the central and eastern Transverse Ranges, southern California [Ph.D. thesis]: Santa Barbara, University of California, 325 p.

Terres, R. R., and Luyendyk, B. P., 1985, Neogene tectonic rotation of the San Gabriel region, California, suggested by paleomagnetic vectors: Journal of Geophysical Research, v. 90, p. 12467–12484.

Thompson, R. N., Morrison, M. A., Hendry, G. L., and Parry, S. J., 1984, An assessment of the relative roles of crust and mantle in magma genesis; An elemental approach: Philosophical Transactions of the Royal Society of London, v. A310, p. 549–590.

Weigand, P. W., 1982, Middle Cenozoic volcanism of the western Transverse Ranges, *in* Fife, D. L., and Minch, J. A., eds., Geology and mineral wealth of the California Transverse Ranges: Los Angeles, California, South Coast Geological Society, p. 170–188.

—— , 1991, Geochemistry and origin of middle Miocene volcanic rocks from Santa Cruz and Anacapa Islands, southern California borderland, *in* Hochberg, F. G., ed., Recent advances in California island research: Santa Barbara Museum of Natural History (in press).

Weigand, P. W., and Frizzell, V. A., Jr., 1986, Age and geochemical data from volcanic rocks in the Plush Ranch, Vasquez, and Diligencia Formations, southern California; Implications for reconstruction of the San Andreas fault: Geological Society of America Abstracts with Programs, v. 18, p. 196.

Weigand, P. W., and Thomas, J. M., 1989, Middle Cenozoic volcanic fields adjacent to the San Andreas fault, central California; Correlation and petrogenesis, *in* Baldwin, E. J., Foster, J. A., Lewis, W. L., and Hardy, J. K., eds., San Andreas fault—Cajon Pass to Wallace Creek: South Coast Geological Society Guidebook 17, v. 1, p. 207–222.

Whistler, D. P., 1967, Oreodonts of the Tick Canyon Formation, southern California: PaleoBios, v. 1, p. 1–14.

Woodburne, M. O., 1975, Cenozoic Stratigraphy of the Transverse Ranges and adjacent areas, southern California: Geological Society of America Special Paper 162, 91 p.

Woodburne, M. O., and Whistler, D. P., 1973, An early Miocene oreodont (Merychyinae, mammalia) from the Orocopia Mountains, southern California: Journal of Paleontology, v. 47, p. 908–912.

Woodburne, M. O., Miller, S. T., and Tedford, R. H., 1982, Stratigraphy and geochronology of Miocene strata in the central Mojave Desert, California, *in* Cooper, J. D., ed., Geologic excursions in the California Desert: Anaheim, California, Cordilleran Section, Geological Society of America Guidebook, p. 47–64.

Ziony, J. I., 1958, Geology of the Abel Mountain area, Kern and Ventura Counties, California [M.A. thesis]: Los Angeles, University of California, 99 p.

MANUSCRIPT RECEIVED BY THE SOCIETY SEPTEMBER 6, 1990

Geological Society of America
Memoir 178
1993

Chapter 9

The Mill Creek Basin, the Potato Sandstone, and fault strands in the San Andreas fault zone south of the San Bernardino Mountains

Peter M. Sadler
Department of Earth Sciences, University of California, Riverside, California 92521
Ali Demirer
Turkish Petroleum Company, Mudafaa cad No. 22, Bakanlikar-Ankara, Turkey
David West and John M. Hillenbrand
Department of Earth Sciences, University of California, Riverside, California 92521

ABSTRACT

Well-lithified Tertiary sedimentary rocks crop out within the San Andreas fault zone south of the San Bernardino Mountains. At the east end of the outcrop is the pre-Pliocene Mill Creek Formation. The diverse compositional facies of the Mill Creek Formation can be explained in terms of a strike-slip basin model. The northern and southern flanks of the basin are characterized by sediments of quite different provevance: garnet- and muscovite-bearing granitoids from the north and Pelona grayschists from the south. Sediments transported into the basin from the southeast are characterized by clasts of volcanic rocks. Conglomeratic sandstone with hornblende- and biotite-bearing granitoid and gneiss clasts entered the basin from the northwest and dominate the axis of the basin.

All of the Tertiary outcrop east of the Mill Creek Formation is assigned to the Potato Sandstone, which has much less compositional variety. The two units are separated by a fault that is probably a major strand of the San Andreas fault, the Wilson Creek strand. The composition and paleocurrents of the Potato Sandstone do resemble the axial deposits in the Mill Creek basin that were derived from the northwest, but the rapid facies changes in strike-slip basins make lithostratigraphic correlations rather unreliable.

In order to account for the garnet- and muscovite-bearing granitoids on the northern flank of the Mill Creek Basin, we suggest that the basin formed in the active Clemens Well–Fenner–San Francisquito fault zone. This is consistent with the pre-Pliocene age of the basin. The Clemens Well fault formed the southern margin. The fault on the northern margin may have been a very early strand in the San Andreas fault zone. The basement clasts in the Potato Sandstone have affinities with the Little San Bernardino Mountains. This suggests that the Potato Sandstone was deposited to the northwest of the Mill Creek basin, perhaps at a later time.

INTRODUCTION

Tertiary sedimentary rocks crop out in much of the San Andreas fault zone along the southern margin of the San Bernardino Mountains (Fig. 1). Aspects of the stratigraphy and prove- nance of these rocks, which constrain the interpretation of the history of the fault, are described below.

Important advances in understanding the San Andreas fault system in the Transverse Ranges have come from a progressive differentiation of the histories of its constituent faults and their

Sadler, P. M., Demirer, A., West, D., and Hillenbrand, J. M., 1993, The Mill Creek Basin, the Potato Sandstone, and fault strands in the San Andreas fault zone south of the San Bernardino Mountains, *in* Powell, R. E., Weldon, R. J., II, and Matti, J. C., eds., The San Andreas Fault System: Displacement, Palinspastic Reconstruction, and Geologic Evolution: Boulder, Colorado, Geological Society of America Memoir 178.

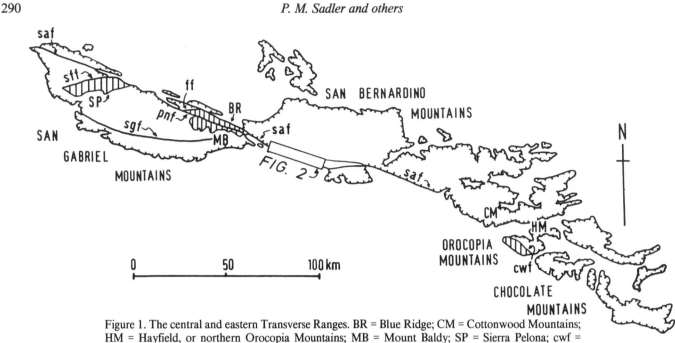

Figure 1. The central and eastern Transverse Ranges. BR = Blue Ridge; CM = Cottonwood Mountains; HM = Hayfield, or northern Orocopia Mountains; MB = Mount Baldy; SP = Sierra Pelona; cwf = Clemens Well fault; ff = Fenner fault; pnf = Punchbowl-Nadeau fault; saf = San Andreas fault; sff = San Francisquito fault; sgf = San Gabriel fault. Vertical ruling indicates Pelona and Orocopia Schists.

separate branches and strands. The trace of the San Andreas fault zone along the southern flank of the San Bernardino Mountains was for many years described in terms of two branches (Dibblee, 1964, 1968). This approach left conflicting evidence for the total displacement and how it should be partitioned between the branches (Woodburne, 1975). Much of the conflict was resolved when Matti and others (1985) explained this segment of the San Andreas fault in terms of four strands braided into a single narrow zone in some places and forming two distinct branches elsewhere.

The realization that relict strands of the San Andreas fault switch back and forth between the north and south branches makes it possible to accept that the total slip across a branch really does vary abruptly from place to place. It also follows that the narrow tract of land between the two branches should include at least two disparate terranes. Yet the well-lithified, terrigenous Tertiary sedimentary rocks that crop out in much of the tract are sufficiently uniform to have been mapped together as the Potato Sandstone (Dibblee, 1970, 1973) and lend some appearance of unity to the fault zone. Careful examination of these sedimentary rocks reveals differences in stratigraphy and provenance; these differences are evidence for distinguishing the major fault blocks and a guide to their former locations.

In this chapter we show that the southeasternmost portion of the Tertiary sedimentary rocks was deposited in a small strike-slip basin, the Mill Creek basin. This part of the outcrop is characterized by rapid lateral changes in clast composition, whereas to the northwest the rocks show considerably more lateral uniformity of facies. The close association of several facies of different provenance in the former strike-slip basin places many useful constraints on the possible position of the basin at the time of deposition. In

the following sections we make the case that the basin formed close to the northern Orocopia Mountains, before initiation of the San Gabriel and San Andreas faults. We also show that the small-scale variability of facies in the strike-slip basin frustrates the attempt to distinguish fault blocks on the basis of lithologic similarity alone. The boundary between laterally uniform and laterally variable facies coincides with one major block boundary that seems to be required by basement rock relationships, but we demonstrate that provenance relationships make it difficult to delimit the blocks elsewhere.

First, it is useful to clarify some aspects of the nomenclature used in this chapter for parts of a fault zone and for the Tertiary sedimentary rocks.

Fault Nomenclature

Following what we understand to be the intent of Matti and others (1985), we use the term strand for those slip surfaces in a fault zone that were active, and essentially continuous, at the same time. A new strand arises when active slip takes a new route through the fault zone and abandons some, though not necessarily all, of the formerly active slip surface. Some parts of inactive, or relict, strands are likely to be found as disjunct fault segments, separated by slip on younger strands. Since we are primarily concerned with a relict strand, the Wilson Creek strand of Matti and others (1985), the term strand is used to refer collectively to all the fault segments that are interpreted to have been part of the same formerly active trace. A relict strand is thus interpretive. We use the term fault, not strand, for individual fault segments that have been recognized on the ground.

We now spell out this distinction for the Wilson Creek strand. Matti and others (1985, p. 6, 10) recognized that the

Wilson Creek fault zone of Smith (1959) and Gibson (1964, 1971) juxtaposes contrasting basement terranes. They presented the case that the Wilson Creek fault is part of an early strand of the San Andreas fault with considerable offset—the Wilson Creek strand. They suggest that this strand also includes the Yucaipa Ridge fault (Smith, 1959), and a previously unnamed fault, which we term the Elder Gulch fault (Fig. 2).

The terms north and south branch are convenient for reference to the margins of the fault zone, when no particular strand is implied.

Mill Creek Formation and Potato Sandstone

Well-lithified, superficially similar, terrigenous Tertiary sedimentary rocks are exposed in the San Andreas fault zone from Waterman Canyon to Wilson Creek (Fig. 2). Stratigraphic relationships are best exposed near City Creek and Mill Creek. In the Mill Creek area it is useful to distinguish two structural provinces:

1. A relatively large, lens-shaped block, dominated by the Yucaipa Ridge syncline, occupies the southern and central portion of the outcrop. The syncline is a simple, open, doubly plunging structural basin, uninterrupted by major faults.

2. Sediments with more complex deformation crop out in a narrow strip that is in fault contact with the east and north margins of the Yucaipa Ridge syncline. This northern fault zone is a strip of smaller tilted blocks, separated by faults. The same structural style continues northwest beyond the Mill Creek area to Waterman Canyon.

Outcrops of Tertiary sedimentary rocks in the San Andreas fault zone near Mill Creek were named the Potato Sandstone by Vaughan (1922). Smith (1959) applied the same name throughout the Yucaipa Ridge Syncline. Subsequent usage has made two partially conflicting changes. Dibblee (1964, 1970, 1973) used the name Potato Sandstone for all the superficially similar Tertiary sedimentary rocks in the fault zone from Wilson Creek to Waterman Canyon. But Owens (1959), Gibson (1964, 1971), Dibblee (1982), Demirer (1985), Sadler and Demirer (1986), and West (1987) used the name Mill Creek Formation for the rocks in the Yucaipa Ridge syncline.

Just how much of the Potato Sandstone any author assigned to the Mill Creek Formation depended on his interpretation of the fault zone on the north side of the Yucaipa Ridge Syncline (summary *in* West, 1987). Owens (1959) and Demirer (1985) have suggested that different formations occupy the syncline and the fault zone. Matti and others (1985) used a similar argument to place their Wilson Creek strand within the sedimentary outcrop, at the southern edge of the northern fault zone. But Smith (1959),

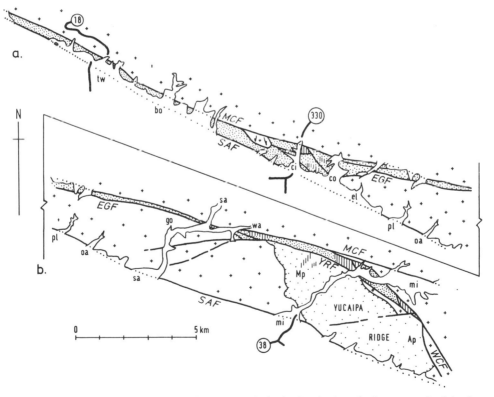

Figure 2. Simplified map of Tertiary sedimentary rocks in the San Andreas fault zone south of the San Bernardino Mountains. Sparse stipple indicates Mill Creek Formation; dense stipple, Potato Sandstone; vertical ruling, red sandstones and siltstones; crosses, crystalline basement rocks. MCF = Mill Creek Fault; EGF = Elder Gulch fault; SAF = south branch of the San Andreas fault; YRF = Yucaipa Ridge fault; WCF = Wilson Creek fault; Ap = Allen Peak; Mp = Morton Peak; bo = Borea Canyon; ci = City Creek; co = Cook Creek; el = Elder Gulch; go = Government Canyon; mi = Mill Creek; oa = Oak Creek; pl = Plunge Creek; sa = Santa Ana River; tw = Twin Creek; wa = Warm Springs Canyon.

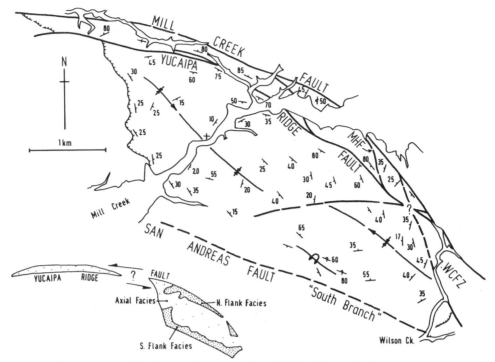

Figure 3. Structure map of the Yucaipa Ridge syncline. WCFZ = Wilson Creek Fault zone. Inset shows distribution of axial and lateral facies; question mark indicates uncertainty concerning the amount of offset on the Yucaipa Ridge fault.

using the name Potato Sandstone, and Dibblee (1982), using the name Mill Creek Formation, managed to fit a single set of units to outcrops in both structural provinces. Gibson (1964, 1971) dismissed the distinctive appearance of the rocks in the northern fault zone as due to intense brecciation and staining.

We regard only the sedimentary rocks of the Yucaipa Ridge syncline as formally attributable to the Mill Creek Formation. The name Potato Sandstone is applied informally to the remainder of the Tertiary outcrop, following Dibblee (1964, 1970, 1973). Whether more of the Potato Sandstone should be included in the Mill Creek Formation is an important open question. The answer influences the concept of the Wilson Creek strand, and is addressed later in this chapter, after descriptions of the Mill Creek Formation and two outcrops of the Potato Sandstone.

MILL CREEK FORMATION IN THE YUCAIPA RIDGE SYNCLINE

The early studies (Smith, 1959; Owens, 1959; Gibson, 1964) differ considerably in their finer stratigraphic subdivision of the sedimentary sequence in the Yucaipa Ridge syncline (summarized *in* Demirer, 1985; West, 1987). The differences are not simply a matter of historical accident; they reflect varying success at recognizing the importance of lateral facies changes. We have identified a consistent pattern of steep interfingering contacts be-

tween the axial facies and several marginal facies (Figs. 3, 4). The best model for this facies architecture is the fill of a strike-slip basin.

The deeply incised drainages of lower Mill Creek and along the San Andreas fault provide the vertical relief necessary to reconstruct the three-dimensional facies relationships in the Yucaipa Ridge syncline. Landsliding is extensive but does not obscure the facies distribution. In the core of the fold, the Mill Creek Formation is essentially an upward thickening and coarsening sequence of sandstone and conglomerate. The amount of detrital chlorite and interbedded shale decreases upward, producing a gradational contact between two mappable units—a green sandstone facies and a buff sandstone facies. These two facies appear to occupy the depositional axis of what we call the Mill Creek basin; their transition is the only facies change in the Yucaipa Ridge syncline that is essentially parallel to bedding and represents a succession in time. In all directions except the northwest, the two axial facies are replaced laterally, by basin-margin facies that are readily distinguished by composition and color (Figs. 3, 5). The marginal facies are typically coarser. Four marginal facies have been recognized: (1) gray, Pelona Schist–bearing breccia; (2) purple, andesite–bearing conglomerate; (3) white, monzonite, and diorite-bearing breccia; and (4) muscovite-rich sandstone and conglomerate. Red sandstone and siltstone occur near the contact between the muscovite-rich sandstones and the axial facies; they are a volumetrically very minor component of the formation, but they suggest a link with the Potato sandstone.

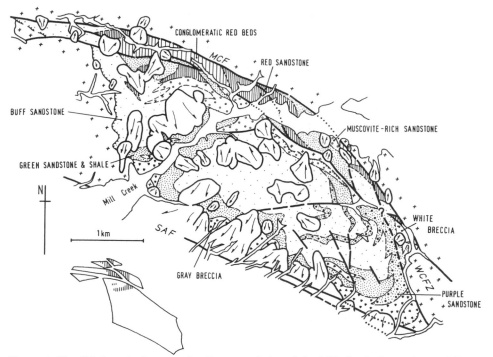

Figure 4. Simplified geologic map of sedimentary facies of the Mill Creek Formation and Potato Sandstone in the Mill Creek area. Arrows indicate landslides; crosses, crystalline basement rocks. Inset shows the most conservative strike-slip restoration of the Yucaipa Ridge and Wilson Creek fault zones; it collects together outcrops of silty redbeds (ruling).

Axial Facies

The sedimentary sequence in the basin axis is described in terms of two facies, beginning with the older.

Green Sandstone Facies. Thin-bedded gray-green sandstone and shale form the base of the fill, or the oldest exposed facies, in the center of the basin. This basal facies is characterized by several tens of meters of interbedded shale and dark gray-green, micaceous wacke with subordinate calcareous shale, and algal limestone. The sandstone beds are typically bounded by well-defined, parallel, bedding surfaces, but packets of several beds are locally deformed into isoclinal slump folds. Some of the thinner sandstones (less than 15 cm) show complete or partial sets of the internal structure sequence described by Bouma (1962) from turbidites. These beds are rarely conglomeratic. The thicker sandstone beds have sharp bases and a pervasive parallel lamination, that is often contorted by soft-sediment deformation. Bed thickness generally increases up-section, where thick, cross-bedded packets of conglomeratic sandstone interfinger with the sandstone and shale. The proportion of interbedded shale decreases up-section. Where conglomeratic sandstone dominates in the upper part of the section, parallel bedding surfaces are replaced by less regular, scoured contacts. Plant remains are abundant in the sandstone; ostracods occur in the shale and limestone.

The granule- and pebble-sized clasts in the conglomerates are not readily compared with the coarser clasts in the conglomerates of other facies. As a result of the fine grain size, quartz and feldspar clasts are abundant. Some conglomeratic layers contain Pelona Schist fragments.

The basal, green sandstone facies is interpreted to be of lacustrine origin, representing the bottomset and lower foreset deposits of a southeastward prograding, "Gilbert-type" delta (Stanley and Surdam, 1978). The bottomset beds, which have incomplete Bouma sequences of internal structure, resemble lacustrine turbidites described elsewhere (Ludlam, 1974; Sturm and Matter, 1978). Ripple lamination, grain orientation, and sole marks indicate transport from the north and northwest (Gibson, 1964; Demirer, 1985). The uppermost part of the sequence exhibits the fluvial character of topset deltaic deposits, and passes conformably upward into a buff sandstone facies that fills much of the basin.

Buff Sandstone Facies. The younger and more extensive axial facies is dominated by light buff–colored, massive, and thick-bedded conglomeratic sandstone, interbedded with greenish brown shaly siltstone. Thick sets of tabular cross-beds occur near the transition to the basal green sandstones. The sandstone in this younger axial facies is still micaceous, but biotite is much more abundant and chlorite is less abundant than in the underlying green sandstones. The facies may exceed 1,000 m in thickness, but probably consists of thinner imbricated packets. The most abundant conglomeratic beds have clasts of biotite quartz monzonite, leucocratic granitoids, biotite gneiss, and schist. The proportion of gneissic clasts is rather variable. Hornblende-bearing granodiorite and diorite clasts are present in some beds. These characteristic conglomerate beds are intercalated with subordinate conglomerates of different composition. Near the base of the unit there are conglomeratic beds with Pelona Schist clasts, and others with andesite clasts. Near the top of the unit in the northwest, conglomeratic beds appear that are dominated by muscovite- and garnet-bearing granitoid clasts. To the southeast, andesite clasts appear in an increasing proportion of conglomerate beds. All these minor, but compositionally distinct, conglomerate beds are distal intercalations of facies from the margins of the outcrop.

The buff sandstone facies is interpreted as braided stream deposits atop a shallow delta that prograded into the basin from the north-northwest.

Age and Provenance. The axial facies provide the only evidence for the age of the Mill Creek Formation. Fossil plants from the green sandstone and buff sandstone members of the formation, collected by Owens (1959), Smith (1959), and Gibson (1964), were identified by D. I. Axelrod, who inferred an "early Pliocene" age to be most likely. He reported to Gibson (1964, p. 9) that the flora might be "as old as Late Miocene or as young as mid-Pliocene," but the boundary of these time-stratigraphic units had not then been re-calibrated to the currently accepted age of 5 Ma. Using more recent equations of floras with land mammal faunas, Woodburne (1975, p. 69) translated the age determination to Clarendonian; this implies that age of the Mill Creek Formation is more likely 10 to 13 Ma. Even for this age the sandstones might be considered unusually well indurated. But induration would have been accelerated by the high content of detrital phyllosilicates and perhaps a high heat flow in the fault zone.

Abundant sedimentary structures indicate that most of the axial deposits entered the basin from the northwest (Gibson, 1964, 1971; Demirer, 1985). The clasts in the conglomerate beds are rounded, so the sources were not necessarily close. Also, since paleoflow directions in the center of a basin probably reflect redirection by the axial topography, the position of sources is not accurately indicated. The rather nondistinctive granitoid and gneissic clasts in the buff sandstones might be derived from the Little San Bernardino Mountains. We show below that the conglomerates of the Potato Sandstone are similar.

Smith (1959, p. 30) reported a limestone clast with well-preserved Permian fusulinids from the buff sandstone facies. The source of this clast is not known, but we suggest that it was not the Furnace Limestone of the San Bernardino Mountains, north of Big Bear Lake. We have found fusulinids in marbles at localities where Richmond (1960) reported corals and brachiopods from the Furnace Limestone. Unlike Smith's find, nearly all of these fusulinids were too recrystallized to preserve diagnostic wall structures. Furthermore, Furnace Limestone crops out in close association with very resistant Cambrian and Pre-Cambrian quartzites; the limestone is unlikely to become part of a detrital assemblage such as the Mill Creek Formation that lacks quartzite clasts.

Marginal Facies

Three marginal facies can be distinguished from the sediments of the basin axis by major differences in composition that impart distinctly different colors (Figs. 5, 6). Where exposed, the northern and southern flanks of the basin are each occupied by a single alluvial facies. A fluvial facies occupies the southeast corner. These facies and possible sources for them are described below in anticlockwise sequence, beginning on the south flank.

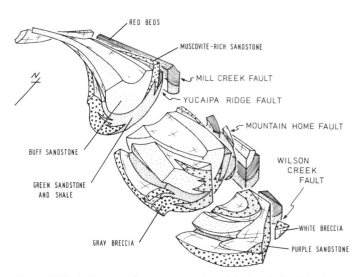

Figure 5. Block diagram of structure and lateral facies relationships in the Yucaipa Ridge syncline and adjoining fault zones. Simplified from Figures 3 and 4.

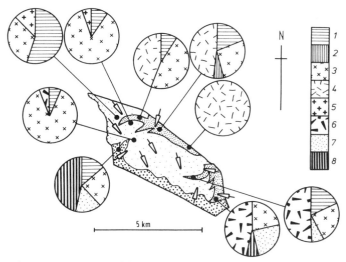

Figure 6. Provenance of the Mill Creek Formation. Arrows summarize paleocurrents determined from sedimentary structure. Proportionally divided circles show composition of clast suites in conglomerates. Key to numbers: 1, biotite-gneiss; 2, foliated biotite-bearing granitoid; 3, biotite quartz-monzonite; 4, muscovite-garnet-bearing granitoid; 5, hornblende diorite and granodiorite; 6, volcanic rocks, mostly andesite; 7, sandstones; 8, Pelona grayschist.

Gray Breccia Facies. Pelona Schist clasts characterize a gray breccia facies, which occurs along the southern edge of the outcrop of the Mill Creek Formation, where it is now truncated by the south branch of the San Andreas fault. The facies is at least 200 m thick and in depositional contact with the basement, but it fingers out abruptly northward into the facies of the basin axis. At the upper limit of this facies the sandstones and conglomerates of the basin axis overstep onto basement rocks.

The gray breccia facies consists of very coarse grained, gray-green, breccio-conglomerate, cobble- and boulder-conglomerate beds, and coarse-grained, chloritic sandstone lenses. In most beds the maximum clast size is 10 to 20 cm, although clasts larger than 50 cm can be found. The matrix of the breccia and conglomerate is a coarse-grained lithic arenite.

At the northwest limit of its outcrop, the gray breccia is dominated by shingle-shaped clasts of Pelona Schist (47 percent), with subordinate mafic gneiss (19 percent), quartz monzonite (12 percent), red and gray fine-grained sandstone (16 percent), and vein quartz (4 percent). The Pelona Schist clasts are essentially all a muscovite-quartz-albite grayschist variety (87 to 96 percent), with subordinate epidote-chlorite-albite greenschist (1 to 6 percent) and metachert (4 to 7 percent). Garnet-bearing schist was not found. The clast composition varies up-section and to the southeast. Plutonic rock clasts become more frequent up-section in the northwest part of the outcrop. To the southeast there is a progressive admixture of better rounded clasts of altered rhyodacite, dark purple andesite, and purple sandstones, whereas the proportion of Pelona Schist clasts falls to 10 to 15 percent.

Imbricated clasts and cross-bedding, measured at 12 localities, indicate transport to the northeast. The coarseness and textural immaturity indicate a nearby source, along the south flank of the basin. The sedimentary structure and lithologic association of the gray breccia facies meet the criteria given by Nilsen (1982) for alluvial fan deposits. It is probably the product of a series of small coalesced fans along an active fault scarp, perhaps a small-scale analog of the Violin breccia in the Ridge basin (Crowell, 1982b).

The source area for the Pelona Schist clasts in the gray breccia facies must provide the correct proportions of grayschist to metachert and greenschist, but it must not provide the garnet-bearing schists. The source should also lie next to a fault that could have been active during deposition in the Mill Creek Basin.

Orocopia Schist, a lithologic equivalent of the Pelona Schist, in the Orocopia Mountains, now southeast of the basin, was suggested by Gibson (1964, 1971) to be the source of the gray breccia. Gibson's argument was incorrectly based on southeastward paleocurrents determined from the axial buff sandstone facies where it does not contain Pelona Schist. The Orocopia Schist outcrop is rather small and heterogeneous; we found an unacceptably high content of greenschist there and distinctly different grayschists with albite porphyroblasts.

On the southwest side of the San Andreas fault, Pelona Schist occurs in the Crafton Hills and scattered hills in the San Bernardino Valley, along Blue Ridge, near Mount Baldy, in the Sierra Pelona, and at Mount Pinos. The Pelona Schist of Blue Ridge, between the Punchbowl and San Andreas faults, is eliminated as a possible source: it is of high metamorphic grade and includes garnet-bearing schist. The grayschists in Crafton Hills and San Bernardino Valley exposures are more siliceous (D. M. Morton, personal communication). Furthermore, they are rejected as sources because that would imply an unacceptably small displacement across the San Andreas fault.

Grayschist and greenschist occur without garnet-bearing schists in the Mount Baldy area (Ehlig, 1981). Morton (in Woodburne, 1975, p. 34) has noted that this is a suitable source for the clasts in the gray breccia. We characterized this source by clast counts at Stockton Flat. The Pelona Schist proportions there (87 to 91 percent grayschist, 3 to 4 percent metachert) are very similar to those in the gray breccia. The clasts would have to be transported across a strand of the San Andreas fault *before* the slip on the Punchbowl fault introduced garnet-bearing schists to the east. This reconstruction is at odds with evidence for very young uplift of the Mount Baldy source (Ray Weldon, written communication, 1986); this Pelona Schist source had probably not been unroofed prior to accumulation of the gray breccia facies.

The Sierra Pelona includes a large outcrop of suitable Pelona Schist, bordered by the San Andreas fault to the east and the San Francisquito fault to the north. For these schists to have been shed into the Mill Creek basin across the San Andreas fault, deposition must occur *after* slip on the Punchbowl fault has removed the garnet-bearing schists. But this event is most likely younger than deposition in the Mill Creek basin (Woodburne, 1975). We favor a model in which the San Francisquito fault forms the southern margin of the Mill Creek basin; this is also the only reconstruction of the southern flank that readily accounts for garnet muscovite granitoid sources on the north flank.

In the southeast corner of the basin, the Pelona Schist–bearing gray breccia contains purple volcanic clasts. These clasts are a better rounded admixture from a separate clast suite (Purple conglomerate facies) without Pelona Schist clasts, so they do not need to be available in the same source terrain as the Pelona Schist.

Purple Conglomerate Facies. Toward the southeast corner of the modern outcrop, the buff conglomeratic sandstone of the basin axis interfingers with sandstone and conglomerate of different provenance. This facies consists of similar bed forms and sedimentary structures and is probably also a braided stream deposit, but it is differentiated because its clast suite includes volcanic rocks and reworked sandstones, and its paleoflow was from the east and northeast (Gibson, 1964, 1971). The facies is about 400 m thick and is typically purple or red. Maximum clast size is about 15 cm and the clasts are rounded. The volcanic clasts include basaltic andesite, porphyritic hornblende andesite, trachyandesite, vesicular basalt, and porphyritic rhyolite.

The rounding of the volcanic clasts suggests a relatively long, fluvial transport path. From the paleocurrent directions it appears likely that this path entered the basin via the fault zone that formed its south flank. Gibson (1964) pointed out that the volcanic clasts could be derived from the Chocolate Mountains; the Mint Canyon Formation of the Soledad basin is an alternate, second-cycle source.

Muscovite-rich Facies. Biotite and chlorite are the dominant detrital micas in the sandstones and conglomerates of the basin axis, but we have distinguished a superficially similar, muscovite-rich conglomeratic sandstone facies near the northern edge of the outcrop. This facies is dominated by very thick-bedded, often cross-bedded, white and buff sandstone and conglomerate with abundant detrital muscovite. The clasts, which are not easily distinguished from the matrix, are muscovite-rich, garnetiferous granite, biotite- and two-mica quartz-monzonites, and gneiss. The facies includes coarse, monolithologic breccio-conglomerate with angular and subangular clasts of the muscovite garnet granite set in a compositionally similar matrix of coarse arkosic arenite. The matrix is more rarely calcareous and resembles lenses of algal, ostracod-bearing limestones found near the base of the Mill Creek Formation (Gibson, 1964; Demirer, 1985). These limestones are contaminated by muscovite-rich arkosic sand and gravel.

This facies is contemporaneous with most of the axial facies. It crops out just south of the Yucaipa Ridge fault, next to exposed crystalline basement. On the flanks of Morton Peak, conglomerate beds of this facies are intercalated with buff conglomeratic sandstone in the younger parts of the axial deposits. There is relatively little mixing with the typical clast suites of the buff sandstone. The muscovite-rich facies becomes finer grained when traced toward the northeast. It is interbedded with both green and buff sandstones of the axial facies, but pinches out rapidly toward the basin center. At their northeastern limit the muscovite-rich sandstones occur with thin red sandstones and green siltstones that are also intercalated with the axial buff sandstones.

Muscovite-rich granitoid sources occur sporadically throughout the eastern Transverse Ranges, where they are often among the youngest of the Mesozoic plutons. Most of the occurrences are minor dikes and far from faults on which they might be restored to the northern flank of the Mill Creek Basin. There are major outcrops across the eastern ramp of the San Bernardino Mountains (Sadler, 1981), in the Little San Bernardino Mountains (Rogers, 1961), and in the Hayfield Mountains north of the Orocopia Mountains (Powell, 1981). The Hayfield Mountains are situated north of the Clemens Well fault and a parallel fault at the northern limit of the Tertiary units in the Orocopia Mountains (Arthur, 1974). The Clemens Well fault can be used to restore the Sierra Pelona to a former position south of the Hayfield Mountains (Powell, 1981).

We have found the large, modern alluvial fan that issues southward from the central Hayfield Mountains and fills much of the basin north of the Orocopia Mountains to be a very good analog for the muscovite-rich facies. The gravels on this fan are dominated by clasts of muscovite-bearing granitoids. There is

coarse, white muscovite-garnet-granite and medium-grained muscovite–biotite–quartz monzonite. Mafic clasts are conspicuously absent. To the east the fan has a small admixture of locally derived biotite-rich schist, gneiss, and muscovite-garnet–bearing calc-silicates. At Salt Wash, which forms the eastern edge of the fan, there is an influx of sand and better rounded cobbles from the Chocolate Mountains. Clasts of vesicular basalts, purple andesites, and layered red tuffs are abundant and reminiscent of the purple andesite-bearing conglomerates of the Mill Creek Formation. Parts of Salt Wash are actively eroding and recycling older red sediments.

Red Sandstone and Siltstone

At the west end of the north flank of the Yucaipa Ridge syncline, the buff sandstones of the basin axis become intercalated with redbeds. The intercalated portion of the section is about 70 m thick, it contains red sandstone and siltstone, green siltstone, reddish or purple conglomerate, white muscovite-rich conglomerate sandstone (as described above) and buff conglomeratic sandstone typical of the basin axis.

Red and purple beds with sharp contacts range from 6 to 60 cm in thickness. They consist of medium- to coarse-grained, silty sandstones. Their sedimentary structures include cross-bedding and convolute and parallel lamination. The beds usually have irregular bottoms and shaley to silty tops. They occur in packets as thick as 12 m and persist laterally at least 100 m. The paler reddish conglomerates contain a clast assemblage that is dominated by olive-green gneisses, hornblende monzonites, and a variety of granitoids. When traced westward, the redbeds lose the conglomeratic component, become very thin, and finally occur only as isolated, indistinct layers of red staining.

This facies is of interest primarily because it resembles redbeds in the Potato Sandstone that crop out in the fault zone to the north.

POTATO SANDSTONE AT MILL CREEK

Before attempting a reconstruction of the Mill Creek basin, we need to describe the Potato Sandstone that crops out in the fault zone north of the Yucaipa Ridge syncline and along the San Andreas fault zone to the northwest. The crucial question is whether these rocks are also part of the Mill Creek basin.

The dominant and most distinctive sedimentary rocks of the northern fault zone at Mill Creek are redbeds. In addition there are buff and green conglomeratic sandstones, which are similar to facies described from the axis of the Yucaipa Ridge syncline.

Redbed Facies

Redbeds crop out on both sides of the Mountain Home fault (Fig. 3). West of this fault, from Mill Creek to Warm Springs Canyon, redbeds dominate the outcrop and only their lower con-

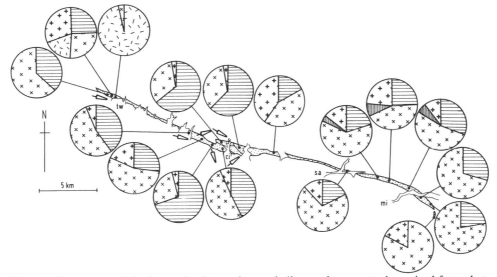

Figure 7. Provenance of the Potato Sandstone. Arrows indicate paleocurrents determined from clast imbrication; proportionally divided circles, composition of clast suites in conglomerates. Symbols as in Figure 6. Abbreviations as in Figure 2.

tact is seen. To the east on Yucaipa Ridge, redbeds are less extensive and only their upper contact is seen. At both their upper and lower contacts, the redbeds are intercalated with sandstone beds that are comparable with the axial facies of the Mill Creek basin. The transitional sections between the redbeds and the buff and green sandstones resemble the occurrence of redbeds on the north flank of the Yucaipa Ridge syncline.

On both sides of the Mountain Home fault, the redbeds include a deeper red to purple silty facies and a paler, pink, conglomeratic facies. In both fault slices the conglomeratic facies is west of the silty facies. The characteristic red to purple color is mostly imparted by iron oxides in the silty portion of the matrix. There is no correlation between pigmentation and degree of shearing (cf. Gibson, 1964, 1971).

The darker, silty facies are alternations of mottled medium-grained sandstone with red and green siltstone beds and less abundant conglomeratic sandstone. The sandstone includes poorly sorted arkosic arenite and feldspathic graywacke, with low quartz content and abundant fresh, detrital biotite. The feldspars are extensively altered to clay minerals and in some cases replaced carbonates. Many sandstone beds are graded or include convolute lamination. Sparse bioturbation (meniscus-filled burrows) is present.

In the paler conglomeratic facies, two conglomerate bed types may be distinguished. Ungraded, or reversed graded, matrix-supported beds with irregular tops are interpreted as debris flow deposits. The clasts are subrounded and lack preferred orientation. Some beds reach 2 m in thickness; the largest clasts exceed 0.5 m, and some protrude into the overlying bed. Lenticular, normally graded, sandy conglomerates, with erosional bases, are attributed to braided channel filling. Many have large-scale cross-bedding or a crude parallel stratification, and rarely include clasts coarser than 30 cm.

In the older redbeds, exposed to the west of the Mountain Home fault, the clast assemblage is characterized by abundant gneiss, hornblende quartz monzonite, crudely foliated biotite granitoid, biotite quartz monzonite, and biotite-muscovite quartz-monzonite (Fig. 7). The hornblende quartz monzonite and an olive-green variety of gneiss become more abundant toward the west; they are absent from the reddish conglomerates exposed on the east side of the Mountain Home fault near the top of the redbed sequence. The younger, eastern clast assemblage includes abundant pegmatite, biotite-rich gneiss, and the biotite- and biotite-muscovite-quartz monzonite.

The clast assemblages in the redbeds contain no rocks that pinpoint a source. The mix of gneisses and granitoids could be derived from the eastern Little San Bernardino Mountains or the Cottonwood Mountains (see descriptions *in* Powell, 1981); it also resembles some of the basement rocks that crop out between Cook Creek and Plunge Creek (Fig. 2), both in the San Bernardino Mountains and within the San Andreas fault zone. The clast assemblages and their range of variation match those from other Potato Sandstone exposures northeast to Twin Creek (Fig. 7).

Green and Buff Conglomeratic Sandstone Facies

Thinner, green, silty sandstone beds and thicker, buff, conglomeratic sandstone beds found within the northern fault zone resemble the green sandstone and buff sandstone facies of the Mill Creek Formation in the Yucaipa Ridge Syncline. Detailed examination of the clast suites in the conglomeratic beds reveals three distinct clast suites. Most beds resemble the buff sandstone facies of the Mill Creek Formation; these contain gneiss, quartz monzonite, and pegmatite. Some beds have a muscovite-rich matrix and include reworked pieces of micritic limestone; they resemble the muscovite-rich facies of the Mill Creek Formation. Other beds contain hornblende-bearing granitoids and olive-green gneisses in various proportions, which is characteristic of the redbeds.

White Breccia Facies

This white breccia is a coarse unit with a very limited outcrop, near the eastern edge of the Yucaipa Ridge syncline in Wilson Creek. It has a maximum thickness of 150 m. Although previously mapped as plutonic basement and/or fault breccia (Smith, 1959; Gibson, 1964), these rocks are clearly sedimentary; they are intercalated with thin, dark, sandy beds and interfinger with the purple conglomerate facies. The clasts are porphyritic biotite quartz monzonite, and subordinate hornblende-biotite quartz diorite. Most clasts are 3 to 8 cm in diameter, but a few are 25 cm in diameter. The white breccia is interpreted to be a very proximal alluvial fan or talus facies. It cannot be derived from basement rocks now immediately east of the Wilson Creek fault zone, since they include quartzite, marble, and schist.

We once sought to include this unit in the Mill Creek Formation (Sadler and Demirer, 1986). Subsequent mapping east of City Creek (see next section) identified such a breccia as the base of the Potato Sandstone.

Summary

The Potato Sandstone in Mill Creek is characterized by an abundance of redbeds. The upper and lower contacts of the redbeds are marked by intercalation with buff or green sandstone facies that resemble the axial facies of the Mill Creek Formation. Although redbeds are quite rare in the Mill Creek Formation, their limited occurrence is in transitional sequences very similar to the Potato Sandstone.

A simple lithostratigraphic correlation can thus be made between the Mill Creek Formation and the Potato Sandstone in the Mill Creek area. There is no proof that the rocks are of the same age, but the similarity of texture and provenance suggests that the Potato Sandstone could be an offset portion of the axial facies of the Mill Creek Formation. It follows from this interpretation that the Yucaipa Ridge fault does not necessarily have large displacement. The Potato Sandstone can be restored close to the red beds in the Mill Creek basin by removing as little as 2 to 5 km of right slip from the Yucaipa Ridge fault (Fig. 4, inset). This is clearly a minimum slip and is based on a simplistic lithostratigraphic interpretation, but it makes the point that the Yucaipa Ridge fault does not necessarily qualify for inclusion in the Wilson Creek strand of the San Andreas fault on lithostratigraphic evidence alone.

Although this lithostratigraphic correlation between the Potato Sandstone and a part of the Mill Creek Formation can be defended as the simplest interpretation of the local facts, it is less attractive in a broader setting. The facies architecture of the Mill Creek Formation is characterized by extreme and small-scale lateral variability. The Potato Sandstone resembles only a small part of the range of lithology in the Mill Creek Formation, and has great lateral uniformity. We now characterize the Potato Sandstone west of Mill Creek by reference to the relatively wide outcrop near City Creek.

POTATO SANDSTONE FROM CITY CREEK TO ELDER GULCH

We have mapped five lithologic units in the Potato Sandstone near City Creek (Fig. 8): coarse basal conglomerate (Tps1), pale conglomeratic sandstone (Tps2), interbedded shale and buff sandstone (Tps3), and a calcareous unit (Tps4) that is transitional to a second pale conglomeratic sandstone (Tps5).

The base of the Potato Sandstone is a coarse breccio-conglomerate (Tps1) with a strong resemblance to the basement rocks beneath. The clasts, which range up to 3 m in size, are

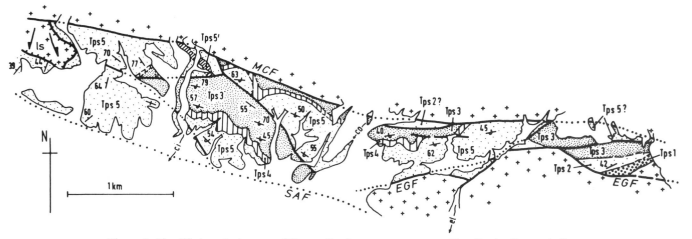

Figure 8. Simplified geologic map of Potato Sandstone outcrops near City Creek. Crosses indicate crystalline basement rocks; Tps1, coarse basal conglomerate; Tps2 and Tps5, pale colored conglomerates; Tps5′: sandy red beds; Tps3, buff and green sandstones interbedded with shale; Tps4, siltstone and shale-rich interval with redbeds and micritic carbonate; ls, landslide deposit. EGF = Elder Gulch fault; MCF = Mill Creek fault; SAF = south branch of San Andreas fault; ci = City Creek; co = Cook Creek; el = Elder Gulch.

dominated by hornblende-bearing diorite and granodiorite, biotite-rich granodiorite, and biotite quartz monzonite (Fig. 7). Some of these rocks have a distinct foliation. The distinctive "megaporphyry" that crops out north of Mill Creek (Matti and others, 1985) was not seen. Subordinate clast types include biotite gneiss and leucogranite.

The coarse basal unit grades upward into interbedded conglomerate and pale arkosic sandstone (Tps2). The sandstone beds have a crude parallel lamination and very low angle cross-lamination. Their pale color results from the coarse grain size and the predominantly white feldspar grains. Siltstone and shale are notably lacking. The clast types in the conglomerate resemble the basal breccio-congloemrate, but the proportions are different. Biotite- and quartz-rich gneiss is usually more abundant and the hornblende granitoid is relatively rare. The clast proportions vary from bed to bed. The maximum grain size in the conglomerates is typically less than 15 cm, but may reach 75 cm. The conglomerate matrix is a pebbly, white granule sand. Many of the clasts, particularly the biotite quartz monzonite clasts, have a red altered rind up to 5 mm thick. Since this red color is not pervasive in the matrix, we speculate that it was acquired before deposition. Where abundant, in the clast suites with a low content of hornblende granitoids, the red rims give the conglomerates a faintly pink appearance.

The succeeding unit, Tps4, consists of distinctive coarse greenish or yellow sandstone, interbedded with siltstone and shale. The sandstone beds have sharp bases and are indistinctly laminated; some have a reversed grading indicative of debris flow. Sedimentary slump structures are common. Large lenses of micritic carbonate are found sporadically in this unit. The interbedded conglomerate is pale colored and dominated by clasts of leucocratic granitoid and biotite- and quartz-rich gneiss; it resembles the hornblende-poor conglomerate beds of the unit below. This unit has a conformable, intercalated upper contact with a second unit of pale sandstones and conglomerates (Tps5). The transition is 20 to 50 m thick and is marked by biotite-rich siltstone, thin red siltstone beds, and frequent horizons and nodules of caliche.

At City Creek the conglomerate and sandstone of unit Tps5 resembles Tps2. The two units are readily distinguishable only by their stratigraphic position relative to Tps3 or Tps1. Farther northwest, however, these beds are intercalated with purple sandstones, and the color contrast reveals intense bioturbation. Just east of City Creek near the north branch of the San Andreas fault, there is an interbedded unit of silty red sandstone (Tps5' in Fig. 8). Tps5 is the only Potato Sandstone unit recognized northwest of the area in Figure 8 and is presumably considerably thicker than Tps2.

Age

Three varieties of fossil leaves, found in Tps3 on the east bank of City Creek, have been identified as *Persea* sp., *Salix* sp., and *Quercus* g. *pliopalmeri* by Daniel Axelrod. He reports (*in* Hillenbrand, 1990, p. 59) that "if the Mill Creek Formation is

Miocene, and about 13 m.y., this collection [Potato Sandstone] could be somewhat younger as judged from the very few fossils that are closely identifiable."

Provenance

Imbricated cobbles in the conglomerates of the Potato Sandstone at City Creek and farther west record paleocurrents from the northwest, subparallel with the fault zone. This is the same as the paleocurrent direction in the axial facies of the Mill Creek Formation.

Possible sources for the conglomerate clasts are very close at hand. Leucocratic and biotite-rich gneiss with pods of weakly foliated, nonporphyritic, biotite quartz monzonite, and hornblende-rich granitoid are characteristic of the basement beneath the basal conglomerate, in the fault block south of the Elder Gulch fault.

Comparison with Other Outcrops

None of the units described from the City Creek outcrop serves to distinguish it decisively from the Potato Sandstone in Mill Creek. Only the proportions of the major lithologic units and the proportions of clast types in the conglomerates change. West of City Creek, the outcrop is dominated by pale conglomerate and sandstone. Locally there are sequences of thin-bedded, burrow-mottled, purple and white sandstone beds. West of Twin Creek there are conglomerate beds that are dominated by a muscovite-biotite-granitoid, with pink feldspars (Fig. 7), but, otherwise we have found essentially the same rock types and clast suites throughout a 30-km strip of outcrop. Ray Weldon has informed us (written communication, 1988) that very similar rocks, with consistent internal stratigraphy and a quartz monzonite basement, reappear farther northwest in the San Andreas fault zone (between Cable Canyon and Cajon Creek). This extends the outcrop to more than 40 km.

RECONSTRUCTION OF THE MILL CREEK BASIN

Almost every major feature of the Mill Creek Formation may be considered characteristic for the exhumed fill of a nonmarine, strike-slip basin, and can be interpreted in terms of a Mill Creek basin that formed at an en echelon step in a strike-slip fault zone. We list below the features of the basin that match the strike-slip basin models given by Crowell (1974a, b, 1982a), Steel (1976), Link and Osborne (1978), Steel and Gloppen (1980), Crowell and Link (1982), Hempton and others (1983), and Dunne and Hempton (1984).

Mill Creek Basin as a Strike-Slip Basin

1. The basin is located in a strike-slip fault zone. Of course, the San Andreas fault is largely a post-Miocene structure, so the currently active faults in the zone postdate the deposition of the Mill Creek Formation.

2. The basin has a fault-bounded, elongate lens shape. The marginal breccia and conglomerate facies suggest active fault scarps during deposition, and indicate that little of the north and south margins has been removed by the younger faulting.

3. The basin fill has been deformed into a simple, doubly plunging syncline. The synclinal hinge coincides with the basin axis, as indicated by facies distributions.

4. The sedimentary fill is unusually thick relative to basin size; this usually implies high accumulation rates.

5. The thickness and facies architecture of the sedimentary fill are asymmetrical.

6. Lateral facies changes are frequent in a pattern comprising an axial facies sequence that interfingers with coarser, marginal facies of different composition (Figs. 2 through 4). The basin axis was occupied by a flood basin or lake. The marginal facies were deposited as deltas and/or alluvial fans.

7. A narrow zone of the coarsest, least mature sediments, including breccias, occupies one of the long margins of the lens-shaped basin. This is characteristically a steep, active fault margin in other examples. The sediments are generated by a series of small, steep drainage basins that are largely confined to the fault-line scarp. Drainage disruption by fault motion is sufficiently frequent to prevent the basins from growing to the point where they could capture streams originating far from the fault zone. Thus, the margin may be very active, and yet not yield a large volume of sediment, or contribute significantly to the axial deposition.

8. A wider zone of less coarse, more mature, but locally conglomeratic sediments is associated with the opposite long margin. It is indicative of less steep marginal slopes and perhaps of less active faulting. The topography of active, strike-slip fault traces may, of course, be determined by the contrast between older, unrelated, landscapes that the fault has juxtaposed. But around basins associated with en echelon steps, dip slip is significant, and the relief of fault line scarps may offer some evidence of fault activity.

9. The largest, most mature, and frequently finest grained depositional systems originate at the ends of the basin and dominate sedimentation in the basin axis. In modern examples this results where the strike-slip fault zone is a valley that channels regional drainage and introduces the farthest traveled sediments to the basin.

10. Paleoflow directions for the whole basin tend to box the compass (Fig. 6). Flow parallel to the basin axis was dominant, but is a poor guide to the location of sediment sources.

11. During the fill of a strike-slip basin, the marginal sources may move relative to the basin center. In a large strike-slip basin with good exposure, the resulting diachronism and imbrication of the marginal facies can sometimes be proven. The Mill Creek Formation is too poorly exposed and imprecisely dated for this.

The Mill Creek basin is accordingly reconstructed (Fig. 9) as a basin, where two strike-slip faults overlap. Both the northern and southern margins were sufficiently active that they were the site of relatively small, steep drainage basins. Larger fluvial systems followed the fault zone into the northwest and southeast ends of the basin. The river that entered from the northwest dominates the fill of the basin axis, but clasts from the southeast extended far into the basin in the earlier part of its history. The age of the Mill Creek basin is not well confined, but it is certainly evidence of strike slip before the Pliocene San Andreas fault, and it may predate the San Gabriel fault.

Sediment Sources

The clast suites derived from the basement that stood on the northern side of the basin are the best guide to its paleogeography. The compositional changes from west to east and the need for a very proximal, prolific source of muscovite-rich granitoid clasts can best be matched by the modern Cottonwood and Hayfield Mountains, north of the Clemens Well fault. The Chocolate and Little Chuckwalla Mountains and the Diligencia Formation (Crowell, 1975) provide sources of volcanic clasts to the east. Widespread late Oligocene and early Miocene redbeds crop out here and in the San Gabriel Mountains; they are part of the classical argument used to link the San Francisquito, Fenner, and Clemens Well faults (Bohannon, 1975). They serve here as evidence that red soils and older sediments were probably available as a source for the red facies in the Mill Creek Basin. The Diligencia Basin includes the closest redbeds in space and time; its interpretation (Arthur, 1974) suggests that a parallel fault was active a few kilometers north of the later Clemens Well fault by 20 to 22 Ma. Thus, it is feasible that the Clemens Well fault developed en echelon steps in the vicinity of Hayfield Mountain.

The most plausible arrangement of sources for all facies in the Mill Creek Basin (Fig. 10) occurs during slip on the San Francisquito–Fenner–Clemens Well fault, as reconstructed by Bohannon (1975), Powell (1981), and others. Powell (1981) has used basement rock assemblages to show that the San Francisquito–Fenner–Clemens Well fault was active before the San Gabriel and San Andreas faults and that it caused the northern margin of the Sierra Pelona to pass south of the granitoid and gneiss terrane of the modern Cottonwood and Hayfield Mountains.

The southern fault-line scarp must have been dominated by Pelona Schists, without the garnet-bearing variety. The north margin of the modern Sierra Pelona, along the San Francisquito fault is a suitable source terrane. We are unable to arrange any of the alternate sources of Pelona Schist (discussed in an earlier section) in a plausible paleogeography that can satisfy the very restricting source requirements of the northern basin margin, and the pre-Pliocene age of the Mill Creek Formation.

Better dating of the Mill Creek Formation may eventually test this interpretation of the original location of the Mill Creek basin. Powell has pointed out to us that the Fenner and San Francisquito faults are overstepped by the Clarendonian lower Punchbowl and upper Mint Canyon Formations respectively; the Mill Creek Formation would have to be older. A second test concerns the basement beneath the Mill Creek basin.

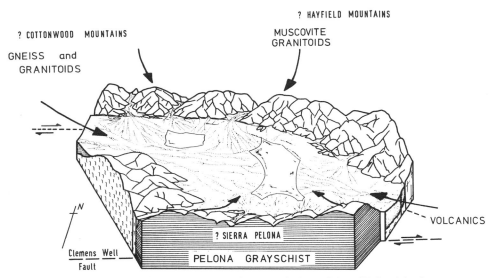

Figure 9. Speculative reconstruction of the morphology of the Mill Creek basin.

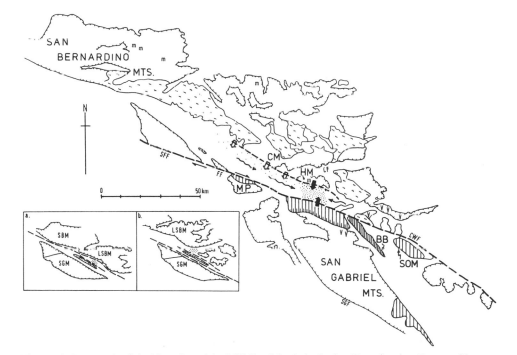

Figure 10. Suggested original location of the Mill Creek basin in the San Francisquito–Fenner–Clemens Well fault zone (SFF-FF-CWF). Outlines of modern bedrock ranges have been repositioned by restoring slip on the San Andreas fault, the San Gabriel fault (SGF), and the Chiriaco fault (CF) after Powell (1981). Vertical ruling indicates Pelona and Orocopia schist; wavy dashes, major gneiss outcrops close to basin; m's, large outcrops of muscovite- and garnet-bearing granitoids; v's, volcanic rocks; dense stipple and black arrows, deposition of Mill Creek Formation; sparse stipple and open arrows, possible deposition of Potato Sandstone. BB = Blue Ridge; CM = Cottonwood Mountains; HM = Hayfield Mountains; MP = Mount Pinos; SOM = southern Orocopia Mountains. Inset shows two stages of restoration of Mill Creek Formation (stippled) to original location. a, Slip relative to San Bernardino Mountains (SBM) and Little San Bernardino Mountains (LSBM), accomplished on Mission Creek strand of San Andreas fault. b, Slip relative to San Gabriel Mountains (sgm) accomplished by Punchbowl-Nadeau fault and Wilson Creek strand of San Andreas fault.

Basement Relationships

The basement to the south of the Mill Creek basin is inferred to have been dominated by Pelona Schists. The basement to the north was quite different; it is inferred to have been dominated by plutonic rocks including a large muscovite- and garnet-bearing body.

The floor of the basin is exposed beneath the Mill Creek Formation; it is not quite like the basement thought to lie to the north or the south. Most of it is a banded quartzo-feldspathic gneiss. Consideration of just the major named strands of the San Andreas fault zone indicates that the same basement block may extend west to Cook Creek ("wcb" of Matti and others, 1985). From Cook Creek to the Santa Ana River, the basement block comprises leucocratic and biotite-rich gneiss with pods of weakly foliated, nonporphyritic biotite quartz monzonite and hornblende-rich diorite to granodiorite. Just west of the Santa Ana River, there are small porphyritic hornblende and biotite-bearing igneous bodies with large pink orthoclase phenocrysts that closely resemble the "megaporphyry" described by Matti and others (1985) from the block north of the Mill Creek fault. We do not find this basement terrane to be obviously different from the gneissic complex immediately north of the north branch of the San Andreas fault in the same area. Dibblee (1970, 1973) and Morton and Miller (1975) did not distinguish the two gneissic terranes.

These rather nondiagnostic hornblende- and biotite-bearing granitoid and gneissic basement rocks thus permit comparison with the San Bernardino Mountains terrane, but Smith (1959) and Matti and others (1985) cited local evidence that the Mill Creek basin is floored, at least in part, by crystalline rocks with affinities to the San Gabriel Mountains or the Orocopia Mountains. An isolated outcrop of Pelona Schist was reported by Smith near the southern margin of the basin; it might be a huge landslide block. Matti and others (1985, p. 10) described a small anorthosite-like body in the basement outcrops at the southeast end of the Yucaipa Ridge fault.

These tentative basement relationships suggest that, before deposition began, the floor of the Mill Creek basin was already a separate fault sliver, offset from the neighboring blocks. Our mapping is not adequate to preclude the possibility that the basement beneath the Mill Creek Formation is a faulted composite of two different terrains.

Relationship to the Potato Sandstone

The relationship between the Mill Creek Formation and the Potato Sandstone is particularly troublesome; it is not easily reconciled with either the concept of a small, deep strike-slip basin or the original concept (Matti and others, 1985) of the Wilson Creek strand of the San Andreas fault. Let us spell out the ambiguities as we see them.

The Potato Sandstone and the axial facies of the Mill Creek Formation both contain red sandstone and siltstone associated with pale conglomerate and darker, buff sandstone, which is interbedded with shale. In both formations the conglomerate and sandstone were deposited by current flow from the northwest. The clast suites in the two areas are similar. These observations might ordinarily be enough to justify a lithologic correlation of two neighboring formations. The inset in Figure 4 shows how the redbeds north and south of the Yucaipa Ridge fault can be aligned by restoring limited right slip. In this reconstruction the Potato Sandstone becomes the northwestward extension of the youngest part of the axial facies in the Mill Creek basin. It is possible that the Potato Sandstone is the upstream portion of the alluvial-fluvial system that fed into the northwest end of the Mill Creek basin. The model thus changes from an isolated strike-slip basin to long faulted rift. In this scenario the Mill Creek Formation accumulates at a low point where streams converge from the northwest and southeast.

But the hallmark of the Mill Creek basin is the extreme lateral variability of its marginal facies. The Potato Sandstone shares none of these distinctive facies, with the possible exception of muscovite-rich sandstone. The Potato Sandstone is aligned parallel to the basin axis, which is the direction likely to show the fewest facies changes, but the modern outcrop of the Potato Sandstone has rather uniform character for 30 to 50 km. Faults within the modern outcrop may have strike-slip character, but the most extreme palinspastic rearrangement still leaves a strip of outcrop that is longer than the Mill Creek basin. A very elongate depositional basin with so much variability concentrated in one part seems less plausible than the basin reconstructed to explain the Mill Creek Formation alone. This is the essence of the first argument against correlation.

If the Potato Sandstone and Mill Creek Formation have a common depositional history, then the Wilson Creek strand of the San Andreas fault cannot include the Yucaipa Ridge fault that separates them. The Wilson Creek strand would be required to run directly from the Wilson Creek fault to the Mill Creek fault. Even if we assume that the resemblances between the Potato Sandstone and Mill Creek Formation are coincidental, it is difficult to make a case for inclusion of the Elder Gulch fault in the Wilson Creek strand. The basal breccio-conglomerates of the Potato Sandstone were derived from a terrane that is indistinguishable from the block south of this fault, so it is not evident that the Elder Gulch fault has a large displacement. If the Elder Gulch and Yucaipa Ridge faults are linked into the same strand, then the Potato Sandstone and Mill Creek Formation appear to be deposited on the same basement block. The basement south of Elder Gulch fault and the basement beneath the west end of the Mill Creek Formation are separated by faults. It is possible that Yucaipa Ridge fault runs west through basement rocks to the south branch of the San Andreas fault zone.

TECTONIC IMPLICATIONS

The Mill Creek basin is evidence of early strike slip in southern California. A consideration of its age and the simplest restoration of sources for its sediment places the active Mill Creek basin in the San Francisquito–Fenner–Clemens Well fault zone. This is sedimentary corroboration of a fault history reconstructed by Powell (1981) from a consideration of crystalline basement rocks. Other authors who sought to determine the slip history of the San Andreas fault (Gibson, 1964, 1971; Matti and others, 1985) have proposed different locations for the Mill Creek Basin; they were unaware of the muscovite-bearing facies on its north side, and their reconstructions are unacceptable. We briefly examine the implications of our interpretation for the San Andreas fault.

The modern outcrop of the Mill Creek Formation lies within the San Andreas fault zone. Thus entrained, it need not have experienced the total slip of the fault, relative to a given source terrain. It is the separation of the inferred northern and southern sources that gives the slip on the whole fault zone.

In order to reach its modern position from a depositional setting south of the Hayfield Mountains, the Mill Creek Formation first needs to travel west within the Clemens Well fault zone about 35 km, to reach the southeast corner of the Cottonwood Mountains. The Sierra Pelona would slip 10 to 20 km farther west. It is not clear when deposition ceased, relative to this movement. About 50 km of right slip is required on the Clemens Well fault zone, after initiation of the Mill Creek basin. Our model for the northern margin of the basin shows a fault that may have been a precursor of strike slip in the San Andreas fault near the Little San Bernardino Mountains. Proper age control for the Potato Sandstone is necessary to substantiate this idea.

The San Gabriel fault has no influence on the path of the Mill Creek basin, as we reconstruct it. The basin is next moved with the San Andreas fault strands. The Mill Creek Formation is now *south* of the Sierra Pelona and the Blue Ridge. In our reconstruction of the Mill Creek basin, it began to the north of these blocks. The total modern separation of the Mill Creek Formation from Sierra Pelona is about 120 km, and from Blue Ridge about 75 km. The difference is the slip on a strand of the Punchbowl-Nadeau fault zone. The Mill Creek Formation is 105 km from the southeast corner of the Cottonwood Mountains. Thus about 225 km of offset is implied for the San Andreas fault. The initial, Wilson Creek, strand of the San Andreas fault was essentially proposed by Matti and others (1985) to account for the exotic location of the Mill Creek basin relative to the San Bernardino Mountains. It would account for most of the separation from the northern sources (ca. 105 km). Matti and others (1985) interpreted the younger Mission Creek strand to pass south of Yucaipa Ridge; it would be the strand of choice to separate the Sierra Pelona and the Mill Creek Formation (ca. 120 km). The Mill Creek basin is a guide to apportionment of slip between these two strands. Also, if slip on the Mission Creek strand entirely post-dates slip on the Wilson Creek strand, then Yucaipa Ridge may once have stood between Blue Ridge and the central San Bernardino Mountains. Gravels shed from Blue Ridge into the central San Bernardino Mountains (see Sadler, this volume) may have acquired andesite and arkosic sandstone clasts from the Mill Creek Formation.

If the Potato Sandstone is part of the axial facies of the Mill Creek Basin, as simple lithologic similarities might suggest, then the preceding interpretation has serious difficulties. The difficulties arise from the fact that the Potato Sandstone between Cook Creek and Oak Creek seems to be derived from a complex of gneisses with hornblende- and biotite-bearing granitoids that still crops out immediately to the south. The outcrop to the south is south of the Elder Gulch fault, in the Wilson Creek block of Matti and others (1985). Thus we would not have evidence for great displacement on either the Elder Gulch fault or the Yucaipa Ridge fault. One answer to these difficulties is to assume that the lithologic similarities between the Potato Sandstone and the axial facies of the Mill Creek Formation, and between the Potato Sandstone and the basement to the south, are coincidental. The clast suites in the Potato Sandstone (Fig. 7) are sufficiently nondiagnostic that they might fit several sources along the 80-km southern margin of the Little San Bernardino Mountains (Fig. 11).

SUMMARY

The Mill Creek Formation was deposited in a very early strike-slip basin that probably developed in the San Francisquito–Fenner–Clemens Wall fault zone. This reconstruction is based on the pre-Pliocene age of the formation and the need to arrange a large source of garnet-muscovite granitoids immediately north of the basin and garnet-free Pelona Schist to the south. The reconstruction also permits clasts of volcanic rocks to be derived from the Chocolate Mountains to the southeast. Only the last two sources have been considered in earlier reconstructions of this basin.

The northern margin of the Mill Creek basin may have been a precursor of the San Andreas fault, but the basin predates all the known strands of that fault. Strands of the San Andreas pass north and south of the Mill Creek Formation, so any reconstruction of the original basin requires a certain apportionment of slip between the strands to the north and south. Strands that now pass north of the basin (especially the Wilson Creek strand) have separated it from the region of the Cottonwood Mountains, about 105 km. Strands that pass south of the basin (especially the Mission Creek Strand) have separated it from the Sierra Pelona, about 120 km.

Because the Mill Creek Formation accumulated in a small, but deep, strike-slip basin, it must be used with caution to interpret the history of the younger San Andreas fault zone where the formation now crops out. The formation includes abrupt lateral changes of lithology, and its northern and southern margins are quite dissimilar. For this reason any offset portions of the Mill Creek Formation might be difficult to recognize. Conversely, other units cannot be assumed to have originated in a different basin simply because they are lithologically distinct. Notice that the southern flank facies of the Mill Creek basin gives no guide to the appearance of its northern flank facies.

The Wilson Creek fault separates the east end of the Mill

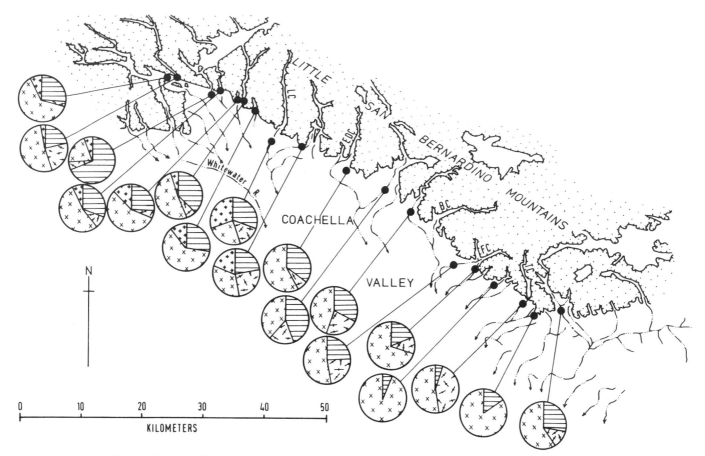

Figure 11. Proportions of major clast types in modern gravels derived from the Little San Bernardino Mountains. Symbols as in Figure 6, with addition of asymmetric crosses to differentiate within category 3 those biotite quartz-monzonite varieties with pink orthoclase phenocrysts. BC = Berdoo Canyon; EDC = East Deception Canyon; FC = Fargo Canyon; LC = Long Canyon.

Creek basin from basement rocks that have affinities with the San Bernardino Mountains. Matti and others (1985), who made this observation, concluded that the fault is part of a regionally important strand of the San Andreas fault. Their Wilson Creek strand includes other faults that separate the Mill Creek Formation from the Potato Sandstone: the Yucaipa Ridge and Elder Gulch faults. In spite of our point about facies changes, the Potato Sandstone consists of lithologies that *can* be matched with the axial facies of the Mill Creek basin. The paleocurrent directions are parallel. The clast suites, although not very specific indicators of their source, are both similar in many ways to gravels from the Little San Bernardino Mountains. Thus, the simplest lithostratigraphic conclusion consistent with the intrinsic evidence from these two units is that they were deposited close together, not separated by a major strand of the San Andreas fault.

Between Cook Creek and Oak Creek the clast suites in the

Potato Sandstone closely resemble a gneissic basement complex that crops out to the south of the Elder Gulch fault, and, thus, in the Wilson Creek Block of Matti and others (1985). Clearly, the simple lithologic correlations between the Mill Creek Formation, the Potato Sandstone, and basement rocks near Elder Gulch are hard to reconcile with our model for Mill Creek basin or with the concept of a Wilson Creek strand. The Wilson Creek strand may include unnamed faults south of the Elder Gulch fault. Both models remain intact if the similarities of the Potato Sandstone and the Mill Creek Formation are assumed to be fortuitous. The Potato Sandstone may be derived from the Little San Bernardino Mountains (Fig. 11). Are the sedimentary and basement rock types sufficiently nondiagnostic to let models hold sway over observations? Better evidence of age seems needed to resolve the questions.

ACKNOWLEDGMENTS

Doug Morton, Jonathon Matti, Ray Weldon, Kris Meisling, Chuck Stelting, Andrew Miall, Mike Murphy, and Mike Woodburne visited the outcrops with us and helped sharpen our wits. William Hubbard granted generous access to his private property. Anne Lilje assisted in the analysis of turbidite-like beds. Dave Lukesh, Rick Garrison, and Dana Prevost shared the results of their mapping for the U.S. Army Corps of Engineers. We thank Doug Morton, Bob Powell, Ray Weldon, and especially Brett Cox, for thorough and insightful reviews of an earlier draft of this chapter.

REFERENCES CITED

Arthur, M. A., 1974, Stratigraphy and sedimentation of lower Miocene non-marine strata of the Orocopia Mountains; Constraints for late Tertiary slip on the San Andreas fault system in southern California [M.S. thesis]: Riverside, University of California, 200 p.

Bohannon, R. G., 1975, Mid-Tertiary conglomerates and their bearing on Transverse Range tectonics, southern California, *in* Crowell, J. C., ed., San Andreas fault in southern California: California Division of Mines and Geology Special Report 118, p. 75–82.

Bouma, A. H., 1962, Sedimentology of some flysch deposits: Amsterdam, Elsevier, 169 p.

Crowell, J. C., 1974a, Sedimentation along the San Andreas fault, California, *in* Dott, R. H., and Shaver, R. H., eds., Modern and ancient geosynclinal sedimentation: Society of Economic Paleontologists and Mineralogists Special Publication 19, p. 292 303.

—— , 1974b, Origin of late Cenozoic basins in southern California, *in* Dickinson, W. R., ed., Tectonics and sedimentation: Society of Economic Paleontologists and Mineralogists Special Publication 22, p. 191–204.

—— , 1975, Geologic sketch of the Orocopia Mountains, *in* Crowell, J. C., ed., San Andreas fault in southern California: California Division of Mines and Geology Special Report 118, p. 99–110.

—— , 1982a, The tectonics of Ridge basin, *in* Crowell, J. C., and Link, M. H., eds., Geologic history of Ridge basin, southern California: Pacific Section, Society of Economic Paleontologists and Mineralogists, p. 25–42.

—— , 1982b, The Violin breccia, Ridge basin, southern California, *in* Crowell, J. C., and Link, M. H., eds., Geologic history of Ridge basin, southern California: Pacific Section, Society of Economic Paleontologists and Mineralogists, p. 89 98.

Crowell, J. C., and Link, M. H., 1982, Ridge basin, southern California, *in* Crowell, J. C., and Link, M. H., eds., Geologic history of Ridge basin, southern California: Pacific Section, Society of Economic Paleontologists and Mineralogists, p. 1–4.

Demirer, A., 1985, The Mill Creek Formation; A strike-slip basin filling in the San Andreas fault zone, San Bernardino County, California [M.S. thesis]: Riverside, University of California, 108 p.

Dibblee, T. W., 1964, Geologic map of the San Gorgonio Mountain Quadrangle, San Bernardino and Riverside Counties, California: U.S. Geological Survey Miscellaneous Geologic Investigations Map I-431, scale 1:62,500.

—— , 1968, Displacements on the San Andreas fault systems in the San Gabriel, San Bernardino, and San Jacinto Mountains, southern California, *in* Dickinson, W. A., and Grantz, A., eds., Proceedings of a Conference on Geologic Problems of the San Andreas Fault System: Stanford, California, Stanford University Publications in Geological Sciences, v. 11, p. 260–276.

—— , 1970, Regional geologic map of the San Andreas and related faults in the eastern San Gabriel Mountains, San Bernardino Mountains and western San Jacinto Mountains, California: U.S. Geological Survey Open-File Map 71-8, 1:250,000.

—— , 1973, Geologic map of the Redlands Quadrangle, California: U.S. Geological Survey Open-File Report 74–1022, scale 1:62,500.

—— , 1982, Geology of the San Bernardino Mountains, southern California, *in* Fife, D. L. and Minch, J. A., eds., Geology and mineral wealth of the Transverse Ranges: Santa Ana, California, South Coast Geological Society Guidebook 10, p. 148–169.

Dunne, L. A., and Hempton, M. R., 1984, Deltaic sedimentation in the Lake Hazar pull-apart basin, southeastern Turkey: Sedimentology, v. 31, p. 401–412.

Ehlig, P. L., 1981, Origin and tectonic history of the basement terrane of the San Gabriel Mountains, central Transverse Ranges, *in* Ernst, W. G., ed., The geotectonic development of California; Rubey Volume 1: Englewood Cliffs, New Jersey, Prentice-Hall, p. 253–283.

Gibson, R. C., 1964, Geology of a portion of the Mill Creek area, San Bernardino county, California [M.S. thesis]: Riverside, University of California, 50 p.

—— , 1971, Non-marine turbidites and the San Andreas fault, San Bernardino Mountains, California, *in* Elders, W. A., ed., Geological excursions in southern California: Riverside, University of California Campus Museum Contributions, v. 1, p. 167–181.

Hempton, M. R., Dunne, L. A., and Dewey, J. F., 1983, Sedimentation in a modern strike-slip basin, southeastern Turkey: Journal of Geology, v. 91, p. 315–330.

Hillenbrand, J. M., 1990, The Potato Sandstone between the Santa Ana River and Badger Canyon, San Bernardino County, southern California [M.S. thesis]: Riverside, University of California, 163 p.

Link, M. H. and Osborne, R. H., 1978, Lacustrine facies in the Pliocene Ridge Basin Group; Ridge Basin, California, *in* Matter, A., and Tucker, M. E., eds., Modern and ancient lake sediments: International Association of Sedimentologists Special Publication 2, p. 169–187.

Ludlam, S. D., 1974, Fayetteville Green Lake, New York; 6, The role of turbidity currents in lake sedimentation: Limnology and Oceanography, v. 19, p. 656–664.

Matti, J. C., Morton, D. M., and Cox, B. F., 1985, Distribution and geologic relations of fault systems in the vicinity of the central Transverse Ranges, southern California: U.S. Geological Survey Open-File Report 85-365, 23 p.

Morton, D. M., and Miller, F. K., 1975, Geology of the San Andreas fault zone north of San Bernardino between Cajon Canyon and Santa Ana Wash, *in* Crowell, J. C., ed., San Andreas fault in southern California: California Division of Mines and Geology Special Report 118, p. 136–146.

Nilsen, T. H., 1982, Alluvial fan deposits, *in* Scholle, P. A., and Spearing, D., eds., Sandstone depositional environments: American Association of Petroleum Geologists Memoir 31, p. 49–87.

Owens, G. V., 1959, Sedimentary rocks of lower Mill Creek, San Bernardino Mountains, California [M.A. thesis]: Claremont, California, Claremont College, 50 p.

Powell, R. E., 1981, Geology of the crystalline basement complex, eastern Transverse Ranges, southern California; Constraints on regional tectonic interpretation [Ph.D. thesis]: Pasadena, California Institute of Technology, 441 p.

Richmond, J. F., 1960, Geology of the San Bernardino Mountains north of Big Bear Lake, California: California Division of Mines and Geology Special Report 65, 68 p.

Rogers, J.J.W., 1961, Igneous and metamorphic rocks of the western portion of Joshua Tree National Monument, Riverside and San Bernardino Counties: California Division of Mines and Geology Special Report 68, 26 p.

Sadler, P. M., 1981, The structure of the northeast San Bernardino Mountains; Notes to accompany 7.5-minute Quadrangle maps submitted for compilation onto the San Bernardino 1 by 2 degree Quadrangle: California Division of Mines and Geology Final Techical Report, Contract 5-1104, 26 p., 12 maps.

Sadler, P. M., and Demirer, A., 1986, Pelona Schist clasts in the Cenozoic of the San Bernardino Mountains, southern California, *in* Ehlig, P. L., (ed.), Neotectonics and faulting in southern California: Geological Society of America Cordilleran Section 82nd Annual Meeting Guidebook and Volume, p. 129–146.

Smith, R. E., 1959, Geology of the Mill Creek area [M.A. thesis]: Los Angeles, University of California, 95 p.

Stanley, K. O., and Surdam, R. C., 1978, Sedimentation on the front of Eocene Gilbert-type deltas, Washakie basin, Wyoming: Journal of Sedimentary Petrology, v. 48, p. 557–573.

Steel, R. J., 1976, Devonian basins of Norway; Sedimentary response to tectonism and varying tectonic context: Tectonophysics, v. 48, p. 207–224.

Steel, R. J., and Gloppen, T. G., 1980, Late Caledonian (Devonian) basin formation, western Norway; Signs of strike-slip tectonics during infilling, *in* Ballance, P. F., and Reading, H. G., eds., Oblique-slip mobile zones: International Association of Sedimentologists Special Publication 4, p. 79–103.

Sturm, M., and Matter, A., 1978, Turbidites and varves in Lake Brienz (Switzerland); Deposition of clastic detritus by density currents, *in* Matter, A., and Tucker, M. E., eds., Modern and ancient lake sediments: International Association of Sedimentologists Special Publication 2, p. 147–168.

Vaughan, F. E., 1922, Geology of San Bernardino Mountains north of San Gorgonio Pass: Berkeley, University of California Publications in the Geological Sciences, v. 13, no. 9, p. 319–341.

West, D., 1987, Geology of the Wilson Creek–Mill Creek fault zone; The north flank of the former Mill Creek basin, San Bernardino County, California [M.S. thesis]: Riverside, University of California, 95 p.

Woodburne, M. O., 1975, Cenozoic stratigraphy of the Transverse Ranges and adjacent areas, southern California: Geological Society of America, Special Paper 162, 91 p.

Manuscript Accepted by the Society September 6, 1990

Geological Society of America
Memoir 178
1993

Chapter 10

The Santa Ana basin of the central San Bernardino Mountains: Evidence of the timing of uplift and strike slip relative to the San Gabriel Mountains

Peter M. Sadler
Department of Earth Sciences, University of California, Riverside, California 92521

ABSTRACT

The Cenozoic Santa Ana basin lies between the San Gorgonio massif and the northern plateau of the San Bernardino Mountains. Cenozoic sediments are considerably thinner on these two upland areas, which are obvious sources only for the Quaternary portion of the basin fill. The Tertiary fill is the Santa Ana Sandstone—an alluvial to lacustrine, pre-orogenic deposit that includes at least four conglomerate facies with different provenance. Only two of these facies can simply be derived from local basement terrain, and of these only one is compatible with the modern relief.

A third facies requires a source of garnet-bearing Pelona Schist, Pelona greenschists and grayschists, greenstones, arkose, and "polka-dot granite" clasts. The distribution of clast sizes suggests a source that lay about 5 km to the south, just across the San Andreas fault. Such a source could have been provided by the Sierra Pelona of the northern San Gabriel Mountains, prior to major offset on the Punchbowl fault zone. The clast suite of the fourth facies was also transported northward, but bears superficial resemblance to the San Gorgonio basement rocks.

The reconstruction of the Santa Ana basin requires that the rocks of the San Gabriel Mountains drew alongside before the uplift of the San Bernardino Mountains. At that time the San Gabriel area was relatively high and stood close to the present position of San Gorgonio Mountain. The modern configuration of the Santa Ana basin was acquired during compression of the Santa Ana basin and thrust faulting of its local sources over the northern margin. The fault at the southern margin is obscured by landsliding and superficial deposits, but may deserve inclusion with the San Andreas fault system.

The Pelona Schist–bearing facies in the Santa Ana basin is now about 120 km from its inferred sources, separated by the San Andreas fault zone. Unfortunately, the age of that facies is poorly constrained. It is certainly older than the uplift of the northern plateau of the San Bernardino Mountains. In some areas beyond the Santa Ana basin, the uplift appears to have begun by 4.2 Ma; in others it is still undetected by 2.5 Ma. The Pelona Schist–bearing facies is apparently younger than the 15-Ma sediments near the base of the basin fill, and may be younger than 6.2-Ma basalts. This remaining range of age includes possibilities that do not fit well with published reconstructions of the San Andreas fault history.

Sadler, P. M., 1993, The Santa Ana basin of the central San Bernardino Mountains: Evidence of the timing of uplift and strike slip relative to the San Gabriel Mountains, *in* Powell, R. E., Weldon, R. J., II, and Matti, J. C., eds., The San Andreas Fault System: Displacement, Palinspastic Reconstruction, and Geologic Evolution: Boulder, Colorado, Geological Society of America Memoir 178.

INTRODUCTION

The San Gabriel Mountains and the San Bernardino Mountains are neighboring Transverse Ranges separated only by the San Andreas fault zone (Fig. 1). Since the total slip on that fault is widely believed to exceed the distance between the two ranges, the rocks they contain surely once lay close alongside each other. So we should not be surprised to find evidence of some exchange of sediment between the two terranes. The Santa Ana basin, in the central San Bernardino Mountains, contains sediment that will be shown to have its source in the San Gabriel Mountains. This sediment transfer is discussed as evidence of the relative relief of the two terranes and the time at which they passed alongside.

In this chapter, the history of the Santa Ana basin is shown to involve two quite distinct basinal episodes, separated by a phase of rapid uplift in the San Bernardino Mountains. This uplift has long been attributed to the development of the bend in the San Andreas fault at San Gorgonio Pass (Allen, 1957; Dibblee, 1975, 1984). Uplift of the San Bernardino Mountains provides key sediment sources for two basins south of the San Andreas fault—the Ridge basin (Ramirez, 1983) and the San Timoteo Badlands (Matti and Morton, 1975; May and Repenning, 1982b). Thus this account of the Santa Ana basin concludes by considering implications for the history of the San Andreas fault.

QUATERNARY SANTA ANA BASIN

The upper reach of the modern valley of the Santa Ana River, in the central San Bernardino Mountains, has anomalous morphology. The river has cut a narrow gorge and deposited limited gravel terraces on the floor of an elongate intramontane basin. The north and south margins of the basin parallel the river course, but were not simply cut by the river; they are the steep bedrock flanks of tectonically elevated blocks (Figs. 1, 2). The northern plateau of the San Bernardino Mountains has been elevated on the Santa Ana thrust, whose trace follows the foot of the escarpment at the north margin of the basin. San Gorgonio Mountain borders the basin to the south; it appears to be a tilted block that is locally in fault contact with rocks of the basin floor.

Relatively small, steep drainage basins that are largely confined to the steep, tectonically controlled flanks of the basin have delivered enormous volumes of coarse alluvial, fluvial, and fluvioglacial sediments to the basin in Quaternary times. Like their unstable source slopes, these deposits have been modified by landsliding (Sadler and Morton, 1989), but they postdate the tectonic activity that blocked out the form of the Quaternary basin. The texture and composition (Fig. 3) of the Quaternary deposits relate very simply to the composition and configuration of the current basin margins. These properties provide a stark contrast with the Santa Ana Sandstone, which lies unconformably below the surficial sediments and landslide deposits; we must turn to it for evidence of the Tertiary precursor basin.

TERTIARY SEDIMENTS OF THE SANTA ANA BASIN

The older deposits in the Santa Ana basin are often still poorly consolidated; but they have compositions (Fig. 3) and textures that cannot be derived from the steep, unstable slopes that flank the modern drainage basin. They record deposition in a Tertiary basin that predates the uplift of the northern plateau and

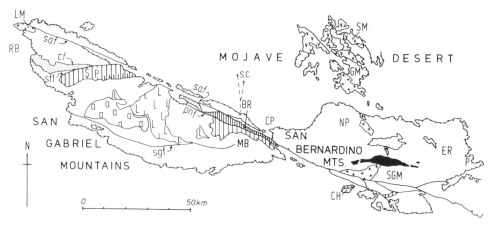

Figure 1. Location of the Santa Ana Sandstone (black) and potential sources of some clast types. Vertical ruling indicates Pelona Schist, garnetiferous varieties indicated by closer ruling; L's = Mount Lowe Granodiorite; a's, anorthosite; v's, Sidewinder Volcanics; crosses, early Mesozoic, hornblende quartz monzonite, and granodiorite; stipple, San Francisquito Formation. BR = Blue Ridge; CH = Crafton Hills; CP = Cajon Pass; ER = eastern ramp of the San Bernardino Mountains; GM = Granite Mountain; LM = Liebre Mountain; MB = Mount Baldy; NP = northern plateau of the San Bernardino Mountains; P = Pioneertown; RB = Ridge Basin; SGM = San Gorgonio Mountain; SM = Sidewinder Mountain; SP = Sierra Pelona; saf = San Andreas fault; cf = Clearwater fault; sff = San Francisquito fault; pnf = Punchbowl-Nadeau fault; s.c. = Sheep Creek; sgf = San Gabriel fault.

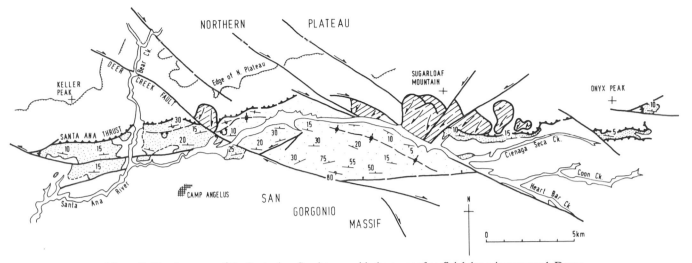

Figure 2. Structure map of the Santa Ana Sandstone, with the cover of surficial deposits removed. Dense stipple indicates fault blocks where base of formation is exposed; light stipple, base not seen. Diagonal ruling indicates large bedrock landslides.

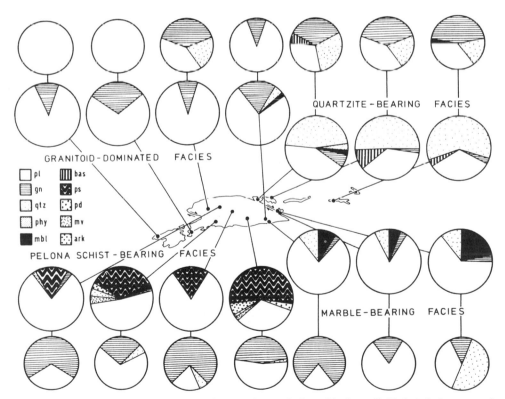

Figure 3. Clast compositions in the Santa Ana Sandstone (indicated by large divided circles) compared with overlying surficial deposits (small circles). The location of samples is shown on an outline version of Figure 4, with restoration of strike slip. Key: pl = plutonic rocks; gn = gneiss; qtz = quartzite; phy = phyllite; mbl = marble; bas = basalt; ps = Pelona Schist; pd = "Polka-dot Granite"; mv = volcanic and metavolcanic rock; ark = arkose.

San Gorgonio Mountain. From 200 to 1,500 m of sandstone, conglomerate, and clay can be found in tilted and folded (Fig. 2) sequences exposed beneath the surficial deposits. The surficial sediments are typically darker and contain more clay.

The older sediments were recognized by Vaughan (1922), who coined the term Santa Ana Sandstone for the conglomerate, sandstone, and shale exposed beneath the surficial gravels near the center of the basin. Dibblee (1964) mapped more of the unit to the east and west, and included sediments of different composition, some of them interbedded with basalt. Subsequent, more detailed, mapping (Sadler, 1981, 1982; Jacobs, 1982; Strathouse, 1982, 1983; Powell and others, 1983) has followed Dibblee's extended usage. Since the name has been applied to all the deposits that predate tilting and uplift in the basin, the Santa Ana Sandstone is essentially a tectono-stratigraphic unit. Not surprisingly, the practice creates difficulties at the east and west ends of the basin where the tectonic setting changes. The westernmost outcrops that Dibblee attributed to the Santa Ana Sandstone lie unconformably beneath Mud Flat. Thin-bedded silty sandstone is exposed there, together with conglomerate in which R. J. Weldon (written communication, 1988) has recognized clast suites that are diagnostic for the Crowder Formation. The easternmost outcrops that might deserve to be included are sandstones interbedded with basalt flows at Pioneertown (Fig. 1); these were originally assigned to the Old Woman Sandstone (Dibblee, 1967).

The outcrop of the Santa Ana Sandstone is a narrow, nearly continuous, east-west strip more than 30 km long and at most about 2 km wide (Fig. 2). Weakly indurated, pale colored boulder conglomerates and coarse pebbly sands, with limited lateral continuity of bedding, dominate the exposures. Immature paleosoils, cut-and-fill lenses, and very poorly sorted debris-flow beds are common. The overall character is indicative of alluvial deposition. Well-bedded sandstone, silt, and clay with a lacustrine character are subordinate except at the center of the basin. The conglomerates of the Santa Ana Sandstone include four significantly different clast suites, which define four areally discrete (Fig. 4) compositional facies.

The key to the Tertiary history of the Santa Ana basin is the age and provenance of the four compositional facies. The rest of this section describes the facies from west to east, presents the evidence for their ages, and finally suggests possible sources.

Granitoid-dominated facies

The conglomerates in the western third of the main outcrop are characterized by clasts of granitoids and granitoid-gneiss. There is a paucity of distinctive clasts in this suite. The granitoids include biotite monzonite and a hornblende-rich diorite. A distinctive pre-Cretaceous, megaporphyritic hornblende granodiorite to monzonite (Morton and others, 1980, Matti and others, 1985) locally forms the basement to this part of the Santa Ana Sandstone and crops out through Angelus Oaks to the south (Fig. 1). It is rarely seen as clasts except at the very base of the formation. The granitoid gneisses bear superficial resemblance to the Baldwin Gneiss of San Gorgonio Mountain. But they may be distinguished from this local source by the paucity of coarsely foliated, biotite-rich varieties, and by the presence of a high percentage of gneiss with a mylonitic fabric. Weakly imbricated cobble fabrics in the conglomerates indicate transport from the south and south-southwest.

The base of the Santa Ana Sandstone in this western area is marked by a thick, red paleosoil succeeded by fossiliferous red siltstones and sandstones that thicken eastward and become intercalated with pale pink, pebbly sandstone and green clay (Fig. 5).

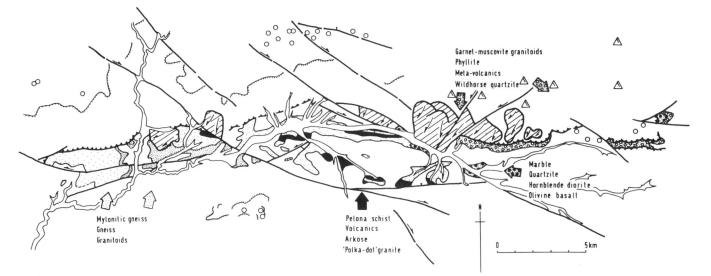

Figure 4. Facies and provenance of the Santa Ana basin. Arrows indicate transport directions and key clast types. Solid black indicates Pelona schist–bearing facies; dense stipple, basal red facies; sparse stipple, mylonitic gneiss-bearing facies; solid circles with stipple, marble-bearing facies; open circles with stipple, quartzite-bearing facies; large open circles, remnant of the upland veneer of quartzite gravels; triangles, main quartzite sources for the veneer.

Pelona Schist–bearing facies

The central portion of the Santa Ana Sandstone includes a higher proportion of sandstone and siltstone. Many of the intercalated conglomerate beds contain distinctive Pelona Schist clast suites, but the proportions of schist types and associated clasts vary appreciably (Fig. 3). Where several conglomerate beds are seen in sequence, each may have its own distinctive provenance. Pelona Schist–bearing conglomerates are found interbedded with others that have no such clasts.

The Pelona Schist clast suite, which accounts for 8 to 51 percent of the clast assemblages in which it occurs, includes varying proportions of muscovite-rich grayschists (25 to 74 percent), greenschists and greenstones with coarse, white albite porphyroblasts (15 to 36 percent), coarse garnet-bearing grayschists (1 to 50 percent), metacherts (0 to 3 percent) and metaserpentinites (0 to 2 percent). The associated clasts include arkosic sandstone that resembles the San Francisquito and the Mill Creek Formations (0 to 5 percent), purple vesicular andesite (0 to 1 percent), porphyritic metavolcanics (0 to 5 percent), quartzite (0 to 14 percent), the "polka-dot granite" (0 to 2 percent) of Pelka (1971) and Ehlert and Ehlig (1977), and a variety of monzonites and diorites (35 to 81 percent).

I have examined this clast assemblage at several localities and on separate occasions with help from Perry Ehlig (California State University, Los Angeles) and Douglas Morton, both of whom are well acquainted with Pelona Schist terranes (e.g., Ehlig, 1981). At one locality we uncovered three small clasts of a chalky white plagioclase with coarse polysynthetic twinning. Although such clasts might be derived from an anorthosite, we were unable to find large clasts of anorthosite. Further-

more, clasts of the distinctive syenites and Mount Lowe Granodiorite, which accompany anorthosite outcrops in the San Gabriel Mountains, were not found in the Santa Ana Sandstone despite a deliberately focused effort.

The largest Pelona Schist clasts are nearly of half-meter diameter (garnet schist, 48 cm; albite-porphyroblast greenschist, 45 cm; grayschist, 45 cm) and are from the southernmost outcrops of the Santa Ana Sandstone, close to the faulted north side of the San Gorgonio massif. The largest arkosic sandstone (28 cm) and "polka-dot granite" (29 cm) clasts occur at the same locality. The areal distribution of the maximum sizes of Pelona Schist clasts (Fig. 6) indicates a northward-transport direction.

The quartzite and metavolcanic clasts are the most durable ones, yet are typically better rounded and often significantly smaller than the softer clasts. No metavolcanic clasts coarser than 6 cm were seen; most are 2 to 3 cm in diameter. The quartzite clasts have a bimodal size distribution; most are as small as the metavolcanic clasts, but, at two locations near the western limit of the facies, white and pale pink quartzite clasts of 15 to 25 cm were present. These larger clasts are indistinguishable from quartzite clasts in the eastern third of the Santa Ana Sandstone, and in a Miocene gravel veneer on the northern plateau of the San Bernardino Mountains (Sadler and Reeder, 1983). By contrast, the smaller quartzite clasts are brown and have surface pits and shatter marks more typical of clasts from better indurated conglomerates. The unexpectedly high roundness of these durable clasts can be explained if they are recycled from older sediments. One such source, with a maximum grain size of less than 10 cm probably supplied all the metavolcanic clasts and the clasts of smaller, pitted quartzites.

The "polka-dot granite," vesicular andesite, and arkose

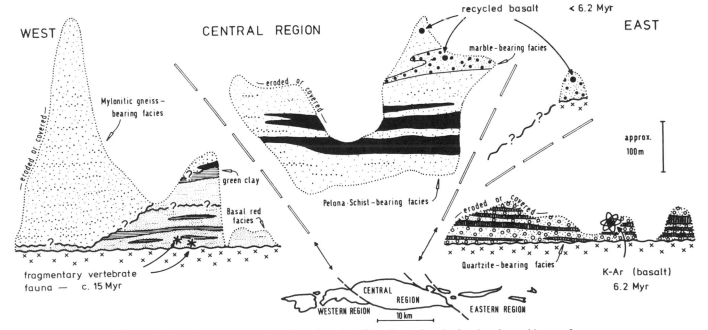

Figure 5. Tentative reconstruction of stratigraphy of the Santa Ana basin, showing evidence of age. Facies symbols as in Figure 4.

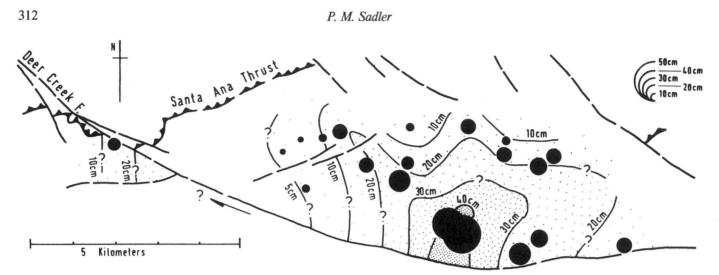

Figure 6. Maximum grain sizes of Pelona Schist clasts in the Santa Ana basin. Proportional symbols indicate location of data, which are not necessarily at precisely equivalent stratigraphic levels.

clasts are found only in the coarsest conglomerates, and appear to be stratigraphically the lowest of the Pelona Schist–bearing conglomerates (Figs. 3, 6). Metavolcanic clasts are most abundant at the same localities but persist farther up-section and northward into the basin. The larger quartzite clasts are found exclusively at two localities that appear to be at the very base of the facies. Up-section and in the direction of transport, the clast suite is less diverse and more uniform.

Marble-bearing facies

Immediately east of the Pelona Schist–bearing facies, the Santa Ana Sandstone is represented by coarse, marble-bearing conglomerates. The latter are tentatively interpreted to lie stratigraphically above the Pelona Schist–bearing facies; they occupy the core of a syncline flanked by the latter facies, but the outcrops are too few to be certain that this is a conformable superposition. The clast suite containing marble includes biotite-rich quartz-monzonite and diorite, foliated biotite quartz monzonite, hornblende granodiorite, calc-silicate, quartzite, and olivine basalt.

Quartzite-bearing facies

The easternmost Santa Ana Sandstone is very poorly indurated, characteristically contains quartzite clasts, and is interbedded with basalt flows. The clast assemblage includes quartzite, phyllite, coarse quartz-pebble metaconglomerate, biotite quartz-monzonite and diorite, muscovite-garnet-granite, olivine basalt, jasper, and a banded metavolcanic rock.

Stratigraphic relations and age

Two northwest-trending fault zones divide the outcrop of the Santa Ana Sandstone into three subequal portions. In the eastern and western thirds, the base of the Santa Ana Sandstone is exposed, but in the central portion no base is seen (Figs. 4, 5).

The central portion is further characterized by landsliding and a thick cover of surficial deposits (Sadler and Morton, 1989). Exposures of the Santa Ana Sandstone are consequently poor there and confined to the steep, lower flanks of incised drainages. Lateral continuity is difficult to establish, but the relationships are tentatively summarized in Figure 5.

The Santa Ana sandstone is everywhere older than the influx of coarse surficial gravels that resulted from the uplift of San Gorgonio Mountain and the northern plateau of the San Bernardino Mountains. Several lines of evidence constrain this uplift to Pliocene or early Pleistocene times. Glacial cirques on the north side of San Gorgonio Mountain prove that it had been uplifted before Wisconsin times. The history of deposition in the San Timoteo Badlands appears to show that the south flank of the mountain began to deliver large volumes of gravel at 5 to 5.4 Ma (Matti and Morton, 1975; May and Repenning, 1982b). The influx of marble clasts into the Old Woman Sandstone Basin during uplift of the northern edge of the San Bernardino Mountains is younger than 2 to 3 Ma (Sadler, 1982; May and Repenning, 1982a). In the western San Bernardino Mountains, Meisling and Weldon (1989) have documented early tectonic activity between 9.5 and 4.1 Ma, which they believe might be correlative with movement on the Santa Ana thrust, but they place the uplift of the modern northern plateau between 2.0 and 1.5 Ma.

The age of the basal red siltstone and soils of the western portion of the Santa Ana Sandstone is apparently between 20 and 12 Ma old, and has been tentatively placed at 15 Ma on the basis of fragmentary microvertebrates (Carlton, in Sadler, 1985, p. 72):

The Santa Ana Sandstone west of Barton Flats may be Hemingfordian or Barstovian in age, based on the tentative identification of two rodent tooth fragments, both assigned to the family Heteromyidae. The first fragment is interpreted to belong to the genus *Proheteromys* and has a crown height similar to *Proheteromys sulculus*. . . .The second is interpreted as belonging to the genus *Cupidinimus*. . . .[The same genera] have been found in the Crowder Formation (Reynolds, 1984).

Basalt interbedded with the quartzite-bearing facies of the Santa Ana Sandstone has a 6.2-Ma K/Ar age (Woodburne, 1975, p. 83). This is the youngest radiometric age from a compositionally homogeneous province of alkali olivine basalts in the northeast San Bernardino Mountains and adjacent Mojave Desert (Neville and others, 1985). Most of the basalt flows yield ages between 7 and 8 Ma (Neville, 1982); the oldest is 8.9 Ma (Oberlander, 1972). It is unsafe to assume that this late Miocene age can be extrapolated westward to the other facies of the Santa Ana Sandstone. The marble-bearing facies includes reworked clasts of the same olivine basalt and is probably younger. This facies appears to occur up-section from the Pelona Schist clasts; it may have overstepped the basalts to rest directly on plutonic basement.

There are no exposed contacts between the quartzite- and marble-bearing facies, but the latter appears to be younger; it contains basalt clasts derivable from the flows interbedded with the quartzite-bearing facies. Furthermore, since both facies locally rest on basement, their mutual contact is probably a disconformable overstep (Fig. 5).

The age relations of the critical Pelona Schist–bearing facies are the most obscure. The facies seems certainly younger than 20 Ma and older than 2 Ma. Best guesses for the age of the basal fauna of the Santa Ana Sandstone and the uplift of the basin margins tentatively narrow the bracket to between 15 and 4 Ma.

Sediment sources

Only the marble-bearing eastern facies of the Santa Ana Sandstone can simply be locally derived. The clast sizes, types, and proportions closely match modern gravels in Coon Creek and Heart Bar Creek, 5 to 8 km to the east. Note that there is no steep, well-defined margin to the modern Santa Ana basin in this direction. The position of the deposits need not have changed relative to this source. Why then is this facies attributed to the Tertiary Santa Ana basin? The marble-bearing facies lacks clasts of the biotite-rich Baldwin gneiss that makes up the steep, young slopes of higher ground immediately to the south and north (Sadler, 1981). Very coarse boulders of this gneiss dominate the overlying surficial deposits (Fig. 3). Thus the marble-bearing facies is quite easy to distinguish from local deposits of the Quaternary basin, and is interpreted to predate establishment of the modern mountainous relief.

Sources for the quartzite-bearing eastern facies are deceptively close. The clast assemblage includes metaconglomerate and spotted phyllite (coarse basal Wildhorse Quartzite and Lightning Gulch Metasiltstone: Cameron, 1981; Sadler, 1981) and banded alkali rhyolites that, taken together, are diagnostic for the Sugarloaf Mountain exposures of prebatholithic metasediments immediately to the north. White quartzites may be derived from other parts of the San Bernardino Mountains, but not this particular assemblage. The problem for the eastern facies is that quartzite clasts reach the same locations today, following the active creeks and talus slopes from the same source, but they are much more angular than their counterparts in the Santa Ana Sandstone (Sadler and Reeder, 1983). Furthermore, the Quaternary clast suites have an admixture of Baldwin Gneiss clasts, from the lower flanks of Sugarloaf Mountain, that is missing from the Santa Ana Sandstone. The trace of the Santa Ana Thrust runs between the quartzite-bearing facies and its inferred source; presumably, slip on the thrust brought the two closer together in the process of configuring the Quaternary Santa Ana basin.

The center of the Tertiary Santa Ana basin requires a nearby source of Pelona Schist to the south. This source must include the garnet-bearing varieties that may be derived in abundance only from the southern edge of Sierra Pelona, west of the Punchbowl fault, or from Blue Ridge, between the San Andreas and the Punchbowl faults (Ehlig, 1981; D. M. Morton, personal communication, 1984). The arkosic wacke and "polka-dot granite" clasts may also be derived from either of these two areas. San Francisquito sandstones are exposed north of both Pelona Schist outcrops. "Polka-dot granite" occurs as dikes in the pre–San Francisquito basement north of Sierra Pelona (Kooser, 1980); it is also abundant, as recyclable clasts in the Punchbowl Formation, above the San Francisquito Sandstones, east of the Punchbowl fault. I have found large, fresh angular blocks of "polka-dot granite" in the superficial deposits at the west edge of the Punchbowl Formation; this suggests that a bedrock source exists on Pleasantview Ridge. Recycled volcanic and quartzite clasts would be available from the conglomerate beds in the San Francisquito Formation (Kooser, 1980).

Pelona Schist exposed on Mount Baldy, and in a series of low ridges southeast to the Crafton Hills, does not have the necessary lithologic variety to qualify as sources for the Santa Ana Sandstone. There are no suitable sources west of the San Gabriel fault. Since anorthosite, syenite, and Mount Lowe Granodiorite crop out extensively in the San Gabriel Mountains south of the Sierra Pelona, it is necessary that sediment from this area be blocked from the Santa Ana basin.

I suggest that the Pelona Schist clasts entered the Santa Ana basin before the Sierra Pelona and Blue Ridge source areas were significantly separated by the Punchbowl fault. This allows Blue Ridge to act as a barrier to the ready passage of anorthosite and Lowe Granodiorite clasts into the basin. If the gravels entered the basin through gaps in a shutter-ridge system produced by the Punchbowl-Nadeau fault zone, then the bed-to-bed changes in composition may be explained.

Sheep Creek near Wrightwood drains steep slopes in Pelona Schist; it then flows north across the San Andreas fault and through a gap in a shutter ridge to deposit a large fan of micaceous sediment that is distinct even on satellite images of the southern Mojave Desert. Since this is not unlike my reconstruction for deposition in the Santa Ana basin, Figure 7 risks a simple quantitative comparison of the two transport systems. Maximum observed grain size is very sensitive to the field techniques. In Sheep Creek, large areas could be searched, and a distinction was made between the largest clast and the size of commonly encountered large clasts (Fig. 8). In extreme cases 75-cm clasts were

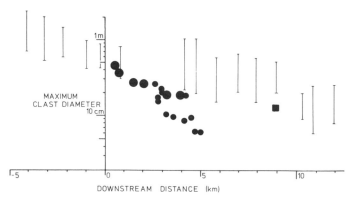

Figure 7. Pelona Schist clast sizes as a function of transport distance. Circles indicate Santa Ana basin; large circles, data points aligned along strike and thus free of shortening caused by folds; square, westernmost locality (Fig. 6), which is apparently offset by right slip. The origin for the distance axis is arbitrarily set at the southern boundary fault, immediately south of the most proximal outcrop. Vertical bars indicate coarsest Pelona Schist clasts encountered in traverses of the active Sheep Creek (see Fig. 1) downstream from the Blue Ridge (see Fig. 8). The origin for the Sheep Creek data is placed at –5 km; this allows the two regressions to converge at the largest clast size seen in Sheep Creek (2 m).

found nearly 20 km from source, and with more field time this distance would likely be extended. The poor exposures of Santa Ana Sandstone provided relatively very small areas to search; this is somewhat compensated for by the concentration of clasts in surficial lags. From Figure 7 we might tentatively conclude that the Pelona Schist source lay on the order of 5 km south of the most proximal outcrop now exposed in the Santa Ana basin. Where it crosses the San Andreas fault, Sheep Creek is joined by other creeks with Pelona Schist sources farther northwest. These tributaries have flowed southeast along the fault zone for up to 4 mi. Thus, the fan and its sources are not symmetrically aligned across the fault.

The simple fan-like pattern in the maximum grain sizes is surprising, considering the lack of stratigraphic control in the data points, the variation in the clast suites, and the likelihood of active strike slip between the source and the basin. The generation of a simple "unsmeared" fan pattern is favored by an entry point for the gravels that is fixed relative to the basin; a canyon or topographic gap north of any active faults would suffice. Second, the influx should probably be short-lived. A strand of the Punchbowl fault may have been active during deposition of the Pelona Schist clasts in the Santa Ana basin.

The Sierra Pelona and Blue Ridge sources can be restored to a position south of the Pelona Schist–bearing conglomerates by restoring some strike slip on the San Andreas and Punchbowl faults. But San Gorgonio Mountain now stands in that position, and obviously had not assumed its modern configuration at that time; it would have blocked the transport path of Pelona Schist clasts. Since clasts of the gneisses, monzonites, and megaporphyritic granodiorite of San Gorgonio Mountain are absent in the

clast suites derived from the south, it is unlikely that they were exposed near the transport path.

Clasts of the western facies are not particularly distinctive. Considering the north and north-northwest paleocurrents, and the inferred source of the Pelona Schist–bearing facies, basement terranes north of the Clearwater fault are a possible source.

LIMITS OF THE TERTIARY BASIN

Since the Tertiary and Quaternary Santa Ana basins have quite separate origins, they need not have comparable size and shape. This section attempts to establish the limits of the Tertiary basin, first by examining the edges of the modern basin, and then by comparing the thickness and composition of Tertiary sediments on the neighboring uplands.

Santa Ana thrust and northern margin

At the foot of the steep north flank of the modern basin is the Santa Ana thrust system: a zone of north-dipping reverse and thrust faults that has carried crystalline rocks of the northern plateau over the older basin fill. Numerous, large bedrock slides obscure the original geometry of the fault, but close to the fault zone the Santa Ana Sandstone steepens and is locally overturned. The thrusting predates the surficial sediments, which were derived primarily from the overriding terrane. Clearly, the overthrusting took place before the modern Santa Ana basin was established.

In the northeast the lower levels of the overriding crystalline terrain expose Baldwin Gneisses, which are conspicuously absent as clasts in the Santa Ana basin. Higher levels carry the sources of quartzite clasts that are now too close to the basin to account for the rounding of the clasts. The overthrust terrain must have been farther north, with its lower levels probably unexposed, during the deposition of the Santa Ana Sandstone (Fig. 9). Sadler and Reeder (1983) concluded from the clast roundness (Fig. 10) that as much as 10 km of the Tertiary Santa Ana basin may have been overridden by the thrust. Although the source of the quartzite clasts is well established, their results must still be regarded as very approximate; it is not clear that overthrusting is the sole cause of the difference in roundness between the Tertiary and Recent clasts.

The thickest lacustrine clay sequences in the Tertiary Santa Ana basin are close to the Santa Ana thrust. There is a lack of southwardly transported sediments in the center of the basin. Quite possibly the original northern flank of the Tertiary basin has been hidden beneath the thrust.

Southern margin

Much of the southern limit of the Santa Ana Sandstone is faulted. A steep but poorly exposed fault zone separates the crystalline basement of San Gorgonio Mountain from steep and locally overturned beds of the Pelona Schist–bearing facies. The

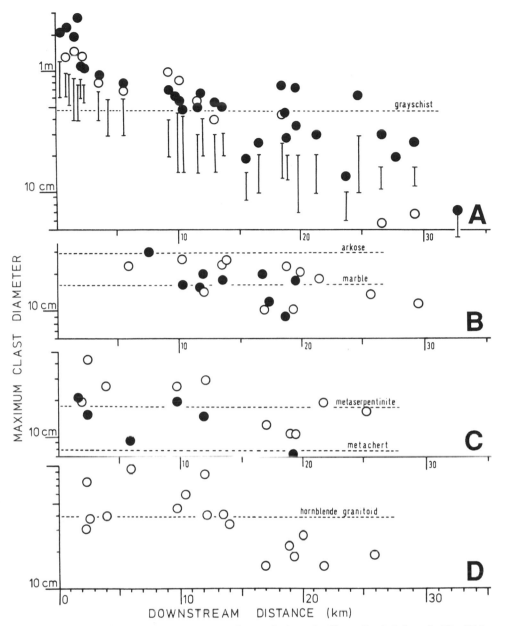

Figure 8. Relationship of clast size to transport distance in the active Sheep Creek, below the Blue Ridge near Wrightwood. Dashed lines indicate the coarsest clasts seen in the Pelona Schist–bearing conglomerates of the Santa Ana Sandstone. A, Pelona Schist clasts; circles plot the coarsest clast seen in several traverses at a given distance downstream; filled circles are grayschist clasts; open circles are albite-porphyroblast greenschists; vertical bars represent the range of sizes that appear to be coarsest in at least 20 percent of the channel traverses at a given distance. B, Arkose (filled circles) and marble clasts. C, Metachert (open circles) and metaserpentinite clasts. D, Granitoid clasts.

westward trace of this fault separates the central and the western portions of the Santa Ana Sandstone. Dip separation is down to the north, but strike slip is significant. The displacements of the lower parts of the Pelona Schist–bearing facies (Fig. 6), the syncline in the Santa Ana Sandstone and the Santa Ana thrust zone fit a right-lateral strike slip of 3 to 6 km. The younger surficial deposits are offset 1 to 2 km. Another northwest-trending splay

from this fault zone separates the central from the eastern portion of the Santa Ana Sandstone. Dip separations are inconsistent, but right-lateral strike slip can account for the displacement of the Santa Ana thrust, the juxtaposition of southwardly transported quartzite-bearing facies with Pelona Schist–bearing facies, and a 3-km right-lateral separation of the toes of bedrock slides on the south flank of Sugarloaf Mountain.

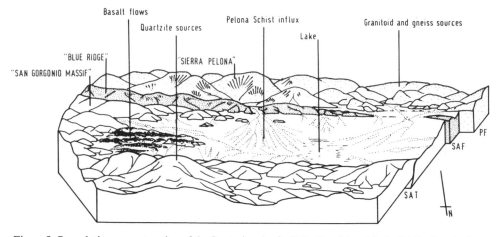

Figure 9. Speculative reconstruction of the Santa Ana basin. PF = Punchbowl fault; SAF = San Andreas fault; SAT = Santa Ana thrust.

Steep zones of crushed basement rock follow the southern limit of the granitoid-dominated facies of the Santa Ana Sandstone. But these beds are tilted northward approaching San Gorgonio Mountain; the north-dipping basal contact with weathered crystalline basement becomes approximately parallel with the steep bedrock slopes to the south. The impression remains that San Gorgonio Mountain is to some extent simply tilted, or warped up.

Upland Tertiary sediments

Only the quartzite-bearing facies of the Tertiary Santa Ana basin appears to have any continuation atop the uplifted basement that bounds the younger basin. There are no Tertiary sections on the northern plateau of the San Bernardino Mountains, or on San Gorgonio Mountain, with thicknesses comparable to the Santa Ana basin. But both uplands have remnants of a veneer of quartzite gravels that were derived from residual peaks of prebatholithic rocks, like Sugarloaf Mountain (Sadler and Reeder, 1983). The remnants are sufficiently widespread to preclude the possibility that a major portion of the fill of the Tertiary Santa Ana basin might once have existed beyond the limits of the younger basin.

The age of the veneer is certainly composite and not easy to specify. Some patches of the upland quartzite gravels are still active. Many are probably contemporaneous with the Tertiary Santa Ana basin, and tenuous lines of evidence suggest that others are older than the quartzite-bearing conglomerates of the Santa Ana basin.

Sadler and Reeder (1983) presented permissive evidence that the upland quartzite gravels include a westward transport path that led into Foster's (1980) unit three of the Crowder Formation in Cajon Pass. Subsequently published vertebrate faunas and magnetostratigraphy (Reynolds, 1984; Weldon, 1984)

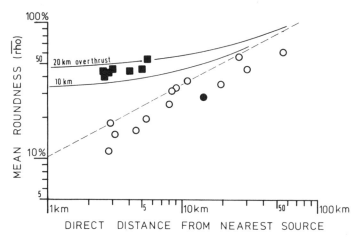

Figure 10. Roundness of quartzite clasts in the upland veneer (circles) and the Santa Ana Sandstone (squares). Solid circle indicates upland veneer remnant south of the Santa Ana basin; dashed regression, modern quartzite clasts in active Santa Ana river basin (Sadler and Reeder, 1983). Roundness determined by the Sames (1966) method, using clasts of 20- to 80-mm intermediate diameter.

show that unit three is at least 11 m.y. old (mid/late Miocene). Gravels of the upland veneer may have been recycled into the Crowder Formation long after their deposition. Indeed, K. E. Meisling and R. J. Weldon (personal communication, 1987) have preferred to explain the changing provenance of the Crowder units solely in terms of the evolution of Mojave Desert sources.

On the eastern ramp of the San Bernardino Mountains, the upland quartzite gravels include an admixture of volcanic and metavolcanic clasts that originated in the Mojave Desert. These portions of the veneer underlie and therefore predate 9 to 6-Ma alkali olivine basalt flows (Sadler, 1982). At Pioneertown, near the eastern limit of the San Bernardino Mountains, coarse arkosic sediments with no quartzite clasts are also found beneath, and interbedded with, late Miocene olivine basalts (Sadler, 1982). The Pioneertown sediments are less than 2 km from basalt flow remnants that cap the upland gravel veneers. This might be an abrupt facies change, but it is quite reasonable to suppose that the arkosic sands and the quartzite gravel veneer are of different age. The arkosic sands thicken where the pile of basalt flows thickens and also interbed with the flows, whereas the quartzite gravel veneer does neither. We must accept that the arkosic sands at Pioneertown are contemporaneous with the basalt flows; the quartzite veneer may be appreciably older.

The relationships described from Pioneertown indicate that the eastward transport of the upland gravels was established before deposition of the eastern facies of the Santa Ana Sandstone. The quartzite-bearing Santa Ana gravels are interbedded with the alkali olivine basalt flows, and must be nearly contemporaneous with the arkosic sands at Pioneertown. The unconsolidated upland veneer of quartzite cobbles and boulders also spread southward onto San Gorgonio Mountain (Sadler and Reeder, 1983). This creates several difficulties:

1. If San Gorgonio Mountain simply did not rise until after the influx of Pelona Schist clasts across it, there must have been a protective sediment mantle to prevent entrainment of San Gorgonio basement clasts into the Santa Ana basin. But survival of upland quartzite gravels suggests that there never was a mantle of the Pelona Schist–bearing sediments. The quartzite gravel itself could not serve as the protective mantle without wholesale mixing into the Pelona schist–bearing clast assemblages. Yet, in the Santa Ana basin, quartzites arrived with only the very first flush of Pelona Schist clasts from the south. Subsequently, there are no quartzite clasts and no San Gorgonio basement clasts.

2. On the mdoern map the transport path of the veneer boulders appears to have crossed the center of the Santa Ana basin. This is impossible after basin formation. Any earlier transport path should leave its trace in the form of quartzite-bearing conglomerates near the base of the Santa Ana Sandstone. They are present to the east, but not in the west; the basement is not exposed beneath the center of the basin.

3. The quartzite clasts of the eastern Santa Ana Sandstone were probably derived from a previously established upland veneer. Sadler and Reeder (1983) showed that the well-rounded clasts have been brought anomalously close to source by the Santa Ana thrust. But the restoration of the thrusting, which eliminates that anomaly, leaves the remnant of the veneer on San Gorgonio Mountain too far from this source above the thrust (Fig. 10).

The earlier observation that there has been strike slip between the Santa Ana basin and San Gorgonio Mountain can eliminate all three incongruencies. Restoration of right-lateral strike slip can take the San Gorgonio basement rocks out of the Pelona Schist transport path. It also places the quartzite-bearing facies of the Santa Ana basin between the northern plateau sources and the remnant of quartzite veneer on San Gorgonio

Mountain. A total of about 10 km of right slip distributed across the northwest-trending faults that traverse the Santa Ana basin places the quartzite veneer on San Gorgonio Mountain at a distance from its source that is more consistent with the roundness of the clasts. This same right slip, if restored, also brings the early Mesozoic, megaporphyritic granodiorite pluton of Angelus Oaks into the northwest-trending line of comparable plutons that crop out from Fairview Valley (Miller, 1978), through the Granite Mountains (Miller, 1977) and across Bear Valley (Richmond, 1960; Sadler, 1981) to the northern edge of the Santa Ana Valley (Sadler, 1985).

The strike-slip solution given above is quite convenient, but without proof. The strike-slip faults have no obvious continuation to the southeast of the Pelona Schist–bearing facies. This does not pose a geometric problem if the slip on the Santa Ana thrust zone is greater southeast of Sugarloaf Mountain than to the west. The relatively thin, quartzte-bearing facies of the Santa Ana Sandstone has affinities with the upland quartzite veneer. In the Tertiary Santa Ana basin the distinction between basin and upland was probably far more subtle than in the modern basin.

UPLIFT OF THE SAN BERNARDINO MOUNTAINS

The Tertiary Santa Ana basin has been shown to predate the uplift of San Gorgonio Mountain and the northern plateau of the San Bernardino Mountains; these uplift events created the Quaternary Santa Ana basin. The deposits of the Tertiary basin—the Santa Ana Sandstone—are seen throughout the floor of the modern basin. In the central portion they have been folded about east-west axes; at the faulted, east-west contacts with adjacent uplifted blocks, bedding is often vertical and overturned. The northern margin of the Tertiary basin is apparently hidden beneath the Santa Ana Thrust. The folding and marginal faulting appear to be closely related.

Evidently, the Tertiary basin was wider than the modern separation between the northern plateau and San Gorgonio Mountain. These two basement blocks have several rock types in common, but once stood farther apart; they must have converged during uplift as a result of compression and overthrusting of the intervening Tertiary Santa Ana basin.

When the Tertiary basin was active, the Blue Ridge region of the San Gabriel Mountains stood close to the south. There is no evidence that rocks of San Gorgonio Mountain were exposed at that time. This block was still low, and probably to the southeast of its present position relative to the Santa Ana basin. Uplift of San Gorgonio Mountain, relative to the Tertiary Santa Ana basin was accomplished via three components: northward tilting, combined with dip slip and right-lateral slip on the bounding faults.

The northern plateau of the San Bernardino Mountains had only modest relief when the Tertiary basin was active. It served to prevent transport of Pelona Schist clasts into the basins of the southern Mojave Desert. Uplift of the plateau was achieved on the Santa Ana Thrust, on which it overrode northern parts of the

Tertiary Santa Ana basin. The uplift postdates the 6.2-Ma basalt flows (9 to 6 Ma, if one prefers to use the range of ages in the larger alkali basalt province) that are overridden by the thrust. In the western San Bernardino Mountains, Meisling and Weldon (1990) have constrained what they believe to be contemporaneous thrusting to the interval between 9.5 and 4.1 Ma.

The northern plateau of the San Bernardino Mountains is highest north of San Gorgonio Mountain, where the Santa Ana thrust is also best developed. To the east and west the plateau surface descends. To the west the Santa Ana thrust becomes less distinct. The Santa Ana Sandstone is folded north of San Gorgonio Mountain, but to the east and west the folds give way to a simple northward tilt. The uplift of the plateau is interpreted to be a response to the northwestward drive of San Gorgonio Mountain into the Tertiary Santa Ana basin. Thus, the folding and tilting of the basin floor, the slip on right-lateral faults and the slip on the Santa Ana thrust work together to accommodate the invading block.

Crystalline rocks of the south flank of San Gorgonio Mountain had become exposed to erosion by 5 to 5.4 Ma (Matti and Morton, 1975; May and Repenning, 1982b). The northern edge of the San Bernardino Mountains is not uplifted until later than 2 to 3 Ma (Sadler, 1982; May and Repenning, 1982a). Perhaps the deformation propagated northward from its presumed origins at the restraining bend in the San Andreas fault system in San Gorgonio Pass. None of the dates for these events is tightly constrained, but everything fits together most easily if the Tertiary Santa Ana basin is eliminated at about 4 to 5 Ma.

The modern Santa Ana drainage developed later, when the southeast San Gabriel Mountains had slipped past the San Gorgonio massif. This allowed a south-flowing drainage basin to develop on the San Andreas fault line scarp and erode headward until it captured streams that drained the intramontane remnant of the Tertiary basin.

IMPLICATIONS FOR THE HISTORY OF THE SAN ANDREAS FAULT

We have seen that the Tertiary Santa Ana basin received sediment from the San Gabriel Mountains, across the trace San Andreas fault system. This phase of the basin history probably terminated at about 4 to 5 Ma. The implications for the history of the fault system emerge when these events are placed into sequence with two other exchanges of sediment between the two areas. They are recorded in the Miocene stratigraphy of the Ridge basin and the Quaternary sediments of the Victorville fan.

Ridge basin

The high San Gabriel Mountains have not been reported to include any sediments of San Bernardino provenance. Farther north, at the top of the fill of the Ridge basin, the Hungry Valley Formation (Crowell, 1982; Ramirez, 1983) includes clasts attributable to Mojave Desert and, perhaps, San Bernardino Mountain sources.

Meta-agglomerates in a suite of siliceous metavolcanic clasts found throughout the Hungry Valley Formation indicate a Mojave Desert source north of Victorville. Marble and quartzite clasts are also common (Ramirez, 1983). I have examined Ramirez' collections and find the clast suite reminiscent of the 17- to 10-m.y. Crowder Formation at the west end of the San Bernardino Mountains. A vertebrate fauna indicates that the Hungry Valley deposits are only 4 to 5 m.y. old (Crowell, 1982). Thus, the Crowder Formation is not correlative, but probably served as a second cycle source for the Hungry Valley Formation.

Victorville Fan

Clasts from Blue Ridge contribute to the Pleistocene "Victorville Fan" (Meisling and Weldon, 1982; Weldon, 1984, 1985). Pelona Schist clasts, with Lowe Granodiorite and San Francisquito Sandstone clasts, are well known from the Harold and Shoemaker Formations of the Cajon Pass area at the west end of the San Bernardino Mountains. The older Phelan Peak Formation (4 to 1.5 m.y.; Weldon, 1984, 1985) has mixed paleocurrent directions, but was apparently dominated by southwestward transport from the Mojave Desert (Foster, 1980). It has not yielded Pelona Schist clasts. The initiation of the Victorville Fan marks the passage of the Pelona Schist terrains of the San Gabriel Mountains beyond the catchment of the Santa Ana basin. The deposition of the Victorville fan is in some respects a younger analog for the Pelona Schist clasts in the Santa Ana basin.

Synthesis

Figure 11 attempts to synthesize evidence from the Santa Ana basin with the histories of sedimentation in the Ridge basin, the Cajon Pass area, and the San Timoteo Badlands. Four frames are used; they represent different spans of time, but seem to be a necessary sequence. The total ranges of possible age for each frame include appreciable overlap. The Tertiary Santa Ana basin is active during the first three frames. It is uplifted before the fourth frame, which includes the Quaternary Santa Ana basin. It is instructive to consider two permissible guesses at the correct ages, one that maximizes the time span involved, and one that minimizes it.

It is possible to compress most of the history of the Santa Ana basin into the 5 Ma conventionally allotted to the San Andreas fault, sensu stricto. This allows the Wilson Creek strand of the San Andreas fault (Matti and others, 1985) to be active during deposition of Pelona Schist clasts in the Santa Ana Basin (at about 4 Ma). In this compressed time scale, the uplift of the north and south margins of the northern plateau of the San Bernardino Mountains is simultaneous. On the other hand, the northwestward progression of uplift from the structural knot in San Gorgonio Pass, and all correlations with deformation in the Cajon Pass area, are lost.

In order to match the deformation and thrusting episodes in Cajon Pass and the Santa Ana basin (as suggested by Meisling

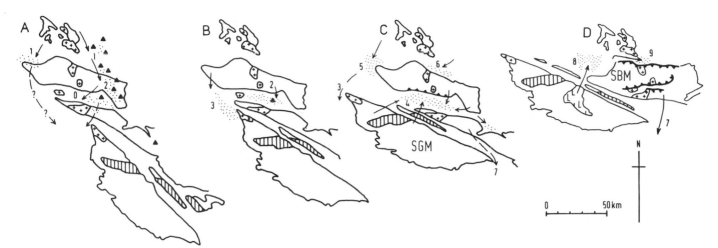

Figure 11. Possible regional restorations of deposits and their sources. The modern margins of San Gabriel Mountains, San Gorgonio massif, and northern plateau of the San Bernardino Mountains are included for reference. The four frames A-D represent different time spans. They follow in necessary sequence, but the total ranges of possible age for each include overlap: A, ?15 to 6 m.y.; B: 9 to 4 m.y.; C; 9 to 3 m.y.; D: 2 to 1 m.y. L: Lowe Granodiorite. Crosses indicate megaporphyry of Matti and others (1985); triangles, 6- to 9-m.y. olivine basalt flows; vertical ruling, Pelona Schist; stipple, active deposition. Sequence key to numbers: 0, Paleosoils form in western Santa Ana basin. 1, Sidewinder Volcanics clasts transported from Mojave Desert; two routes into the Ridge basin are tentatively suggested. 2, Quartzite clasts, derived principally from Sugarloaf Mountain, enter eastern Santa Ana basin; some reach west end of future San Gorgonio Mountain block, others mix with 1 and travel at least to Pioneertown. 3, Ridge basin in position to deposit Hungry Valley Formation by reworking from Crowder Formation of Cajon Pass. 4, Clast assemblage with Pelona Schist can now enter Santa Ana basin around west end of future San Gorgonio Mountain. 5, Phelan Formation. 6, Earliest deposition in Old Woman Sandstone basin. 7, Clasts from San Gorgonio Mountain available to San Timoteo Badlands. 8, Victorville Fan. 9, Youngest, synorogenic deposition in Old Woman Sandstone basin.

and Weldon, 1990), the history may be allowed to expand up to about 9 Ma. The Pelona Schist clasts may have entered the Santa Ana basin before 6 Ma, but only if the San Andreas fault was initiated earlier than is popularly supposed. Perhaps it is time to question this conventional wisdom. Certainly, Weber (1986) has written that there are grounds to reexamine the relationships between the San Andreas fault and the older San Gabriel fault that are said to constrain their ages.

SUMMARY

The Santa Ana basin has a three-phase history. There was a Tertiary Santa Ana basin that is quite distinct from the Quaternary basin; the two are separated by a phase of considerable uplift in the San Bernardino Mountains. The Santa Ana Sandstone was deposited in the Tertiary basin.

The Tertiary Santa Ana basin received sediment from high ground to the south as already uplifted rocks, now seen in the Sierra Pelona and Blue Ridge portions of the San Gabriel Mountains slipped along the southern margin of the basin. At the time of this sediment transfer, the crystalline basement rocks now exposed on San Gorgonio Mountain were probably buried. They were certainly located farther from the basement rocks of the

northern plateau of the San Bernardino Mountains than today, probably to the southwest.

The source of the Pelona Schist clasts in the Santa Ana basin now lies about 120 km to the northwest, separated from the basin by all the strands of the Punchbowl and San Andreas fault zones. From the discussion above and the strand sequence proposed by Matti and others (1985), it seems most likely that one strand of the Punchbowl fault and all strands of the San Andreas fault have been active since deposition of the Pelona Schist clasts. The Tertiary history of Santa Ana basin offers no evidence for the apportionment of the 120 km between these strands.

Presumably as a result of the restraining bend in the San Andreas fault system, San Gorgonio Mountain was uplifted and forced northwest into the Santa Ana basin. The basin folded immediately north of the San Gorgonio block; to the east and west the basinal deposits were simply tilted northward. The deformation extended to basement rocks north of the basin, which rose as a thrust-bound plateau, and partially overrode the Tertiary basin fill. The elongate form of the Quaternary Santa Ana basin thus developed as a welt in a region of general uplift and compression. Streams that drained the slopes of the basin were integrated into the modern Santa Ana River by capture.

ACKNOWLEDGMENTS

Careful reviews of an earlier draft by Perry Ehlig and Ray Weldon allowed substantial improvements in this chapter. Perry Ehlig, Douglas Morton, and Jonathon Matti, gave invaluable advice on the identity and possible sources of several clast types. Vince Ramirez showed me the clast suites that he had collected from the Hungry Valley Formation. Bob Reynolds and Cleet Carlton prepared and identified the vertebrate fossils. Marilyn Kooser, Wes Reeder, and David West assisted in the field.

REFERENCES CITED

Allen, C. R., 1957, San Andreas fault zone in San Gorgonio Pass: Geological Society of America Bulletin, v. 68, p. 315–349.

Cameron, C. S., 1981, Geology of the Sugarloaf and Delamar Mountain areas, San Bernardino Mountains, California [Ph.D. thesis]: Cambridge, Massachusetts Institute of Technology, 399 p.

Crowell, J. C., 1982, Pliocene Hungry Valley Formation, Ridge basin, southern California, in Crowell, J. C., and Link, M. H., eds., Geologic history of Ridge basin, southern California: Pacific Section, Society of Economic Paleontologists and Mineralogists, p. 143–150.

Dibblee, T. W., 1964, Geologic map of the San Gorgonio Mountain Quadrangle, San Bernardino and Riverside Counties, California: U.S. Geological Survey Miscellaneous Geologic Investigations Map I-431, scale 1:62,500.

——— , 1967, Geologic map of the Morongo Valley Quadrangle, San Bernardino and Riverside Counties, California: U.S. Geological Survey Miscellaneous Geologic Investigations Map I-517, scale 1:62,500.

——— , 1975, Late Quaternary uplift of the San Bernardino Mountains on the San Andreas and related faults, in Crowell, J. C., ed., The San Andreas fault in southern California: California Division of Mines and Geology Special Report 118, p. 127–135.

——— , 1984, Geology of the San Bernardino Mountains, southern California, in Fife, D. L. and Minch, J. A., eds., Geology and mineral wealth of the California Transverse Ranges: Santa Ana, California, South Coast Geological Society, p. 149–169.

Ehlert, K. W., and Ehlig, P. L., 1977, The "polka-dot" granite and the rate of displacement on the San Andreas fault in southern California: Geological Society of America Abstracts with Programs, v. 9, p. 415–416.

Ehlig, P. L., 1981, Origin and tectonic history of the basement terrane of the San Gabriel Mountains, central Transverse Ranges, in Ernst, W. G., ed., The geotectonic development of California; Rubey Volume 1: Englewood Cliffs, New Jersey, Prentice-Hall, p. 253–283.

Foster, J. H., 1980, Late Cenozoic tectonic evolution of Cajon Valley, southern California [Ph.D. thesis]: Riverside, University of California, 242 p.

Jacobs, S. E., 1982, Geology of a part of the upper Santa Ana River valley, San Bernardino Mountains, San Bernardino County, California [M.S. thesis]: Los Angeles, California State University, 107 p.

Kooser, M. A., 1980, Stratigraphy and sedimentology of the San Francisquito Formation, Transverse Ranges, California [Ph.D. thesis]: Riverside, University of California, 201 p.

Matti, J. C., and Morton, D. M., 1975, Geologic history of the San Timoteo Badlands, southern California: Geological Society of America Abstracts with Programs, v. 7, p. 344.

Matti, J. C., Morton, D. M., and Cox, B. F., 1985, Distribution and geologic relations of fault systems in the vicinity of the central Transverse Ranges, southern California: U.S. Geological Survey Open-File Report 85-365, 23 p.

May, S. R., and Repenning, C. A., 1982a, New evidence for the age of the Old Woman Sandstone, Mojave Desert, southern California, in Cooper, J. D., compiler, Geologic excursions in the Transverse Ranges: Geological Society of America Cordilleran Section Field Trip Guidebook 6, p. 93–96.

——— , 1982b, New evidence for the age of the Mount Eden fauna, southern California: Journal of Vertebrate Paleontology, v. 2, no. 1, p. 109–113.

Meisling, K. E., and Weldon, R. J., 1982, The late Cenozoic structure and stratigraphy of the western San Bernardino Mountains, in Cooper, J. D., compiler, Geologic excursions in the Transverse Ranges: Geological Society of America Cordilleran Section Field Trip Guidebook 6, p. 75–81.

——— , 1989, The Late Cenozoic tectonics of the northwestern San Bernardino Mountains, southern California: Geological Society of America, Bulletin, v. 101, p. 106–128.

Miller, C. F., 1977, Alkali-rich monzonites, California; Origin of near silica-saturated alkaline rocks and their significance in a calc-alkaline batholithic belt [Ph.D. thesis]: Los Angeles, University of California, 283 p.

Miller, E. L., 1978, Geology of the Victorville region, California [Ph.D. thesis]: Houston, Texas, Rice University, 226 p.

Morton, D. M., Cox, B. F., and Matti, J. C., 1980, Geologic map of the San Gorgonio wilderness, San Bernardino County, California: U.S. Geological Survey Miscellaneous Field Studies Map MF-1161-A, scale 1:62,500.

Neville, S. L., 1982, Late Miocene alkaline volcanism, south-central Mojave Desert and northeast San Bernardino Mountains, California [M.S. thesis]: Riverside, University of California, 156 p.

Neville, S. L., Schiffman, P., and Sadler, P. M., 1985, Ultramafic inclusions in late Miocene alkaline basalts from Fry and Ruby Mountains, San Bernardino County, California: American Mineralogist, v. 70, p. 668–677.

Oberlander, T. M., 1972, Morphogenesis of granitic boulder slopes in the Mojave Desert, California: Journal of Geology, v. 80, p. 1–19.

Pelka, G. J., 1971, Paleocurrents of the Punchbowl Formation and their interpretation: Geological Society of America Abstracts with Programs, v. 3, p. 176.

Powell, R. E., Matti, J. C., Cox, B. F., Oliver, H. W., Wagini, A., and Campbell, H. W., 1983, Mineral resource potential map of the Sugarloaf roadless area, San Bernardino County, California: U.S. Geological Survey Miscellaneous Field Studies Map MF-1606-A, scale 1:24,000.

Ramirez, V. R., 1983, Hungry Valley Formation; Evidence for 220 km of post-Miocene offset on the San Andreas fault, in Andersen, D. W., and Rymer, M. J., eds., Tectonics and sedimentation along faults of the San Andreas system: Pacific Section, Society of Economic Paleontologists and Mineralogists, p. 33–44.

Reynolds, R. E., 1984, Miocene faunas in the lower Crowder Formation, Cajon Pass, California; A preliminary discussion, in Hester, R. L., and Hallinger, D. E., eds., San Andreas Fault–Cajon Pass to Wrightwood: Pacific Section, American Association of Petroleum Geologists Guidebook 55, p. 17–20.

Richmond, J. F., 1960, Geology of the San Bernardino Mountains north of Big Bear Lake, California: California Division of Mines and Geology Special Report, v. 65, 68 p.

Sadler, P. M., 1981, The structure of the northeast San Bernardino Mountains; Notes to accompany 7.5-minute quadrangle maps submitted for compilation onto the San Bernardino 1 by 2 degree quadrangle: California Division of Mines and Geology Final Technical Report Contract 5-1104, 26 p., 12 maps.

——— , 1982, Provenance and structure of late Cenozoic sediments in the northeast San Bernardino Mountains, in Cooper, J. D., compiler, Geologic excursions in the Transverse Ranges: Geological Society of America Cordilleran Section Field Trip Guidebook 6, p. 83–91.

——— , 1985, Santa Ana Sandstone; its provenance and significance for the late Cenozoic history of the Transverse Ranges, in Reynolds, R. E., ed., Geologic investigations along Interstate 15, Cajon Pass to Manix Lake, California: Redlands, California, San Bernardino County Museum, p. 69–78.

Sadler, P. M., and Morton, D. M., 1989, Landslides of the uppermost Santa Ana river basin and the adjacent San Bernardino Mountains of southern California, in Sadler, P. M., and Morton, D. M., eds., Landslides in a semi-arid environment: Publications of the Inland Geological Society, v. 2, p. 356–386.

Sadler, P. M., and Reeder, W. A., 1983, Upper Cenozoic, quartzite-bearing gravels of the San Bernardino Mountains, southern California; Recycling and mixing as a result of transpressional uplift, *in* Andersen, D. W., and Rymer, M. J., eds., Tectonics and sedimentation along faults of the San Andreas system: Pacific Section, Society of Economic Paleontologists and Mineralogists, p. 45–57.

Sames, C. W., 1966, Morphometric data of some recent pebble associations and their application to ancient deposits: Journal of Sedimentary Petrology, v. 36, p. 126–142.

Strathouse, E. C., 1982, The Santa Ana Sandstone (Miocene, in part) and evidence for late Cenozoic orogenesis in the San Bernardino Mountains, *in* Cooper, J. D., compiler, Geologic excursions in the Transverse Ranges: Geological Society of America Cordilleran Section Field Trip Guidebook 6, p. 97–102.

—— , 1983, Late Cenozoic fluvial sediments in Barton Flats and evidence for orogenesis, San Bernardino Mountains, San Bernardino County, southern California [M.S. thesis]: Riverside, University of California, 127 p.

Vaughan, F. E., 1922, Geology of San Bernardino Mountains north of San Gorgonio Pass, Berkeley, University of California Publications in the Geological Sciences, v. 13, no. 9, p. 319–341.

Weber, F. H., 1986, Geologic relationships between the San Gabriel and San Andreas faults, Kern, Los Angeles and Ventura Counties: California Geology, v. 39, p. 5–14.

Weldon, R. J., 1984, Implication of the age and distribution of the late Cenozoic stratigraphy in Cajon Pass, southern California, *in* Hester, R. L., and Hallinger, D. E., eds., San Andreas Fault–Cajon Pass to Wrightwood: Pacific Section, American Association of Petroleum Geologists Guidebook 55, p. 9–16.

—— , 1985, Implications of the age and distribution of the late Cenozoic stratigraphy in Cajon Pass, southern California, *in* Reynolds, R. E., ed., Geologic investigations along Interstate 15, Cajon Pass to Manix Lake, California: Redlands, California, San Bernardino County Museum, p. 59–68.

Woodburne, M. O., 1975, Cenozoic stratigraphy of the Transverse Ranges and adjacent areas, southern California: Geological Society of America, Special Paper 162, 91 p.

MANUSCRIPT RECEIVED BY THE SOCIETY SEPTEMBER 6, 1990

Index

[Italic page numbers indicate major references]

Typeset by WESType Publishing Services, Inc., Boulder, Colorado
Printed in U.S.A. by Malloy Lithographing, Inc., Ann Arbor, Michigan